DIGITAL TWIN TECHNOLOGY AND SMART GRID

Advances in Digital Twin Computing and Sensor Networks

DIGITAL TWIN TECHNOLOGY AND SMART GRID

IRAKLIS VARLAMIS
Professor of Data Management, Department of Informatics and Telematics, Harokopio University of Athens, Greece

ANTOINE BAGULA
Professor, University of the Western Cape, South Africa

HOSSEIN HASSANI
Professor, Webster University in Vienna, Austria

ELSEVIER

Elsevier
Radarweg 29, PO Box 211, 1000 AE Amsterdam, Netherlands
50 Hampshire Street, 5th Floor, Cambridge, MA 02139, United States

Publisher's note: Elsevier takes a neutral position with respect to territorial disputes or jurisdictional claims in its published content, including in maps and institutional affiliations.

Books and Journals published by Elsevier comply with applicable product safety requirements. For any product safety concerns or queries, please contact our authorised representative, Elsevier B.V., at productsafety@elsevier.com.

For accessibility purposes, images in electronic versions of this book are accompanied by alt text descriptions provided by Elsevier. For more information, see https://www.elsevier.com/about/accessibility.

Notices

Knowledge and best practice in this field are constantly changing. As new research and experience broaden our understanding, changes in research methods, professional practices, or medical treatment may become necessary.

Practitioners and researchers must always rely on their own experience and knowledge in evaluating and using any information, methods, compounds, or experiments described herein. In using such information or methods they should be mindful of their own safety and the safety of others, including parties for whom they have a professional responsibility.

To the fullest extent of the law, neither the Publisher nor the authors, contributors, or editors, assume any liability for any injury and/or damage to persons or property as a matter of products liability, negligence or otherwise, or from any use or operation of any methods, products, instructions, or ideas contained in the material herein.

ISBN: 978-0-443-36664-2

Publisher: Mara E. Conner
Acquisitions Editor: Neil Craig Smith
Editorial Project Manager: Bhaviya Koul
Production Project Manager: Kiruthigadevi
Cover Designer: Raman Kumar

Typeset by Aptara, New Delhi, India

Contents

14. Cloud computing for Digital Twin power grids

SANA SALEH, ZOHAIB SAADAT AND MUHAMMAD TARIQ

15. Fog computing for Digital Twin power grids

AMIT KUMAR TYAGI, SAURABH SINGHAL, AJEET KUMAR SHARMA AND AVINASH KUMAR SHARMA

16. Edge computing for Digital Twin-enabled future electrical grid systems

ABUBAKAR AHMAD MUSA, HENGSHUO LIANG, QINGQING LI, MILON BISWAS, CHENG QIAN, SIPING LI AND WEI YU

17. Quantum computing with Digital Twins and power grids

S. CHIDAMBARANATHAN

18. Integrating generative AI and LLMs with digital twins in smart grids

NARGES FATEMI, SIAVASH HOSSEINI AND JAVAD FATTAHI

19. Harnessing generative AI and large language models for Digital Twin-enabled smart grids

ABIODUN EMMANUEL ONILE, HOSSEIN NOUROLLAHI HOKMABAD, EDUARD PETLENKOV AND JURI BELIKOV

20. Integration of generative artificial intelligence and large language models into Digital Twin-based smart grids

SAVERIO IEVA, DAVIDE LOCONTE, GIUSEPPE LOSETO, MICHELE RUTA, FLORIANO SCIOSCIA, DAVIDE MARCHE AND MARIANNA NOTARNICOLA

Contributors

Helder Aranha Independent researcher, Lisbon, Portugal

Felipe Arraño-Vargas UNSW Sydney, School of Electrical Engineering and Telecommunications, Sydney, NSW, Australia

Juri Belikov Software Science Department, Tallinn University of Technology, Tallinn, Estonia

V.C. Bharathi School of Computer Science and Engineering, VIT-AP University, Amaravati, Andhra Pradesh, India

Federico Bianchi Ricerca sul Sistema Energetico, Milano, Italy

Milon Biswas Department of Computer and Information Sciences, Towson University, Towson, MD, United States

S. Chidambaranathan Department of Computer Applications, St. Xavier's College, Palayamkottai, Tamil Nadu, India

Istvan David McMaster University, Hamilton, ON, Canada; McMaster's Centre for Software Certification (McSCert), McMaster University, Hamilton, ON, Canada

Narges Fatemi School of Electrical Engineering and Computer Science, Faculty of Engineering, University of Ottawa, Ottawa, ON, Canada

Javad Fattahi School of Electrical Engineering and Computer Science, Faculty of Engineering, University of Ottawa, Ottawa, ON, Canada

Philippe Glass Computer Science Center, University of Geneva, Geneva, Switzerland

Eduardo Gomez-Luna High-Voltage Research Group – GRALTA, School of Electrical and Electronic Engineering, Faculty of Engineering, University of Valle, Santiago de Cali, Colombia

Kaladi Govindaraj Department of Computer Science and Engineering, Aditya University, Surampalem, Andhra Pradesh, India

Guillaume Guerard De Vinci Higher Education, De Vinci Research Center, Paris, France

Hossein Nourollahi Hokmabad Software Science Department, Tallinn University of Technology, Tallinn, Estonia

Siavash Hosseini School of Electrical Engineering and Computer Science, Faculty of Engineering, University of Ottawa, Ottawa, ON, Canada

Mira Jiang Harvard College, Harvard University, MA, United States

Pooja Kapoor Noida Institute of Engineering & Technology, Greater Noida, Uttar Pradesh, India

Chitranjanjit Kaur GNA University, Phagwara, Punjab, India

Georgios Konstantinou UNSW Sydney, School of Electrical Engineering and Telecommunications, Sydney, NSW, Australia

Burgess Lauren Department of Computer and Information Sciences, Towson University, Towson, MD, United States

Saverio Ieva Department of Electrical and Information Engineering, Polytechnic University of Bari, Bari, Italy; donkeyPower S.r.l., Bari, Italy

Qingqing Li Department of Computer and Information Sciences, Towson University, Towson, MD, United States

Siping Li Department of Computer and Information Sciences, Towson University, Towson, MD, United States

Hengshuo Liang School of Engineering & Technology, University of Washington Tacoma, WA, United States

Xiaoran Liu McMaster University, Hamilton, ON, Canada

Davide Loconte Department of Electrical and Information Engineering, Polytechnic University of Bari, Bari, Italy

Giuseppe Loseto donkeyPower S.r.l., Bari, Italy; Department of Engineering, LUM Giuseppe Degennaro University, Casamassima, Italy

Chao Lu Department of Computer and Information Sciences, Towson University, Towson, MD, United States

Kandan Mariadoss Department of Computing Technologies, School of Computing, Faculty of Engineering and Technology, SRM Institute of Science and Technology, Kattankulathur, Tamil Nadu, India

Massimiliano Masi Autostrade per L'Italia SpA, Rome, Italy

Davide Marche Lutech S.p.A., Bari, Italy

Joan S. Moreano-Gaviria High-Voltage Research Group – GRALTA, School of Electrical and Electronic Engineering, Faculty of Engineering, University of Valle, Santiago de Cali, Colombia

Abubakar Ahmad Musa Department of Computer and Information Sciences, Towson University, Towson, MD, United States

Marianna Notarnicola Lutech S.p.A., Bari, Italy

Abiodun Emmanuel Onile Software Science Department, Tallinn University of Technology, Tallinn, Estonia

Tanja Pavleska Laboratory for Open Systems and Networks, Jozef Stefan Institute, Ljubljana, Slovenia

Eduard Petlenkov Department of Computer Systems, Tallinn University of Technology, Tallinn, Estonia

Cheng Qian Department of Computer Science and Information Technology, Hood College, MD, United States

Jeni Gracia R C Saveetha School of Engineering, Saveetha Institute of Medical and Technical Sciences(SIMATS), Saveetha University, Chennai, Tamil Nadu, India

Marcello Restelli Department of Electronic, Information, and Bioengineering, Politecnico di Milano, Milano, Italy

Michele Ruta Department of Electrical and Information Engineering, Polytechnic University of Bari, Bari, Italy; donkeyPower S.r.l., Bari, Italy

Zohaib Saadat Smart Technologies and Applied Intelligence Research (STAIR) Lab, Department of Computer Engineering, School of Engineering, National University of Computer and Emerging Sciences (NUCES), Islamabad, Pakistan

Davide Salaorni Department of Electronic, Information, and Bioengineering, Politecnico di Milano, Milano, Italy

Sana Saleh Smart Technologies and Applied Intelligence Research (STAIR) Lab, Department of Computer Engineering, School of Engineering, National University of Computer and Emerging Sciences (NUCES), Islamabad, Pakistan

Gabriel Rodrigues Santos Production Engineering Department, Polytechnic School of the University of São Paulo, São Paulo, Brazil

Floriano Scioscia Department of Electrical and Information Engineering, Polytechnic University of Bari, Bari, Italy; donkeyPower S.r.l., Bari, Italy

Giovanni Paolo Sellitto Independent researcher, Rome, Italy

Giovanna Di Marzo Serugendo Computer Science Center, University of Geneva, Geneva, Switzerland

Ajeet Kumar Sharma School of Computer Science & Engineering, IILM University, Greater Noida, Uttar Pradesh, India

Avinash Kumar Sharma Department of Computer Science & Engineering (CSE), Sharda School of Computing Science & Engineering (SSCSE), Sharda University, Greater Noida, Uttar Pradesh, India

Zhiwei Shen UNSW Sydney, School of Electrical Engineering and Telecommunications, Sydney, NSW, Australia

Saurabh Singhal Department of CSE, Greater Noida Institute of Technology (Engg. Institute), Greater Noida, Uttar Pradesh, India

Muhammad Tariq Smart Technologies and Applied Intelligence Research (STAIR) Lab, Department of Computer Engineering, School of Engineering, National University of Computer and Emerging Sciences (NUCES), Islamabad, Pakistan

Francesco Trovò Department of Electronic, Information, and Bioengineering, Politecnico di Milano, Milano, Italy

Amit Kumar Tyagi School of Law Forensic Justice & Policy Studies, National Forensic Sciences University, Gandhinagar, Gujarat, India

Iraklis Varlamis Department of Informatics and Telematics, Harokopio University of Athens, Athens, Greece

Qianlong Wang Department of Electrical and Computer Engineering, Old Dominion University, VA, United States

Wei Yu Department of Computer and Information Sciences, Towson University, Towson, MD, United States

Saima Zafar Department of Electrical Engineering, National University of Computer and Emerging Sciences, Lahore, Pakistan

Eduardo Zancul Production Engineering Department, Polytechnic School of the University of São Paulo, São Paulo, Brazil

Preface

The evolution of our energy infrastructure is one of the defining technical challenges of the 21st century. As we transition from the centralized, unidirectional power systems of the past to the distributed, dynamic, and decarbonized grids of the future, the complexity of managing these networks has grown exponentially. The modern smart grid is no longer just a system of copper and steel; it is a cyber-physical ecosystem pulsating with data, defined by intermittent renewable generation, and driven by the active participation of prosumers.

In this volatile landscape, the margin for error is increasingly narrow. Traditional methods of simulation and control—effective for the static grids of yesterday—struggle to keep pace with the real-time demands of today's energy transition. This is where the concept of the Digital Twin emerges, not merely as a novelty, but as a necessity.

This book captures the convergence of the two powerful domains, namely Digital Twin Technology and Smart Grid. It articulates a vision where high-fidelity virtual replicas do not just mirror the physical world but interact with it—predicting failures before they occur, optimizing flows in real time, and securing critical infrastructure against increasingly sophisticated threats.

This volume brings together a diverse chorus of global experts to dissect the "how" and "why" of this technology. From the fundamental definitions and architectural pillars to advanced applications in machine learning, fault detection, and social acceptance, this book offers a comprehensive roadmap. It challenges us to look beyond the hype and understand the rigorous engineering and data science required to build a grid that is not only smart but also resilient, secure, and human-centric.

For researchers, engineers, and policymakers, this work serves as both a foundation and a guidepost. It reminds us that the Digital Twin is more than a model; it is the bridge to a sustainable energy future.

Iraklis Varlamis
Antoine Bagula
Hossein Hassani

Introduction by the editors

The rapid digitization of the energy sector has given rise to unprecedented opportunities and challenges. As the smart grid becomes the backbone of modern society, ensuring its reliability, efficiency, and security is paramount. Digital Twin Technology and Smart Grid were conceived to explore the transformative potential of pairing advanced digital modeling with critical power infrastructure.

This book is designed to guide the reader through the lifecycle of this technology—from its conceptual roots to its most advanced applications in artificial intelligence and cybersecurity. We have structured the content to build a narrative that is both educational for newcomers and technically rigorous for experts.

The journey begins with the fundamentals. In Chapter 1: Introduction to Digital Twins for the Smart Grid, we establish the baseline definitions, distinguishing between digital twins, digital shadows, and digital models, while exploring the technology's history and its critical role across the grid's lifecycle—from design to predictive maintenance. Chapter 2: Introduction to Digital Twin: Applications expands on this by detailing the three fundamental pillars—real space, virtual space, and the bidirectional link—and explaining how these models achieve self-regulation and self-diagnosis.

To understand the environment in which these twins operate, Chapter 3: Introduction to Smart Grid provides a comprehensive overview of the smart grid's evolution. It details the shift from legacy infrastructure to a system defined by advanced metering infrastructure (AMI), distributed energy resources (DERs), and self-healing capabilities.

With the foundations laid, the book moves into specific technological implementations. Chapter 4: Introduction to Digital Twin technology for power grids offers a state-of-the-art analysis of current research, identifying key clusters such as enterprise strategy, cybersecurity, and control systems, and proposing a novel cybersecurity-oriented architecture.

Data analytics is the engine of any digital twin, and this is the focus of the next two chapters. Chapter 5: Using Machine Learning algorithms for prediction and diagnosis problems introduces the spectrum of machine learning techniques—from supervised and unsupervised learning to reinforcement learning—demonstrating their utility in fault detection and energy optimization. Chapter 6: Machine Learning Algorithms for Forecasting and Diagnosis Applications in Smart Grids dives deeper into deep learning architectures, specifically convolutional (CNN), recurrent (RNN), and graph neural networks (GNN), and presents a case study on using Federated Learning to preserve user privacy.

The human element is often overlooked in technical literature, but it is addressed here in Chapter 7: Intelligent Digital Twins for Social Acceptance by Design in Smart Grids. This chapter proposes a novel framework where "intelligent" digital twins act as agents that negotiate energy contracts while adhering to social values, privacy rules, and fairness to prevent the "tragedy of the commons."

As grids become more inverter-dominated, reliability becomes harder to maintain. Chapter 8: Fault Detection in Modern Power Systems tackles this by examining how data-driven methods are replacing traditional impedance-based protection, particularly in the challenging contexts of microgrids and renewable generation.

Validation is a critical step before deployment, and Chapter 9: Digital Twin Testing Platforms for Power Grids addresses this by introducing the "Virtual Physical Twin"—a specialized testbed that allows operators to validate strategies in a safe environment without risk ing the actual grid. Moving to control strategies, Chapter 10: Automatic Control and Reinforcement Learning Techniques for Microgrids contrasts traditional model predictive control with data-driven reinforcement learning, offering robust solutions for managing decentralized microgrids.

Specific grid assets require tailored digital strategies. Chapter 11: Towards Digital and Smart Substations: A Digital Twin Approach outlines a service-oriented architecture for automating these critical nodes. When major outages do occur, Chapter 12: Machine learning for power system restoration demonstrates how ML-augmented digital twins can optimize the complex and delicate "black-start" process, whereas Chapter 13: Algorithm and Simulation for a Digital Twin-Based Protection System proposes adaptive mechanisms that allow the grid to self-heal and isolate faults dynamically.

The computational infrastructure required to run these twins is immense and varied. We explore the computing continuum across three dedicated chapters: Chapter 14: Cloud Computing for Digital Twin Power Grids discusses the scalability provided by centralized cloud architectures; Chapter 15: Fog-Computing for Digital Twin Power Grids examines the middle layer essential for reducing latency; and Chapter 16: Edge computing for Digital Twin-enabled future electrical grid systems focuses on decentralized processing directly at the source of data. Pushing the boundaries of computation further, Chapter 17: Quantum Computing With Digital Twins and Power Grids investigates how quantum algorithms can resolve high-dimensional optimization problems that remain intractable for classical computers.

As the grid becomes more decentralized, the need for flexible management systems grows. Chapter 18: Integrating generative AI and LLMs with digital twins in smart grids addresses this by exploring software-defined microgrids (SDMG) and examines how LLM-based interfaces can solve critical interoperability challenges by acting as a semantic bridge between diverse devices and protocols, utilizing retrieval-augmented generation (RAG) to streamline communication and coordination. Moving from coordination to core modeling fidelity, Chapter 19: Harnessing generative AI and large language models for Digital Twin-enabled smart grids introduces the novel "Physics-Informed Transformer Architecture" (PITA). This approach merges the data-driven power of transformer models with fundamental physical laws to create hybrid digital twins that are both accurate and physically consistent, while also deploying conversational chatbots to enhance consumer engagement in demand-side management. Addressing the usability of these complex systems, Chapter 20: Integration of generative artificial intelligence and large language models into Digital Twin-based smart grids introduces a framework where GenAI and large language models act as semantic interfaces, allowing operators to query and interact with the digital twin using natural language.

Security is a recurring theme, culminating in Chapter 21: Artificial Intelligence, Digital Twin, and Blockchain Technology for Secure Smart Grids. This chapter presents a unified, multilayered security architecture that combines the predictive power of AI, the simulation capabilities of digital twins, and the immutable integrity of Blockchain to defend against cyber-physical threats.

Finally, Chapter 22: Future Perspectives for Digital Twin-Based Power Grids looks ahead. It synthesizes the potential of integrating technologies such as fog computing and IoT to create a decentralized, user-driven electricity market, offering a vision of what the next generation of power systems will look like.

We hope this book serves as a vital resource for understanding how the convergence of physical and digital worlds will power our future.

Iraklis Varlamis
Antoine Bagula
Hossein Hassani

PART I

Introduction

CHAPTER

1

Introduction to Digital Twins for the Smart Grid

Xiaoran Liu[a], *Istvan David*[a,b]

[a]McMaster University, Hamilton, ON, Canada [b]McMaster's Centre for Software Certification (McSCert), McMaster University, Hamilton, ON, Canada

1.1 Foundations and definitions

Digital Twins (DT) are real-time and high-fidelity virtual representations of physical assets (Rasheed et al., 2020), referred to as the *physical twin* (Kritzinger et al., 2018) or *actual twin* (Tao et al., 2019). Digital Twins reflect the current state of the physical twin and offer cost-effective, safe alternatives for interacting with it. To avoid confusion, and whenever necessary, collections of digital and physical twins, collectively, are referred to as *digital twin systems*.

A unique trait of a digital twin system is the strong bidirectional coupling between the digital and physical twin in which the digital twin controls the physical twin through computational reflection—the behavior exhibited by a reflective system based on a causal connection to the reflected system (Maes, 1987). In digital twin systems, such a causal connection is typically achieved through the digital twin continuously processing sensor data from the physical twin. Computational reflection of the physical twin is what allows the digital twin to reason about the current and future states of the physical twin and control it in accordance with optimality, safety, and other goals of interest.

These concepts are shown in Fig. 1.1. The digital twin contains various models of the grid, capturing, for example, network dynamics, resource utilization, and environmental conditions (e.g., temperature, humidity, light conditions). These models are maintained through real-time data acquisition. Services make use of models, e.g., to simulate future states of the grid, and to present the results of such simulations to end-users through graphical user interfaces. Models in the digital twin can be inclusive of the power distribution network, the consumer and prosumer (power consumers who also produce and contribute power) points, as well as the primary power generators, e.g., power plants, wind farms, and photovoltaic devices. The more subsystems the digital twin covers, the more control it may have over the grid.

1.1.1 Brief history of Digital Twins

The concept of the digital twin is rooted in NASA's Apollo programs, tracing back to the 1960s. In response to spacecraft failures, NASA envisioned "living models" to support its Apollo missions (Allen, 2021), i.e., simulators that evaluate the health of the spacecraft based on continuously ingested data. One could consider this one of the first Digital Twins. A few decades later, in the early 2000s, driven by advanced product lifecycle management (PLM) ambitions, Grieves and Vickers (2017) suggested a similar mechanism to develop the "conceptual ideal for PLM," later labeled as "the information mirroring model." The notion of a digital twin has been eventually defined in a 2010 NASA roadmap document (Shafto et al., 2010) as follows: "an integrated multiphysics, multiscale, probabilistic simulation of an as-built vehicle or system that uses the best available physical models, sensor updates, fleet history, etc., to mirror the life of its corresponding flying twin." The digital twin is also considered "ultra-realistic," thanks to high-fidelity physical models, real-time sensor data, and the ability to aggregate maintenance history and fleet data. In their hype cycle report for emerging technologies, Gartner reported in 2017 that Digital Twins entered the innovation trigger phase, i.e., a rapid growth of interest in the concept and subsequent adoption is imminent in the

Digital Twin Technology and Smart Grid.
DOI: https://doi.org/10.1016/B978-0-443-36664-2.00001-0

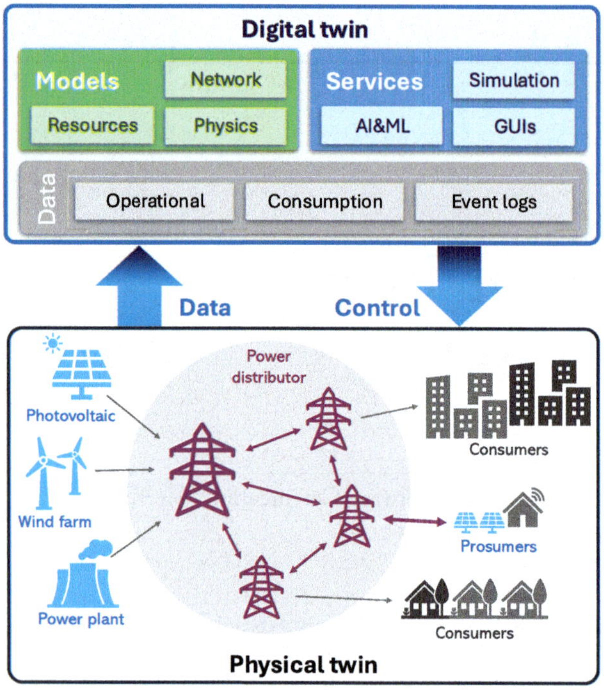

FIGURE 1.1 A typical digital twin of a power grid.

next 5–10 years. Due to an unusually rapid progress, the 2018 edition of the same report already placed Digital Twins at the peak level of (inflated) expectations (https://www.gartner.com/en/documents/3885468). Later reports do no longer mention Digital Twins, and it is widely assumed that Digital Twins have reached the steady level of maturity, i.e., the so-called plateau of productivity (Paredis & Vangheluwe, 2022). Indeed, today, Digital Twins are drivers to complex and mature solutions. Some of the key applications of Digital Twins include automated optimization, real-time reconfiguration, and intelligent adaptation of physical systems. These capabilities and the simplicity of the digital twin concept render Digital Twins particularly appealing in settings where cyber-physical systems have to be controlled in a computer-automated fashion, such as manufacturing (Touhid et al., 2023), uncrewed avionics (Lei et al., 2021), smart ecosystems (Michael et al., 2024), and smart grids (Jafari et al., 2023).

1.1.2 Why Digital Twins matter?

Digital Twins provide a substantial competitive advantage across industries. Capgemini reports an average of 13% decrease in costs and a 15% increase in operational efficiency across organizations implementing Digital Twins (https://www.capgemini.com/be-en/insights/research-library/digital-twins/). McKinsey reports these numbers being in the 20%–25% range in warehouse operations (https://www.mckinsey.com/capabilities/operations/our-insights/improving-warehouse-operations-digitally) with additional 30%–50% reductions in machine downtime (https://www.mckinsey.com/capabilities/operations/our-insights/capturing-the-true-value-of-industry-four-point-zero). Thus, the business case for Digital Twins is rather clear. In many domains, Digital Twins are key drivers of digital transformation—the process of organizations adapting to changes in their business through utilizing digital technology (Vial, 2019). As companies progress through their digital transformation journeys, they adjust their value creation processes and strategies, largely driven by the opportunities in digitalization (Kraus et al., 2021). Not only do Digital Twins enable digital transformation in many domains, but the endeavor itself through which companies develop and deploy Digital Twins often also enforces the digital transformation of the company. This is because advanced digitalization requires the acquisition of new skills, tools, licenses, and best practices. Even domains that have been least amenable to digitalization—such as agronomy and healthcare—adopt digital twin technology at a rapid pace (David & Syriani, 2023).

Another foundational enabler of digital transformation, the digital thread (Margaria & Schieweck, 2019) makes extensive use of digital twin technology, too. The digital thread is a record of a product's or system's lifetime from its creation to its retirement. Digital Twins act as real-time data collection facilities across the entire lifecycle and as

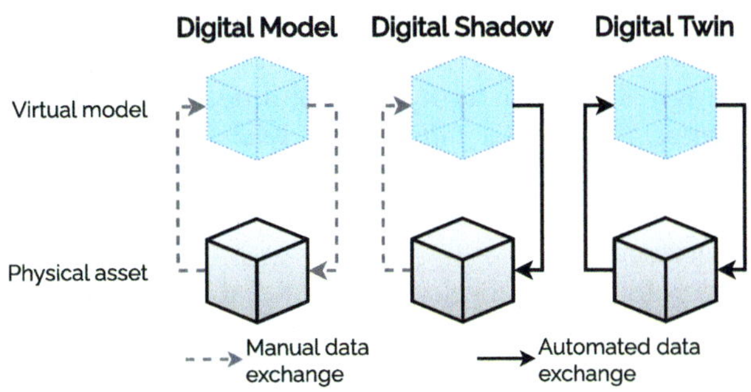

FIGURE 1.2 Digital assets classified by the level of automation in data exchange with the physical asset, as per Kritzinger et al. (2018).

contextualized representations that continuously reflect the current state and behavior of physical assets. This real-time mirroring allows the digital thread to stay up-to-date and allows for informed decision-making throughout the value chain. For example, in a smart grid, the digital thread might capture the operational history of a wind turbine, ranging from its design parameters, through its manufacturing and installation records, to real-time performance data and maintenance logs. When the digital twin detects physical indications of wear and tear, this information is recorded in the digital thread, allowing for predictive maintenance and repair. This rich information contained in the digital thread allows quicker repair and maintenance operations as technicians can access and correlate error diagnostics with specifications.

In addition to driving digital transformation, Digital Twins excel in driving sustainability transition journeys, too. Today, we see an increasing interest in aligning digital transformation principles with corporate sustainability ambitions (Lazazzara et al., 2024). Industry 4.0 is being gradually replaced by Industry 5.0, an improvement that puts research and innovation at the service of the transition to a sustainable, human-centric and resilient industry (https://research-and-innovation.ec.europa.eu/research-area/industrial-research-and-innovation/industry-50_en). By integrating artificial intelligence (AI), automation, and robotic technology into manufacturing production, while enabling workers to play an active role in decision-making (i.e., promoting on human creativity), Industry 5.0 concentrates on value-added tasks, while automating repetitive and monotonous tasks. Digital Twins serve as the platform for the integration of various technological components and allow humans to interface with them in real-time. By that, Digital Twins also foster circularity in systems engineering (David et al., 2024).

1.1.3 Definitions: What is and what is not a Digital Twin?

A common typology of digital twin systems is due to Kritzinger et al. (2018), who distinguish between three types of systems, depending on the level of automation in data exchange and control (Fig. 1.2). *Digital Twins* process data from the physical twin in an autonomous fashion, and possess the ability to control them. A system is called a *digital shadow* when automated control cannot be guaranteed, but still processes data from the physical system continuously. Finally, a *digital model* is updated with data manually while having no control over the physical system.

Quite often, digital shadows are generously labeled as Digital Twins, even though this is not entirely correct. In a somewhat more detailed view, Tao et al. (2019) suggest *services* and *data* to be promoted to first-class components in digital twin systems. In their model, data takes the central role through which physical and virtual entities are connected, and based on which services are provided by the digital twin. In reaction to the often-overlooked human aspects of Digital Twins, David and Bork (2024) suggest the separation of the ability and liberty of a digital twin to act autonomously. In this classification, *human-supervised Digital Twins* can control physical systems, but they are not allowed to. To shift toward fully automated Digital Twins, human factors, especially trust in the digital twin, need to improve. In contrast, *human-actuated Digital Twins* are unable to control physical twins in an automated fashion, even though this capability would be needed, and instead, they rely on human actuation, e.g., by issuing work commands to humans.

More industry-attuned definitions are provided by companies and suppliers that provide digital twin platforms or use them for innovative services. Some of these are well-aligned with the usual definitions of Digital Twins. For example, NVIDIA defines Digital Twins as "*virtual representations of products, processes, and facilities that enterprises use to design, simulate, and operate their physical counterparts*. (https://www.nvidia.com/en-eu/glossary/digital-twin/)" An interesting element in this definition is processes which, apparently, can be twinned, too. This is not as surprising as it seems at first. When twinning a process, a digital twin reflects a real-life process based on its observable and measurable outputs, and controls the process based on simulations about its future states. Other industry actors focus more on the extensions of the digital twin concept through their own unique capabilities. For example, Dassault Systèmes defines the notion of a *virtual twin* as "*the representation of the full lifecycle, behavior and evolution of a product or system, starting with a 3D model that captures the shape, dimensions and properties of a physical object.* (https://www.3ds.com/virtual-twin)" Dassault Systèmes achieves this through advanced 3D experience, enabled by its 3DEXPERIENCE platform. (https://www.3ds.com/3dexperience/)

With the uptick in the adoption of Digital Twins, their general typology has experienced a noticeable shift. Reports both from industry (Muctadir et al., 2024) and academia (Dalibor et al., 2022) hint at an overall lack of understanding of Digital Twins and the surrounding ecosystems. Today, Digital Twins are used as an umbrella term that covers real Digital Twins and some limited variants, too, e.g., digital shadows and even digital models. While there is often no point in arguing against such umbrella concepts, we provide a sufficiently clear, fit-for-purpose definition of Digital Twins.

A digital twin is a real-time, sufficiently realistic computational reflection of a system with the ability to control it.

This high-level definition turns out to be adequate for the vast majority of situations, and there are some notable properties it implies and that merit discussion.

Computational reflection is provided by a model that is continuously updated with data from the physical twin to reflect the prevalent state of the twinned system. This model is used by computer machinery for the automated reasoning about the physical system's state, and by humans to observe, inspect, and operate physical systems.

The model is an abstraction of reality, capturing only the essential properties of the system (Minsky, 1965; Stachowiak, 1973), and it is valid under a limited set of conditions (Zeigler et al., 2018).

The model is digital. This is contrasted with the actual system, which is typically of a physical nature. The model is, therefore, amenable to computer-automated analysis and simulation. *Examples of digital twin systems.* To illustrate this definition, here are some examples of systems that rely on Digital Twins.

Robotic assembly system for smart manufacturing, in which a manufacturing system is the physical twin, and it is observed and controlled through a cloud-based digital twin (Touhid et al., 2023). The digital twin contains models of the geometry of robot arms, models of kinematics to understand and control its movement and states (e.g., safe states, unstable states, etc). Such solutions constitute the foundations of Industry 4.0 and Industry 5.0.

A smart building, in which the building is the physical twin, and it is monitored and controlled for optimal temperature (El Mokhtari et al., 2022). In the digital twin, 3D models and models of physics aid automated decision-making, integrated into a unified building information model. Such solutions constitute the foundations of Construction 4.0.

Controlled environment agriculture, in which a smart agriculture enclosure is the physical twin, and it is monitored and controlled for optimal crop growth conditions, including temperature, humidity, and lighting conditions (David et al., 2023). Discrete event models may capture states of machines (e.g., on/off), and continuous (e.g., Simulink-based) energy models may capture the energy states of plants. Equipment (e.g., irrigation and air conditioning) is controlled based on insights from high-performance simulators. *Examples of systems that are not Digital Twins,* at least not in the strict sense.

A monitoring software is not a digital twin in the strict sense. A monitoring software has no control over the physical system, and typically, monitoring software has no central computational model to reflect the system's state.

Closed-loop control is not a digital twin. In some research communities, closed-loop control is considered a form of a digital twin. This is not entirely correct. While the control element is clearly there, a controller does not facilitate the computational reflection of the system.

Models@run.time are not Digital Twins. Models@run.time (Blair et al., 2009) support computational reflection of physical systems but are not capable of controlling them.

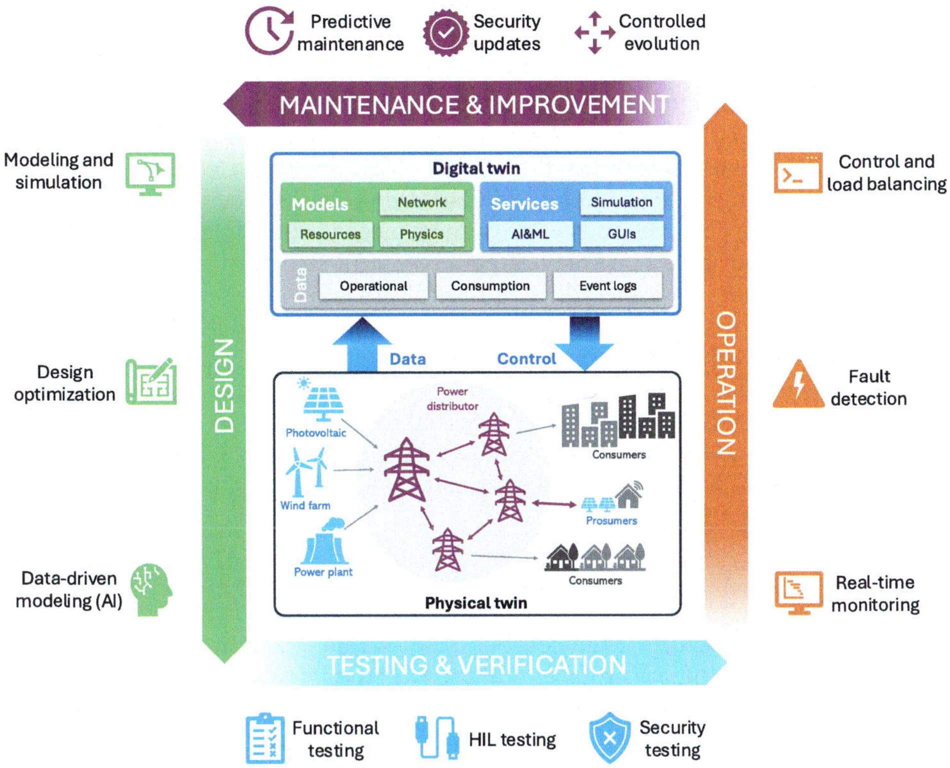

FIGURE 1.3 Key digital twin responsibilities along the lifecycle of smart grids.

1.2 The case for Digital Twins in smart grids

Power grids are a critical infrastructure to modern society. They enable nowadays digitalized world and economy, and directly affect our daily lives. Providing safe, stable, and sufficient electricity is essential for sustainable development worldwide. However, the power industry is facing serious challenges: growing energy demand, limited resources, and the looming threat of climate change render traditional power generation and distribution practices insufficient (Mchirgui et al., 2024). The growing electricity demand from electric vehicles and data centers requires more flexible and responsive grid management that supports a sustainable energy transition. However, traditional centralized power generation cannot efficiently integrate variable renewable energy sources essential for carbon emission reduction, such as solar and wind energy. These challenges create an urgent need to modernize power infrastructures. Smart grids tackle these challenges by using two-way flows of electricity and information to create a widely distributed automated energy delivery network (Fang et al., 2011). In smart grids, electricity can flow bi-directionally, and information flows multi-directionally (Fig. 1.3). This enables, for example, transforming consumers into *prosumers* (producer-consumers) who generate power through small-scale power generators, such as rooftop solar panels and small wind turbines, and feed excess electricity back to the grid (Sarwar & Asad, 2016). Prosumers continuously share consumption and generation data with grid operators, who respond with dynamic pricing signals and demand management requests. This creates a dynamic system where smart grids constantly adjust power supply and distribution based on real-time data from thousands of distributed energy sources.

Apart from being highly dynamic systems, the topology and organizational structure of smart grids change rapidly too, for example, due to prosumers adding and removing distributed generation sources, system overloads, and occasional outages (Pan et al., 2020). Meanwhile, advanced technologies such as Internet of Things (IoT) sensors for real-time monitoring, advanced metering infrastructure (AMI), large-scale battery-energy-storage systems, and vehicle-to-grid (V2G) chargers are being introduced to the power sector, making it more complex. These new smart technologies greatly enrich computing and communication capabilities and make power management more efficient, reliable, and sustainable. However, they create new operational challenges that cannot be addressed through physical infrastructure alone (Jafari et al., 2023). Together, dynamic prosumer behavior and advanced technological complexity of smart grids create a control problem: real-time load balancing must now reconcile thousands of bidirectional prosumers whose injections rise and fall with weather, tariffs, and local storage schedules.

Digital Twins offer a powerful solution to address these challenges across the entire lifecycle of smart grid systems. The digital twin keeps a synchronized virtual replica of each asset and of the grid as a whole, so that planners and cyber-defenders can all act on shared, up-to-date data. This replica underpins day-to-day monitoring, optimization, security analytics, and predictive maintenance (Das et al., 2024), and it becomes a real-time decision support tool when contingencies strike, such as equipment failure and coordinated cyberattacks (Zheng et al., 2022). Furthermore, Digital Twins enable real-time coordination of thousands of prosumers by continuously modeling their bidirectional power flows and predicting their behavior based on weather patterns, tariff structures, and storage schedules, creating a closed-loop control system that automatically adjusts physical grid operations to maintain optimal performance. When smart grids and Digital Twins work together, energy providers can achieve both operational efficiency and environmental sustainability goals through optimized renewable energy integration (Irfan et al., 2023; Sleiti et al., 2022).

In the following, we look at how Digital Twins play a crucial role in four lifecycle phases of a smart grid: design, testing & verification, operation, and maintenance & improvement. Through real-world examples and use cases, we discuss how the unique capabilities of Digital Twins enable new ways to enhance grid reliability, resilience, and performance.

1.2.1 Design

In the design of smart grid infrastructures, engineers build various analytical and simulation models of the physical system. These models constitute the primary artifacts later used in a digital twin. Digital Twins in the design phase can be used, for example, to simulate the impact of adding a new solar farm or wind park to the grid before construction, explore the associated costs and risks, and visualize recommended design decisions for key stakeholders.

1.2.1.1 Modeling and simulation

During the design phase, Digital Twins serve as foundational tools built on two core capabilities: comprehensive modeling and simulation of smart grid systems, and the connected nature between digital and physical twins that enables continuous synchronization. These digital representations allow engineers to create virtual smart grid systems, addressing various aspects of design complexity through different modeling approaches. Several distinct modeling approaches emerge to address specific aspects of smart grid design complexity. One comprehensive approach involves modeling both physical and cyber layers simultaneously. For example, the ANGEL digital twin employs such a dual-layer design methodology, in which integrated models represent both the physical grid infrastructure and its cybersecurity framework (Danilczyk et al., 2019). This design approach ensures that security considerations are embedded into the system architecture, enabling the digital twin to validate data integrity and evaluate grid performance within a unified modeling framework. For real-time applications, probabilistic modeling offers advantages. Milton et al. (2020) propose an approach for online diagnostic analysis of power electronic converters using real-time, probabilistic digital twinning.

The approach involves developing real-time, probabilistic simulation models with stochastic variables and embedding them into power converter controllers running on field programmable gate array (FPGA) computing devices—specialized hardware that can be reconfigured for specific computational tasks.

1.2.1.2 Design optimization

Digital Twins also support design optimization by leveraging operational insights through a feedback loop from operational monitoring back into the design phase. For example, rotor speed data collected from an operating wind turbine can be analyzed to determine the most efficient speed for power generation. These resulting insights can be fed back into the design phase to inform future designs or modifications of wind turbine systems (Wang et al., 2021). At a broader system level, Digital Twins enable strategic infrastructure planning and optimization. Advanced analytics applied to the digital twin can help discover ideal load distribution patterns, reduce peak load pressures, and recommend modifications to improve system resilience and efficiency (Jafari et al., 2023).

1.2.1.3 Data- and AI-driven modeling

Building upon these foundational modeling capabilities, advanced techniques like AI can further enhance digital twin functionalities in smart grid design. The convergence of AI, IoT, and cyber-physical systems with smart grid infrastructure enables the continuous gathering, processing, and analysis of real-time data to accurately assess current and future states (Jafari et al., 2023). One particularly important application of AI-enhanced Digital Twins is climate resilience planning. Climate resilience modeling integrates physical hazard assessment with data-driven capabilities. An innovative approach introduces modeling and predicting the performance of electric power networks (EPNs) when exposed to extreme weather events (Braik & Koliou, 2023). The suggested digital twin framework integrates physical models, such as hazard and vulnerability assessments, with data-centric methodologies, specifically, a dynamic Bayesian network (DBN), to create a high-fidelity hybrid model that can receive updates in near-real-time through sensor data.

1.2.2 Testing and verification

Once comprehensive digital twin models are established during the design phase, they provide the foundation for rigorous testing and validation before system deployment. Digital Twins provide a safe and cost-efficient environment for testing extreme scenarios that would be impossible or too risky to perform on actual grid infrastructure (Mansour et al., 2023). Key examples of such scenarios pertain to the safety of the grid, e.g., testing the impact of extreme weather conditions or electromagnetic pulse (EMP) attacks; as well as the security of the grid, e.g., testing cybersecurity attacks through a coordinated false data injection attack on multiple substations.

1.2.2.1 Functional testing

By creating a dynamic digital replica of the physical grid, operators can simulate various load scenarios in controlled environments before they occur in the real-world. This simulation capability allows teams to test how the system responds to potential imbalances, identify weak points, and verify that protection systems respond appropriately. For example, engineers can simulate scenarios with rapid load fluctuations or unexpected demand spikes to verify system stability (Mchirgui et al., 2024). These tests help validate the effectiveness of existing control systems and protection mechanisms, ensuring they perform as designed under stress conditions. This capability allows grid operators to explore worst-case scenarios and boundary conditions that would otherwise remain theoretical, generating practical knowledge that helps prevent costly failures and outages. By testing various dangerous scenarios on the digital copy first, engineers can predict potential problems before they ever happen in the real-world. A practical example of this safe testing environment is when Digital Twins are used for wind turbines (Wang et al., 2021). Engineers can create a physics-based model—a digital twin—of an actual wind turbine to test how it would perform in extreme conditions. For instance, they can simulate what happens when the turbine faces dangerously high wind speeds or when it operates below its normal cut-off speed.

1.2.2.2 Hardware-in-the-loop testing

Digital Twins allow for advanced hardware-in-the-loop (HIL) testing of critical power grid components, too (Nguyen et al., 2022). Integrating experimental hardware components into the actual grid would be too risky. Digital Twins can be used, for example, in the pre-integration phase to assess the impact of hardware elements—for example, relays, controllers, and storage—to be deployed on the grid. Digital Twins, in such a setup, simulate the behavior of the overall grid, while power-hardware-in-the-loop (PHIL) components render test scenarios more realistic. This allows for detecting rare edge cases (so-called "black swan events") in a simulated grid that pure in-silico models may miss.

1.2.2.3 Security testing

Digital Twins also allow for security testing of smart grids. Traditional security testing on active power grids is problematic because it can disrupt service to thousands of customers. With Digital Twins, security experts can accurately model how the physical grid functions and run comprehensive security tests without causing any service disruptions. This approach establishes a common framework for developing standardized models that allow for continuous and thorough security testing (Atalay & Angin, 2020). Security teams can repeatedly test for vulnerabilities, update their testing procedures as new threats emerge, and ensure that the grid remains protected, all without affecting the day-to-day operation of the actual power system. Another critical application is testing protection systems, the safeguards designed to prevent cascading failures in the grid (Kuber et al., 2022). A digital twin of a relay contains all the same protection functions and algorithms as the physical device. This virtual approach goes beyond just verifying individual components; it allows engineers to validate how the entire protection system works together without needing physical access to substations. This saves time, reduces costs, and improves safety for maintenance crews.

1.2.3 Operation

During the operational phase, Digital Twins are live-synchronized systems that provide real-time insights and control capabilities. These virtual replicas continuously stream data and analytics to support maintenance decisions, user interactions, and immediate fault detection across the entire grid infrastructure. A pertinent example is real-time voltage regulation and congestion management. By maintaining a synchronized virtual replica of the grid, the digital twin continuously monitors node voltages and line loadings. When demand peaks or renewable generation fluctuates, the twin can simulate corrective actions such as reactive power dispatch from distributed energy resources or reconfiguration of network topology. These simulations allow operators to identify the most effective intervention in real-time, preventing voltage collapse or line overloading.

1.2.3.1 *Real-time monitoring, control, and load balancing*

Digital Twins provide operators with comprehensive, up-to-date information about all grid components and systems. This continuous monitoring capability extends across various grid assets, from individual components to entire subsystems, enabling precise operational control and decision-making (Mchirgui et al., 2024). For wind turbines (Wang et al., 2021), Digital Twins collect comprehensive operational data including output power, wind conditions, temperature, electrical currents, and mechanical parameters. This data can be used to analyze rotor speed performance and ensure safe operation within established limits. For energy storage systems, Digital Twins are used to visualize, process, and assess the performance in the virtual environment. This technology is used to determine the scheduling programs for the operational process (Steindl & Kastner, 2021).

1.2.3.2 *Real-time fault detection*

Digital Twins can process real-time data from both cyber and physical domains to detect anomalies, and their multidimensional characteristics enable the real-time detection of faults that improve overall grid reliability and stability (Khan et al., 2023). In a typical case, researchers implemented a digital twin for 110 kV oil-filled transformers that integrated multiple diagnostic sensors, including ultraviolet monitoring, thermal imaging, and discharge measurements to create a comprehensive digital model (Khalyasmaa et al., 2020). This multi-sensor digital twin provided instant feedback on transformer condition and detected performance degradation over time, enabling operators to identify potential issues before they escalated into costly failures. For distributed energy resources, researchers develop neural network-based fault detection systems based on a digital twin. These systems use two-stage machine learning tools to process massive data streams from smart meters, identify potential fault locations, and reduce the complexity of monitoring distributed grid components (Tzanis et al., 2020).

1.2.3.3 *Cybersecurity and real-time threat detection*

Digital Twins enhance cybersecurity through real-time data analysis capabilities that improve cyber-attack detection performance while reducing response times (Khan et al., 2023). Advanced security frameworks like ANGEL demonstrate sophisticated protection capabilities by modeling both cyber and physical grid layers. The system provides real-time data visualization and health assessment across various operating conditions, enabling evaluation of both current system status and potential security threats. Using physics-based modeling combined with machine learning algorithms, ANGEL create comprehensive defense mechanisms that reduce grid vulnerabilities and accelerate threat detection (Danilczyk et al., 2019). For complex grids like China's ultra-high voltage systems, Digital Twins provide essential decision support with sub-second response times. These real-time analysis systems handle unprecedented operational complexity while maintaining rapid responsiveness for critical security decisions (Zhou et al., 2020).

1.2.3.4 *Customer engagement and demand response*

Digital Twins transform customer participation in grid operations by providing unprecedented visibility into energy usage patterns. Instead of waiting for monthly bills, customers receive real-time information about their electricity consumption, timing, and costs. Smart meters serve as the primary interface for this enhanced engagement (Das et al., 2024), enabling customers to understand their energy usage patterns and make informed decisions. When combined with real-time pricing, this visibility allows customers to reduce costs by shifting consumption away from peak-price periods, contributing to overall grid efficiency.

1.2.4 Maintenance and improvement

The wealth of operational data collected through digital twin monitoring creates opportunities beyond day-to-day operations. As Digital Twins collect operational data over time, they become increasingly valuable for planning future upgrades and optimizations. The historical information captured by Digital Twins provides insights that would be impossible to gather from the physical system alone, creating a foundation for continuous improvement (Jafari et al., 2023). For example, by analyzing vibration and temperature data in a digital twin, operators can predict a bearing failure in a grid-scale transformer weeks in advance, allowing for planned maintenance instead of an emergency outage.

1.2.4.1 *Predictive maintenance*

By predicting maintenance needs based on actual asset conditions rather than predetermined schedules, digital twin technologies reduce unnecessary maintenance, extend equipment life, and ensure optimal performance (Coppolino et al., 2023). By deploying Industrial Internet of Things (IIoT) networks and computational resources at the edge of the system, it becomes possible to collect and process both historical and real-time data directly at the source. This enables data-driven machine learning algorithms with the support of digital twin analytics to generate optimized maintenance schedules for each subsystem (Khan et al., 2023). The dynamic nature of Digital Twins allows operators to model the grid's evolution over extended periods, simulating how load patterns might change over months and years (Jafari et al., 2023). This long-term perspective allows utilities to predict and mitigate potential imbalances that might emerge as consumption patterns change, renewable generation increases, and new technologies like electric vehicles become more prevalent. Additionally, predicting the remaining useful life (RUL) of components in the smart grid becomes more accurate and efficient. Digital twin can forecast when components are likely to fail, allowing operators to schedule maintenance before problems occur (Aizpurua et al., 2017). This predictive approach helps avoid unexpected failures and costly service disruptions.

1.2.4.2 *Security updates*

Unlike traditional security approaches that remain static until breaches occur, Digital Twins enable an adaptive security posture that evolves based on emerging threats. Coppolino et al. (2023) implement a digital twin in a real-world case study on an operational smart grid in Kropa, Slovenia, where the twin provided a comprehensive view of the grid's health and enabled quick detection and communication of potential threats. As threats evolve, the digital twin captures new attack patterns and automatically updates its security protocols, creating a continuously improving security posture that stays ahead of emerging risks.

1.2.4.3 *Controlled evolution*

Digital Twins enable controlled system evolution through operational optimization. Wind turbine operations provide an illustration (Wang et al., 2021). Wind turbine blade pitch (the angle of the blades relative to the wind) significantly affects both energy production and equipment longevity, and finding the perfect pitch angle for varying wind conditions has traditionally been challenging. With Digital Twins, an optimization program can use the digital twin to test hundreds of pitch angle settings for various wind speeds, directions, and turbine locations in order to find the best configuration. Then, such optimization can be used to close the control loop that controls the wind turbine's blade rotation. As power grids incorporate more diverse energy sources, Digital Twins provide a platform for continuous optimization across these hybrid resources. For example, the system can learn seasonal patterns in renewable generation, gradually adjusting control algorithms to maximize renewable utilization, reduce operational costs while maintaining grid stability (Khan et al., 2023). This knowledge can then inform future infrastructure investments.

1.3 Engineering Digital Twins

The tight coupling of Digital Twins with physical systems merits special considerations in their development. Here, we review some of the key software organization and architectural guidelines that are used in digital twin engineering.

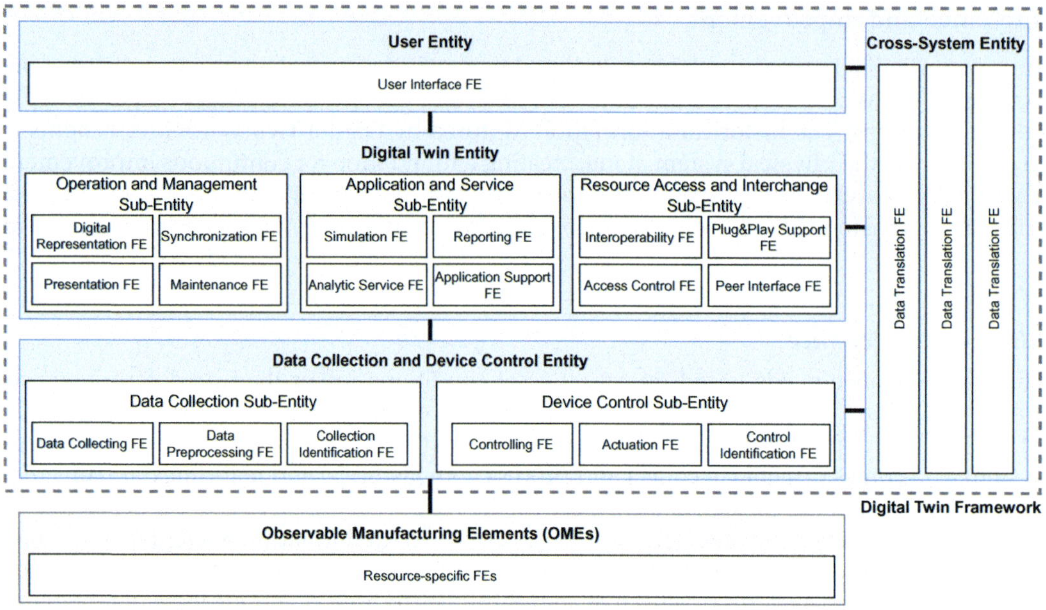

FIGURE 1.4 Functional view of the ISO 23247 digital twin reference architecture. *Reproduced from Shao (2021).*

1.3.1 The ISO 23247 reference architecture

Despite the wide adoption of Digital Twins, their engineering methods are still in the early stages of maturity. This, in turn, gives rise to some pressing software engineering problems, such as the relative scarcity of architectural guidelines and standards. Especially in regulated domains—such as smart grids—software certification is a typical need (Song et al., 2019). However, in the absence of standards, certifying software is a costly endeavor.

The *ISO 23247–Digital twin manufacturing framework* standard is a relatively new one, specifically tailored for Digital Twins (https://www.iso.org/standard/75066.html). Published in 2001 by the International Organization for Standardization (ISO), the standard defines a development framework for Digital Twins, particularly in the manufacturing domain. It consists of four parts: Part 1 – Overview and general principles; Part 2 – Reference architecture; Part 3 – Digital representation of manufacturing elements; Part 4 – Information exchange. For software architecture purposes, Part 2 provides a detailed, layered reference architecture (https://www.iso.org/standard/78743.html). As shown in Fig. 1.4, the reference architecture consists of domains, refined into domain entities, sub-entities, and, eventually, functional entities (FEs). A functional entity is a unit of coherent functionality that ISO-compliant Digital Twins shall implement. For example, the *simulation FE* predicts the behavior of physical entities, the *analytic service FE* manages and analyzes data collected from physical entities, and the result of simulations, the *reporting FE* generates reports about production results, simulation predictions, and the results of data analytics, and the *application support FE* provides services for implementing applications such as predictive and reactive maintenance, open and closed-loop applications. These four FEs together constitute the *application and service sub-entity* whose role is to provide functionality related to applications and services. This sub-entity is situated within the *digital twin entity*, which is the (set of) system(s) providing functionalities for Digital Twins, such as realization, management, synchronization, and simulation. The reader is referred to Shao (2021) and Shao et al. (2023) for detailed elaboration on the ISO 23247 standard.

Standardization of and for Digital Twins is subject to active research and development. Since ISO 23247 is currently the only digital twin-focused standard, its limitations are often on display in domains outside of manufacturing. For example, the reference architecture does not provide a specific functional entity for enhanced data storage, despite acknowledging the need for data exchange via databases or cloud systems (Ferko et al., 2023). Advanced analytics and deep-learning FEs are also missing (Kang et al., 2024). The need for such FEs has been identified, e.g., in digital twin-enabled AI simulation (Liu & David, 2025). The standard does not address verification, validation, and uncertainty quantification (VVUQ) tasks, which are fundamental in regulated domains (Shao et al., 2024). Finally, while there exists an *interoperability support FE* in the standard, it is seldom implemented in practical applications (Ferko et al., 2023), leaving digital twin aggregation and composition efforts without actionable guidelines (Adesanya et al., 2025). Fortunately, there are some important ongoing standardization efforts for Digital Twins, including two

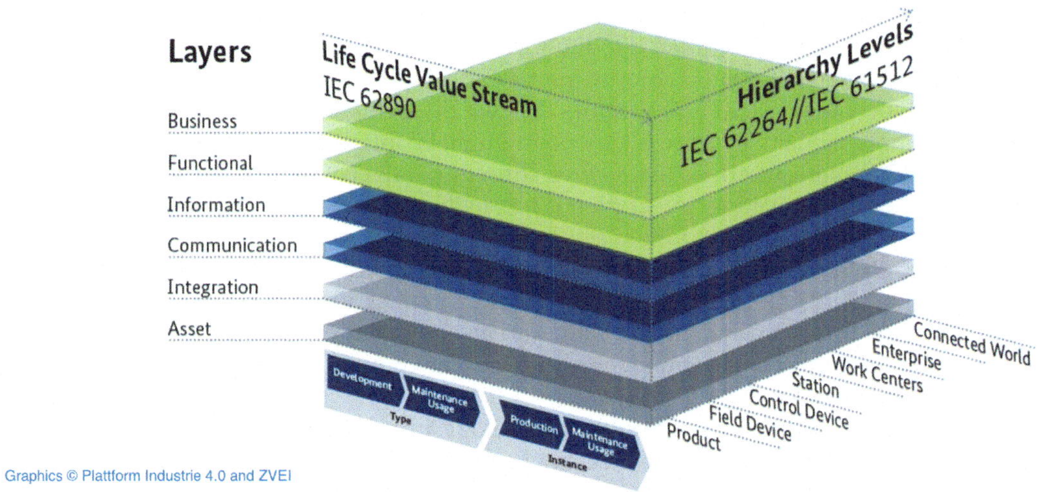

FIGURE 1.5 Overview of the RAMI 4.0 reference model. *From Schweichhart (2016).*

new parts to the ISO 23247 standard – Part 5: digital thread for Digital Twins, and Part 6: digital twin composition (Shao, 2024). The joint technical committee of the ISO and the International Electrotechnical Commission (IEC) – ISO/IEC JTC 1 – is currently working on refining definitions and concepts related to Digital Twins, as well as developing maturity models and reference architectures. The Digital Twin Consortium of the Object Management Group (OMG) is another consortium that works on standardization, with a focus on cross-domain industry standards. In the interim, more purpose-specific standards can be used to augment Digital Twins, particularly for data manage-ment, data security, and information modeling. For example, the OPC Unified Architecture (OPC UA) (https://opcfoundation.org/developer-tools/documents/?type=Specification) is a frequently used standard in complex digital twin applications where integration is a concern (Adesanya et al., 2025). It defines a standardized framework for reliable and secure communication among diverse subsystems.

The reader is referred to Shao (2024), Ferko et al. (2023), and David et al. (2025) for further pointers on standardization.

1.3.2 Reference architectural model industrie 4.0 and the asset administration shell

Another important reference model is the Reference Architectural Model Industrie 4.0 (RAMI 4.0). Compared to the ISO 23247 standard, RAMI 4.0 defines higher-level guidelines, focusing on how to situate Industry 4.0 systems, such as Digital Twins in the overall corporate information technology (IT) infrastructure. As shown in Fig. 1.5, RAMI 4.0 defines the following three dimensions to approach Industry 4.0 system development in a structured fashion.

Hierarchy levels concern the value creation systems that create the product, ranging from the *product* itself through work *station*s to the external, *connected world*. This hierarchy is compliant with the IEC 62264 (Enterprise-control system integration) (https://www.iso.org/standard/57308.html) and IEC 61512 (Batch control) (https://webstore.iec.ch/en/publication/107554) standards.

Life cycle value stream allows for describing the lifecycle of the product from early conceptualization to retirement. The lifecycle model is compliant with the IEC 62890 (Industrial-process measurement, control and automation - Life-cycle-management for systems and components) (https://webstore.iec.ch/en/publication/30583) standard, which establishes basic principles for lifecycle management of systems and components used for industrial-process measurement, control and automation.

Layers are IT layers, specifically, and allow for describing products and systems in terms of, e.g., physical components (*asset*), the data they operate with (*information*), and the functionality such systems provide (*functional*). It is easy to see how these layers correspond to the concept of digital twin.

The three-dimensional model of RAMI 4.0 establishes a common vocabulary for stakeholders and fosters a better understanding of the role and implementation of Digital Twins in an Industry 4.0 context. The key benefit of RAMI 4.0 is that it provides a unified framework across three orthogonal dimensions, within which processes, component hierarchies and IT concerns are properly aligned. By relying on standards in each dimension, the RAMI 4.0 framework actively supports independence from vendors, products, and technologies.

Subject matter experts and stakeholders typically interact with the RAMI 4.0 framework through domain-specific architectural views and viewpoints. These viewpoints slice the three dimensions in a way that is the most useful for their purposes. For example, a business process viewpoint might be interested in business, functional, informational layers in the station–work center–enterprise hierarchies, throughout the entire lifecycle. Or, an integration viewpoint might be interested in the communication–integration layers in the development–maintenance phases of the product. In general, views are viewpoints applied to a specific problem, model, or architecture. A more formal definition is given by the ISO/IEC/IEEE 42010:2022 (Software, systems and enterprise — Architecture description) standard (https://www.iso.org/standard/74393.html), which defines an architecture viewpoint as a set of conventions for the creation, interpretation and use of an architecture view; and an architecture view as the information part comprising portion of an architecture description. We remark here that in some communities and industries, "view specification" and "viewpoint" are synonymous.

The RAMI 4.0 framework supports digital twin technology in numerous ways. First, it defines a high-level, standard-ized model for Digital Twins to organize the key components of Digital Twins (physical assets, digital representations) along with their integration points (data and control) in a clear lifecycle model. Digital Twins are typically realized at the upper layers of RAMI 4.0. While the framework allows for detailed modeling and operationalization of Digital Twins, its richness also increases the complexity of implementing Digital Twins. To alleviate this issue, the Asset Administration Shell (AAS) has been developed by the German Electrical and Electronic Manufacturers' Association for the standardized digital representation of assets of the Reference Architecture Model Industrie 4.0 (RAMI4.0) (Oakes et al., 2023). The AAS provides a machine-readable, device-independent, hierarchical standard language (metamodel) for describing the properties of assets. As such, the AAS is a key enabler of digital twin technology within the RAMI 4.0 framework.

1.3.3 Open challenges in implementing Digital Twins for the smart grid

As technological capabilities increase, the industry is increasingly recognizing the disruptive potential of digital twin technologies. However, the development of Digital Twins for smart grids is hindered by several challenges due to the complex nature of the power infrastructure. Here, we review some of the most pressing challenges.

Data infrastructure and management: The foundation of any successful smart grid digital twin implementation rests on a robust technical infrastructure capable of handling massive data volumes from smart meters, supervisory control and data acquisition (SCADA) systems, and grid sensors. Current IT systems often prove inadequate for supporting the intensive real-time data analysis demands required for smart grid operations, creating a fundamental barrier to implementation (Jafari et al., 2023). Also, smart grid digital twin systems generate enormous amounts of data from distributed energy resources, smart meters, and grid monitoring devices, which must be processed and analyzed in real-time (Mchirgui et al., 2024). To address these infrastructure limitations, organizations can invest in high-performance computing solutions, including advanced graphics processing units (GPUs) and cloud-based services from providers like Amazon, Microsoft, and Google (Jafari et al., 2023). Edge computing emerges as a particularly valuable approach, reducing data transmission delays while increasing bandwidth efficiency (Favour, 2024). For effective data management, implementing distributed storage solutions becomes essential, e.g., using Hadoop Distributed File System (HDFS) and cloud-based platforms like Amazon S3, Microsoft Azure Blob Storage, or Google Cloud Storage (Yassin et al., 2023). These technologies provide the scalable and reliable storage and retrieval capabilities necessary for handling the vast datasets generated by digital twin applications.
In smart grids, an additional challenge lies in integrating heterogeneous legacy systems, such as legacy SCADA architectures, analog meters, and proprietary communication protocols with modern IoT-based data streams. Without proper middleware, data adapters, and semantic alignment, these legacy systems may create information silos or introduce latency and inconsistency into the digital twin. Therefore, interoperability standards and data integration frameworks—such as the Common Information Model (CIM) of the Distributed Management Task Force (DMTF) (https://www.dmtf.org/standards/cim) and the earlier mentioned OPC UA—become critical for achieving a unified, coherent view of the grid across both old and new infrastructure.

Connectivity and real-time processing: Smart grid Digital Twins depend on seamless connectivity between physical grid infrastructure and their virtual counterparts, yet this remains one of the most challenging aspects for utility operators. Despite advances in smart grid communication technologies, significant connectivity issues persist in substation networks, smart meter communications, and distribution automation systems, including software errors, power outages, and missing data that can disrupt critical grid monitoring capabilities (Das et al., 2024, Jafari et al., 2023). In addition, the latency in data collection from IoT devices such as smart meters and grid sensors limits the efficiency of Digital Twins (Almasan et al., 2022). Given that the quality and timeliness of IoT data streams fundamentally determine digital twin performance, smart grid operators should prioritize robust communication protocols and implement backup data collection methods to ensure continuous, reliable connections between physical and Digital Twins (Ebrahimi, 2019). To meet these demands, grid providers often adopt hybrid communication architectures that combine multiple technologies, e.g., fiber-optic links for critical backbone infrastructure, 5G and cellular networks for distributed assets, and radio-frequency or low-power wide-area networks for remote sensors in areas that are hard to reach. Such a hybrid approach improves coverage and redundancy, but at the same time, it introduces latency management challenges, as data from heterogeneous networks needs to be synchronized, time-stamped, and fused into a coherent stream for real-time decision-making. These challenges can be addressed, e.g., through intelligent edge processing and time-sensitive networking (TSN) (Zhang et al., 2024) which may help minimize delays.

Security: The extensive data exchange inherent in smart grid digital twin systems creates substantial security vul-nerabilities that pose risks to power infrastructure (Jafari et al., 2023). These systems collect and process sensitive grid operational data from multiple sources, including IoT devices, SCADA systems, and grid sensors, making them high-value targets for cyberattacks. The interconnected nature of smart grid networks amplifies security risks, as vulnerabilities in one component can potentially compromise the entire system (Casola, 2023). Ensuring data security requires implementing multiple layers of protection, starting with advanced encryption methods, where blockchain technology provides a particularly effective solution through its decentralized and tamper-resistant architecture (Ebrahimi, 2019). Organizations can also adopt privacy-enhancing technologies and systematic data governance practices to minimize exposure risks. These protective measures include strategic data minimization to limit the volume of sensitive information collected, robust anonymization techniques to obscure individual identities, and comprehensive identity protection protocols to prevent unauthorized access and potential data breaches (Botín-Sanabria et al., 2022). Ensuring security requires implementing multiple layers of protection through advanced encryption methods. Privacy-enhancing technologies and systematic data governance practices help further minimize exposure risks. Beyond simple data breaches, an exposed digital twin may influence the grid in adverse and even catastrophic ways. Successful cyberattacks may potentially propagate through the physical grid, causing service disruptions or equipment damage. Thus, a digital twin of a smart grid must be treated as a critical attack surface and protected with the same rigor as, e.g., SCADA and control systems.

1.4 Advanced topics in digital twin technology

There are some advanced topics in digital twin technology that merit further discussion in relation to smart grids.

1.4.1 Digital Twins and artificial intelligence

The success of digital twin technology hinges on the quality and fidelity of the models that capture essential properties of the physical twin. Complex physical systems, however, are not trivial to model manually. Smart grids are pertinent examples of such systems, as they require modeling of physics, power, production-consumption dynamics, environmental conditions, as well as fault modes and security concerns. AI can alleviate the complexity of modeling in Digital Twins in at least two ways.

Surrogate models: Surrogate models are simplified or reduced models that are used in place of more complex models. Surrogate models approximate the behavior of more complex, expensive, or computationally intensive models, but often, these approximations are sufficient, and the reduced computation time and costs compensate for the lost precision. AI models are prime examples of surrogate models in modern systems engineering. Instead of manually modeling a complex system (e.g., the physics of a smart grid), a surrogate AI model can be trained on operational data and used in a digital twin to replace resource-intensive numeric calculations with approximations the AI model has been trained for. Clearly, a surrogate AI model will provide reliable results only for the cases it has been trained for, while a general numeric model is more generally applicable. Thus, a surrogate AI model finds trade-offs between generality and speed, too. Some of the typical examples of AI models used in modern Digital Twins are the following.

Deep neural networks (DNNs) can learn complex, hierarchical patterns from data with a sufficiently high number of dimensions. DNNs are structured into multiple hidden layers, which progressively transform data into increasingly complex structures. DNNs are often used, e.g., for energy forecasting in a smart building, a DNNs might take environmental conditions (temperature, humidity) and occupancy information (hour of day) as input, detects correlations in the first layer (e.g., high temperature → more AC usage), combines these correlations in the second layer (e.g., high temperature outside + the building is crowded → maximum AC usage and energy consumption), and output the energy prediction in the output layer.

Physics-informed neural networks (PINNs) improve over DNNs by integrating physical laws (e.g., differential equations) directly into the learning process. DNNs may produce output that is not possible in the real-world. For example, a DNN trained to predict voltage in a smart grid might give plausible outputs, but violate power flow equations during unusual demand spikes. A PINN, on the other hand, is trained to respect the underlying physics, ensuring predictions remain realistic. PINNs achieve this by building laws of physics directly into the loss function during training. A by-product of these physics-based hints during training is the reduced data requirements of training PINNs.

Gaussian process regression (GPR) is a non-parametric Bayesian approach for interpolation in Gaussian processes. GPR is a probabilistic machine learning method that predicts functions based on observed data, such as wind flows over a turbine. GPR is uncertainty-aware, i.e., the variance of predictions is known. GPR also works well with limited datasets. These two properties are the strengths of GPR over DNNs.

The choice between a manually-built model and an AI surrogate depends on the objective of the digital twin. For safety-critical applications, such as real-time grid stability control, fault detection, or protection coordination, manually-built models are preferred because they are transparent and easier to verify and certify. In contrast, AI surrogate models act as black-boxes and are harder to verify and certify, and as a consequence, less suitable for safety-critical applications. However, AI surrogates are excellent for tasks where rapid scenario exploration is more valuable than provable correctness and safety, such as long-term capacity planning, demand forecasting, or policy evaluation. In such scenarios, AI surrogates offer an advantage by enabling faster-than-real-time simulation of a high number of scenarios. In some cases, a hybrid approach is used where manually-built models act as a reference model, and surrogates are deployed for fast approximation, with occasional recalibration to ensure validity as the system evolves.

Automated inference of digital twin components: AI and machine learning can be used in the development phase of Digital Twins, too. Developing simulators for Digital Twins is a pertinent example. Simulators are computer programs that encode an in-silico probabilistic mechanism and enact its changes over a sufficiently long period of time (Ross, 2012). As such, they are key enablers of Digital Twins. However, simulator engineering is hindered by systems complexity (Spiege et al., 2005). Reusing and composing existing simulator components has been the traditional approach to building complex simulators. However, such techniques are hindered by vertical challenges stemming from inappropriate abstraction mechanisms, horizontal challenges stemming from different points of view, and increased search friction due to the abundance of information (Page & Opper, 1999). To overcome the challenge of manual simulator construction, automated inference methods, e.g., based on reinforcement learning (RL), have been suggested. RL is a machine learning mechanism that works by trial-and-error. The RL agent observes the workings of its environment in reaction to the agent's action. By that, the agent learns beneficial actions in specific situations. In simulator engineering, an RL agent can learn simulator input-output mappings and encode them in its policy, which subsequently can be used as a surrogate model. Integrated into a digital twin, an RL agent can actively experiment with its surroundings with reasonable boundaries of safety. For example, an RL agent in a digital twin of a smart grid might change the behavior of the grid as it optimizes it, and learn whenever an optimization action has turned out to be beneficial. RL can augment a digital twin and implement a continuous learning and calibration mechanism to keep simulators up-to-date (David & Syriani, 2023).

AI simulation: AI simulation is the technique of applying AI and simulation technology to develop AI agents and the simulated environments in which they can be trained, tested, and sometimes deployed (Gartner Research, 2023). A fundamental challenge in modern AI development is the scarcity of high-quality training data, particularly in domains where real-world data collection is expensive, dangerous, or impractical (Farahani & Monsefi, 2023; Zhou et al., 2017). To overcome this issue, high-fidelity simulators can be used to generate synthetic training data for AI components.

Digital Twins are particularly well-suited for AI simulation due to their unique characteristics (Liu & David, 2025). First, the emergence of DTs elevated the quality, fidelity, faithfulness, and performance of simulators. These well-performing and high-fidelity simulators align well with the goals of AI simulation. Second, Digital Twins can support both passive observation and active experimentation with their physical counterparts, allowing for purposeful data collection that addresses specific AI training needs. The sophisticated simulation capabilities inherent in Digital Twins can generate diverse scenarios and edge cases that would be difficult or impossible to encounter in real-world operations. The primary advantage of AI simulation lies in the ability to generate large volumes of high-quality, labeled training data under controlled conditions, while preserving the fidelity required for effective sim-to-real transfer. Digital Twins support this process through either rapid, small-batch interactions or large-scale data generation. Moreover, digital twin-enabled AI simulation extends beyond simple data generation and labeling, enabling the creation of rich, interactive virtual training environments. Notably, AI simulation applications are predominantly built on genuine Digital Twins—those with bidirectional data and control flows. This underscores the importance of fully realized digital twin capabilities for effective AI training.

Digital twin-enabled AI simulation has proven to be particularly useful in complex domains such as wireless networks and robotics, where the combination of system complexity and sparse real-world data makes traditional AI training approaches challenging. For smart grid applications, this approach can enable safe training of control algorithms and optimization strategies without risking disruption to critical power infrastructure.

1.4.2 Digital Twins and sustainability

Sustainability is the ability to endure (Lago et al., 2015) and maintain the function of a system over an extended period of time (Hilty et al., 2006). Driven by the increasingly more pronounced preference of individuals and society at large for human-centered, ecological, and economically viable systems, sustainability is becoming a key characteristic and driver of value in modern engineered systems, motivating companies to incorporate its various forms in their value propositions and operations, fostering circularity (David et al., 2024) and value retention across value chains (Stucki et al., 2024). Sustainability of our technical and socio-technical systems has now become a strategic goal for governments as well, stimulating an array of international undertakings and innovation programs, e.g., the Industry 5.0 initiative of the European Commission. (https://researchand-innovation.ec.europa.eu/research-area/industrial-research-and-innovation/industry-50en) Digital Twins have opened new frontiers for systems sustainability and excel at implementing various sustainability ambitions. Digital Twins improve on traditional modeling capabilities by putting models into action (Bork et al., 2024). The central tenet of this book is that Digital Twins enable more sustainable power generation as key components of smart grids. In addition, Digital Twins can control physical systems for various sustainability goals, rendering the system itself sustainable.

Digital Twins contribute to sustainability in smart grids along environmental, economic, and social dimensions. From an *environmental* perspective, Digital Twins allow grid operators to dynamically and automatically optimize the energy mix by simulating different dispatch strategies and enacting the optimal one. By that, Digital Twins help maximize the use of renewable energy sources and minimize the overall carbon intensity of the grid. For instance, by predicting solar and wind generation profiles across the grid and aligning them with real-time demand, Digital Twins help ensure that clean energy is consumed when available, thus reducing dependency on fossil-based balancing mechanisms. From an *economic* perspective, Digital Twins enable the predictive maintenance of critical assets, such as transformers, storage units, and transmission lines. Through the continuous monitoring of equipment health and failure forecasting, Digital Twins reduce unplanned downtime and associated costs. Optimized asset utilization extends the operational lifetime of expensive infrastructure elements, too. Finally, from a *social* perspective, Digital Twins contribute to the resilience and reliability of the grid, which reduces outage times and stabilizes energy access for communities. Digital Twins enable the simulation of fault scenarios and stress conditions, which help operators prepare mitigation strategies in advance. In addition, Digital Twins can be used to evaluate and enact policies for equitable energy distribution, ensuring that vulnerable or remote populations benefit equally from sustainable energy.

Apart from implementing sustainability by Digital Twins, it is also important to implement Digital Twins in a sustainable way. In terms of a digital technology, such as Digital Twins, primary sustainability considerations include technical sustainability (e.g., evolution (David et al., 2023)), environmental sustainability (e.g., energy-efficiency (Bellis & Denil, 2022)), and social sustainability (e.g., situating the human in the loop (David & Bork, 2024)). Digital Twins are prime enablers of high-level legislative frameworks aiming at sustainable development. For example, twin transition (Shajari & David, 2025) advocates identifying joint digital and sustainability transformation pathways that mutually amplify each other. Digital Twins improve information flow across organizational silos and support decision-making about sustainability goals. Circular systems engineering (David et al., 2024) situates Digital Twins as a key enabler in the design, operation, and retirement of complex systems, and also puts forward that at the same time, Digital Twins must become sustainable in order to avoid defeating their very purpose (i.e., to improve the sustainability of complex systems). In response to such emerging needs, the intensity of research both in Digital

Twins for sustainability (Heithoff et al., 2023; Xu et al., 2022) and sustainability of Digital Twins (Bellis & Denil, 2022; David et al., 2023; Fur et al., 2023) has increased recently.

1.5 Concluding remarks: The future of digital twin-enabled smart grids

This chapter provided an introduction to the foundations of Digital Twins and made the case for employing them in smart grids. As engineered systems become more complex and autonomous, digital twin technology gains importance as the unified technological platform for design, testing, operation, and maintenance. Smart grids are prime examples of such complex systems, in which unique design and operation challenges arise from the combination of physical and software components. As high-fidelity in-silico replicas of physical components, Digital Twins provide safe and cost-efficient experimentation facilities in the design and verification phase of smart grids. In the operation phase of smart grids, Digital Twins enable automated load balancing of grids through real-time simulation and decision-making. These, and an array of similar benefits, position Digital Twins as crucial technological components in smart grids.

There are some key technological and methodological trends that are expected to shape the digital twin landscape in smart grids. First, the increasing adoption and deployment of edge computing infrastructure, along with the increasing computing power of edge nodes, is expected to allow Digital Twins to be hosted closer to physical assets. This, in turn, will enable localized and distributed decision-making – a critical capability for fast-changing grid conditions. Second, trends in Digital Twins point toward aggregated and integrated Digital Twins (Adesanya et al., 2025), which are expected to give rise to a full system-of-systems grid twin composed of individual Digital Twins of assets. System-of-system principles, such as subsystem independence and ad-hoc dynamic cooperation of constituent systems, align exceptionally well with the need to understand hard-to-model emergent behaviors of grids. Finally, cognitive Digital Twins, i.e., Digital Twins that learn to self-optimize, are expected to become key enablers of the next generation of smart grids. With accessible AI solutions and affordable and mature hardware technology to support it, cognitive Digital Twins will contribute to more predictable, stable, and efficient grids.

Acknowledgment

We acknowledge the support of the Natural Sciences and Engineering Research Council of Canada (NSERC), DGECR-2024-00293 (End-to-end Sustainable Systems Engineering).

References

Adesanya, F., Silva, K. C., Valdemar, V., Neto, G., and David, I. 2025. Systems of twinned systems: A systematic review. Technical Report. https://arxiv.org/abs/2505.19916

Aizpurua, J. I., Catterson, V. M., Stewart, B. G., McArthur, S. D. J, Lambert, B., Ampofo, B., Pereira, G., & Cross, J. G. (2017). Determining appropriate data analytics for transformer health monitoring. In *10th international topical meeting on nuclear plant instrumentation, control, and human-machine interface technologies, NPIC and HMIT 2017* (pp. 1–11). American Nuclear Society.

Allen, B. A. 2021. Digital Twins and living models at NASA. https://ntrs.nasa.gov/citations/20210023699.

Atalay, M., & Angin, P. (2020). A Digital Twins approach to smart grid security testing and standardization. In *IEEE international workshop on metrology for Industry 4.0 & IoT* (pp. 435–440). IEEE.

Bellis, S., & Denil, J. (2022). Challenges and possible approaches for sustainable digital twinning. In *Proceedings of the 25th international conference on model driven engineering languages and systems: Companion proceedings (Montreal, Quebec, Canada) (MODELS '22)* (pp. 643–648). Association for Computing Machinery. doi:10.1145/3550356.3561551.

Blair, G., Bencomo, N., & France, R. B. (2009). Models@ run.time. *Computer, 42*(10), 22–27. doi:10.1109/MC.2009.326.

Bork, D., David, I., Reinhartz-Berger, I., España, S., Guizzardi, G., & Proper, H. (2024). The role of modeling in the analysis and the design of sustainable systems. *Communications of the Association for Information Systems*. doi:10.17705/1CAIS.05434.

Botín-Sanabria, D. M., Mihaita, A.-S., Peimbert-García, R. E., Ramírez-Moreno, M. A., Ramírez-Mendoza, R. A., & Lozoya-Santos, J. J. (2022). Digital twin technology challenges and applications: A comprehensive review. *Remote Sensing, 14*(6), 1335.

Braik, A. M., & Koliou, M. (2023). A novel digital twin framework of electric power infrastructure systems subjected to hurricanes. *International Journal of Disaster Risk Reduction, 97*(2023), 104020.

Casola, L. (2023). *Opportunities and challenges for Digital Twins in engineering. Proceedings of a Workshop–in Brief.* Washington, D.C., US: National Academies Press.

Coppolino, L., Nardone, R., Petruolo, A., & Romano, L. (2023). Building cyber-resilient smart grids with Digital Twins and data spaces. *Applied Sciences, 13*(24), 13060.

Dalibor, M., Jansen, N., Rumpe, B., Schmalzing, D., Wachtmeister, L., Wimmer, M., & Wortmann, A. (2022). A cross-domain systematic mapping study on software engineering for Digital Twins. *Journal of Systems and Software, 193*(2022), 111361. doi:10.1016/j.jss.2022.111361.

Danilczyk, W., Sun, Y., & He, H. (2019). *ANGEL: An intelligent digital twin framework for microgrid security. In 2019 North American Power Symposium (NAPS)* (pp. 1–6). Wichita, KS, USA: IEEE. https://doi.org/10.1109/NAPS46351.2019.9000371.

Das, O., Hamza Zafar, M., Sanfilippo, F., Rudra, S., & Lal Kolhe, M. (2024). Advancements in digital twin technology and machine learning for energy systems: A comprehensive review of applications in smart grids, renewable energy, and electric vehicle optimisation. *Energy Conversion and Management: X, 24,* 100715.

David, I., Archambault, P., Wolak, Q., Vu, C. V., Lalonde, T., Riaz, K., Syriani, E., & Sahraoui, H. (2023). Digital Twins for cyber-biophysical systems: Challenges and lessons learned. In *ACM/IEEE 26th international conference on model driven engineering languages and systems (MODELS)* (pp. 1–12). IEEE. doi:10.1109/MODELS58315.2023.00014.

David, I., Archambault, P., Wolak, Q., Vu, C. V., Lalonde, T., Riaz, K., … Sahraoui, H. (2023). *Towards a taxonomy of digital twin evolution for technical sustainability. 2023 ACM/IEEE International Conference on Model Driven Engineering Languages and Systems Companion (MODELS-C)* (pp. 934–938). Västerås, Sweden: IEEE. https://doi.org/10.1109/MODELS-C59198.2023.00147.

David, I., & Bork, D. (2024). Infonomics of autonomous Digital Twins. *Advanced information systems engineering* (pp. 563–578). Springer. doi:10.1007/978-3-031-61057-8_33.

David, I., Bork, D., & Kappel, G. (2024). Circular systems engineering. *Software and Systems Modeling, 23*(2), 269–283. doi:10.1007/s10270-024-01154-4.

David, I., Shao, G., Gomes, C., Tilbury, D., & Zarkout, B. (2025). Interoperability of Digital Twins: Challenges, success factors, and future research directions. *Leveraging applications of formal methods, verification, and validation. Application areas* (pp. 27–46). Springer. doi:10.1007/978-3-031-75390-9_3.

David, I., & Syriani, E. (2023). *Automated inference of simulators in Digital Twins*. CRC Press. doi:10.1201/9781003425724-11.

Ebrahimi, A. (2019). Challenges of developing a digital twin model of renewable energy generators. In *IEEE 28th international symposium on industrial electronics (ISIE)* (pp. 1059–1066). IEEE.

El Mokhtari, K., Panushev, I., & McArthur, J. J. (2022). Development of a cognitive digital twin for building management and operations. *Frontiers in Built Environment, 8*. doi:10.3389/fbuil.2022.856873.

Fang, X., Misra, S., Xue, G., & Yang, D. (2011). Smart grid–The new and improved power grid: A survey. *IEEE Communications Surveys & Tutorials, 14*(4), 944–980.

Farahani, B., & Monsefi, A. K. (2023). Smart and collaborative industrial IoT: A federated learning and data space approach. *Digital Communications and Networks, 9*(2), 436–447.

Favour, A. E. (2024). Evaluating the integration of edge computing and serverless architectures for enhancing scalability and sustainability in Cloud-based Big Data Management. *Journal of Engineering Research and Reports, 26*(7), 347–365.

Ferko, E., Bucaioni, A., Pelliccione, P., & Behnam, M. (2023). *Standardisation in digital twin architectures in manufacturing* (pp. 70–81). L'Aquila, Italy: IEEE. https://doi.org/10.1109/ICSA56044.2023.00015.

Fur, S., Heithoff, M., Michael, J., Netz, L., Pfeiffer, J., Rumpe, B., & Wortmann, A. (2023). *Sustainable digital twin engineering for the internet of production* (pp. 101–121). Springer.

Almasan, P., Ferriol-Galmés, M., Paillisse, J., Suárez-Varela, J., Perino, D., López, D., … Barlet-Ros, P. (2022). *Digital twin network: Opportunities and challenges*. arXiv preprint.

Gartner Research. 2023. Innovation insight: AI simulation. https://www.gartner.com/en/documents/4037399.

Grieves, M., & Vickers, J. (2017). *Digital twin: Mitigating unpredictable, undesirable emergent behavior in complex systems* (pp. 85 113). Springer. doi:10.1007/978-3-319-38756-7_4.

Heithoff, M., Hellwig, A., Michael, J., & Rumpe, B. (2023). Digital Twins for sustainable software systems. In *IEEE/ACM 7th international workshop on green and sustainable software (GREENS)* (pp. 19 23). doi:10.1109/CREENS59328.2023.00010.

Hilty, L. M., Arnfalk, P., Erdmann, L., Goodman, J., Lehmann, M., & Wäger, P. A. (2006). The relevance of information and communication technologies for environmental sustainability – A prospective simulation study. *Environmental Modelling & Software, 21*(11), 1618–1629. doi:10.1016/j.envsoft.2006.05.007.

Irfan, M., Niaz, A., Habib, M. Q., Shoukat, M. U., Hussain Atta, S., & Ali, A. (2023). Digital twin concept, method and technical framework for smart meters. *European Journal of Theoretical and Applied Sciences, 1*(3), 105–117.

Jafari, M., Kavousi-Fard, A., Chen, T., & Karimi, M. (2023). A review on digital twin technology in smart grid, transportation system and smart city: Challenges and future. *IEEE Access, 11*(2023), 17471–17484.

Kang, M.-S., Lee, D.-H., Bajestani, M. S., Kim, D. B., & Do Noh, S. (2024). Edge computing-based digital twin framework based on ISO 23247 for enhancing data processing capabilities. *Machines, 13*(1), 19.

Khalyasmaa, A. I., Eroshenko, S. A., Shatunova, D. V., Larionova, A. A., & Egorov, A. O. (2020). Digital twin technology as an instrument for increasing electrical equipment reliability. *IOP Conference Series: Materials Science and Engineering*: 836. Bristol, UK: IOP Publishing https://doi.org/10.1088/1757-899X/836/1/012005.

Khan, S. A., Rehman, H. Z. U., Waqar, A., Khan, Z. H., Hussain, M., & Masud, U. (2023). *Digital twin for advanced automation of future smart grid. In 2023 1st International Conference on Advanced Innovations in Smart Cities (ICAISC)* (pp. 1–6). Jeddah, Saudi Arabia: IEEE. https://doi.org/10.1109/ICAISC56366.2023.10085428.

Kraus, S., Jones, P., Kailer, N., Weinmann, A., Chaparro-Banegas, N., & Roig-Tierno, N. (2021). Digital transformation: An overview of the current state of the art of research. *Sage, 11*(3), 21582440211047576.

Kritzinger, W., Karner, M., Traar, G., Henjes, J., & Sihn, W. (2018). Digital Twin in manufacturing: A categorical literature review and classification. *IFAC-PapersOnLine, 51*(11), 1016–1022.

Kuber, S., Sharma, M., Bonetti, A., Harispuru, C., & Soroush, A. (2022). *Virtual testing of protection systems using digital twin technology* (pp. 1–8). College Station, TX, USA: IEEE. https://doi.org/10.1109/CPRE55809.2022.9776572.

Lago, P., Koçak, S. A., Crnkovic, I., & Penzenstadler, B. (2015). Framing sustainability as a property of software quality. *Communication of the ACM, 58*(10), 70–78. doi:10.1145/2714560.

Lazazzara, A., Reina, R., & Za, S. (2024). Digital transformation and sustainability goals: Advancing the "Twin Transition". *Toward digital and sustainable organisations* (pp. 1–7). Springer.

Lei, L., Shen, G., Zhang, L., & Li, Z. (2021). Toward intelligent cooperation of UAV swarms: When machine learning meets digital twin. *IEEE Network, 35*(1), 386–392. doi:10.1109/MNET.011.2000388.

Liu, X., & David, I. (2025). AI simulation by Digital Twins: Systematic survey, reference framework, and mapping to a standardized architecture. *Software and Systems Modeling*, (2025). doi:10.1007/s10270-025-01306-0.

Maes, P. (1987). Concepts and experiments in computational reflection. In Conference proceedings on object-oriented programming systems, languages and applications *(Orlando, Florida, USA)* (OOPSLA '87) (pp. 147–155). ACM. doi:10.1145/38765.38821.

Mansour, D.-E. A., Numair, M., Zalhaf, A. S., Ramadan, R., Darwish, M. M. F, Huang, Q., Hussien, M. G., & Abdel-Rahim, O. (2023). Applications of IoT and digital twin in electrical power systems: A comprehensive survey. *IET Generation, Transmission & Distribution, 17*(20), 4457–4479.

Margaria, T., & Schieweck, A. (2019). The digital thread in Industry 4. 0. In W Ahrendt, S Lizeth, & T Tarifa (Eds.), *Integrated formal methods* (pp. 3–24). Springer.

Mchirgui, N., Quadar, N., Kraiem, H., & Lakhssassi, A. (2024). The applications and challenges of digital twin technology in smart grids: A comprehensive review. *Applied Sciences, 14*(23), 10933.

Michael, J., David, I., & Bork, D. (2024). Digital twin evolution for sustainable smart ecosystems. In *Proceedings of the ACM/IEEE 27th international conference on model driven engineering languages and systems (Linz, Austria) (MODELS Companion '24)* (pp. 1061–1065). ACM. doi:10.1145/3652620.3688343.

Milton, M., De La O, C., Ginn, H. L., & Benigni, A. (2020). Controller-embeddable probabilistic real-time Digital Twins for power electronic converter diagnostics. *IEEE Transactions on Power Electronics, 35*(9), 9850–9864.

Minsky, M. (1965), *Matter, mind and models* (Vol. 1, pp. 45–49). Washington, DC, USA: Spartan Books.

Muctadir, H. M., Manrique Negrin, D. A., Gunasekaran, R., Cleophas, L., van den Brand, M., & Haverkort, B. R. (2024). Current trends in digital twin development, maintenance, and operation: An interview study. *Software and System Modelling, 23*. doi:10.1007/s10270-024-01167-z.

Nguyen, V. H., Tran, Q. T., Besanger, Y., Jung, M., & Nguyen, T. L. (2022). Digital twin integrated power-hardware-in-the-loop for the assessment of distributed renewable energy resources. *Electrical Engineering, 104*(2), 377–388. doi:10.1007/s00202-021-01246-0.

Oakes, B. J., Parsai, A., Meyers, B., David, I., Mierlo, S. V., Demeyer, S., Denil, J., Meulenaere, P. D., & Vangheluwe, H. (2023). A digital twin description framework and its mapping to Asset Administration Shell. In *Model-driven engineering and software development* (pp. 1–24). Springer. doi:10.1007/978-3-031-38821-7_1.

Page, E. H., & Opper, J. M. (1999). *Observations on the complexity of composable simulation* (pp. 553–560). Phoenix, AZ, USA: ACM. https://doi.org/10.1145/324138.324433.

Pan, H., Dou, Z., Cai, Y., Li, W., Lei, X., & Han, D. (2020). *Digital twin and its application in power system* (pp. 21–26). Shanghai, China: IEEE. https://doi.org/10.1109/ICPRE51194.2020.9233278.

Paredis, R., & Vangheluwe, H. (2022). Toward a digital Z framework based on a family of architectures and a virtual knowledge graph. *Proceedings of the 25th international conference on model driven engineering languages and systems (MODELS-C): Companion proceedings* (pp. 491–496). ACM.

Rasheed, A., San, O., & Kvamsdal, T. (2020). Digital twin: Values, challenges and enablers from a modeling perspective. *IEEE Access, 8*(2020), 21980–22012.

Ross, S. M. (2012). *Simulation (5. ed)*. Academic Press.

Schweichhart, K. 2016. Reference architectural model Industrie 4.0 (RAMI 4.0) – An introduction. https://ec.europa.eu/futurium/en/system/files/ged/a2-schweichhart-reference_architectural_model_industrie_4.0_rami_4.0.pdf.

Sarwar, M., & Asad, B. (2016). *A review on future power systems; technologies and research for smart grids* (pp. 1–6). Islamabad, Pakistan: IEEE. https://doi.org/10.1109/ICET.2016.7813247.

Shafto, M., Conroy, M., Doyle, R., Glaessgen, E., Kemp, C., LeMoigne, J., & Wang, L. (2010). Draft modeling, simulation, information technology & processing roadmap. *Technology Area, 11*(2010), 1–32.

Shajari, B., & David, I. (2025). *Bridging the silos of digitalization and sustainability by twin transition: A multivocal literature review* (pp. 23–34). Dublin, Ireland: IEEE. https://doi.org/10.1109/ICT4S68164.2025.00012.

Shao, G. (2021). *Use case scenarios for digital twin implementation based on ISO 23247 (NIST Advanced Manufacturing Series 400-2)*. Gaithersburg, MD, USA: National Institute of Standards and Technology https://doi.org/10.6028/NIST.AMS.400-2.

Shao, G. (2024). Manufacturing digital twin standards. In *Proceedings of the ACM/IEEE 27th international conference on model driven engineering languages and systems companion* (pp. 370–377). ACM. doi:10.1145/3652620.3688250.

Shao, G., Frechette, S., & Srinivasan, V. (2023). An analysis of the new ISO 23247 series of standards on digital twin framework for manufacturing. In *International manufacturing science and engineering conference: 87240*. American Society of Mechanical Engineers V002T07A001.

Shao, G., Kibira, D., & Frechette, S. (2024). Digital Twins for advanced manufacturing: The standardized approach. *Digital Twins, simulation, and the metaverse: Driving efficiency and effectiveness in the physical world through simulation in the virtual worlds* (pp. 145–169). Springer.

Sleiti, A. K., Kapat, J. S., & Vesely, L. (2022). Digital twin in energy industry: Proposed robust digital twin for power plant and other complex capital-intensive large engineering systems. *Energy Reports, 8*(2022), 3704–3726.

Song, E. Y., Nguyen, C., & Gopstein, A. (2019). *Review of smart grid standards for testing and certification landscape analysis (NIST Technical Note 2042)*. Gaithersburg, MD, USA: National Institute of Standards and Technology https://doi.org/10.6028/NIST.TN.2042.

Spiegel, M., Reynolds, P. F., Jr., & Brogan, D. C. (2005). *A case study of model context for simulation composability and reusability* (pp. 437–444). Orlando, FL, USA: IEEE Computer Society. https://doi.org/10.5555/1162708.1162784.

Stachowiak, H. (1973). *Allgemeine modelltheorie. Springer*.

Steindl, G., & Kastner, W. (2021). Semantic microservice framework for Digital Twins. *Applied Sciences, 11*(12), 5633.

Stucki, M., Meierhofer, J., Gal, B., Gallina, V., & Eisl, S. (2024). Data driven value creation in industrial services including remanufacturing. *Procedia Computer Science, 232*(2024), 2240–2248. doi:10.1016/j.procs.2024.02.043.

Tao, F., Zhang, H., Liu, A., & Nee, A. Y. C. (2019). Digital Twin in industry: State-of-the-art. *IEEE Transactions on Industrial Informatics, 15*(4), 2405–2415. doi:10.1109/TII.2018.2873186.

Touhid, M. T. B., Marne, M., Oskroba, T., Mirahmadi, S. A., Zhu, E., Mehrabian, A., Defersha, F., & Yang, S. (2023). Building a cloud-based digital twin for remote monitoring and control of a robotic assembly system. *The International Journal of Advanced Manufacturing Technology, 129*(9), 4045–4057. doi:10.1007/s00170-023-12611-7.

Tzanis, N., Andriopoulos, N., Magklaras, A., Mylonas, E., Birbas, M., & Birbas, A. (2020), *A hybrid cyber physical digital twin approach for smart grid fault prediction* (Vol. 1, pp. 393–397). Tampere, Finland: IEEE. https://doi.org/10.1109/ICPS48405.2020.9274723.

Vial, G. (2019). Understanding digital transformation: A review and a research agenda. *The Journal of Strategic Information Systems, 28*(2), 118–144.

Wang, M., Wang, C., Hnydiuk-Stefan, A., Feng, S., Atilla, I., & Li, Z. (2021). Recent progress on reliability analysis of offshore wind turbine support structures considering digital twin solutions. *Ocean Engineering, 232*(2021), 109168.

Yassin, M. A. M., Shrestha, A., Salahel, S., & Ahmed, R. (2023). Digital twin in power system research and development: Principle, scope, and challenges. *Energy Reviews, 2*, 100039.

Zeigler, B. P., Muzy, A., & Kofman, E. (2018). *Theory of modeling and simulation: Discrete event & iterative system computational foundations*. Academic press.

Zhang, T., Wang, G., Xue, C., Wang, J., Nixon, M., & Han, S. (2024). Time-sensitive networking (TSN) for industrial automation: Current advances and future directions. *ACM Computing Surveys, 57*(2), 38. doi:10.1145/3695248.

Xu, L., de Vrieze, P., Lu, X., & Wang, W. (2022). Digital twins approach for sustainable industry. In Advanced Information Systems Engineering Workshops: CAiSE 2022 International Workshops, Leuven, Belgium, June 6–10, 2022, Proceedings (pp. 126–134). Cham, Switzerland: Springer International Publishing. https://doi.org/10.1007/978-3-031-07478-3_11

Zheng, T., Liu, M., Puthal, D., Yi, P., Wu, Y., and He, X. 2022. Smart grid: Cyber attacks, critical defense approaches, and digital twin. arXiv preprint.

Zhou, L., Pan, S., Wang, J., & Vasilakos, A. V. (2017). Machine learning on big data: Opportunities and challenges. *Neurocomputing, 237*(2017), 350–361. doi:10.1016/j.neucom.2017.01.026.

Zhou, M., Yan, J., & Zhou, X. (2020). Real-time online analysis of power grid. *CSEE Journal of Power and Energy Systems, 6*(1), 236–238.

CHAPTER

2

Introduction to Digital Twin: Applications

Eduardo Gomez-Luna, Joan S. Moreano-Gaviria

High-Voltage Research Group – GRALTA, School of Electrical and Electronic Engineering, Faculty of Engineering, University of Valle, Santiago de Cali, Colombia

2.1 Definition of Digital Twins

Digital Twins (DTs) are exact virtual models of physical objects, processes, or systems, such as generators, motors, transformers, and electrical networks, that enable real-time analysis, monitoring, and optimization of the performance of these physical assets. These models can run multiple simulations of any scenario to study how the physical object will behave under certain conditions. This is highly useful for predicting future failures and making decisions before they occur, such as scheduling preventive maintenance or modifying input or output parameters (voltage, current, torque, etc.) without directly intervening in the real equipment. This is possible thanks to bidirectional data transmission between the digital model and the real model, integrating Internet of Things (IoT) sensor technologies, historical data analysis, advanced algorithms, and artificial intelligence (AI), technologies that are part of Industry 4.0. Unlike a simple simulation, a DT is based on three fundamental pillars (Soares et al., 2023) shown in Fig. 2.1:

- **Real Space**, where the real-world devices, processes, or systems under study are located and generate data.
- **Virtual Space**, which replicates the physical system in a digital environment, based on real-time data and computational models.
- **Bidirectional Communication**, which enables data synchronization between the physical entity and the virtual model.

While traditional simulators typically focus on specific scenarios or somewhat conceptual conditions, DTs provide a more comprehensive and dynamic representation. Due to real-time data updates, they enable precise monitoring of the current state of the physical object, facilitating more effective decision-making (Latorre et al., 2024).

Additionally, DTs possess self-regulation and self-diagnosis capabilities (Soares et al., 2023):

- **Self-regulation:** Ensures that the changes made in the DT comply with the physical limitations of the real object, thus optimizing its performance. For example, a DT of a wind turbine would prevent a simulation from increasing the rotor speed beyond the mechanical limits of the physical alloy that models it, thereby avoiding structural damage and extending the useful life of the equipment.
- **Self-diagnosis:** Evaluates the state of the system based on current and historical data, identifying operational problems. For example, in the case of an industrial motor, the DT can detect anomalous patterns in vibrations or temperatures that indicate premature bearing wear, alerting to the need for preventive maintenance before a failure occurs.

DTs are composed of several elements that enable the behavior of a physical system to be reflected in a virtual environment. These key components are shown in Table 2.1.

To provide a comprehensive understanding of DTs in the context of electrical power systems, this article is structured around four key aspects:

- **Fundamentals and architecture of DTs (physical/virtual/bidirectional):** DTs consist of three interconnected components: the actual physical equipment (such as a transformer or generator installed in the field), its digital

Digital Twin Technology and Smart Grid.
DOI: https://doi.org/10.1016/B978-0-443-36664-2.00025-3

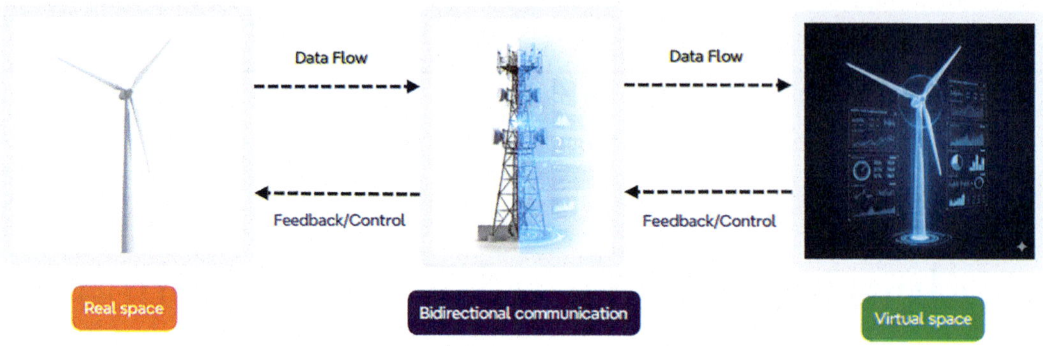

FIGURE 2.1 Fundamental pillars of digital twins. *Source: Own creation using* xmind.app.

TABLE 2.1 Summary of the key components of Digital Twins.

Component	Purpose	Key technologies	Example Tools
Data acquisition	Collect real-time information from physical assets, including voltages, currents, temperature, vibrations, and environmental conditions.	Internet of Things (IoT) devices, sensors, actuators, and communication protocols.	SCADA systems, PLCs, smart meters, RTUs.
Data processing and modeling	Clean, integrate, and process the collected data to create accurate digital representations of physical system behavior.	High-performance computers. Real-time simulators.	MATLAB/SIMULINK, Python (Pandas, NumPy), offline, real-time, and online software, and artificial intelligence.
Simulation and analysis	Execute scenario tests and predict system behavior under various operating conditions.	Computational clusters, offline, real-time and online simulation engines.	MATLAB/SIMULINK, Python (Pandas, NumPy), offline, real-time and online software, artificial intelligence.
Visualization and interaction	Present analysis results through interactive interfaces to facilitate online and real-time decision-making.	Business intelligence tools, AI-enhanced analytics, and dashboard platforms.	Power BI, Tableau, Grafana, Apache Superset.

copy (a virtual model that reproduces its behavior), and the Bidirectional Communication between the two. This structure allows physical systems to be monitored and controlled in real time through their digital replicas. This is very useful because operators can make adjustments and tests on the virtual model without having to touch the actual equipment, reducing risks and operating costs.

- **Basic functionalities and modeling approaches:** DTs offer several important functions for managing electrical systems efficiently: they can simulate real-time operation, predict potential failures before they occur, optimize performance, and test different operating scenarios. There are different approaches to creating these accurate digital models. Some are based on the laws of physics and mathematical equations that describe how equipment works, others use AI and learn from historical data analysis, and there are also methods that combine both techniques to obtain better results.
- **Cross-sector applications:** Although DT technology was initially developed for the manufacturing and aerospace industries, in recent years it has been applied with great success in the energy sector. Today, DT is used to model and optimize all types of electrical equipment and systems: from individual generators and transformers to entire distribution networks. This allows electric utilities to improve service reliability, increase operational efficiency, and make their systems more resilient to failures. It also makes it easier to schedule preventive maintenance and make informed decisions before problems occur, without the need to intervene directly in the actual equipment.
- **Integration into smart grids and current research trends:** The incorporation of DT into smart grids represents an important step toward the modernization of electrical systems, in line with the digital transformation that characterizes Industry 4.0. Currently, research in this field focuses on several objectives: improving the ability to process large amounts of data in real time, developing more accurate AI algorithms to predict failures and optimize operations, creating standards that allow different systems and platforms to work together seamlessly,

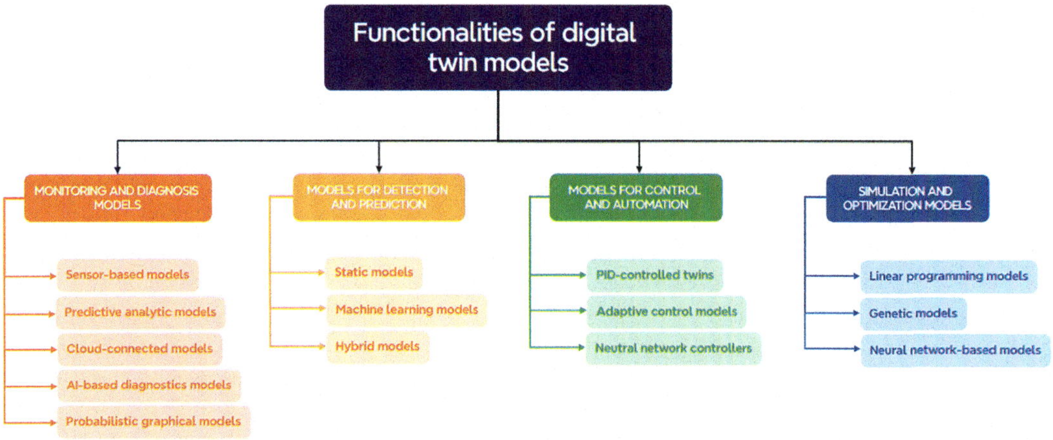

FIGURE 2.2 Summary of digital twin functionalities. The functionalities of the digital twin models are shown, divided into the four most important sections. Bidirectional coupling between real and virtual entities underpins all DT intelligence. *Source: Own creation using* xmind.app.

and exploring new applications to better integrate renewable energies and manage electricity demand more flexibly and efficiently.

2.2 Importance of Digital Twins: Why and what are DTs used for?

DTs have become a key technology in the digital transformation of Industry 4.0, thanks to their ability to recreate, monitor, analyze, predict, and optimize physical systems in real time. To understand their role in digital transformation, Fig. 2.2 shows a summary of the key functionalities of DTs, each of which is detailed further.

1. **Monitoring and diagnostic models:** These models enable real-time observation and analysis of physical assets to detect behavioral changes and anomalies. By continuously collecting data through sensor networks, they facilitate preventive and corrective maintenance planning, extending equipment life and improving system performance. The taxonomy includes six types of submodels (Rayhana et al., 2024):
   a. **Sensor-based models:** Use distributed physical sensor networks for continuous real-time data collection and transmission.
   b. **Predictive analytics models:** Anticipate future system behavior and identify potential failures using historical and real-time data.
   c. **Cloud-connected models:** Leverage cloud infrastructure for distributed data processing, storage, and analysis without location constraints.
   d. **System dynamics models:** Represent complex systems through feedback loops and stock flow structures to analyze the root causes of system behavior.
   e. **AI-based diagnostic models:** Use neural networks, decision trees, or deep learning to identify patterns and diagnose problems efficiently.
   f. **Probabilistic graphical models:** These use Bayesian networks to estimate the probabilities of specific situations based on the available information.
2. **Models for detection and prediction:** These models have the critical function of identifying deviations between current physical states and virtual simulations, enabling early detection of anomalies and prediction of future system behavior. Accuracy at this forecasting stage is what enables the system's proactive response. By anticipating problems before they manifest, detection and prediction models provide the trigger signal for control models. There are three main approaches to this function (Rayhana et al., 2024):

   a. **Static models:** Use predefined thresholds and deviation analysis; their effectiveness depends on accurate calibration.
   b. **Machine learning (ML) models:** Analyze historical data to detect hidden patterns and dynamically predict anomalies in complex environments.
   c. **Hybrid models:** Combine traditional physics-based methods with AI techniques, such as deep learning, to improve robustness and accuracy.

TABLE 2.2 Summary of DT functionalities: Tools and results.

Function	Representative tools	Main Result
Monitoring and diagnosis	SCADA systems, Power BI, Tableau, Azure.	Online visualization, real-time, and anomaly detection.
Detection and prediction	TensorFlow, PyTorch, statistical analysis, and artificial intelligence tools.	Early anomaly alerts and failure prediction.
Control and automation	PID controllers, MATLAB/SIMULINK, adaptive controller systems, neural network controllers.	System stabilization and autonomous operation.
Simulation and Optimization	Genetic algorithms, LP solutions (such as CPLEX, Gurobi), neural networks, and offline, real-time and online simulation software.	Maximization of efficiency and reliability, minimization of costs and risks.

DT, Digital Twin.

3. **Models for control and automation:** These models leverage AI and real-time data to maximize efficiency, profitability, and performance through iterative simulations. They optimize resource allocation, reduce costs, and improve overall efficiency. As shown in Fig. 2.2, three main approaches are used (Rayhana et al., 2024):

   a. **PID-controlled twins:** These use Proportional-Integral-Derivative (PID) controllers to regulate and adjust system output in real time. These models are used in applications such as temperature control or driving systems, where the system response needs to be quickly modified based on changing conditions. The PID controller adjusts the system output based on three terms: proportional (to correct the current error), integral (to correct accumulated errors), and derivative (to anticipate future errors). (Rayhana et al., 2024)
   b. **Adaptive control models:** These are designed to dynamically adjust to any changes in the system, allowing the model to adapt and reconfigure in real time based on new conditions or disturbances. These models are known for their ability to learn and modify their behavior without constant manual intervention, making them ideal for complex and dynamic environments. (Rayhana et al., 2024)
   c. **Neural network controllers:** These are control models based on deep learning that use neural networks to manage complex and nonlinear systems. These controllers are designed to optimize system responses and facilitate advanced decision-making processes, especially in environments where system behavior is difficult to model or highly dynamic. (Rayhana et al., 2024).

4. **Simulation and optimization models:** These models leverage AI and real-time data to maximize efficiency, profitability, and performance through iterative simulations. They optimize resource allocation, reduce costs, and improve overall efficiency. As shown in Fig. 2.2, three main approaches are used (Rayhana et al., 2024):

   a. **Linear programming models:** Solve optimization problems with linear relationships to maximize or minimize objective functions subject to constraints.
   b. **Genetic models:** Apply evolution-inspired algorithms to explore large solution spaces and adapt to different systems.
   c. **Neural network-based models:** Learn complex patterns and perform adaptive optimization in environments with large data volumes and nonlinear relationships.

This taxonomy (Fig. 2.2 and Table 2.2) provides a comprehensive framework for understanding how different DT models are applied in electrical power systems, facilitating the selection of appropriate approaches based on specific operational requirements.

2.3 Applications of Digital Twins: Where are DTs used? Fields of application of DTs

2.3.1 Manufacturing

DTs enable the optimization of processes, cost reduction, and improvement of product quality in various industries, including food processing, automobile manufacturing, and electronic device production, among others. Through detailed predictive analytics and simulations, DTs enhance efficiency and performance.

In this sector, DTs collect sensor data (such as temperature, speed, and vibration) from machines to analyze their real-time behavior. By integrating this data into a virtual model, continuous monitoring of machines and manufacturing processes is achieved, allowing for—failure prediction and downtime reduction, optimization of production processes, faster adaptation to demand changes, reduction of raw material waste, lower energy consumption, and improved final product quality.

2.3.2 Health

In this sector, DTs enable more personalized monitoring and analysis of patients by integrating their medical history, real-time health parameters (such as pulse, breathing, blood glucose levels, etc.), and genetic information. With this data, DTs, through AI and ML algorithms, help predict diseases in patients (tendency to develop diabetes, bone diseases, cancer, among others) and reduce risks by allowing the early initiation of the corresponding treatments.

For example, tools such as Philips' HeartModel perform the simulation and analysis of cardiac function through mathematical models of the heart, while HeartNavigator facilitates visualization and interaction for surgical planning through interactive 3D interfaces, demonstrating a complete DT workflow in medicine that integrates from the acquisition of clinical data to faster and more accurate medical decision-making (Soares et al., 2023).

The article titled "Digital twin technology and applications in the medical field: Connecting the virtual with precision medicine" analyzes the impact of digital twin technology in the medical sector, specifically in precision medicine. To this end, data from patients' organs are collected in real time through clinical sensors, processed, and modeled to create a virtual model of the organ. Subsequently, scenario simulations and tests are carried out to evaluate the effectiveness and possible risks of different medical treatments, allowing for more precise and reliable results (Chen & Zhu, 2024).

2.3.3 Infrastructure and smart cities

DTs used in infrastructures such as large buildings or bridges collect data from vibration sensors, structural stresses, and weather conditions, among others, to monitor structural integrity, predict and prevent failures, simulate hypothetical seismic scenarios, and optimize maintenance, while reducing repair costs.

Beyond individual structures, urban-scale DTs integrate data from traffic sensors, power grids, and emergency services to simulate the impact of new policies, urban planning decisions, or large-scale events. This allows the planning of new buildings by analyzing their impact on traffic, public safety, emergencies, the environment, and energy management. Likewise, they enable the simulation of situations under specific economic, social, environmental, and technological conditions, influencing the development and future of anything from a city to an entire country.

For example, in Singapore, DTs have modernized the monitoring of urban water systems through the Smart Water Grid model, which optimizes water distribution and quality in real time (Soares et al., 2023).

2.3.4 Energy and sustainability

DTs in the electric power sector analyze everything from individual components such as transformers, generators, or wind turbines to large-scale transmission and distribution networks. They enable:

- Monitoring of equipment status using network analyzers or sensors that measure temperature, vibrations, current, and voltage.
- Planning preventive and corrective maintenance.
- Failure simulation (e.g., service interruptions, lightning strikes, line-to-line faults, and line-to-ground faults).
- Adjustment in power generation and distribution, responding quickly to demand fluctuations.
- Improvement of the energy efficiency of the grid.

In cities like Singapore, Amsterdam, and Barcelona, DTs have been integrated for efficient energy consumption management, achieving (Soares et al., 2023):

- Reduction of energy waste.
- Optimization of traffic and infrastructure maintenance.
- Development of more sustainable urban ecosystems.

TABLE 2.3 Cross-sector applications of DT technology.

Industry	Main objective	Representative example	DTs impacts	References
Manufacturing	Process optimization and improvement of product and service quality.	Monitoring of industrial production lines.	Failure prediction, downtime reduction, waste minimization, and improvement of energy efficiency.	Rayhana et al. (2024)
Health	Personalized patient monitoring and treatment planning.	Philips HeartModel for cardiac simulation and HeartNavigator for surgical planning.	Early disease prediction, personalized treatment and improvement in surgical precision and results.	Soares et al. (2023), Chen and Zhu (2024)
Infrastructure and smart cities	Structural integrity monitoring and urban planning optimization.	Singapore Smart Water Grid (SWG) for real-time water distribution management.	Preventive failure detection, seismic scenario simulation, traffic optimization, and maintenance cost reduction.	Soares et al. (2023)
Energy and sustainability	Electrical grid optimization, renewable energy connection, and integration of new technologies.	Monitoring of electrical grids in Singapore, Amsterdam, and Barcelona.	Real-time equipment monitoring and management, predictive maintenance, demand response optimization, and energy loss reduction.	Soares et al. (2023)

DT, Digital Twin.

Table 2.3 provides an overview of how DT technology has been implemented in various industrial sectors beyond its origins in manufacturing. Each industry has adapted this technology to its particular objectives, achieving significant impacts ranging from fault prediction and process optimization to improved accuracy in medical treatments and efficient management of urban infrastructure (Soares et al., 2023).

2.4 Implementation of Digital Twins in the field of smart grids: How are DTs being used in smart grids?

In smart grids, DTs are being widely used to enhance the efficiency, stability, and security of the electrical grid. Their implementation enables real-time monitoring, simulations, and optimizations of both individual components and the overall system. Following are the main ways in which DTs are being applied in smart electrical grids (Gao et al., 2022):

1 **Transmission:** DTs offer an accurate representation of transmission systems for real-time monitoring and failure prediction in high-voltage assets, such as transmission lines, power transformers, and protection equipment.
   a. **Digital modeling of transmission lines and electrical pipelines:** Virtual models are built to replicate the physical and electrical behavior of the lines through data processing and modeling, applying physics-based models such as differential equations that consider factors such as impedance, loads, and losses. They improve the real-time panoramic monitoring capability of transmission lines (Gao et al., 2022).
   b. **Failure prediction and maintenance optimization:** Thanks to the self-diagnostic capabilities of the DT, historical and real-time data are analyzed to identify patterns that allow for the anticipation of failures in structures, insulators, or conductors, facilitating predictive maintenance. They contribute to the reduction of maintenance costs and time. In addition, they facilitate line maintenance management and line operation management, moving from reactive to proactive maintenance (Zhang et al., 2022).
   c. **Real-time monitoring of weather and structural conditions:** Through continuous data acquisition with IoT sensors and SCADA systems, meteorological data and physical sensors (temperature, wind, mechanical stress) are integrated to adjust network operation and prevent failures due to extreme events. DTs are used to

improve the response time of the communication network and the real-time panoramic monitoring capacity of transmission lines (Rayhana et al., 2024).

2 **Smart substations:** DTs provide a comprehensive view of the status and performance of all substation equipment.

a. **Integration of equipment and sensor data:** The data acquisition component enables continuous collection of electrical and mechanical variables from transformers, switches, relays, and protection systems, among other substation components. In physical scenario replication tests (IEC 61850 substations), the DT achieved an average delay between Sampled Values (SV) messages of 249.59 μs (±0.91 μs), which closely resembles the expected time window of 205 μs. The minimum delta time between SV frames was 18.00 μs in the DT (Soares et al., 2023). Simulation of an overcurrent scenario showed that the average difference for current between the physical and DTs was 0.004 amperes, and for voltage it was 0.006 volts (Soares et al., 2023).

b. **Event simulation and maintenance optimization**: Using specialized software such as MATLAB/SIMULINK for simulation and analysis, different scenarios are recreated, such as equipment failures or disconnections, to assess their impact and plan corrective actions without compromising the actual system. DT models help diagnose and optimize maintenance costs. The financial management level of the substation is improved (Zhang et al., 2022).

c. **Virtual inspection and remote monitoring of assets:** Using visual interfaces and specialized control panels, it is possible to access the operational status of substation components without physical intervention, improving safety and operational efficiency.

3 **Power distribution:** DTs in electric power distribution facilitate demand management, integration of distributed energy resources (DERs), and improvement of service quality.

a. **Distribution network modeling to improve stability:** Using data processing and modeling techniques with physics-based models, representations are constructed that reflect the low- and medium-voltage network, enabling the identification of critical points and the simulation of solutions. Verification of the accuracy of the network model in memory (in-memory computing) takes only 0.021 s, underscoring the speed of the data and calculation service (Wu et al., 2023).

b. **Fault restoration optimization through simulations:** The simulation and analysis component allows different reconfiguration strategies to be evaluated in the event of interruptions, achieving faster and more effective service restoration. DT allows the real-time failure rate of distribution equipment to be calculated and the level of risk to be quantified, enabling the mode of operation and maintenance to be transformed from "retrospective" to "prospective" (Zhang et al., 2022).

c. **Reduction of technical losses and service quality improvement:** The self-regulating capabilities of the DT allow real-time analysis of energy flows and loads to make adjustments that reduce Joule losses and improve voltage profiles for users. High accuracy is achieved in the model: the voltage magnitude error is basically 10^{-6} (in p.u.) and the angle error is up to −0.000609376 degrees (Wu et al., 2023). The self-regulating capabilities of the DT allow real-time adjustments that reduce Joule effect losses (Wu et al., 2023).

4 **Renewable energies:** DTs enable the optimization of the connection and operation of intermittent renewable sources, such as solar and wind energy, within the power grid.

a. **Simulation of wind farms, photovoltaic, and hydroelectric plants:** Through simulation and analysis using specialized mathematical models, DTs model the dynamic behavior of these sources according to operating conditions and specific design. DTs inspect wind turbine blades for wear and tear, contributing to the longevity of the turbine blades (Rayhana et al., 2024).

b. **Energy generation forecasting based on weather conditions:** Using data-driven models, such as ML, trained with historical operational information, climate data (solar radiation, wind speed, and water level) is integrated to accurately estimate future energy production. They help maximize energy distribution and production and achieve the best possible energy outcome.

c. **Optimization of energy consumption and storage:** DTs play a crucial role in predicting maintenance needs, enabling informed decisions to be made about battery use, electric vehicle charging, or grid injection, by knowing the expected generation and demand in real time. The use of DT has significantly reduced energy waste (Latorre et al., 2024).

5 **Data centers:** Data centers are massive energy consumers, and their reliability is critical. DTs are used to optimize their energy efficiency and ensure operational continuity.

a. **Intelligent monitoring of energy consumption in servers and data centers:** Through continuous data acquisition using sensors and IoT devices, DTs monitor the energy performance of IT equipment and cooling systems in real time.

b. **Performance optimization through real-time data analysis:** Data processing and modeling using ML techniques allow the operation and ventilation of servers to be adjusted to maximize performance and minimize consumption. They promote the intelligent updating of the operation and maintenance of the machine room. The automation of lighting and climate control in buildings reduces the carbon footprint (Zhang et al., 2022).
c. **Integration with energy management systems:** Through graphical interfaces and specialized control panels, they connect to energy management platforms (EMS) to coordinate consumption with energy availability, reducing costs and improving sustainability.

In addition to sector-specific benefits, DTs offer general operational advantages applicable to all areas of smart grids:

- **Reduction in operating costs:** DTs reduce unplanned downtime and associated costs by enabling predictive maintenance. Early detection of potential problems minimizes maintenance costs and system downtime, resulting in significant savings (Rayhana et al., 2024).
- **Improved efficiency:** They ensure that systems operate at their maximum potential, improving overall performance through continuous monitoring.
- **Advanced prediction:** They enable industries to move from reactive to proactive maintenance by predicting potential failures and preventing catastrophic events.
- **Prediction accuracy:** In network optimization contexts, DT models have been shown to predict critical service level agreement metrics with a Mean Absolute Percentage Error of 3.8% in unseen topologies (Rayhana et al., 2024).
- **Sustainability:** DTs act as enablers for achieving United Nations Sustainable Development Goals (SDGs), such as Responsible Consumption and Production (SDG-12), Affordable and Clean Energy (SDG-7), and Climate Action (SDG-13) (Rayhana et al., 2024).

2.5 Literature review on Digital Twins in energy management

Having established the fundamental functionalities and applications of DTs in the energy sector, it is essential to examine how recent research has addressed specific implementations and technological advances in this field. The scientific literature reveals a growing interest in the development of DT-based solutions to optimize energy management, from individual substations to complete power grids and urban integration. Some of the relevant works are as follows:

1. **Protection and digital substations:** The study called "Digital twin application for the evaluation of protection performance in IEC-61850-based digital substations" (Lak et al., 2023) stands out for its focus on the validation of electrical protections through DTs based on the IEC-61850 standard. Its unique contribution lies in the integration of the Power Systems Computer Aided Design (PSCAD) simulator with advanced communication protocols such as Transmission Control Protocol/Internet Protocol (TCP/IP) and Inter-Process Communication (IPC) to create a cost-effective testing environment that avoids the use of expensive physical equipment. The connection with energy management is established through the optimization of substation reliability, critical components in the distribution of electric power.
2. Another study was conducted in the field of "Enabling digital twins for smart substation networks" (Soares et al., 2023), which complements the previous work by delving into practical implementation using SCD files, COMTRADE, and GOOSE/SV protocols. Its distinctive value lies in the use of simulators such as Mininet to replicate IED behaviors and communication networks, managed by a digital twin orchestrator. This research connects directly with energy management by enabling the simulation and optimization of critical communications in intelligent substations.
3. **Adaptive protection and distributed resources:** In this section, the article "A new digital twins-based overcurrent protection scheme for DERs integrated distribution networks" (Gómez-Luna et al., 2023) provides an innovative perspective by integrating physical protection with real-time simulation through Power Hardware-in-the-Loop configuration. Its uniqueness lies in the evaluation of the behavior of real protections under the integration of DERs. The connection with energy management is direct: it addresses one of the greatest current challenges of power grids, which is the safe and efficient integration of distributed renewable sources.
4. **Urban applications and systemic integration:** In the field of urban applications and systemic integration, "Intelligent transportation systems: exploring DT technologies in smart grid, transportation systems, and smart cities"

(Kanimozhi et al., 2024), broadens the perspective to multisectoral applications that integrate smart grids, micro-grids, and transportation systems. Their distinctive contribution lies in the use of 3D modeling (Autodesk 3ds Max) and simulation environments (Siemens Prescan) for the creation of comprehensive DTs, where energy management is approached from the perspective of optimizing urban systems, taking into account the interaction between different consumption sectors. This vision is complemented by the work of "Application and development of digital twins in smart cities"(Cai, 2024), who incorporates emerging technologies such as 5G, blockchain, AI, and extended realities (3R), offering a comprehensive framework for urban digitalization in which energy management is coordinated with other city systems to optimize resource efficiency and strengthen governance.

5. **Architectures and advanced models:** Along these lines, "Power distribution safety assessment based on the digital twin" (Sun et al., 2023), introduces the use of the TOPSIS algorithm as a quantitative evaluation tool, supported by cloud architectures for real-time processing. Their contribution lies in the application of multicriteria methods for security assessment, which is linked to energy management through predictive monitoring and operational risk prevention. Complementarily, "Application of digital twins in smart grids" (Zhang et al., 2022) propose the Digital Twin Power Grid Model, based on multidimensional mapping that encompasses form, data, business, energy flow, and temporal dimensions. Its distinctive contribution is to overcome the limitations of previous models that did not comprehensively consider the characteristics of energy flow, thus enabling a complete perception of the state of the grid and the deduction of its temporal states. Finally, "Digital deduction application of power grid operation mode based on digital twin model" (Wu et al., 2023) focuses on operational adaptation in the face of the growing penetration of renewable energies. Their differential value lies in the digital deduction methodology that uses QS files, in-memory data networks (Hazelcast), and microservices (Spring Cloud), directly linking energy management with the efficient coordination of renewable resources and real-time adaptation of operating modes.

The comparative analysis reveals three main levels of application: substation studies [A1, A2] focus on localized systems with IEC-61850 standards, complete network research [D2, D3] addresses large-scale systemic integration, while urban approaches [C1, C2] integrate multiple sectors with emerging technologies such as 5G and blockchain. Simulation technologies vary from PSCAD [A1] for traditional power systems, Mininet [D2] for communication emulation, to PHIL configurations [B1] that combine real hardware with digital simulation. Analysis methods evolve from multicriteria techniques such as TOPSIS [D1] to artificial intelligence [C1] and multidimensional mapping models [D2], while architectures migrate toward cloud-distributed platforms and microservices [D1, D3] for greater scalability.

The analyzed works demonstrate that DTs in energy management have evolved from specific simulation tools to integrated platforms that address systemic challenges. The trend toward the integration of renewable energies, the digitalization of power grids, and the development of smart cities positions DTs as an essential enabling technology for the energy transition and urban sustainability.

The synthesis of the reviewed works reveals fundamental common themes that consolidate the field of DTs in energy management. The IEC-61850 standard emerges as a fundamental pillar that facilitates interoperability and communication in digital substations, providing a common language for intelligent devices. Simulation tools, especially MATLAB/Simulink and PSCAD, constitute the technological basis for power system modeling and analysis, while real-world data captured in formats such as COMTRADE enable the validation and calibration of digital models. All these elements converge toward the common goal of creating safer, more efficient, and self-healing power grid systems, capable of dynamically adapting to changing operating conditions and anticipating failures before they occur.

2.6 Conclusion

This chapter has established a comprehensive understanding of digital twins as a transformative technology that transcends traditional simulations to create dynamic digital ecosystems with continuous bidirectional connectivity with their physical counterparts.

- **Definition and nature of DTs:** DTs are fundamentally distinguished from conventional simulations by their ability to maintain an exact virtual representation of physical objects, processes, or systems that constantly evolves through the integration of real-time data. This characteristic makes them tools for analysis, monitoring, and optimization that allow studying the behavior of physical systems under various conditions without directly intervening in the real equipment.

- **Fundamental pillars:** The architecture of DTs is based on three essential pillars: the Real Space, where the devices, processes, or physical systems that generate data are located; the Virtual Space, which replicates the physical system in a digital environment based on real-time data and computational models; and Bidirectional Communication, which enables data synchronization between the physical entity and the virtual model. This tripartite structure, complemented by self-regulation and self-diagnosis capabilities, ensures that changes made in the DT respect the physical limitations of the real object while continuously assessing the system's state.
- **Key technological components for industry 4.0:** DTs are composed of four integrated technological elements: data acquisition through IoT sensors, actuators, and SCADA systems that capture real-time variables; processing and modeling that use data engineering, statistics, and ML techniques to represent the behavior of the physical system; simulation and analysis through specialized software that allow exploring operational scenarios without risk; and visualization and interaction through dashboards and graphical interfaces that facilitate informed decision-making.
- **Transformative core functions:** DTs perform four categories of specialized functions that redefine the management of complex systems. Monitoring and diagnostic models enable continuous observation of the state of physical objects through sensors, predictive analytics, cloud architectures, system dynamics models, and AI techniques. Detection and prediction models identify anomalies and forecast future behaviors using static models, ML, and hybrid implementations. Control and automation models supervise and manage automated processes through PID controllers, adaptive control systems, and neural networks. Finally, simulation and optimization models use linear programming, genetic algorithms, and neural networks to maximize efficiency and performance.
- **Sectoral impact and applications:** The versatility of DTs is manifested in their successful application across critical sectors. In manufacturing, they optimize processes and reduce costs through continuous monitoring of machinery. In healthcare, they personalize treatments by integrating medical histories with real-time parameters. In infrastructure and smart cities, they monitor structural integrity and facilitate urban planning. In energy and sustainability, they revolutionize the management of power grids, from individual components to complete transmission and distribution systems.
- **Transformation of smart power grids:** In the specific context of smart power grids, DTs demonstrate their transformative potential through applications in transmission (line modeling, fault prediction, and climate monitoring), smart substations (equipment data integration, event simulation, and virtual inspection), power distribution (network modeling, restoration optimization, and loss reduction), renewable energy (plant simulation, generation forecasting, and storage optimization), and data centers (intelligent monitoring, performance optimization, and integration with energy management systems).

While the concept of DT is no longer new, its large-scale application in power grids and integration with urban systems is still in the early stages of exploration. The future trend points toward fully autonomous and self-repairing systems, driven by the amalgamation of technologies such as AI, IoT, and edge computing. To achieve this future, it is crucial to address several research gaps and technical and regulatory challenges:

- **Interoperability and standard alignment:** Interoperability and standard harmonization are fundamental challenges, given that DT models currently often lack a consistent framework that facilitates large-scale deployment. The IEC 61850 standard is a crucial pillar that facilitates communication and interoperability in digital substations, defining the model and facilitating communication between devices (Lak et al., 2023). However, the lack of uniform standards and regulations hinders scalability and adaptability, affecting the ability of DTs to interact across multiple sectors, including smart grids and urban systems. This lack of standardization not only complicates integration but can also compromise data accessibility and security, underscoring the need to develop consistent regulations to mitigate the risk of cyberattacks (Soares et al., 2023).
- **Distributed architectures that preserve privacy:** The future of the DTs requires modular and scalable distributed architectures capable of managing the enormous volumes of data generated. Cloud computing provides the necessary storage capacity and scalability, but edge computing is vital for real-time processing and reducing latency in critical systems. For example, edge nodes can train customized prediction models for nearby areas and store the information locally, which helps alleviate the overload in the cloud center (Zhang et al., 2022). Despite the architectural benefits, data security and privacy remain important concerns, as DTs handle sensitive and confidential information. The constant flow of data between the physical and DTs can expose vulnerabilities. Therefore, it is imperative to implement end-to-end data encryption techniques and robust cybersecurity measures to protect information from potential threats (Latorre et al., 2024).

- **Integration with AI agents, federated learning (FL), and DT governance frameworks:** To exploit the full potential of the DT, its integration with AI and ML is essential, as these technologies not only have the potential to simplify DT implementation and reduce costs, but are also crucial for the analysis of nonlinear and uncertain parameters (Gómez-Luna et al., 2023). AI is fundamental for intelligent fault diagnosis and adaptive process optimization. A key trend is FL, a promising approach for industrial IoT environments and smart cities, which allows AI models to be trained in a distributed manner, improving communication efficiency while preserving the privacy of local datasets. Finally, it is crucial that governance frameworks be developed to ensure the ethical and legal use of DTs, addressing concerns such as system autonomy, fairness and equity, as well as transparency in data handling to ensure user trust (Latorre et al., 2024).

In summary, the future requires multidisciplinary collaboration to overcome implementation barriers, from the integration of rigorous technical standards (such as IEC 61850) to the creation of governance frameworks that facilitate the adoption of advanced AI and distributed architectures in a secure and ethical manner.

AI disclosure

During the preparation of this work the author(s) used ChatGPT to improve the writing of some paragraphs and to improve the use of technical terms, without affecting the originality of the submitted document. After using this tool/service, the author(s) reviewed and edited the content as needed and take(s) full responsibility for the content of the publication.

References

Cai, Z. (2024). Application and development of digital twins in smart cities. In *Proceedings - 2024 4th International Conference on Big Data Engineering and Education, BDEE 2024*. Thailand: Institute of Electrical and Electronics Engineers Inc http://ieeexplore.ieee.org/xpl/mostRecentIssue.jsp?punumber=10723339. doi:10.1109/BDEE63226.2024.00012.

Chen, S., & Zhu, P. (2024). Digital twin technology and applications in the medical field: Bridging the virtual for precision medicine. In *Proceedings–2024 IEEE 4th International Conference on Digital Twins and Parallel Intelligence, DTPI 2024*. China: Institute of Electrical and Electronics Engineers Inc http://ieeexplore.ieee.org/xpl/mostRecentIssue.jsp?punumber=10778636. doi:10.1109/DTPI61353.2024.10778770.

Gao, B., Liu, H., Su, H., Hong, Z., Ji, C., & Zhao, M. (2022). Design and typical application of digital twin architecture in Smart Grid. In *Proceedings - 2022 7th Asia Conference on Power and Electrical Engineering (ACPEE)*. China: Institute of Electrical and Electronics Engineers Inc http://ieeexplore.ieee.org/xpl/mostRecentIssue.jsp?punumber=9783222. doi:10.1109/ACPEE53904.2022.9783642.

Gómez-Luna, E., Candelo-Becerra, J. E., & Vasquez, J. C. (2023). A new digital twins-based overcurrent protection scheme for distributed energy resources integrated distribution networks. *Energies, 16*(14). http://www.mdpi.com/journal/energies/. doi:10.3390/en16145545.

Kanimozhi, N., Sumathy, G., Maheshwari, A., Arunarani, A. R., & Sherin Shibi, C. (2024). Intelligent transportation systems: Exploring digital twin technologies in smart grid, transportation systems and smart cities. *International Conference on Advances in Data Engineering and Intelligent Computing Systems (ADICS)*. doi:10.1109/ADICS58448.2024.10533581.

Lak, P. Y., Key, S., Yoon, S. M., & Nam, S. R. (2023). Digital twin application for the evaluation of protection performance in IEC-61850-based digital substations. *International Conference on Advanced Power System Automation and Protection, APAP 2023*. South Korea: Institute of Electrical and Electronics Engineers Inc http://ieeexplore.ieee.org/xpl/mostRecentIssue.jsp?punumber=10347561. doi:10.1109/APAP59666.2023.10348444.

Latorre, L., Rego, E., & De Leo, L. (2024). *Reporte de Tecnología*. Gemelos Digitales.

Rayhana, R., Bai, L., Xiao, G., Liao, M., & Liu, Z. (2024). Digital Twin models: Functions, challenges, and industry applications. *IEEE Journal of Radio Frequency Identification, 8*, 282–321. http://ieeexplore.ieee.org/xpl/mostRecentIssue.jsp?punumber=7433271. doi:10.1109/JRFID.2024.3387996.

Soares, A. A. Z., Lopes, Y., Fernandes, N. C., Muchaluat-Saade, D. C., & Mosse, D. (2023). Enabling digital twins for smart substation networks. In *IEEE International Conference on Communications, Control, and Computing Technologies for Smart Grids, SmartGridComm 2023 - Proceedings*. Brazil: Institute of Electrical and Electronics Engineers Inc http://ieeexplore.ieee.org/xpl/mostRecentIssue.jsp?punumber=10333305. doi:10.1109/SmartGridComm57358.2023.10333905.

Sun, L., Shang, Y., Ma, Z., Zhou, H., & Cui, H. (2023). Power distribution safety assessment based on the digital twin. *4th International Conference on Smart Grid and Energy Engineering, SGEE 2023*. China: Institute of Electrical and Electronics Engineers Inc http://ieeexplore.ieee.org/xpl/mostRecentIssue.jsp?punumber=10481519. doi:10.1109/SGEE60678.2023.10481776.

Wu, Q., Zhou, M., Yan, J., & Feng, D. (2023). Digital deduction application of power grid operation mode based on digital twin model. *IEEE Joint International Information Technology and Artificial Intelligence Conference (ITAIC)*. China: Institute of Electrical and Electronics Engineers Inc https://ieeexplore.ieee.org/xpl/conhome/1800501/all-proceedings. doi:10.1109/ITAIC58329.2023.10409069.

Zhang, Y., Ma, C., Luo, J., Zhu, W., Wu, Y., & Zhang, X. (2022). Application of digital twins in smart grids. *IEEE 4th International Conference on Power, Intelligent Computing and Systems, ICPICS 2022*. China: Institute of Electrical and Electronics Engineers Inc http://ieeexplore.ieee.org/xpl/mostRecentIssue.jsp?punumber=9872525. doi:10.1109/ICPICS55264.2022.9873758.

C H A P T E R

3

Introduction to smart grid

Saima Zafar

Department of Electrical Engineering, National University of Computer and Emerging Sciences, Lahore, Pakistan

Introduction and Scope of the Chapter

Digital Twin technology, when integrated with smart grid systems, creates a virtual replica of physical grid components and subsystems, reflecting their real-time state and functioning through continuous updates with live data. These capabilities enable simulation, monitoring, and optimization of the grid's operations and performance, which in turn support improved decision-making, enhanced reliability, and proactive maintenance. This chapter provides an introduction to the smart grid, outlining its architecture, key features, enabling technologies, implementation challenges, major applications, and future outlook within the context of modern power infrastructure (Box 3.1).

BOX 3.1

"What This Chapter Delivers"

- Understanding of smart grid fundamentals and architecture.
- Global historical evolution and policy lessons.
- Enabling technologies and implementation challenges.
- Applications and future trends.

3.1 What is a smart grid?

The smart grid can be defined as an advanced electric grid system that integrates sophisticated sensing, digital communication, and automation technologies to optimize the production, distribution, and consumption of electricity. It represents an enhancement of the legacy 20th-century power grid by incorporating advanced metering infrastructure (AMI), sensing technologies, digital communication, automated control, security, and artificial intelligence (AI). Together, these innovations make the grid more resilient, autonomous, and intelligent, resulting in significantly improved overall efficiency. The smart grid enables automated electricity management by continuously monitoring grid performance in real-time, seamlessly integrating renewable energy sources, and enhancing energy efficiency for both utilities and consumers. As shown in Fig. 3.1, the smart grid functions as a centralized intelligent network that integrates generation, distribution, and consumption, enabling efficient control, reliability, and active consumer engagement.

The layered structure of the smart grid can be summarized through its major operational levels, along with the key components and functions associated with each, as shown in Table 3.1. These integrated components and systems work collaboratively to execute the core functions of the smart grid effectively (Khan et al., 2020).

The key elements that form the basic smart grid architecture include:

- **Distributed energy resources:** Distributed energy resources (DER) are decentralized energy generation and storage units such as solar photovoltaic panels, wind turbines, battery storage systems, combined heat and power

Digital Twin Technology and Smart Grid.
DOI: https://doi.org/10.1016/B978-0-443-36664-2.00018-6

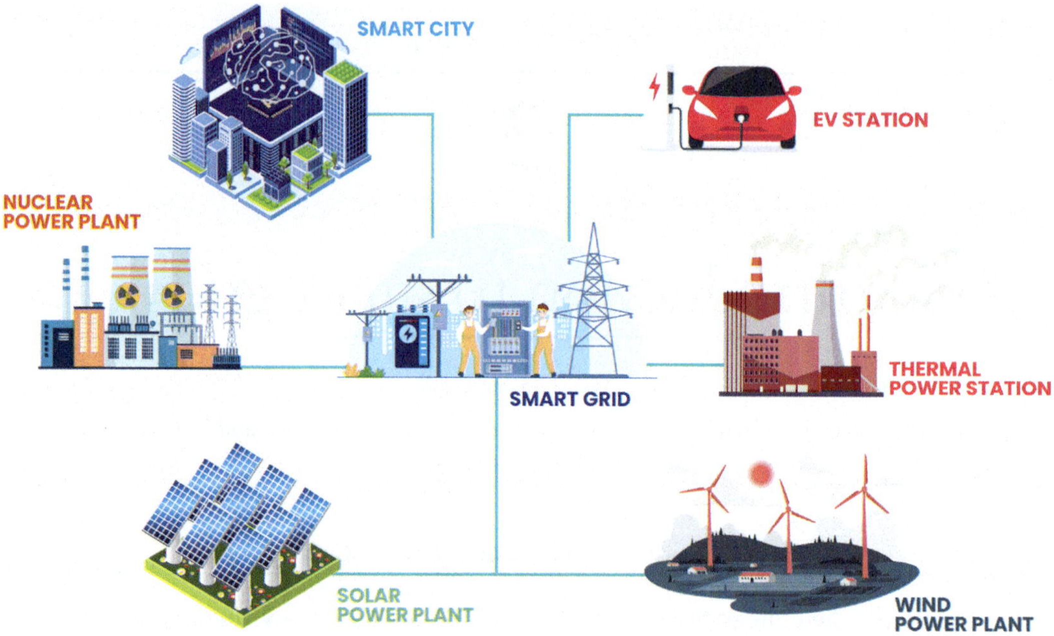

FIGURE 3.1 Smart grid.

TABLE 3.1 Smart grid architecture layers, components, and functions.

Layer	Core function	Representative components
Generation	Decentralized production	Distributed energy resources (DERs), renewable power plants, energy storage systems (ESSs)
Transmission	Grid stability and monitoring	Smart transformers, phasor measurement units (PMUs), supervisory control and data acquisition (SCADA) systems
Distribution	Automation and control	Distribution automation system (DAS), advanced metering infrastructure (AMI), IoT sensors and devices, DER integration equipment, digital communication systems
Consumption	User-side intelligence	Smart meters, electric vehicles (EVs), demand response (DR) systems

(CHP) units, and electric vehicles (EVs), located near the point of consumption. They are integrated with the grid to enhance its resilience and reduce reliance on centralized power plants.

- **Energy storage system:** The energy storage system (ESS) comprises technologies such as batteries, pumped hydro storage, thermal energy storage, and others, each with unique features suited to specific applications. They store excess energy during periods of low demand and release it when demand is high, thereby balancing supply and demand, enhancing grid reliability, and reducing strain on main power plants.
- **Distribution automation system:** The distribution automation system (DAS) manages electricity distribution networks using sensors, control devices, and software to monitor and operate the system. In the event of an outage or disruption, power can be automatically rerouted to maintain service continuity.
- **Smart meters and AMI:** The most common central component of the smart grid is the smart meter, which primarily functions to record and transmit energy consumption data to utilities. AMI is a system comprising advanced digital devices, including smart meters and sensors, along with bidirectional communication networks and meter data management systems (MDMS). It collects, stores, and analyzes data from these devices, providing detailed usage analytics and insights for utilities and consumers.
- **Advanced sensors and monitoring systems:** These devices monitor the generation, transmission, and distribution subsystems of the grid, detecting and reporting faults or discrepancies to utilities, thereby enabling faster responses to outages and maintenance needs.
- **Digital communication systems:** Advanced digital communication systems enable real-time, secure, and reliable bidirectional communication between various grid components and subsystems, ensuring effective coordination and control.

- **DR and load management system:** The demand response (DR) and load management system are designed to balance electricity demand and supply, especially during periods of peak load. DR refers to the adjustment of electricity usage by end-users in response to signals from utilities. A load management system is a digital platform used by utilities to coordinate with smart devices, monitor consumption, and automatically adjust loads as needed.

3.2 Historical context and evolution

The development of the traditional electric grid has been a continuous process that started in the late 19th century with the advent of electrical power generation. In the early days, electricity was generated by small, localized power plants, commonly powered by coal or water, to supply energy to nearby communities. As demand grew, these plants were connected through an expanding network of transmission lines, enabling electricity to be distributed over wider areas. A breakthrough came with the installation of the first electrical distribution system in New York City by Thomas Edison in 1882, followed by the introduction of alternating current (AC) systems by Nikola Tesla and George Westinghouse, which enabled long-distance transmission of electricity. In the mid-20th century, the evolution of supervisory control and data acquisition (SCADA) systems became a transformative step in the transition to automated processes across industrial and utility sectors. The expansion and increasing complexity of power grids and other critical infrastructure underscored the growing need for centralized control and real-time monitoring. Early SCADA systems addressed this need by integrating mainframe computing, remote terminal units (RTUs), and basic telemetry, allowing operators to oversee and manage equipment from a centralized location. This transition from manual, labor-intensive methods to automated control greatly enhanced operational efficiency, reliability, and scalability, enabling faster responses to system disruptions and reducing downtime. These developments established the technological foundation for the modern smart grid, which builds on early innovations through advanced digital communication, real-time analytics, distributed control, and integration of renewable energy sources (Momoh, 2012).

Over the years, scientific advancements and technological breakthroughs, including the construction of large-scale power plants, the installation of substations, and improvements in transformer and transmission line efficiency, supported the expansion of the electric grid to a nationwide scale. Today, the grid has evolved into a sophisticated system that balances supply and demand, integrates diverse generation sources, ensures reliable electricity delivery across vast areas, and supports emerging energy technologies (Ponce-Jara et al., 2017).

Amid escalating global energy demands and the push to reduce carbon emissions, there is a growing urgency to modernize the aging electric grid to meet modern requirements. Traditional power grids, designed decades ago for centralized and unidirectional energy flow, are inadequate for addressing the complexities of contemporary energy systems. These legacy systems face significant challenges in integrating variable and decentralized renewable energy sources such as solar panels, wind turbines, and battery storage, all of which require real-time, dynamic management of supply and demand. Additionally, the need for a more responsive grid architecture has become even more critical due to increasing concerns over grid reliability, environmental sustainability, cybersecurity threats, and the ongoing transition to electric transportation (Ponce-Jara et al., 2017; The smart grid – history, present and future, 2025).

To overcome these challenges, the smart grid has emerged as a transformational innovation, reshaping the conventional electric grid into an intelligent and adaptable network. The modern grid integrates communications infrastructure with the electric grid, incorporating advanced sensors, communication technologies, automation, and data analytics to monitor and efficiently manage electricity flows and grid operations. These technologies enable utilities to detect outages promptly, optimize energy consumption, accurately forecast demand, and integrate diverse energy sources without compromising grid stability. The smart grid also empowers consumers by providing real-time information and insights into energy usage patterns, allowing them to make informed energy choices and participate in the DR programs. Fundamentally, the transformation of the grid through advanced technologies represents not just an upgrade but a progressive shift toward a more sustainable, resilient, and user-centric energy ecosystem.

The smart grid we see today has evolved through several significant milestones. Table 3.2 presents an overview of substantial smart grid developments and milestones in leading countries worldwide, highlighting the contributions of the United States, Japan, South Korea, Germany, and China.

The United States was among the first countries to adopt these technologies, with early trials for implementing AMI for real-time energy consumption data beginning in the late 1990s and early 2000s. Subsequent efforts, particularly in states like California, focused on testing the DR programs to manage grid load during peak periods. These pilot projects provided valuable insights into how digital technologies could transform grid operations by improving reliability, optimizing energy distribution, and enabling consumers to play a more active role in managing energy usage. Such initiatives paved the way for broader smart grid deployment and highlighted the potential for improved

TABLE 3.2 Global smart grid milestones.

Country	Year	Initiative	Focus	Outcome
United States	Late 1990s–early 2000s	Early AMI trials	Real-time energy consumption monitoring	Provided insights on digital technologies, improved reliability, optimized energy distribution, and enabled consumer participation
Japan	Early 2000s	Pilot smart grid projects (Yokohama, Kitakyushu)	Energy reliability, efficiency, security	Modernized grid infrastructure, aligned with decarbonization and energy diversification goals
United States	2009	American Recovery and Reinvestment Act (ARRA)	DOE $4.5B investment in smart grid	Nationwide AMI roll-out, automated substations, enhanced communication networks
South Korea	2010	Jeju smart grid test-bed	Integration of renewables, AMI, EVs, ESS, DR	Identified critical components for smart grid, enabled real-time consumer energy management
Germany	2012	Smart Grid 2.0 (Baden-Württemberg)	Renewable integration, storage, and DR programs	Enhanced grid efficiency, strengthened stability in urban and rural areas
South Korea	2012	Smart grid roadmap	Nationwide deployment strategy	Nationwide smart grid deployment plan by 2030, accelerated implementation in urban/industrial areas
China	2010s–present	State-owned smart grid projects, UHV lines, AMI	Grid modernization, renewable integration	Strengthened energy network, reduced emissions, supported clean energy targets, promoted international smart grid development via BRI

AMI, *advanced metering infrastructure;* DOE, *department of energy;* DR, *demand response;* ESS, *energy storage system; EVs, electric vehicles; UHV, ultra-high-voltage; BRI, Belt and Road Initiative.*

efficiency and greater integration of renewable energy sources. The U.S. government played a pivotal role in accelerating the deployment and expansion of smart grid technology. In 2009, the U.S. Department of Energy (DOE) allocated approximately $4.5 billion under the American Recovery and Reinvestment Act (ARRA) to support smart grid initiatives nationwide. This funding, along with projects such as the installation of smart meters and the development of advanced grid technologies, including automated substations and enhanced communication networks, marked significant steps toward a more automated grid. Internationally, Japan, South Korea, Germany, and China also made substantial investments to modernize their grids. These government-backed programs spurred innovation and provided financial and regulatory support to utilities, enabling broader implementation. Pilot projects quickly evolved into large-scale programs aimed at enhancing grid reliability, reducing emissions, and integrating renewable energy sources (Fatfinger, 2024).

Several countries have made remarkable progress in smart grid development and deployment. Japan began its transition toward a smart grid in the early 2000s, driven by the need for greater energy reliability, efficiency, and security. The government and private sector collaborated to modernize the electric grid, launching initial pilot projects in cities such as Yokohama and Kitakyushu. Over time, Japan's smart grid initiatives evolved to align with broader national goals of decarbonization, energy diversification, and disaster resilience, positioning the country as a leader in smart energy innovation.

South Korea has also emerged as a global leader in smart grid technology, supported by its commitment to technological innovation and energy conservation. In 2010, it launched the Jeju Smart Grid Test-Bed on Jeju Island, one of the world's most advanced pilot sites. The project integrated renewable energy sources along with AMI, EVs, ESSs, and DR, enabling consumers to adjust electricity usage based on real-time pricing. This comprehensive approach helped South Korea identify the critical components needed for a fully functional smart grid, including grid management software and advanced data analytics. The success of the Jeju project paved the way for the 2012 Smart Grid Roadmap, which outlined plans for nationwide deployment by 2030. Strong government support and substantial private investment have accelerated implementation in urban and industrial regions, marking major progress in integrating digital technologies with the national energy grid (Kang et al., 2023).

Germany's Energiewende played a crucial role in advancing smart grid innovations, focusing heavily on integrating renewable sources, particularly wind and solar, into the energy system alongside smart technologies. The German

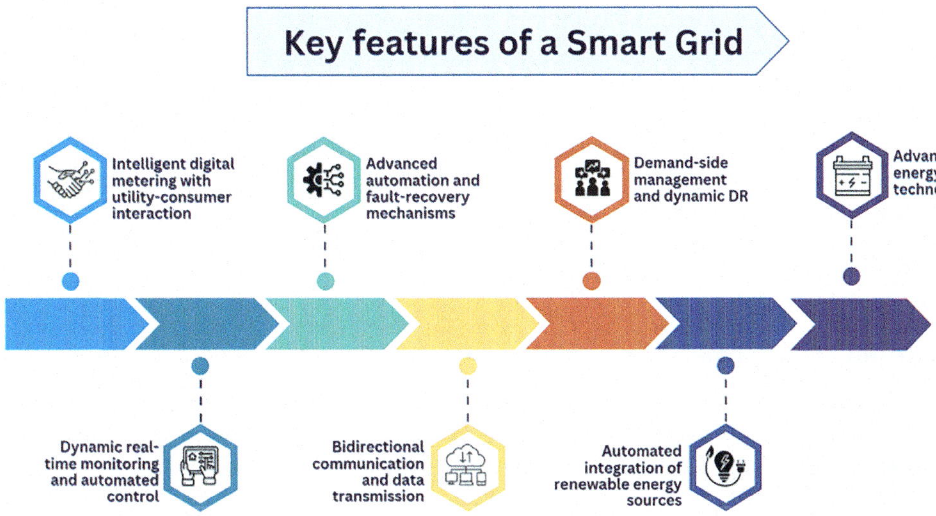

FIGURE 3.2 Key smart grid features that optimize system performance, operational reliability, and consumer participation.

government invested substantial resources in these initiatives, emphasizing AMI deployment and implementing DR strategies. In 2012, the country launched its first large-scale project, "Smart Grid 2.0," in Baden-Württemberg. The project combined renewable integration, storage systems, and DR to enhance grid efficiency. As the share of renewables continues to grow, Germany has steadily expanded its smart grid across both urban and rural regions to strengthen grid stability.

China has made substantial investments in smart grid technology, positioning itself as a global leader in the sector. Rapid urbanization, the pursuit of efficiency, and the need to integrate a growing share of renewable energy have made modernizing the nation's power infrastructure a top priority. Through major state-owned enterprises, China has installed large-scale smart grid systems featuring ultra-high-voltage (UHV) transmission lines, AMI, and sophisticated grid automation technologies. Guided by long-term strategies, including the 5-year plans, Made in China, 2025, and the goal of achieving carbon neutrality by 2060, China has strengthened its energy network, reduced emissions, and supported ambitious clean energy targets. Through initiatives such as the Belt and Road Initiative, China is also supporting the development of smart grid infrastructure in other countries (Ngar-yin Mah et al., 2017).

These international milestones emphasize the crucial role government policies and innovative pilot projects have played in shaping the global smart grid movement. They show how early investments laid the foundation for widespread adoption of intelligent energy infrastructure and highlight the smart grid's growing importance in improving grid performance and supporting the worldwide transition toward a more sustainable, low-emission energy landscape.

3.3 Key features of a smart grid

The key features that distinguish the smart grid from the traditional electric grid, as summarized in Fig. 3.2, are elaborated in the following sections (Feng et al., 2021; Ohanu et al., 2024; Ugwu et al., 2023).

3.3.1 Intelligent digital metering with utility-consumer interaction

The smart grid replaces conventional meters with smart meters, which are part of an intelligent digital metering infrastructure called AMI. This infrastructure is a fundamental component of smart grid technology, enabling instantaneous bidirectional communication between utilities and consumers, efficient energy management, and enhanced grid performance. In contrast to conventional metering systems, which provide periodic energy consumption readings over time, digital metering infrastructure comprises an integrated network of smart meters, communication technologies, and data management systems that enable bidirectional information exchange between energy providers and consumers. Using this infrastructure, the utilities collect extensive instantaneous electricity consumption data for in-depth insight into individual and collective electricity usage patterns.

These insights help utilities to perform accurate load forecasting, optimize distribution, minimize operational costs, and manage power outages. Similarly, consumers utilize insights from metering infrastructure to make smarter, data-driven choices regarding their electricity consumption, adapt their usage patterns, and benefit from flexible pricing schemes, helping to decrease overall electricity expenses. The digital metering infrastructure enables the integration of the power grid and DERs and EVs. It also supports the DR with two-way communication between the utilities infrastructure and consumers, where consumers adjust their consumption behavior based on incentives. Overall, intelligent digital metering with advanced capabilities is instrumental in modernizing the electric grid, paving the way toward a smarter and more sustainable energy future.

3.3.2 Dynamic real-time monitoring and automated control

A core feature of the smart grid is the real-time dynamic monitoring and automated control, which enables the utilities to accurately and efficiently supervise and manage the flow of electricity in the grid's generation, transmission, and distribution subsystems. Utilizing sensors, smart meters, and intelligent devices across the grid, the utilities collect real-time electricity flow and consumption data to detect outages, voltage fluctuations, or malfunctioning equipment, enabling them to respond swiftly. This feature also supports seamless integration of renewable energy sources, such as solar and wind, thereby amplifying operational efficiency and reliability. Furthermore, real-time monitoring and control are crucial for effective DR programs, enabling coordination between electricity demand and supply. This leads to reduced energy wastage, electricity cost savings, and system stability. Consequently, real-time dynamic monitoring and automated control play a pivotal role in shaping the future of energy systems.

3.3.3 Advanced automation and fault recovery mechanisms

Advanced automation and fault recovery mechanisms, which give self-healing capabilities to the smart grid, are the most transformative features of these systems. Using sensors, actuators, and advanced communication mechanisms, faults are automatically detected, and the affected sections are isolated, thus restoring operations without human intervention most of the time. This automated fault recovery significantly impacts the Quality of Service (QoS) by reducing the duration and impact of outages and downtime. During Hurricane Irma in Florida, intelligent devices on the grid helped isolate faults, and around 50% of outages were restored without human intervention. A huge number of outages were prevented due to the self-healing capabilities of the smart grid. This not only saved significant business revenue by minimizing operational disruptions but also improved public safety by ensuring continuous access to critical services like healthcare, emergency response, and transportation. In the 2020 wildfire season in California, some power lines failed due to excessive heat and disruptions. The automated switches isolated the affected sections of the grid and rerouted power, resulting in power being restored to 92% of consumers within one minute. These examples show how self-healing capabilities have shifted the power grid from reactive to proactive, minimizing downtime, avoiding large-scale blackouts, and enabling faster detection and localization of faults. Furthermore, these mechanisms also support reduced equipment stress, safer and quicker restoration without overloading the system (Erenoğlu et al., 2024).

3.3.4 Bidirectional communication and data transmission

Seamless collaboration between utilities and consumers, supported by the grid infrastructure, is imperative for the operational efficiency of the smart grid, which requires continuous bidirectional communication and data exchange. Unlike the conventional power grid, which supports unidirectional flows from utility to consumer, the smart grid leverages advanced communication networks to enable real-time, bidirectional data exchange. This way, utilities can monitor grid operations, identify faults, and dynamically adjust energy distribution, while consumers can receive real-time updates on their energy consumption, peak load hours, pricing, and available incentives. This bidirectional communication over advanced wired and wireless technologies is also vital to the success of the DR programs, where consumers receive real-time updates to adjust their energy usage during periods of high demand, enabling effective alignment of supply and demand. Due to the advancements in wired and wireless communication technologies, high-speed and reliable real-time communication is possible in today's smart grid. This feature offers several benefits, including incentivizing consumers who own renewable energy systems or EVs to sell excess energy back to the grid, reducing power outages and electricity losses, improving grid reliability, and lowering energy costs (Liu et al., 2024).

3.3.5 Demand side management and dynamic demand response

Demand side management (DSM) is a broader term used to represent all activities aimed at regulating consumers' energy usage to increase system efficiency and reduce peak demand. It includes DR, which is one component of the broader DSM toolkit. The DSM includes long-term and ongoing energy efficiency initiatives, load shifting strategies, and behavioral change efforts. Some examples of DSM efforts include programs endorsing energy-efficient appliances and lighting, time-of-use electricity pricing to encourage off-peak consumption, and support for home energy management systems (EMSs) that monitor and report usage patterns. The implementation of DSM relies on the capabilities provided by smart grid technologies. Likewise, the DR, which refers to short-term, instantaneous changes in consumers' electricity usage in response to utility signals such as price changes or reliability issues, is also supported by the advanced functionalities of the smart grid systems. Contrary to the conventional grid, where energy consumption is generally passive, the DR systems provide a mechanism to balance energy demand and supply in real-time by actively engaging consumers in managing their energy usage, particularly during times of peak demand or grid stress. This is possible due to two-way communication technologies, which enable the grid to send signals to the connected devices, appliances, or smart meters to adjust power consumption. This way, it becomes possible for utilities to curtail or defer electricity consumption in non-critical areas, such as temperature control systems in commercial buildings or reschedule industrial processes. For utilities, this lowers dependence on expensive, high-emission peak power plants and decreases the likelihood of blackouts. For consumers, it leads to lower electricity bills and increased awareness of energy usage. The DR schemes are event-driven, short-term, incentive-based, and grid-focused. These schemes support seamless integration of renewable energy sources into the electric grid by helping to maintain grid stability, shifting loads to match renewable output, and encouraging decentralized participation (Bakare et al., 2023).

3.3.6 Automated integration of renewable energy sources

A transformative characteristic of a smart grid is the integration of a wide range of intermittent and decentralized renewable energy sources with the conventional electric grid. The legacy grid primarily relies on centralized, fossil-fuel-based generation, such as coal-fired, oil-fired, natural gas-fired, and diesel generators, while the smart grid supports the incorporation of diverse renewable energy sources, such as solar, wind, hydroelectric, and biomass into the power system. This reduces reliance on fossil-fuel-based sources, resulting in benefits such as reduced greenhouse gas emissions, lower air and water pollution, conservation of natural resources, and preservation of ecosystems. This integration would not have been possible without the advanced sensing, monitoring, communication, and control technologies of the smart grid that facilitate real-time grid responsiveness to the intermittent output of renewable sources. The challenges associated with integrating renewable energy sources to the grid, such as grid stability, voltage regulation, and frequency control, are managed by the real-time data and automated control features of the smart grid. Besides, when wind or solar output fluctuates due to changing weather conditions, the grid can quickly adapt by redirecting power flows, maintaining supply-demand balance, and storing surplus energy in batteries or other storage systems for later use. The smart grid also enables greater participation by decentralized renewable sources, allowing consumers with rooftop solar panels or home battery systems to generate and store their electricity, while also having the option to feed surplus energy back into the grid. This contributes to a more robust, flexible, and sustainable energy system, supporting the transition to a low-carbon future (Kiasari et al., 2024).

3.3.7 Advanced energy storage technologies

Energy storage plays a crucial role in the functionality of a smart grid. These technologies, such as electrochemical storage in the form of batteries, mechanical storage in the form of pumped hydro storage, and thermal and chemical storage, allow excess energy produced during times of low demand or abundant renewable generation to be stored and dispatched when demand rises or renewable output declines. This stabilizes the grid by moderating the variability of renewable energy sources, which fluctuate with weather and time of day. Besides reliability and conservation, these systems help lower reliance on fossil-fuel-based peaking power plants. They contribute to the grid's overall stability by providing ancillary services, such as voltage regulation, frequency control, and black-start capabilities, in conjunction with grid management systems and load shifting, which support successful DR programs. The most widely used energy storage in the smart grid is lithium-ion batteries, pumped hydro storage, flow batteries, compressed air energy storage, and the emerging technology of solid-state batteries. Lithium-ion batteries are the leading energy storage technology for the smart grid due to their high energy density, fast response, long battery

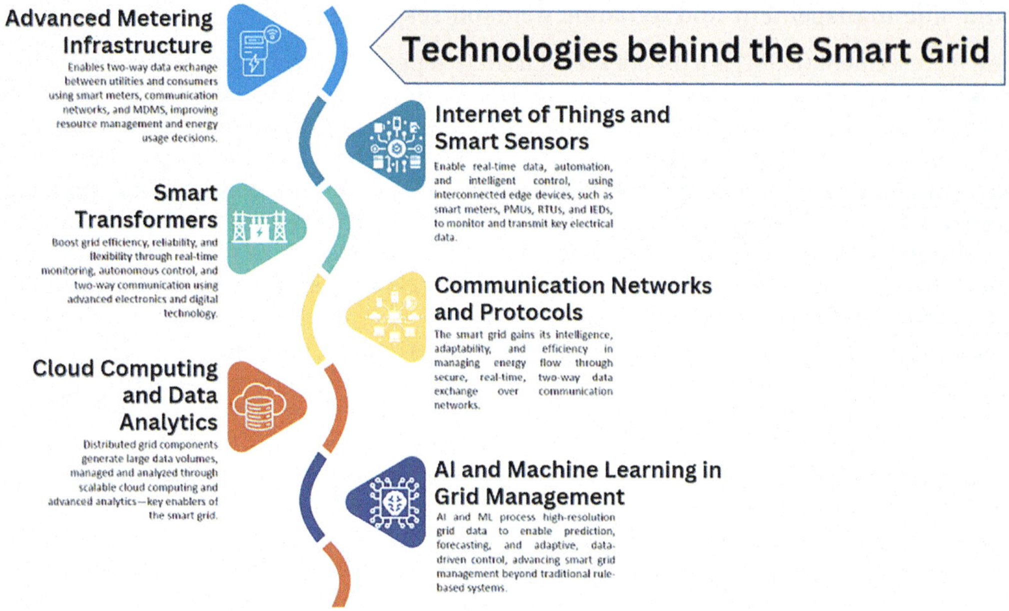

FIGURE 3.3 Smart grid technologies. Technologies that underpin smart grid functionality, facilitating intelligent management, system coordination, adaptability, and operational efficiency.

life, and support for integrating renewable energy sources. They are primarily used for grid-scale energy storage, residential energy storage, and EV charging stations. Pumped hydro is a mature and high-capacity energy storage solution often used in areas with appropriate geography that supports massive water storage. Flow batteries such as Vanadium Redox are gaining traction, particularly for large-scale and long-duration energy storage in the smart grid. Some of their benefits include long life cycles, scalable capacity, and the ability to release energy over extended durations, suitable for stabilizing the grid during extended shortages of renewable energy (Fang et al., 2025).

3.4 Technologies behind the smart grid

The key features of a smart grid discussed in the previous sections, such as automation, bidirectional energy flow, and enhanced monitoring, are realized through a set of enabling technologies. AMI, Internet of Things (IoT) sensors, robust communication protocols, and AI-driven analytics collectively allow the grid to operate intelligently and efficiently. By integrating these technologies, the theoretical capabilities of the smart grid are translated into practical, real-world functionalities, enabling improved energy management, seamless communication, automation, reliability, and consumer engagement. This section focuses on these technologies as illustrated in Fig. 3.3 and their role in implementing the previously defined smart grid features (Ayub Khan et al., 2023; Hui et al., 2020; Mazhar et al., 2023a).

3.4.1 Advanced metering infrastructure

AMI is a core component of the smart grid that employs advanced technologies to collect and transfer data between utilities and consumers, enabling utilities to efficiently manage resources and consumers to make informed decisions about their energy usage. It comprises smart meters, data concentrators, communication systems, and MDMS, collectively functioning to support the collection, transmission, and interpretation of detailed real-time system data. Unlike traditional analog meters, which are passive, manual, and limited in data collection and functionality, smart meters are active, digital, and intelligent devices featuring real-time monitoring, remote management, and dynamic pricing. They are capable of recording periodic interval data, tracking voltage levels, frequency, and power quality in addition to energy consumption. Smart meters utilize digital sensors and microprocessors to measure electrical parameters and are connected to the AMI for remote access and control. They are also equipped with communication modules that facilitate data transmission to a central system for analysis.

TABLE 3.3 Communication technologies in smart grid.

Technology	Range	Data rate	Power consumption	Typical use cases in smart grid
NB-IoT	Up to 10–15 km	~20–250 kbps	Low	WAN – smart metering, grid monitoring
LoRaWAN	Up to 15–20 km (rural); 2–5 km (urban)	~0.3–50 kbps	Very low	WAN/NAN – metering, grid sensors, AMI
Zigbee	10–100 m	~250 kbps	Low	HAN – home energy management, smart appliances
Wi-SUN	Up to 1–2 km	~300 kbps–1 Mbps	Low	NAN – smart metering, street lighting
Cellular (4G/5G)	10–50 km (depending on cell size)	Mbps–Gbps (esp. 5G)	Medium to high	WAN – distribution automation, real-time monitoring
PLC (power line communication)	Local (up to several km) via power lines	~2–500 kbps (varies by type: NB-PLC or Broadband Power-Line Communication (BB-PLC))	Medium	HAN/NAN – smart metering, load control
Wi-Fi	~50–100 m (indoors)	~10–600 Mbps (802.11 n/ac)	High	HAN – in-home displays, energy apps
Bluetooth low energy (BLE)	~10–100 m	~125 kbps–2 Mbps	Very low	HAN – wearable energy monitors, smart plugs

HAN, *home area network;* LoRaWAN, *long range wide area network;* NAN, *neighborhood area network;* NB-IoT, *narrowband IoT;* NB-PLC, *narrowband power line communication;* WAN, *wide area network.*

Technically speaking, precise, high-resolution, and time-synchronized energy consumption data, including real-time voltage, current, and power quality metrics from AMI, support the implementation of DR programs, real-time and accurate short-term, medium-term, and long-term load forecasting, and timely outage detection and restoration. Yet high-resolution consumption data collected by AMI raises privacy concerns, as it can potentially reveal detailed information about individuals' daily routines, appliance usage, and occupancy patterns. For system integration of distributed renewable energy sources, AMI enables automated meter reading, remote service connectivity management, time-of-use pricing, and net metering. When interfaced with a distribution management system (DMS) and EMS, this data enables state estimation, fault location isolation and service restoration (FLISR), voltage/VAR optimization (VVO), voltage regulation, and load distribution. In summary, AMI has a transformational role in today's smart grid, as it provides bidirectional data-driven intelligence to utilities and consumers. This turns consumers into active and prudent members of the energy ecosystem and empowers utilities to implement diverse programs for efficient grid operation.

3.4.2 Internet of Things and smart sensors

The IoT and smart sensor technologies are indispensable to the modern smart grid operations, supporting intelligent distributed architecture, real-time data collection and dissemination, and swift automated control across the grid infrastructure. In the context of the smart grid, IoT comprises an extensive network of interconnected edge devices that gather and transfer high-resolution data on key electrical parameters. The edge devices include smart meters (which are part of the AMI), phasor measurement units (PMUs), RTUs, and intelligent electronic devices (IEDs), which report parameters such as voltage, current, phase angle, frequency, harmonics, and power factor. These edge devices communicate over various suitable wired and wireless communication technologies like narrowband IoT (NB-IoT), long range wide area network (LoRaWAN), global system for mobile communications/general packet radio service (GSM/GPRS), radio frequency (RF), and power line communication (PLC) to transfer data over heterogeneous networks, frequently connected via cloud platforms of edge computing nodes to support distributed processing (Avdaković et al., 2025). These communication technologies are summarized in Table 3.3, highlighting their key characteristics in terms of technology, range, data rate, power consumption, and typical use cases in the smart grid. This provides a comparative overview of how different technologies support smart grid applications, from short-range IoT sensor networks to long range communication for wide area monitoring and control.

Smart sensors are integrated across the entire system for condition-based, risk-based, and threshold-based monitoring and diagnostics of assets in the transmission and distribution networks of the grid infrastructure. These assets include transformers, switchgear, substations, cybersecurity devices, and ESSs. Their predictive maintenance is essential for minimizing equipment failure, which can cause unplanned outages. Additionally, real-time data from IoT devices supports DR programs, voltage optimization, control and automation, fault detection, isolation, and restoration (FDIR), EVs integration, cybersecurity, data analytics, and dynamic load balancing. This is especially critical in power systems that incorporate a large proportion of DERs and rely heavily on variable renewable generation. The IoT-based systems incorporate operational resilience and self-healing grid capabilities along with AI and machine learning (ML) algorithms. IoT technologies and smart sensors underpin the cyber-physical power system, enabling the shift to a more intelligent and resilient grid.

3.4.3 Smart transformers

Smart transformers are an important component of modern power systems, contributing to the efficiency, reliability, and flexibility of the grid. They utilize advanced technologies of power electronics, digital control, and communication systems to enhance their functionality beyond traditional transformers. These transformers are equipped with intelligent sensors, microprocessors, and communication systems supporting real-time monitoring, autonomous regulation, and two-way communication. They can continuously monitor voltage, current, frequency, harmonics, power factor, temperature, and tap position and disseminate this data to a central control center for data analytics and autonomous control (Rahbarimagham & Gharehpetian, 2025).

In contrast to traditional transformers meant for stepping up or stepping down voltages in the power system, the technical role of smart transformers is significantly enhanced, encompassing a broader range of functions beyond traditional power regulation. They are equipped with technologies such as on-load tap changing (OLTC) for voltage regulation under load, solid-state switches, power electronics interfaces in case of solid-state transformers (SSTs), and insulation systems. They play their role in dynamic voltage regulation, automated diagnostics, voltage stability, frequency regulation, and grid robustness. We discuss these roles in sequence.

Dynamic voltage regulation by smart transformers involves the continuous, real-time monitoring and control of voltage levels to offset fluctuations and ensure grid stability. SSTs are specifically designed for this purpose, with the integration of digital control systems and advanced power electronics. They employ power electronic converters, solid-state tap changers, sensors, and feedback mechanisms, and digital embedded control system. They support self-diagnostics, which refers to the transformer's ability to monitor its condition and operating status in real-time through embedded sensors and microcontrollers. They also support fault detection and isolation, which is the transformer's ability to detect electrical faults and respond swiftly to minimize damage, using real-time monitoring, high-speed sampling, and signal analysis. For fault isolation, selective tripping is used, so that only the faulty section is isolated, minimizing the impact on the rest of the grid, which continues to operate.

When distributed energy sources are integrated into the power grid, they introduce issues such as voltage fluctuations, bidirectional power flow, frequency instability, harmonics, and other power quality concerns, which must be managed to ensure smooth operation. Smart transformers support the seamless integration of distributed energy sources by providing dynamic voltage support, bidirectional power flow, reactive power compensation, and grid fault ride-through capabilities.

3.4.4 Communication networks and protocols

The smart grid derives its intelligence, adaptability, and efficiency in managing energy generation, transmission, distribution, and consumption from secure, real-time, bidirectional data transfer over communication networks. These networks must comply with strict QoS prioritization, be designed based on fault-tolerant and resilient architectures, and implement end-to-end data encryption to meet the stringent requirements of the smart grid. Efficient communication networks also support the integration of edge computing and cloud platforms into the grid. Edge computing brings data processing closer to edge devices, enabling faster decision-making and a swift response to issues, while cloud platforms, often accessed over the internet, provide centralized computing power and large-scale storage capabilities. Overall, advanced communication networks are crucial to realizing the robust, fault-tolerant, and adaptive nature of the smart grid, supporting distributed energy management, real-time grid optimization, and integration of renewable energy sources.

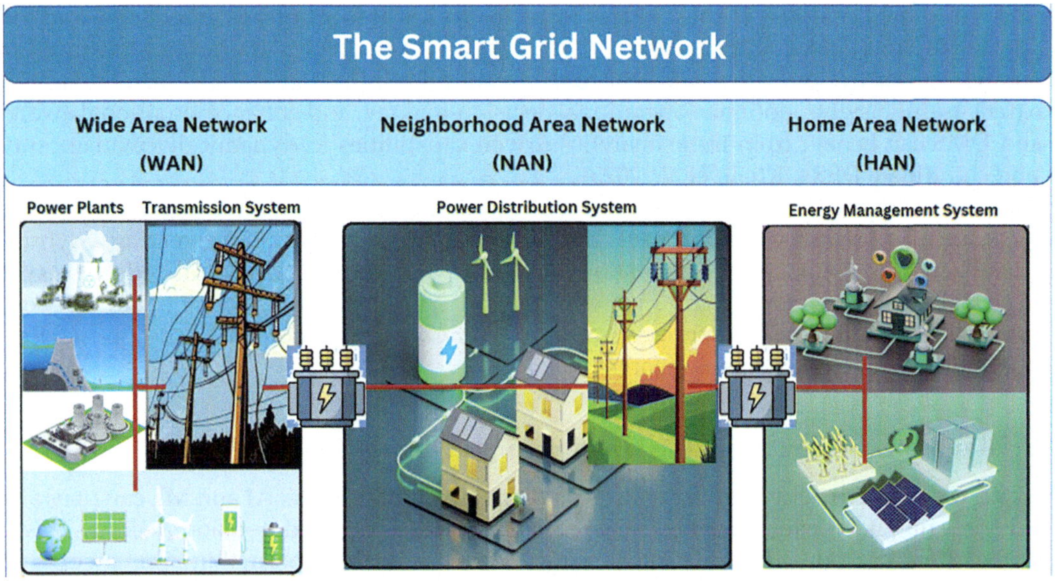

FIGURE 3.4 Hierarchical smart grid communication network showing WAN, NAN, and HAN for efficient monitoring, control, and data flow. *HAN*, Home area network; *NAN*, neighborhood area network; *WAN*, wide area network.

Smart grid communication networks are structured into tiers, namely home area network (HAN), neighborhood area network (NAN), and wide area network (WAN), based on the geographical coverage, data transmission requirements, and functional roles within the grid infrastructure. These networks, shown in Fig. 3.4, provide bidirectional communication between the various components of the grid infrastructure, such as AMI, EMSs, IEDs, grid central control units, DERs, and EVs, enabling real-time monitoring, control, and automated operations. The three tiers of grid communication networks utilize various wired and wireless communication technologies for high-speed, secure, and reliable communication between grid components. Some wired communication technologies used in the smart grid include PLC, fiber optics, wired local area network (LAN), and serial communication, and wireless technologies include wireless LAN, wireless personal area network (PAN), cellular, LoRaWAN, NB-IoT, and satellite communication.

The choice among these communication systems is based on the range and geographical coverage, data rate and latency requirements, reliability, availability, scalability, security, cost, interoperability, power, and regulatory constraints. The communication protocols utilized in the HAN include ZigBee, Wi-Fi, Z-Wave, and Bluetooth, all of which are designed for short-range communication. In the NAN, protocols such as 6LoWPAN, Wi-SUN, PLC, and LoRaWAN are used, while in the WAN, fiber optic communication, Long Term Evolution (LTE)/5G, microwave, radio, and Internet Protocol/Multi-Protocol Label Switching (IP/MPLS) are employed. The choice of communication protocols, including International Electrotechnical Commission (IEC) 61850, DNP3, Modbus, and Message Queuing Telemetry Transport (MQTT), guarantees interoperability and standardized data exchange between diverse systems (Qays et al., 2023).

3.4.5 Cloud computing and data analytics

Distributed grid components generate a massive amount of data, which is aggregated, stored, processed, and analyzed using cloud computing and advanced data analytics. Scalable cloud infrastructure and intelligent data processing are the key technologies behind the smart grid. Edge computing is used alongside the cloud to overcome latency challenges in real-time control applications, enabling faster local processing while leveraging the cloud's scalability and data management.

In the smart grid, AMI, PMUs, SCADA systems, and IoT-enabled IEDs generate uninterrupted, time-synchronized data at high temporal resolution. Cloud computing platforms commonly deployed through service models such as Infrastructure as a Service (IaaS), Platform as a Service (PaaS), and Software as a Service (SaaS), deliver the computational elasticity and horizontally scalable storage architectures essential for processing and managing heterogeneous data types, including structured, semi-structured, and unstructured datasets.

Real-time advanced data analytics enable utilities to perform critical functions such as short- and long-term load forecasting, predictive fault detection, prognostics, DSM, and asset condition monitoring. Advanced ML and deep learning techniques, predictive analytics, and real-time stream processing frameworks (e.g., Apache Kafka, Spark Streaming) extract useful insights, optimize operational decision-making, and enhance situational awareness. Cloud-based EMS and DMS can benefit from these analytics-driven capabilities to dynamically manage supply-demand equilibrium and coordinate DERs (Khalif et al., 2025).

Additionally, cloud computing supports edge cloud integration by providing centralized infrastructure, management tools, and scalability that edge environments utilize to function effectively, thus creating a hybrid architecture where the centralized cloud manages computationally intensive tasks and decentralized edge systems handle latency-sensitive tasks. In short, cloud computing and data analytics are the backbone of smart grid computation and intelligence, enabling a multitude of services such as real-time insights, automated decision-making, and flexible control over complex, distributed energy networks.

3.4.6 Artificial intelligence and machine learning in grid management

As various grid components generate a large amount of high-resolution data, AI and ML can play a pivotal role in processing and analyzing it to develop models for predictions, forecasting, and decision-making. This facilitates the evolution of conventional rule-based control into adaptive, data-driven decision-making frameworks and advanced smart grid management systems. By utilizing large datasets generated by AMI, PMUs, SCADA systems, and IEDs across the grid, these technologies can discover complex patterns and behaviors, predict trends, equipment and systems failures, optimize load distribution, and assist dynamic strategic planning. This results in a responsive, resilient, and efficient grid that continuously learns from historical and real-time data to adapt to changing conditions.

The most widely used ML techniques in a smart grid include algorithms such as support vector machines (SVMs), random forests, neural networks, and paradigms like reinforcement learning. These are trained to identify faults and equipment failures, predict short-term and long-term load patterns, and improve grid efficiency. Short-term, medium-term, and long-term load forecasting and predictive modeling of renewable energy sources utilize time-series prediction models such as recurrent neural networks (RNNs), transformers, XGBoost, and hybrid models for efficient demand forecasting, operational planning, and energy dispatch techniques. The capability of AI and ML algorithms to analyze vast datasets and patterns has made them instrumental in DR optimization. These algorithms identify effective load curtailment strategies by learning from load consumption and dynamic tariff signals. AI-driven models are also used to ensure grid reliability by implementing predictive maintenance models that analyze temperature, vibration, partial discharge, oil quality, electrical stress, ambient conditions, electrical signatures, and age and maintenance records to estimate the remaining useful life of grid assets. These assets include transformers, circuit breakers, switchgear, substation equipment, ESSs, smart meters, and sensors, etc. AI algorithms, such as reinforcement learning and multi-agent systems, play a pivotal role in the integration of smart grids and DERs by enabling real-time coordination, decentralized control, and optimization within microgrid EMS (Khedkar & Ramesh, 2021).

AI plays a crucial role in the smart grid's security by reducing false positives, identifying threats in real-time, and classifying intrusion attempts. AI-based intrusion detection systems, predictive analytics, automated response, and vulnerability management form a comprehensive cybersecurity framework that enables proactive threat detection, rapid attack mitigation, and robust protection of systems and infrastructure. Cloud and edge computing are also enhanced by AI models that use large datasets in the cloud for training purposes and enable high-speed inference at the grid edge.

The growing integration of renewable energy sources, EVs, and active energy prosumers has increased the complexity of the smart grid. The various grid components generate large amounts of data, enabling revolutionary technologies of AI and ML to facilitate optimization, predictions, resilience, security, and autonomous decision-making, thereby shaping the foundation of next-generation EMSs.

3.5 Challenges and barriers to implementation

The most important challenges associated with implementing a fully automated smart grid to its full potential are discussed in this section, with an overview provided in Fig. 3.5 (Chenine et al., 2014; Colak et al., 2016; Fan et al., 2021; Khalid, 2024; Mazhar et al., 2023b). The challenges and barriers are grouped under three main categories: technical integration challenges, economic and policy constraints, and cybersecurity and privacy risks. Table 3.4 links each identified challenge to its corresponding emerging mitigation solution.

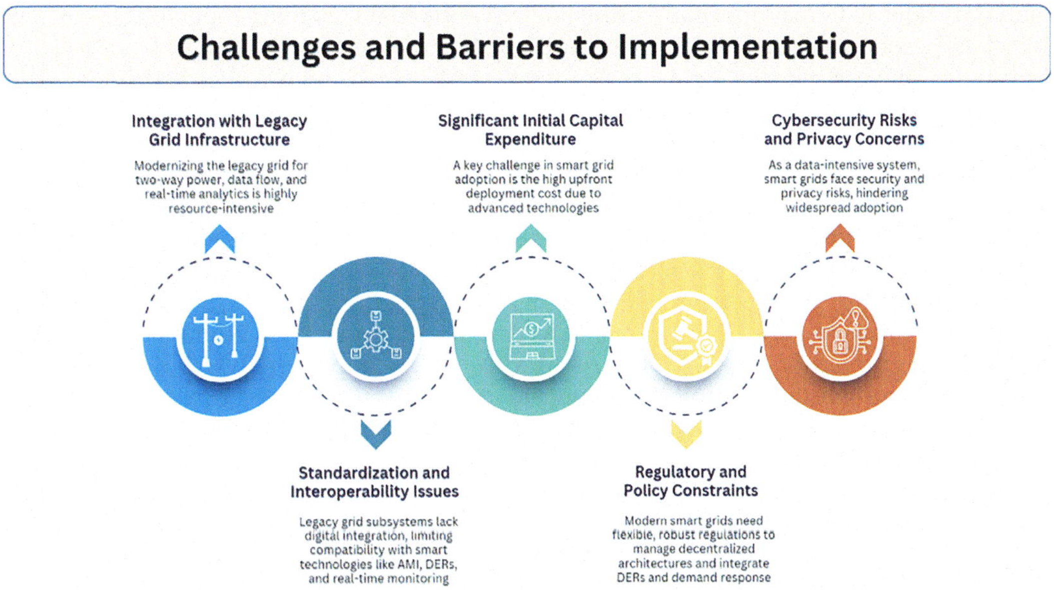

FIGURE 3.5 Challenges and barriers to implementation of the smart grid.

TABLE 3.4 Smart grid challenges and corresponding mitigation strategies.

Challenge category	Specific challenge	Mitigation / Emerging solution
Technical integration challenges	Legacy grid infrastructure	Protocol gateways, retrofitting interfaces, hybrid communication architectures
	Lack of interoperability	Adoption of IEC 61850, open standards, unified data models, interoperable middleware
	Integration of heterogeneous devices	IoT platforms, standardized application programming interfaces (APIs), and device certification frameworks
Economic and policy constraints	High upfront investment costs	Cost-sharing models, government incentives, public-private partnerships (PPP)
	Regulatory uncertainty	Updated regulatory frameworks, performance-based regulation, and harmonized policies
	Slow utility adoption	Pilot programs, demonstration projects, and economic impact assessments
Cybersecurity and privacy risks	Vulnerabilities in grid communication networks	Zero trust architecture, continuous authentication, secure key management
	Data privacy concerns	Privacy-preserving data analytics, encryption, and anonymization
	Increased attack surface due to IoT integration	Intrusion detection systems (IDS), network segmentation, and security-by-design engineering

IoT, *Internet of Things*.

3.5.1 Technical integration challenges

3.5.1.1 Integration with legacy grid infrastructure

A significant challenge to the smart grid is achieving seamless integration with the aging and complex traditional grid infrastructure, much of which was developed decades ago. Traditionally, the electric grid was deployed for unidirectional power flow from remotely located large centralized power plants to end consumers, lacking in sensing and communication capabilities. Upgrading the huge infrastructure of the legacy electric grid to support bidirectional power flow and data flow, with real-time monitoring and advanced data analytics capabilities, is resource-intensive in both technical and economic terms. This transition is challenging due to the technical complexity of achieving seamless connectivity between modern and legacy systems, as well as the high cost of retrofitting grid components and subsystems. Another significant challenge is the need for advanced control, coordination, and management due to system complexity. This includes integration with DERs, DR, EV charging networks, and managing a large amount of data generated by smart meters and IEDs. The DERs comprising solar panels, wind turbines, and ESSs, with intermittent and fluctuating generation tendencies, require advanced control and management to ensure grid stability. The DR programs require dynamic two-way interactions, necessitating significant changes and advancements that are nonexistent in the legacy grid. The integration of EV charging networks also requires expensive upgrades, management of grid overloading, mitigation of consumption spikes, and improved coordination. Hence, for a smooth transition to a next-generation intelligent electric grid, seamless and economically efficient integration between modern smart grid technologies and legacy infrastructure is important, which mandates smart planning, technical development, and substantial capital investment.

Integrating legacy protocols like Modbus RTU and DNP3 into modern IP-based systems presents a range of technical challenges. Designed for serial communication over RS-232 or RS-485, these protocols lack native support for IP networking, making seamless interaction with ethernet or cloud-based systems difficult. Key obstacles include incompatible data formats, stringent timing requirements, restricted addressing schemes, and a lack of built-in security features. Protocol gateways play a vital role in enabling this integration by translating legacy communications into IP-compatible formats, encapsulating serial data into network packets, and often introducing capabilities such as buffering, data normalization, and basic security improvements. However, successful deployment and maintenance of these gateways requires deep expertise in both legacy protocol operation and modern network architecture to ensure dependable and secure integration.

3.5.1.2 *Standardization and interoperability issues*

The transition from the legacy electric grid to a smart grid critically depends on standardization to ensure compatibility and interoperability. Standardization refers to the practice of aligning data and processes with a uniform set of rules or standards. This ensures consistency and compatibility across the various subsystems and processes (Box 3.2).

BOX 3.2

"Key Standards for Interoperability and Security."

- **IEC 61850 – Utility automation Open Systems Interconnection (OSI):** Application–Session; extends to transport/network via Manufacturing Message Specification/Generic Object Oriented Substation Event/Sampled Values (MMS/GOOSE/SV). **Smart Grid Architecture Model (SGAM):** Process/Field zones: communication, information, component layers.
- **IEEE 2030.5 – DER & DR OSI:** Application, transport, network (REST over Transmission Control Protocol/Internet Protocol (TCP/IP)). **SGAM:** Customer–DER domains; communication, information layers.
- **CIM (IEC 61970/61968) – Semantic data model OSI:** Application, presentation. **SGAM:** Information, business layers (transmission, distribution, market).
- **NISTIR 7628 – Cybersecurity framework OSI:** Cross-cutting (all layers). **SGAM:** Cross-layer (component, communication, information, function, business).

Legacy grid subsystems rely on proprietary protocols and have limited digital integration, which hinders their ability to seamlessly connect with smart grid components such as AMI, DERs, and real-time monitoring systems. There is a need for universally adopted standards for interconnection and communication across components, vendors, and technologies. These include communication standards for information exchange between devices and subsystems in the smart grid, interconnection standards for connecting power-generating systems to the grid, data and information standards to support consistent data formatting and transmission across systems, smart metering standards to ensure interoperability among metering systems and back-end infrastructure systems, and cybersecurity standards to protect the authenticity and confidentiality of smart energy workflows. The interconnectivity of diverse devices across multiple layers of smart grid architecture becomes challenging due to differences in communication protocols, varying data formats, and standards, necessitating a cohesive framework for interoperability through standardized protocols. Much effort has been invested in this task, as demonstrated by standards such as IEC 61850, Institute of Electrical and Electronics Engineers (IEEE) 2030.5, IEEE 1547, Common Information Model (CIM), Device Language Message Specification/Companion Specification for energy Metering (DLMS/COSEM), and NISTIR 7628, among others. IEC 61850 serves as a foundational standard for communication in electrical substations, enabling seamless interoperability between IEDs through standardized data models and communication protocols. Supporting this framework, IEC 62351 outlines key cybersecurity measures for power system communications, helping to protect critical infrastructure from cyber threats and ensuring the confidentiality and integrity of transmitted data. More effort is needed to accelerate the transition to a fully automated grid for sustainable energy innovation (Strasser et al., 2024).

3.5.2 Economic and policy constraints

3.5.2.1 *Significant initial capital expenditure*

A primary challenge in transitioning toward a fully automated smart grid lies in the high preliminary cost and expenditure requirements associated with its deployment. All the subsystems that make up a smart grid architecture and its operations require significant capital investment due to their reliance on advanced technologies. The implementation of AMI, smart sensing systems, communication networks, cloud data centers, ESSs, automated substations, and cybersecurity frameworks requires a substantial monetary commitment. These components utilize state-of-the-art hardware and sophisticated software, demanding continuous monitoring, updates, and maintenance to function effectively. Governments must also allocate substantial funding to train the workforce to operate and maintain high-tech equipment, software platforms such as SCADA and EMS, and to support the long-term management of these complex systems. Moreover, this expenditure is further exacerbated by the additional requirements of interoperability, scalability, and robust security. The integration of DERs, such as solar panels and EVs, also requires costly upgrades to the existing grid infrastructure, adding to the overall system cost.

Modernizing electric grids represents a significant financial challenge for developing countries, where limited resources must be carefully allocated among competing national needs. Paradoxically, many of these nations rely on outdated grid infrastructure that urgently requires upgrades to improve reliability, boost efficiency, and accommodate renewable energy sources. Yet, the substantial upfront costs of modernization remain a major obstacle. India, for example, which is home to one of the world's largest and most complex power systems, faces serious difficulties in revamping its aging infrastructure amid rapid urbanization and growing energy demand. To address these issues, the Indian government has introduced programs such as the Smart Cities Mission and the Ujwal Distribution Company (DISCOM) Assurance Yojana (UDAY), which aim to enhance grid performance and strengthen the financial position of utilities. However, the success of these initiatives often hinges on support from international financial institutions like the World Bank, the Asian Development Bank, and other multilateral agencies. These organizations play a crucial role by offering not only low-interest funding but also technical expertise and policy support to enable sustainable and resilient grid development in emerging economies.

Challenges also increase when multiple funding parties, beyond just the government, are involved in grid management and maintenance. While financial challenges can be mitigated by innovative funding models such as public-private partnerships and other incentives that guarantee return on investment, long payback periods, and uncertainty regarding regulatory support and cost recovery often discourage private investors.

3.5.2.2 *Regulatory and policy constraints*

The legacy regulatory and policy frameworks were tailored to a centralized and less complex grid infrastructure. In contrast, modern smart grid systems require robust and flexible complex regulatory structures to manage decentralized autonomous architecture and integration of DERs and DR capabilities. These regulatory and policy-related barriers hinder large-scale acceptance and successful implementation of smart grid technologies. There is a lack of standardized regulations for interoperability among smart grid components, technologies, and retailers, leading to fragmented systems and operational inefficiencies. Another challenge to innovations in these technologies is inconsistency in rules and regulations related to cybersecurity, privacy, and consumer rights enforcement in different countries. It serves as a barrier to international cooperation and innovation. Additionally, market barriers may exist for new participants, such as aggregators, microgrid operators, and EV owners, seeking to participate in energy trading (Jørgensen et al., 2025).

The uncertain regulatory support also affects the cost recovery mechanisms for utilities, resulting in a reluctance to make investments. The policy frameworks concerning market and pricing mechanisms, grid and infrastructure access, regulatory and consumer protections, operational and market innovation, and clean energy incentives often fall behind technological progress, potentially hindering innovation and delaying the shift toward a more sustainable energy system. Overcoming these challenges requires comprehensive, future-focused regulations and policies that establish clear guidelines, encourage investment, and foster collaboration among participants, including governments, utilities, technology vendors, and consumers.

3.5.3 Cybersecurity risks and privacy concerns

Similar to all highly interconnected and data-intensive digital systems, smart grid networks are vulnerable to security attacks and privacy concerns, posing a major obstacle to their widespread adoption. Technologies enabling real-time, data-driven operations, including AMI, IoT, cloud computing, and communication infrastructure, amplify the grid's exposure to cyber threats by expanding potential points of attack. Smart grid networks are vulnerable to various cyberattacks, including distributed denial-of-service, man-in-the-middle attacks, malware infections, data manipulation (including false data injection), phishing and social engineering, unauthorized access to control systems, supply chain attacks, insider attacks, and firmware and software exploits. Successful cyberattacks on smart grid networks can have severe and far-reaching consequences, disrupting the energy sector and society at large. There can be interruptions to the electricity supply, causing blackouts, damage to critical infrastructure, an impact on grid stability, and manipulation of pricing mechanisms. Besides operational turmoil, such breaches can trigger financial losses, dent consumer trust, and jeopardize confidential user data and information. In highly orchestrated, advanced cases, attackers could control grid operations, causing damage to assets and system breakdown. At the same time, privacy concerns also become challenging due to the large amount of data collection involved. This data includes consumers' energy consumption data, appliance performance data, and occupancy details. If accessed without authorization or as a result of a breach, this data can be misused, leading to identity theft, intrusive surveillance, and other types of data exploitation (Chenine et al., 2014; Mazhar et al., 2023b). Along with authentication, authorization, and non-repudiation, it is crucial to ensure the core security principles of confidentiality, integrity, and availability (CIA) of data. Techniques such as data encryption, multi-factor authentication, intrusion detection, firewalls, antivirus software, data loss prevention, patch management, zero trust architecture, network segmentation, and regular vulnerability assessments are required. Administrative and policy measures must also be implemented, including security awareness training, an incident response plan, access control policies, backup and disaster recovery procedures, a security governance framework, and strategies for third-party risk management. To protect user privacy and build public trust, transparent data governance policies are required, along with regulatory compliance and robust consumer consent mechanisms. As smart grid technologies continue to develop, cybersecurity must be integrated as a core design principle, not an addendum, requiring ongoing investment in resilient security frameworks, incident handling strategy, and workforce cybersecurity awareness programs.

3.6 Smart grid applications

Smart grid applications encompass a range of technologies and software solutions designed to enhance the monitoring, regulation, productivity, and consistency of the electric power grid. These applications utilize advanced technologies in automation, communication, and data analytics to monitor and control the generation, transmission, distribution, and consumption systems. These include real-time grid monitoring, fault detection, the DSM, peak load management, decentralized power generation, DER management, and EV integration, as shown in Fig. 3.6. These applications make the power grid more intelligent, flexible, and efficient.

Real-time grid monitoring and fault detection is a critical application that examines data from sensors and smart meters to identify component malfunctions, abnormalities, and faults. The goal is to enable faster response times and prevent power outages. A key application of smart grid technology aimed at optimizing user energy consumption is DSM and peak load reduction. Based on real-time energy consumption patterns and grid conditions, utilities encourage consumers to manage their energy usage, thus improving energy efficiency and grid stability. An important grid application to enhance energy resilience and sustainability is microgrids and decentralized power generation. Microgrids are localized grid systems that often integrate renewable energy resources and can operate independently or in coordination with the main grid. By enabling decentralized power generation closer to the point of use, these systems minimize transmission losses and increase grid reliability.

An important smart grid application is DER management, which assists in managing renewable energy resources such as solar panels, wind turbines, and battery storage, as a part of the grid infrastructure. It is vital to efficiently

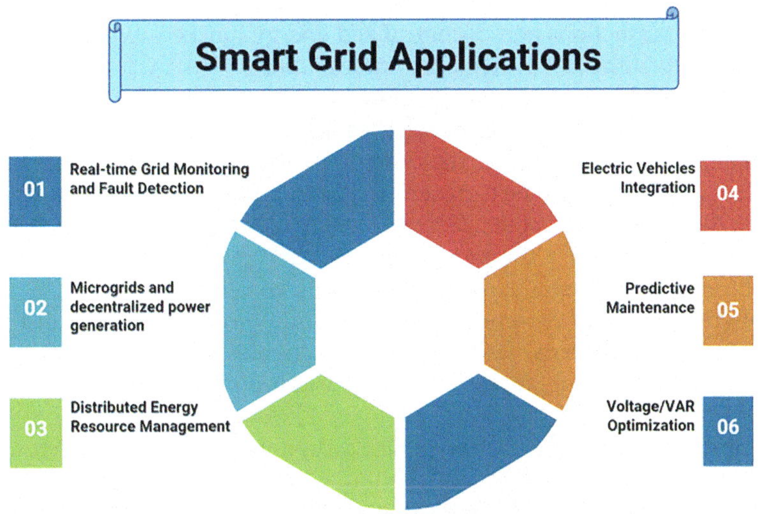

FIGURE 3.6 Illustration of some important applications of the smart grid.

integrate and manage these resources to balance demand and supply in real-time, thus maximizing the benefits of renewable resources for a low-carbon energy future. A breakthrough application of the smart grid is EV integration, which has transformed EVs from energy consumers to dynamic grid assets. Vehicle-to-Grid (V2G) technology supports EVs to return excess stored energy to the grid, helping to balance supply and demand, support grid stability, and enhance overall energy efficiency (Khan et al., 2025; Nafi et al., 2016; Powell et al., 2024). Another critical application is predictive maintenance, which utilizes sensor data and advanced analytics to identify potential equipment failures before they happen. This proactive approach enables utilities to perform timely maintenance, minimize unplanned downtime, and prolong the lifespan of assets. Likewise, VVO is essential for the automatic regulation of voltage levels and reactive power throughout the distribution network. By continuously fine-tuning these parameters, VVO helps reduce energy losses, enhance system efficiency, and ensure power quality remains within regulatory standards.

3.7 Future of the smart grid

The smart grid is envisioned to play a key role in achieving a sustainable energy future. This pertains to a future where global energy consumption aligns with environmentally viable, economically feasible, and socially responsible ways. Such a future can be achieved by transitioning from fossil fuels and harmful practices to cleaner, safer, and more reliable energy systems that safeguard the planet and improve the quality of life globally. The smart grid forms the foundation for reforming the energy system to be cleaner, more efficient, and capable of meeting environmental and economic goals. This is because the smart grid facilitates the integration of renewable energy resources into the existing grid infrastructure, thus supporting and promoting the wide-scale use of such resources. It represents a significant advancement toward an energy system that is not only efficient and reliable but also resilient and increasingly autonomous, capable of adapting intelligently to changing demands and conditions while minimizing environmental impact (Ethirajan & Mangaiyarkarasi, 2025)

Over the last decade, the rapid evolution and refinement of several technologies have significantly impacted advancements in the smart grid, providing the underlying framework and acting as the bedrock supporting its development. Some of the revolutionary and impactful technologies include cybersecurity and blockchain, 5G wireless networks, IoT, AI, ML, cloud and edge computing, big data analytics, digital twin technology, and renewable energy sources. Continued advancements in these technologies will play a pivotal role in shaping the future of the smart grid.

Advancements in cybersecurity are crucial to protecting highly sensitive grid equipment and operations against threats and intrusions. Without strong protection against such threats, the fully automated grid, now increasingly driven by AI-integrated grid automation, would be vulnerable to compromise, resulting in vast-scale disruptions. As the grid evolves toward a more interconnected Internet of Energy (IoE), blockchain technology aids secure and transparent peer-to-peer energy transactions and decentralized energy markets, where consumers can trade surplus electricity with peers or the grid without the need for intermediaries. At the same time, 5G wireless networks provide the ultra-low latency and high-speed data communication required among millions of interconnected devices throughout the grid, while resilience engineering in smart grids ensures that these complex, digitalized energy systems can anticipate, withstand, and recover from disturbances. Together, these innovations represent the core future technologies that will define the next-generation of smart grid infrastructure (Akagic & Džafić, 2024; Alam et al., 2025; Butt et al., 2021; Gonçalves et al., 2025; Kiliç et al., 2024; Laroussi et al., 2023; Tuan et al., 2025).

The following key takeaways summarize the most important insights from this section regarding the future of smart grids, highlighting technological, sustainability, and security considerations.

Convergence of technologies: Smart grids represent the integration of information and communication technology (ICT), AI, and renewable energy systems. This convergence enables real-time monitoring, predictive analytics, automation, and intelligent decision-making, transforming the traditional grid into a more responsive and adaptive energy network.

Sustainability through innovation: The integration of DERs, digital twins, and IoT-driven analytics allows utilities to optimize energy production, distribution, and consumption, while reducing carbon footprints. These innovations support the creation of a resilient and sustainable energy ecosystem, where renewable energy, energy storage, and consumer assets interact seamlessly.

Security and governance: As smart grids scale, robust cybersecurity measures and governance frameworks become critical. Techniques such as AI/ML-based behavioral analytics, deception technology, secure DevOps, cloud security posture management, and blockchain help protect data, prevent intrusions, and ensure reliable, safe, and transparent grid operations. Strong governance ensures compliance with regulations and builds public trust in the evolving energy landscape.

In the future, the smart grid will rely on IoT to enable real-time monitoring, control, and automation. AI, ML, and big data analytics will analyze large amounts of data from these IoT devices for forecasting, fault identification, and optimization. Digital twin technology will create real-time virtual simulations of the grid, enabling utilities to model, predict, and optimize operations to prevent disruptions. As these technologies continue to advance, they will enable the smart grid to become increasingly efficient and robust. The IoE harnesses these advanced technologies to realize a sophisticated future smart grid, where energy systems, storage units, EVs, and IEDs seamlessly interact to create a smarter, more efficient, and sustainable energy ecosystem (Box 3.3).

BOX 3.3

"Future Directions: Toward IoE and Digital Twin Integration"

- **Internet of Energy (IoE):** Fully connect DERs, EVs, storage, and intelligent devices for **real-time energy management and adaptive load balancing.**
- **Digital Twins:** Simulate and optimize grid operations for **proactive fault prevention, predictive maintenance, and dynamic resource allocation.**
- **IoE–Digital Twin integration:** Enable **automated, data-driven decision-making,** enhancing **resilience, sustainability, and efficiency.**
- **Supporting technologies:** Advanced **IoT, AI/ML, and big data,** combined with robust **cybersecurity**, will drive safe and reliable future smart grids.

Future innovations in grid management will also focus on enhancing cybersecurity. Some emerging and advanced security measures include behavioral analytics that uses AI/ML to detect anomalies in user or system behavior, deception technology, where fake assets are deployed to mislead intruders and detect breaches, cloud security posture management, to ensure security in cloud management, secure DevOps which integrates security throughout the software development lifecycle, and blockchain for integrity verification, which ensures tamper-proof records and audit trails. These improvements will allow grids to operate more efficiently (Butt et al., 2021; Ethirajan & Mangaiyarkarasi, 2025).

References

Akagic, A., & Džafić, I. (2024). Enhancing smart grid resilience with deep learning anomaly detection prior to state estimation. *Engineering Applications of Artificial Intelligence, 127*. doi:10.1016/j.engappai.2023.107368.

Alam, Md. M., Hossain, M. J., Habib, Md. A., Arafat, M. Y., & Hannan, M. A. (2025). Artificial intelligence integrated grid systems: Technologies, potential frameworks, challenges, and research directions. *Renewable and Sustainable Energy Reviews, 211*. doi:10.1016/j.rser.2024.115251.

Avdaković, S., Dedović, M. M., Hasković, E., & Jašarevic, Z. (2025). Communication technologies for the smart grid—A short review and a case-study analysis. *Electrotechnical Review /Elektrotehniski Vestnik, 92*(3), 86–96. https://ev.fe.uni-lj.si/3-2025/Avdakovic.pdf.

Ayub Khan, A., Ali Laghari, A., Rashid, M., Li, H., Javed, A. R., & Gadekallu, T. R. (2023). Artificial intelligence and blockchain technology for secure smart grid and power distribution automation: A state-of-the-art review. *Sustainable Energy Technologies and Assessments, 57*, 103282. doi:10.1016/j.seta.2023.103282.

Bakare, M. S., Abdulkarim, A., Zeeshan, M., & Shuaibu, A. N. (2023). A comprehensive overview on demand side energy management towards smart grids: Challenges, solutions, and future direction. *Energy Informatics, 6*(1). doi:10.1186/s42162-023-00262-7.

Butt, O. M., Zulqarnain, M., & Butt, T. M. (2021). Recent advancement in smart grid technology: Future prospects in the electrical power network. *Ain Shams Engineering Journal, 12*(1), 687–695. http://www.elsevier.com/wps/find/journaldescription.cws_home/724208/description#description. doi:10.1016/j.asej.2020.05.004.

Chenine, M., Ullberg, J., Nordstrom, L., Wu, Y., & Ericsson, G. N. (2014). A framework for wide-area monitoring and control systems interoperability and cybersecurity analysis. *IEEE Transactions on Power Delivery, 29*(2), 633–641. doi:10.1109/tpwrd.2013.2279182.

Colak, I., Sagiroglu, S., Fulli, G., Yesilbudak, M., & Covrig, C.-F. (2016). A survey on the critical issues in smart grid technologies. *Renewable and Sustainable Energy Reviews, 54*, 396–405. doi:10.1016/j.rser.2015.10.036.

Erenoğlu, A. yşeK. ü, Sengor, I., & Erdinç, O. (2024). Power system resiliency: A comprehensive overview from implementation aspects and innovative concepts. *Energy Nexus, 15*. doi:10.1016/j.nexus.2024.100311.

Ethirajan, V., & Mangaiyarkarasi, S. P. (2025). An in-depth survey of latest progress in smart grids: paving the way for a sustainable future through renewable energy resources. *Journal of Electrical Systems and Information Technology, 12*(1). doi:10.1186/s43067-025-00195-z.

Fan, D., Ren, Y. i, Feng, Q., Liu, Y., Wang, Z., & Lin, J. (2021). Restoration of smart grids: Current status, challenges, and opportunities. *Renewable and Sustainable Energy Reviews, 143*, 110909. doi:10.1016/j.rser.2021.110909.

Fang, Z., Shek, J. K. H., & Sun, W. (2025). A review of grid-connected hybrid energy storage systems: Sizing configurations, control strategies, and future directions. *Journal of Energy Storage, 118*. doi:10.1016/j.est.2025.116226.

Feng, C., Wang, Y. i, Chen, Q., Ding, Y. i, Strbac, G., & Kang, C. (2021). Smart grid encounters edge computing: Opportunities and applications. *Advances in Applied Energy, 1*, 100006. doi:10.1016/j.adapen.2020.100006.

Gonçalves, V. fP., Serrano, A. L. M., Rodrigues, G. A. P., de Oliveira, M. N., Meneguette, R. I., Bispo, G. D., Peixoto, M. G. M., & Filho, G. P. R. (2025). An empirically validated framework for automated and personalized residential energy-management integrating large language models and the internet of energy. *Energies, 18*(14). doi:10.3390/en18143744.

Hui, H., Ding, Y. i, Shi, Q., Li, F., Song, Y., & Yan, J. (2020). 5G network-based Internet of Things for demand response in smart grid: A survey on application potential. *Applied Energy, 257*, 113972. doi:10.1016/j.apenergy.2019.113972.

Jørgensen, B. N., Gunasekaran, S. S., & Ma, Z. G. (2025). Impact of EU Laws on AI adoption in smart grids: A review of regulatory barriers, technological challenges, and stakeholder benefits. *Energies, 18*(12). doi:10.3390/en18123002.

Kang, C., Kirschen, D., & Green, T. C. (2023). The evolution of smart grids. *Proceedings of the IEEE, 111*(7), 691–693. doi:10.1109/jproc.2023.3284213.

Khalid, M. (2024). Smart grids and renewable energy systems: Perspectives and grid integration challenges. *Energy Strategy Reviews, 51*, 101299. doi:10.1016/j.esr.2024.101299.

Khalif, A. A., Mohamoud, H. M., Hirsi, A., Jama, M., Mohamud, A. H., Audah, L., Ahmed, M., Abdirahman, A. M., & Adi, P. D. P. (2025). Smart grid data management and analytics using cloud computing: trends and future directions. In *International conference on new trends in computing sciences, ICTCS 2025* (pp. 472–477). IEEE. doi:10.1109/ICTCS65341.2025.10989444.

Khan, B., Getachew, H., & Alhelou, H. H. (2020). Components of the smart-grid system. In *Solving urban infrastructure problems using smart city technologies: Handbook on planning, design, development, and regulation* (pp. 385–397). Elsevier. doi:10.1016/B978-0-12-816816-5.00017-6.

Khan, S., Mazhar, T., Shahzad, T., Amir khan, M., Ur Rehman, A., & Hamam, H. (2025). Integration of smart grid with Industry 5.0: Applications, challenges and solutions. *Measurement: Energy, 5*, 100031. doi:10.1016/j.meaene.2024.100031.

Khedkar, M. K., & Ramesh, B. (2021). AI and ML for the smart grid. In *Intelligent renewable energy systems: Integrating artificial intelligence techniques and optimization algorithms* (pp. 287–306). Wiley. doi:10.1002/9781119786306.ch11.

Kiasari, M., Ghaffari, M., & Aly, H. H. (2024). A Comprehensive review of the current status of smart grid technologies for renewable energies integration and future trends: The role of machine learning and energy storage systems. *Energies, 17*(16). doi:10.3390/en17164128.

Kiliç, S. K., Özdemir, K., Yavanoğlu, U., & Özdemir, S. (2024). Enhancing smart grid efficiency through AI technologies. In *Proceedings IEEE international conference on Big Data, BigData 2024* (pp. 7068–7074). IEEE. doi:10.1109/BigData62323.2024.10825117.

Laroussi, I., Huan, L., & Xiusheng, Z. (2023). How will the internet of energy (IoE) revolutionize the electricity sector? A techno-economic review. *Materials Today: Proceedings, 72*, 3297–3311. doi:10.1016/j.matpr.2022.07.323.

Liu, Y., Liu, C., Tao, J., Liu, S., Wang, X., & Zhang, X. (2024). Corrective evaluation of response capabilities of flexible demand-side resources considering communication delay in smart grids. *Electronics, 13*(14). doi:10.3390/electronics13142795.

Mazhar, T., Irfan, H. M., Haq, I., Ullah, I., Ashraf, M., Shloul, T. A., Ghadi, Y. Y., & Imran, D. H. E. (2023a). Analysis of challenges and solutions of IoT in smart grids using AI and machine learning techniques: A review. *Electronics, 12*(1). www.mdpi.com/journal/electronics. doi:10.3390/electronics12010242.

Mazhar, T., Muhammad Irfan, H., Khan, S., Haq, I., Ullah, I., Iqbal, M., & Hamam, H. (2023b). Analysis of cyber security attacks and its solutions for the smart grid using machine learning and blockchain methods. *Future Internet, 15*(2), 83. doi:10.3390/fi15020083.

Momoh, J. (2012). *Smart grid: Fundamentals of design and analysis*. Wiley. doi:10.1002/9781118156117.

Nafi, N. S., Ahmed, K., Gregory, M. A., & Datta, M. (2016). A survey of smart grid architectures, applications, benefits and standardization. *Journal of Network and Computer Applications, 76*, 23–36. https://www.sciencedirect.com/science/article/pii/S1084804516302314. doi:10.1016/j.jnca.2016.10.003.

Ngar-yin Mah, D., Wu, Y.-Y., & Hills, P. R. (2017). Explaining the role of incumbent utilities in sustainable energy transitions: A case study of the smart grid development in China. *Energy Policy, 109*, 794–806. doi:10.1016/j.enpol.2017.06.059.

Ohanu, C. P., Rufai, S. A., & Oluchi, U. C. (2024). A comprehensive review of recent developments in smart grid through renewable energy resources integration. *Heliyon, 10*(3). https://www.sciencedirect.com/science/journal/24058440. doi:10.1016/j.heliyon.2024.e25705.

Ponce-Jara, M. A., Ruiz, E., Gil, R., Sancristóbal, E., Pérez-Molina, C., & Castro, M. (2017). Smart Grid: Assessment of the past and present in developed and developing countries. *Energy Strategy Reviews, 18*, 38–52. doi:10.1016/j.esr.2017.09.011.

Powell, J., McCafferty-Leroux, A., Hilal, W., & Gadsden, S. A. (2024). Smart grids: A comprehensive survey of challenges, industry applications, and future trends. *Energy Reports, 11*, 5760–5785. https://www.sciencedirect.com/science/journal/23524847. doi:10.1016/j.egyr.2024.05.051.

Qays, M. d. O., Ahmad, I., Abu-Siada, A., Hossain, M. d. L., & Yasmin, F. (2023). Key communication technologies, applications, protocols and future guides for IoT-assisted smart grid systems: A review. *Energy Reports, 9*, 2440–2452. doi:10.1016/j.egyr.2023.01.085.

Rahbarimagham, H., & Gharehpetian, G. B. (2025). The effect of smart transformers on the optimal management of a microgrid. *Electric Power Systems Research, 238*. doi:10.1016/j.epsr.2024.111044.

Strasser, T. I., Widl, E., Kuchenbuch, R. éA., Lázaro-Elorriaga, L., Laraudogoitia, B. T., Ginocchi, M., Penthong, T., Ponci, F., Gyrard, A., Kung, A., Mac Gregor, C. A., Montero, C. G., & Algaba, E. R. (2024). Towards interoperability testing of smart energy systems—An overview and discussion of possibilities. *IET Conference Proceedings, 2024*(29), 263–268. doi:10.1049/icp.2024.4670.

Fatfinger (2024). The evolution of smart grids and workforce management, https://fatfinger.io/the-evolution-of-smart-grids-and-workforce-management/

N/A (2025). The smart grid – history, present and future. *Smart grids for the many Linköping University*. Linköping University https://smartgridsforthe
many.se/en/research-focus/smartgrids-history-present-future.

Tuan, L. A., Bouloumpasis, I., & Steen, D. (2025). Resilience-driven planning and real-time control strategies: Challenges and solutions. *Current Sustainable/Renewable Energy Reports, 12*(1). doi:10.1007/s40518-025-00273-w.

Ugwu, J., Odo, K. C., Ohanu, C. P., García, J., & Georgious, R. (2023). Comprehensive review of renewable energy communication modeling for smart systems. *Energies, 16*(1). http://www.mdpi.com/journal/energies/. doi:10.3390/en16010409.

CHAPTER

4

Introduction to Digital Twin technology for power grids

Tanja Pavleska[a], *Massimiliano Masi*[b], *Helder Aranha*[c], *Giovanni Paolo Sellitto*[d]

[a]Laboratory for Open Systems and Networks, Jozef Stefan Institute, Ljubljana, Slovenia [b]Autostrade per L'Italia SpA, Rome, Italy [c]Independent researcher, Lisbon, Portugal [d]Independent researcher, Rome, Italy

4.1 Introduction

The evolution of electrical power systems toward increased complexity, decentralization, and digitalization has brought about unprecedented challenges in grid planning, operation, and maintenance. In response to these challenges, Digital Twin (DT) technology has emerged as a transformative solution, offering dynamic, data-driven virtual models that mirror the real-time behavior and state of physical systems. DTs offer virtual representations of physical assets or systems that can be continuously updated through sensor data and computational models, enabling predictive, diagnostic, and prescriptive analytics across the grid lifecycle (Grieves & Vickers, 2016).

While DTs were initially developed for aerospace and manufacturing applications, their adoption within the energy sector is rapidly expanding. In the context of power grids, DTs replicate subsystems such as substations, transformers, or entire network topologies, capturing their real-time behavior and facilitating continuous monitoring, simulation, and optimization. Their implementation in grid environments has shown measurable impact in enabling condition-based maintenance, improving asset utilization, enhancing resiliency, and informing strategic planning (Abo-Khalil, 2023; Bazmohammadi et al., 2022).

At a technical level, DTs are structured in multilayered architectures that include physical, data, model, simulation, and application layers (Kritzinger et al., 2018). These layers collectively enable the integration of field data with advanced simulation and decision-support tools. Moreover, DTs exhibit varying levels of fidelity, from static digital models and real-time shadows to fully interactive DTs with bidirectional control capabilities, depending on the degree of connectivity, model accuracy, and computational capability (Bragatto et al., 2023; Brosinsky et al., 2020). Hence, their implementation in power grids is underpinned by a convergence of enabling technologies such as artificial intelligence (AI), edge/cloud computing, data analytics, and advanced visualization platforms. These technologies support the core DT functionalities, facilitating a shift from reactive to proactive and autonomous grid operations.

Recent research highlights the growing role of DTs in supporting decentralized grid management, integrating renewable generation, and ensuring cybersecurity. Particularly, DTs are being used to simulate and evaluate cyber-physical threats, providing a safe testbed for intrusion detection, anomaly classification, and vulnerability assessment (Ayyalusamy et al., 2022; Coppolino et al., 2023). However, the integration of DTs into real-world power systems remains complex, constrained by issues like interoperability with legacy infrastructure, data silos, and limited standardization. Moreover, as distributed energy resources (DERs) and prosumers proliferate, the grid's architecture is evolving from a centralized "hub-and-spoke" model to a more distributed and adaptive configuration. In such environments, DTs must not only reflect operational dynamics but also support lifecycle governance.

Despite their promise, DTs also face challenges in real-world deployment, including data integration complexity, model validation, computational demands, and cybersecurity risks. Ongoing advancements in AI, communication networks, and simulation platforms should aim at addressing these issues and paving the way for wider adoption and more sophisticated applications.

Digital Twin Technology and Smart Grid.
DOI: https://doi.org/10.1016/B978-0-443-36664-2.00016-2

This chapter responds to these challenges by offering a comprehensive analysis of the state of research on DTs in power grids. Through a systematic semantic and bibliometric review, we identify key research trajectories, technological enablers, and implementation gaps. Based on this analysis, we propose a novel cybersecurity-oriented DT architecture that addresses lifecycle governance and interoperability, and demonstrate its application through a detailed proof-of-concept use case. Our contribution seeks to bridge the gap between conceptual modeling and operational deployment, and to inform future development of scalable, secure, and standards-aligned DT frameworks for energy systems.

4.2 Comprehensive analysis of the state of the art

4.2.1 Research methodology

The methodology for constructing the visual and analytical map of research on DT technology for power grids is grounded in a multi-phase process that combines semantic bibliometric analysis with systematic screening and thematic synthesis. This approach enables both a macro-level understanding of the field's intellectual structure and a micro-level extraction of actionable insights. For this purpose, we use VOSviewer[a], a specialized software tool that uses advanced text mining and natural language processing techniques to generate a co-occurrence network, where nodes represent terms and edges represent their frequency of co-mention within the same documents (van Eck & Waltman, 2010). VOSviewer was selected due to its methodological alignment with the research objective of uncovering thematic structures in the DT literature, but also because it offers a rigorous approach for extraction and visualization, which is essential for identifying dominant research clusters and semantic patterns (van Eck & Waltman, 2010). Other tools, like BibExcel, are primarily designed for data preprocessing and require external tools for network visualization (Persson et al., 2009). Similarly, CitNetExplorer specializes in citation network exploration but lacks integrated term co-occurrence features (van Eck & Waltman, 2014). While CiteSpace (Chen, 2006) and Bibliometrix/Biblioshiny (Aria & Cuccurullo, 2017) do offer complementary functionalities (such as burst detection or thematic evolution), they are less optimized for the specific task of generating detailed co-occurrence maps based on textual metadata. VOSviewer provides a cohesive environment for both data processing and interactive visual analysis. Given the focus on mapping conceptual proximities and identifying latent research gaps, the clustering algorithm and distance-based layout of VOSviewer proved to be the most suitable choice for supporting the semantic exploration required in this work.

The first step involves obtaining a suitable dataset of publications for constructing and analyzing the co-occurrence of scientific terms within the field of interest in a given period. To begin, a targeted search query is constructed to reflect the scope of interest, focusing on the intersection of DTs and power grids, as well as the architectural (system) considerations of their application in the domain. This query is run across major scientific publication databases (Scopus, Web of Science, Google Scholar, and ACM) through the VOSviewer interface or data import features. The output is a rich corpus of thousands of documents, forming the initial dataset for semantic mapping. The visualization helps to reveal distinct clusters of research based on term co-occurrence patterns, which are color-coded and spatially organized to indicate thematic proximity. These clusters allow identifying major areas of research concentration, interconnected themes across disciplines, emerging topics, and peripheral or underexplored domains. In addition, the visual layout provides immediate insights into how research communities emerge and how concepts interrelate, serving as a valuable heuristic for identifying maturity zones and knowledge gaps in the field.

To enhance the precision and relevance of the subsequent analysis, the semantic mapping was refined, filtering the dataset to remove duplicated or low-quality records and excluding studies that do not offer technical depth or direct relevance to DT applications in power systems. In addition, the scope was limited to peer-reviewed journal articles and high-quality conference proceedings. This screening narrows the focus to a smaller subset of documents that are closely aligned with the identified clusters and research questions. This curated set forms the basis for deeper qualitative and quantitative investigation. Hence, the final phase represents a deep-dive analysis of the refined set of studies, where each selected document is reviewed to extract specific information across dimensions such as modeling approaches, application domains, technological enablers, validation methods, and implementation technology.

This approach yields a comprehensive synthesis of the current state of the art, identifying strategic insights and uncovering research gaps. The outcome is not only a clearer view of how DTs are shaping the future of power grids, but also a set of targeted recommendations for advancing research, development, and deployment.

[a] https://www.vosviewer.com/

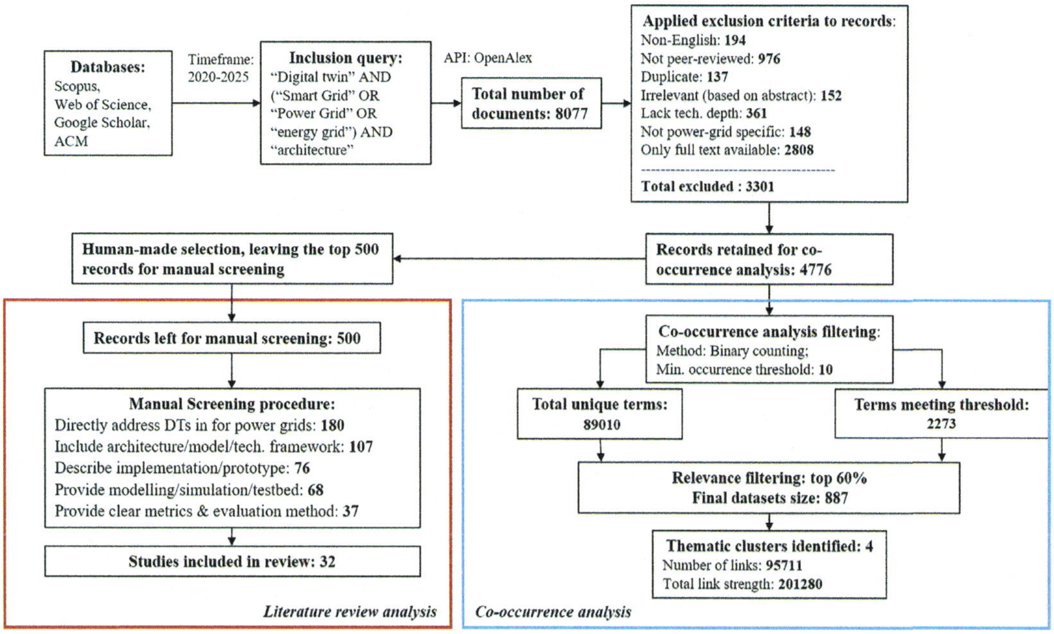

FIGURE 4.1 Joint workflow for the literature review and the co-occurrence analysis.

4.2.2 Analysis and visualization of the research literature

A combination of semantic analysis and bibliometric filtering guided the systematic modeling process, which was carried out using the VOSviewer tool. To construct a robust dataset, we queried the largest scientific databases with the following search string:

```
"Digital Twin" AND ("Smart Grid" OR "Power Grid" OR "energy grid") AND "architecture"
```

The scope of the search was limited to publications from 2020 to 2025, ensuring the inclusion of only the most recent and relevant research. This initial query yielded a corpus of 8077 documents (records), which then underwent further refinement based on semantic criteria and predefined exclusion filters to enhance dataset quality and focus. After applying the exclusion criteria, 4776 documents were left for the term co-occurrence analysis. The overall procedure for processing these documents for both the co-occurrence and the literature review analysis is shown in the diagram in Fig. 4.1. The figure illustrates the decision for term filtering/document exclusion made at each step, together with the results of that decision.

For this analysis, we applied a binary counting method, where only the presence or absence of a term in a document is considered, regardless of how frequently it appears. A minimum occurrence threshold was set to filter out low-frequency noise; in our case, terms must appear in at least 10 documents to be included. Out of a total of 89,010 terms, 2273 met this threshold. To further refine the semantic map, a relevance score was calculated, selecting the top 60% of terms based on their distribution and connectivity. This resulted in a final visualization dataset comprising 887 terms, forming 95,711 links with a total link strength of 201,280, which is a strong indication of the semantic richness and interconnectedness of the research field. This method enabled us to identify core thematic clusters, uncover emerging research areas, and establish a quantitative foundation for further in-depth analysis.

After applying the exclusion criteria, 4776 documents were left for the term co-occurrence analysis. For this analysis, we applied a binary counting method, where only the presence or absence of a term in a document is considered, regardless of how frequently it appears. A minimum occurrence threshold is also set to filter out low-frequency noise; in our case, terms must appear in at least 10 documents to be included. Out of a total of 89,010 terms, 2273 met this threshold. To further refine the semantic map, a relevance score is calculated, selecting the top 60% of terms based on their distribution and connectivity. This results in a final visualization dataset comprising 887 terms, forming 95,711 links with a total link strength of 201,280, which is a strong indication of the semantic richness and interconnectedness of the research field. This method enables us to identify core thematic clusters, uncover emerging research areas, and establish a quantitative foundation for further in-depth analysis.

The final visualization represents a scientific co-occurrence network visualization depicting four major clusters, each corresponding to a different research focus. This is shown in Fig. 4.2.

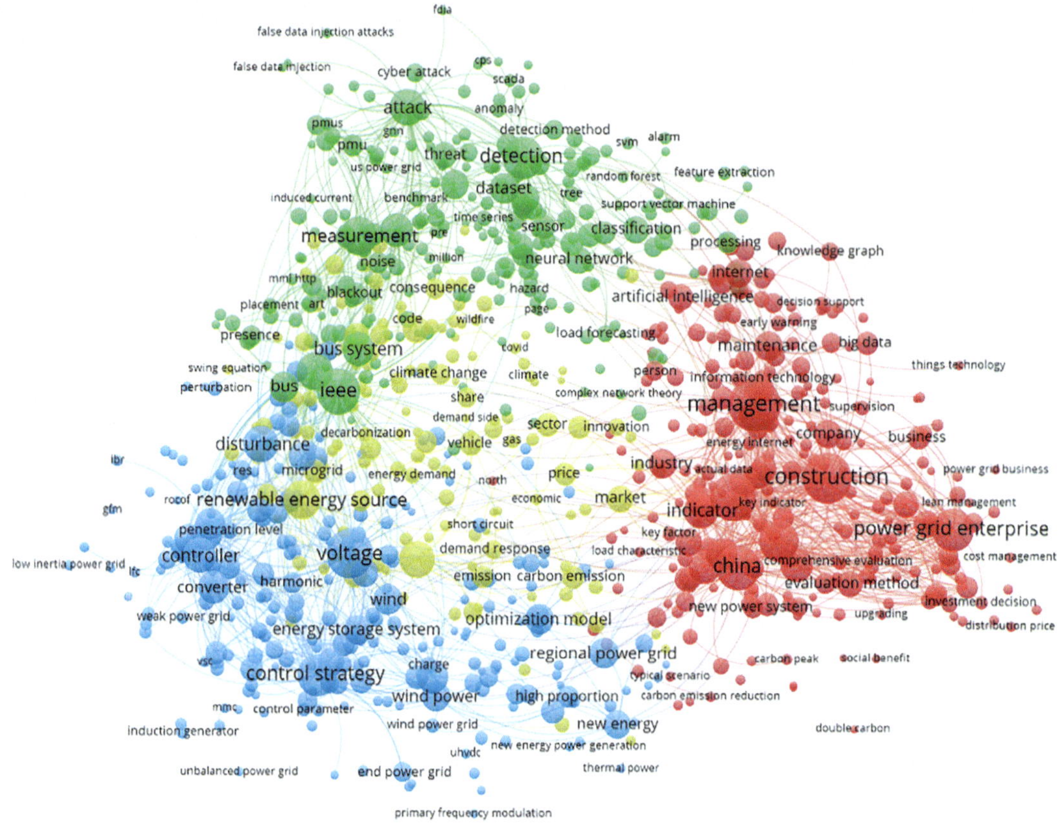

FIGURE 4.2 Co-occurrence network visualization of the research literature.

In addition to revealing keyword clusters, the figure shows how research topics group and relate within the DT ecosystem in energy systems. Each node represents a keyword, whereas node size reflects keyword frequency. Edge thickness indicates co-occurrence strength, the colors separate the clusters (themes), and the proximity of objects reflects their semantic relatedness.

The red cluster in the co-occurrence network reflects a strong focus on enterprise strategy, management, and evaluation, particularly in the context of large-scale DT adoption. Central keywords suggest institutional and strategic deployment of DT technologies, with a notable geographic emphasis on China, where there is a significant concentration of research and practical implementation in policy-driven and state-supported environments. The themes emerging from this cluster highlight the role of DTs in enhancing operational efficiency, enabling predictive and preventive maintenance, and supporting lean management practices. Organizations are increasingly using DTs not only as technical tools but as integral components of business intelligence and asset management. Other terms like indicator, evaluation, and digital economy point to a rising demand for quantifiable performance metrics and structured cost-benefit analysis to justify and optimize DT investments. Moreover, the presence of big data and information technology underscores the convergence of DTs with enterprise-level IT systems, facilitating data-driven decision-making at both the operational and strategic levels.

The green cluster represents a concentrated body of research on cybersecurity, detection, and measurement within the context of DT technology for power grids. The central themes here reflect the importance of DTs as tools for identifying and mitigating cyber-physical threats. Techniques such as anomaly detection and defense against false data injection attacks are prominent, supported by the use of advanced machine learning (ML) algorithms, including neural networks and feature extraction methods. A defining feature of this cluster is its emphasis on real-time monitoring and analysis. Technologies like phasor measurement units (PMUs) play a critical role, enabling the collection of high-frequency data necessary for fast and accurate detection of grid anomalies. The most common feature extraction techniques used in that context are:

- Principal component analysis (PCA), used to reduce dimensionality by transforming correlated variables into a smaller set of uncorrelated components. PCA helps detect voltage instability, oscillations, or event onset, serving as a precursor to anomaly classifiers.

- Neural network-based autoencoders, which learn compressed latent representations of PMU signals and are particularly useful in unsupervised learning settings where labels (e.g., fault types) are scarce.
- Wavelet transform, capturing both time and frequency characteristics, is useful for detecting transient faults, harmonic distortions, and signal irregularities. It is critical in event detection tasks.
- Statistical time-series features, with descriptive statistics like mean, variance, skewness, kurtosis, or zero-crossing rates, are computationally simple, but very effective for detecting abnormalities in voltage or frequency signals.
- Spectral and entropy-based features (spectral entropy, approximate entropy, and power spectral density) are used to quantify the disorder or variability in PMU streams, which is particularly helpful in dynamic stability analysis.

Feature extraction directly impacts DT fidelity, as it determines how well the virtual model reflects the operational state of the physical grid. In DT implementations aimed at real-time grid health monitoring or cyber-attack detection, feature extraction pipelines are embedded in the data ingestion layer, allowing the DT to remain responsive and predictive without being overloaded by raw data volume.

In addition to the technical terms, the appearance of *IEEE* in the green cluster as a frequently associated term reflects ongoing efforts to align DT implementations with recognized industry standards, ensuring interoperability and reliability. This alignment is particularly important in power grid contexts, where frameworks like IEC 62443 (for industrial cybersecurity), IEEE 2030.5 (for communication with DERs), and IEC 61850 (for substation automation protocols) offer critical guidance for structuring secure and interoperable systems. Similarly, standards like IEEE 1547 govern DER interconnection, while the emerging IEEE P2805 and ISO/IEC 30186 frameworks address the formal representation and maturity evaluation of DTs. Their integration into DT architectures enhances not only technical compatibility across systems but also regulatory compliance and operational resilience. Clearly, the green cluster illustrates a technically advanced domain where DTs serve not only as digital mirrors of physical infrastructure, but also as proactive defense systems used to simulate attack scenarios, test resilience, and support the design of secure smart grid architectures.

The blue cluster represents the technical and engineering core of DT research in power grids, emphasizing areas such as control systems, optimization, and grid dynamics. The dominant keywords here reflect a deep engagement with the mechanisms that maintain grid stability and performance. This cluster is particularly focused on simulating and managing the behavior of dynamic components in the grid, including wind power, energy storage systems, and unbalanced or low-inertia power systems. At the core is the use of DTs as real-time simulation and actuation tools. DT technology is leveraged to model the interactions between converters, storage devices, and fluctuating energy sources, enabling predictive control and fine-tuning of grid parameters such as voltage and frequency. The attention to components like harmonic distortion, voltage source converters, and power output showcases the precision with which DTs are being applied to tackle emerging engineering challenges in decentralized and renewable-heavy grid environments. This reinforces the concept of DTs as active participants in power system control loops, rather than static digital replicas.

Finally, *the yellow cluster* centers on the role of DT technology in advancing sustainability, emission reduction, and the global energy transition. The prominent themes in the cluster highlight a research focus that joins environmental goals with economic modeling. This cluster captures DTs' application as strategic tools to model, simulate, and optimize low-carbon energy systems. In this context, DTs are used for forecasting emissions, analyzing renewable energy penetration, and exploring climate scenarios, enabling energy planners and policymakers to make informed, data-driven decisions. Terms like *optimization model, price,* and *energy demand* reflect a growing integration of energy economics into the design of DT systems, supporting evaluations of market responses, cost-benefit dynamics, and investment risks associated with clean energy adoption. The presence of demand response strategies suggests a move toward adaptive grid behavior, where DTs enable simulation of consumer behavior and flexible load management under varying environmental and market conditions.

4.2.3 Cross-cluster integration

The cross-cluster integration themes in the DT research landscape reveal the increasing interdisciplinarity of the field, where technical, strategic, and environmental aspects converge to create more holistic and responsive energy systems. These thematic intersections highlight the synergistic roles of DTs across different domains and point toward the emergence of comprehensive, multi-functional platforms for grid innovation.

The intersection between the red and green clusters (enterprise management and cybersecurity) illustrates the integration of organizational oversight with real-time threat monitoring. This reflects how DTs are being embedded into enterprise systems not just for asset management and operational analytics, but also for incorporating cybersecurity dashboards, enabling companies to align performance management with risk detection and mitigation strategies. The connection between the green and blue clusters (security and control systems) demonstrates the convergence of intrusion detection mechanisms with dynamic control architectures. Here, DTs serve as both observers and actors, detecting anomalies and simultaneously recalibrating control strategies in response to detected threats. This enables resilient grid operations, particularly in the face of sophisticated cyber-physical attacks. The bridge between the blue and yellow clusters (control and sustainability) shows how DTs are central to managing low-carbon, distributed energy systems. In this integrated view, DTs are not only stabilizing systems through real-time control but also optimizing performance against environmental objectives like emissions reduction and renewable integration. This positions DTs as key enablers of climate-aligned grid operations. Finally, the link between the yellow and red clusters (sustainability and enterprise strategy) demonstrates the growing importance of carbon accounting, Environmental, Social, and Governance (ESG) metrics, and sustainability indicators in corporate energy strategies. DTs help organizations simulate, track, and report environmental performance, making them essential tools for strategic investment, regulatory compliance, and stakeholder engagement.

These cross-cluster themes reinforce the notion that DTs are not isolated technological tools, but integrated systems that operate across technical, managerial, economic, and environmental dimensions, supporting a system-of-systems approach to grid modernization.

4.2.4 Discussion and research opportunities

The insights derived from these analyses offer a high-level understanding of where DT research in the energy sector is currently concentrated, as well as where it may evolve in the near future. One notable pattern is China's dominance in the red cluster. This may reflect a substantial national-scale deployment and research investments in DT technologies by Chinese power enterprises, highlighting China as both a prominent market player and a research hub. Another key insight is the multidisciplinary expansion of the field. While the blue cluster reflects DT's engineering roots in control systems and dynamic modeling, the yellow cluster incorporates economic modeling, sustainability, and policy evaluation, and the red cluster anchors DT in corporate strategy and transformation. This broad spread across clusters illustrates how DTs have evolved from niche simulation tools into complex sociotechnical systems, enabling decision-making at both operational and strategic levels. The green cluster's focus indicates that cybersecurity has emerged as a standalone research pillar within DT applications (Alcaraz & Lopez, 2024). With its dedicated cluster centered around anomaly detection, classification, and threat mitigation, cybersecurity is treated not as an auxiliary feature but as an integral function of DT ecosystems, particularly for critical infrastructure protection. The role of AI and Internet of Things (IoT) is also prominent, reflecting the infusion of AI and ML into DT architectures (Antonesi et al., 2025).

However, the analysis also reveals several critical research gaps and future opportunities. Concepts like *human-in-the-loop* systems, which integrate human decision-makers into DT feedback loops, are noticeably underrepresented (Yan et al., 2025). Similarly, there is a lack of research explicitly addressing *interoperability*, which is vital for integrating DTs with legacy infrastructure and across vendor ecosystems (Bjørnskov et al., 2025). Additionally, *ethical and regulatory considerations*, such as data privacy, accountability in autonomous decisions, and algorithmic transparency, are largely missing from current discussions. Although *IEEE* appears as a keyword, there is little depth in the exploration of *standardization frameworks*, indicating a need for more structured guidelines to support scalable and secure deployments. Finally, there is an absence of emphasis on *cross-border and regional power trading scenarios*, a crucial consideration as grids become increasingly interconnected and decarbonization efforts demand transnational coordination. These gaps present rich opportunities for future exploration, innovation, and policy development in the rapidly expanding field of DT technology for power systems.

4.3 In-depth analysis of the most relevant research approaches

While the co-occurrence analysis with VOSviewer provided valuable insight into the structure of the research field, it did not enable a granular assessment of methodological rigor, practical implementation, or technical contributions. Therefore a second phase of analysis was carried out to identify and examine the most relevant and impactful studies in greater depth. To move from broad mapping to focused insight, the initial dataset underwent progressive refinement using a set of predefined exclusion criteria. The aim was to isolate studies that offered substantive and technically detailed contributions to the field of DTs in power grid systems.

From the initial 4776 studies, the top 500 most relevant documents were selected for manual screening by all authors. Through this filtering process, we arrived at a final curated set of 32 studies that met all or most of the following inclusion criteria:

- The study must directly address DT applications within the context of power grids.
- DTs must be applied specifically to grid systems.
- Studies must include a proposed architecture, model, or technical framework.
- They must go beyond theory, describing system-level implementation or prototype development.
- Detailed modeling approaches, simulation environments, or testbeds must be provided.
- Clear metrics or evaluation methods must be reported.

The studies were assessed holistically, and inclusion was based on whether they satisfied a critical mass of these criteria, rather than any single dimension. Each of the 32 selected studies was analyzed using a standardized data extraction protocol designed to align with the clusters and density trends identified in the VOSviewer analysis. The extraction focused on five analytical dimensions:

- Study focus: Determines the general area within the power grid research, and where the specific approach falls. This helps position the study in some of the previously defined clusters.
- Implementation type: Classifies the modeling paradigm employed by each study as physics-based, data-driven, or hybrid, and documents specific modeling techniques and computational frameworks.
- Grid application area: Maps the study to one or more grid domains, including generation, transmission, distribution, microgrids, and DER integration, identifying specific use cases like state estimation, stability analysis, or renewable coordination.
- Key technology: Records the technologies employed, such as IoT sensors, Supervisory Control And Data Acquisition (SCADA) systems, AI/ML tools, and cloud/edge computing platforms, also noting the level of integration with existing infrastructure and protocols (e.g., IEC 61850, Modbus).
- Performance metrics: Documents the evaluation framework used to validate the DT system. This included the type of validation (e.g., real-time simulation, hardware-in-the-loop) and potential performance metrics reported (accuracy, latency, and simulation fidelity).

This layered approach enabled us to move from broad thematic insight to detailed technical synthesis. Moreover, it allowed us to identify specific gaps and inconsistencies in evaluation methodologies and highlight opportunities for cross-cluster innovation and interdisciplinary integration.

The remainder of this review presents the synthesized findings from these 32 studies, as shown in Table 4.1, reflecting their contributions across the core dimensions described earlier. Each of the table entries is also categorized according to the cluster from the co-occurrence map to which they belong.

Based on this analysis, we can observe that, while research on DT technology in the power and energy sector has grown significantly in recent years, there is a strong need for more intelligent, adaptive, and secure management of the increasingly complex grid systems. The analyzed studies demonstrate a wide variety of applications, ranging from component-level simulations to enterprise-wide planning tools, with varying levels of implementation maturity, technological integration, and performance evaluation metrics.

Abo-Khalil (2023) presents a real-time hybrid simulation platform designed to analyze voltage stability and dynamics, reflecting the increasing importance of DTs in operational decision support. Similarly, Bassey (2024) focuses on renewable energy systems, using DTs for component-level simulation and testing. These studies exemplify the trend of applying DTs to specific technical challenges within the power system lifecycle, particularly in generation and control. They focus on system-level modeling, reflecting the research concentrated within the blue cluster of the co-occurrence analysis and semantic mapping in the previous section. These contributions are pivotal in validating DTs as real-time, closed-loop tools rather than static digital replicas. However, while component and system-level control are well represented, the literature shows fewer examples of multiscale or federated architectures that integrate control functions across layers (component ↔ system ↔ enterprise).

In contrast, Bai and Wang (2022) adopt a more conceptual approach, proposing an architectural framework for intelligent power grid development using DTs. Their analysis incorporates technologies like big data, IoT, and AI, positioning DTs as foundational to broader smart grid initiatives. This reflects an important direction in the literature where DTs are not merely reactive monitoring tools, but strategic enablers of long-term system intelligence and interoperability. These studies are representative of the red cluster from our co-occurrence analysis, structured around the decision-making and enterprise-level problems for using DT technology. However, the analysis shows limited

TABLE 4.1 Detailed analysis of the most relevant studies.

Study	Study focus	Implementation type	Grid application area	Key technologies	Performance metrics
Bai and Wang (2022)	**DT structure and key technologies for the intelligent power grid**	**Hybrid**	**Enterprise-level intelligent operation of the physical grid**	**Big data, cloud computing, IoT, AI**	**Conceptual and structural analysis**
Gutmann et al. (2023)	Power consumption and productivity balance	Hybrid	Data flow between the factory and its DT	Siemens Tecnomatix plant simulation v2201, Shelly 3EM energy sensors, Mosquitto MQTT broker v2.0.15, Telegraf agent v1.24.2, InfluxDB v2.4.0, Plantsim Python	Monte Carlo cross-validation; mean, standard deviation; two-sample t-test, real-time data management toolchain
Thoma et al. (2023)	Microgrid management of energy demand	Hybrid	Microgrid, peak shaving	Nissan electric vehicles, bidirectional charging stations, stationary battery system	Integration of the vehicles into a car-sharing system
Ayyalusamy et al. (2022)	**Cybersecurity analysis**	**Hybrid**	**Smart Grid cyber-physical system**	**MATLAB/Simulink, Raspberry Pi, communication protocols (Modbus, IEC61850, MQTT)**	**Real-time simulation, hardware-in-the-loop**
Coppolino et al. (2023)	Cybersecurity monitoring	Data-driven	Smart Grids	FIWARE, Security Information and Event Management (SIEM)	Real-world case study
Danilczyk et al. (2019)	Microgrid security	Physics-based	Microgrid security	Legacy-comm. techniques and WAMs	Simulation example
Danilczyk et al. (2020)	Grid anomaly detection	Data-driven	Transmission systems	WAMs, PMUs, deep learning Convolutional Neural Network (CNN)	Validation on IEEE 9- and IEEE 39-bus standard benchmarks
Kandasamy et al. (2022)	Cybersecurity testing and research	Hybrid	Microgrid systems	SPEEDGOAT, MATLAB-Simulink, virtual machines, Docker containers	Real-time simulation
Saad et al. (2020)	Energy cyber-physical systems	Hybrid	Grid state estimation, stability analysis	AWS, IoT	Real-time implementation
Sen et al. (2023)	Replication and analysis of realistic multistage cyberattacks on Smart Grids	Hybrid	Cybersecurity in active distribution grids	Pandapower, Docker containers, Containernet, IEC 60870-5-104 and Modbus protocols	Real-time simulation, hardware-in-the-loop
Siraj Khan et al. (2022)	Real-time cyber-physical analysis	Physics-based	Power distribution systems	Hardware-in-the-loop	Hardware-in-the-loop testbed
Abo-Khalil (2023)	**Power system stability**	**Hybrid**	**Voltage/dynamic stability, predictive maintenance, renewable energy integration**	**Real-time hybrid simulation platform**	**Voltage and stability analysis**

(*continued on next page*)

TABLE 4.1 Detailed analysis of the most relevant studies—cont'd

Study	Study focus	Implementation type	Grid application area	Key technologies	Performance metrics
Bassey (2024)	Enhanced design and testing of renewable energy systems	Hybrid	Renewable energy sources – solar and hydroelectric power	MATLAB/Simulink, ANSYS, COMSOL Multiphysics, CAD software, Microsoft Azure IoT, and AWS IoT Core	Simulation-based validation of component performance
Brosinsky et al. (2020)	Energy management systems	Hybrid	Grid state estimation, stability analysis	SCADA, Wide Area Monitoring Systems (WAMS), Digital Dynamic Mirror	Case study
Cai et al. (2021)	Power grid, State estimation	Data-driven	Medium and low-voltage distribution networks	Artificial neural network (ANN), harmonic spectra	Field test
Darbali-Zamora et al. (2021)	Distribution voltage regulation	Hybrid	Distribution grid management	ProDROMOS, Particle Swarm Optimization, OpenDSS	Real-time simulation, field demonstration
Han et al. (2022)	DT prototyping	Physics-based	Renewable energy integration, grid stability	PMUs, Real-Time Digital Simulator, GTNET-SKT card	Real-time simulation, hardware-in-the-loop
Ruhe et al. (2023)	Hierarchical DT architecture	Hybrid	Distribution grids	Real-time interface, RT simulation environment, Opal-RT	Real-time simulation, hardware-in-the-loop
Sundhu et al. (2024)	Islanding control in distribution systems	Hybrid	Grid state estimation, islanding control	SCADA	Evaluation of the actual energy distribution system
Tran et al. (2024)	Voltage control in distribution networks	Data-driven	Voltage control, stability analysis	AI/ML	Validation on a real low-voltage network
Wileman et al. (2023)	Degradation monitoring and fault detection	Physics-based	Power electronic converters	Matlab/Simulink, Simscape toolbox, Wavelet scattering, SVM	DT simulation of the power electronic converter
Bazmohammadi et al. (2022)	**Microgrid concepts and applications**	**Hybrid**	**Microgrids**	**Sensor networks, IoT, high-fidelity models and simulation platforms**	**Comparative analysis of DT applications in microgrids**
Bragatto et al. (2023)	Active distribution networks	Data-driven	Distribution network management	Real-time monitoring, digital simulation frameworks	Continuous operation over a month
Hong et al. (2024)	Networked microgrids	Data-driven	Networked microgrids	Blockchain, LSTM-based controller	Hardware-in-the-loop simulations
Meyer et al. (2022)	Radio-mesh smart meter network	Data-driven	Outage management in a low-voltage area	DT of radio-mesh smart meter network	Experimental data comparison
Nasiri and Kavousi-Fard (2023)	Energy hub management and grid resiliency	Data-driven	Grid stability and resilience, water-power system	Amazon Cloud Service, Bat-based optimization algorithm	Real-time simulation, hardware-in-the-loop
Paldino et al. (2022)	Dynamic thermal rating of transmission lines	Hybrid	Transmission lines of the power system	Python, scikit-learn library	Prediction accuracy, dimensionality reduction with feature selection, error correlation analysis

(continued on next page)

TABLE 4.1 Detailed analysis of the most relevant studies—cont'd

Study	Study focus	Implementation type	Grid application area	Key technologies	Performance metrics
Pelekis et al. (2023)	Promote hydrogen and accelerate the energy transition through Power-to-Gas (P2G) technologies	Hybrid	P2G tech. and regenerative hydrogen fuel cells into the electricity transmission system	PyPSA, OSeMOSYS, Pandapipes, Python-based MLOps framework with XGBoost, Random Forest, NBEATS, TCN, and LSTM. Streamlit, Apache Superset, MATRYCS Visualization Engine, Keycloak	Multi-resolution simulations and optimization
Rana et al. (2024)	Smart Grid estimation	Physics-based	Grid state estimation, renewable energy integration	IoT sensors, computer-aided tools	Real-time simulation, hardware-in-the-loop
Sahoo et al. (2025)	DT-enabled smart microgrids (DT-SMGs)	Hybrid	SMG system, substation, and feeder automation, distributed energy management	N/A	Comparative study on situational awareness, security, and resilience
Somda and Liu (2024)	Resilient power distribution systems	Hybrid	Renewable energy integration, grid stability	10 MW battery energy storage system, 2 MW photovoltaic plant, 5 MW conventional synchronous generator, NS-3 network simulator, GTNETx2 card	Electromagnetic transient simulations using a real-world test case
Özkan et al. (2024)	Microgrids	Data-driven	Microgrid operations	Deep reinforcement learning	Optimization scenario

*Each **bolded** row denotes the start of the cluster, starting from red, to green, to blue, and yellow.*

work that integrates DTs with financial modeling, ESG metrics, and regulatory compliance in the sector. These domains are expected to grow as utilities adopt DTs for broader organizational strategy.

Several studies address the integration of DTs into distributed and decentralized grid contexts. Bazmohammadi et al. (2022), for instance, explore hybrid implementations of DTs in microgrids, comparing various architectures and their operational performance. Moreover, they highlight the role of DTs in carbon accounting and demand response. Bragatto et al. (2023) extend this further by exploring real-time monitoring applications, emphasizing the role of DTs in system resilience and safety assurance, as well as in environmental and economic modeling. These hybrid approaches, combining physical testbeds with digital simulations, are crucial in validating DT frameworks for environments that demand both real-time responsiveness and cyber-physical coordination. This aligns with the yellow cluster, confirming the DTs' growing use for supporting sustainability metrics and demand-side flexibility. However, there remains a lack of standardization in how performance is evaluated, and very few studies explicitly incorporate environmental impact as a core metric.

Finally, a growing subset of DT research focuses on the integration of cyber-physical security measures. Several studies highlight the use of DTs for passive monitoring and active detection of system anomalies. For instance, the work by Han et al. (2022) explores anomaly detection frameworks within DT environments, underlining their importance in safeguarding grid assets from cyber threats such as false data injection. These contributions align with the green cluster in the co-occurrence network, emphasizing the increasing importance of cybersecurity measures. However, the literature review also reveals that, while security is recognized as a critical feature, very few studies implement real-time intrusion detection or resilience simulations as part of their DT architectures, as well as the inclusion of the DT as part of the security-by-design practices. Moreover, security features are rarely integrated into enterprise-level DTs. This marks an important gap in the current body of applied research.

From a technological standpoint, most studies integrate a mix of simulation environments, communication protocols, and analytics platforms. Key technologies such as AI/ML, cloud-edge computing, and co-simulation platforms, although frequently mentioned, are unevenly adopted. Moreover, performance metrics vary widely, with some studies providing quantitative evaluation based on system stability, latency, or power flow accuracy, while others rely on conceptual validation or comparative analysis. This confirms the earlier findings about the need for standardized metrics and benchmarks to evaluate DT implementations and ensure replicability and cross-compatibility. Additionally, the literature also reveals a fragmented landscape and disparities in implementation types, with many DT implementations remaining simulation-based or conceptual, and few operational deployments or validation in industrial settings.

The overall analysis of the research field is indicative of the fact that future DT research should lie in cross-domain solutions that are not only technically robust but also secure, economically viable, and environmentally aligned. The following sections introduce such an approach.

4.4 Use case: Cybersecurity Digital Twin for lifecycle governance of power grids

This section applies the findings of our literature review and focuses on addressing key research gaps identified across the red (enterprise strategy and evaluation) and green (cybersecurity and monitoring) clusters. We present a cybersecurity DT use case designed for power grid governance across the full system lifecycle. This use case specifically targets three underexplored, but critical dimensions in DT deployments for power grids: human-in-the-loop decision-making, regulatory alignment, and interoperability frameworks. Drawing inspiration from Masi et al. (2023), the use case outlines a methodology for designing and operationalizing a cybersecurity DT that extends beyond simulation to become a continuously evolving architectural element of the smart grid system. It adopts a security-by-design approach and integrates threat modeling into the system architecture, ensuring resilience throughout the entire lifecycle.

We first describe the key concepts of the methodology and their relationship within the approach. Then, we illustrate its application to a real-world smart grid scenario. It is important to note that the intermediate models and results are less focused on accuracy and more on coherence.

4.4.1 Key concepts

4.4.1.1 Reference model for information assurance and security

The reference model for information assurance and security (RMIAS) framework defines four key dimensions for embedding security into system design and operation (Cherdantseva & Hilton, 2013). At its core is a generic *Security Development Lifecycle (SDLC)*, designed to be superimposed onto the system's own lifecycle, ensuring that information assurance is addressed from design through operation to retirement. The *Taxonomy* dimension provides a structured classification of cyber-physical assets, facilitating coherent security analysis. The *Security Goals* dimension formalizes desired system behaviors under threat, including objectives such as *Accountability, Auditability, Authenticity, Availability, Confidentiality, Integrity, Nonrepudiation,* and *Privacy*. These dimensions collectively enable a unified framework for risk and *countermeasure* analysis across technical and organizational domains. Crucially, RMIAS assumes human-in-the-loop interaction, promoting collaboration between business leaders, who define what must be protected, and technical experts, who determine how to protect it. This alignment supports cost-effective, adaptive security strategies tailored to evolving operational and threat environments.

4.4.1.2 Enterprise architecture

Modern power grids are complex systems of systems, often incorporating Industrial Automation Control Systems. To manage this complexity, enterprise architecture frameworks such as the Smart Grid Architecture Model (SGAM) provide an abstraction framework of system behavior and inter-component relationships (Cen-Cenelec-Etsi, 2014; Masi et al., 2018). SGAM defines five interoperability layers: (1) Business, (2) Function, (3) Information, (4) Communication, and (5) Component. These layers span across two structural dimensions: *Zones*, representing hierarchical management levels (from Market to Process), and *Domains*, representing the electrical energy conversion chain (from Bulk Generation to Customer Premises). Together, Layers, Zones, and Domains form architectural viewpoints, each addressing stakeholder-specific concerns (Harrison, 2016; Group, 2012). Each SGAM layer can be traced back to a specific standard, as shown in Table 4.2, allowing the design of a robust model for grid architectures. Cybersecurity, in

TABLE 4.2 Alignment between SGAM layers and standards.

SGAM layer	Role	Relevant standards (examples)
Business	Governance, risk, policy, audit.	ISO/IEC 27001; NIST CSF
Function	Security processes, continuity, and SOC playbooks.	IEC 62443-2-x; ISO 22301
Information	Information models; data object security.	CIM (IEC 61970/61968); IEC 62351
Communication	Protocol and communication security.	IEC 61850; IEEE 2030.5; DNP3-SA; IEC 60870-5-104; TLS
Component	Component-level security; DER interconnection.	IEC 62443-3-x/-4-x; IEEE 1547

DER, distributed energy resources.

and of itself a cross-cutting concern, is expressed through a dedicated architectural view. This cybersecurity view, built on abstracted components with intended security behaviors, serves as a crucial intermediate construct for designing concrete DTs, ensuring that risk management is integrated into system design from the outset.

4.4.1.3 *Threat modeling*

Unlike conventional threat-oriented approaches that rely on historical system behavior, the RMIAS framework adopts a goal-oriented model applicable at any stage of the system lifecycle. Once security goals are defined, for example, in an Industrial Automation Control System, organizations can apply threat-based modeling by leveraging existing knowledge, tools, and threat intelligence to identify effective countermeasures. These countermeasures, by design, support the achievement of security goals and the mitigation of vulnerabilities. Their implementation is influenced by multiple factors: the nature of the assets to protect, available organizational resources (e.g., domain expertise, technical capabilities, budget), and the evolving threat landscape. Importantly, human-in-the-loop interaction remains essential to systematically assess and adjust solutions over time, ensuring responses remain contextually and economically viable. A simulation-driven threat modeling construct, capable of rapid experimentation, actionable output, and real-time feedback, is particularly well-suited for cybersecurity DTs. In this work, we adopt the Meta-Attack Language (MAL) framework (Johnson et al., 2018) to model the DT and enable experimentation with threat scenarios.

The following section outlines how these concepts are operationalized step by step, enabling coherent, adaptive, and standards-compliant DT implementation across the full system lifecycle.

4.4.2 Methodology for use case operationalization

STEP 1: The process begins with the definition of a cybersecurity taxonomy, rooted in the RMIAS framework. This taxonomy provides a formalized and multidimensional classification scheme for cyber-physical assets to protect based on attributes such as *Form* (physical or digital), *Sensitivity* (impact of compromise), *Location* (degree of organization's control over the asset), and lifecycle *State* (e.g., deployed, retired). This taxonomy enables precise security reasoning across different system stages, supports modularity in modeling, and serves as a reusable lens through which to structure future updates and risk evaluations.

STEP 2: Next, this taxonomy is mapped onto the SGAM reference architecture, which enables a structured, interoperable view of the system across five layers: business, function, information, communication, and component. Using SGAM's *Zones* and *Domains*, a cybersecurity viewpoint is created to guide the modeling of how assets are exposed to threats at different architectural levels. This step ensures alignment with grid standards and fosters cross-stakeholder transparency in system design, critical for regulatory reporting and enterprise governance.

STEP 3: To embed security into the system's evolution, the RMIAS SDLC is adapted to the organization's development lifecycle. This alignment ensures that security planning, implementation, validation, and monitoring are continuously embedded from design through operation and decommissioning. It also ensures that updates to the system can be assessed proactively using the DT, rather than treated as ad hoc security tasks.

STEP 4: With these architectural foundations in place, the system-wide cybersecurity view (specified in the cybersecurity viewpoint defined in STEP 2) is now built, and the actual construction of the DT begins. A full dissection of assets to protect is carried out across and over SGAM layers, zones, and domains, exposing the security-relevant aspects of the former. In legacy systems where architectural documentation is unavailable, the NIST Cybersecurity

Framework (NIST, 2024) provides a structured path to discover assets and evaluate protections using a technology-agnostic control library tailored to industrial automation systems.

STEP 5: Once assets are cataloged and dissected, a risk assessment is performed. Here, the organization defines RMIAS security goals for each asset or cluster, prioritizing based on sensitivity and business value. This step directly connects business expectations to security responses, allowing for balanced trade-offs between risk appetite, regulatory burden, and cost of countermeasures.

STEP 6: The DT is then instantiated using the MAL, modeling cyber-physical assets, their interconnections, and security countermeasures. The model is run in a simulation environment, e.g., MALSIM (Johnson et al., 2018), allowing analysts to simulate realistic multistage cyberattacks. Vulnerabilities, threat vectors, and attacker profiles are encoded in the model, which is iteratively refined to simulate the impact of countermeasures on key security goals. This simulation becomes a space for experimentation and dialogue between security architects and business decision-makers, where mitigation strategies can be tested and debated without impacting live systems.

STEP 7: Finally, the DT becomes a persistent governance tool. As the organization evolves (through system upgrades, policy changes, or emerging threats), the DT can be updated and reused. It supports continuous decision-making by simulating the impact of proposed changes and validating whether security posture remains within acceptable thresholds. This capability is essential for lifecycle security governance, especially in Smart Grids, where systems are long-lived and exposed to rapidly shifting regulatory and technological pressures.

4.5 Use case contextualization: Predictive cybersecurity maintenance for DER operations

Building upon the use case "Control Reactive Power of a DER Unit" as described in Annex B, paragraph B.2.4 of the SGAM documentation (Group, 2012), we demonstrate the applicability of the previously introduced methodology by illustrating how a cybersecurity DT can support predictive security maintenance in a real-world DER operational scenario. In this scenario, we consider a Distributed Energy Resources (DER) operator responsible for managing a network of rooftop photovoltaic (PV) panels and residential battery storage systems, sited at the Customer Premises attached to the distribution grid. Reflecting the evolution of energy system architectures, the DER was adopted into the Smart Grid Conceptual Model (Group, 2012) as a distinct domain, and later formally recognized in the NIST Framework through NISTIR 7628 (NIST, 2014). This is also an acknowledgment of their growing importance in enabling resilient, decentralized, and sustainable energy systems.

In the original SGAM-based use case, the business objective of the distribution grid operator (DSO) is to control the reactive power output of DER installations in order to stabilize voltage levels within the distribution grid. This supports the broader responsibility put upon DSOs of ensuring power quality and system reliability, enabling energy delivery to end-users under defined economic, technical, and regulatory constraints. To extend the value of this use case, which is essentially mapping existing devices, infrastructures, functions, communication and information standards, its business objective and constraints over the SGAM model, we embed it within the methodology for cybersecurity DT design; it stands transformed from a static operational objective into a living security governance tool. The DT of the system-of-systems described according to SGAM is constructed not only according to its underlying 'electrical behavior' but also to perform continuous threat analysis and vulnerability assessment, allowing both DSO and DER operators to move from reactive responses to predictive maintenance and security planning.

We assume a scenario in which a routine inspection report from field technicians identifies vulnerabilities in both physical protection mechanisms (tamper seals, physical access controls) and cybersecurity infrastructure (outdated firmware, unsecured communication channels to the DER Controller). These findings serve as the trigger for engaging the cybersecurity DT, which is already integrated into the DSO and DER operators' asset and risk management ecosystem.

The DT enables several key functions:

- Asset classification using RMIAS, establishing the sensitivity and exposure level of each DSO and DER component.
- Architectural mapping of intra-DSO and DSO-DER communication pathways and control interfaces onto the SGAM framework to visualize interdependencies.
- Simulation of attack scenarios to assess the impact of cyber-physical intrusions or control logic manipulation on voltage stability and grid safety.
- Prioritization of countermeasures through security goal modeling, balancing mitigation strategies with business constraints.

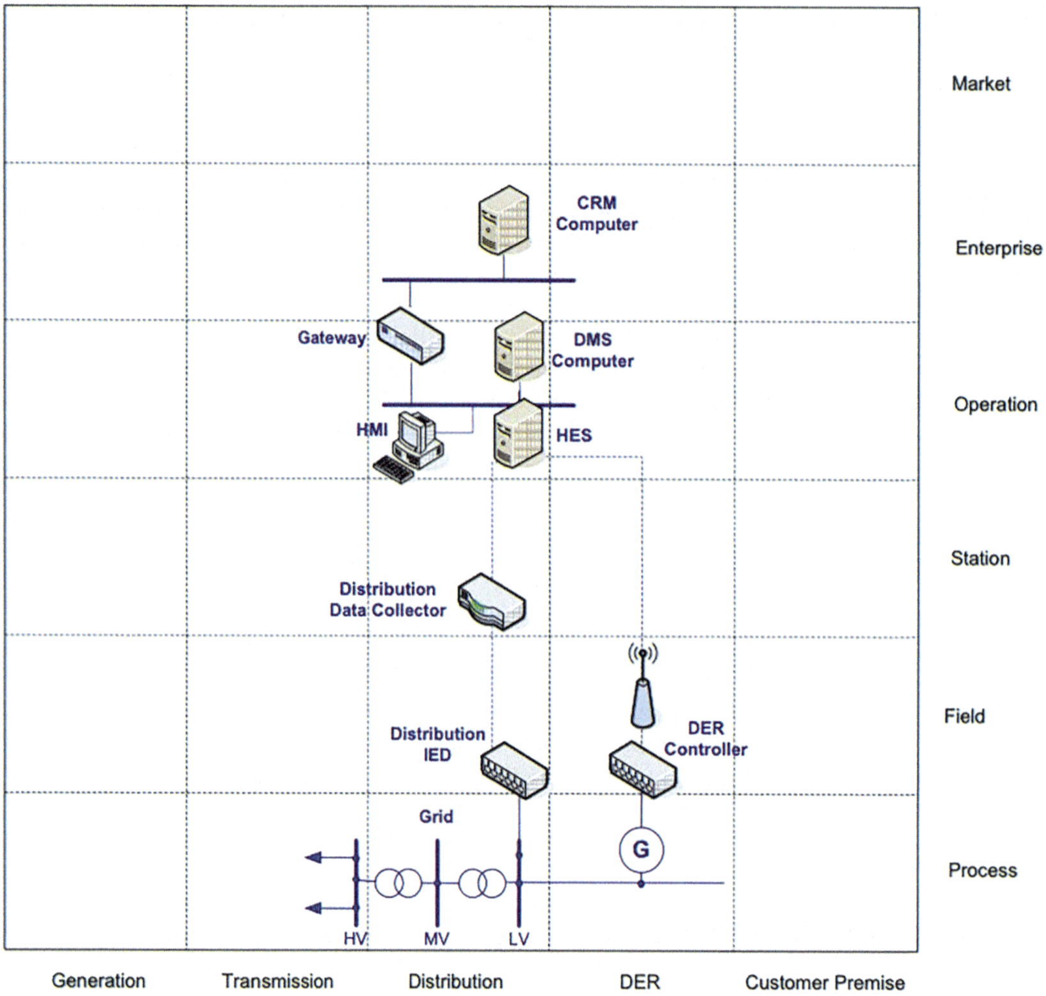

FIGURE 4.3 A component-layer SGAM view of the use case. *SGAM*, Smart Grid Architecture Model.

Fig. 4.3 provides a component-layer SGAM view of the use case. Key actors include Intelligent Electronic Devices (IEDs) in the field, which are responsible for measuring and communicating local voltage and current, as well as actuation upon some measurements in a localized area.

SCADA systems at the distribution operator level provide centralized monitoring and command functionalities, whereas the Distributed Generation injects reactive or active power into the distribution grid, controlled by the DER Controller, interfacing with SCADA.

Next, we show how the DT can become a security-aware system asset that links routine field operations with enterprise-level risk management by employing the introduced methodology.

4.5.1 Extending the SGAM use case through the methodology

Following the methodology, all assets defined in the SGAM-based technology view are translated into a DT model using concepts from the MAL coreLang library (Wideł et al., 2023). This translation is the basis for constructing a simulation-capable, cybersecurity-aware DT that reflects not only functional system behavior but also the underlying vulnerabilities and attack vectors associated with each asset. With that, we are able to extend the utility of SGAM-based planning into the operational phase and ensure that cybersecurity becomes an integrated function of day-to-day grid management, not an afterthought. The specific system model is shown in Fig. 4.4, illustrating the simulated DT model architecture. The boxes represent computational nodes corresponding to key fields and control components within the power grid infrastructure. These include the Distribution *IED*, the DER Controller, and higher-level supervisory systems such as the *Head-End System (HES)* and *Human-Machine Interface (HMI)*. The arrows indicate

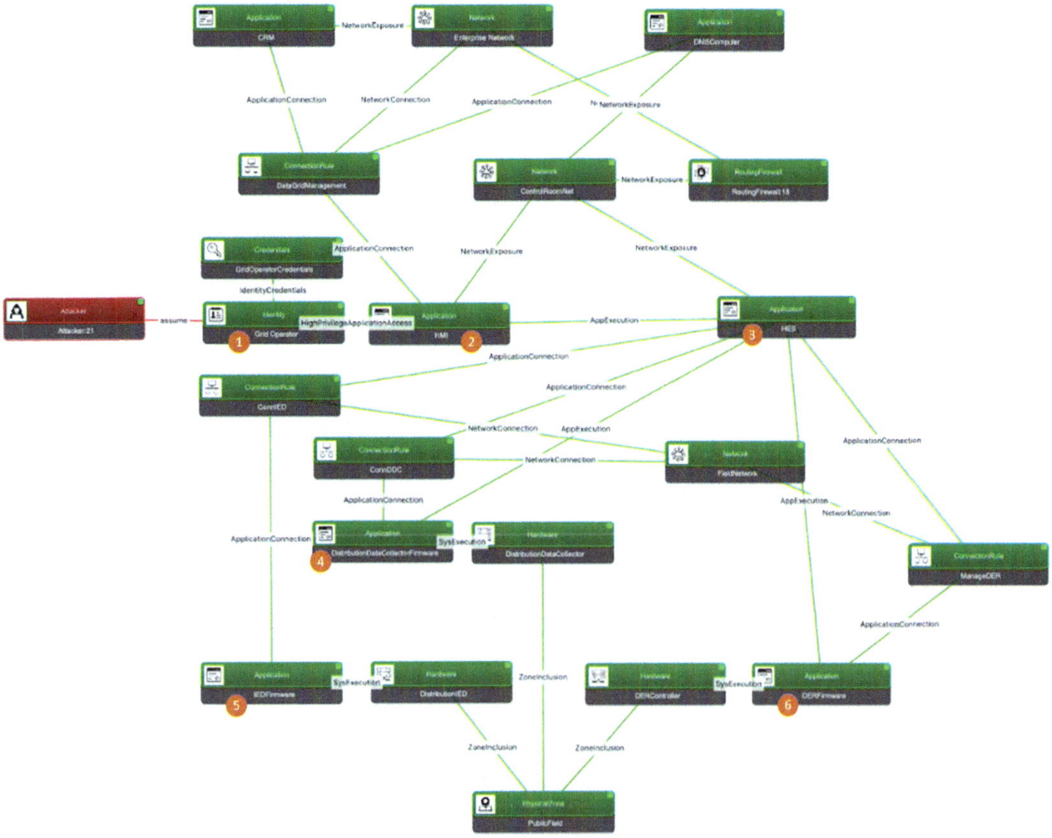

FIGURE 4.4 The system model for simulating the use case with methodology steps marked.

logical communication links over designated network segments. Notably, field devices interact via public or less-trusted network environments, whereas control systems are isolated within a protected Control Room Network. This structural segmentation reflects typical real-world deployment and forms the basis for modeling attack surfaces and defensive containment within the simulation.

The source files of the design and simulation of the use case can be found on GitHub[b]. The MALSIM simulation was executed on a workstation equipped with an Intel Core i9 6-Core CPU @ 2.9 GHz, with 32 GB of RAM, running macOS Sequoia. The attack-defense scenario presented in this use case was completed in approximately 3.2 seconds of wall-clock time, with all DT operations performed in a single-threaded simulation mode.

Within this model, hardware assets (*Distribution IED, Distributed Data Connector*, and *DER Controller*) are instantiated as computational nodes executing firmware. These devices are susceptible to a variety of physical threats, including manipulation (manumission), theft, and operating system compromise (OS poisoning). While known software vulnerabilities specific to these components are not explicitly modeled in this simulation instance, the framework supports future extension with vulnerability databases for increased threat fidelity.

The DT also incorporates two key control applications: the *HES* and the *HMI*. Due to its operator-facing role, the HMI is assumed to require user authentication credentials. In many operational technology (OT) contexts, such credentials are commonly shared among users, poorly maintained, or use weak defaults (e.g., "admin/admin"), thereby posing a significant attack vector. The HMI is also modeled to execute privileged actions on the HES, creating a pathway for lateral privilege escalation. The *HES* itself maintains three connection rules with the *Distribution IED, DER Controller*, and *Distributed Data Connector*, respectively. These communication links are modeled as network-exposed channels, which may be implemented via field-level 4G/5G cellular networks or proprietary fiber-optic infrastructures. Such connectivity reflects real-world deployment practices, but also introduces exposure to remote intrusion attempts and protocol-based attacks.

Both the HES and HMI are dual-hosted within a control room network, often configured as a dedicated virtual local area network (VLAN) for operational control systems. This VLAN is modeled with a firewall boundary, isolating

[b] https://github.com/mascanc/dtwin/tree/main

it from the broader enterprise network, which hosts critical corporate applications such as *Customer Relationship Management* and *Distribution Management Systems*. These enterprise systems access data from the HMI through a defined connection rule, reflecting a common architectural pattern where operational and enterprise domains intersect through shared Application Programming Interfaces or middleware. This connection becomes another choke point for possible data leakage or intrusion if improperly secured.

This system exposes a multilayered attack surface, which is simulated and assessed using the DT. In alignment with IEC 62443-2-1, the model does not quantify the attacker's assumed capabilities directly (e.g., skill level, resource availability). Instead, it focuses on the exploitability of system components, distinguishing between trivially exploitable weaknesses and those requiring advanced skills or prolonged dwell time, which inherently increase the likelihood of attacker detection.

4.5.2 Attack scenario simulation through the methodology

With the modeled system, we can now instantiate a threat-aware DT of the system architecture introduced in the SGAM use case. This DT not only functions as a virtual replica of the operational flows but also as a risk simulation platform capable of forecasting threat propagation, estimating mitigation efficacy, and informing strategic security decisions across the asset lifecycle.

We first revisit each of the steps defined within the methodology, and then present an attack scenario testing different defensive strategies.

STEP 1: This use case assumes that a formal security taxonomy has been established based on the RMIAS framework (Step 1); that all system assets have been classified according to the four dimensions; and that the organization has adopted the core set of security goals defined in RMIAS. These serve as the guiding objectives for all downstream threat modeling, simulation, and governance decisions within the DT environment.

STEP 2: A cybersecurity viewpoint based on the SGAM model has been defined and the corresponding view generated, allowing security architects to visualize dependencies, flows, and potential attack surfaces. This viewpoint ensures interoperability with existing grid infrastructure and regulatory frameworks while enabling traceable analysis across domains and zones, such as Customer, DER, and Distribution.

STEP 3: The organization has integrated the Security Development Lifecycle derived from RMIAS into its broader system development process. This ensures that security is embedded as a foundational requirement from the earliest phases of design through deployment and maintenance. As a result, security assessments, DT updates, and simulation-based threat evaluations become part of routine development and operational governance, rather than isolated or reactive events.

STEP 4: At the time of this analysis, the SGAM-based cybersecurity view remains structurally unchanged, as no new assets requiring protection were identified during the most recent inspection. The architectural model is deemed current, and all existing devices and data flows are accounted for. The organization did not identify any missing design documentation or legacy systems outside the existing architectural model. All assets and their interdependencies have already been captured through prior SGAM mapping and RMIAS lifecycle integration.

STEP 5: However, the attack surface of the system to protect has increased according to the field report findings: during inspection, two previously unrecorded vulnerabilities were identified. The first concerns weak credentials for the grid operator in the HMI, potentially allowing unauthorized access. The second involves a known tampering technique applicable to a specific model of Distribution-level IEDs, which allows adversaries to inject false voltage readings. These false measurements could propagate through the SCADA system, inducing grid-level instability. A targeted risk analysis using the existing DT model confirmed that, while the security goals remain unchanged, the likelihood of intrusion and operational disruption has increased, introducing a need for immediate mitigation.

STEP 6: With the DT instantiated and all system components modeled using the MAL coreLang framework, we are able to emulate the attacker behavior and test different defensive strategies. This stage is central to DT's predictive value, as it enables reasoning about system vulnerabilities, potential attack paths, and the effectiveness of various countermeasures, without exposing the live infrastructure to risk.

STEP 7: As the cybersecurity DT is operational, the simulation model was updated to reflect the two newly identified exploits. These were encoded as distinct threat vectors in the DT simulation environment, allowing for predictive testing of attack impacts. The simulation demonstrated potential effects on grid-voltage stability and unauthorized system access. Countermeasures, including password-hardening protocols and tamper-detection firmware updates, were iteratively simulated and evaluated in terms of cost, feasibility, and effectiveness. A balanced mitigation strategy was jointly selected by the security architect and business stakeholder, ensuring operational resilience and alignment with business constraints.

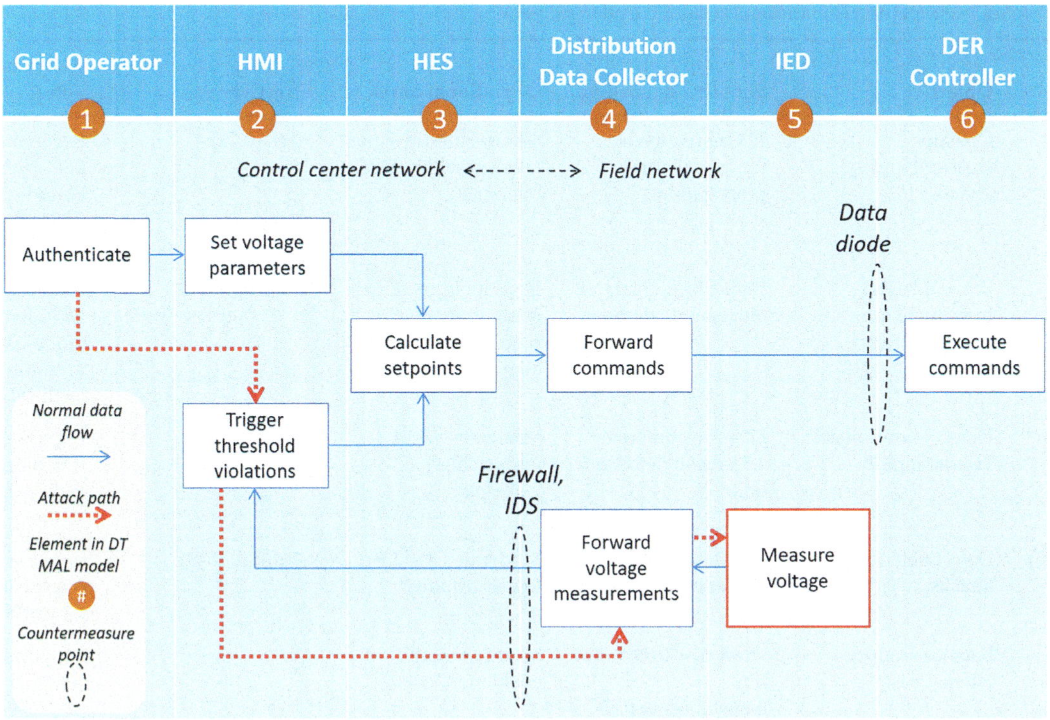

FIGURE 4.5 Threat analysis storyboard with MAL mappings, attack path, and countermeasure points. *MAL*, Meta-Attack Language.

To illustrate this, in the simulation scenario, we represent an attacker exploiting a common vulnerability in the HMI. The simulated vulnerability could be a basic cross-site scripting (XSS) flaw, the presence of a webshell, or weak user credentials, all of which are prevalent in many OT environments. Once the attacker gains access to the HMI, the simulation models their progression through the system toward compromising IEDs, a class of assets identified earlier as critical to preserve for ensuring grid stability and operational integrity.

This scenario is illustrated in Fig. 4.5, and is encoded in MAL using the configuration as follows:

```
rewards:
'DERFirmware:denyFromNetworkingAsset': 10
agents:
'Attacker1':
type: attacker
agent_class: BreadthFirstAttacker
entry_points:
- 'SoftwareVulnerability:attemptExploit'
'Defender1':
type: defender
agent_class: PassiveAgent
```

In this setup, the attacker agent attempts to exploit any reachable vulnerability, beginning with a low-effort entry such as the software weakness in the HMI. A reward of 10 is assigned when the attacker successfully denies or disables the DER firmware through a network-based compromise, emphasizing this objective as the simulation's primary threat concern. As the simulation unfolds in MALSIM, the attacker is able to progress through the modeled system. The event log excerpt given further demonstrates a successful compromise:

```
INFO:malsim.__main__:Iteration 346
INFO:malsim.__main__:Agent "Attacker1" chose action: IEDFirmware:deny
INFO:malsim.mal_simulator:Stepping through iteration 345/1000000000
INFO:malsim.mal_simulator:Attacker agent "Attacker1" stepping through "IEDFirmware:deny"(192).
INFO:malsim.mal_simulator:Attacker agent "Attacker1" compromised "IEDFirmware:deny"(192).
INFO:malsim.mal_simulator:All attackers are terminated
INFO:malsim.mal_simulator:Removing agent Attacker1
```

TABLE 4.3 Smart grid security evaluation mini suite (condensed view).

Measure adopted	Objective	Simulation outcome	Operational impact	Effectiveness	Remarks/technician note
Hardened firewall	Contain cross-network attacks	72% reduction in successful attack propagations	+18 ms control signal latency (within DER tolerance)	High (0.72)	Recommended for field/substation boundaries; validate per IEC 62443 Zones & Conduits
Data diode	Prevent lateral movement into control centers	100% block of lateral movement attempts	Loss of bidirectional communication; requires protocol adaptation	Very High (1.00)	Ideal for telemetry-only DER; hybrid diode-gateway advised for limited control feedback
IDS (Substation Zone)	Detect known attack vectors rapidly	87% detection rate, 2.5 s mean detection time	Minimal system impact; high visibility	High (0.87)	Deploy anomaly-based IDS near switch/aggregation layers
IDS (Field Zone)	Detect field-level attacks	49% detection rate, slower correlation	Moderate CPU load; lower visibility	Medium (0.49)	Complement with centralized correlation for full coverage
Optimization layer	Balance security, latency, cost	Firewall = 0.59 score; IDS (Substation) = 0.65 score	Multicriteria DT evaluation loop	—	Supports iterative configuration tuning (security-latency-cost equilibrium)

```
INFO:malsim.mal_simulator:Removing agent Defender1
–
INFO:malsim.__main__:Game Over.
Total reward "Attacker1" 90
Total reward "Defender1" -90
```

This outcome confirms that, under current configurations, an attacker can traverse the system from a compromised HMI to the DER control layer. Notably, the event log entry "IEDFirmware:deny" indicates that the attacker successfully executed a firmware-level denial action on the IED. In real-world terms, this could render the IED inoperative or unresponsive, effectively disabling its ability to measure or transmit voltage and current data. Such a compromise can lead to a critical loss of situational awareness for the grid operator, potentially delaying or preventing appropriate control responses during abnormal grid conditions. With this insight, the DT becomes a decision support environment for exploring alternative security configurations. Stakeholders can now pose strategic questions such as:

- *What if we configure a hardened firewall at the field network boundary?*
  For example, a simulation of the hardened firewall configuration in our scenario showed a 72% reduction in successful attack propagations from the enterprise network to operational assets. It did introduce an average latency of 18 ms in control signal transmission, which remained within acceptable bounds for DER management operations.
- *Would inserting a data diode on DER communication channels prevent lateral movement?*
  The simulation indicated that the one-way communication constraint imposed by the data diode successfully blocked all lateral movement attempts into the control center infrastructure. However, the diode-limited bidirectional command-and-status communication requires protocol adaptation for critical control loops.
- *How effective would an Intrusion Detection System (IDS) be, and where should it be deployed?*
  Placing an anomaly-based IDS at the substation switch zone, it detected 87% of known attack vectors within 2.5 seconds on average. When placed at the DER field zone, its detection rate dropped to 49%, showing the importance of IDS positioning for both visibility and detection latency.

Table 4.3 shows a condensed view of the simulation choices and results.

Other defensive enhancements can also be added to the model, which would require simulations to be rerun to evaluate impact, trade-offs, and cost-effectiveness. The reward structures in the scenario file can also be adjusted to reflect not just security objectives but also operational constraints, such as the cost of deploying a technology, its maintenance burden, or the increased complexity introduced into system operations. In this way, the cybersecurity DT evolves from a passive monitoring tool to an active architecture optimization platform, capable of supporting iterative, evidence-based decisions. By simulating threats, testing mitigations, and optimizing configurations in a closed virtual loop, grid operators can proactively manage their risk posture while balancing technical, financial, and human resource considerations.

4.5.3 Discussion

The use case illustrates how conceptual frameworks can be effectively operationalized to address key gaps in DT deployment for power grids. By extending the SGAM-based reactive power control scenario with cybersecurity modeling and simulation, we showed how DTs can evolve from passive replicas into proactive, adaptive governance tools. A central contribution is the demonstration of *lifecycle integration*, addressing a critical gap identified in the bibliometric analysis, presented in Section 4.2.4. Unlike approaches that treat security as a postdeployment add-on, the proposed methodology in the use case embeds threat modeling and security goals throughout the system's lifecycle, aligning with the RMIAS. This enables dynamic risk mitigation and fosters organizational readiness without disrupting system operations.

The use case also addressed the research gaps concerning *interoperability and standardization*. By modeling system entities based on the SGAM architecture model and aligning communication and security behavior with relevant standards (e.g., IEC 62443, IEEE 2030.5), the proposed methodology enables stakeholders to reason about security across layers and vendor ecosystems. This reflects the complexity of modern power grids, where components must remain interoperable despite heterogeneity in protocols, vendors, and management domains. It also provides a formal pathway for integrating regulatory concerns into the DT design process, a dimension often overlooked in technology-centric DT implementations.

Another notable contribution to bridging the gaps identified in Section 4.2.4 is the integration of *human-in-the-loop* mechanisms by demonstrating how the simulation environment fosters collaboration between cybersecurity architects and business stakeholders. Through interactive threat-response modeling, the DT enables strategic co-decision making, aligning with the sociotechnical complexity of critical infrastructure governance.

Methodologically, the use of MAL and MALSIM shows the value of simulation platforms for performing dynamic and data-driven threat analysis. The attacker–defender simulation demonstrates how DTs can quantify risks and prioritize mitigations based on both impact and business value, capabilities not commonly found in traditional DT research.

While the use case provides a valuable proof-of-concept, it also presents areas for further refinement. Currently, the methodology illustrated in the use case relies on predefined threat vectors and simplified behavioral assumptions about attackers. Future work will focus on integrating real-time threat intelligence, adaptive learning, and federated DT networks (Abdessadak et al., 2025), as well as incorporating automated policy enforcement and compliance monitoring to support evolving cyber-governance needs (Sinneh & Yanxia, 2026).

4.6 Conclusion

This chapter presented a dual contribution to the evolving field of DTs for power grids by bridging analytical insight and practical implementation.

The first contribution was a bibliometric and semantic co-occurrence analysis, which mapped the academic literature and revealed four dominant research clusters. This revealed conceptual fragmentation and highlighted critical gaps, notably in lifecycle integration, human-in-the-loop systems, cybersecurity modeling, and standards alignment. These findings underscore the need for more integrated, standards-aware, and interdisciplinary DT approaches that embed sociotechnical considerations and regulatory readiness into system design from the outset.

The second contribution addressed these gaps through a cybersecurity-oriented DT architecture and simulation use case. Using a model extended with threat simulation, lifecycle-spanning integration, and human decision loops, the use case demonstrated how DTs can transition from passive observatories to active resilience-enabling systems. Technical feasibility was illustrated using MALSIM, with scenario-specific interactions reflecting real-world threat dynamics. The alignment with IEC 62443 and IEEE 2030.5 further showcased the system's regulatory compliance and standards-aware design.

Together, these contributions substantiate the central claim that DTs for power grids must evolve into lifecycle-aware, adaptive, and multistakeholder platforms, combining technical fidelity with strategic foresight. This synthesis enables a more secure, flexible, and intelligent digital infrastructure for critical energy systems.

To advance this trajectory, further research should:

- **Develop standardized maturity metrics and benchmarks**, including testbeds and performance indices, to evaluate DT functionality across architectural, operational, and security dimensions.
- **Incorporate federated and hierarchical modeling**, enabling multiscale coordination across DERs, microgrids, and national infrastructure while preserving data sovereignty.
- **Embed advanced AI techniques**, including generative AI and context-aware reasoning, to support intelligent diagnostics, adaptive planning, and human-machine collaboration in high-stakes grid scenarios.
- **Leverage edge-cloud orchestration architectures** for latency-sensitive control functions and large-scale strategic simulations.
- **Align with emerging frameworks,** such as the Industrial Metaverse and Industry 5.0, positioning DTs as cognitive, immersive, and participatory platforms rather than static system mirrors.

In this way, the research proposed here offers not only a descriptive framework but a prescriptive foundation for building intelligent, trustworthy, and interoperable DTs for the energy sector.

Key Points: What this chapter delivers:

Method: A semantic+bibliometric pipeline (2020–25) to map DT research for power grids (Scopus, WoS, Google Scholar, and ACM).
Co-occurrence map: Four research clusters (enterprise/strategy; cybersecurity/measurement; control/optimization; sustainability/economics) with cross-cluster links.
Curated review: A focused synthesis of key studies (scope, implementation type, grid area, tech stack, metrics).
Cybersecurity DT methodology: A Lifecycle approach, combining RMIAS (goals/SDLC), SGAM (views), and MAL/MALSIM threat modeling.
Use case (DER/SGAM): Predictive security maintenance for voltage/reactive-power control with attacker–defender simulation.
Gaps & next steps: Human-in-the-loop, standardized metrics/benchmarks, deeper interoperability, and federated DTs.

Key Points: Clusters' properties and takeaways:

Red – strong focus on enterprise strategy: roadmap, asset management, ROI, governance. **Takeaway: DTs as strategic enablers for grid modernization.**
Green – body of research on Cybersecurity/Measurement research on cybersecurity, detection, and measurement within the context of DT technology for power grids. **Takeaway: Security telemetry dominates DT-security overlap.**
Blue – technical and engineering core of DT research focused on control systems, optimization, and grid dynamics. **Takeaway: Focus on real-time constraints and stability.**
Yellow – role of DT technology in advancing Sustainability / Economics: evaluations of market responses, cost-benefit dynamics, and investment risks associated with clean energy adoption. **Takeaway: Linking technical performance with economics and sustainability.**

References

Abdessadak, A., Ghennioui, H., Thirion-Moreau, N. è, Elbhiri, B., Abraim, M., & Merzouk, S. (2025). Digital twin technology and artificial intelligence in energy transition: A comprehensive systematic review of applications. *Energy Reports, 13*, 5196–5218. doi:10.1016/j.egyr.2025.04.060.

Abo-Khalil, A. G. (2023). Digital twin real-time hybrid simulation platform for power system stability. *Case Studies in Thermal Engineering, 49*, 103237. doi:10.1016/j.csite.2023.103237.

Alcaraz, C., & Lopez, J. (2024). Digital Twin-assisted anomaly detection for industrial scenarios. *International Journal of Critical Infrastructure Protection, 47*, 100721. doi:10.1016/j.ijcip.2024.100721.

Antonesi, G., Cioara, T., Anghel, I., Michalakopoulos, V., Sarmas, E., & Toderean, L. (2025). A systematic review of transformers and large language models in the energy sector: Towards agentic digital twins. *Applied Energy, 401*, 126670. doi:10.1016/j.apenergy.2025.126670.

Aria, M., & Cuccurullo, C. (2017). Bibliometrix: An R-tool for comprehensive science mapping analysis. *Journal of Informetrics, 11*(4), 959–975. https://doi.org/10.1016/j.joi.2017.08.007.

Ayyalusamy, V., Sivaneasan, B., Kandasamy, N. K., Xiao, J. F., Abidi, K., & Chandra, A. (2022). Hybrid Digital Twin architecture for power system cyber security analysis. In *Proceedings of the 11th International Conference on Innovative Smart Grid Technologies - Asia, ISGT-Asia 2022. 9798350399660* (pp. 270–274). http://ieeexplore.ieee.org/xpl/mostRecentIssue.jsp?punumber=10003265. doi:10.1109/ISGTAsia54193.2022.10003563.

Bai, H., & Wang, Y. (2022). Digital power grid based on digital twin: Definition, structure and key technologies. *Energy Reports, 8*, 390–397. doi:10.1016/j.egyr.2022.10.328.

Bassey, K. E. (2024). Enhanced design and development simulation and testing. *Engineering Science & Technology Journal, 3*(2), 18–31. doi:10.51594/estj.v3i2.1283.

Bazmohammadi, N., Madary, A., Vasquez, J. C., Mohammadi, H. B., Khan, B., Wu, Y., & Guerrero, J. M. (2022). Microgrid Digital Twins: Concepts, applications, and future trends. *IEEE Access, 10*, 2284–2302. http://ieeexplore.ieee.org/xpl/RecentIssue.jsp?punumber=6287639. doi:10.1109/ACCESS.2021.3138990.

Bjørnskov, J., Thomsen, A., & Jradi, M. (2025). Large-scale field demonstration of an interoperable and ontology-based energy modeling framework for building digital twins. *Applied Energy, 387*, 125597. doi:10.1016/j.apenergy.2025.125597.

Bragatto, T., Bucarelli, M. A., Carere, F., Cresta, M., Gatta, F. M., Geri, A., Maccioni, M., Paulucci, M., Poursoltan, P., & Santori, F. (2023). Near real-time analysis of active distribution networks in a Digital Twin framework: A real case study. *Sustainable Energy, Grids and Networks, 35*, 101128. doi:10.1016/j.segan.2023.101128.

Brosinsky, C., Krebs, R., & Westermann, D. (2020). Embedded Digital Twins in future energy management systems: Paving the way for automated grid control. *At - Automatisierungstechnik, 68*(9), 750–764. doi:10.1515/auto-2020-0086.

Cai, H., Song, X., Zeng, Y., Jiang, T., Schlegel, S., & Westermann, D. (2021). A practical approach to construct a digital twin of a power grid using harmonic spectra. In *56th International Universities Power Engineering Conference: Powering Net Zero Emissions, UPEC 2021 - Proceedings. 9781665443890. (2021)* http://ieeexplore.ieee.org/xpl/mostRecentIssue.jsp?punumber=9548148. doi:10.1109/UPEC50034.2021.9548199.

Chen, C. (2006). CiteSpace II: Detecting and visualizing emerging trends and transient patterns in scientific literature. *Journal of the American Society for Information Science and Technology, 57*(3), 359–377. doi:10.1002/asi.20317.

Cherdantseva, Y., & Hilton, J. (2013). A reference model of information assurance & security. In *Proceedings - 2013 International Conference on Availability, Reliability and Security, ARES 2013* (pp. 546–555). doi:10.1109/ARES.2013.72.

Coppolino, L., Nardone, R., Petruolo, A., Romano, L., & Souvent, A. (2023). Exploiting Digital Twin technology for cybersecurity monitoring in smart grids. *ACM International Conference Proceeding Series* http://portal.acm.org/. doi:10.1145/3600160.3605043.

Danilczyk, W., Sun, Y., & He, H. (2019). ANGEL: An Intelligent Digital Twin framework for microgrid security. *51st North American Power Symposium, NAPS 2019*. United States: Institute of Electrical and Electronics Engineers Inc http://ieeexplore.ieee.org/xpl/mostRecentIssue.jsp?punumber=8974143. doi:10.1109/NAPS46351.2019.9000371.

Danilczyk, W., Sun, Y. L., & He, H. (2020). Smart Grid anomaly detection using a deep learning Digital Twin. In *52nd North American Power Symposium (NAPS)* (pp. 1–6). doi:10.1109/NAPS50074.2021.9449682.

Darbali-Zamora, R., Johnson, J., Summers, A., Jones, C. B., Hansen, C., & Showalter, C. (2021). State estimation-based distributed energy resource optimization for distribution voltage regulation in telemetry-sparse environments using a real-time Digital Twin. *Energies, 14*(3), 774. doi:10.3390/en14030774.

Grieves, M., & Vickers, J. (2016). Digital twin: Mitigating unpredictable, undesirable emergent behavior in complex systems. In *Transdisciplinary perspectives on complex systems: New findings and approaches* (pp. 85–113). Springer International Publishing. doi:10.1007/978-3-319-38756-7_4.

Cen-Cenelec-Etsi. (2014). SG-CG/M490/K_ SGAM usage and examples. Secretariat CCMC. https://www.cencenelec.eu/media/CEN-CENELEC/AreasOfWork/CEN-CENELEC_Topics/Smart%20Grids%20and%20Meters/Smart%20Grids/4_sgcg_methodology_sgamusermanual.pdf.

Cen-Cenelec-Etsi Smart Grid Coordination Group. (2012). *Smart grid reference architecture*. CEN-CENELEC-ETSI.

Gutmann, T., Nyffenegger, F., Pellegrini, M., Cabrucci, A., & Guzzini, A. (2023). A Digital Twin-based approach for the optimization of floor-ball manufacturing. *Electronics, 12*(24), 4979. doi:10.3390/electronics12244979.

Han, J., Hong, Q., Feng, Z., Syed, M. H., Burt, G. M., & Booth, C. D. (2022). Design and implementation of a real-time hardware-in-the-loop platform for prototyping and testing digital twins of distributed energy resources. *Energies, 15*(18). http://www.mdpi.com/journal/energies/. doi:10.3390/en15186629.

Harrison, R. (2016). *TOGAF 9 Foundation study guide—3rd Edition*. Van Haren Publishing.

Hong, Y. Y., Alano, F. I., Lee, Y. D., Jiang, J. L., & Yeh, J. N. (2024). Digital Twin-based blockchain for power support in networked microgrids. *Taiwan IEEE Access, 12*, 86675–86689. http://ieeexplore.ieee.org/xpl/RecentIssue.jsp?punumber=6287639. doi:10.1109/ACCESS.2024.3416672.

Johnson, P., Lagerström, R., & Ekstedt, M. (2018). A meta language for threat modeling and attack simulations. *ACM International Conference Proceeding Series*. Association for Computing Machinery Sweden http://portal.acm.org/. doi:10.1145/3230833.3232799.

Kandasamy, N. K., Venugopalan, S., Wong, T. K., & Leu, N. J. (2022). An electric power digital twin for cyber security testing, research and education. *Computers and Electrical Engineering, 101*. https://www.journals.elsevier.com/computers-and-electrical-engineering. doi:10.1016/j.compeleceng.2022.108061.

Kritzinger, W., Karner, M., Traar, G., Henjes, J., & Sihn, W. (2018). Digital Twin in manufacturing: A categorical literature review and classification. *IFAC-PapersOnLine, 51*(11), 1016–1022. doi:10.1016/j.ifacol.2018.08.474.

Masi, M., Pavleska, T., & Aranha, H. (2018). Automating smart grid solution architecture design. *IEEE International Conference on Communications, Control, and Computing Technologies for Smart Grids, SmartGridComm 2018*. Austria: Institute of Electrical and Electronics Engineers Inc http://ieeexplore.ieee.org/xpl/mostRecentIssue.jsp?punumber=8580358. doi:10.1109/SmartGridComm.2018.8587457.

Masi, M., Sellitto, G. P., Aranha, H., & Pavleska, T. (2023). Securing critical infrastructures with a cybersecurity digital twin. *Software and Systems Modeling, 22*(2), 689–707. https://www.springer.com/journal/10270. doi:10.1007/s10270-022-01075-0.

Meyer, L., Jorstad, K. L., Mortensen, L. K., Santos, A. Q., Handrup, K., & Shaker, H. R. (2022). Digital Twin of radio-mesh smart meter communication for outage management using a multi-agent system. *10th International Conference on Smart Energy Grid Engineering, SEGE 2022*. Denmark: Institute of Electrical and Electronics Engineers Inc http://ieeexplore.ieee.org/xpl/mostRecentIssue.jsp?punumber=9889733. doi:10.1109/SEGE55279.2022.9889773.

Nasiri, G., & Kavousi-Fard, A. (2023). A Digital Twin-based system to manage the energy hub and enhance the electrical grid resiliency. *Machines, 11*(3), 392. doi:10.3390/machines11030392.

NIST. (2014). *Guidelines for Smart Grid Cybersecurity (No. NIST Internal or Interagency Report (NISTIR) 7628 Rev. 1)*. National Institute of Standards and Technology. doi:10.6028/NIST.IR.7628r1.

NIST. (2024). *The NIST Cybersecurity Framework (CSF) 2.0 (No. NIST CSWP 29)*. U.S. Department of Commerce. doi:10.6028/NIST.CSWP.29.

Özkan, E., Kök, İı., & Özdemır, S. (2024). DeepTwin: A deep reinforcement learning supported digital twin model for micro-grids. *IEEE Access, 12*, 196432–196441. doi:10.1109/access.2024.3521124.

Paldino, G. M., De Caro, F., De Stefani, J., Vaccaro, A., Villacci, D., & Bontempi, G. (2022). A Digital Twin approach for improving estimation accuracy in dynamic thermal rating of transmission lines. *Energies, 15*(6). https://www.mdpi.com/1996-1073/15/6/2254/pdf. doi:10.3390/en15062254.

Pelekis, S., Sarmas, E., Georgiadou, A., Karakolis, E., Ntanos, C., Dimitropoulos, N., Kormpakis, G., & Doukas, H. (2023). TWINP2G: A digital twin architecture for optimal power-to-gas planning. In *Proceedings of the International Conferences on e-Society 2023, ES 2023 and Mobile Learning 2023, ML 2023*. IADIS Press Greece.

Persson, O., Danell, R., & Wiborg Schneider, J. (2009). (PDF) How to use Bibexcel for various types of bibliometric analysis. https://www.researchgate.net/publication/285473885_How_to_use_Bibexcel_for_various_types_of_bibliometric_analysis.

Rana, M., Shetty, S., Aved, A., Cruz, E. A., Ferris, D., & Morrone, P. (2024). Attack-resilient cyber-physical system state estimation for smart grid digital twin design. In E. Blasch, F. Darema, & A. Aved (Eds.), *Dynamic Data Driven Applications Systems. DDDAS 2022. Lecture Notes in Computer Science*. Vol. 13984. Springer, Cham. https://www.springer.com/series/558. doi:10.1007/978-3-031-52670-1_31.

Ruhe, S., Schaefer, K., Branz, S., Nicolai, S., Bretschneider, P., & Westermann, D. (2023). Design and implementation of a hierarchical digital twin for power systems using real-time simulation. *Electronics, 12*(12), 2747. doi:10.3390/electronics12122747.

Saad, A., Faddel, S., & Mohammed, O. (2020). IoT-based digital twin for energy cyber-physical systems: Design and implementation. *Energies, 13*(18), 4762. doi:10.3390/en13184762.

Sahoo, B., Panda, S., Kumar Rout, P., Bajaj, M., & Blazek, V. (2025). Digital twin enabled smart microgrid system for complete automation: An overview. *Results in Engineering, 25*, 104010. doi:10.1016/j.rineng.2025.104010.

Sen, Ö., Bleser, N., Henze, M., & Ulbig, A. (2023). A cyber-physical digital twin approach to replicating realistic multi-stage cyberattacks on smart grids. *IET Conference Proceedings, 2023*(6), 995–999. doi:10.1049/icp.2023.0614.

Sinneh, I. S, & Yanxia, S. (2026). Federated deep MPC-enabled digital twin and multiagent learning framework for secure and scalable smart nano grid energy management. *Renewable Energy Focus, 56*, 100762. doi:10.1016/j.ref.2025.100762.

Siraj Khan, M. M., Giraldo, J., & Parvania, M. (2022). Real-time cyber-physical analysis of distribution systems using digital twins. *IEEE International Conference on Communications, Control, and Computing Technologies for Smart Grids, SmartGridComm 2022*. United States: Institute of Electrical and Electronics Engineers Inc http://ieeexplore.ieee.org/xpl/mostRecentIssue.jsp?punumber=9960863. doi:10.1109/SmartGridComm52983.2022.9961022.

Somda, B., & Liu, C. C. (2024). Digital Twin for resilient power distribution systems. *22nd International Conference on Intelligent Systems Applications to Power Systems, ISAP 2024*. United States: Institute of Electrical and Electronics Engineers Inc http://ieeexplore.ieee.org/xpl/mostRecentIssue.jsp?punumber=10744224. doi:10.1109/ISAP63260.2024.10744374.

Sundhu, A. A., Saadatmandi, S., & Arnone, M. (2024). Enhancing grid resilience: Islanding control in a digital twin of real electrical distribution system. In *Proceedings - 2024 IEEE 6th Global Power, Energy and Communication Conference, GPECOM 2024*. Italy: Institute of Electrical and Electronics Engineers Inc http://ieeexplore.ieee.org/xpl/mostRecentIssue.jsp?punumber=10582548. doi:10.1109/GPECOM61896.2024.

Thoma, D., Sachenbacher, M., Leucker, M., & Ali, A. T. (2023). A Digital Twin for coupling mobility and energy optimization: The ReNuBiL Living Lab. In *CEUR Workshop Proceedings 16130073 CEUR-WS Germany*. Lübeck, Germany: CEUR-WS.org https://ceur-ws.org/Vol-3507/paper5.pdf.

Tran, T. P., Tuan Tran, Q., Le, M. T., & Caire, R. (2024). Voltage control using optimization-based method in a Digital Twin. In *France 2024 IEEE 22nd Mediterranean Electrotechnical Conference, MELECON 2024* (pp. 138–143). Institute of Electrical and Electronics Engineers Inc. http://ieeexplore.ieee.org/xpl/mostRecentIssue.jsp?punumber=10608458. doi:10.1109/MELECON56669.2024.10608689.

van Eck, N. J., & Waltman, L. (2010). Software survey: VOSviewer, a computer program for bibliometric mapping. *Scientometrics, 84*(2), 523–538. http://www.springerlink.com/content/0138-9130. doi:10.1007/s11192-009-0146-3.

van Eck, N. J., & Waltman, L. (2014). CitNetExplorer: A new software tool for analyzing and visualizing citation networks. *Journal of Informetrics, 8*(4), 802–823. http://www.journals.elsevier.com/journal-of-informetrics/. doi:10.1016/j.joi.2014.07.006.

Wideł, W., Hacks, S., Ekstedt, M., Johnson, P., & Lagerström, R. (2023). The meta attack language - a formal description. *Computers & Security, 130*, 103284. https://doi.org/10.1016/j.cose.2023.103284.

Wileman, A. J., Aslam, S., & Perinpanayagam, S. (2023). A component level digital twin model for power converter health monitoring. *IEEE Access, 11*, 54143–54164. http://ieeexplore.ieee.org/xpl/RecentIssue.jsp?punumber=6287639. doi:10.1109/ACCESS.2023.3243432.

Yan, J., Liu, Z., Leng, J., Zhao, J. L., Chen, C., Zhang, D., Tao, Y., Wang, Y., Liu, T., Zhang, C., Tong, Y., Mourtzis, D., & Wang, L. (2025). Human-centric artificial intelligence towards Industry 5.0: Retrospect and prospect. *Journal of Industrial Information Integration, 47*, 100903. doi:10.1016/j.jii.2025.100903.

PART II

Digital Twin computing for microgrids

CHAPTER

5

Using Macine Learning algorithms for prediction and diagnosis problems

Zohaib Saadat, Sana Saleh, Muhammad Tariq

Smart Technologies and Applied Intelligence Research (STAIR) Lab, Department of Computer Engineering, School of Engineering, National University of Computer and Emerging Sciences (NUCES), Islamabad, Pakistan

5.1 Introduction: Overview of machine learning in Digital Twin for smart grids

The incorporation of machine learning (ML) into Digital Twin (DT) frameworks has markedly enhanced the functionalities of smart grid systems, facilitating real-time decision-making, predictive analytics, and autonomous control. ML, a subset of artificial intelligence (AI) encompassing a wide array of data-driven methodologies, includes supervised learning, unsupervised learning, deep learning (DL), and reinforcement learning (RL). ML methodologies, including decision trees, neural networks (NN), and clustering techniques, enable systems to examine extensive, diverse datasets from the grid to identify significant patterns, predict system behavior, and facilitate proactive actions. The choice of a certain ML method is contingent upon the task type, classification, regression, clustering, or control, along with the availability of labeled data and the intricacy of the system dynamics involved. Supervised learning is proficient in fault classification and load forecasting, whereas RL is more effective for adaptive energy management and policy optimization.

Since the emergence of Industry 4.0, DT technology has seen widespread adoption across various sectors, including manufacturing, healthcare, transportation, and energy, for its ability to replicate physical systems and enable data-driven decision-making. In the context of smart grids, DTs serve as dynamic, real-time models that mirror the condition and behavior of physical grid assets using continuous sensor data. These virtual counterparts not only facilitate real-time monitoring but also support predictive analysis through "what-if" simulations, allowing operators to anticipate faults, optimize system performance, and enhance grid resilience. This ability to transition from reactive to proactive management makes DTs a foundational technology for the next generation of intelligent, sustainable power systems. (Opy Das, 2024). The realization of these capabilities is enabled by the convergence of advanced communication technologies and the Internet of Things (IoT). These technologies enable continuous data collection from a multitude of sensors embedded within the grid infrastructure, ensuring that DTs remain synchronized with their physical counterparts. High-speed data transmission and edge-to-cloud integration further enhance the responsiveness and scalability of smart grid DT implementations, laying the groundwork for real-time visibility and control.

However, to unlock the full potential of DTs, beyond mere real-time visualization, an intelligent layer is needed to analyze and interpret the vast volumes of data being collected. ML plays a crucial role in this regard, offering powerful tools to extract meaningful insights, detect patterns, and make data-driven predictions. ML is particularly indispensable in dynamic environments like microgrids, where system behavior is highly variable and often non-linear (Bassey, 2024). By integrating ML algorithms into DT frameworks, smart grids can not only monitor but also learn from historical and real-time data to forecast equipment failures, diagnose system anomalies, and optimize energy distribution with unprecedented precision. Moreover, ML empowers DTs to proactively detect cyber threats and operational irregularities that traditional grid monitoring systems struggle to deliver effectively.

Digital Twin Technology and Smart Grid.
DOI: https://doi.org/10.1016/B978-0-443-36664-2.00004-6

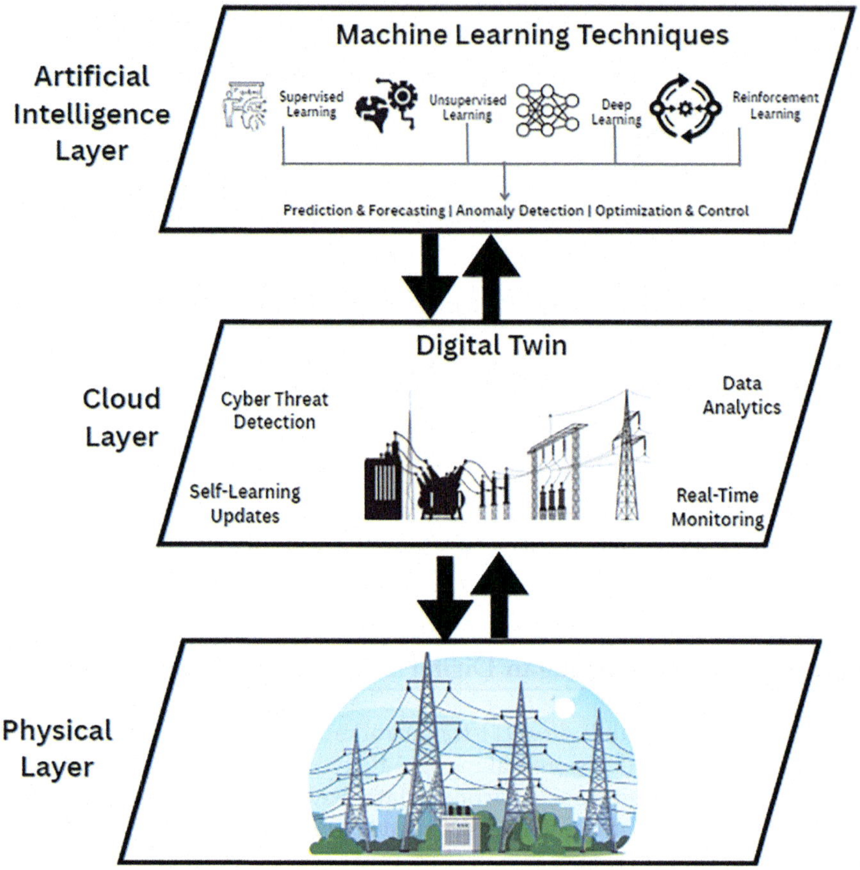

FIGURE 5.1 ML-DT integration architecture for smart grids: illustrates the architecture integrating ML techniques within a DT framework, showcasing the data flow across the physical, cloud, and AI layers of a smart grid. *AI*, aritificial intelligence; *DT*, Digital Twins; *ML*, machine learning.

This introductory section highlights the synergy between DTs and ML in smart grids to lay the groundwork for a more thorough examination of use cases, algorithms, challenges, and emerging trends (Nabil Mchirgui, 2024). The multilayered architecture of ML-enhanced DTs in smart grids is illustrated in Fig. 5.1, showcasing how real-time data from the physical layer is processed by DT and analyzed through various ML techniques. This layered integration enables functionalities like prediction, optimization, and cyber threat detection. For instance, an ML-enhanced DT can learn the typical operating behavior of a transformer and promptly alert operators when sensor readings deviate from the norm, signaling a potential fault. It may also analyze years of historical data to accurately forecast future demand. This predictive and diagnostic capability is critical for modern power systems, which must navigate the complexities of distributed renewable generation, fluctuating demand, and increasingly intricate cyber-physical interactions.

Traditional grid management techniques struggle to process the vast volume and complexity of data produced by smart grids. In contrast, data-driven models excel at identifying trends, anomalies, and correlations within such datasets (Wangyong Guo, 2025). Put simply, ML transforms a DT from a passive, real-time replica into a proactive decision-support tool. By drawing actionable insights from both historical and real-time data, ML-enhanced DTs improve simulation accuracy and forecasting capability. While a conventional DT offers a synchronized virtual model of grid assets, an AI-enabled DT evolves through continuous learning, adapting its behavior to match real-world dynamics.

Conventional control and fault-monitoring systems are often limited in their ability to fully capture the interdependencies within smart grid infrastructures or to offer reliable real-time insights. In contrast, DTs provide a robust framework for advanced functionalities such as predictive maintenance (PdM), fault detection, and energy optimization. A notable example is NASA's DT initiative, which integrates high-fidelity simulations with onboard health management systems, maintenance logs, and operational telemetry to perform PdM, demonstrating a high standard of safety and reliability. Given the intricate web of interconnected components and the integration of renewable energy sources, smart grids require advanced management techniques to maintain system stability, reliability, and efficiency. ML algorithms are indispensable in this context, helping address several critical challenges:

- **PdM:** Anticipating potential component failures to minimize downtime and optimize maintenance schedules.

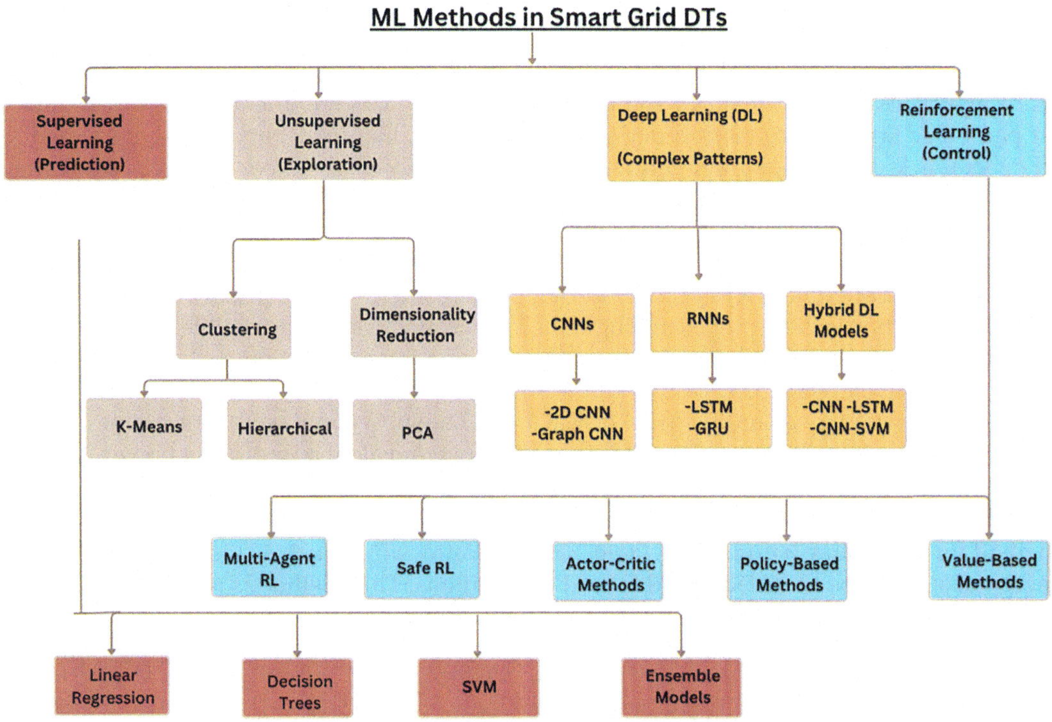

FIGURE 5.2 Taxonomy of ML methods for smart grid DTs: shows a hierarchical taxonomy of ML methods for smart grid DTs, organizing supervised, unsupervised, deep, and reinforcement learning techniques alongside their representative algorithms and primary functional roles.

- **Fault detection:** Identifying and analyzing operational anomalies to prevent disruptions and ensure system stability.
- **Energy optimization:** Balancing electricity production and consumption by leveraging distributed energy resources (DERs) such as solar and wind, thereby improving overall energy efficiency.

Importantly, these advancements address pressing industry needs. As global power networks grow older and more complex, particularly with the addition of consumer-producers (prosumers), microgrids, and bidirectional energy flows, traditional top-down management models become insufficient. AI integration enables smart grids to remain robust and adaptive in the face of this growing complexity (Nabil Mchirgui, 2024).

The sections that follow provide a structured overview of the application of ML in DT-based smart grids. We start by exploring the primary ML techniques, ranging from basic regression models to advanced DL and RL approaches, and then move to classifying the main use-case categories, namely prediction and diagnosis, where ML adds substantial value. We also examine key challenges, such as data quality issues and real-time processing constraints, while discussing the transformative potential of emerging paradigms like explainable AI (XAI) and federated learning (FL) in shaping the future of intelligent, AI-driven DTs.

5.2 Methods and algorithms for machine learning in smart grid Digital Twins

Gaining a thorough grasp of the fundamental ML methods and algorithms that make these capabilities possible is crucial before delving into the particular uses of ML in DT-based smart grids. The problem's nature, the availability of labeled data, and the intricacy of the patterns encoded in the data all influence the ML strategy that is used, whether it is for control, classification, or regression tasks. The four main categories of ML techniques, supervised learning, unsupervised learning, DL, and RL, are outlined in this section in an organized manner (see Fig. 5.2). Every method's working principle, shared algorithms, and applicability to smart grid DT scenarios are evaluated.

5.2.1 Supervised learning methods

In smart grid DT systems, supervised learning is one of the most popular ML techniques. This paradigm allows models to learn mappings between features and target variables by training them on historical data that contains input-output pairs. For tasks where labeled datasets are available, like load forecasting, equipment defect classification, and cyberattack detection, supervised learning is especially well-suited. Supervised learning algorithms are trained by labeled examples, including historical data with established outcomes (target variables). Regression methods (for continuous outcomes) and classification methods (for discrete categorical outcomes) are encompassed under this domain. Supervised learning is commonly employed in smart grids for diagnostics (e.g., classifying fault types or equipment conditions) and predictions (e.g., forecasting loads or prices). Supervised models excel in generating precise predictions when ample training data with ground truth is accessible.

These ML categories offer a versatile toolkit for addressing various prediction and control challenges in smart grid DTs. Understanding the strengths, assumptions, and customary applications of each method is crucial for choosing the appropriate strategy according to the task at hand. The subsequent sections examine each category of ML techniques in detail, beginning with supervised learning, the most prevalent paradigm in power system applications.

5.2.1.1 Linear regression

A basic method for forecasting a numerical value. Simple relationships, such as the relationship between temperature and energy usage, can be modeled by LR in grid applications. This method is restricted to linear correlations, but it is simple to understand. For day-ahead load forecasting, which refers to predicting electrical demand for each hour of the following day, LR could be employed as a baseline; however, nonlinear ML models often perform better for complex load patterns.

5.2.1.2 Decision trees and Random Forests

Decision trees split data based on feature thresholds to predict an outcome (e.g., "if voltage drops and current spikes, classify as fault"). They are intuitive (mimicking rule-based decision-making) and handle nonlinearity well in many cases.

Random Forests technique, an ensemble of many decision trees, offers improved accuracy and robustness by averaging many decision paths to determine an outcome. In a microgrid fault diagnosis scenario, a Random Forest might take inputs like feeder measurements and output the fault category with high accuracy, as it can capture complex interactions between features without much parameter tuning. Additionally, these models offer a certain amount of explainability (feature importance), which is beneficial for vital infrastructure (Ojetola, 2022).

5.2.1.3 Support vector machines

Support vector machines (SVMs) are powerful supervised learning techniques that are frequently used for classification problems. Support vector regression (SVR), its regression equivalent, applies the same ideas to predictive modeling. Either fitting a regression line that best captures the data or optimizing the area between decision boundaries is how both strategies operate. These techniques are well-suited for grid applications with complex, high-dimensional feature spaces. While SVRs can be used to represent continuous relationships like load forecasting or voltage estimates, SVMs perform better for classification, especially in high-dimensional feature spaces, by identifying the optimal hyperplane to split classes with the largest margin.

By distinguishing between attack and normal data, SVM-based methods have been applied in grid security to classify power system interruptions or detect intrusions. In a similar vein, SVRs can offer insights based on regression for evaluating grid stability. Both approaches frequently work well on smaller to medium-sized datasets and can handle nonlinear patterns using kernel functions. However, training can be computationally taxing for very big datasets. In order to provide operators with a rapid, automated diagnostic, an SVM may be trained in smart grids to classify power quality events (voltage sags, swells, and outages) based on waveform characteristics.

5.2.1.4 Ensemble methods

A durable predictive model can be produced by combining several weak learners using ensemble approaches, especially boosting algorithms. Techniques like gradient boosting and XGBoost, which go beyond Random Forests, combine a number of weak predictive models, typically decision tree stumps, to create a better model. These techniques have demonstrated strong performance. For example, XGBoost has been applied to short-term load forecasting and has demonstrated accuracy comparable to NNs despite being quicker to train and, in certain situations, simpler to adjust. Given the multifaceted nature of energy consumption (weather, human activities, etc.), ensembles are advantageous because they efficiently manage nonlinear linkages and variable interactions. They improve accuracy and resilience in activities like fault classification and load forecasting (discussed later) in smart grid situations.

5.2.1.5 Shallow neural networks

Shallow neural networks (SNNs) and DL differ in both depth and architectural complexity. The purpose of SNNs, which usually include only one or two hidden layers, is to simulate nonlinear input-output interactions in tasks that are only moderately difficult. On the other hand, DL typically refers to multilayered NNs or those that use sophisticated architectures like recurrent neural networks (RNNs) or convolutional neural networks (CNNs), which are intended to extract sequential or hierarchical characteristics from data.

SNNs can work effectively in smart grids for tasks with moderate data complexity, like simple load forecasting, classifying device statuses, or identifying patterns in smart meter data. The most popular type of SNNs are multi-layer perceptrons (MLPs), which approximate complex relations using nonlinear activation functions. MLP-based networks showed early success in applications such as energy time-series forecasting and grid event classification, even before the advent of DL. SNNs are reasonably easy to create with modern software libraries and, with the right training, can outperform more direct methods, even if they demand a considerable amount of training data and processing resources. Shallow architectures continue to be advantageous in applications where DL is not necessarily required, such as anomaly detection and solar power forecasting.

In contrasting these supervised approaches, tree-based models are common in power system applications because they can handle nonlinearity and categorical features effectively and are reasonably interpretable (rules); linear models are the most straightforward and interpretable but may underfit complex trends; NNs (particularly deep ones, which are covered later) frequently provide the best accuracy for extremely complicated patterns at the expense of interpretability and requiring more data. SVMs provide robust categorization, but they can be black-box models and do not scale as well to vast amounts of data. Fortunately, a lot of grid ML tasks use somewhat big data (for example, a utility may have a few years' worth of hourly data for a specific feeder's load, which is manageable by most algorithms, or a few tens of thousands of points). When there are few labeled examples of errors or assaults, supervised learning may be supplemented by unsupervised techniques, as we will see next, or by simulation (using DT to create synthetic labeled data).

5.2.2 Unsupervised and semisupervised learning

5.2.2.1 Unsupervised learning

Unsupervised learning focuses strongly on finding patterns in unlabeled data. This ML technique is crucial for exploratory analysis, anomaly identification, and grouping related behaviors in smart grid DTs. The capacity for unsupervised learning to find "interesting" patterns without labels is essential since many problems (particularly those pertaining to system health and security) lack comprehensive labeled datasets of all potential circumstances.

5.2.2.1.1 Clustering algorithms

These algorithms cluster data points according to how similar they are. Using historical load profiles, for instance, clustering might be used to find common daily trends (weekday vs. weekend, seasonally high vs. low demand days, etc.). One unsupervised learning algorithm in particular, K-means clustering, divides data into k different clusters by reducing the variance within each cluster. Each data point is assigned to the closest cluster center, or centroids, which are then iteratively updated to better reflect the data's structure (Yang, 2020).

Clustering in a microgrid allows for the division of users or devices into groups based on shared usage patterns. This aids in the design of targeted demand responses and the identification of anomalies, or unexpected deviations in a device's behavior from its cluster. Power quality events could be grouped using hierarchical clustering to view the taxonomy of disruptions without any prior classification. Data's innate structure can be revealed through clustering, which helps direct additional research.

5.2.2.1.2 Principal component analysis

Principal component analysis (PCA) is a dimensionality reduction technique that determines which linear feature combinations, or principal components, capture the most variance in the data. In grid applications, PCA can reduce high-dimensional sensor datasets (such as data streams from phasor measurement units, or PMUs) to a small number of principal components that represent the state of the system. In addition to aiding in visualization, this can also act as an anomaly detector: an out-of-bound situation is indicated if the real-time data point in PCA space is outside of the typical range, such as outside a specific distance from the cluster of previous points. Through the observation of anomalous trajectories in reduced-dimensional subspaces of system measurements, operators have employed PCA to identify occurrences such as voltage instability or line interruptions.

5.2.2.1.3 Anomaly detection and outlier methods

There are specific techniques (many of which are unsupervised or one-class classifiers) for identifying outliers in addition to clustering and PCA, which are both useful for anomaly detection. Only "normal" data is used to train One-Class SVM, which then marks anything that does not fit into that class as abnormal. Data and points that are easier to identify as more likely anomalies are scored using Isolation Forests, which randomly sample and partition data. These techniques have been used in power grid intrusion detection systems, which mimic typical network traffic and sound an alarm at any variation (maybe an attack). An anomaly detector used in equipment monitoring may be able to identify a wind turbine's typical vibration frequency signature before spotting the emergence of a new frequency component, which could be a symptom of an emerging gearbox malfunction.

5.2.2.1.4 Association rule learning

Large datasets can reveal hidden patterns and connections using association rule learning algorithms like Apriori, despite their less frequent use in smart grid applications. For Apriori, finding frequent item-sets in a dataset and generating rules based on the idea that any subset of a frequent itemset must likewise be frequent is the primary goal. By examining the frequency with which items or conditions occur together in transactional or sequential data, it is frequently used to find significant "if-then" patterns, such as event co-occurrences. These algorithms could extract association rules from event logs, such as "if frequency drop and voltage drop occur together, then a generator trip event likely happened." Unsupervised relationship discovery like this can improve a DT's reasoning engine's knowledge base.

5.2.2.2 Semisupervised learning

Semisupervised learning is also pertinent in many situations, such as when a large amount of data may be unlabeled, but just a tiny percentage is identified. Semisupervised learning is an ML technique that enhances model performance by using both labeled and unlabeled data during training, particularly in situations when labeled data is hard to come by or prohibitively expensive. Leveraging the enormous unlabeled corpus is made possible by methods like clustering with known class prototypes or training an autoencoder on all data and then fine-tuning on labeled data. An excellent example would be anomalies in the smart grid; it is possible that only a limited labeled dataset is available for a few known faults or attacks, but the goal is to detect new ones. In order to sharpen its boundaries, semisupervised anomaly detection may use a small number of known anomalous cases in addition to training a model on primarily normal data.

5.2.2.3 Contrasting the two against each other

Less accurate results are usually obtained from unsupervised techniques than from supervised ones (for example, classifiers supply a specific label, but clustering does not provide the "name" of a cluster; one must interpret it). On the other hand, unsupervised learning can reveal surprising things. Supervised models deal with expected scenarios in a DT, whereas unsupervised learning may operate continually in the background to identify any odd patterns, acting as an early warning system. When an unsupervised model flags an odd occurrence, engineers can classify it as a new kind of fault. This combination can then be added to the training set of a supervised model to enhance classification in the future. With its emphasis on iterative development and real-time adaptation, this continuous learning loop is highly compatible with DT philosophy.

Within the DT framework, this complementary relationship creates an ongoing learning cycle. The first line of defense is provided by unsupervised models, which search through massive unlabeled data streams for new anomalies and hidden patterns. Supervised models are trained to correctly classify these anomalies for future detection after domain experts have identified and labeled them. By acting independently on these predictions and classifications, putting control measures into place, optimizing setpoints, or initiating mitigation protocols, RL agents ultimately complete the feedback cycle. The self-healing adaptive intelligence that ML-driven DTs promise for smart grids is embodied in this smooth transition from discovery to diagnosis to independent action.

5.2.3 Deep learning techniques

DL refers to multilayered neural network models that are capable of learning complex, hierarchical data representations. DL's cutting-edge results in problems involving speech, pictures, and sequences have made it popular in many domains.

DL techniques have begun to play a significant role in the context of DTs and smart grids, particularly as the amount of data (from high-frequency sensors, high-resolution weather forecasts, etc.) increases. Deep models eliminate the need for human feature engineering by automatically learning features from raw data. Certain crucial DL architectures, such as CNNs and RNNs, have enabled great strides to be made in creating incredibly powerful AI agents.

5.2.3.1 Convolutional neural networks

CNNs are well-known for their ability to analyze images, but they can also be used to analyze spatial or spatiotemporal data in power systems. CNNs can be used to identify patterns (such as oscillations or faults) in 2D "images" of power system data, such as frequency versus time spectrograms from disturbance recordings or 2D heatmaps of voltage phase angles throughout a grid. CNNs have been used to uncover features that may be challenging to manually code by detecting and classifying transients in waveform data from grid sensors. A CNN could recognize complicated system states in a DT situation by processing data from multiple sensors simultaneously (think of sensor array data as comparable to pixels).

Applying CNNs to graph-structured data (using graph convolution) to learn from the grid's topology is another new application. This is more complex, but studies indicate that graph CNNs can assist with tasks like stability evaluation and state estimation by utilizing the network structure. Approaches like PE-MAGCN, SAFE-TCS, and Hierarchical Grouping (HGCNN) can significantly improve a system's robustness, accuracy, and tolerance to missing data (Zhong, 2021). Emerging transformer-based visual architectures extend this approach; Vision Transformers have been used for anomaly detection in PMU data, offering improved sensitivity to subtle waveform deviations compared to conventional CNNs (Liu et al., 2025). Graph-based transformer frameworks such as SmartFormer demonstrate the adaptability of attention mechanisms to spatially distributed DT models, enhancing transferability across network configurations (Saeed et al., 2025).

5.2.3.2 Recurrent neural networks

RNNs are a type of neural network designed for sequential data. They use recurrent connections to maintain a 'memory' of prior inputs. RNNs are perfect for time-series tasks in smart grids, including load forecasting, equipment condition monitoring, and anomaly detection, because they can capture temporal dependencies, unlike feedforward networks.

Advanced versions of standard RNNs, such as long short-term memory (LSTM) and gated recurrent unit (GRU) networks, are made to get around problems like vanishing gradients that make it difficult to learn long-term dependencies. LSTMs give the network the ability to selectively keep or discard data over time by including memory cells and gating mechanisms, such as input, output, and forget gates. On the other hand, GRUs streamline this design by integrating the input and forget gates into a single update gate, which leads to faster training and fewer parameters while preserving high performance on sequential data. LSTM and GRU models work especially well in smart grid DTs for jobs where it is crucial to capture intricate temporal patterns, like multistep demand forecasting, renewable generation prediction, and condition-based equipment monitoring.

An extension of conventional LSTM networks, bidirectional LSTMs (BiLSTMs), process input sequences both forward and backward. In this way, they simultaneously record data from the past (prior time steps) and the future (upcoming time steps). BiLSTMs are particularly helpful in applications like event detection or fault classification in smart grids, where both historical and future trends around a point in time can provide crucial insight. This dual perspective improves the model's capacity to learn dependencies that standard unidirectional LSTMs might miss. RNNs have memory for past inputs and are made for sequence data. Long-term dependencies can be captured by LSTM and GRU networks, which solve the vanishing gradient problems of RNNs. Both sequence anomaly detection and time-series forecasting (load, generation, and prices) substantially rely on these.

An LSTM in a microgrid DT might use the last N hours of measurements (load, solar output, and battery condition) as input to forecast the numbers for the following hour and perhaps the deviation from normal for anomaly identification. In fact, LSTMs have demonstrated outstanding performance in energy system forecasting tasks. Interestingly, GRU networks beat LSTM and BiLSTM models in terms of accuracy and computational efficiency when it comes to photovoltaic (PV) power forecasting (Abu-Rub, 2023). As they can track the development of wear and tear, they have also been used to forecast equipment health indicators over time. For multistep forecasting, predicting an entire future trajectory rather than simply one step, stacked LSTMs or encoder-decoder LSTMs are employed. Because daily loads and seasonal trends are cyclical, LSTMs frequently automatically encode them.

5.2.3.3 *Hybrid deep model techniques*

To tackle complicated jobs, various deep architectures are frequently integrated. A CNN–LSTM hybrid for spatiotemporal forecasting is a typical example in smart grids. For instance, a CNN may extract features from spatial data (such as weather maps or grid topology influence), and an LSTM would then process these features over time to anticipate load or renewable production.

Another hybrid strategy is to integrate NNs with physics-based models into a coherent framework (also known as gray-box models or physics-informed NNs). The hybrid DT model that we previously saw was an example: it integrated an RNN that learns the data-driven components of the behavior with differential equation models that encode the physical rules of a battery and HVAC system. As a result, DERs could be accurately modeled, and tailored energy efficiency recommendations could be made. Recent architectures combine sequence modeling and attention mechanisms. For instance, a hybrid LSTM–Transformer model has demonstrated superior performance for short-term power load forecasting by capturing both temporal dependencies and long-range correlations in grid dynamics (Pentsos et al., 2025). These hybrid models, which combine domain expertise and data learning, are highly promising because they frequently use less data for training and provide more reliable extrapolations.

5.2.3.4 *Principal insights*

The ability to handle large-scale multidimensional data (such as hundreds of sensors or high-resolution time series), automation of feature extraction, and accuracy (capturing highly nonlinear and complicated correlations) are some of the benefits of DL in the context of smart grid DT. Nevertheless, they have drawbacks: deep models are often less interpretable ("black boxes"), necessitating explainability techniques (covered later); they require a large amount of computational power (although developments in edge computing and GPUs/TPUs aid); and they typically require a large amount of training data, which could be problematic if we only have a small number of historical examples of, say, blackouts or uncommon faults. However, DL is expected to become even more important as data increases and pretraining on simulations becomes more widespread. In pilot projects, grid operators are starting to use these. For example, DL is being used to identify complex events in PMU data that threshold-based detectors were previously unable to identify. Transformer architectures are increasingly used to capture multiscale temporal patterns; the MFformer model aggregates features across multiple frequencies to enhance forecasting accuracy in electricity load prediction (Tong & Liu, 2025).

5.2.4 Reinforcement learning methods

RL, which basically learns what to do under different conditions to accomplish an objective, is the preferred method for sequential decision-making issues, even though supervised and unsupervised learning are excellent at prediction and pattern recognition. Controlling a microgrid or optimizing energy use, as described in the use cases, is naturally formulated as an RL problem: at each time step, the controller (agent) observes the current state (loads, generation, prices, etc.) and must select control actions (dispatch levels, battery charging rate, demand-response signals) that affect the system's future state and immediate reward (e.g., improved stability or negative cost).

5.2.4.1 *Policy gradient and actor-critic techniques*

A subclass of RL approaches known as policy gradient methods computes gradients of expected rewards in order to directly optimize the policy, which is a mapping from states to actions. In contrast to value-based methods like Q-learning, examples include the policy gradient and actor-critic approaches, which are especially successful for complex control tasks and continuous action spaces.

Deep-Q network (DQN) uses a deep neural network to approximate the Q-function in complicated situations with huge state spaces, such as a power system with several continuous variables. Q-learning is a traditional RL technique that determines a Q-value for every state-activity combination (the predicted cumulative reward from performing that action in that state). In real life, a DQN can learn how to use a battery, for instance: Actions may be charge, discharge, or idle; states could include the battery's current charge, the time-of-day, and the power price; and the reward could be negative cost, meaning that maximizing reward entails reducing cost. Research has demonstrated that DQN agents can effectively learn peak shaving techniques. For example, in Belgium, an installation resulted in cost savings and increased efficiency for a home microgrid.

To enable steadier and sample-efficient learning, actor-critic designs integrate two elements: the critic, which assesses the proposed actions, and the actor, which suggests them. These techniques work especially well for multi-objective energy optimization, voltage regulation, and fine-grained control of DERs in smart grid DTs. Instead of using value estimates, these methods (such as REINFORCE, A3C, deep deterministic policy gradient [DDPG], and proximal policy optimization [PPO]) directly modify the policy (the decision strategy). Furthermore, for continuous action spaces, such as managing volt/volt-ampere reactive (VAR) devices with continual settings, DDPG is often used (Dong, 2021). Research on microgrid energy management has made use of the stable actor-critic technique known as PPO. Through reward shaping, an agent with PPO is better able to manage several goals and restrictions. Centralized PPO has been shown to achieve 12.7% higher economic benefit by optimizing renewable energy utilization and storage systems, all while reducing carbon emissions by 15.3% through better coordination of wind turbines, photovoltaics, and pumped storage (Tan, 2024). An RL agent managing an industrial microgrid in Norway, for example, successfully learned to lower both peak demand charges and overall energy costs by combining PPO with a forecasting model to manage energy under time-of-use pricing.

5.2.4.2 Safe reinforcement learning

Safe RL focuses on implementing autonomous control because it involves training agents that maximize performance while adhering to stringent safety constraints. Making sure the RL agent does not do anything risky (such as overloading a line or voltage out of range) is an essential part of real-world deployment. Methods include employing a supervisory layer that overrides the agent if it is going to breach a known constraint or including penalty terms in the reward for constraint violations. In order to keep the agent's recommendations within a safe range, several researchers have merged model-predictive control and RL, also known as hybrid model-predictive control RL (RL-based MPC). There is a lot of emphasis on safe exploration and safe policy adherence for RL because grids are important infrastructure (Liu, 2023). For instance, an agent may learn to maximize energy usage, but a safety feature makes sure that it never pushes the battery over its limit or makes the grid unstable (Ye, 2023).

5.2.4.3 Multiagent reinforcement learning

Diverse agents that learn to cooperate or compete may control diverse entities (generators, storage, and flexible loads) in a power grid. In situations like transactive energy or decentralized control, where each agent (e.g., every home with solar & storage) maximizes its profit while simultaneously advancing the goals of the grid as a whole, multiagent RL (MARL) might be helpful. It is possible to learn coordination methods, such as consensus procedures or the sharing of a global reward component. MARL is a developing field of study because, despite its complexity, it mimics the actual structure of power systems (for example, numerous microgrids in a network learning to trade electricity among themselves optimally).

5.2.4.4 Principal insights

RL differs from previous ML techniques as it creates data through interaction. Since you would not want an agent arbitrarily cutting off electricity in a real grid while training, a DT environment is almost a requirement for teaching RL for grid control. To make the agent resilient, DT can model hundreds of scenarios, such as various load profiles and outages. The ability of RL controllers to adjust to shifting circumstances without requiring explicit reprogramming is one of the main benefits noted. An RL agent, for instance, can progressively learn a new optimal policy through ongoing training in the event that fuel prices fluctuate, or a new PV farm is introduced, while a traditional rule-based controller would require human redesign. When tested on a microgrid benchmark, researchers found that all RL algorithms developed efficient energy management techniques. One method, an upgraded actor-critic, performed the best and reduced operating expenses by more than 20% when compared to the baseline (Mihet-Popa, 2024). In extremely dynamic scenarios, such as islanded microgrid operation during an outage, this demonstrates how RL might find counterintuitive control rules that a human operator might not consider.

5.3 Taxonomy of machine learning applications in DT-enabled microgrids

The operational goal of the grid is closely related to the selection of ML models within the DT framework. While control-oriented and optimization tasks prioritize low-latency interpretability and robustness under dynamic conditions, forecasting-oriented applications (like load and renewable prediction) usually focus on accuracy and long-term generalization. As a result, each algorithm category, supervised, unsupervised, deep, or RL, depends on

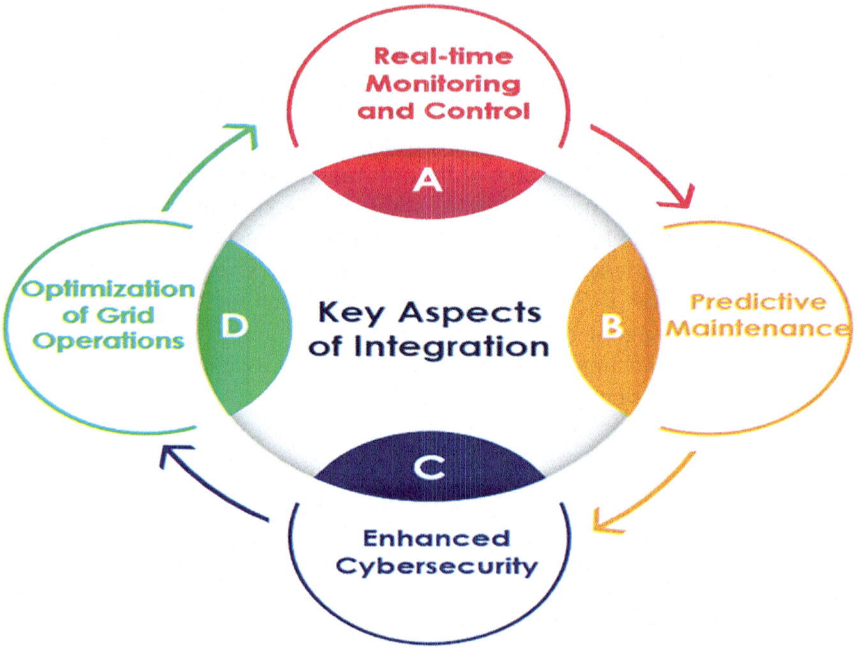

FIGURE 5.3 Key features of ML-driven DTs in smart grids: depicts the core capabilities enabled by ML and DT integration: (A) real-time monitoring, (B) PdM, (C) cybersecurity, and (D) grid optimization.

how well it strikes a balance between data accessibility, computational cost, and real-time responsiveness. The main issues that come up when implementing these models in real-world smart grid settings, such as data quality, time constraints, and system integration, are examined in the section that follows.

This section's taxonomy of ML applications is organized according to a methodological framework that gives operational goals and data properties essential to smart grid DTs top priority. Three interrelated criteria are used to guide the selection and classification of algorithms: (1) the main operational goal, differentiating between prediction (e.g., load forecasting) diagnosis (e.g., fault detection) and control (e.g., energy optimization) (2) the type of data that is available including its volume labeling status and format (time-series spatial or tabular) and (3) the tasks temporal and robustness requirements such as the need for real-time inference explainability or adaptability to non-stationary grid conditions. This methodical approach guarantees a practical mapping of ML techniques to the most pertinent and significant use cases in DT-enabled grids, going beyond a purely algorithmic classification to one based on engineering utility.

This taxonomy mirrors the smart grid's core operational priorities: anticipate future conditions, detect and diagnose anomalies, respond through PdM, and optimize control actions. It thus follows a logical progression from prediction to diagnosis to intelligent response, aligning ML techniques with real-world grid management workflows. The incorporation of ML into DT frameworks for microgrids can potentially tap into powerful capabilities in real-time monitoring and adaptive control. As microgrids evolve to expand into variable renewable energy sources, distributed generation, and dynamic load profiles, old rule-based control and diagnostics methods grow increasingly insufficient. Data gathered and reflected in a DT framework enables, through ML, smart and proactive decision-making across multiple operational layers. It is crucial to group ML applications according to the fundamental grid functions they enhance in order to fully utilize these advantages.

ML can address a variety of prediction and diagnosis issues in smart grids and microgrids, particularly when integrated with DT frameworks. In general, the application cases fall into four primary categories: energy optimization & control, cybersecurity/anomaly detection, fault diagnosis & PdM, and load forecasting. These major ML application areas in smart grid DTs, as determined by recent research, are depicted in Fig. 5.3. A crucial component of grid management is covered by each category:

With an emphasis on four core areas, load forecasting, cybersecurity and anomaly detection, fault diagnosis and PdM, and energy optimization and control, this section offers a structured taxonomy of ML application cases in DT-enabled microgrids. By classifying these applications based on their diagnostic and predictive capabilities, we look for a thorough framework that demonstrates the concrete ways in which ML enhances current smart grid operations. To provide a concise operational overview, Table 5.1 summarizes representative ML algorithms mapped to their corresponding smart grid functions, highlighting key strengths and limitations.

TABLE 5.1 ML algorithm-to-use-case map for DTs in smart grids.

Task	Representative models	Strength	Limitation
Load forecasting	LR, LSTM, CNN–LSTM hybrids, ANFIS	Captures nonlinear and temporal patterns; supports short- and medium-term forecasts	Requires extensive high-quality data; sensitive to concept drift
Fault detection	SVM, RF, CNN, Decision Tree	High classification accuracy; adaptable to new fault types	Needs labeled fault data; retraining required for evolving conditions
Predictive maintenance (PdM)	LSTM, Autoencoder, Isolation Forest, One-Class SVM	Predicts Remaining Useful Life (RUL); detects early degradation trends	Demands long historical datasets; model complexity
Cybersecurity/Anomaly detection	PCA, One-Class SVM, Autoencoder, Clustering (k-means, DBSCAN)	Identifies unseen cyber threats and abnormal activities; works with unlabeled data	Threshold selection affects false alarm rate; scalability issues
Energy optimization and control	DQN, PPO, DDPG, Multiagent RL	Learns optimal control policies via interaction; enables real-time adaptation	Safety assurance and convergence stability remain challenges

CNN, convolutional neural networks; DDPG, Deep Deterministic Policy Gradient; DQN, *deep-Q network;* LR, *linear regression;* DT, *Digital Twin; LSTM, long short-term memory;* ML, *machine leanring;* RF, *Random Forest;* RL, *reinforcement learning; SVM, support vector machine.*

5.3.1 Load forecasting

A fundamental prerequisite for dependable and economical microgrid operation is accurate load forecasting. This relates to making useful predictions regarding future electricity demand across short-term (minutes to days) or long-term horizons, and it is one of the most important responsibilities in power systems. Traditional forecasting methods find it difficult to remain accurate as patterns of energy use become more dynamic due to elements like prosumer behavior, renewable energy, and electric car charging.

ML models can better forecast future demand by utilizing high-resolution historical and real-time data inside a DT framework. These projections can then support grid stability and energy efficiency by facilitating more intelligent scheduling of generation, storage, and demand-response measures. To predict the load for each future period, ML models are trained on historical load data, weather data, time-of-day patterns, and other pertinent factors in a DT-based smart grid. Operators and automated control systems can better arrange resources and balance generation and demand with the aid of accurate load predictions. The nonlinear and variable behavior of contemporary loads driven by distributed solar/wind power and electric vehicles is difficult for traditional load forecasting approaches like time-series analysis (ARIMA) and simple regression to handle. By identifying intricate patterns in the data, ML techniques have proven to perform better (Wangyong Guo, 2025).

S, LR (as a baseline for load patterns), and more complex learning models like LSTM are common techniques. Because LSTM networks can handle sequences with seasonality or trends and learn temporal connections, they are very useful for load forecasting. According to recent research, LSTM-based models predict microgrids with renewable energy variability more accurately than standard models (Silva-Rodriguez, 2024). For instance, one study developed a hybrid forecaster that handled both linear and nonlinear load patterns by combining an LSTM with a statistical model, producing daily load predictions with excellent accuracy.

An alternative method is weighted least adaptive neuro-fuzzy inference system (WLANFIS), a hybrid strategy that blends NNs, weighted least squares (WLS), and an adaptive neuro-fuzzy inference system (ANFIS). The ANFIS framework itself combines fuzzy logic's reasoning powers with NNs' capacity for learning. WLANFIS uses WLS and NNs to produce a fine-tuned dataset that aids in identifying the best fuzzy set ranges within the fuzzy system, as well as the ideal amount and kinds of membership functions. The accuracy and flexibility of ANFIS are enhanced by this connection. The findings demonstrate that WLANFIS outperforms other methods in terms of predicting accuracy and good generalization ability, especially for medium-term load forecasting, with a mean absolute percentage error as low as 2.66%. (Ali et al., 2021).

In practice, a DT may utilize an LSTM model to continually forecast a microgrid's demand for the upcoming hours. This would enable the controller, or human operator, to proactively send out generators or storage units. Demand-response initiatives and market operations (bidding tactics) both benefit from accurate load projections. In conclusion, ML-driven load forecasting in DTs lowers costs and blackout risks by ensuring that energy supply efficiently satisfies demand.

5.3.2 Fault detection and predictive maintenance

5.3.2.1 Fault detection

In the field of fault detection and diagnosis, smart grids need to quickly identify and diagnose equipment failures or malfunctions. Additionally, in order to undertake preventive maintenance, operators are striving toward PdM, which foresees failures before they occur. By identifying patterns of typical vs anomalous behavior in electrical equipment, ML algorithms built into a grid's DT are highly effective at both jobs. ML models can continually assess sensor streams from transformers, inverters, lines, and other assets, including currents, voltages, temperatures, vibrations, and more, in a microgrid DT. The model either initiates automated mitigation or flags anomalies or fault signatures for review.

Supervised learning techniques are commonly used for fault detection and classification. For example, SVM can be used to differentiate between normal transients and faulty conditions, and decision trees and random forests are used to classify the type of fault (e.g., short-circuit, open-circuit) based on measured parameters. Either simulated fault scenarios or historical fault event data are used to train these models. Five distinct ML classifiers, including SVM, k-nearest neighbors, decision trees, random forests, and NNs, were assessed for determining fault type and location in a single case study on a low-voltage DC microgrid; the results demonstrated high accuracy, proving the viability of data-driven fault diagnosis in microgrids (Ojetola, 2022).

The speed and versatility of ML-based fault detection in DTs are two of its main advantages. Near real-time reaction is made possible by models' ability to detect faults frequently milliseconds after they occur, which is significantly faster than human operators. Furthermore, these models may adjust to new fault patterns through retraining with fresh data, guaranteeing that the DT will continue to function well even when system conditions change. The foundation of robust DT applications is ML-driven fault detection because of its quick detection and ongoing learning capabilities.

5.3.2.2 Predictive maintenance techniques

PdM, as a data-driven maintenance approach, makes use of sensor data and analytical models to forecast when equipment will break down, enabling prompt and economical repairs. By evaluating the health and remaining usable life of an asset, PdM goes beyond defect identification. For instance, a DT enhanced with ML could forecast that a transformer has a high chance of failing over the next ten days because of slow trends in insulation degradation. Prognostic modeling and anomaly detection are frequently combined in PdM techniques. Here, unsupervised or semisupervised ML techniques are useful: Even in the absence of labeled "fault" samples for training, anomaly detection algorithms (such as one-class SVM, isolation forests, or clustering techniques) can detect minute variations in equipment sensor data that presage a failure.

Degradation patterns are also utilized to predict future conditions or time-to-failure using time-series models and NNs (including LSTM and autoencoders). According to a comprehensive literature analysis, research on the use of DTs for PdM has accelerated since 2018, emphasizing goals like asset health forecasting and significant obstacles like model complexity and a lack of failure data . Reducing unplanned downtime is a noteworthy result of ML-driven PdM in smart grids; utilities can prevent both catastrophic failures and needless maintenance by precisely replacing or repairing components as needed.

An energy distribution network's DT study, for instance, demonstrated that by modeling different grid scenarios and employing ML to forecast component failures, maintenance could be planned as efficiently as possible, reducing downtime . Essentially, ML-enabled PdM transforms reactive maintenance into proactive maintenance by transforming the grid's massive data into foresight.

5.3.2.3 Cybersecurity and anomaly detection

Smart grids are handling data flows in addition to power flows as a result of growing digitization, making them susceptible to cyberattacks. In the crucial area of cybersecurity, ML can help with anomaly and intrusion detection. When combined with AI, a DT of the grid can function as a cyber-defense platform that keeps an eye on device activity, control orders, and network traffic for indications of abnormalities or attacks. Since many cyberattacks seem like odd patterns that diverge from regular activities, anomaly detection techniques (unsupervised ML) are especially crucial in this context. For instance, anomaly detection models that have learned the typical patterns can detect an unauthorized control signal or an abrupt increase in data traffic on a substation's network. A baseline of typical multidimensional behavior can be established using statistical models or clustering techniques, and any departure from this sets off an alert.

To test and improve defenses, DTs can even replicate cyberattack scenarios. The DT can serve as a training environment for intrusion detection by modeling both acceptable behavior and possible attack behaviors using ML. If labeled datasets of known attacks vs typical occurrences are available, supervised learning can also be used. For example, a Random Forest or NN classifier can be trained to identify the hallmarks of malware or phishing attempts in grid control systems. The grid's virtual replica serves as both a tool for operations and a sentinel for network security thanks to the combination of DT and AI for cybersecurity.

Benign anomalies (equipment miscalibrations, communication faults, etc.) must be detected in addition to intentional attacks. As previously said, ML-based DTs are skilled in fault anomaly identification, which is similar to cybersecurity anomaly detection in that both require identifying the odd one out of a stream of data. Anomaly detection at scale becomes difficult as the smart grid develops into an IoT system with millions of connected devices; ML techniques (such as deep autoencoders that can recognize intricate normal patterns and identify reconstruction errors) are being used more and more in this field.

In conclusion, ML significantly increases grid resilience against outages and attacks by enabling DTs to serve as real-time monitoring and alert systems for both physical problems and cyber incursions. In order to preserve stability and service, the system can swiftly isolate a malfunction or a compromised component and then reconfigure or start defense mechanisms.

5.3.2.4 Energy optimization and control

Energy optimization and control are significant areas of application for ML in DT-based microgrids. Microgrids frequently function in intricate settings, juggling dynamic loads and a variety of dispersed energy sources, including solar panels, wind turbines, batteries, and diesel generators. The purpose is to respect limits (e.g., battery capacity, power balance) while optimizing goals such as cost, energy efficiency, or dependability. While conventional model-predictive or rule-based control techniques are helpful, specifically, RL provides a potent means of achieving adaptive, nearly optimum control policies in real time. Since a DT can mimic the microgrid's reaction to events without endangering the actual system, it offers the ideal environment for training and implementing such learning-based controllers.

By interacting with the environment (the microgrid DT model) and receiving feedback in the form of rewards or penalties (for instance, a reward could be higher when operation costs are lower, and penalties for constraint violations or unmet demand), an agent (the control algorithm) learns in an RL approach. The agent eventually finds a control strategy (which maps states to actions) that maximizes cumulative reward, successfully learning tactics like the best way to charge or discharge batteries, dispatch generators, or reduce demand during peak hours. To prevent peak charges, a DQN agent, for example, can learn to do peak shaving by utilizing stored energy during times of high demand. Likewise, sophisticated algorithms like PPO and DDPG are used for continuous management of inverters and voltage levels, when actions are real-valued (e.g., setpoints) as opposed to discrete. In fact, when an RL agent utilizing the PPO algorithm was trained to regulate a battery energy storage system in conjunction with load forecasting in one industrial microgrid research, it significantly reduced operational costs (by more than 20%) when compared to a baseline that did not learn (Upadhyay et al., 2024). This was achieved by the agent learning the best energy arbitrage and peak demand limiting strategies (charging when electricity is cheap or plentiful, discharging when it is expensive or scarce).

Other ML techniques, including RL, also help with optimization. For example, supervised ML models can predict energy or solar power prices ahead of time, and these predictions are then entered into an optimization solver for scheduling. A DT might use a demand forecasting model to forecast tomorrow's electricity consumption, and then feed those predictions into an optimization algorithm that schedules battery charging and discharging at the lowest cost. This hybrid approach, also known as predict-and-optimize, is especially effective. The DT may integrate such pipelines and then run what-if simulations to figure out the best course of action. Additionally, multiagent systems can be employed, where many RL agents (for different assets or microgrid clusters) learn to coordinate with each other, often using game theory or FL techniques (explained further).

An autonomous energy management system that continuously self-tunes is the ultimate goal of ML-driven control in DTs. The results include increased reliability (quick corrective response to disruptions), increased efficiency (lower fuel costs, greater use of renewables), and the capacity to manage complexity (many devices and limitations) that

TABLE 5.2 Key smart grid prediction/diagnosis problems and common ML solutions.

Problem	Goal	Impacts	Typical ML techniques used
Load forecasting	Make short- or long-term predictions about future power demand (load) to help with planning and balancing.	lower reserve margins and optimal generator dispatch.	• NNs (such as ANN and LSTM for time-series) and regression models (ARIMA, LR) (Divisekara, 2020; Victorio, 2021). • Using ensemble techniques (like XGBoost) to increase accuracy (Tripto, 2020).
Fault detection & diagnosis	As issues or anomalous circumstances arise in grid components (lines, transformers, etc.), identify and categorize them.	Quick fault separation, little equipment damage, or downtime.	• Classification algorithms trained on fault signatures, including Random Forest, SVM, k-NN, and Decision Trees. • CNN on high-frequency waveform data for defect type identification is an example of signal processing plus ML (Hasan, 2018). • ML-enhanced rule-based expert systems for pattern identification (Bukhari, 2020).
PdM techniques	Monitor deterioration to predict equipment problems before they occur and schedule maintenance as efficiently as possible.	Prevent disastrous breakdowns, increase asset longevity, and minimize maintenance expenses.	• Using sensor trends and anomaly detection (one-class SVM, clustering) to identify wear early on (Al-amri, 2021). • Time-series health index forecasting (e.g., forecast battery health or transformer oil quality). • Using sequences of condition data in DL techniques (Autoencoders, LSTM) is used to predict the remaining useful life (Serradilla, 2022).
Cyber threat detection (intrusion detection)	Find malware, illegal access, or unusual control activity in the grid's control and communication systems.	Assure operational security by preventing or lessening cyberattacks.	• Network traffic anomaly detection without supervision (using PCA and clustering to identify outliers) (Hwang, 2020). • Supervised classification of known attack patterns (e.g., on intrusion datasets using Random Forest or Naive Bayes) (Liberini, 2022). • Deep Autoencoders that can recognize typical device activity and react to deviations (Meidan, 2018).
Energy optimization & control	For cost, efficiency, or stability, optimize demand-side management, power dispatch, and storage utilization in real-time.	lower energy costs, improved use of renewable energy sources, and steady voltage and frequency.	• RL for adaptive control policies (Q-learning, DQN, DDPG, PPO) (Pallonetto, 2019). • Hybrid ML + optimization, which uses ML model forecasts to provide predictive control (Ali, 2019). • Distributed resource coordination using multiagent learning, where each agent is responsible for a battery or generator (Faheem, 2024).

CNN, *convolutional neural networks;* DDPG, *Deep Deterministic Policy Gradient;* DL, *deep learning;* DQN, *deep-Q network;* LSTM, *long short-term memory;* ML, *machine learning;* NNs, *neural networks;* PCA, *principal component analysis;* PPO, *Proximal Policy Optimization;* SVM, *support vector machine;*

human or classical controls cannot readily handle. Because of this, adaptive ML control is essential to "Energy 4.0" projects and future smart grids.

Table 5.2 lists key smart grid issues that ML can solve, along with the usual techniques that are employed in each instance, to provide a summary of the aforementioned use cases:

Beyond the methodological mapping discussed earlier, quantitative results from recent studies demonstrate the tangible benefits of ML integration in DT frameworks. Table 5.3 consolidates these performance metrics to facilitate comparison and reproducibility:

TABLE 5.3 Performance snapshot of representative ML-DT case studies.

Application area	Model/Method	Key metric	Reported performance	Reference
Load forecasting	WLANFIS (Hybrid ANFIS)	Mean absolute percentage error (MAPE)	2.66%	Ali et al. (2021)
Energy optimization	PPO (Proximal Policy Optimization)	Operational cost reduction	>20%	Upadhyay et al. (2024)
Energy optimization	PPO (Centralized)	Economic benefit increase	12.7%	Tan (2024)
Energy optimization	PPO (Centralized)	Carbon Emission Reduction	15.3%	Tan (2024)
Energy optimization	Upgraded Actor-Critic	Operational Cost Reduction	>20%	Mihet-Popa (2024)

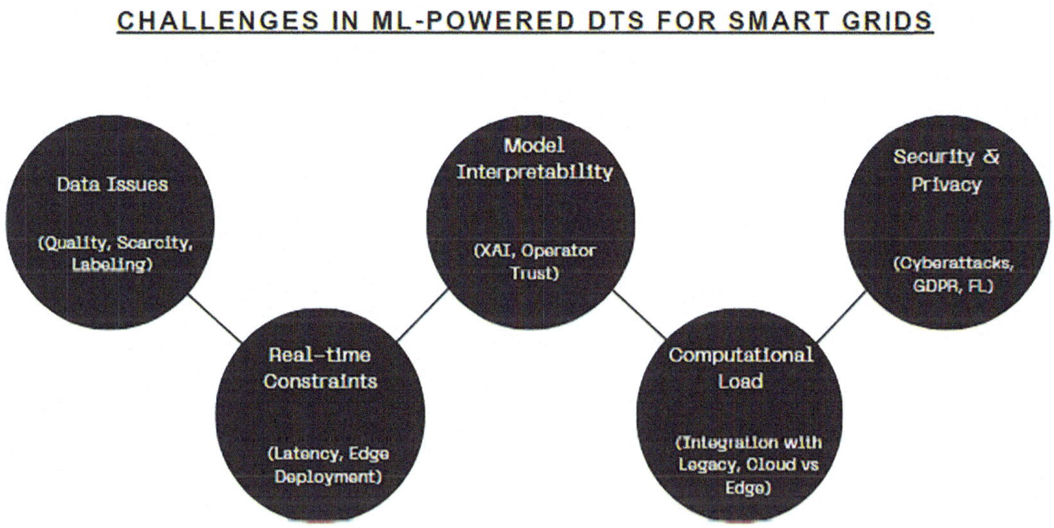

FIGURE 5.4 Key challenges in deploying ML-powered DTs in smart grids: highlights the major deployment challenges: data quality, real-time processing, interpretability, integration, and security.

Note: A lot of applications mix several methods. For example, an LSTM load forecast may be used to feed an RL agent that manages a battery, or an anomaly identified by an unsupervised model could activate a supervised classifier to determine its cause. For these AI-driven processes, the DT serves as the integrating platform.

5.4 Challenges and limitations

Although there is great potential for integrating ML and DT technology into smart grids, there still remain a number of obstacles to overcome. For a successful, large-scale deployment in the real world, mastering these is essential. We address the main obstacles, including data availability and quality, real-time limits, interpretability (explainability) of the model, and computational constraints (e.g., security and integration issues) in this section. The core challenges faced are summarized in Fig. 5.4, include:

5.4.1 Data quality and availability

ML requires a huge amount of data for training. A DT requires high-quality data from the physical grid, such as sensor readings, event logs, maintenance records, etc., in order to make precise predictions and detect problems. In actuality, though, data may be noisy, lacking, or not recorded at the necessary frequency. Since many failure events, such as transformer explosions or cascading blackouts, are fortunately uncommon, there are not many labeled instances available to train prediction algorithms on those crucial occurrences (Dinter et al., 2022). Furthermore, it could be challenging to combine data silos from several utility companies and microgrid operators. Careful data governance and perhaps data-sharing systems are required for this.

To eliminate outliers and fill in gaps, a lot of work is put into data cleaning, validation, and preprocessing before ML is deployed. Since the twin should ideally reflect every pertinent element of the physical system, any missing or subpar data causes a blind spot in the twin; the difficulty is increased in a DT context. In an assessment, Mchirgui et al. point out that it is crucial to guarantee data availability and quality throughout the lifecycle, necessitating procedures to determine what data is crucial, guarantee its accuracy, and get rid of "weak data" (Nabil Mchirgui, 2024). Another factor is that labeling it frequently takes professional assistance to get ground truth labels. For instance, was a reported anomaly actually a cyberattack or merely a sensor glitch? Labeling and validating events can require a significant amount of work, which is sometimes not fully handled in research prototypes. Using simulations to synthesize data, transfer learning from other domains, or active learning, where the model can ask an expert for labeling on unclear cases, are some ways to overcome data issues.

5.4.2 Real-time constraints

Supervising and regulating the smart grid is a real-time endeavor. Any ML model employed in grid operations must satisfy stringent latency and reliability criteria. A highly complex model that requires seconds for computation may prove useless in scenarios where, for instance, identifying a flaw and activating a breaker must occur within milliseconds to avert equipment damage.

DTs may need to continuously generate predictions or choices in addition to continuously ingesting data streams from supervisory control and data acquisition (SCADA), PMUs, and smart meters. It can be difficult to make sure that communication delays, ML inference, and data pipelines all fit within real-time constraints. This occasionally entails deploying algorithms on edge devices nearer the assets or utilizing lighter models. Accuracy and speed are frequently traded off; for example, a DL model may be slower than a simpler model but more accurate.

To deploy bigger models under real-time limitations, methods like model compression, quantization, or the use of specialist hardware (FPGAs, GPUs) in control centers are being investigated. Although an AI-driven DT framework produced remarkable energy-saving outcomes, one study noted that it did not go into great detail on the scalability and computational complexity for large-scale deployment, which is a legitimate worry.

The ML algorithms must endure the increased load without failure if a system is required to expand to thousands of nodes. A real-time constraint also suggests resilience to transient faults or missing data; if one sensor's data is briefly lost, the model should adapt gracefully and perhaps fill it in through prediction. Perhaps mixed architectures (e.g., a rapid rule-based fallback that maintains safety while the ML model delivers optimal recommendations) are needed to achieve real-time reliability. Therefore, using methods from real-time systems engineering, the DT and ML system design must take into consideration the worst-case timing possibilities.

5.4.3 Model interpretability and explainability

Many ML algorithms, particularly DL and ensemble approaches, are "black boxes" since they do not simply explain the reasoning behind a particular prediction or choice. Operators and engineers require knowledge of and confidence in these AI models in vital infrastructure, such as the electrical grid. For instance, the operator will want to know the reasons (which sensor or condition prompted it) if an ML model suggests turning off a feeder to prevent an incident. As a result, smart grids now require XAI. (We shall look a little closer into XAI in a later section.) XAI methods, including rule extraction, SHAP values, and feature importance analysis aim to shed light on the underlying workings or factors that influence complex model decisions.

Regulators may also mandate explainability for AI decisions that affect markets or grid operations. Engineers can validate model performance and detect issues by incorporating explainability into DT dashboards. The implementation of interpretable models, such as decision trees or linear models, in specific system components, or the incorporation of explanation modules into NNs, exemplifies current research focused on enhancing the transparency of AI models for grid management.

Since part of the system's behavior follows known physics, one solution is to embed domain knowledge into the model architecture.

5.4.3.1 *Physics-guided neural networks*

These models guarantee that predictions stay in line with accepted physics principles by directly incorporating physical rules into their structure or loss functions. They are more interpretable by design and are especially useful in engineering settings where dependability and trust are crucial. Finding a balance between the accuracy of more complicated models and the simplicity of interpretable models is challenging. User trust is another issue: if a model occasionally makes mistakes, operators could become unconfident unless its limitations are clearly known. Therefore, validation, visualization of model confidence, and user training are necessary for human–ML collaboration to be successful.

Explainability should be viewed as a crucial necessity rather than an optional feature since it is crucial for diagnosing DT behavior. For instance, system performance can be directly enhanced by identifying whether a persistent false alarm results from an operating condition that was not recorded during training or from a defective training dataset. In conclusion, the requirement that AI-driven choices be transparent and trustworthy is a significant obstacle to wider implementation in the electrical industry.

5.4.4 Computational integration and limitations

It takes a lot of computing power to implement a DT with sophisticated ML for a big smart grid. Memory, computing infrastructure, and interface with older systems are among the problems. Integrating a new DT platform with SCADA systems that were created decades ago without causing any disturbances is a difficult task for many utilities. Computational restrictions can take many different forms. For example, if DT uses high-fidelity simulations (such as electromagnetic transient models of the grid) in conjunction with ML, it might be costly to execute in real-time. If the amount of data pouring from IoT devices is not controlled, it can become overwhelming (the phrase "big data" is definitely applicable; high sampling rate PMUs distributed throughout a grid can produce gigabytes of data every day). A sturdy IT infrastructure is needed to store, process, and secure this data. Cloud computing and distributed computing (edge-cloud coordination) are frequently suggested solutions; while lesser inference runs on local servers or edge devices for speed, heavy computation, such as retraining large models, can take place offline or in the cloud (Ali Aghazadeh Ardebili, 2024).

However, this creates a dependency on network connectivity; the local system should continue to operate even if the cloud connection is lost. Aligning the elements of the DT, the databases, the operator interfaces, the ML models, and the physical simulations presents another integration problem. It is a software engineering issue to make sure they all interact successfully, and that data formats and semantics are consistent. DT standards and interoperability frameworks are just starting to take shape; without them, every deployment may require a one-off engineering effort. There is also the financial consideration: creating and sustaining an advanced DT with AI may necessitate a large expenditure on hardware, software, and human resources. The absence of more accessible options or collaborative initiatives may render it unfeasible for smaller utilities. Costs and development effort can be reduced through the use of open-source initiatives and industry collaborations, such as open standards for DT data interchange.

Last but not least, one must consider the possibility of model errors. Since an ML model is only as good as its training, it is necessary to incorporate suitable fail-safes, such as human-in-the-loop confirmation for high-impact judgments or alarms that a prediction is dubious. In conclusion, DT & ML's technological integration into the current grid control environment is a challenge that involves more than simply algorithms; for it to be genuinely successful, system architecture, standards, and workflow management are all necessary.

5.4.5 Security and privacy

Maintaining the cybersecurity of DT itself is crucial given the rise in connectivity and data sharing (particularly when utilizing cloud or FL, which are covered next). Ironically, the DT creates a new attack surface even though it is designed to improve grid cybersecurity. Protecting data integrity is essential since corrupted data may cause ML models to generate incorrect predictions. It is necessary to use methods like blockchain (for data integrity and logging), encryption, and robust authentication. If consumer energy usage data is included, privacy is also an issue (for example, load profiles can reveal personal activity). When handling such data, laws like the General Data Protection Regulation (GDPR) are relevant. This increases interest in techniques that save raw data locally, such as FL (covered in section 5.5). When designing AI-driven DT systems, security and privacy must be considered from the beginning rather than as an afterthought.

TABLE 5.4 Challenge-mitigation matrix for ML-driven DTs.

Challenge	Mitigation strategies
Data scarcity & quality	Synthetic data generation, transfer learning, active learning, data governance frameworks
Real-time latency	Edge computing, model compression, quantization, lightweight algorithms, specialized hardware (FPGAs/GPUs)
Model interpretability	Explainable AI (XAI), SHAP/LIME techniques, rule extraction, Physics-guided neural networks (PGNNs)
Computational integration	Edge-cloud coordination, open standards, modular architecture, legacy system interfaces
Security & privacy	Blockchain for data integrity, federated learning, encryption, and robust authentication

5.4.6 Principal insights

Despite the challenges that have been presented, progress is being made across the board. Pilot projects are providing light on real-world challenges and answers, and academic and corporate groups are actively studying potential solutions to these difficulties. The following part will provide an overview of some of the recent advancements and research avenues that have been developed with the intention of overcoming these limitations and enhancing the potential of DT technology applications in smart grids.

To clearly link the challenges discussed earlier with potential mitigation strategies, Table 5.4 presents a concise challenge- mitigation matrix, illustrating how each obstacle in deploying ML and DT in smart grids can be addressed.

The ethical and governance aspects of DTs and ML frameworks become just as important as their technical performance as they become integrated into critical grid operations. In accordance with local data-protection laws like the GDPR, data privacy must be guaranteed when sensitive operational or customer information is shared across utilities or cloud environments. In order to ensure that algorithmic choices like load-curtailment scheduling or fault prioritization do not unintentionally disadvantage specific user groups or geographical areas, fairness and bias mitigation are essential. Additionally, transparent model documentation, decision logs, and human-in-the-loop validation should be used to uphold accountability and traceability. Building operator trust and public acceptance of AI-driven DT systems will require the establishment of explicit governance frameworks that specify accountability for model behavior, data stewardship, and security compliance.

5.5 Future research trends

As AI and DT technologies advance, a number of emerging trends are influencing how they are used in microgrids and smart grids. These developments aim to address current challenges and open up new possibilities. These include XAI and transparent models, FL and collaborative analytics, and the use of blockchain and other technologies for increased security and trust. We also discuss other breakthroughs, such as standardization attempts, transfer learning, and AI informed by physics.

5.5.1 Explainable artificial intelligence and transparent models

XAI directly addresses the interpretability and explainability challenge highlighted in Section 5.4.3, ensuring that operators can understand and trust AI-driven decisions in smart grids. Explainability is essential, as mentioned in the challenges. The goal of future research is to improve the interpretability and reliability of AI models for smart grids. Creating XAI techniques specifically for power systems is part of this. One trend, for instance, is the use of hybrid models, in which AI only forecasts the residuals and a core physics model provides a baseline. This allows the influence of AI to be limited by known physics, which makes it easier to understand. Another strategy is rule extraction, in which algorithms try to extract decision trees or rules that roughly represent the behavior of a complicated model in various regimes after it has been trained.

Interactive visualization tools for DT outputs are especially important because they would enable engineers to determine which features or sensors were mostly responsible for a given control action recommendation or anomaly alarm. This would be a significant factor in the overall significance of these tools. An XAI framework for smart grid operations has been proposed over recent years. Each AI recommendation is accompanied by an explanatory graph that shows the data or sequence of events that led to it (Babu, 2025). Prompt-based architectures, such as the dual-agent transformer network, are also emerging for explainable voltage prediction, enabling interpretable coordination between forecasting and control agents (Xu et al., 2025).

This reflects a larger movement toward trustworthy AI, which aims to make sure AI systems are not only efficient but also dependable, equitable, and in line with the expectations of human operators. Explainability is associated with legal accountability and compliance in critical systems; we anticipate that future grid AI would incorporate audit logs and decision explanation archives (particularly if AI ever performs autonomous control operations). Additionally, the DT might demonstrate its reasoning to an operator, who could then provide feedback (e.g., "this recommendation was not appropriate because of X"), which would be used to improve the model (a type of human-guided ML). This is known as user-in-the-loop learning. In conclusion, AI in smart grids of the future will be much more transparent, utilizing methods from XAI research to transform opaque systems into transparent ones so that grid operators can interact with confidence (Kobayashi & Alam, 2024).

5.5.2 Federated learning and collaborative analytics

FL helps mitigate privacy and data-sharing challenges discussed in Section 5.4.5, enabling collaborative model training across utilities without exposing sensitive local data. FL is an ML technique where model training is dispersed over multiple devices or organizations, and only the learned parameters, not the raw data, are shared and aggregated. This is particularly attractive for electric power systems due to concerns around data ownership and privacy. For instance, data about loads, malfunctions, and cyber events may be kept by many utility companies or microgrid operators and may not be directly shared due to privacy or legal constraints. Data from several grids is frequently identically distributed and nonindependent (non-IID), which presents a significant issue in this setting. In reality, utilities might range significantly in terms of event frequency, equipment kinds, and consumption patterns, which means that local statistics are distributed differently. One of the primary challenges for using FL in smart grids is its non-IID character, which makes it more difficult for the global model to converge well and maintain accuracy across all participants.

By using FL, they can work together to train a global ML model (for example, load forecasting or anomaly detection) by training on their own local data and communicating gradient updates to a central coordinator while training on their local data. In this manner, all participants' expertise is included in the final model without disclosing private information. Initial studies on FL in smart grids are being conducted for tasks like load forecasting across regions and intrusion detection, where each substation trains on its network data to improve a global attack detector (Shanker, 2022). Additionally, federated DT frameworks are being envisioned, in which a federation mechanism aligns the models for a system-wide perspective while each sub-system twin learns locally (Zhou, 2024). FL ensures user privacy by transmitting just gradients throughout the training process. Moreover, sophisticated FL architectures that integrate deep RL, DTs, and generative adversarial networks can be employed for the identification of privacy risks (Ali, 2023). A future vision includes a federated smart grid DT network, where each utility's DT applies ML to its data and regularly shares models securely to create a more resilient meta-DT that, for instance, can identify an attack pattern that was observed in one area but is now present in others. For instance, a decentralized False Data Injection Attack (FDIA) detection approach based on deep FL, utilizing attentive aggregation, can perform parallel computation and effectively detect stealthy FDIA across all nodes concurrently (Tahir, 2022).

Additionally, this distributes the processing load and enables edge devices (such as EV chargers or smart meters) to take part in federated edge learning, which is the process of learning local trends. Although there are still difficulties in managing non-IID (nonidentically distributed) data discrepancies between participants, ensuring federated convergence, and communication overhead, these problems are being addressed by continuing research (Mohamed Rafik Aymene Berkani, 2025). FL is anticipated to be a key component of future resilient, privacy-aware grid AI and fits in nicely with the decentralized structure of energy systems.

5.5.3 Blockchain integration and secure management decentralization

Blockchain addresses trust, integrity, and secure integration challenges described in Sections 5.4.4–5.4.5 by providing immutable records and enhancing data provenance in DT-enabled grids. Blockchain and other distributed ledger technologies are being intensively researched to improve DTs in the energy sector for security, transparency, and decentralized coordination. The primary advantage of blockchain technology is its capacity to produce an unchangeable and verifiable record of occurrences. To guarantee that the data utilized by DTs and ML models has not been altered, sensor readings from a power grid, for example, can be hashed and saved on a blockchain. This integrity is crucial since trust in the data directly affects trust in the DT model. The Pearson Correlation Coefficient is determined for each connection, yielding a value of 0.9749 in the absence of an adversary and 0.1282 upon the introduction of an adversary into the network (Kamal, 2021).

Automation is also made possible by blockchain. Without human involvement, smart contracts can validate DT-driven actions or settle energy transactions. For instance, a smart contract might automatically confirm the result and immediately pay participants if a DT predicts energy savings from a demand-response event. Likewise, digital identities for devices (or their twins) stored on a blockchain can stop spoofing attempts because the network would reject harmful or unidentified identities. Blockchain is frequently viewed as a key component of Smart Grid 3.0 (the "Energy Internet"), which aims to achieve decentralized decision-making, prosumer-to-prosumer energy trading, and highly automated microgrids. According to this idea, blockchain offers the layer of trust and verification required for autonomous coordination among dispersed assets and peer-to-peer energy marketplaces. This new energy paradigm has been said to be laid by the combination of DTs, blockchain, and even cutting-edge technologies like the Metaverse (Tariq, 2024). Selecting appropriate convergence models, mathematical frameworks that connect distributed optimization, blockchain, and DTs, is a key research challenge in this field. Realism, computational complexity, and compliance with the developing Smart Grid 3.0 framework must all be balanced in these simulations (Adnan & Tariq, 2024).

Other DT-related technologies can also be supported by blockchain. For FL, for instance, it can act as a secure audit trail, guaranteeing that contributions from various utilities are traceable and even rewarded with tokens. (Tariq, 2021) To improve grid resilience against attacks, blockchain has been integrated with DL, Software-Defined Networking, and DTs in cybersecurity to authenticate control commands and record system status. (Prabhat Kumar, 2024)

Even if issues like scalability and energy-intensive consensus processes still exist, developments like proof-of-stake and permissioned blockchains are helping to resolve them. All things considered, blockchain has great potential to build a secure, open, and self-sufficient energy ecosystem when paired with DTs and AI. In the near future, a DT might use ML to anticipate energy output, use blockchain technology to conduct peer-to-peer energy trading, and automatically settle transactions. All of this would happen without the need for a central authority and with complete participant trust. (Banerji et al., 2024)

5.5.4 Standardization and interoperability of Digital Twin platforms

Standard data models and application programming interfaces (API) for DTs in the energy sector will be pushed as use increases. Initiatives from groups like the IEEE, IEC, and industry alliances are probably going to result in frameworks that guarantee a DT of one system can communicate or exchange data with another. Common evaluation metrics for DT accuracy, standard interfaces for real-time data streaming, and standard schemas for asset data are a few examples of this. Innovation would be accelerated by a standardized approach since models and tools that work with the twins of many utilities may be created.

Standardized assessment criteria, including common datasets for grid anomaly detection, are expected for ML specifically. This will foster competition and promote algorithmic progress, much to the impact of computer vision with ImageNet, among others. Numerous open datasets and competitions now exist, particularly in load forecasting; it would be advantageous to broaden them to encompass comprehensive DT scenarios, including simulated microgrid data and fault conditions.

5.5.5 Transfer learning and domain adaptation

Although every power system or microgrid is different, many of the lessons learned can be applied to other situations. Adapting an ML model that was trained in one setting to another is known as transfer learning. For instance, instead of beginning from scratch, a fault detection model learned on one microgrid may be modified for another by retraining only the final layer. Libraries of pretrained models for common jobs are probably in the future (for example, a vendor's "generic wind turbine fault predictor" model that a wind farm operator's DT can modify with local data).

Models can be adjusted to new data distributions using domain adaptation techniques, which is helpful for transferring a model from a 50 to a 60 Hz grid environment. FL is also related to this, as one participant's model might assist in initializing another's. When implementing AI for a new system, the overall result will be a reduction in development time because we would not constantly start from scratch but rather apply the knowledge that has been gathered from various grid initiatives.

5.5.6 Edge/fog-based Digital Twins and the role of 5G/6G technologies

Distributing the DT across cloud and edge is made possible by the introduction of ultra-fast, low-latency connections (5G and beyond). Parts of the DT (including ML inference) can be located close to the field devices (edge or fog nodes) rather than in a single central system. Because only insights, not raw data, may be transferred upstream, this lowers latency and bandwidth consumption. Utilizing micro-DTs at the device level, for instance, every solar inverter has a twin that performs local ML-based control, and coordinating with the system-level twin is a popular approach.

Synchronization is guaranteed by advanced networking. The hierarchical regulation of power grids is modeled by this hierarchy of twins and AI agents. Additionally, it increases resilience because local agents can continue to operate on their own even in the event that they lose contact with central control. The best way to divide ML jobs between edge and cloud computing (also known as split learning), and how 6G's expected features (such as integrated sensing and connectivity) will improve DT capabilities, is being investigated.

5.5.7 Key observations

DT technology in smart grids possesses a very collaborative and interdisciplinary future. AI will evolve to be more secure (utilizing blockchain and resilient architecture), more reliable (explainable and informed by physics), and more decentralized (federated and edge-based). These advancements suggest the concept of an autonomous, self-optimizing grid, often referred to as the "cognitive grid," wherein digital representations of grid components cooperate to efficiently and reliably manage energy through constant data-driven learning. Researchers and master's students exploring this subject will encounter several unresolved problems and innovative solutions, many of which are now in development.

5.6 Conclusion

ML and DT technology have the potential to revolutionize smart grid and microgrid management. This chapter provided a thorough review of how ML methods enable DTs to address diagnosis and prediction issues in the energy sector. From load forecasting and PdM to fault diagnosis, cybersecurity, and optimal control, the main application areas were examined, showing that data-driven models can significantly enhance foresight and decision-making across these domains. A range of ML techniques was explored, highlighting how supervised learning offers tools for classification and regression for labeled problems, unsupervised learning reveals anomalies and hidden structures in unlabeled data, DL enables robust pattern extraction for complex time-series and spatial data, and RL facilitates adaptive control policies for real-time optimization. Recent literature-backed instances of these techniques in power system DT implementations were also discussed.

Success hinges on tackling important interdisciplinary issues, such as guaranteeing high-quality data access, attaining interpretability and user confidence, offering adequate real-time processing power, and blending in seamlessly with current grid operations. New approaches like blockchain, XAI, and FL suggest strategies to get beyond these obstacles.

The main takeaway is unmistakable: combining DTs and ML is a revolutionary step for energy systems, not only a small improvement. Engineers and data scientists must work together across disciplines to realize this goal, fusing cutting-edge AI with physics-based knowledge. The ultimate objective is challenging but attainable: a self-optimizing, cognitive grid that is safe, robust, and able to independently manage dispersed resources while maintaining constant communication with humans.

5.7 Key takeaways

- **ML families & eoles:** Supervised, unsupervised, deep, and RL each offer unique capabilities, from forecasting and anomaly detection to adaptive control.
- **DT integration:** DTs provide a virtual replica of the grid, enabling real-time monitoring, simulation, and predictive analytics.

- **Applications:** Key domains include load forecasting, PdM, fault detection, cybersecurity, and energy optimization.
- **Challenges & mitigations:** Data quality, real-time constraints, interpretability, computational integration, and security can be addressed via XAI, FL, blockchain, model compression, and edge computing.
- **Future trends:** Physics-informed models, federated learning, blockchain, transfer learning, and edge-based DTs are shaping the next-generation smart grids.
- **Ultimate goal:** Realizing a cognitive, self-optimizing, secure, and resilient energy system requires interdisciplinary collaboration between engineers, data scientists, and operators.

References

Abu-Rub, M. (2023). Harnessing recurrent-based deep learning models for time series photovoltaic power forecasting. *IECON 2023- 49th Annual Conference of the IEEE Industrial Electronics Society*. doi:10.1109/IECON51785.2023.10312298.

Adnan, M., & Tariq, M. (2024). Smart Grid 3.0: Navigating the future—Unleashing the Power of metaverse, blockchain, and Digital Twins in the evolution of smart grids. *SSRN*. doi:10.2139/ssrn.4801458.

Al-amri, R. (2021). A review of machine learning and deep learning techniques for anomaly detection in IoT data. *Applied Sciences, 11*(12). doi:10.3390/app11125320.

Ali, M. (2019). Online remaining useful life prediction for lithium-ion batteries using partial discharge data features. *Energies, 12*(22). doi:10.3390/en12224366.

Ali, M. (2023). Federated learning for privacy preservation in smart healthcare systems: A comprehensive survey. *IEEE Journal of Biomedical and Health Informatics, 27*(2), 778–789. doi:10.1109/JBHI.2022.3181823.

Ali, M. A, Tariq, M., Poor, M., & Vincent, H. (2021). Load forecasting through estimated parametrized based fuzzy inference system in smart grids. *IEEE Transactions on Fuzzy Systems, 29*(1), 156–165. doi:10.1109/TFUZZ.2020.2986982.

Ali Aghazadeh Ardebili, M. (2024). Digital Twins of smart energy systems: A systematic literature review on enablers, design, management and computational challenges. *Energy Informatics, 7*(94). doi:10.1186/s42162-024-00385-5.

Babu, M. A. S. (2025). Applications of Artificial Intelligence in (machine learning /deep learning) smart grid. *Interantional Journal of Scientific Research in Engineering and Management, 09*(03), 1–9. doi:10.55041/IJSREM42202.

Banerji, A., Sharma, K., & Meikandasivam, D. V. (2024). DT technology: A smart grid perspective. *AIP Conference Proceeding, 2966*, 020012.

Bukhari, S. (2020). Convolutional neural network-based intelligent protection strategy for microgrids. *Iet Generation Transmission & Distribution, 14*(7), 1177–1185. doi:10.1049/iet-gtd.2018.7049.

Dinter, R., Tekinerdogan, B., & Catal, C. (2022). Predictive maintenance using digital twins: A systematic literature review. *Information and Software Technology, 151*(107008). https://edepot.wur.nl/574763. doi:10.1016/j.infsof.2022.107008.

Divisekara, R. (2020). Forecasting the red lentils commodity market price using sarima models. *Sn Business & Economics, 1*(1). doi:10.1007/s43546-020-00020-x.

Dong H., (2021). Robust regional coordination of inverter-based Volt/Var control via multi-agent deep reinforcement learning. **12** 5420–5433, doi:10.1109/tsg.2021.3104139.

Faheem, M. (2024). A lightweight smart contracts framework for blockchain-based secure communication in smart grid applications. *Iet Generation Transmission & Distribution, 18*(3), 625–638. doi:10.1049/gtd2.13103.

Hasan, M. (2018). Bearing fault diagnosis under variable rotational speeds using Stockwell transform-based vibration imaging and transfer learning. *Applied Sciences, 8*(12). doi:10.3390/app8122357.

Hwang, R. (2020). An unsupervised deep learning model for early network traffic anomaly detection. *IEEE Access, 8*, 30387–30399. doi:10.1109/access.2020.2973023.

Kamal, M. (2021). Blockchain-based lightweight and secured V2V communication in the internet of vehicles. *IEEE Transactions on Intelligent Transportation Systems, 22*(7), 3997–4004. doi:10.1109/TITS.2020.3002462.

Kobayashi, K., & Alam, S. (2024). Explainable, interpretable, and trustworthy AI for an intelligent digital twin: A case study on remaining useful life. *Engineering Applications of Artificial Intelligence, 129*. https://www.sciencedirect.com/science/article/pii/S0952197623018043. doi:10.1016/j.engappai.2023.107620.

Liberini, V. (2022). Radiomics and artificial intelligence in prostate cancer: new tools for molecular hybrid imaging and theragnostics. *European Radiology Experimental, 6*(1). doi:10.1186/s41747-022-00282.

Liu, D., Zang, C., Zeng, P., Li, W., Wang, X., Liu, Y., & Xu, S. (2023). Deep reinforcement learning for real-time economic energy management of microgrid system considering uncertainties. *Frontiers Energy Research, 11*. https://www.frontiersin.org/journals/energy-research/articles/10.3389/fenrg.2023.1163053/full. doi:10.3389/fenrg.2023.1163053.

Liu, Z., Wang, Y., Wang, Q., & Hu, M. (2025). Vision transformer-based anomaly detection in smart grid phasor measurement units using deep learning models. *IEEE Access, 13*, 44565–44576. doi:10.1109/ACCESS.2025.3549679.

Meidan, Y. (2018). N-baiot—network-based detection of iot botnet attacks using deep autoencoders. *Ieee Pervasive Computing, 17*(3), 12–22. doi:10.1109/mprv.2018.03367731.

Mihet-Popa, S. (2024). Energy management system for an industrial microgrid using optimization algorithms-based reinforcement learning technique. *Energies*. doi:10.3390/en17163898.

Mohamed Rafik Aymene Berkani, A. (2025). Advances in federated learning: Applications and challenges in smart building environments and beyond. *Computers, 14*(4). doi:10.3390/computers14040124.

Nabil Mchirgui, N. (2024). The applications and challenges of digital twin technology in smart grids: A comprehensive review. *MDPI, 14*(23), 10933. https://www.mdpi.com/2076-3417/14/23/10933#:⟨:text=inclusion%20of%20machine%20learning%2C%20it,life%20entity.

Ojetola, S., Reno, M. J., Flicker, J., Bauer, D., & Stoltzfuz, D. (2022). Testing machine learned fault detection and classification on a DC microgrid. *IEEE Power & Energy Society Innovative Smart Grid Technologies Conference (ISGT)* https://ieeexplore.ieee.org/document/9817517. doi:10.1109/ISGT50606.2022.9817517.

Opy Das, M. (2024). Advancements in digital twin technology and machine learning for energy systems: A comprehensive review of applications in smart grids, renewable energy, and electric vehicle optimisation. *Energy Conversion and Management, 24.* https://doaj.org/article/2d34497e4aa840e6ade2c3f5a854ee34#:⟨:text=systems%2C%20including%20forecasting%2C%20anomaly%20detection%2C,The.

Pallonetto, F. (2019). Demand response algorithms for smart-grid ready residential buildings using machine learning models. *Applied Energy, 239,* 1265–1282. doi:10.1016/j.apenergy.2019.02.020.

Pentsos, V., Tragoudas, S., Wibbenmeyer, J., & Khdeer, N. (2025). A hybrid LSTM-transformer model for power load forecasting. *IEEE Transactions on Smart Grid, 16*(3), 2624–2634. doi:10.1109/TSG.2025.3535407.

Prabhat Kumar, R. (2024). Digital twin-driven SDN for smart grid: A deep learning integrated blockchain for cybersecurity. *Solar Energy, 263*(111921). https://doi.org/10.1016/j.solener.2023.111921. doi:10.1016/j.solener.2023.111921.

Saeed, F., Rehman, A., Shah, H. A, Diyan, M., Chen, J., & Kang, J.-M. o (2025). SmartFormer: Graph-based transformer model for energy load forecasting. *Sustainable Energy Technologies and Assessments, 73.* doi:10.1016/j.seta.2024.104133.

Serradilla, O. (2022). Deep learning models for predictive maintenance: A survey, comparison, challenges and prospects. *Applied Intelligence, 52*(10), 10934–10964. doi:10.1007/s10489-021-03004-y.

Shanker, S. (2022). Digital Twins and federated learning for smart cities and their applications. *IGI Global,* 1–9. https://doi.org/10.4018/978-1-6684-3733-9.ch009. doi:10.4018/978-1-6684-3733-9.ch009.

Silva-Rodriguez, J. (2024). LSTM-based net load forecasting for wind and solar power-equipped microgrids. In *2024 56th North American Power Symposium (NAPS)* (pp. 1–6). IEEE. https://ieeexplore.ieee.org/document/10741801. doi:10.1109/NAPS61145.2024.10741801.

Tahir, B. (2022). Experience-driven attack design and federated-learning-based intrusion detection in industry 4.0. *IEEE Transactions on Industrial Informatics, 18*(9), 6398–6405. doi:10.1109/TII.2021.3133384.

Tan X., (2024). Multi-layer energy management and strategy learning for microgrids: A proximal policy optimization approach . doi: 10.3390/en17163990.

Tariq, M. (2021). Integration of blockchain and federated learning for Internet of Things: Recent advances and future challenges. *Computers & Security, 108.* https://doi.org/10.1016/j.cose.2021.102355. doi:10.1016/j.cose.2021.102355.

Tariq, M. (2024). Exploring the convergence of Metaverse, Blockchain, Artificial Intelligence, and digital twin for pioneering the digitization in the envision smart grid 3.0. *Computers and Electrical Engineering, 120.* https://doi.org/10.1016/j.compeleceng.2024.109709. doi:10.1016/j.compeleceng.2024.109709.

Tong, H., & Liu, J. (2025). MFformer: An improved transformer-based multi-frequency feature aggregation model for electricity load forecasting. *Electric Power Systems Research, 243.* doi:10.1016/j.epsr.2025.111492.

Tripto, N. (2020). Evaluation of classification and forecasting methods on time series gene expression data. *Plos One, 15*(11). doi:10.1371/journal.pone.0241686.

Upadhyay, S., Ahmed, I., & Mihet-Popa, L. (2024). Energy management system for an industrial microgrid using optimization algorithms-based reinforcement learning technique. *Energies, 17*(16). https://www.mdpi.com/1996-1073/17/16/3898#:⟨:text=Reinforcement%20learning%20has%20emerged%20as,using%20a%20reward%20signal%20that. doi:10.3390/en17163898.

Victorio, M. (2021). Price forecast methodologies comparison for microgrid control with multi-agent systems. *2021 IEEE Madrid PowerTech.* IEEE. doi:10.1109/powertech46648.2021.9494970.

Wangyong Guo, S. (2025). Power grid load forecasting using a CNN-LSTM network based on a multi-modal attention mechanism. *MDPI, 15*(5). https://www.mdpi.com/2076-3417/15/5/2435#:⟨:text=,Therefore%2C%20leveraging%20advanced%20artificial. doi:10.3390/app15052435.

Xu, W., Kamsin, A., Jiang, L., Mei, Y., Zhou, X., & Huang, J. (2025). Dual-agent transformer network based on prompt engineering for voltage prediction. *IEEE Access, 13,* 105786–105797. doi:10.1109/ACCESS.2025.3580456.

Yang, K. (2020). Unsupervised K means clustering algorithm. *IEEE Access, 8,* 80716 80727. https://doi.org/10.1109/ACCESS.2020.2988796. doi:10.1109/ACCESS.2020.2988796.

Ye, Y. (2023). Safe deep reinforcement learning for microgrid energy management in distribution networks with leveraged spatial–temporal perception. *IEEE Transactions on Smart Grid, 14,* 3759–3775. https://ieeexplore.ieee.org/document/10040615. doi:10.1109/TSG.2023.3243170.

Zhong, J. (2021). System-scale-free transient contingency screening scheme based on steady-state information: A pooling-ensemble multi-graph learning approach. *IEEE Transactions on Power Systems, 37,* 294–305. doi:10.1109/tpwrs.2021.3097331.

Zhou, Y. (2024). Double robust federated Digital Twin modeling in smart grid. *IEEE Internet of Things Journal, 11*(24), 39913–39931. doi:10.1109/JIOT.2024.3450433.

Bassey, K. E., et al. (2024). Digital twin technology for renewable energy microgrids. *Engineering Science & Technology Journal, 5*(7), 2248–2272. https://doi.org/10.51594/estj.v5i7.1319.

CHAPTER

6

Machine learning algorithms for forecasting and diagnosis applications in smart grids

Mira Jiang[a], *Hengshuo Liang*[b], *Cheng Qian*[c], *Wei Yu*[d]

[a]Harvard College, Harvard University, MA, United States [b]School of Engineering & Technology, University of Washington, Tacoma, WA, United States [c]Department of Computer Science and Information Technology, Hood College, MD, United States [d]Department of Computer and Information Sciences, Towson University, Towson, MD, United States

6.1 Introduction

Cyber-physical systems (CPS) have revolutionized the landscape of smart grids, integrating advanced sensing, communication, and control technologies to enhance efficiency, reliability, and sustainability (Lin et al., 2017; Liu et al., 2019; Mallapuram et al., 2017; Philip et al., 2021; Qian et al., 2022, Qian et al., 2024; Rajkumar et al., 2010; Stankovic, 2014; Sun et al., 2016; Xu et al., 2018; Yu et al., 2018). The smart grid significantly improves how energy is generated and distributed (Tuballa & Abundo, 2016). The traditional power grid operates in a one-way electricity flow, whereas the smart grid enables two-way electricity and information flows. Physically deployed sensors commonly embedded or attached within smart meters could gather granular sensing data, say energy consumption measurements, which are processed via integrated data analytics components, offering actionable insights to adjust and optimize energy distribution instantly (Kayastha et al., 2014). Furthermore, these insights empower consumers with real-time feedback, fostering greater human engagement concerning energy-saving behaviors. By "human dimension," we refer to the impact of human actions and decisions on energy demand and usage patterns. This impact can significantly affect the smart grid's effectiveness concerning the management of peak loads and promotion of sustainable energy integration practices, among others (Lin et al., 2016; Mohamed et al., 2019; Zhang et al., 2015).

Integrating federated learning (FL) within the smart grid system further improves operational efficiency and privacy. One of the key features of FL is its decentralized learning nature, which enables model training across a number of distributed edge devices. With FL, a global model can be updated by the server that aggregates local model parameters from individual participants that retrain the model using locally available data. One commonly adopted method in FL is weighted average aggregation (McMahan et al., 2017). It computes the global model parameters as a weighted average of all local models. In such an aggregation, the size of each participant's local dataset is taken into account. This way ensures that clients that contribute more data can have a proportionally greater influence on the aggregated global model, resulting in optimized overall learning performance and robustness.

The ongoing progress of the smart grid significantly impacts energy delivery and management (Fang et al., 2011; Lin et al., 2012; Liang et al., 2024; Xu et al., 2017). Forecasting and diagnostics are two critical functions in the operation of modern smart grids. Forecasts help operators anticipate future grid states, such as load demand, generation output, voltage levels, or equipment stress, enabling proactive decisions that improve efficiency, reduce costs, and maintain stability. Diagnostics, on the other hand, aims to identify and understand faults as they occur, facilitating rapid response and system recovery. Both tasks are becoming more complex due to the increasing integration of distributed energy resources, the variability introduced by integrating renewable generation, and the tightly coupled nature of cyber-physical grid operations. Traditional rule-based or model-driven methods are often limited in their ability to scale, adapt to changing patterns, or provide accurate insights in real-time.

Digital Twin Technology and Smart Grid.
DOI: https://doi.org/10.1016/B978-0-443-36664-2.00002-2

Recent advances in artificial intelligence (AI), particularly deep learning, have transformed the way time-series data, sensor streams, and spatial dependencies are analyzed in CPS, such as power systems (Chen et al., 2024; Hatcher & Yu, 2018; Le et al., 2025; Liang et al., 2019; Liu et al., 2019, 2021; Mohammadi et al., 2018; Musa et al., 2023). Deep learning models can extract patterns directly from data, without requiring explicit feature engineering or prior knowledge of system dynamics. This chapter focuses on three representative neural network architectures useful for smart grid applications. In particular, convolutional neural networks (CNNs) are designed to extract local spatial features from structured inputs, such as weather grids or power signal spectrograms. Recurrent neural networks (RNNs) are ideal for modeling time-dependent sequences and learning temporal correlations in data streams, including load curves or voltage waveforms. A notable variant, the long short-term memory (LSTM) network, improves memory retention and mitigates vanishing gradient problems. Graph neural networks (GNNs) extend deep learning to irregular, non-Euclidean structures (Liang et al., 2022). This makes them highly effective for modeling power grids as graphs, in which nodes represent components and edges represent physical or logical connections. Integrating AI into smart grid operations holds significant promise for enhancing both forecasting and diagnostics (Liang et al., 2024a, 2024b; Liang et al., 2019). By automating the analysis of massive and diverse datasets, AI tools can improve situational awareness, enable predictive maintenance, and support optimal control under uncertainty. These tools offer adaptability to noisy and dynamic environments and the ability to uncover complex spatial-temporal relationships that traditional models cannot easily capture. As power systems continue evolving toward more decentralized, data-driven, and renewable-heavy configurations, deep learning models will play a central role in ensuring reliable, efficient, resilient, and intelligent grid performance. Each section examines use cases across the three core domains of the power system—generation, power transmission and distribution, and end-user consumption—illustrating the comparative applications of CNNs, RNNs, and GNNs in solving domain-specific challenges.

To illustrate the feasibility of deep learning techniques in the smart grid, we conduct a case study concerning leveraging distributed machine learning to address forecast-related issues. An Edge AI-based approach that enables data management and processing at the edge, which is closer to the data source, is a promising method to enhance the efficiency of grid operations. Integrating FL (Chen et al., 2021; Su et al., 2021; Tian et al., 2022), a representative distributed machine learning framework, with Edge AI enables forecasting performance comparable to that of the traditional centralized model. In the meantime, this approach preserves data privacy for two primary reasons. First, Edge AI is capable of carrying out data processing and decision-making in real-time at the user end (close to the data source). This can minimize delays and eliminate the need to transfer data to the central server constantly. Second, FL, with its collaborative model training that shares global model parameters with local models, reassures data privacy. It avoids transferring all data from the user end to a central server for model training, thereby protecting data privacy without directly transmitting or sharing all data to the cloud.

To summarize, our research has two main contributions:

(1) First, we conduct a thorough literature review on deep learning applications in energy forecasting and diagnostics. In particular, we examine existing research efforts to address forecasting problems from two aspects. One is the type of deep learning models consisting of CNNs, RNNs, and GNNs. The other is the smart grid key components, from power generation, transmission, distribution, and consumers. We systematically review existing research efforts in different combinations. We further review examples of existing efforts to apply CNNs, RNNs, and GNNs to address diagnostics-related issues in the smart grid.

(2) Second, we conduct a case study to design an Edge AI-based architecture for the smart grid that integrates FL, reducing the decision-making latency and preserving data privacy. This architecture deploys the lightweight model on remote edge devices to adapt to device heterogeneity (different computing resources, memory, battery, etc.). To comprehensively evaluate the proposed architecture, we define four dimensions to assess performance and design two lightweight machine learning models with similar lightweight models but different structures: a DNN and an LSTM model. We further utilize FL to train the DNN model in a distributed manner, achieving performance comparable to that of a centralized DNN. We use both centrally and distributedly trained models for monthly load forecasting as a test case. The evaluation results confirm that the centrally trained DNN outperforms the LSTM on the baseline in the same constrained environment. At the same time, the FL-based DNN performs better than the centrally trained DNN. This demonstrates that FL can protect users' data privacy while leading to better performance.

The remaining sections are organized as follows. Section 6.2 conducts a comprehensive review of AI applications in smart grids concerning both prediction and diagnosis problems. Section 6.3 presents a case study about engaging

TABLE 6.1 Problem space of deep learning for forecasting in smart grids.

Model type	Generation (X_1)	Transmission & distribution (X_2)	Consumer (X_3)
CNN (Y_1) (spatial patterns)	Spatial forecasting based on weather grids or satellite imagery; extract localized patterns in solar and wind inputs.	Capture spatial signal patterns for system state estimation and anomaly prediction; applied to grid flow and waveform data.	Learn spatial usage patterns tied to building layout, appliance location, or occupancy-driven variation.
RNN (Y_2) (temporal sequences)	Model time-dependent generation patterns influenced by weather variability; useful for wind and solar prediction.	Learn the temporal evolution of grid states like voltage and frequency, applied to short-term stability and operational forecasting.	Forecast user demand and behavior over time; support scheduling, load prediction, and peak demand management.
GNN (Y_3) (structural topology)	Model spatial dependencies between distributed generators; incorporates relationships between sensors or sources.	Leverage network topology for node-level state forecasting (e.g., optimal power flow (OPF), voltage); robust to reconfiguration and line changes.	Capture inter-node correlations across users in decentralized grids; models load propagation through the distribution network.

distributed learning to carry out energy forecasts. Section 6.4 shows some related future research problems. Finally, Section 6.5 concludes the chapter.

6.2 Deep learning in energy prediction and diagnosis

6.2.1 Overview

Deep learning techniques have been applied across a wide range of forecasting tasks in smart grids, with each model architecture offering distinct advantages depending on the problem itself. CNNs, RNNs, and GNNs (Hatcher & Yu, 2018) represent three foundational approaches, each aligning with a different dimension of smart grid data. Specifically, CNNs are especially well-suited for spatially structured inputs, such as gridded weather data or visual representations of load patterns, making them effective for forecasting in generation and consumer contexts. RNNs, particularly LSTM and gated recurrent unit (GRU) variants, excel at modeling sequential dependencies and are commonly used in scenarios that require temporal prediction, such as short-term load forecasting or voltage stability assessments. GNNs, on the other hand, are ideal for systems in which the underlying structure is relational or topological, such as transmission networks or distributed energy systems.

When mapped against different forecasting domains—generation, transmission, and distribution, and end-user consumption—this classification reveals a natural alignment between data modality and model architecture. For instance, power generation forecasting for renewable resources often involves spatial and temporal dynamics, such as cloud movement or wind fluctuations, benefiting from hybrid CNN-RNN approaches. Power transmission and distribution applications that rely on the physical structure of the grid are increasingly leveraging GNNs to predict state variables across nodes and branches. At the consumer level, a combination of CNNs and RNNs has proven effective in capturing appliance-level patterns and daily load cycles. Table 6.1 visualizes this cross-section of techniques and applications, offering a conceptual framework for understanding where different deep learning models are most appropriately deployed in smart grid forecasting tasks from power generation to consumers.

6.2.2 Prediction

Predictive models are used to anticipate a wide range of events—from voltage anomalies and load fluctuations to equipment wear and power generation variability. These capabilities enable proactive decision-making, allowing grid operators to allocate resources, balance loads, and respond to real-time dynamic conditions. Prediction is a cornerstone of smart grid stability and efficiency.

Deep learning has significantly expanded the reach and accuracy of predictive modeling in CPS, such as the smart grid. By learning complex spatial, temporal, and relational patterns from sensor data, weather inputs, and system behavior, deep learning models (e.g., CNNs, RNNs, GNNs) are helping to anticipate grid behavior more effectively than traditional methods. Considering that the smart grid consists of power generation, transmission, distribution,

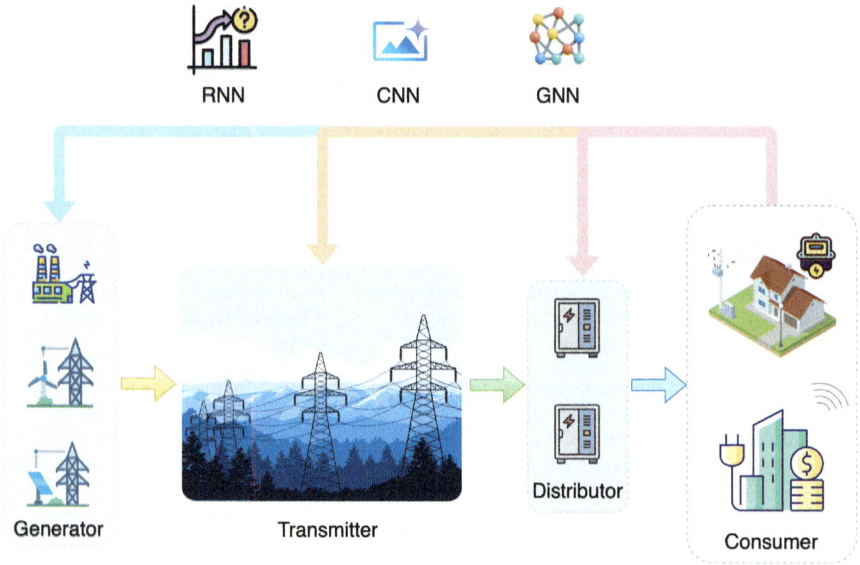

FIGURE 6.1 Deep learning-assisted prediction in smart grid.

TABLE 6.2 Summary of prediction tasks.

Model type	Architecture	Prediction target	Relevance to grid balance
CNN-based	CNN + CatBoost (Kim et al., 2023) CNN-LSTM (Abedinia et al., 2024)	Solar radiation Wind power output	Solar power production management: Uncertainty-aware production planning
	CNN + GCN (Li et al., 2024)	PV and wind generation	Integrated energy production optimization
RNN-based	DWT-LSTM (Nabavi et al., 2021) LSTM (Wu et al., 2023)	Building energy demand, load, and price forecasts	Load balancing and battery scheduling, Grid stability and storage coordination
	MIMO-LSTM (Nelega et al., 2023)	Multi-step PV output	Long-term energy planning
GNN-based	Spatio-temporal GNN (Cheng et al., 2022) GNN + Bi-LSTM (Liao et al., 2023)	Intra-hour PV output, ultra-short-term wind	Real-time solar balancing Variability management in production

Bi-LSTM, *bidirectional long short-term memory;* CNN, *convolutional neural networks;* DWT-LSTM, *discrete wavelet transformation long short-term memory;* GCN, *graph convolutional network;* GNN, *graph neural networks;* LSTM, *long short-term memory;* MIMO, *multiple-input, multiple-output;* PV, *photovoltaic;* RNN, *recurrent neural networks.*

and end-users, we systematically review the deep learning models in those key components. Fig. 6.1 presents the architecture of deep learning-assisted prediction in the smart grid.

Prediction of power generation: Establishing an accurate and cost-effective predictive model is central to maintaining stability in renewable energy-based CPS, such as managing anomalies in high-voltage components. These anomalies often arise from mismatches between energy supply and demand, rapidly changing environmental conditions, or inherent fluctuations in power output from renewable energy sources (e.g., solar, wind). Thus, accurately forecasting energy production helps mitigate these risks by enabling proactive energy scheduling, storage management, and grid load balancing. In recent years, deep learning methods have significantly advanced prediction capabilities. In the following, we review notable applications of each deep learning model architecture in forecasting for renewable energy systems (Table 6.2).

(1) *CNN-based forecasting approaches:* CNNs are particularly effective in extracting spatial features from structured inputs such as gridded weather data or satellite imagery. Their role in renewable energy forecasting is established in several recent studies. One notable example is a hybrid CNN-CatBoost model developed for solar radiation forecasting across various geographical regions (Kim et al., 2023). The architecture combines a two-layer Conv1D

CNN, which extracts spatial features from inputs such as extra-atmospheric solar radiation, temperature, humidity, and cloud volume, with CatBoost, which is a gradient-boosting algorithm optimized for categorical data. While the inclusion of wind speed had a minimal effect on accuracy, the hybrid design demonstrated improved forecasting performance compared to either model used independently.

Another study applies a combined CNN-LSTM architecture to the task of wind power forecasting (Abedinia et al., 2024). The CNN layers process spatially distributed weather and turbine data, while the LSTM component learns temporal dependencies within the wind power time series. This approach produces probabilistic forecasts, which provide both expected values and uncertainty ranges, enhancing operational decision-making in variable-energy environments.

A more complex model expands the utility of CNNs in a multi-cluster renewable energy setting (Li et al., 2024). Specifically, this work uses a 2D CNN with varied filter sizes to represent spatial patterns across multiple photovoltaic (PV) and wind power stations, combined with a lightweight graph convolutional network (GCN) that captures relational dependencies between these stations. The hierarchical learning architecture supports joint prediction of wind and solar generation in geographically dispersed systems, offering a scalable solution for integrated energy forecasting.

(2) *RNN-based forecasting approaches:* RNNs, and in particular LSTM networks, are widely used for time-series prediction. Their capacity to retain information across sequences makes them ideal for modeling energy usage, pricing, and environmental patterns over time. A comprehensive discrete wavelet transformation (DWT)-LSTM model was developed to forecast energy demand and supply at the building level while supporting energy scheduling strategies (Nabavi et al., 2021). In their system, DWT decomposes climate and energy data into frequency components, which enhances the LSTM's ability to learn nuanced temporal patterns. Inputs include dynamic pricing, weather variables, and historical consumption patterns. The model enables predictive control of storage systems and grid import/export decisions, yielding 87% electricity cost savings and an 84% reduction in grid energy imports.

Another LSTM-based approach focuses on optimizing storage in smart grids by forecasting future load and price patterns (Wu et al., 2023). Using a one-shot load decomposition technique and LSTM predictions, this framework helps manage uncertainties related to energy availability and market dynamics, supporting more efficient real-time storage control. In the context of PV output prediction, researchers developed a multiple-input, multiple-output (MIMO) LSTM architecture (Nelega et al., 2023). Their designed model ingests historical sequences and predicts multiple future time points simultaneously, enhancing accuracy over single-step forecast models. It is particularly useful in applications requiring long horizon energy planning.

(3) *GNN-based forecasting approaches:* GNNs offer a viable framework for modeling spatial and topological relationships in interconnected systems. This makes them especially relevant for renewable energy applications involving multiple, spatially dispersed sources and sensor networks. A study using satellite images to predict intra-hour PV power output constructed a spatial-temporal graph from cloud movement patterns and applied a GNN to forecast energy generation (Cheng et al., 2022). By minimizing image redundancy and computational complexity, the model offers fast and flexible short-term forecasts, which are essential in rapidly changing weather conditions.

Another model combines GNN with bidirectional LSTM and an enhanced Bootstrap technique to predict ultra-short-term wind power and its associated uncertainty (Liao et al., 2023). Here, wind farms and meteorological stations are represented as nodes in a graph, with edges capturing spatial correlations. The GNN learns inter-node relationships, while the bidirectional LSTM captures temporal trends in each node's historical data. The Bootstrap method provides prediction intervals, which support more reliable decision-making under uncertainty.

In conclusion, all these models represent a growing trend toward hybrid, data-driven forecasting systems that combine spatial, temporal, and relational learning. CNNs are effective in extracting detailed spatial patterns, RNNs are suited for learning time-dependent behaviors, and GNNs model structural relationships across distributed nodes. As renewable energy generation becomes more prevalent and decentralized, predictive models that unify these capabilities will be central to achieving reliability and operational efficiency in smart grids.

6.2.3 Power system fault detection and system optimization

While forecasting models anticipate future grid states, diagnostic models identify and localize faults arising from prediction uncertainty or physical disturbances, which together form complementary components of the smart grid resilience cycle. Fault detection and system optimization in both high-voltage transmission and low-voltage distribution networks are essential for the reliability and safety operations of modern smart grids. With the integration of renewable energy sources, fault patterns and grid dynamics have become increasingly complex. Predictive modeling using deep learning has enabled significant progress in identifying, classifying, and localizing faults, as well as enhancing control and optimization tasks across the transmission and distribution hierarchy.

(1) *CNN-based approaches:* CNNs are increasingly being used in power systems for their ability to extract meaningful features from complex electrical signals or visual data. A hybrid CNN-LSTM model has been proposed for carrying out fault detection and location estimation across power transmission lines (Alhanaf et al., 2024). This model processes voltage and current data collected from multiple points in the network. The 1D CNN extracts key signal characteristics, which are then used by an LSTM layer to detect fault types and accurately estimate their location. This architecture is particularly beneficial for handling dynamic fault currents from renewable energy sources, thereby ensuring effective grid monitoring.

Another relevant application is the use of CNNs in the visual domain. To overcome limitations in transmission line inspection, such as narrow defect classification, lack of comprehensive datasets, and inconsistent evaluation methods, a deep learning detection system based on faster R-CNN was developed (Liang et al., 2020). This approach effectively identifies defects in transmission line components from inspection images and remains robust under challenging conditions such as complex backgrounds and variable lighting.

Further, a CNN-based method was proposed for single-ended fault location in high-voltage transmission lines using traveling wave signals (Deng et al., 2023). Instead of relying solely on wavehead detection, this method performs full-waveform feature extraction to capture high-frequency components, thereby enhancing fault localization accuracy. This novel technique significantly improves robustness in complex fault scenarios where traditional models struggle, offering practical advantages in real-world transmission networks.

Furthermore, convolutional network-based autoencoders are capable of recognizing unsupervised anomalies in multivariate energy-related data. One example is the robust deep autoencoder (Zhou & Paffenroth, 2017) that utilizes deep CNN layers to capture features and reconstruct data. In this way, it is capable of compressing or consolidating normal energy consumption patterns and also identifying instances with high reconstruction errors as anomalies. By identifying localized spatial patterns in temperature, load, and voltage fluctuations, such a model is beneficial for recognizing abnormal consumption behaviors (e.g., sensor faults, cyberattacks, and equipment failures) without requiring labeled anomaly samples.

(1) *RNN-based approaches:* RNNs are particularly well-suited for time-dependent data in power systems, such as voltage, current, and discharge sequences. A two-layer RNN model, followed by fully connected layers and a classification correction mechanism, was introduced for detecting series arc faults in low-voltage multi-load systems (Li et al., 2020). These faults are difficult to detect due to their chaotic behavior. Still, the RNN model bypasses manual analysis by learning patterns directly from data, enhancing both speed and accuracy in real-time detection. Concerning power distribution, detecting partial discharge (PD) events is vital for maintaining grid health. A learning-based multiscale feature engineering framework uses RNNs to address this issue (Huang et al., 2024). The approach extracts both global and local signal features and processes them through an RNN to identify PD patterns, delivering higher performance in both accuracy and interpretability compared to traditional techniques.

For voltage stability assessment in smart grids, an RNN model incorporating spatial attention mechanisms has been developed (Zhu et al., 2021). This model accounts for spatial correlations among buses and uses spatial attention factors to rectify voltage trajectories before feeding them into an LSTM-based sequential learning pipeline. The result is a robust and scalable system for online short-term voltage stability assessment, which performs reliably even with noisy or incomplete phasor measurement unit data.

In addition, the LSTM-based variational autoencoder, also referred to as LSTM-VAE, represents an enhanced version of RNN-based architectures for anomaly detection (Guha et al., 2023). The key idea of LSTM-VAE is that it utilizes LSTM layers to capture temporal features, while the VAE is used to learn normal pattern distributions in the latent space. It is capable of recognizing anomalies in smart grid sensor data by analyzing loss in multivariate time-series reconstruction. One unique benefit is its unsupervised nature, which enables the recognition of both known and unknown fault types, even without labeled data.

(1) *GNN-based approaches:* GNNs are ideally suited for power systems because they can model the complex topologies inherent in transmission and distribution networks. A comprehensive study comparing deep learning models for solving optimal power flow (OPF) problems showed that GNNs outperform CNNs and fully connected networks, particularly in scenarios involving variable grid topologies or line contingencies (Falconer & Mones, 2023).

GNNs excel at leveraging structural information, making them ideal for real-time system optimization under dynamic conditions. For topology detection in power distribution networks, a novel dual-graph structure graph neural network (DGS-GNN) model was introduced (Ebtia et al., 2025). This approach transforms topology detection into a link prediction task, using only node attributes, such as phasor measurement unit (PMU) data, at inference time. By aligning a similarity graph with the actual network topology during training, the model achieves high accuracy even under noisy conditions and varying grid configurations, making it suitable for large-scale distribution systems.

Moreover, the spatio-temporal graph autoencoder (STGAE) (Chen et al., 2021) belongs to a GNN-based method. It can be used for conducting anomaly detection by utilizing its ability to model both spatial relationships between grid components and their temporal changes. The STGAE utilizes graph convolutional layers combined with temporal attention to result in its capability of discovering normal, benign patterns in dynamic power grid networks. The model is capable of recognizing anomalies through its ability to measure reconstruction errors across time-evolving graph snapshots. With its unique characteristics, STGAE can be utilized to perform tasks such as real-time detection of faults and cyberattacks in grid monitoring systems.

A final GNN application, PowerFlowMultiNet, addresses the challenges of modeling unbalanced three-phase power flow in distribution grids (Ghamizi et al., 2025). By representing each phase as a separate graph in a multigraph structure and applying message-passing mechanisms, this model captures asymmetrical dependencies across phases. It outperforms both traditional and deep learning methods in computational speed and accuracy across systems ranging from 13 to over 900 buses. This marks a significant advancement in real-time unbalanced power flow analysis in distribution grids.

In conclusion, CNN, RNN, and GNN models offer comprehensive tools for fault detection, location, classification, and system optimization in both transmission and distribution domains. Their ability to process multiscale, time-series, and topological data positions them as essential components of intelligent monitoring and control in next-generation smart grids.

6.2.4 Predictive modeling for end-user energy consumption

At the consumer level—spanning individual homes, commercial buildings, and other smart spaces—predictive modeling plays a central role in improving energy efficiency, managing load profiles, and integrating renewable generation and storage. The increasing penetration of decentralized energy resources and smart appliances adds complexity to forecasting user demand, requiring advanced machine learning models to account for diverse usage behaviors, fluctuating weather, and dynamic pricing. CNNs, RNNs, and GNNs each offer valuable tools to improve short- and long-term forecasts for energy consumption and management at the end-user level.

(1) *CNN-based approaches:* CNNs are commonly applied to energy consumption forecasting tasks at the residential and building levels because of their ability to extract key features from multi-dimensional datasets. One study developed a hybrid model that integrates CNNs with a multi-layer bidirectional long short-term memory network M-(M-BDLSTM) to address the prediction issue of short-term energy consumption within residential buildings (Ullah et al., 2020). The CNN layers handle spatial and contextual data—such as temperature and historical consumption—while the M-BDLSTM captures sequential patterns in forward and backward directions. This hybrid model outperformed traditional methods by better accounting for irregular occupant behavior and changing weather patterns.

In another smart home application, a CNN-RNN model was combined with a Gray Wolf Optimization (GWO) algorithm to predict household energy use and optimize appliance scheduling (Singh et al., 2023). The CNN extracts energy usage features from raw time-series data, while the LSTM model learns temporal dynamics. Adding GWO enhances the model's parameter tuning and scheduling efficiency, demonstrating effective load balancing and energy cost savings. CNNs have also been applied to public building environments. One study used a CNN paired with a genetic algorithm to conduct the prediction of energy usage clusters in public infrastructure (Abdelaziz et al., 2023). By identifying consumption pattern groups more accurately than CNNs alone, the method supports more efficient energy supply and demand coordination in large shared spaces. A more comprehensive CNN-based system also appears in smart grid load forecasting, where CNNs were integrated into a hybrid model alongside a bidirectional LSTM encoder and the Beluga Whale Optimization (BWO) algorithm (Asiri et al., 2024). The model, known as load forecasting system based on hybrid deep learning with Beluga Whale Optimization (LFS-HDLBWO), leverages CNNs for carrying out feature extraction, LSTM for sequential learning, and BWO for hyperparameter optimization.

The system demonstrated superior accuracy in capturing dynamic load profiles, significantly improving resource allocation and grid reliability.

(1) *RNN-based approaches:* RNNs, especially LSTM and GRU architectures, are widely used for forecasting consumer energy use because of their ability to model sequential and time-varying data. In residential applications, RNNs have been used to develop storage control policies that respond to renewable energy availability and real-time pricing. One such model employed LSTM networks in combination with one-shot load decomposition to predict both energy consumption and pricing fluctuations (Wu et al., 2023). The approach enables consumers to minimize grid imports and maximize energy cost savings through smarter storage scheduling.

Another model addresses both load forecasting and renewable energy generation using a stacked GRU-RNN framework (Xia et al., 2021). This model integrates correlation-based feature selection and an adaptive training strategy, allowing it to capture nonlinear patterns in user behavior and grid performance. The model proved effective in both univariate and multivariate forecasting settings, improving the operational efficiency of home energy systems. RNNs are also central to the LFS-HDLBWO model (Asiri et al., 2024), where bidirectional LSTM encoders play an important role in modeling long-term dependencies in user load data, further demonstrating the versatility of RNNs in end-user applications.

(1) *GNN-based approaches:* GNNs are particularly effective for decentralized energy systems, where spatial dependencies and network topology play a key role in energy flow and load behavior. A recent study applied GNNs to forecast the behavior of three-phase inverters, which are essential components in residential and small commercial PV systems (Khamis & Agamy, 2023). Using bond graph modeling to convert electrical circuits into graphs, the model applied graph convolutional layers to capture voltage and current dynamics. This method enabled highly accurate predictions of inverter behavior, offering valuable insights for real-time control and fault detection in distributed systems.

Another GNN-based approach focuses on electricity load forecasting in decentralized grids, using synthetic and real-world datasets to validate performance (Campagne et al., 2024). The study demonstrates that GNNs, including GCNs and GraphSAGE, significantly outperform traditional time-series methods by capturing spatial dependencies between consumers and integrating relational information across nodes. This is particularly useful in renewable-heavy, decentralized environments where energy consumption patterns shift rapidly based on local conditions and peer behavior.

Summary: These CNN, RNN, and GNN-based models enable more accurate, context-aware energy predictions at the end-user level. Whether in homes, buildings, or other smart environments, deep learning approaches facilitate improved energy scheduling, dynamic pricing responses, demand forecasting, and renewable energy resource integration. As consumer level energy systems grow more complex and data-rich, hybrid and graph-based models will play an increasingly central role in driving the development and deployment of intelligent energy ecosystems of the future.

6.2.5 Diagnostics in power systems

While predictive modeling in CPS aims to forecast future behavior, diagnostics focuses on identifying and classifying faults once they occur. Accurate and timely fault diagnosis is critical for maintaining grid reliability and resilience, minimizing downtime, and preventing cascading and/or other types of failures. Unlike prediction, diagnostics often rely on real-time signal analysis to determine the nature, location, and severity of faults such as short circuits, overcurrents, capacitor failures, and phase imbalances. Deep learning techniques have emerged as viable tools in automating diagnostic processes, offering improved accuracy, adaptability, and deployment potential in edge or low-resource environments. Fig. 6.2 presents the architecture of deep learning-assisted diagnosis in the smart grid.

CNN-based diagnostic approaches: CNNs have proven effective in fault classification tasks, especially when combined with advanced signal processing techniques. A recent study introduced a lightweight CNN architecture for fault diagnosis in grid-connected PV systems, targeting implementation on edge devices (Almahfouz et al., 2024). The model uses two-dimensional scalograms, generated from raw multi-channel time-series data via continuous wavelet transform (CWT), as input to the CNN. This hybrid of CWT-based preprocessing and CNN-based classification enables efficient feature extraction from noisy signal data. Additionally, the energy valley optimizer (EVO) algorithm is used to fine-tune the CNN's hyperparameters, improving performance without extensive manual tuning. The

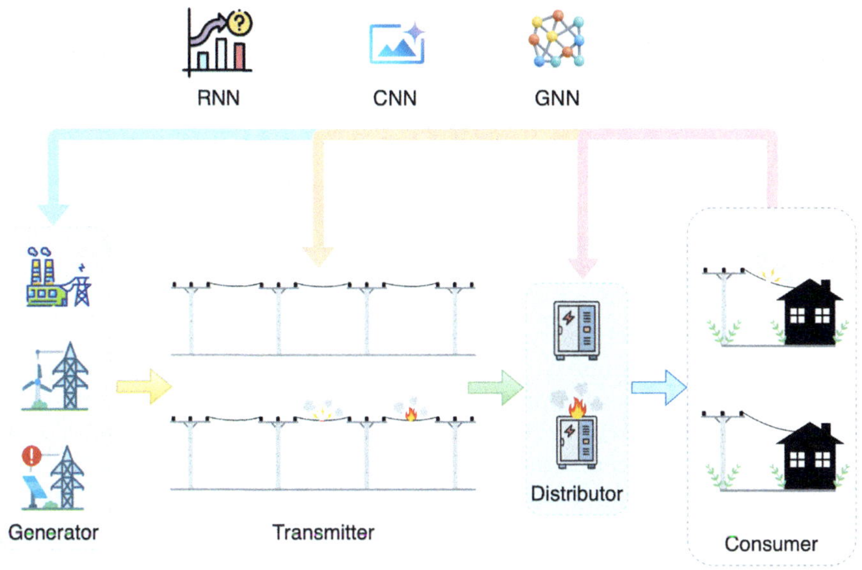

FIGURE 6.2 Deep learning-assisted diagnosis in smart grid.

resulting system supports real-time, resource-efficient fault diagnosis in solar installations, enhancing grid stability and system resilience at the edge.

Additionally, recent advancements include the integration of transformer architectures with CNNs for fault detection tasks in power systems. For example, Thomas et al. (2023) proposed a hybrid deep learning model, which combines 1-D CNNs with transformer attention blocks to classify fault currents in power distribution systems. The CNN layers rapidly extract local temporal patterns from long current-signal sequences. At the same time, the transformer captures their long-range dependencies, enabling accurate identification of both conventional shunt faults and elusive high-impedance faults. The evaluation results demonstrate that the CNN-based transformer outperforms regular CNN-based systems on fault detection. Likewise, Shen et al. (2021) investigated a physics-informed CNN model for fault detection, integrating physical system knowledge into the learning process to achieve improved accuracy. The authors employed a custom loss function combining data fidelity with physics-based constraints to enhance performance, thus ensuring physically meaningful predictions and superior fault detection outcomes.

RNN-based diagnostic approaches: RNNs, known for their capability of capturing temporal patterns, are well-suited for diagnosing faults that develop or manifest over time. One study presents a hybrid wild horse optimizer (WHO)-RNN model for diagnosing faults in capacitor voltage transformers (CVTs), which are critical components in high-voltage transmission systems (K et al., 2024). Traditional diagnosis methods for CVTs require physical inspections and downtime, making them inefficient and labor-intensive. The WHO-RNN model automates this process using 30-day sequences of ratio and angle difference data—key indicators of CVT health—to classify faults such as capacitor breakdown, oil leakage, and inter-turn short circuits. The WHO algorithm optimizes hyperparameters such as learning rate and hidden node count, enabling the model to achieve 93% accuracy. Importantly, the system operates in real-time without requiring equipment shutdowns, providing utilities with a powerful tool for proactive maintenance and enhanced grid safety.

Meanwhile, Lee et al. (2021) proposed an attention-based RNN for diagnosing inter-turn short-circuit faults in permanent magnet synchronous motors (PMSM), which are critical components frequently used in power system generators. The network comprises a bidirectional long short-term memory (BI-LSTM) encoder and an LSTM decoder, which are combined with an attention mechanism to exploit specific encoder states effectively. It takes negative-sequence current (NSC), positive-sequence current (PSC), and rotational speed as inputs and directly outputs a measure of fault severity. Experimental results confirm the algorithm's accuracy in assessing fault severity in a variety of operating and fault scenarios. Likewise, Lee et al. (2025) proposed a bidirectional gated recurrent unit (Bi-GRU)-based model for carrying out fault diagnosis in medium-voltage DC systems. The model simultaneously diagnoses multiple fault categories, including fault types and locations, achieving high accuracy (95.54%) and robust performance under noisy conditions. Additionally, the authors incorporated a sliding window that enables real-time fault detection within milliseconds.

GNN-based diagnostic approaches: GNNs have recently been applied to power system diagnostics, particularly for spatially distributed grids where structural information plays a key role in fault localization. A diagnostic framework combines recurrent graph neural networks (R-GNNs) with transfer learning to enable scalable and accurate fault diagnosis using only voltage measurements (Nguyen et al., 2023). Unlike traditional methods that rely on current data or relay-based systems, this approach improves generalizability and reduces the need for hardware installation. The model is trained in stages: first, an R-GNN is trained to detect fault events; its learned representations are then transferred to additional R-GNNs for fault type classification and fault localization. By leveraging transfer learning, the framework achieves high diagnostic accuracy with reduced training data and faster convergence. The approach shows strong potential for scalable grid-wide implementation, especially in systems where current sensing is not feasible or cost-effective.

Also, Wang and Hu (2025) proposed a scheme for accurately carrying out fault diagnosis and localizing faults for power systems using the graph sample and aggregate (GraphSAGE) algorithm. The designed algorithm leverages inductive learning capabilities and effectively utilizes topological graph data. Compared to GCN and graph attention networks (GAT), the designed algorithm exhibits superior accuracy, adaptability, and practical effectiveness. Its key advantages include effective identification of critical nodes and strong robustness to topology variations, achieving approximately 3% higher accuracy than GCN and 2% higher accuracy than GAT. Additionally, the proposed method offers greater economic feasibility than traditional traveling wave approaches for power grid lines up to 7.5 km in length. Likewise, Ren et al. (2025) proposed a defense mechanism for power grid faults under physically coordinated cyber-physical attacks. They introduced a GAT-based fault localization (GAT-FL) algorithm that can effectively identify faults by extracting high-level features from graph data. Additionally, the authors formulated the line state identification under PCPA as a linear programming problem, integrating parameters derived from the GAT-FL algorithm into the linear programming (LP) objective function to improve accuracy and computational efficiency.

In conclusion, CNN, RNN, and GNN approaches highlight the growing capabilities of machine learning in power system diagnostics. CNNs excel at localized fault classification using spectral data, RNNs provide powerful tools for interpreting fault progression over time, and GNNs introduce a structural lens for diagnosing and localizing faults across complex grid topologies. As grid complexity increases with renewable penetration and decentralization, such intelligent diagnostic systems will become vital for maintaining safe and reliable energy infrastructure.

6.3 Case study: Distributed AI for load forecasting in energy CPS

This section introduces a case study that engages the Edge AI-based architecture to forecast load in the smart grid. In particular, we first review the concept of Edge AI and FL. To validate the practical feasibility and effectiveness of the distributed learning approaches discussed in the literature, we now design a case study that leverages an Edge AI-based architecture for carrying out load forecasting in smart grids. We then present our proposed architecture, explain the dataset used for evaluation, and describe the designed scenarios, followed by the performance evaluation results.

6.3.1 Overview

The smart grid significantly improves how energy is generated and distributed (Tuballa & Abundo, 2016). The traditional power grid operates in a one-way electricity flow, whereas the smart grid allows for two-way electricity and information flows. This smart grid system enables more efficient management of data feedback from the sub-grid meters and improves decision-making performance (Mohamed et al., 2019). Moreover, the smart grid integrates sensors, network systems, and data analytics into the meters to achieve real-time response (Kayastha et al., 2014).

Edge Intelligence, also known as Edge AI, involves processing data at the user's end, which can be deployed much closer to the data source rather than on a remote cloud-based, centralized server (Deng et al., 2020). Such a design has several benefits, particularly helpful for systems like the smart grid. The rationale behind this approach is that processing data on the user's side is capable of significantly reducing latency without transmitting data to a remote, centralized server. Thus, not only can response times be reduced, but also energy and network resources for data transmission can be further saved. Furthermore, Edge AI can assist in enhancing data privacy by allowing sensitive data to be processed locally, rather than transmitted to a remote central server (Amin & Hossain, 2020).

FL is a representative decentralized machine learning architecture. Within it, the global model is trained by the process of model sharing and updating with local models (reflected by model parameters) trained on local

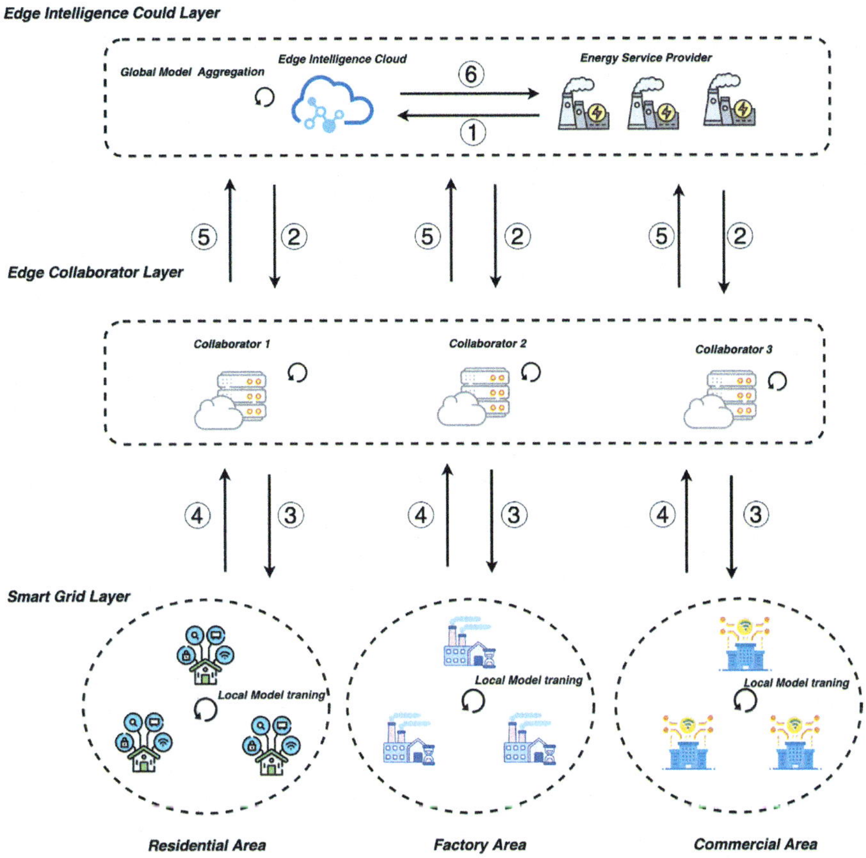

FIGURE 6.3 Federated Edge AI for energy consumption forecast in the smart grid.

datasets, rather than collecting data from all local nodes to a central server that performs model training as a whole (Kairouz et al., 2021; Pene et al., 2024; Wang et al., 2024). As FL keeps data local on edge devices and only needs to share the parameters of the updated model with the edge server, it can reduce the risk of data breaches. It can also improve robust learning efficacy as more edge devices can be invited to join the collaborative learning environment within the smart grid. In this way, the grid operations can significantly benefit from data collected across multiple locations while utilizing the computing resources on individual devices.

6.3.2 Architecture

Fig. 6.3 shows an Edge AI-based architecture in the smart grid for load forecasting. It consists of three layers:

(1) *Edge intelligence cloud layer:* This layer includes two major parts: The energy service provider will generate electricity and deliver it to the user, and the Edge intelligence cloud contains the global model to be shared and can also help make some decisions.
(2) *Edge collaborator layer:* This layer is a gateway for sending the global model parameter to lower layers and uploading locally trained updates to upper layers.
(3) *Smart grid layer:* This layer contains all the sub-grids with meters and edge devices in different areas, such as residential, factory, and commercial areas.

There are several key steps in FL as shown in Fig. 6.3: (1) Sending the raw model parameters and the prediction task to the cloud server. (2) Sending global model parameters to each collaborator. (3) Collaborators send parameters to their local models and train them with their local datasets. (4) Uploading local model parameters to the collaborator. (5) The global model aggregator collects model parameters from all the collaborators. (6) The aggregated global model performs the task and returns the result.

After showing the architecture, we now consider the performance of FL for load forecasting from the following aspects:

- *Prediction time windows:* The forecasting process becomes more challenging as the predicted time span increases. For instance, predicting energy consumption for the upcoming month is much more complicated than predicting it for the upcoming week.
- *FL configuration:* This is about the deployment of FL. It comprises the number of clients participating in the learning process and the number of epochs to be allocated for each client with local model training. It also includes the number of rounds for the global model to update its parameters, the aggregation method to be used by the global model, as well as the security mechanism, such as encryption, to secure the transmission of parameters.
- *Model complexity:* This is about the global model regarding the number of layers and the core mechanism used to build the model. Generally, having more layers and a more complex design leads to better performance. However, our study focuses on lightweight models, where complexity concerns the core mechanism used to build the model.
- *Dataset similarity in energy consumption profile*: The concept of "similarity" refers to whether all the participating FL clients have similar energy consumption profiles. If they are similar, it will significantly benefit the global model's performance in predicting the same energy profiles and vice versa.

6.3.3 Data collection and scenarios

The dataset (UK power networks, 2014) provides a comprehensive collection of energy usage data from London households. This dataset, gathered in 2014, offers valuable insights into energy consumption patterns and trends within residential properties across the city from 2011 to 2014, covering 5566 smart meters. It includes smart meter IDs to capture energy usage at regular intervals, along with timestamps.

To make the most of the dataset, we employ the *K*-means method (Kodinariya & Makwana, 2013) to group smart meters according to their interval energy usage. We then select the cluster with the largest number of meter records and randomly pick one meter with a year's data for training. Subsequently, we randomly selected another meter with a different year's worth of data from the same cluster for testing.

Based on Section 6.3.2, four aspects could impact the performance. We design our scenarios by configuring those four aspects as follows:

(1) We set the prediction windows as a month-to-month task, which may be moderately difficult for the model to forecast.

(2) Since we intend to use a lightweight model, we fix a range for the model size (parameter numbers) with different core mechanisms (DNN and LSTM).

(3) Regarding dataset similarity, as mentioned in Section 6.3.3, we group smart meters based on the similarity of their energy consumption profiles.

(4) We use weighted average aggregation (McMahan et al., 2017) for the FL configuration. For the local model, we set the epochs to 10, 20, and 30. We also fix the number of rounds at 20 for the global models. We set a fixed client number at 15.

With this configuration, we split the evaluation into two parts: (a) *Centralized model training as baseline:* In this baseline, we compare those two lightweight models in central training mode, with 20 epochs to evaluate the training performance and to test the month-to-month prediction performance. (b) *FL model comparing with centralized model:* We train the model that outperforms the baseline using the FL framework to test if the model can perform similarly to the central model under lightweight, constrained conditions.

6.3.4 Performance evaluation

We thoroughly evaluate the efficacy of deep learning models through a comprehensive performance assessment. In the following, we first present our evaluation methodology and then explain the results.

(1) *Methodology:* For the experiment, we use TensorFlow (Abadi et al., 2016) and TensorFlow Federated (TTF, 2019) as our platform. We use a server with an Intel I9-9980XE CPU and a high-end NVIDIA RTX 4090 GPU for model training. As for designing those two lightweight models: (a) *DNN* (Goodfellow et al., 2016): We design a DNN model with four layers in total, containing 54177 parameters with a size of 211.63 KB. (b) *LSTM* (Hochreiter & Schmidhuber, 1997): We design a model of four LSTM layers with a dropout rate of 0.2, in total 128,929 parameters

TABLE 6.3 Experimental parameters for federated learning case study.

Parameter	Value
Dataset source	UK Power Networks (London households)
Dataset size	One year of smart meter readings
Train/Test split	One meter for training; one for testing (same cluster)
Number of clients	15
Prediction window	Month-to-month load forecasting
Local epochs	10, 20, 30
Global rounds	20
Aggregation method	Weighted average aggregation
Model types	DNN, LSTM
Evaluation metrics	MAE, MSE, RMSE

DNN, *dense neural network;* LSTM, *long short-term memory;* MAE, *mean absolute error;* MSE, *mean squared error;* RMSE, *root mean squared error.*

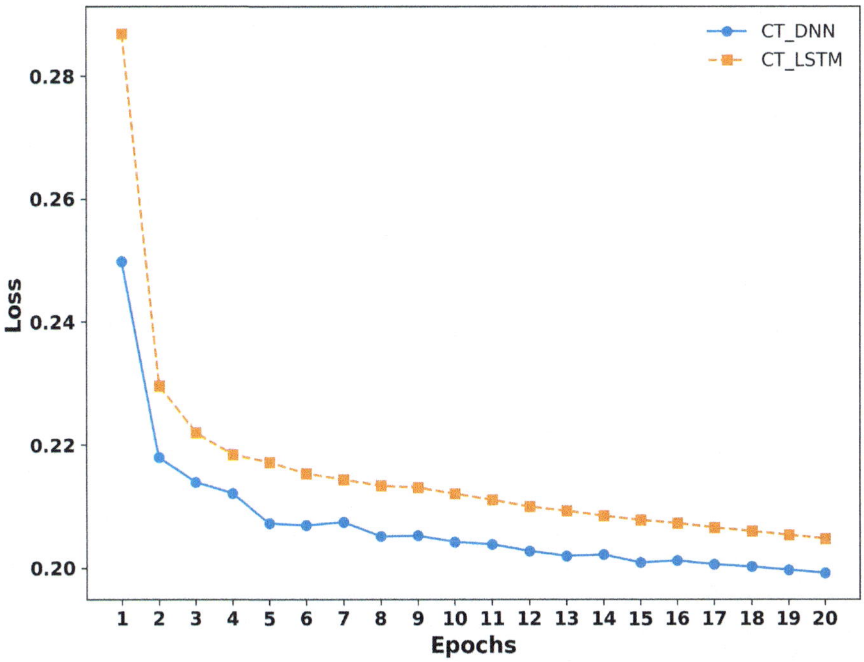

FIGURE 6.4 Loss value update for 20 epochs centralized training.

with a size of 503.63 KB. To improve training efficiency, we also apply the early stopping method (Prechelt, 1998) to both models.

We utilize root mean squared error (RMSE), mean squared error (MSE), and mean absolute error (MAE) as loss functions to measure the accuracy of model training and validation performance. Additionally, we use these three metrics (i.e., RMSE, MAE, and MSE) to assess the efficacy of learning models. We also create plots to track the loss value updates during 20 epochs for both centralized and distributed training. Table 6.3 summarizes the dataset characteristics and experimental configuration used in the FL case study.

(2) *Results:* We now show and discuss the performance results.

Centralized training performance: In Fig. 6.4, during 20 epochs, the centrally trained dense neural network (CT-DNN) updates its loss value faster than the centrally trained long short-term memory (CT-LSTM). However, both models have similar loss values at the end of the last epoch. Despite the LSTM having more parameters than the DNN, the fixed total epoch of 20 and the use of four layers hinder the LSTM from delivering better performance.

TABLE 6.4 Comparison of model performance metrics for centralized training.

Model name	MAE	MSE	RMSE
CT-DNN	0.2072	0.1184	0.3416
CT-LSTM	0.2160	0.1332	0.3610

MAE, *mean absolute error;* MSE, *mean squared error;* RMSE, *root mean squared error*

TABLE 6.5 Comparison of model performance metrics for month to month forecasting in centralized mode.

	CT-DNN			CT-LSTM		
Month	**MAE**	**MSE**	**RMSE**	**MAE**	**MSE**	**RMSE**
January	0.0563	0.0079	0.0838	0.1304	0.0210	0.1443
February	0.0528	0.0082	0.0823	0.1290	0.0208	0.1434
March	0.0587	0.0087	0.0883	0.1293	0.0209	0.1440
April	0.0544	0.0081	0.0835	0.1309	0.0212	0.1444
May	0.0485	0.0063	0.0763	0.1284	0.0196	0.1394
June	0.0485	0.0064	0.0755	0.1302	0.0203	0.1419
July	0.0502	0.0062	0.0753	0.1290	0.0200	0.1407
August	0.0514	0.0069	0.0792	0.1287	0.0202	0.1416
September	0.0550	0.0084	0.0840	0.1307	0.0215	0.1455
October	0.0412	0.0054	0.0652	0.1307	0.0199	0.1402
November	0.0521	0.0072	0.0788	0.1287	0.0205	0.1424
December	0.0532	0.0076	0.0803	0.1311	0.0213	0.1451

MAE, *mean absolute error;* MSE, *mean squared error;* RMSE, *root mean squared error.*

After calculating the average MAE, MSE, and RMSE over 20 epochs as shown in Table 6.4, it is evident that CT-DNN has lower error across all three metrics than CT-LSTM. Then, we use those two trained models to forecast month-to-month energy consumption for 12 months. As illustrated in Table 6.5, CT-DNN achieves lower errors across the three metrics (MAE, MSE, and RMSE) than CT-LSTM in all 12 months. It confirms that the CT-DNN outperforms the CT-LSTM in the baseline comparison and can be trained with an FL approach for further comparison.

FL training performance: Based on the baseline comparison, we consider the same DNN model as the global model to conduct the FL training with different local epoch times (10, 20, and 30). As shown in Fig. 6.5, the FL-based DNN model achieves a faster-updating loss than the CT-DNN, and as the local epoch number increases, the update becomes more rapid. Because the FL approach divides the entire dataset into 15 parts per client, each client trains the model on a local, small-scale dataset. Thus, the more epochs used for training the local dataset, the faster the model will update its loss.

As shown in Table 6.6, after calculating the average MAE, MSE, and RMSE over 20 epochs, it shows that the FL-based DNN experiences lower errors than the centralized DNN across these three metrics. This confirms that the FL approach can outperform centralized models. Additionally, with more local epochs, the errors decreased, leading to improved performance. With 15 clients training locally, the global model can learn more about data patterns from each client during each local epoch training. This improves the global model's generalization. There are two main reasons why centralized training is not better than the FL approach, even when the entire dataset is used for global optimization. First, the centralized lightweight model may struggle to understand the data pattern due to the sheer volume of records (the entire year's records). This makes it hard for the model to learn effectively. Second, in our restricted environments, the centralized approach involves training with only a 4-layer model and 20 epochs for training, resulting in a poorer understanding of the data pattern than the FL approach.

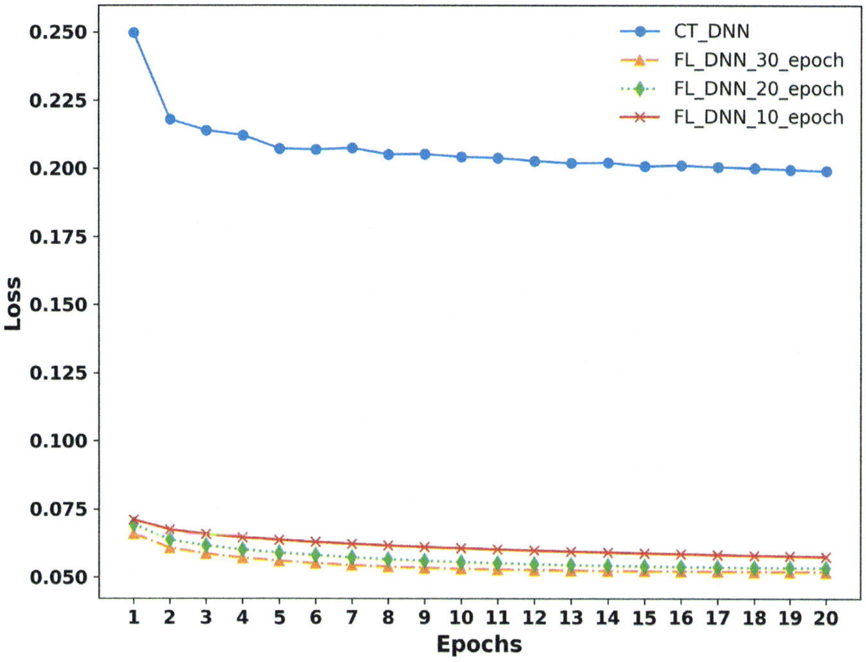

FIGURE 6.5 Loss value update for 20 epochs FL training. *FL,* Federated learning.

TABLE 6.6 Comparison of model performance metrics for FL training.

Model name	MAE	MSE	RMSE
CT-DNN	0.2072	0.1184	0.3416
FL-DNN-10 epochs	0.061	0.0132	0.1146
FL-DNN-20 epochs	0.0570	0.0113	0.1061
FL-DNN-30 epochs	0.0546	0.0110	0.1047

FL, *federated learning;* FL-DNN, *federated learning dense neural network;* MAE, *mean absolute error;* MSE, *mean squared error;* RMSE, *root mean squared error.*

It is worth mentioning that although no statistical significance tests (e.g., *p*-values or confidence intervals) are included, the performance improvements, as seen in our data, are substantial and consistent across all reported metrics. The errors are averaged over the final training epochs to reduce the effect of fluctuations, and the large differences observed (e.g., MAE reduced from 0.2072 to 0.0546) provide strong practical evidence of the FL model's benefits. While we acknowledge the absence of formal statistical testing, the consistent and significant reduction in error across all evaluation metrics supports the validity of our conclusion.

Regarding the month-to-month forecast task, the result is presented in Table 6.7. We observe that the FL-DNN model outperforms the CT-DNN model in most months, achieving lower errors for MAE, MSE, and RMSE. In some cases, such as in May, CT-DNN has a smaller MAE than FL-DNN-10. Nonetheless, as the local epoch increases, the effectiveness of FL-DNN improves and surpasses that of CT-DNN. In most cases, they increase the local epoch, which results in better performance for FL-DNN.

In summary, in the baseline with the constrained environment (same training epoch count, same-range model size), CT-DNN performs better than CT-LSTM, even though the DNN model has a simpler structure. The FL setup shows that the FL approach is capable of realizing better performance, and with more local epochs, performance improves further. This confirms that FL can not only achieve better performance than centralized learning but also preserve data privacy. To complement the qualitative discussion on privacy and communication efficiency, we highlight the key differences between centralized learning and FL in Table 6.8. As shown in the table, FL enhances privacy by keeping raw data local and transmitting only model parameters, reduces bandwidth consumption by avoiding large-scale data transfers, and achieves comparable or improved training efficiency through parallel local updates. These

TABLE 6.7 Comparison of model performance metrics for month-month forecasting in FL mode.

	CT-DNN			FL-DNN-10 epochs			FL-DNN-20 epochs			FL-DNN-30 epochs		
Month	MAE	MSE	RMSE	MAE	MSE	RMSE	MAE	MSE	RMSE	MAE	MSE	RMSE
January	0.0563	0.0079	0.0838	0.0525	0.0076	0.0873	0.0526	0.0073	0.0856	0.0525	0.0074	0.0858
February	0.0528	0.0082	0.0823	0.0515	0.0080	0.0892	0.0515	0.0078	0.0885	0.0520	0.0079	0.0889
March	0.0587	0.0087	0.0883	0.0568	0.0084	0.0916	0.0564	0.0082	0.0903	0.0565	0.0082	0.0904
April	0.0544	0.0081	0.0835	0.0529	0.0078	0.0886	0.0545	0.0078	0.0886	0.0546	0.0078	0.0882
May	0.0485	0.0063	0.0763	0.0489	0.0063	0.0794	0.0474	0.0060	0.0772	0.0480	0.0060	0.0777
June	0.0485	0.0064	0.0755	0.0488	0.0061	0.0784	0.0509	0.0065	0.0808	0.0513	0.0065	0.0806
July	0.0502	0.0062	0.0753	0.0490	0.0060	0.0778	0.0493	0.0061	0.0780	0.0497	0.0062	0.0787
August	0.0514	0.0069	0.0792	0.0490	0.0066	0.0813	0.0498	0.0066	0.0812	0.0502	0.0066	0.0812
September	0.0550	0.0084	0.0840	0.0526	0.0084	0.0915	0.0524	0.0082	0.0906	0.0533	0.0083	0.0914
October	0.0412	0.0054	0.0652	0.0393	0.0050	0.0707	0.0398	0.0050	0.0708	0.0400	0.0049	0.0702
November	0.0521	0.0072	0.0788	0.0503	0.0071	0.0843	0.0494	0.0069	0.0831	0.0504	0.0070	0.0834
December	0.0532	0.0076	0.0803	0.0504	0.0075	0.0863	0.0500	0.0072	0.0847	0.0507	0.0072	0.0850

FL, *federated learning;* FL-DNN, *federated learning dense neural network;* MAE, *mean absolute error;* MSE, *mean squared error;* RMSE, *root mean squared error.*

TABLE 6.8 Comparison of centralized learning and FL.

Metric	Centralized	Federated	Benefit
Data shared	Raw data	Gradients/weights	Enhanced privacy (no raw data sent)
Bandwidth usage	High (large data uploads)	Moderate (model updates only)	Reduced communication load
Training time	Baseline single-model training	Comparable; often improved via parallel local updates	Faster or similar convergence

characteristics make FL particularly suitable for smart grid environments, where data sensitivity, heterogeneous devices, and communication constraints are prevalent.

6.4 Future research directions

This section discusses several of the most important future research directions, focusing on their implications for the smart grid.

6.4.1 Deep learning in forecast and diagnostics

Recent advances have shown that hybrid deep learning models—such as combinations of CNNs, RNNs, and GNNs—can effectively leverage spatial, temporal, and topological features to improve forecasting of renewable energy generation, system load, and overall grid states. Despite these advancements, most existing models remain tailored to specific data types, prediction horizons, or narrowly defined tasks, such as short-term solar output forecasting or wind power prediction.

A promising direction for future research lies in the development of unified, multi-modal, and multiscale deep learning frameworks that can integrate heterogeneous data sources—including weather forecasts, satellite imagery, and sensor streams—into a single cohesive modeling pipeline. These frameworks should be capable of simultaneously predicting multiple grid variables, such as generation, demand, and voltage, across a range of temporal and spatial resolutions. Furthermore, to remain effective in evolving grid environments, such models shall adapt dynamically to both changing system conditions and data quality, potentially through techniques such as transfer learning, domain adaptation, or continual learning. Ultimately, these integrated forecasting systems would provide a more holistic, flexible, and robust foundation for operational and planning decisions in increasingly complex, data-driven, and renewable-heavy smart grids.

A critical direction for future research in deep learning-based diagnostics is the development of models that are adaptable and robust across a wide range of operating conditions, topologies, and data environments. As smart grids become more dynamic and heterogeneous, diagnostic systems must be able to generalize beyond the specific scenarios they were trained on and maintain high performance when dealing with noise with varying strengths, missing data, cyber threats, and evolving grid configurations. Techniques, including transfer learning, domain adaptation, meta-learning, and continual learning, offer promising pathways toward building models that can adapt rapidly to new conditions with minimal retraining. By prioritizing adaptability, future diagnostic systems can become more scalable, resilient, and deployment-ready, enabling reliable real-time fault detection and response across increasingly complex smart grid infrastructures.

6.4.2 Handling not identically and independently distributed data distributions

In real-world smart grid environments, the data collected by different edge devices are often not identically and independently distributed (non-IID) due to varying energy consumption patterns across households, industries, and geographical locations. This diversity presents significant challenges to FL algorithms, which typically assume identically and independently distributed data for optimal performance. The research efforts should focus on developing FL methods that can effectively handle the diverse data types in smart grids. Researchers could explore the idea of designing customized FL models that can adjust to each client's data distribution while still benefiting from shared insights as a solution. One promising strategy involves using learning and multitask learning frameworks to customize models based on various users' or regions' distinct consumption habits.

Furthermore, to enhance model convergence, it would be beneficial to design aggregation methods that consider the statistical properties of local models. For example, using weighted averages to improve the learning process based on similarities in data distributions or performance metrics can be beneficial. Exploring methods to categorize clients with data distributions and training them with individual models can help reduce the impact of non-IID data. Implementing FL could boost performance in smart grid systems by capturing specific consumption patterns. It is crucial to conduct analyses to grasp FL algorithms' convergence behavior and limitations in non-IID scenarios. Studies like these help shape the creation of customized techniques to handle the intricacies of energy usage data.

6.4.3 Optimizing model complexity for edge devices

Edge devices in the smart grid (e.g., smart meters, sensors, actuators) often have limited computational resources, memory, and energy supply. Deploying machine learning models on these devices requires careful optimization to balance performance and resource consumption. Future research can focus on optimizing model complexity to suit the constraints of edge devices within the smart grid. Applying model compression techniques (pruning, quantization, low-rank factorization, etc.) can reduce model size without significantly affecting accuracy. This leads to the feasibility of deploying complex, efficient learning models on resource-constrained devices. Knowledge distillation can be utilized by generating smaller "student" models based on large/complex "teacher" models. This approach enables complex models to be approximated by lightweight versions suitable for deployment on smart grid edge devices. Employing neural architecture search can automate the design of efficient model architectures tailored to the specific resource constraints and performance requirements of edge devices. Considering the specific hardware characteristics of edge devices during model design is also important. Utilizing hardware-aware optimization, including the use of accelerators like GPUs or specialized AI chips when available, can enhance computational efficiency. Developing energy-efficient algorithms for training and inference is essential to prolong device battery life and reduce operational costs in the smart grid.

6.4.4 Enhancing privacy guarantees

While FL inherently provides data privacy by keeping raw data on local devices, privacy risks may not be fully immunized. Adversaries can infer sensitive information from model updates shared during the FL process. Future work should explore advanced privacy-preserving techniques to strengthen privacy guarantees in the smart grid. Integrating differential privacy mechanisms can insert controlled noise into model updates, providing mathematical guarantees against individual data point inference while balancing the trade-off with model utility. Implementing secure aggregation protocols, including homomorphic encryption or secure multi-party computation, can ensure that the central server cannot access individual model updates in plaintext, protecting against both external and internal threats. Developing tools and frameworks for privacy auditing and verification can help assess and verify the privacy properties of FL systems, enabling stakeholders to make informed deployment decisions. How to make sure that the design is compliant with data protection regulations General Data Protection Regulation (GDPR), California Consumer Privacy Act (CCPA) is important, as these may have implications for data handling and user consent within the smart grid context.

FL supports GDPR compliance by holding sensitive data locally on user-owned devices, thereby decreasing the disclosure of personally identifiable information (PII) and fulfilling the GDPR's principle of data minimization. Additionally, FL can improve GDPR's accountability and transparency requirements by tracing data handling processes and assuring that data remains contained within user devices. Such transparency can make sure that users are fully informed before giving their consent, thereby facilitating effective data governance and privacy protection. Furthermore, the integration of FL with Edge AI offers robust security protection for smart grid infrastructure by following the National Institute of Standards and Technology (NIST) Cybersecurity Framework and NIST Special Publication 800-53 (NIST SP800-53). FL and Edge AI further enhance compliance with data protection and cybersecurity regulations by considering mechanisms (e.g., adaptive communication, proper client selection, and data and model aggregation at different levels, among others). Studying potential attack vectors (e.g., model inversion, membership inference attacks) and developing defenses to safeguard against them can enhance the system's robustness.

6.4.5 Improving communication efficiency

FL requires more communication and interactions between edge devices and the central server, which is used to exchange model updates and parameters. Communication overhead can become a bottleneck in smart grid environments with limited bandwidth or unreliable network connections. Future research directions include improving communication efficiency to address scalability-related issues and evaluating the practicality of FL in the smart grid. Reducing the size of model updates through techniques like sparsification, quantization, or encoding can reduce the amount of data transmitted per communication round. Adjusting the frequency and timing of communication based on network conditions, model convergence rates, or energy constraints—through adaptive communication strategies—can optimize efficiency. Implementing client selection mechanisms, in which a subset or group of clients participates in individual training rounds. The selection of the client group can consider data quality, network connectivity, or resource availability, which is beneficial to communication load reduction. Exploring decentralized learning architectures, such as peer-to-peer FL models, in which clients communicate directly with each other, can potentially reduce reliance on central servers and distribute communication demands. Utilizing edge caching and aggregation, where intermediate edge nodes aggregate model updates from nearby devices before transmitting them to the central server, can reduce the number of transmissions over long distances in the smart grid.

6.5 Final remark

This chapter explored the dual challenges of improving forecasting performance and ensuring data privacy in modern smart grids. We began with a comprehensive review of deep learning applications—particularly CNNs, RNNs, and GNNs—across three critical components of the power system: generation, transmission, distribution, and end-users. We introduced a two-dimensional problem space to categorize where each deep learning model is most effective, highlighting their alignment with spatial, temporal, and structural data characteristics within the grid. To demonstrate the feasibility of deploying deep learning in real-world settings, we presented a case study featuring an Edge AI-based architecture for load forecasting. This architecture integrates FL with lightweight neural networks to support distributed learning at the edge while preserving user data privacy. We evaluated the system along four key dimensions: data similarity, FL configuration, prediction window size, and model complexity. Two baseline models, DNN and LSTM, were tested under centralized training, and the better-performing model (DNN) was then trained using the FL framework. Our results confirmed that FL-DNN not only outperformed the centralized DNN but also improved further as the number of local training epochs increased.

These findings demonstrate that deep learning—especially when paired with FL and edge computing strategies—offers a promising path toward scalable, privacy-preserving, and high performance forecasting in smart grids. As grid systems continue to evolve, the integration of intelligent learning frameworks like these will play a pivotal role in supporting both resilient operation and user trust.

References

Abadi, M., Barham, P., Chen, J., Chen, Z., Davis, A., Dean, J., Devin, M., Ghemawat, S., Irving, G., Isard, M., Kudlur, M., Levenberg, J., Monga, R., Moore, S., Murray, D. G., Steiner, B., Tucker, P., Vasudevan, V., Warden, P., … Zheng, X. (2016). Tensorflow: A system for large-scale machine learning. In *Proceedings of the 12th USENIX Conference on Operating Systems Design and Implementation, ser. OSDI'16. USA* (pp. 265–283). USENIX Association.

Abdelaziz, A., Santos, V., & Dias, M. S. (2023). Convolutional neural network with genetic algorithm for predicting energy consumption in public buildings. *IEEE Access, 11*, 64049–64069.

Abedinia, O., Ghasemi-Marzbali, A., Shafiei, M., Sobhani, B., Gharehpetian, G. B., & Bagheri, M. (2024). Wind power forecasting enhancement utilizing adaptive quantile function and CNN-LSTM: A probabilistic approach. *IEEE Transactions on Industry Applications, 60*(3), 4446–4457.

Alhanaf, A. S., Farsadi, M., & Balik, H. H. (2024). Fault detection and classification in ring power system with dg penetration using hybrid CNN-LSTM. *IEEE Access, 12*, 59953–59975.

Almahfouz, M., Alhussein, A., & Khader, A. A. (2024). Lightweight deep learning framework for real-time fault diagnosis in grid-connected PV systems. *Scientific Reports, 14*(1), 1–12.

Amin, S. U., & Hossain, M. S. (2020). Edge intelligence and internet of things in healthcare: A survey. *IEEE Access, 9*, 45–59.

Asiri, M. M., Aldehim, G., Alotaibi, F. A., Alnfiai, M. M., Assiri, M., & Mahmud, A. (2024). Short-term load forecasting in smart grids using hybrid deep learning. *IEEE Access, 12*, 23504–23513.

Campagne, E., Amara-Ouali, Y., Goude, Y., and Kalogeratos, A., "Leveraging graph neural networks to forecast electricity consumption," 2024. https://arxiv.org/abs/2408.17366

Chen, C., Li, P., Chen, G., & Yao, L. (2021). Learning graph representations with temporal attention for anomaly detection in dynamic graphs. *IEEE Transactions on Neural Networks and Learning Systems, 32*(9), 4030–4042.

Chen, Z., Tian, P., Liao, W., & Yu, W. (2021). Zero knowledge clustering based adversarial mitigation in heterogeneous federated learning. *IEEE Transactions on Network Science and Engineering, 8*(2), 1070–1083.

Chen, Z., Tian, P., Qian, C., Liao, W., Hussaini, A., & Yu, W. (2024). Machine learning frameworks in IoT systems: A survey, case study, and future research directions. In Z. Lyu (Ed.), Smart spaces, *ser. intelligent data-centric systems* (pp. 197–218). Academic Press. https://www.sciencedirect.com/science/article/pii/B9780443134623000157.

Cheng, L., Zang, H., Wei, Z., Ding, T., & Sun, G. (2022). Solar power prediction based on satellite measurements – A graphical learning method for tracking cloud motion. *IEEE Transactions on Power Systems, 37*(3), 2335–2345.

Deng, F., Zhong, Y., Zeng, Z., Zu, Y., Zhang, Z., Huang, Y., Feng, S., & Zeng, X. (2023). A single-ended fault location method for transmission line based on full waveform features extractions of traveling waves. *IEEE Transactions on Power Delivery, 38*(4), 2585–2595.

Deng, S., Zhao, H., Fang, W., Yin, J., Dustdar, S., & Zomaya, A. Y. (2020). Edge intelligence: The confluence of edge computing and artificial intelligence. *IEEE Internet of Things Journal, 7*(8), 7457–7469.

Ebtia, A., Ghafouri, M., Debbabi, M., Kassouf, M., & Mohammadi, A. (2025). Power distribution network topology detection using dual-graph structure graph neural network model. *IEEE Transactions on Smart Grid, 16*(2), 1833–1850.

Falconer, T., & Mones, L. (2023). Leveraging power grid topology in machine learning assisted optimal power flow. *IEEE Transactions on Power Systems, 38*(3), 2234–2246.

Fang, X., Misra, S., Xue, G., & Yang, D. (2011). Smart grid—The new and improved power grid: A survey. *IEEE Communications Surveys & Tutorials, 14*(4), 944–980.

Ghamizi, S., Cao, J., Ma, A., & Rodriguez, P. (2025). PowerFlowMultiNet: Multigraph neural networks for unbalanced three-phase distribution systems. *IEEE Transactions on Power Systems, 40*(1), 1148–1151.

Goodfellow, I., Bengio, Y., & Courville, A. (2016). *Deep learning*. MIT Press http://www.deeplearningbook.org.

Guha, D., Chatterjee, R., & Sikdar, B. (2023). Anomaly detection using lstm-based variational autoencoder in unsupervised data in power grid. *IEEE Systems Journal, 17*(3), 4313–4323.

Hatcher, W. G., & Yu, W. (2018). A survey of deep learning: Platforms, applications and emerging research trends. *IEEE Access, 6*, 24411–24432.

Hochreiter, S., & Schmidhuber, J. (1997). Long short-term memory. *Neural Computation, 9*(8), 1735–1780.

Huang, C., Ding, S., Li, S., & Liu, R. (2024). Lmfe: Learning-based multiscale feature engineering in partial discharge detection. *IEEE Transactions on Neural Networks and Learning Systems, 35*(5), 5848–5856.

K, V., Vijay, P., R, S., & B, S. (2024). Enhancing fault detection and diagnosis through an integrated chaos-rnn model. In *International conference on intelligent innovative practices in engineering management (IIPEM)* (pp. 1–6).

Kairouz, P., McMahan, H. B., Avent, B., Bellet, A., Bennis, M., Bhagoji, A. N., Bonawitz, K., Charles, Z., Cormode, G., Cummings, R., et al. (2021). Advances and open problems in federated learning. *Foundations and Trends in Machine Learning, 14*(1–2), 1–210.

Kayastha, N., Niyato, D., Hossain, E., & Han, Z. (2014). Smart grid sensor data collection, communication, and networking: a tutorial. *Wireless Communications and Mobile Computing, 14*(11), 1055–1087.

Khamis, A. K., & Agamy, M. (2023). Three-phase inverter dynamics predictions using GNN-based regression model. In *IEEE energy conversion congress and exposition (ECCE)* (pp. 6289–6294). Nashville, TN, USA: IEEE.

Kim, H., Park, S., Park, H.-J., Son, H.-G., & Kim, S. (2023). Solar radiation forecasting based on the hybrid CNN-catboost model. *IEEE Access, 11*, 13492–13500.

Kodinariya, T. M., & Makwana, P. R. (2013). Review on determining number of cluster in k-means clustering. *International Journal, 1*(6), 90–95.

Le, M., Huynh-The, T., Do-Duy, T., Vu, T.-H., Hwang, W.-J., & Pham, Q.-V. (2025). Applications of distributed machine learning for the internet-of-things: A comprehensive survey. *IEEE Communications Surveys Tutorials, 27*(2), 1053–1100.

Lee, B., Kim, Y., Lee, H., & Kang, C. (2025). Bidirectional gated recurrent unit neural network for fault diagnosis and rapid maintenance in medium-voltage direct current systems. *Sensors, 25*(3), 693.

Lee, H., Jeong, H., Koo, G., Ban, J., & Kim, S. W. (2021). Attention recurrent neural network-based severity estimation method for interturn short-circuit fault in permanent magnet synchronous machines. *IEEE Transactions on Industrial Electronics, 68*(4), 3445–3453.

Li, W., Liu, Y., Li, Y., & Guo, F. (2020). Series arc fault diagnosis and line selection method based on recurrent neural network. *IEEE Access, 8*, 177815–177822.

Li, Z., Ye, L., Song, X., Luo, Y., Pei, M., Wang, K., Yu, Y., & Tang, Y. (2024). Heterogeneous spatiotemporal graph convolution network for multi-modal wind-pv power collaborative prediction. *IEEE Transactions on Power Systems, 39*(4), 5591–5608.

Liang, F., Hatcher, W. G., Liao, W., Gao, W., & Yu, W. (2019). Machine learning for security and the internet of things: The good, the bad, and the ugly. *IEEE Access, 7*, 158126–158147.

Liang, F., Hatcher, W. G., Xu, G., Nguyen, J., Liao, W., & Yu, W. (2019). Towards online deep learning-based energy forecasting. In *28th International Conference on Computer Communication and Networks (ICCCN)* (pp. 1–9). Valencia, Spain: IEEE.

Liang, F., Qian, C., Yu, W., Griffith, D., & Golmie, N. (2022). Survey of graph neural networks and applications. *Wireless Communications and Mobile Computing, 2022*(1), 9261537.

Liang, H., Qian, C., Yu, W., Griffith, D., & Golmie, N. (2024a). Assessing deep learning performance in power demand forecasting for smart grid. *International Journal of Sensor Networks, 44*(1), 36–48.

Liang, H., Yu, W., Qian, C., Guo, Y., Griffith, D., & Golmie, N. (2024b). Deep learning-based power quality disturbance detection and classification in smart grid. *International Journal of Sensor Networks, 46*(3), 161–175.

Liang, H., Zuo, C., & Wei, W. (2020). Detection and evaluation method of transmission line defects based on deep learning. *IEEE Access, 8*, 38448–38458.

Liao, W., Wang, S., Bak-Jensen, B., Pillai, J. R., Yang, Z., & Liu, K. (2023). Ultra-short-term interval prediction of wind power based on graph neural network and improved bootstrap technique. *Journal of Modern Power Systems and Clean Energy, 11*(4), 1100–1114.

Lin, J., Yu, W., & Yang, X. (2016). Towards multistep electricity prices in smart grid electricity markets. *IEEE Transactions on Parallel and Distributed Systems, 27*(1), 286–302.

Lin, J., Yu, W., Yang, X., Xu, G., & Zhao, W. (2012). On false data injection attacks against distributed energy routing in smart grid. In *IEEE/ACM third international conference on cyber-physical systems* (pp. 183–192). Beijing, China: IEEE.

Lin, J., Yu, W., Zhang, N., Yang, X., Zhang, H., & Zhao, W. (2017). A survey on internet of things: Architecture, enabling technologies, security and privacy, and applications. *IEEE Internet of Things Journal, 4*(5), 1125–1142.

Liu, X., Qian, C., Hatcher, W. G., Xu, H., Liao, W., & Yu, W. (2019a). Secure internet of things (iot)-based smart-world critical infrastructures: Survey, case study and research opportunities. *IEEE Access, 7*, 79523–79544.

Liu, X., Xu, H., Liao, W., & Yu, W. (2019b). Reinforcement learning for cyber-physical systems. In *IEEE international conference on industrial internet (ICII)* (pp. 318–327). IEEE, Orlando, FL, USA.

Liu, X., Yu, W., Liang, F., Griffith, D., & Golmie, N. (2021). Toward deep transfer learning in industrial internet of things. *IEEE Internet of Things Journal, 8*(15), 12163–12175.

Mallapuram, S., Ngwum, N., Yuan, F., Lu, C., & Yu, W. (2017). Smart city: The state of the art, datasets, and evaluation platforms. In *IEEE/ACIS 16th international conference on computer and information science (ICIS)* (pp. 447–452). Wuhan, China: IEEE.

McMahan, B., Moore, E., Ramage, D., Hampson, S., & y Arcas, B. A. (2017). Communication-efficient learning of deep networks from decentralized data. *Artificial intelligence and statistics* (pp. 1273–1282). PMLR.

Mohamed, A., Refaat, S. S., & Abu-Rub, H. (2019). A review on big data management and decision-making in smart grid. *Power Electronics and Drives, 4*(1), 1–13.

Mohammadi, M., Al-Fuqaha, A., Sorour, S., & Guizani, M. (2018). Deep learning for iot big data and streaming analytics: A survey. *IEEE Communications Surveys Tutorials, 20*(4), 2923–2960.

Musa, A. A., Hussaini, A., Liao, W., Liang, F., & Yu, W. (2023). Deep neural networks for spatial-temporal cyber-physical systems: A survey. *Future Internet, 15*(6). [Online]. Available https://www.mdpi.com/1999-5903/15/6/199 .

Nabavi, S. A., Motlagh, N. H., Zaidan, M. A., Aslani, A., & Zakeri, B. (2021). Deep learning in energy modeling: Application in smart buildings with distributed energy generation. *IEEE Access, 9*, 125439–125461.

Nelega, R., Greu, D. I., Jecan, E., Rednic, V., Zamfirescu, C., Puschita, E., & Turcu, R. V. F. (2023). Prediction of power generation of a photovoltaic power plant based on neural networks. *IEEE Access, 11*, 20713–20724.

Nguyen, B. L. H., Vu, T. V., Nguyen, T.-T., Panwar, M., & Hovsapian, R. (2023). Spatial-temporal recurrent graph neural networks for fault diagnostics in power distribution systems. *IEEE Access, 11*, 46039–46050.

Pene, P., Liao, W., & Yu, W. (2024). Incentive design for heterogeneous client selection: A robust federated learning approach. *IEEE Internet of Things Journal, 11*(4), 5939–5950.

Philip, N. Y., Rodrigues, J. J. P. C., Wang, H., Fong, S. J., & Chen, J. (2021). Internet of things for in-home health monitoring systems: Current advances, challenges and future directions. *IEEE Journal on Selected Areas in Communications, 39*(2), 300–310.

Prechelt, L. (1998). Early stopping — but when? *Neural networks: Tricks of the trade* (pp. 55–69). Springer.

Qian, C., Liu, X., Ripley, C., Qian, M., Liang, F., & Yu, W. (2022). Digital twin—cyber replica of physical things: Architecture, applications and future research directions. *Future Internet, 14*(2). https://www.mdpi.com/1999-5903/14/2/64.

Qian, M., Qian, C., Xu, G., Tian, P., & Yu, W. (2024). Smart irrigation systems from cyber–physical perspective: State of art and future directions. *Future Internet, 16*(7). https://www.mdpi.com/1999-5903/16/7/234.

Rajkumar, R. R., Lee, I., Sha, L., & Stankovic, J. (2010). Cyber-physical systems: The next computing revolution. In *Proceedings of the 47th design automation conference, ser. DAC '10* (pp. 731–736). New York, NY, USA: Association for Computing Machinery. doi:10.1145/1837274.1837461.

Ren, J., Zhao, K., Zhang, G., Liu, X., Zhai, C., and Xiao, G., (2025). Fault localization and state estimation of power grid under parallel cyber-physical attacks, arXiv e-prints, arXiv-2503.

Shen, S., Lu, H., Sadoughi, M., Hu, C., Nemani, V., Thelen, A., Webster, K., Darr, M., Sidon, J., & Kenny, S. (2021). A physics-informed deep learning approach for bearing fault detection. *Engineering Applications of Artificial Intelligence, 103*, 104295.

Singh, T., Solanki, A., Sharma, S. K., Jhanjhi, N. Z., & Ghoniem, R. M. (2023). Grey wolf optimization-based CNN-LSTM network for the prediction of energy consumption in smart home environment. *IEEE Access, 11*, 114917–114935.

Stankovic, J. A. (2014). Research directions for the internet of things. *IEEE Internet of Things Journal, 1*(1), 3–9.

Su, Z., Wang, Y., Luan, T. H., Zhang, N., Li, F., Chen, T., & Cao, H. (2021). Secure and efficient federated learning for smart grid with edge-cloud collaboration. *IEEE Transactions on Industrial Informatics, 18*(2), 1333–1344.

Sun, Y., Song, H., Jara, A. J., & Bie, R. (2016). Internet of things and big data analytics for smart and connected communities. *IEEE Access, 4*, 766–773.

Thomas, J. B., Chaudhari, S. G., S. , K. V., & Verma, N. K. (2023). Cnn-based transformer model for fault detection in power system networks. *IEEE Transactions on Instrumentation and Measurement, 72*, 1–10.
Tian, P., Liao, W., Yu, W., & Blasch, E. (2022). WSCC: A weight-similarity-based client clustering approach for non-IID federated learning. *IEEE Internet of Things Journal, 9*(20), 20243–20256.
TTF, "Tensorflow federated: Machine learning on decentralized data," https://www.tensorflow.org/federated, 2019.
Tuballa, M. L., & Abundo, M. L. (2016). A review of the development of smart grid technologies. *Renewable and Sustainable Energy Reviews, 59*, 710–725.
"UK power networks: Smartmeter energy consumption data in london households," 2014, Accessed: 6 February 2024. https://data.london.gov.uk/dataset/smartmeter-energy-use-data-in-london-households
Ullah, F. U. M., Ullah, A., Haq, I. U., Rho, S., & Baik, S. W. (2020). Short-term prediction of residential power energy consumption via CNN and multi-layer bi-directional LSTM networks. *IEEE Access, 8*, 123369–123380.
Wang, F., & Hu, Z. (2025). A novel fault diagnosis and accurate localization method for a power system based on graphsage algorithm. *Electronics, 14*(6), 1219.
Wang, Q., Liao, W., Guo, Y., McGuire, M., & Yu, W. (2024). Blockchain-empowered federated learning through model and feature calibration. *IEEE Internet of Things Journal, 11*(4), 5770–5780.
Wu, J., Lu, C., & Wu, C. (2023). Learning-aided framework for storage control facing renewable energy. *IEEE Systems Journal, 17*(1), 652–663.
Xia, M., Shao, H., Ma, X., & de Silva, C. W. (2021). A stacked GRU-RNN-based approach for predicting renewable energy and electricity load for smart grid operation. *IEEE Transactions on Industrial Informatics, 17*(10), 7050–7059.
Xu, G., Yu, W., Griffith, D., Golmie, N., & Moulema, P. (2017). Toward integrating distributed energy resources and storage devices in smart grid. *IEEE Internet of Things Journal, 4*(1), 192–204.
Xu, H., Yu, W., Griffith, D., & Golmie, N. (2018). A survey on industrial internet of things: A cyber-physical systems perspective. *IEEE Access, 6*, 78238–78259.
Yu, W., Liang, F., He, X., Hatcher, W. G., Lu, C., Lin, J., & Yang, X. (2018). A survey on the edge computing for the internet of things. *IEEE Access, 6*, 6900–6919.
Zhang, X., Yang, X., Lin, J., Xu, G., & Yu, W. (2015). Towards efficient and secured real-time pricing in the smart grid. In *IEEE global communications conference (GLOBECOM)* (pp. 1–6). San Diego, CA, USA: IEEE.
Zhou, C., & Paffenroth, R. C. (2017). Anomaly detection with robust deep autoencoders. In *Proceedings of the 23rd ACM SIGKDD international conference on knowledge discovery and data mining* (pp. 665–674). Halifax NS Canada: Association for Computing Machinery.
Zhu, L., Hill, D. J., & Lu, C. (2021). Intelligent short-term voltage stability assessment via spatial attention rectified RNN learning. *IEEE Transactions on Industrial Informatics, 17*(10), 7005–7016.

CHAPTER

7

Intelligent Digital Twins for social acceptance by design in smart grids

Giovanna Di Marzo Serugendo, Philippe Glass
Computer Science Center, University of Geneva, Geneva, Switzerland

7.1 Introduction

The global transition toward sustainable energy systems has accelerated the development and deployment of *smart grids*—intelligent, adaptive infrastructures that integrate digital technologies with traditional power networks. These systems promise to enhance energy efficiency, reliability, and resilience by enabling real-time monitoring, decentralized energy production, and dynamic demand-response mechanisms. However, the success of smart grids depends not only on technological innovation but also on *social acceptance,* which remains a critical yet often underestimated factor in their widespread adoption.

As energy consumers evolve into *prosumers*—simultaneously producing and consuming energy—the complexity of managing distributed energy resources increases. This shift necessitates the introduction of *autonomous decision-making capabilities* within the grid to coordinate energy flows, prevent overloads, and optimize storage and consumption. In this context, *intelligent digital twins*—virtual replicas of physical grid components enhanced with AI—emerge as a powerful tool. These Digital Twins enable predictive control, facilitate peer-to-peer energy exchange, and support grid stability by anticipating and mitigating peak loads.

Beyond operational efficiency, Digital Twins also offer a novel avenue to *embed social acceptance by design*. Through the integration of algorithms that address key societal concerns—such as *privacy protection, mitigation of the tragedy of the commons, prevention of free riding behaviors,* and *strategic use of energy storage during blackouts*—these systems can align technological functionality with human values and expectations. Our main contribution is to bridge the gap between technical efficiency and human-centered design by integrating social acceptance criteria into the configuration and behavior of intelligent digital twins representing prosumers in a microgrid.

7.1.1 Context of study

This work is part of a larger research project, called LASAGNE, involving various partners (academics, companies, and living labs communities). As part of this project, social sciences researchers at Haute École spécialisée de Suisse occidentale (*HES-SO) Valais* conducted a qualitative study involving *focus groups and semi-structured interviews* with both potential users of microgrids and stakeholders engaged in microgrid initiatives. The findings revealed a rich and nuanced understanding of how microgrids are perceived—not only in terms of their *technical and economic features,* but also through *logistical, organizational, and relational dimensions.*

A key insight from this study is that the success of a microgrid project is not solely determined by the deployment of infrastructure and technology. Rather, it critically depends on the *social and economic context*, as well as on the *perceptions, expectations, and concerns* of its potential users. Negative perceptions—such as fears around data privacy, lack of transparency, or unclear responsibilities—act as *restraining forces* that can significantly hinder the progress of a microgrid, even when technical conditions are favorable, and some participants are already engaged.

The study also identified a set of *salient attributes*—core characteristics that users associate with microgrids. These attributes were used to construct *profile maps* that describe different types of microgrid communities

Digital Twin Technology and Smart Grid.
DOI: https://doi.org/10.1016/B978-0-443-36664-2.00008-3

(Fragnière et al., 2024). Drawing from social psychology, the concept of *salience* refers to beliefs or attitudes that are particularly prominent in a person's cognitive field and thus more likely to influence behavior (Krech & Crutchfield, 1948).

In the context of microgrids, salient attributes span both *technical* and *mental* dimensions. Technical attributes include the *robustness of infrastructure, emergency preparedness, energy storage capabilities*, and *support for new members*. Mental attributes encompass *team spirit, shared values, participant engagement*, and *identification with sustainability goals* such as energy sobriety and renewable energy adoption (Fragniere et al., 2023).

These attributes were grouped into four main categories to build community profiles: *Energy network* (e.g., grid connection preferences, renewable energy mix), *infrastructure management* (e.g., internal vs. external control), *data management* (e.g., data sharing policies, anonymous consumption comparisons), and *organizational structure* (e.g., meeting frequency, self-regulation practices).

Based on these insights, our collaboration with HES-SO Valais informed the *design of adaptive behaviors* for Digital Twins acting within a microgrid. Rather than prescribing fixed behaviors for specific microgrid instances, our approach demonstrates how *Digital Twin behavior can be tailored* to reflect the unique profile and preferences of a given microgrid community.

In this chapter, we present a proposal for *embedding social acceptance attributes into Digital Twin behavior*, enabling them to respond to community-specific requirements. We discuss the framework we designed and implemented, as well as a suite of algorithms enacting social acceptance. These algorithms cover privacy and personal data protection, dynamic pricing, free rider tracking and penalization, and storage management. We evaluate this approach through a series of scenarios, testing how well Digital Twins can incorporate selected features derived from the salient attributes identified in the study.

7.2 Background on Digital Twins

Barricelli et al. (2019) define the Digital Twin as a computer-based model that simulates, emulates, mirrors, or "twinning" the life of a physical entity, which can be an object, a process, a human, or a human-related feature. Di Marzo Serugendo et al. (2022) define four levels of Digital Twins: the *static* Digital Twin, which provides a static model or representation, the *functional* twin, which enriches the static twin with functional capabilities, the *self-adaptive or real-time* twin, which further integrates real-time data, and finally the *intelligent* Digital Twin, which works alongside the physical twin and can act on behalf of the physical twin. The intelligent Digital Twins may then possess cognitive abilities such as self-adaptation, communication, negotiation, reasoning, or learning.

According to Marah and Challenger (2022), autonomous agents have powerful capabilities that enable them to meet the requirements of integration into a Digital Twin. Indeed, Digital Twins, especially intelligent ones, benefit from the autonomy (sensing and acting) available to agents in multi-agent systems (MAS) (Wooldridge, 2009). Therefore, MAS provides a powerful implementation paradigm for programming Digital Twins.

In the context of this contribution, we developed **intelligent Digital Twins** as autonomous software agents that interact on the one hand with end-user applications and consumption/production edge devices and their data, and on the other hand, coordinate their respective actions through the coordination platform. The qualifier "intelligent" for the Digital Twins corresponds to the definition by Di Marzo Serugendo et al. (2022) and means that such an agent (Marah & Challenger, 2022; Wooldridge, 2009) possesses capabilities of making autonomous decisions, reasoning, learning, knowledge acquisition, processing real-time data, setting up or revoking contracts, and communication with other twins.

7.3 Related works

7.3.1 Digital Twins in the energy domain

Tzanis et al. (2020) present an approach based on cybernetic Digital Twins to anticipate future power failures in a smart grid within a very short time frame. In this approach, each twin combines a machine learning model and a discrete deterministic mathematical model. Experiments have shown promising results on low-latency networks. Ebrahimi (2019) designed a Digital Twin model of complex renewable energy generators, such as wind or hydro power plants. They present a comprehensive modeling strategy that contains different modeling levels and applies uncertainty and smart algorithm tools.

7.3.2 Social acceptance by design in smart grids

As part of an early work on literature review of social acceptance in smart grids, Wolsink (2012) argues for a socially constructed, institutional approach, and introduces the idea of common-pool resource management, stressing the need for self-governance, flexible regulation, and community cooperation in microgrids. More recently, Senyapar and Bayindir (2023) provide a comprehensive analysis of the literature on smart grids, highlighting the fact that economic benefits are often more influential than environmental concerns in driving social acceptance. It also emphasizes the importance of user roles, safety, and artificial intelligence integration, and calls for more attention to the social dimensions of smart grid adoption. Valkering et al. (2014) investigate enablers and barriers to social acceptance for smart energy systems. Besides enablers such as financial incentives, we find barriers provided by loss of control in the smart energy system, privacy and security concerns, as well as free rider behavior.

7.4 Intelligent Digital Twins and coordination in smart grids

Intelligent digital twins for smart grids, implemented as agents or MAS, are central to enabling autonomy, adaptability, and resilience in modern energy infrastructures. These Digital Twins interact and communicate through a structured coordination mechanism that ensures coherent behavior across distributed components. To support this, we employ a specific coordination model known as *LASAGNE*, which provides built-in features for supporting interactions and coordination of activities among Digital Twins via a shared virtual environment. The *LASAGNE coordination model* is an evolution of the *SAPERE coordination model* architecture (Zambonelli et al., 2015) and middleware (Castelli et al., 2015). We base our work on the recent extensions proposed by Ben Mahfoudh et al. (2020) and complete it with Digital Twins, acting on behalf of prosumers, specifically tailored for smart grids. The LASAGNE coordination model draws inspiration from biochemical processes and operates through a *tuple space*—a shared, decentralized data environment where Digital Twins exchange information indirectly. Within this tuple space, Digital Twins publish and retrieve *Live Semantic Annotations (LSAs)* - structured data tuples that encapsulate their properties, intentions, or observations. These LSAs evolve over time, influenced by *coordination laws* (also called *eco-laws*) that govern the dynamics of the system. Unlike traditional publish/subscribe models, this approach enables *indirect, context-aware coordination* through a blackboard-like mechanism, where agents interact by modifying and reacting to shared data rather than sending direct messages. Fig. 7.1 shows the sequence diagram with the LSAs flows between a producer and a consumer Digital Twin in order to reach an energy contract. Consumer 1 sends a request for 75 W energy, followed by a matching offer from producer 1. The process then continues with a contract formation by consumer 1 and a contract confirmation from producer 1. The process ends with a final acknowledgment sent to producer 1, which will trigger the actual provisioning of energy.

The coordination platform, instantiated on a computational node, executes a set of eco-laws that simulate chemical reactions to transform and propagate LSAs. The currently implemented laws include:

- *Bonding*: Matches and links LSAs from different Digital Twins based on compatibility or relevance. For example, all producers bond with properties labeled as "request": consequently, the Bonding law will transmit to the producer's LSA the requests that consumers submit through their LSA to signal a lack of energy.
- *Decay*: Gradually reduces the significance of outdated LSAs, effectively cleaning obsolete data. For example, a stop regulation order has an expiry counter that is decremented at each decay cycle, so that they are cleaned up regularly. This expiry time must be sufficient for this order to be propagated to all other prosumer LSAs across the network before being deleted.
- *Spreading*: Disseminates LSAs across the network to ensure information availability and redundancy. For example, energy request properties are distributed among neighboring nodes so that they can also be retrieved by producers located in neighboring nodes.
- *Aggregation*: Combines LSAs from multiple sources to synthesize higher-level insights or summaries. For example, the produced energy values announced by producers are aggregated at the network level using a summation operator, allowing the computation of the total energy produced across the entire network.

In this framework, *Digital Twins* are not merely passive replicas of physical systems but are *active software agents* (Wooldridge, 2009) capable of sensing, acting, learning, and negotiating (Di Marzo Serugendo et al., 2022). Each twin represents a physical device or user and operates autonomously within the coordination model. By submitting and retrieving LSAs, these agents dynamically adapt to environmental changes, collaborate with peers, and contribute to the overall intelligence of the smart grid.

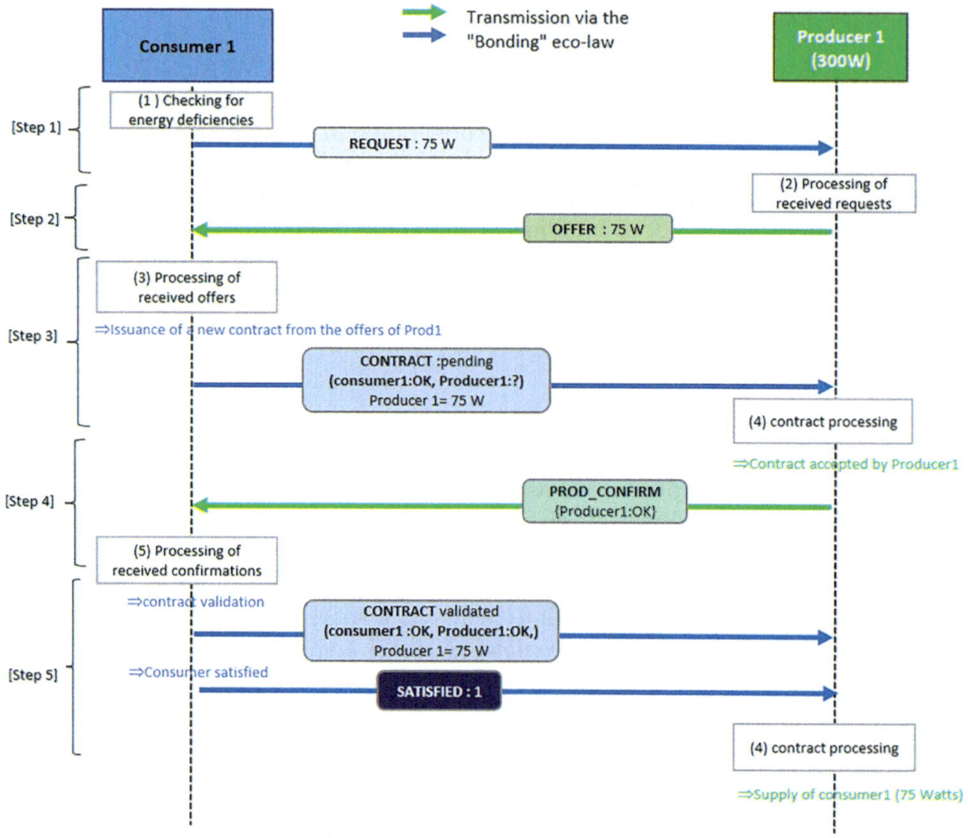

FIGURE 7.1 LSA flows: sequence diagram of data exchanged between two prosumer twins to generate an energy contract.

The coordination model that we implemented uses the various mechanisms to enable the Digital Twins to *negotiate contracts* related to energy exchanges, to *forecast energy consumption or production,* and to *avoid peaks of energy* consumption or production. It also supports the Digital Twins in providing *elements of social acceptance* by design, such as ensuring privacy or overcoming free riding behavior.

7.4.1 Intelligent Digital Twins for smart grids applications

We defined four distinct types of Digital Twins, each serving a specific function in the coordination and regulation of energy exchanges:

- Energy Digital Twins:
  - **Consumers**: Represent devices that consume electricity.
  - **Producers**: Represent devices that generate electricity.
  - These two roles can also be merged into a single entity known as the **prosumer**, capable of both consuming and producing energy.
- Supervisory Digital Twins:
  - **Learning** twin: Predicts energy consumption and production at the node level using historical and real-time data.
  - **Regulator** twin: Manages critical situations such as peak shaving and ensures system stability.

Energy Digital Twins actively participate in negotiation processes for energy exchange. Consumers submit energy demands, while producers respond with offers. The coordination model and its implementation, as well as the negotiation process, including contract formation and validation, are fully detailed in our previous research (Glass & Di Marzo Serugendo, 2023). Each node hosts as many energy Digital Twin instances as there are producers and consumers. In contrast, supervisory Digital Twins are instantiated once per node and are launched with the node's

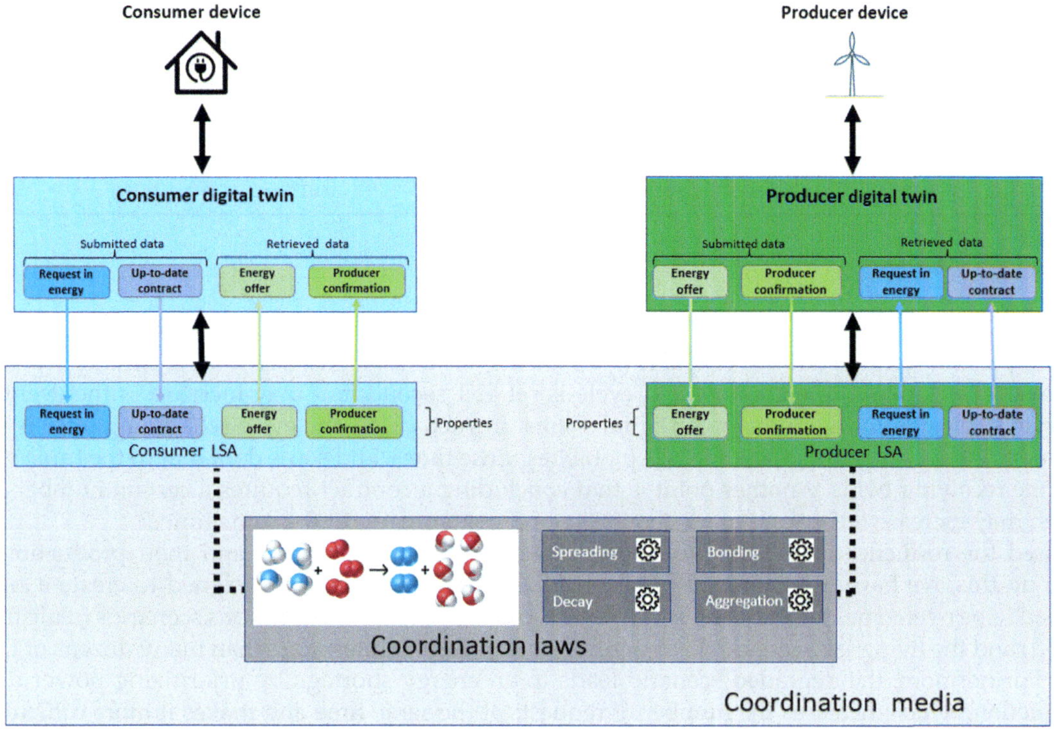

FIGURE 7.2 Flow of information exchanged between producers and consumer agents.

coordination platform. Fig. 7.2 shows consumers and producers' Digital Twins coordinating their activities through the coordination platform.

Implementation: To operationalize our coordination model, we implemented it as a *middleware layer*, enabling seamless interaction between the Digital Twins acting in the smart grid. The Digital Twins—representing consumers, producers, prosumers, and supervisory agents—are realized as *autonomous software agents* and interact through the middleware. The whole system is deployed on actual computational nodes. This architecture allows for decentralized decision-making and real-time coordination and enacts the various algorithms described in this paper. All the software layers mentioned are developed in Java, except for certain classifiers, which are developed in Python using the Keras library. To support development and facilitate monitoring, we also built a *web-based application* integrated with the middleware, providing a user-friendly interface for observing and debugging the behavior of Digital Twins during runtime. The web application uses Angular 9, Node.js, and D3JS, and the coordination service functions as a REST server: all data exchanges between this service and client processes use the JSON format. We have made the sources of the implementation available for free access from a GitHub repository (address: https://github.com/philippe-glass/energy). The operating mode explains how to install and configure the coordination platform, the web client, the test simulators, and the MariaDB used to store the scenario data. The "run_configs" folder contains platform configurations of four different nodes that we used in our experiment. Due to excessive volumes, the log files generated by the executions have not been saved in this GitHub repository.

Scenarios: The scenarios we developed and evaluated are rooted in *real-world data* collected from a *living-lab in Geneva*, ensuring that our platform reflects realistic energy usage patterns and operational conditions. Since May 2022, the electricity consumption history has been collected every minute by smart meters in the living laboratory for 12 entities distributed across four buildings: household appliances, solar panels, and external supply from the canton's main supplier. However, the dataset remains private as it belongs to the municipality of Meyrin, where the living laboratory is located.

To *reproduce the specific conditions* described in these scenarios, we *synthesized targeted datasets* derived from the original real-world data. These synthetic datasets allowed us to simulate diverse and controlled situations while maintaining fidelity to actual usage behaviors. Importantly, all scenarios were executed using our *fully implemented Digital Twin agents*, running the actual coordination algorithms designed to handle the various challenges and interactions described in the scenarios. For example, the realistic scenario attempts to reproduce consumption and production based on actual household statistics. Unlike the living laboratory scenario, it also uses randomness to

TABLE 7.1 Results obtained with the following four scenarios: "living-lab," "realistic," "degraded," and "overload," tested for 60 minutes.

Max response time (sec.)				Ratio of missing energy[b]			
Living-lab (12 DT[a])	Realistic (< 50 DT)	Degraded (< 50 DT)	Overload (>100 DT)	Living-lab (12 DT)	Realistic (< 50 DT)	Degraded (< 50 DT)	Overload (>100 DT)
11	13	29	84	0.27%	0.61%	2.19%	28.14%

[a] *DT: Number of digital twins in the experiment.*
[b] *cumulative missing energy (Wh) / cumulative demanded energy (Wh) during the experiment.*

generate power data by applying a normal distribution to the statistical values that have been collected. We have made an example of a scenario dataset available (see realistic_scenario_dataset.csv on the GitHub repository).

Limitations: The duration of a coordination cycle is set at 1 second, but may increase in the event of network overload or an explosion in the number of Digital Twins. In such cases, the cycle scheduling jitter is greater than zero, and the middleware cannot guarantee acceptable response times, and some data sent to the tuple space may be ignored by the receiving twins. Another point is that concluding a contract requires a certain number of execution cycles, which may increase if there are many requests or offers to process at the same time.

We assessed the resilience of Digital Twins when their number is increased, and their production capacity is reduced. To do this, we have implemented purely fictitious random scenarios designed to create a situation with extensive load to be overcome. Table 7.1 shows the results obtained with three fictitious scenarios (realistic, degraded, and overload) and the living-lab scenario. Each fictitious scenario generates more than many dozens of Digital Twins at random. Furthermore, the degraded scenario leads to an energy shortage by prioritizing power demand over power production, which increases the number of requests at the same time and makes it more difficult to obtain a supply contract. The overload scenario generates more than 100 Digital Twins. The results of maximal response time show good resilience when the number of Digital Twins in a node remains "reasonable." Significant degradation occurs in degraded mode and when the number of Digital Twins per node exceeds 100. In the latter case, we can take advantage of the flexibility of coordination models, which allow us to modify the network topology in order to better distribute Digital Twins across different nodes, thereby reducing the maximum number of Digital Twins per node. Since the computing units of each node operate in parallel, this will eliminate bottlenecks on certain nodes.

7.5 Social acceptance by design

We discuss here how social acceptance is taken into account in the modeling and implementation of Digital Twins interacting via the coordination platform. In general, consideration of the social acceptance criteria defined by potential users is essential to give a project based on AI and digital technologies a serious chance of success. Studies have shown that the absence of this social dimension has led to many project abandonments. In addition to this, microgrid applications are based on the Customer-to-Customer (C2C) model, which means that users have a role to play as producers as well as consumers (and therefore that of a prosumer). Moreover, a microgrid organization requires sharing and distributing resources within the same community, which means that several rules need to be put in place to ensure that the community runs smoothly and that there is mutual respect among stakeholders. Given the importance of the social dimension, we integrated a number of these aspects into the algorithms regulating the Digital Twin's behaviors. This integration is based on the work carried out by the University of HES-SO Valais, which defines user-centered approaches to design a microgrid community combining social aspects and AI technology (Fragniere et al., 2023).

7.5.1 Privacy

To meet social acceptance criteria related to privacy, we enforce confidentiality rules on the data exchanged in the tuple space: a Digital Twin's personal information is hidden from other agents.

These confidentiality rules apply to the various data exchanged, such as requests, offers, contracts, or producer confirmations: they consist of protecting each data submitted through an LSA individually. We use the notion of "wrapper" to protect access to sensitive objects. As an example of application, Fig. 7.3 shows the visibility of an electricity supply contract between consumer number 1 and providers number 2 and 3 (which supplies consumer 1). We can see that:

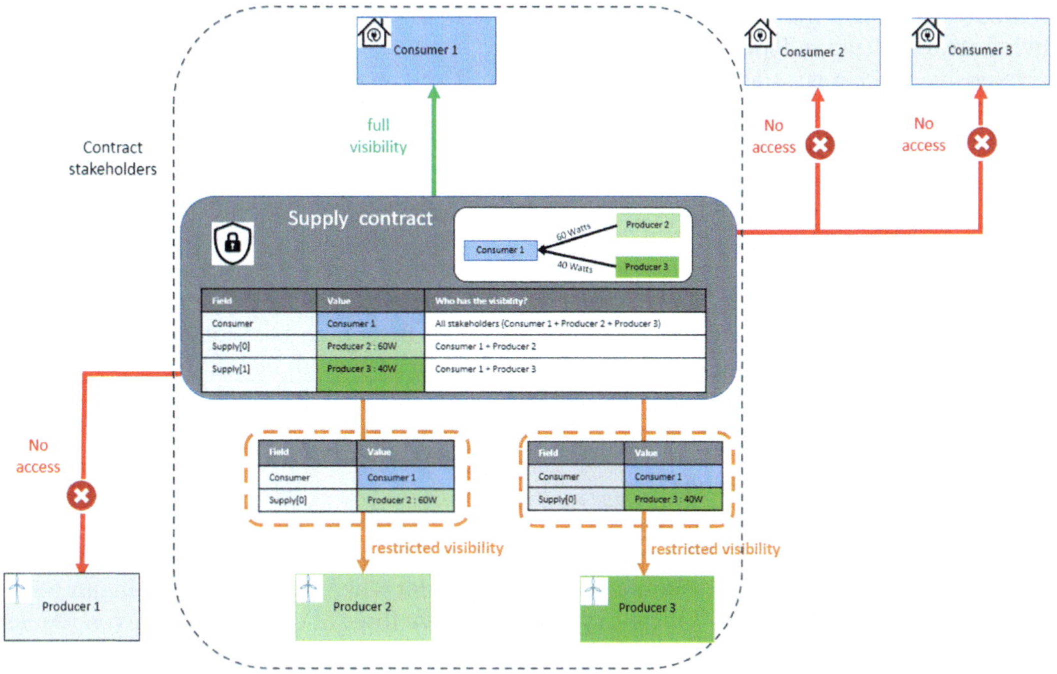

FIGURE 7.3 Visibility of prosumer Digital Twins on an electricity supply contract (approximation of consumer 1 by producer 2 and producer 3).

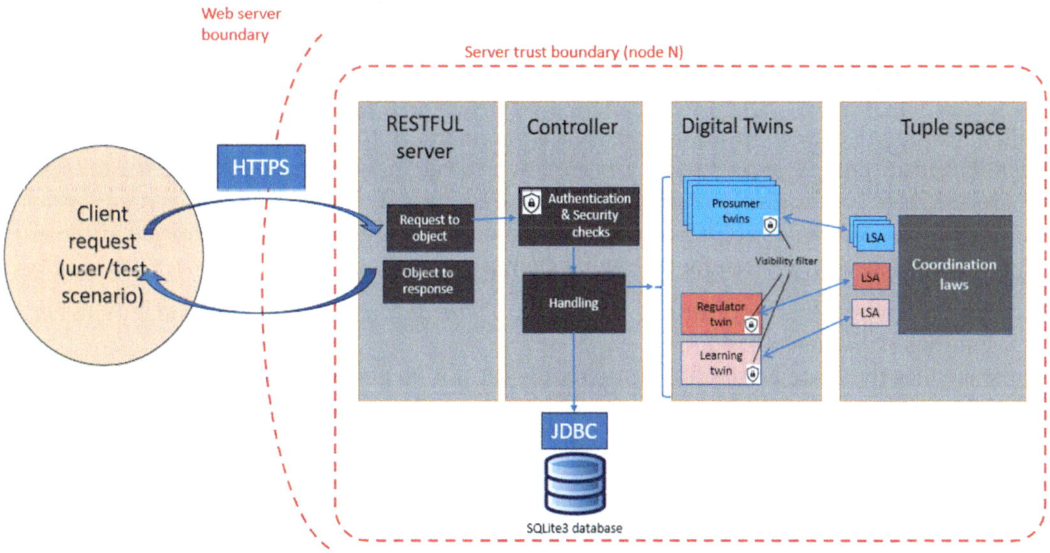

FIGURE 7.4 Threat model including the different layers. The learning and regulator twins have visibility over the prosumer twins, whereas a prosumer twin can only see certain information emitted by other prosumer twins.

- All prosumers not included in the contract stakeholders have no access to the contract data (this is the case for consumers number 2 and 3, as well as producer number 1).
- Consumer number 1, as the contract owner, has full access (it has full visibility of the two supplies it receives).
- The two suppliers (number 2 and 3) have partial access: they only see the partial supply that concerns them. Each producer does not have access to the supply data of the other producers involved in the same contract.

As shown on the threat model (Fig. 7.4), all external client requests are filtered and processed at the REST server level. Digital twins, coordination laws, tuple types, and visibility rules are managed exclusively internally on the server side and cannot be accessed or modified from the outside. A user request cannot initialize a prosumer at its level. All Digital Twins have an identifier, an associated node, and a profile, and this information is stored on the server of each node or LSA that the Digital Twin has already visited. The prosumer cannot have additional visibility. Supervisor twins are initialized independently of any user requests.

However, defensive mechanisms against denial-of-service (DoS) attacks have not yet been implemented. Such attacks can quickly overload the server. Furthermore, Attacks on the server executable have not been investigated: if the executable itself is hacked, the behavior of Digital Twins, eco-laws, and tuple space may be affected, and data may be maliciously modified.

7.5.2 Tragedy of the commons

The "tragedy of the commons" scenario illustrates how individually rational decisions can lead to collective failure. In a microgrid, this occurs when all Digital Twins act in their own best interest, causing a system-wide shortage. For example, if everyone charges their electric vehicles at 11 p.m. to benefit from lower night-time tariff, the resulting demand spike can overwhelm the grid—despite each decision being optimal in isolation. We provided a full description of this case in a previous paper (Glass & Di Marzo Serugendo, 2023).

7.5.2.1 *Tragedy of the common's scenario*

To perform this scenario, we introduced a pricing policy that prompts all Digital Twins to request electricity simultaneously during the lowest price window. This setup stresses the system to test its resilience. The pricing table sets the minimum price one minute ahead, creating a narrow time slot where demand peaks. In our test, 10 consumer twins compete for energy from a single producer that can only serve three at once, leaving seven requests unmet.

7.5.2.2 *Avoiding the tragedy of the commons with dynamic pricing*

To reduce the supply-demand imbalance, we introduced a dynamic pricing strategy that adjusts in real-time based on the producer's remaining capacity. Eq. (7.1) shows the calculation of a producer's dynamic price, based on its generated and supplied power. The price increase factor varies on a scale of 1–10 depending on demand.

$$\text{price} = \text{default\_price} * \min\left(10.0, \frac{\text{produced (W)}}{\text{produced (W)} - \text{provided (W)}}\right) \tag{7.1}$$

As availability drops, prices rise, prompting unserved consumers to shift their demand to less congested time slots. This helps distribute energy requests more evenly over time and alleviates peak pressure on the system. Fig. 7.5 shows the process used to iteratively distribute energy demands over time using a dynamic pricing that favors the least popular periods. A new iteration is carried out when there are unfulfilled requests at the current time. At each iteration, the Digital Twins perform the following operations:

1. Each producer supplies requests according to its capacity. This reduces its available capacity to zero if the overall capacity is not sufficient to meet all requests.
2. Each producer applies dynamic pricing by recalculating all slot tariffs based on updated availability.
3. Each dissatisfied consumer updates the tariff table and revises the time slot of its request according to the new tariffs (it must target the period corresponding to the lowest price).

7.5.2.3 *Results*

In this section, we propose to assess the impact of dynamic pricing on tariffs and on the ability to avoid energy shortages due to excessive demand. For this, we define a key performance indicator, *KPI*1.1: ratio of missing energy.

$$KPI_{1.1} = \frac{\text{cumulative missing energy } (Wh)}{\text{cumulative demanded energy } (Wh)}$$

Given that the objective is to reduce the energy missing, the value of *KPI*1.1 must be minimized. Table 7.2 shows a comparison of energy missing when dynamic pricing is applied or not, with different pricing cycle durations. The results clearly show that dynamic pricing allows for reducing missing energy to almost zero. Considering the experiment conducted, we can explain this result as follows: with the indexed pricing policy in place, unsatisfied Digital Twins automatically defer their requests to later time slots. In our example, three of the seven remaining requests are fulfilled in the second slot, three more in the third, and the final one in the fourth. This staggered response helps smooth out demand and prevents overload during peak times. The rate of energy missing is even lower when the pricing cycle is longer, as the relative time taken to generate a new contract is shorter compared to the price stagnation durations. Our experiment, therefore, demonstrated a solution for addressing the tragedy of the commons

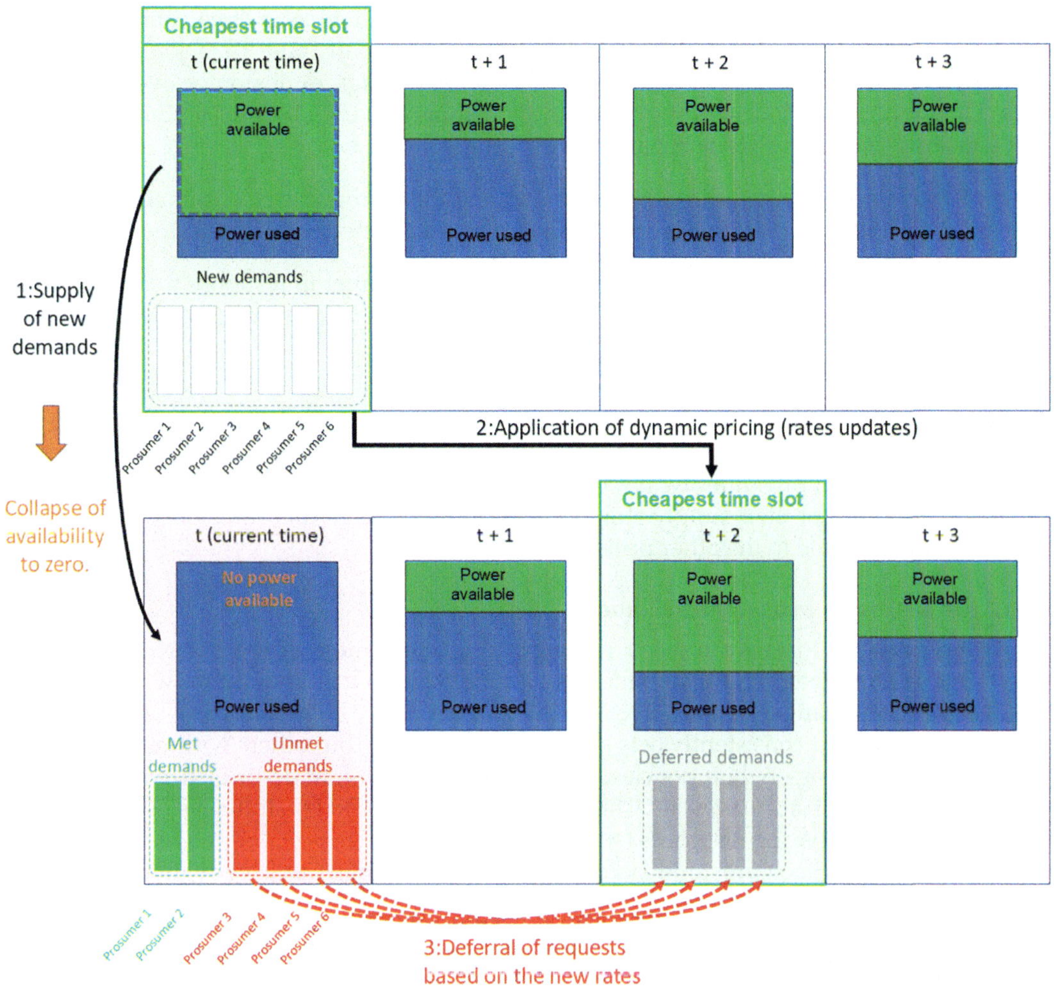

FIGURE 7.5 Using dynamic pricing to spread energy demand over time.

TABLE 7.2 Tragedy of the common scenario: Assessing the energy missing when using (or not using) dynamic pricing.

	Dynamic pricing deactivated (flat pricing)		Dynamic pricing activated	
Price cycle duration (min)	Cumulative missing (Wh)	*KPI*1.1: ratio missing/demanded	Cumulative missing (Wh)	*KPI*1.1: ratio missing/demanded
3	3.28	71.08%	0.07	1.39%
10	10.71	69.12%	0.04	0.27%
30	35.03	70.10%	0.06	0.11%

using dynamic pricing, which we compared to flat pricing by time slot. In this experiment, we did not introduce admission controls or random delays to better quantify the magnitude of the effect. Furthermore, the consumers served first were chosen at random: an improvement would be to distinguish between consumers who are willing to pay more to be served immediately and those who are willing to delay their request in order to pay less.

7.5.3 Overcoming "free riding" behaviors

For a microgrid community to function, the various prosumers are supposed to share energy resources and therefore contribute to supplying their neighbors when they are short of energy, in the fairest possible way. Indeed, a failure to respect this principle of equality of supply leads to frustration and demotivation, and therefore to a reduction in the general well-being of the community. We refer to this type of behavior as "free riding," consisting of taking advantage of the services offered by the community while ceasing to contribute. To tackle these failures of fairness principles, the adopted strategy consists of penalizing prosumers through exclusion or overbilling in the decision-making process. This strategy can be integrated into the algorithms executed by the Digital Twins: they can interact in a way to penalize any "free riding" type behavior:

- On the one hand, by detecting "free riding" behaviors among the surrounding prosumers.
- On the other hand, by adapting their own supply strategy when this behavior is detected.

We propose two variants of a strategy that can be adopted by a prosumer in response to the free riding behaviors:

- One consisting of no longer sending offers to "free riders."
- Another consists of overcharging energy offers sent to "free riders."

For data privacy reasons, we propose that it is the regulator twin that evaluates the behavior of the prosumer twins attached to the same node and assigns them credit scores (positive for rewards and negative for penalties) according to the obtained evaluation. Each prosumer twin will then emit its current credit score in the energy requests, which will allow the producers to adapt their supply strategy according to the credit score. Only the final score shown in the energy requests will be visible to the other peers.

7.5.3.1 Criteria for assessing prosumer behaviors

We defined several criteria to evaluate each participant's contribution to the smart grid community, which influence how energy is allocated. Prosumers take into account this contribution score when deciding whether to accept or prioritize energy exchanges.

For instance, if a Digital Twin consistently acts selfishly - benefiting from others without giving back - it may be penalized by its peers. Depending on their policies, other Digital Twins might de-prioritize the free rider, increase its energy price, or exclude it from future requests. This mechanism is inspired by strategies used to detect and manage free riding in peer-to-peer file-sharing networks (Alotibi et al., 2019).

The regulator twin evaluates each prosumer according to the following criteria of evaluation:

- *Level of altruism* (Eq. (7.2)): distribution of energy surplus to prosumers in the community when the surplus exceeds a certain threshold. For each prosumer, this criterion score considers the energy supplied to other prosumers (denoted by *bonusProvided Wh)* and subtracts the energy that should have been supplied (denoted by *malusNotGiven Wh*):

$$\text{scoreSupplies} = \text{bonusProvided } Wh - \text{malusNotGiven } Wh \tag{7.2}$$

- *Ability to save energy* (Eq. (7.3)): avoid excessive consumption in relation to current availability at the node level. For each prosumer, the real consumption is compared with the computed ideal consumption (with a defined margin between 0 and 1):

$$\text{scoreConsumptions} = (1 + \text{ margin }) * \text{ idealConsumption } Wh - \text{consumed } Wh \tag{7.3}$$

- *Equity in distribution* (Eq. (7.4)): This implies that each prosumer with surplus energy supplies their peers without favoring one over another, by ensuring roughly equivalent response times. For each prosumer that has supplied energy, we consider the distribution of average response times of peer prosumers that have requested energy: this distribution is denoted by X in the equation. For each peer prosumer, we thus evaluate the average time taken to respond to its requests, and an arbitrary average response time is assigned if the peer prosumer has never been supplied:
  X = {average supply response time per inquirer}

$$\text{scoreEquity} = k1 * \text{ supplied } Wh * \frac{\bar{X} - k2 * \sigma_X}{\bar{X}} \tag{7.4}$$

($k1$ and $k2$ are constant coefficients between 0 and 1, X is the average of the X distribution, and σ_X is the standard deviation of the X distribution).

- *Ratio of "green energy" consumption* (Eq. (7.5)). For each prosumer, this score, which increases with renewable energy consumption, corresponds to the amount of renewable energy consumed divided by the total amount of energy consumed:

$$\begin{aligned}\text{scoreGreenEnergy} &= k1' * \text{ idealConsumption } Wh \\ &*(\text{ consumed RenewableW } h / \text{ consumed } Wh)\end{aligned} \tag{7.5}$$

($k1'$ is a constant coefficient between 0 and 1).

The overall credit score is simply equal to the sum of the scores assigned to each of the previous criteria (see Eq. (7.6)). If this score is negative, the prosumer will be penalized at the microgrid level. One improvement could be to weight each criterion score according to its relative importance.

$$\text{totalScore} = \text{scoreSupplies} + \text{ scoreConsumptions} + \text{scoreEquity} + \text{scoreGreenEnergy} \tag{7.6}$$

7.5.3.2 *Credit score computation by the regulator twin*

To assess the criteria and ultimately calculate the credit score (reward or penalty), the regulator twins regularly retrieve the following data:

- The current consumptions and supplies.
- The demanded power energy needs not yet satisfied: we distinguish between requests from consumers that already have penalties, because we consider that a prosumer is not obliged to supply a consumer that has penalties, i.e., a negative credit score.
- The power offered to meet demand (in new offers established to meet consumers' demands).
- The response time of new supplies.

This evaluation is performed at each event of a prosumer twin linked to the node (update or creation of production/demand). For each prosumer, the regulator twin calculates historical data, which is used to evaluate the various criteria assessed. The principle is to establish different evaluation data for each prosumer at regular intervals, then add them together over time to generate credit scores corresponding to the time elapsed since the last calculation.

This historical data in Wh (energy in Watt-Hour) is then aggregated by summing the energy scores in Wh obtained at the different evaluation timestamps (power multiplied by the time differential). Aggregation is performed every 20 assessment cycles (For example, approximately every minute if one assessment is performed every 3 seconds on average). For each prosumer, a credit score instance is created, corresponding to actual rewards or penalties, depending on whether they are positive or negative.

After each credit score calculation, the regulator twin submits the computed credit score table as an LSA property, the content of which is then retrieved by the prosumer twins. Each prosumer retrieves its own credit score, adds it to its cumulative internal credit score, which is then stored in its credit score attribute.

7.5.3.3 *Use of the credit score by prosumers*

A consumer can maintain a positive credit score while being supplied with energy by a producer. Depending on the amount of credit used, the credit score balance may be reset to zero. It should be noted that credit scores (reward or penalties) are modeled by the same attributes, whose value becomes negative in the case of penalties. A prosumer can, for example, absorb part of its penalty in supplying a consumer, or by paying more for electricity for a certain period of time. This way, the prosumer can pay off all its penalties.

7.5.3.4 *Taking credit scores into account in prosumer decision-making*

In processing energy requests, prosumers able to supply energy take specific decisions toward consumers when producing an energy offer. We propose two different alternatives, one consisting of excluding a penalized prosumer (i.e., ignoring its request), the other consisting of increasing the price ($/Wh) in the offer intended for that demanding prosumer.

7.5.3.5 *Scenario including a "free rider" prosumer*

To experiment with free riding behavior and the credit score-based solution, we implemented a microgrid scenario with three prosumers as depicted in Fig. 7.6.

- One of the three prosumers applies a "free riding" policy, which consists of requesting energy while refusing to supply energy to its partners.
- The other two prosumers apply an "altruistic" type policy: they agree to supply their partners, unless these partners have a negative credit score—when the variant that excludes "free riding" behavior is applied.

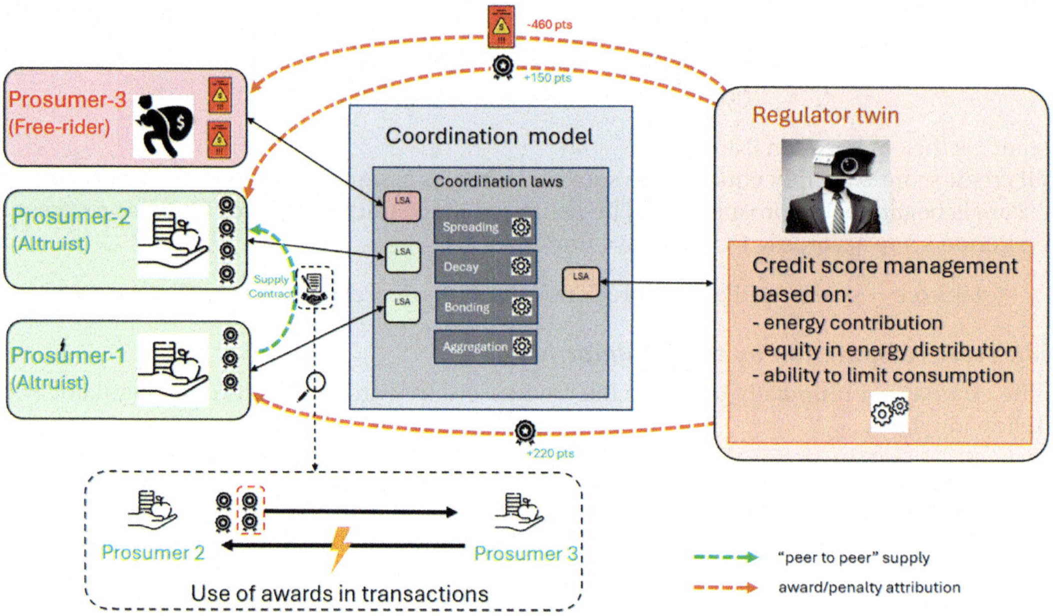

FIGURE 7.6 Reward scenario with two altruistic prosumers and one "free rider" prosumer.

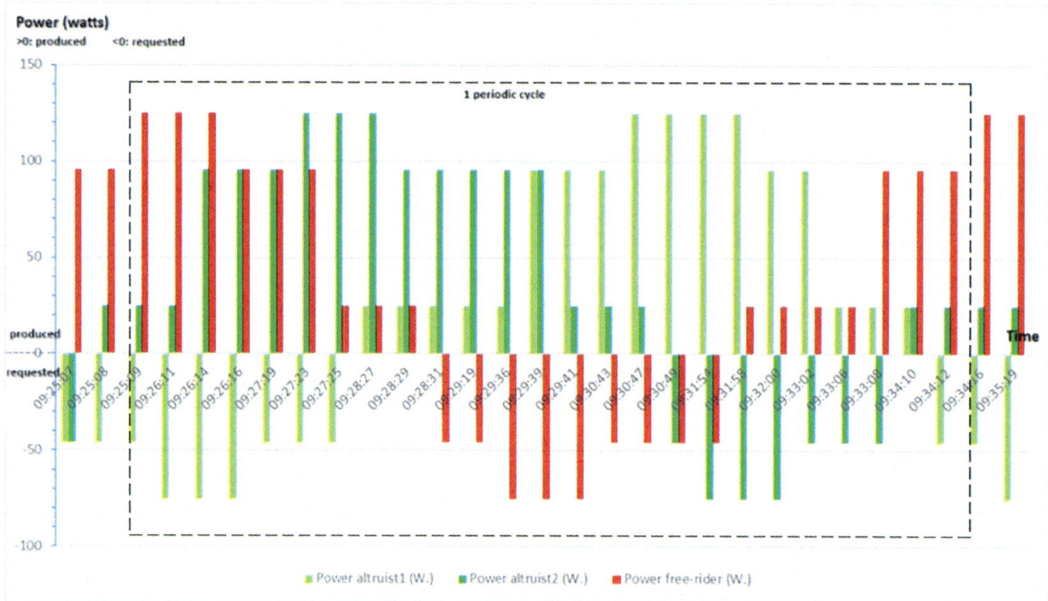

FIGURE 7.7 Hypotheses of the experiment: power produced by each prosumer during the "credit score" scenario.

Fig. 7.7 shows the output of each prosumer, which is positive when it produces energy and negative when it needs energy. The three prosumers have equivalent needs and production, but not at the same time. Each of them switches from consumer to producer and vice versa at different times, which makes it possible to test the fact that each of the three prosumers will be asked by the others to supply energy.

7.5.3.6 Results

Variant based on the exclusion of "free riding" prosumers: Initially, we applied the first variant, which consisted of no longer supplying a prosumer with a negative credit score balance. The first tests very quickly confirmed the exclusion of the "free rider" prosumer as a consumer. In this variant, we propose to assess the energy supplied to free riders

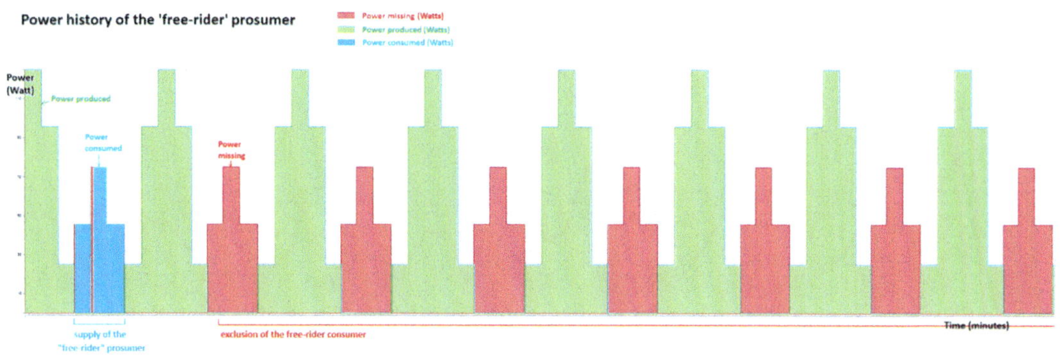

FIGURE 7.8 Consumption history of the "free rider" prosumer.

TABLE 7.3 Free rider scenario: Assessment of the energy missing to a free rider based on activation of credit score management.

	Credit score deactivated		Credit score activated	
Scenario duration (min)	Cumulative missing (Wh)	$KPI_{2.1}$: ratio missing/demanded	Cumulative missing (Wh)	$KPI_{2.1}$: ratio missing/demanded
20	0.08	1.21%	3.08	43.88%
60	0.35	1.42%	19.85	69.12%

despite their "abusive" behavior. For this, we define *KPI*2.1: ratio of missing energy for "free riding" prosumers.

$$KPI_{2.1} = \frac{\text{cumulative missing energy free riders } (Wh)}{\text{cumulative demanded energy free riders } (Wh)}$$

Given that the objective is to reduce "unfair" supplies to free rider prosumers, the value of $KPI_{2.1}$ must be maximized. Fig. 7.8 displays the consumption history of the "free rider" in the first hour of the scenario: rectangles filled in blue correspond to actual supplies, and those filled in red correspond to unsupplied energy demands. History confirms that the "free rider" was only supplied during the first cycle of the scenario. The first supply can be explained by the fact that the "free rider" credit score starts with a null value and falls below the tolerance threshold (which is slightly negative) only after the "free rider" has received its first penalties. We can see that the narrow red vertical stripes that appear in the first supply block are solely due to the time taken to generate a supply contract. The following red blocks confirm that all subsequent requests for supplies are refused by the other prosumers (including during the entire period of the scenario). This result confirms that a prosumer twin can exclude as a potential consumer any prosumer penalized for bad behavior. This could encourage a "free riding" prosumer to cooperate further by supplying energy to the other prosumers requesting energy. Table 7.3, which shows the ratio of energy not supplied for free riding prosumers (*KPI*2.1), confirms the effectiveness of positive credit score management in limiting energy distribution to free riders. Due to the fixed free rider detection time, the rate obtained is higher when the test duration is longer.

*Variant based on price adjustment:*Fig. 7.9 shows the price (\$/Wh) paid by each of the three prosumers during an experiment lasting several hours. The default price is set to 1.0 \$/Wh. These are averages taken over 5-minute periods. We can easily see that the "free rider" prosumer is very quickly penalized and pays 1.5–2 times the default price, whereas the "altruistic" prosumer is very quickly rewarded and pays prices varying between 0 and 0.7 times the default price. The more attractive discount obtained by "altruist 2" compared to "altruist 1" (see Fig. 7.9) can be explained by a more in-depth analysis of the credit utilization data: during the scenario, both collect roughly equivalent amounts of credit but "altruist 2" is supplied over shorter periods (because its supply contracts are interrupted earlier on average), which means that it has more credit left over for each new contract and therefore a better discount. In this variant, we assess the impact of credit score management on the average unit price (\$/Wh). For this, we define *KPI*2.2, which compares the unit price (\$/Wh) paid by free riders and non-free riders in the form of a ratio.

$$KPI_{2.2} = \frac{AVG_{\text{prosumer} \in \text{freeriders}}\{\text{rate}\}}{AVG_{\text{prosumer} \notin \text{freeriders}}\{\text{rate}\}}$$

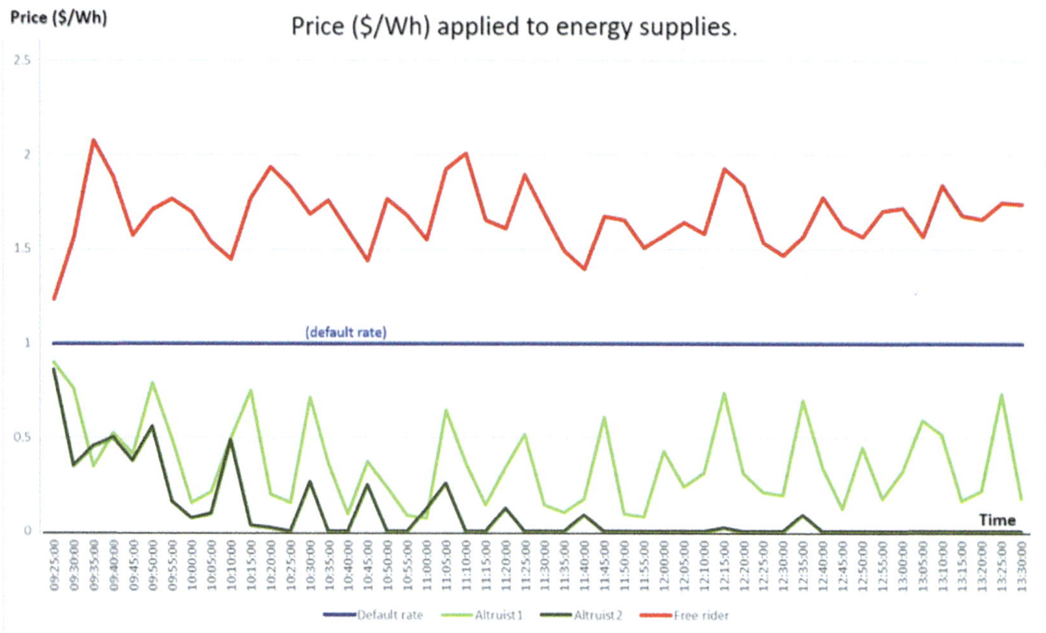

FIGURE 7.9 Price ($/Wh) applied to energy supplies. This is the energy price ($/Wh) paid by each of the three prosumers mentioned. The default price is set to 1.0 $/Wh (represented by the curve in blue). Free rider pays 1.5–2 times the default price; altruists receive discounts proportional to credit.

TABLE 7.4 Free rider scenario: Assessment of the energy price ($/Wh) paid by a free rider, based on activation of credit score management.

	Credit score deactivated			**Credit score activated**		
Scenario duration (min)	**AVG[a](price) non-free rider**	**AVG(price) free rider**	**$KPI_{2.2}$: ratio AVG(price) free riders/non-free riders**	**AVG(price) non-free rider**	**AVG(price) free rider**	**$KPI_{2.2}$: ratio AVG(price) free riders/non-free riders**
240	1.0	1.0	**1.0**	0.236	1.677	**7.106**

[a] *AVG: arithmetic mean*

Given that the objective is to penalize free riders in terms of pricing, the value of $KPI_{2.2}$ must be maximized. Table 7.4 shows a synthetic comparison of the average price in dollars per Watt-Hour ($/Wh) paid by free riders whether credit score management is applied or not: We can observe that when the credit score mechanism is activated, free riders pay on average seven times more per Wh than "altruistic" prosumers, who benefit from a reduction in the tariff. Overall, these results confirm that the use of credit score has a considerable impact on energy prices. This could be an effective way of penalizing "free riding" behavior and encouraging prosumers to improve their involvement in the smooth running of the microgrid community. It should be noted that the experiments were not conducted extensively enough to obtain a sensitivity analysis of the results obtained based on certain parameters, for example, by varying the credit tolerance threshold (below which penalties are applied) or the severity of penalties applied to the unit price (overload intensity).

7.5.4 Use of energy storage during a blackout

In a microgrid, the prevention of energy shortages is one of the main concerns and responds to many sources of worry, particularly when certain supplies are critical. To meet this challenge, we propose to model energy storage at the level of a microgrid and to integrate storage management into the behavior of Digital Twins. We consider local energy storage as a way of coping with a general shortage that may be caused by the temporary shutdown of an external supplier. More concretely, we propose two quite different solutions in terms of organization and implementation: either each prosumer manages its own storage exclusively for its own needs, or all the prosumers

TABLE 7.5 Comparison of private and common storage.

	Private storage	Common storage
Storage twin	Not running	Running
Digital twin that manages storage	All prosumer twins: each has its own battery (with a different capacity).	Only the "storage twin": the capacity of its battery must be sufficient to store the energy received from all the surrounding prosumer twins.
Transfer of energy surplus to storage	internal transfer: each prosumer transfers its surplus of energy exclusively to its private storage.	Peer-to-peer transfer: each prosumer transfers its surplus of energy exclusively to the storage twin, in the form of energy donation.
Transfer of energy from storage	"Internal" transfer: the energy supplied by private storage compensates for the prosumer's need. If the latter is sufficient, it fully compensates for its need: the prosumer is then self-sufficient in energy.	Peer-to-peer transfer. First, the storage twin produces energy using its battery in order to have sufficient power to meet energy demands. Secondly, the storage twin sends an energy offer to the energy requester. Finally, if the offer suits the requester, a new contract is issued, and the storage twin supplies the requester. This stage involves the common mechanism used for any exchange of energy between a consumer and a producer.
Advantage	This solution is simpler and more immediate because it is an internal transfer: there is no protocol and there is no need to transfer energy among the Digital Twins.	This solution enables energy to be stored for the whole community (higher capacity) and is fairer: in the event of a shortage, all consumers should be served in the same way, according to the capacity of the shared battery (there is no disparity in capacity between prosumers).
Disadvantages	Consumers with low-capacity batteries are at a significant disadvantage. Storage is generally less efficient because capacity is limited compared with a shared battery.	Energy transfers are more complex, particularly for supplying an energy demand: several steps are required, and in particular, the generation of an energy contract between the requester and the storage twin. In addition, all the prosumers have to cooperate to supply the joint storage battery.

in the microgrid share a common storage that can be used by all the prosumer members of the microgrid. These two solutions are refered as "private " and "connon" storage, respectively. The following section describes these two approaches.

7.5.4.1 Private and common storage approaches

In both approaches (private storage or shared storage), we propose to integrate energy storage management into the prosumer Digital Twin. The main difference is whether to manage storage in each twin for its own needs or in a single dedicated twin for everyone's needs. We describe how each approach works as follows:

- **Private storage**: At each execution cycle, each prosumer transfers its power surplus to its own storage within the limit of its storage capacity. A prosumer can use its own stored energy later if they need it, when it cannot be met from external supplies (i.e., from external suppliers or from other prosumers located in the microgrid).
- **Common storage**: At start-up, the coordination platform launches an additional prosumer twin which manages the common storage for the whole micro-network: we call it a "storage twin," even though at the implementation level it is also a prosumer twin. This prosumer does not consume energy and can produce energy to supply other prosumers if its battery is charged (i.e., if the energy balance of its battery is sufficient). Its storage battery is discharged at start-up, but it can receive surplus energy from all the prosumers on the microgrid. At each execution cycle governed by the coordination laws, each consumer transfers its surplus energy to the storage twin in the form of an "energy donation" via its LSA via the "DONATION" property: in this way, the storage twin gradually builds up a stock of energy. When certain prosumers need energy that cannot be supplied either by their peers or by an external supplier, the storage twin will produce energy to meet these needs using its storage battery. The storage twin will act as the other producers: the only difference is that it produces energy using its battery.

Table 7.5 summarizes the two approaches. Figs. 7.10 and 7.11 show an example of an application of these two approaches in a microgrid with three prosumers:

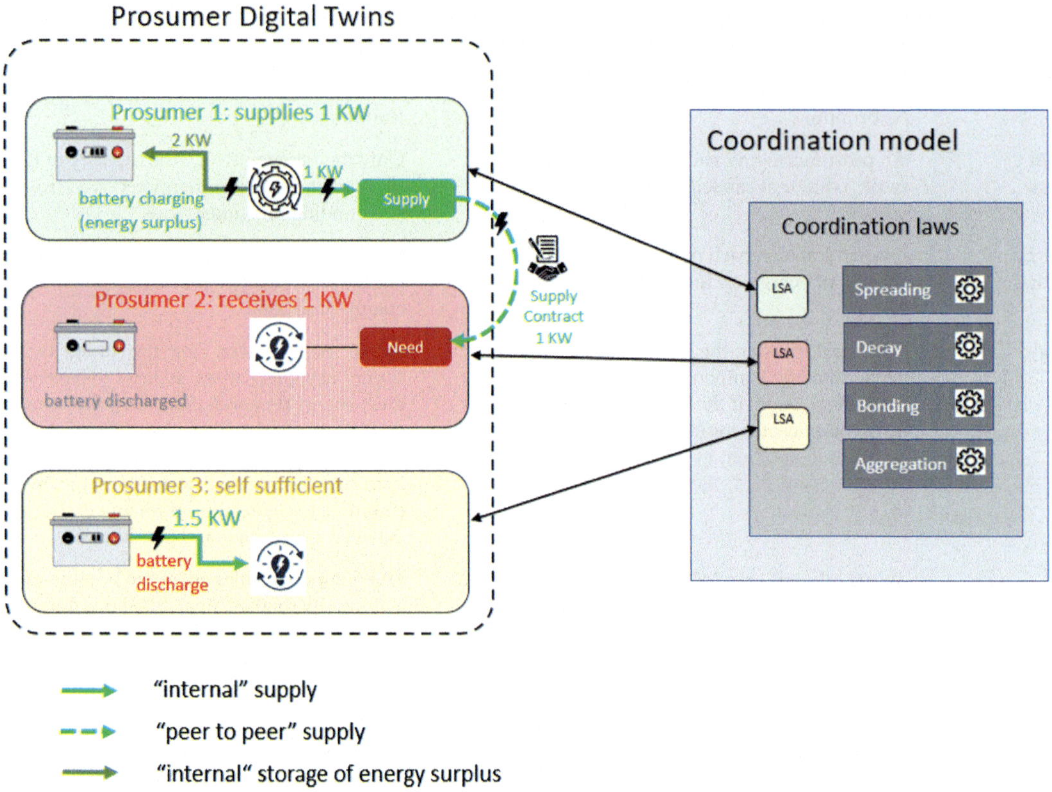

FIGURE 7.10 Private energy storage: Each prosumer manages its own battery for its own needs.

- Prosumer 1, which produces 3 kW, 1 kW of which is used to supply prosumer 2. In the case of private storage, the 2 kW surplus is used to charge its own battery, while in the case of common storage, this surplus is transferred to the "storage twin" in the form of a donation.
- Prosumer 2, which consumes 1 kW (received from prosumer 1). In the case of private storage, it cannot use its battery, which is discharged.
- Prosumer 3, which consumes 1.5 kW. In the case of private storage, consumer 3 is supplied by its own battery, and in the case of shared storage, by the "storage twin."

The "storage twin" is specific to the common storage approach and therefore only appears in Fig. 7.11.

7.5.4.2 Scenario-based assessment

We have implemented a scenario with three prosumer Digital Twins belonging to the same microgrid node, whose production fluctuates cyclically in the same way as for the credit score scenario, but whose average production gradually decreases in stages, until they demand more energy than they produce. Fig. 7.12 shows the production of each of the three local prosumers. An external supplier producing 1000 watts continuously is added: its production stops after 30 minutes, for 20 minutes. The shutdown period is shown in red in Fig. 7.13 (the width of the graph is not sufficient to represent this outage period in its entirety). The principle of the scenario is that the prosumers start from favorable conditions, enabling them to store energy easily, and that this stored energy is then used when they need energy: from a given moment, not only does the external supplier no longer produce, but the prosumers no longer produce enough to meet their needs. The two storage variants (private and common storage) are tested with this same scenario. In assessing the energy shortages, we have allowed a tolerance of 5 seconds when drawing up an energy contract. This time is effectively required for a prosumer to be supplied by the storage twin or another prosumer twin, which plays the role of producer in the supply contract.

$$KPI_{3.1} = KPI_{1.1} = \frac{\text{cumulative missing energy } (Wh)}{\text{cumulative demanded energy } (Wh)}$$
$$KPI_{3.2} = \text{cumulative consumption stored energy } (Wh)$$

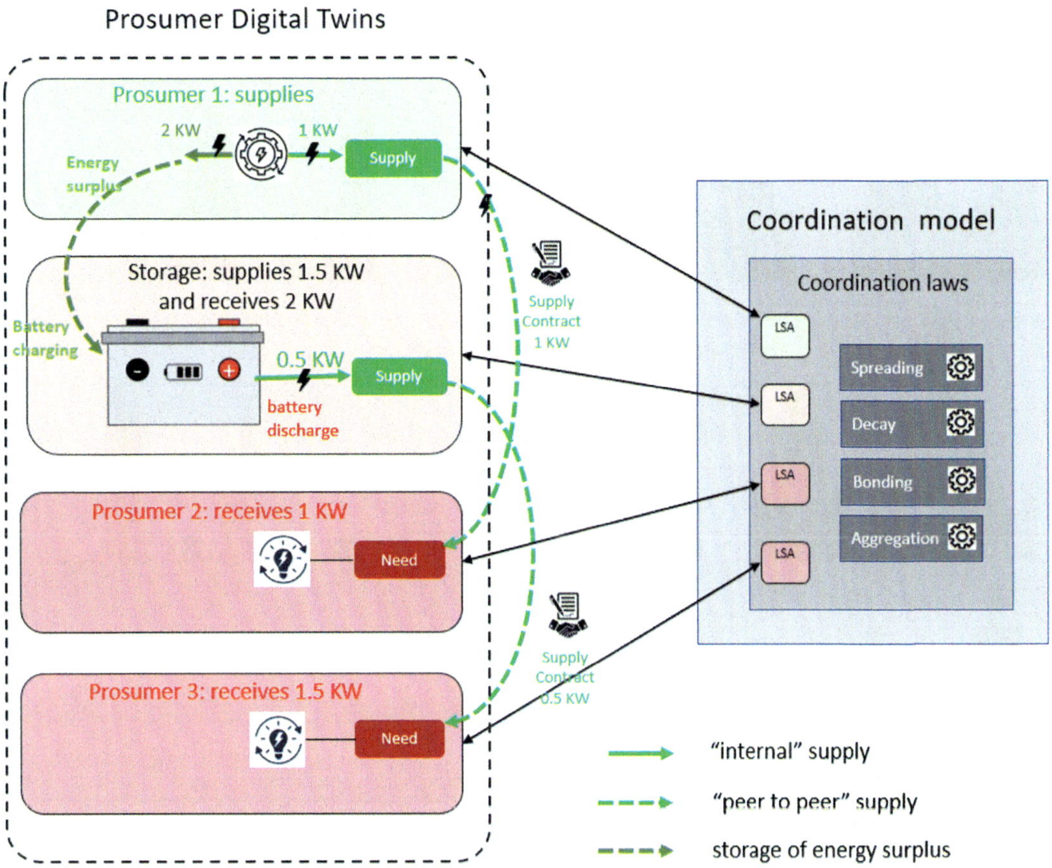

FIGURE 7.11 Common energy storage: The "storage twin" manages energy storage for the whole microgrid and uses the stored energy to meet the energy demands of surrounding prosumers.

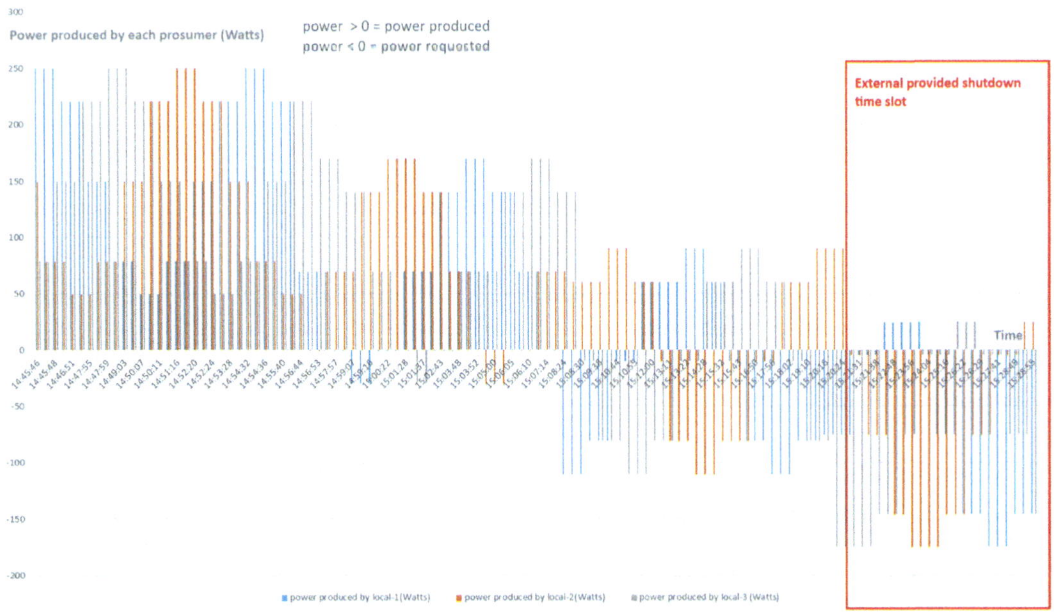

FIGURE 7.12 Hypotheses of the experiment: Power produced by each prosumer during the "storage" scenario.

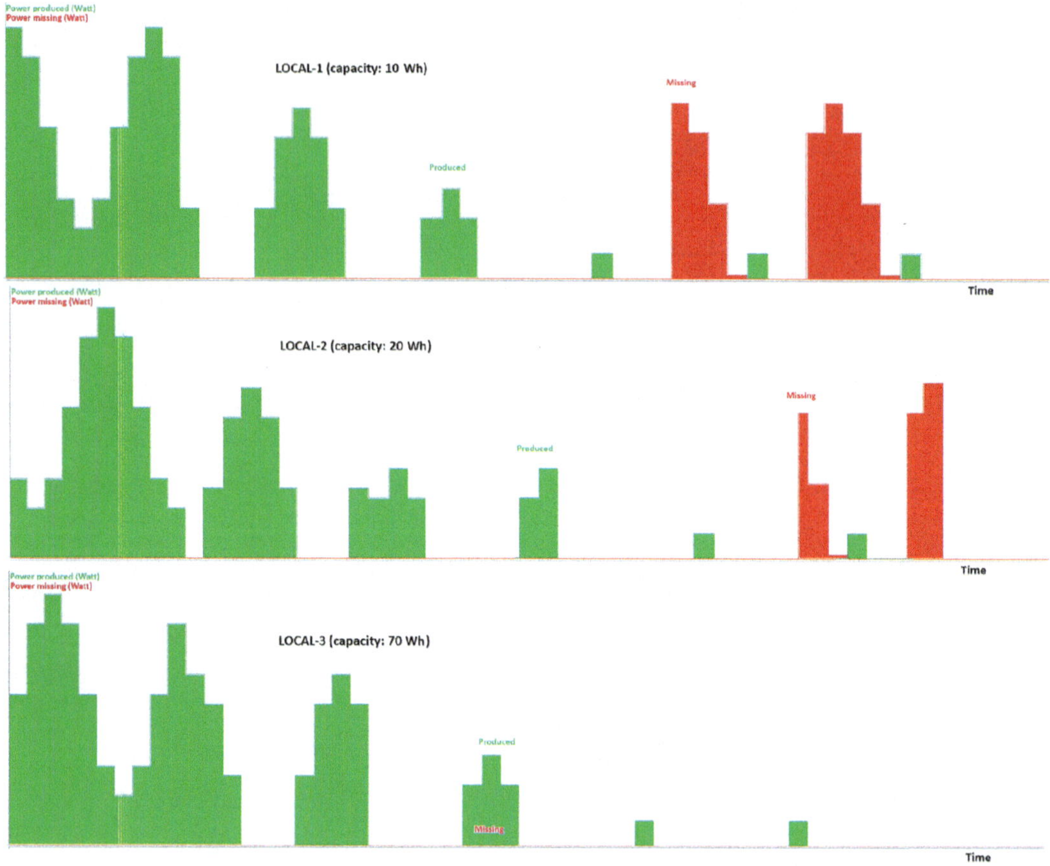

FIGURE 7.13 Private storage: History of the wattage produced and missing during the scenario, for each prosumer.

TABLE 7.6 Private storage: Evaluation of stored and missing energy per prosumer.

Prosumer	Capacity (Wh)	Cumulative missing (Wh)	$KPI_{3.1}$: Ratio missing/demanded	$KPI_{3.2}$: Cumulative storage used (Wh)
Local-1	10	41.91	39.63%	5.14
Local-2	20	15.79	17.52%	16.28
Local-3	70	0.00	0.00%	36.22

In the same way as $KPI_{3.1}$, $KPI_{1.1}$ should be minimized and given that the objective is to store as much energy as possible, the value of $KPI_{3.2}$ must be maximized.

7.5.4.3 Assessment of private storage

In this variant, each prosumer manages its own battery: we have deliberately given them very different capacities: 10, 20, 70 Wh for each of the three. This allows us to test the impact of inequality of storage between the three prosumers.

Fig. 7.13 shows the history of the power produced and missing for each prosumer. The graph clearly indicates that the first two prosumers have energy missing (shown by the staircase steps in the red curve during the last third of the scenario, which corresponds to the period of the external supplier's outage). The third prosumer, which has a much higher storage capacity (70 Wh), has no energy losses: this absence of energy shortfall is a direct consequence of the battery capacity. This also indicates that the battery can avoid or reduce shortages during an outage, within the limits of its storage capacity. The second prosumer, which has a higher capacity than the first one, has relatively less energy missing than the first. Table 7.6 shows the cumulative stored and missing energy scores for each consumer:

TABLE 7.7 No storage: Evaluation of missing energy per prosumer.

Prosumer	Capacity (Wh)	Cumulative missing (Wh)	$KPI_{3.1}$: Ratio missing/demanded	$KPI_{3.2}$: Cumulative storage used (Wh)
Local-1	0	25.30	69.40%	0
Local-2	0	26.44	79.83%	0
Local-3	0	32.33	74.29%	0

TABLE 7.8 Common storage: evaluation of missing energy per prosumer. The storage twin is also listed, as it indicates the total storage energy used to supply the three prosumers.

Prosumer	Cumulative missing (Wh)	$KPI_{3.1}$: Ratio missing/demanded	$KPI_{3.2}$: Cumulative storage used for supply (Wh)
Local-1	0.51	1.41%	0.00
Local-2	0.75	1.88%	0.00
Local-3	0.63	1.56%	0.00
Storage	0.00	0.00%	126.16

this confirms that the amount of storage energy used depends on private storage capacity, and that not all prosumer twins are in the same boat when it comes to avoiding energy shortages.

Table 7.7 shows the cumulative missing energy scores for each consumer when storage is deactivated: the total missing is equal to 84.07 Wh compared with 57.7 Wh when storage is activated. This confirms the overall reduction in missing energy, but it is surprising to see that the cumulative missing energy of the first consumer decreases when storage is deactivated. Actually, when storage is deactivated, prosumers cannot use their energy surplus for their own storage: in this case, they use it to supply other prosumers when they need energy, which tends to even out the scores for missing energy among consumers. The third prosumer, which managed not to run out of energy in the private storage scenario, increases its shortage, while the first prosumer, which experienced the greatest shortage in the private storage scenario, tends to reduce it. The result of the third prosumer is due to the latency introduced by the time needed to generate and validate contracts with the storage twin. This processing time does not exist in the private storage version because the prosumer's personal battery powers the prosumer directly without the need for a contract. This occurs, for example, for the third prosumer ("local 3"), for whom the cumulative missing energy is perfectly zero in the private storage version (see Table 7.6), whereas it is 0.63 Wh in the shared storage version (see Table 7.8). On a more general note, we observe a trade-off between private storage and community sharing. Private storage leads to inequality due to the different storage capabilities of the various prosumers. Instead, community sharing ensures higher overall efficiency, since energy distribution is optimized. It complexifies the overall process and clearly leads to different individual outcomes, from full to partial energy supply.

7.5.4.4 Assessment of common storage

We have run the common storage scenario under the same conditions, with the same three local prosumers and the same external supplier. The storage twin has a 200 Wh battery: this twin is automatically initialized when the coordination platform is started up (only if common storage is activated). In this variant, we consider that the shared storage can be equipped with a battery of much greater capacity, thanks to the shared investment in the microgrid.

Fig. 7.14 shows the history of the power produced and missing for each prosumer in the same way as for the private storage scenario. We can see here that the energy shortages are almost non-existent (we can only see red peaks appearing in places in the zone where the external supplier's outage is located). These energy shortages can be explained by the start-up time of a supply, which can sometimes exceed 5 seconds (this is the time it takes for an energy contract to be made with the storage twin). Table 7.8 shows the cumulated missing energy for each prosumer. The scores are all below 1 Wh, and the ratios are all between 1% and 2%; they confirm that the missing energy is only due to the time needed to start a new supply and that common storage is the most appropriate solution to avoid shortages for all prosumers. In addition, the total energy used from common storage is 126.16 Wh, which is

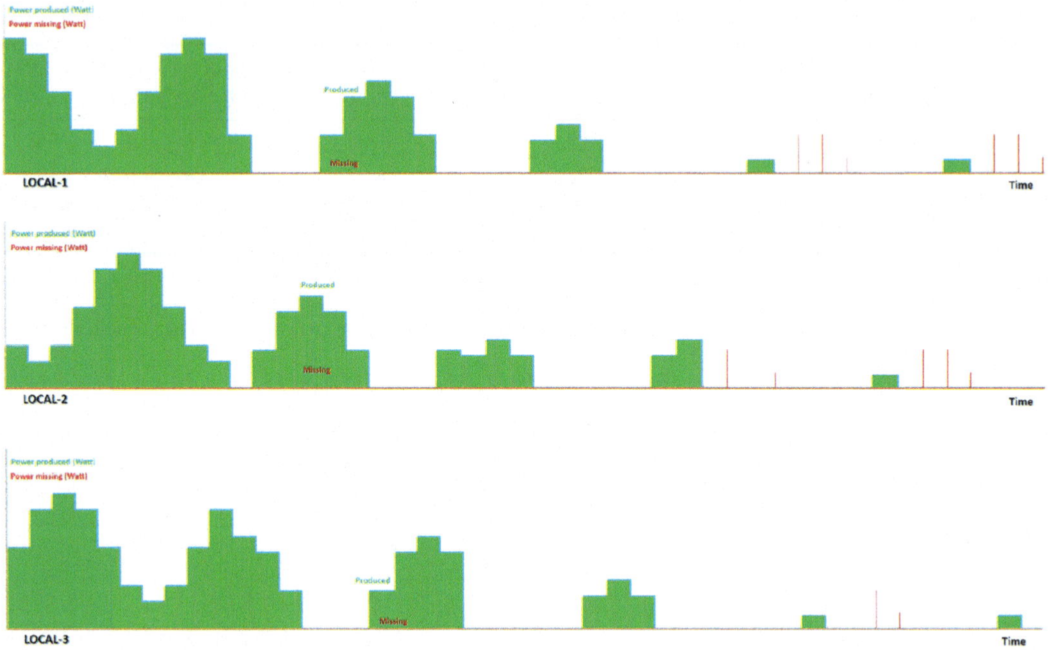

FIGURE 7.14 Common storage: history of the wattage produced and missing during the scenario, for each prosumer. The common storage scenario shows low energy shortage and maximization of energy usage.

TABLE 7.9 Common storage with reduced capacity and critical requests from "Local-1" prosumer: Evaluation of missing energy per prosumer.

Prosumer	Cumulative missing (Wh)	$KPI_{3.1}$: Ratio missing/demanded	$KPI_{3.2}$: Cumulative storage used for supply (Wh)
Local-1	0.11	0.31%	0.00
Local-2	10.02	22.73%	0.00
Local-3	6.18	14.65%	0.00
Storage	0.00	0.00%	110.35

much higher than the total energy used from private storage (57.64 Wh, which corresponds to the sum of cumulative storage used by the three prosumers in Table 7.6).

However, the private storage solution allows the energy shortage to be reduced strictly to zero (provided that the battery capacity of each prosumer is sufficient), which is not the case for shared storage, because there will always be slight shortages due to the time needed to start-up new contracts with the storage twin.

7.5.4.5 *Assessment of supply from storage for critical requests*

This is a more critical variant of the common storage scenario in which the total energy is not sufficient to supply all consumers, but for which certain needs are a priority and must therefore be supplied, whatever the cost. We have used the same conditions as in the previous scenario, but with the following modifications: we have set the priority level for requests from the first prosumer to the highest level, and we have reduced the capacity of the common battery to 70 MH, so that it can no longer store enough energy to meet all needs. High priority requests in LSA are identified by the "priority" attribute, which specifies the priority level and can take one of the values "LOW," "MEDIUM," or "HIGH."

Table 7.9 shows the accumulated missing energy obtained for each prosumer ("Local-1," which has a high priority level, is shown in red). The results confirm that "local-1" is always served with a shortfall of almost zero, while the other prosumers, which have a lower priority, have energy shortfalls, which can be explained by the limits of the

TABLE 7.10 Integration of salient attributes attached to the **"Network"** topic.

Attribute (yes/no value)	Component concerned	Roadmap: Implementation to be done	Expected impact
Connection to the power grid?indicates if the microgrid also accepts to exchange energy with nodes that are located outside the microgrid.	Prosumer Twin	If the value is "no," a prosumer twin will not consider energy objects (demand, offer, contract) coming from a node located outside the microgrid. We also need to implement and configure a node address filter in the configuration of the coordination platform to precisely define the perimeter of the microgrid.	If "no," this increases community self-sufficiency and resilience, but this also reduces flexibility during deficits.
Energy mix? indicates if the microgrid accepts or not to use also non-renewable energies.	Prosumer Twin	If the value is "no," a prosumer twin will not consider offers whose energy source has a high or medium impact on the environment. (The level of impact is already implemented in the objects' modeling of energy flows).	If "no," this reduces the environmental impact but also reduces flexibility during deficits.
Dynamic pricing? It is a question of whether (or not) the price of electricity varies over time according to supply and demand.	Prosumer Twin	When it establishes an energy offer to meet a demand, a prosumer twin will apply a different price calculation depending on this choice of this attribute: either a dynamic tariff which depends on the quantity of energy available and the quantity of energy demanded from the microgrid, or a fixed tariff which depends only on the time slot (day/night). The idea of the dynamic tariff is to be able to adjust the proposed tariff according to supply and demand, which can encourage prosumers to review their range of consumption to optimize costs and avoid better distribution according to time slots.	This encourages users to take responsibility by reducing consumption peaks. It also helps users to save money.
Dynamic pricing and time slots choice Avoiding the tragedy of the commons with dynamic pricing and adaptive time slots for energy consumption.	Prosumer Twin	Prosumer twins apply dynamic pricing strategies based on energy availability. Prosumers decide to shift energy consumption to later time slots based on availability and price (Feature implemented, see Section 7.5.2).	This encourages users to take responsibility by avoiding consumption peaks and spreading the consumption over time. It also helps users to save money.

battery's capacity, which was deliberately reduced for this test. More generally, these results show that the Digital Twins are able to use the available storage energy in such a way as to best serve the most important demands.

7.6 Roadmap: Integration of salient attributes in the decision-making of Digital Twins

In this section, we present the next stage of our work, which will consist of integrating some of the salient attributes resulting from the research carried out by HES-SO Valais. As these attributes have not yet been implemented, we designed the adjustments to be made and provided them in the form of tables (one per topic): for each attribute, we identify the software updates to be made in the Digital Twin (or software components) concerned, as well as the expected impacts for a microgrid community.

7.6.1 Identification of community profile salient attributes by HES-SO Valais

Here, we present some attributes that Fragnière et al. (2024) have defined and then used to design profile maps for card-based surveys that simulate different community configurations. The cards were tested in focus groups and refined to reflect real-world preferences, to identify socially acceptable and sustainable microgrid models. Salient attributes relate to: (1) *Network:* energy infrastructure (self-sufficient smart grid vs grid connected to the regional grid;

TABLE 7.11 Integration of salient attributes attached to the **"Community Life"** topic.

Attribute (yes/no value)	Component concerned	Roadmap: Implementation to be done	Expected impact
Regulation? This consists of imposing quotas based on the number of people living in the household and the size of the flat/house.	Regulator twin	If set to yes: the twin regulator will reinforce the assessment of each prosumer's consumption allocation, taking into account two other characteristics: the surface area of the house/apartment and the number of people living in the house/apartment: the lower these two parameters are, the lower the tolerated consumption limits will be, which will influence the calculation of the credit score allocated to a prosumer according to its consumption.	This encourages users to cooperate more to achieve a better distribution of consumption among prosumers.
Regulation, dynamic pricing, or prosumer exclusion Overcoming "free riding" behavior	Regulator twin prosumer twin	The regulator twin assesses prosumers' behavior and calculates credit scores (reward or penalties). Prosumer twins decide to either exclude a free rider prosumer or to increase the energy price. (Feature implemented, see Section 7.5.3).	This encourages users to cooperate more and to consume energy in a fair way.
Private and common energy storage Ensuring resilience during a blackout	Prosumer Twin	Prosumers twins may have the possibility to store the energy they produce in a private storage for their personal use. Alternatively, they can choose to store that energy in a common storage, used in an optimized manner in case of energy shortage. (Feature implemented, see Section 7.5.4).	This reinforces the resilience of the community in case of an energy shortage.

favoring or not renewable sources of energy); pricing mechanisms (dynamic prices and peak/off-peak pricing or not), (2) *Information*: information sharing (periodic or systematic updates; level of sharing of data among the community), (3) *Community life*: community engagement (frequency of meetings); regulations (imposing fixed or varied quota); and governance (internal or external management). It should be noted that we only list here the attributes that we can integrate into our model.

7.6.2 Integration study of community profile salient attributes

To date, we have not yet translated these attributes at the implementation level and in the Digital Twins' behaviors, but we have identified the adaptations to be made in the software components for each attribute concerned. In the following three tables, we show how implementation should proceed in order to integrate each of the salient attributes selected. For ease of reading, they are separated by topic: Table 7.10 concerns the **"Network,"** while Table 7.11 concerns the **"Community life"** and Table 7.12 the **"Information"**.

It should be noted that in addition to the implementations indicated in each table, all of these attributes will also need to be integrated into the platform's configuration so that they can be easily modified from the platform's web administration.

7.6.3 Management of two levels of flexibility

The principle is that the regulator uses two different levels of flexibility in assessing consumer consumption: a normal level and a more flexible level for holidays or periods of shortage when urgent supplies are needed. In fact, the lack of flexibility can block supply for critical needs or reduce the motivation of members, especially if they have made prolonged efforts over a long time to limit their consumption. Today, the level of flexibility is not yet implemented, but we have also identified the component to be updated. As with the integration of the "regulation"

TABLE 7.12 Integration of salient attributes attached to the **"Information"** topic(*): The REST server is the web server attached to the coordination platform. It provides reports on electricity consumption and production at the level of a prosumer or the microgrid.

Attribute (yes/no value)	Component concerned	Roadmap: Implementation to be done	Expected impact
Data sharing with all members? enables or disables the transfer of production and consumption data to all other microgrid members. The prosumer names are anonymized.	REST server(*)	The visibility restrictions are already implemented on the object exchanged among prosumer twins, but we must adapt it on the server of the web application, which generates a history of consumption. If set to no, each prosumer user of the microgrid will have visibility only on its own consumption and production history, and only the administrator of the web application will have visibility on the consumption/production history of all prosumers. If set to yes, the visibility on all consumption/production will be activated for all members of the microgrid. The names of all prosumers must be anonymized on all reports.	This encourages highly motivated prosumers to cooperate, but it may discourage others who might feel they are being monitored.
Consumption comparison? Enables or disables the display of production and consumption data of "direct" neighbors. The prosumer names are anonymized.	REST server	This attribute refines the previous visibility criterion by adding an intermediate level of visibility, which applies only to the prosumer's neighbors: data is only accessible within a precise list of prosumer neighbors to be defined. This concerns the prosumers attached to the same node, and possibly including those connected to the neighbor nodes that are directly connected to the local prosumer's node.	Disabling the visibility of direct neighbors' data allows the microgrid to mitigate the effects of the previous option and thus achieve better cooperation if some prosumers feel overly monitored.
Privacy Ensuring confidentiality rules in relation to actual data consumed, produced or involved in energy contracts.	REST Server Regulator Twin	Visibility restrictions are implemented in the coordination platform. Information on energy contracts is not visible for all prosumers. Only the regulator twin has access to actual data (Feature implemented, see Section 7.5.1).	Respects prosumers' privacy toward each other.

salient attribute, the integration of flexibility will have to be considered in the calculation of the consumption limit applied in the allocation of the credit score (mentioned in Table 7.11). The flexibility can be simply integrated as another input sent to the function that calculates the acceptable consumption limit (in addition to the two prosumers' home characteristics mentioned: the number of people and the apartment surface). In this way, it will be possible to raise the consumption quotas according to the flexibility level.

7.7 Conclusion

This work was initially motivated by the need to develop a *framework for social acceptance by design* in smart grid systems. As energy infrastructures become increasingly decentralized and reliant on active user participation, it is essential to embed social values-such as trust, fairness, and cooperation - into the core of technological design. Our agent-based coordination model addresses this challenge by enabling advanced functionalities such as *peer-to-peer energy exchange, load balancing*, and *resilience during blackouts*, while also incorporating algorithmic mechanisms to mitigate key social acceptance barriers, including *privacy concerns, free riding*, and the *tragedy of the commons*. The algorithms implemented, which handle data protection, dynamic pricing, free rider penalization, and storage management, are designed within an ongoing interdisciplinary research project, in close collaboration with social scientists, ensuring that the technical solutions align with societal expectations and user behavior. The results of the simulations showed that these algorithms successfully overcame the main obstacles to social acceptance. The model is not only technically robust but also socially informed, aiming to foster long-term engagement and trust within energy communities.

Our coordination model and algorithms are designed to ensure explainability, auditability, and trust. Every decision, whether related to prices, rewards, or penalties, follows well-defined rules, can be systematically logged, and remains fully traceable throughout the process. Beyond social acceptance, we also developed further complementary algorithms supporting predictions among prosumers in a gossip-based, decentralized, and emergent manner (Glass & Di Marzo Serugendo, 2025) and provided a complete, comprehensive description of the whole approach (Glass, 2025).

The salient attributes identified have not all yet been incorporated into Digital Twins. While the model conceptually integrates critical dimensions such as *network configurations, information sharing* protocols, and *regulatory* frameworks, these features are not yet fully implemented in the current prototype. Future work will focus on operationalizing these aspects to further enhance the realism and applicability of the system. This includes the following steps: (1) integrating the salient attributes previously identified, (2) refining data governance mechanisms, (3) formalizing community rules, and (4) expanding the model's adaptability to diverse regulatory environments.

To enable future scalability, it is essential to reflect on how the proposed framework can be generalized beyond the Geneva pilot. Within the LASAGNE project, additional pilots in Sweden and other regions are planned, sharing similar characteristics while introducing new contextual challenges. In the European Union, adaptation will require compliance with EU grid initiatives or heterogeneous regulatory frameworks such as General Data Protection Regulation (GDPR), emphasizing interoperability, consumer rights, and stringent data privacy standards. Cultural factors must also be considered, as salient features identified for Switzerland may differ significantly elsewhere. These variations call for a careful analysis of local practices and expectations, ensuring that technological components and governance mechanisms are aligned with the specific regulatory and cultural context. By bridging the gap between technical innovation and social integration, this work contributes to the development of *resilient, user-centered smart grids* that are not only efficient but also widely acceptable and sustainable.

References

Alotibi, B., Alarifi, N., Abdulghani, M., & Altoaimy, L. (2019). Overcoming free-riding behavior in peer-to-peer networks using points system approach, Procedia Computer Science. In *The 10th international conference on ambient systems, networks and technologies (ANT 2019) /The 2nd international conference on Emerging Data and Industry 4.0 (EDI40 2019) / affiliated workshops: 151* (pp. 1060–1065).

Barricelli, B. R., Casiraghi, E., & Fogli, D. (2019). A survey on digital twin: Definitions, characteristics, applications, and design implications. *IEEE Access, 7*, 167653–167671.

Ben Mahfoudh, H., Di Marzo Serugendo, G., Naja, N., & Abdennadher, N. (2020). Learning-based coordination model for spontaneous self-composition of reliable services in a distributed system. *International Journal on Software Tools for Technology Transfer, 22*, 417–436.

Castelli, G., Mamei, M., Rosi, A., & Zambonelli, F. (2015). Engineering pervasive service ecosystems: The SAPERE approach. *ACM Transactions on Autonomous and Adaptive Systems (TAAS), 10*, 1–27.

Di Marzo Serugendo, G., Cutting-Decelle, A.-F., Guise, L., Corme-nier, T., Khan, I., & Hossenlopp, L. (2022). Digital twins: From conceptual views to industrial applications in the electrical domain. *Computer, 55*, 16–25.

Ebrahimi, A. (2019). Challenges of developing a digital twin model of renewable energy generators. In *IEEE 28th international symposium on industrial electronics (ISIE)* (pp. 1059–1066). Vancouver, Canada: IEEE.

Fragnière, E., Sandoz, S., Abdenadher, N., & Moresino, F. (2024). Developing card-based surveys: Focus group insights for smart grid prosumer optimization. In *12th international conference on smart grid (icSmartGrid)* (pp. 579–581). Setúbal, Portugal: IEEE.

Fragniere, E., Sandoz, S., Abdenadher, N., Moussa, M., Di Marzo Serugendo, G., & Glass, P. (2023). Fostering "energy communities": An ethnographic-seci approach to user-centered residential micro-smart grid adoption. In *11th international conference on smart grid (icSmartGrid)* (pp. 01–05). Paris, France: IEEE.

Fragniere, E., Sandoz, S., Abdenadher, N., Moussa, M., Di Marzo Serugendo, G., & Glass, P. (2023). Improving the social acceptability of microgrids. In *IEEE renewable energy and sustainable e-mobility conference (RESEM)* (pp. 1–5). Bhopal, India: IEEE.

Glass, P. (2025). *Coordination model and Digital Twins for managing energy consumption and production in a smart grid, Doctoral thesis.* University of Geneva. doi:10.13097/archive-ouverte/unige:186221.

Glass, P., & Di Marzo Serugendo, G. (2023). Coordination model and digital twins for managing energy consumption and production in a smart grid. *Energies, 16*.

Glass, P., & Di Marzo Serugendo, G. (2025). Gossip coordination mechanism for decentralised learning. *Energies, 18*, 2116.

Krech, D., & Crutchfield, R. S. (1948). *Theory and problems of social psychol-ogy*: 36. McGraw-Hill.

Marah, H., & Challenger, M. (2022). Intelligent agents and multi agent systems for modeling smart digital twins. In *13th international workshop on engineering multi-agent systems, EMAS*. MI, USA: Springer Nature.

Senyapar, H. N. D., & Bayindir, R. (2023). The research agenda on smart grids: Foresights for social acceptance. *Energies, 16*, 6439.

Tzanis, N., Andriopoulos, N., Magklaras, A., Mylonas, E., Birbas, M., & Birbas, A. (2020). A hybrid cyber physical digital twin approach for smart grid fault prediction. In *IEEE conference on industrial cyber-physical systems (ICPS): 1* (pp. 393–397). Tampere, Finland: IEEE.

Valkering, P., Laes, E., Kessels, K., Uyterlinde, M., & Straver, K. (2014). How to engage end-users in smart energy behaviour? *EPJ Web of Conferences, 79*, 04003.

Wolsink, M. (2012). The research agenda on social acceptance of distributed generation in smart grids: Renewable as common pool resources. *Renewable and Sustainable Energy Reviews, 16*, 822–835.

Wooldridge, M. (2009). *An introduction to multiagent systems* (2 ed.). Wiley.

Zambonelli, F., Omicini, A., Anzengruber, B., Castelli, G., De Angelis, F. L., Serugendo, G. D. M., Dobson, S., Fernandez-Marquez, J. L., Ferscha, A., Mamei, M., Mariani, S., Molesini, A., Montagna, S., Nieminen, J., Pianini, D., Risoldi, M., Rosi, A., Stevenson, G., Viroli, M., … Ye, J. (2015). Developing pervasive multi-agent systems with nature-inspired coordination. *Pervasive and Mobile Computing, 17*, 236–252.

CHAPTER

8

Fault detection in modern power systems

Guillaume Guerard

De Vinci Higher Education, De Vinci Research Center, Paris, France

8.1 Introduction

Conventional power system protection, a discipline built on impedance calculations and overcurrent relays, was engineered for a grid dominated by synchronous generators. This legacy architecture provided high rotational inertia and, critically, delivered predictable, high-magnitude fault currents that formed the unambiguous signal for protection schemes. The global energy transition fundamentally disrupts this paradigm. The widespread integration of inverter-based resources (IBRs), such as solar and wind power, alters fault physics in ways that challenge the foundational assumptions of these legacy systems (Mobashsher et al., 2025).

The core of the problem lies in the protective behavior of power electronic converters. To protect their sensitive internal components, IBRs actively limit fault current contributions, suppressing them from a clear 5–10 per unit (p.u.) signal in traditional systems to a transient-rich, low-level response of around 1.1–1.5 p.u. This behavior not only blinds traditional overcurrent relays but also distorts the voltage and current phase relationships used by more sophisticated impedance-based protection, risking a failure to operate. Compounded by non-sinusoidal waveforms and complex, bidirectional power flows in distributed energy resource (DER) networks, these dynamics compel a transition from static, threshold-based protection to adaptive, data-driven strategies to ensure system integrity (Alcala-Gonzalez et al., 2021).

Historically, grid protection has been evaluated by three key performance indicators (KPIs): detection latency (time to classify a fault), localization error (precision of location), and false trip rate. The challenge today is not the novelty of these metrics, but the severe degradation in performance and the intensified trade-offs between them. Achieving low-latency is harder when the signal is weaker, and minimizing false trips is more difficult when fault signatures resemble normal operational transients. Overcoming this requires a new philosophy built on high-resolution, time-synchronized data from sensors like phasor measurement units (PMUs) and other Internet of Things (IoT) devices (Yang et al., 2024). These technologies provide the rich spatiotemporal data needed to see the subtle, wide-area patterns of modern faults.

Central to this transformation are machine learning (ML) and deep learning (DL) algorithms, which are adept at identifying the complex, non-linear fault signatures that elude traditional models. However, ML models are purely data-driven and lack innate physical understanding. This is where they are complemented by Digital Twin (DT) frameworks—high-fidelity, real-time virtual representations of physical grid assets (Huang et al., 2024). In this paradigm, the DT serves not as a passive simulator, but as an active validation engine. It provides the physics-based context and topological constraints that ground the ML inferences, ensuring that detection and classification results are physically plausible. This synergy—ML for rapid pattern detection and the DT for physical validation and "what-if" analysis—creates a robust framework for proactive maintenance, reliable decision-making, and the safe testing of novel protection schemes before deployment.

As power grids evolve into renewable-dominant, bidirectional ecosystems, this transition from static rules to intelligent, data-driven methodologies is not an optional upgrade but a fundamental requirement for maintaining reliability. This chapter presents a structured framework for designing and implementing these next-generation systems. Its contributions include establishing a clear connection between fault signatures and their application across distinct domains, providing practical deployment checklists, examining ML as a core technology, presenting

Digital Twin Technology and Smart Grid.
DOI: https://doi.org/10.1016/B978-0-443-36664-2.00012-5

an operationalization roadmap using DTs, and addressing the intersecting challenges of cybersecurity and operator trust.

The chapter is organized to follow a logical progression from problem formulation to solution. Section 8.2 establishes the fundamental signals indicative of a fault. Section 8.3 applies this knowledge to the unique challenges within microgrids, transmission and distribution networks, and generation facilities. Section 8.4 presents a unified overview of data-driven methods for interpreting modern fault data. Section 8.5 details the operationalization of these models using DT. Recognizing that advanced systems introduce new vulnerabilities, Section 8.6 addresses the critical concerns of security and trust. Finally, Section 8.7 explores future trajectories, including self-healing grids and the need for industry-wide standardization.

8.2 Fundamentals, fault signals, and indicators

Reliable fault detection is contingent upon identifying electrical and non-electrical anomalies that signal either incipient deterioration or imminent failure. While conventional protection focused on high-magnitude disturbances such as short circuits, contemporary methodologies integrate a wider array of signatures to enable proactive, condition-based intervention (Ibrahim et al., 2024). This section establishes a taxonomy of these fundamental fault signals and maps them to the physical grid components they affect.

8.2.1 A taxonomy of fault signatures

Faults and degradation processes manifest as distinct signatures captured by monitoring systems. These indicators constitute the primary evidence for detection and diagnostic algorithms and are broadly classified by their temporal characteristics. The first set includes high-speed electrical signatures (developing in microseconds to seconds) that are essential for immediate, real-time fault protection. The second comprises slow-evolving physical indicators (developing over weeks to years) that serve as precursors to failure and are foundational for long-term condition monitoring.

- **Voltage sags and swells:** While transient voltage deviations remain principal indicators of faults, their morphology is changing. In traditional systems, an unbalanced fault creates a distinct, asymmetric voltage sag or open circuit (Zhang et al., 2023a). In IBR-rich networks, however, fast-acting inverter controls can cause the fault to appear more symmetric across all three phases, potentially confusing classification algorithms that rely on traditional voltage asymmetry patterns (Farrokhabadi et al., 2019).
- **Current surges and sequence components:** A rapid increase in current magnitude was the hallmark signature of a low-impedance fault. In IBR-dominated systems, this surge is actively limited by converter controls. More importantly, the negative-sequence current—a robust indicator of unbalanced faults in synchronous systems—is often suppressed by inverter control logic. This necessitates more sensitive techniques that analyze not just magnitude, but subtle changes in positive-sequence phase angles or high-frequency components.
- **Waveform distortion and harmonics:** A critical signature unique to modern grids is the distortion of the fundamental 50/60 Hz waveform. IBRs inject a characteristic harmonic spectrum during normal operation. A fault, however, causes a sudden, dramatic shift in this spectrum and can introduce high-frequency transients (in the kHz range). Analyzing the evolution of this harmonic content provides a fast and data-rich signal for detecting and classifying faults that produce minimal fundamental-frequency current change.
- **Frequency deviations:** In a system with low rotational inertia—a direct consequence of displacing synchronous generators with IBRs—grid frequency becomes a more volatile and sensitive indicator. Faults that disconnect even moderate amounts of generation or load can cause a much faster rate of change of frequency (RoCoF), providing a critical wide-area signal of a severe disturbance that threatens system stability (Farrokhabadi et al., 2019).
- **Partial discharge (PDs):** As a slow-evolving indicator, these low-energy dielectric breakdowns in insulation are a critical tool for predicting catastrophic failure in high-voltage equipment like cables, transformers, and switchgear. Advanced analytics now focus on classifying PD patterns to not only detect but also diagnose the specific type and severity of insulation degradation (Ibrahim et al., 2024).
- **Abnormal temperatures and vibrations:** These physical stressors are leading indicators of electrical failure. Thermal anomalies at joints or in windings indicate excessive resistance, while abnormal mechanical vibrations in rotating machinery (e.g., wind turbines) can signal bearing wear or rotor imbalance. Continuous monitoring

of these parameters using IoT-enabled sensors enables the detection of these failure pathways long before a fault occurs, forming the basis of predictive maintenance (Collacott, 2012).

8.2.2 Mapping signatures to asset failure modes

The diagnostic value of signals is maximized when they are directly correlated with specific failure modes in physical assets. This mapping allows a transition from simple fault detection to root-cause analysis, but the nature of this analysis is shifting. While traditional electromechanical assets exhibit slow-moving thermal and physical degradation, modern power electronic assets are defined by faster, more complex failure modes rooted in semiconductor physics and control software.

This evolution is best understood through examples. For a legacy power transformer, assessing health involves monitoring multiple, slow-moving signals. The risk of internal insulation failure is diagnosed by tracking PD activity and analyzing dissolved gas anomalies over months or years. Concurrently, physical indicators such as abnormal temperature gradients, detected via thermal imaging, or shifts in acoustic emissions can signify winding deformation. This is the domain of classic condition monitoring.

In contrast, consider a power electronic converter at the heart of a solar farm. Its primary failure modes are semiconductor insulated-gate bipolar transistor (IGBT) degradation and DC-link capacitor failure. An aging capacitor manifests as an increase in the voltage ripple and a subtle shift in the converter's harmonic output. An impending IGBT failure may produce no advanced thermal signature but can be predicted by analyzing high-frequency (kHz–MHz) emissions and transient voltage overshoots during switching events—signals that are invisible to conventional monitoring.

The integrity of high-value assets is therefore increasingly assessed through sensor fusion that synthetically combines these disparate data streams. For a utility-scale battery system, fusing thermal imaging data that identifies hot cells with battery management system (BMS) data on cell voltage and impedance provides a far more reliable indicator of thermal runaway risk than any single signal alone (Collacott, 2012). This integrated, multi-domain approach is foundational to modern fault diagnostics. Table 8.1 provides a detailed overview of these relationships, linking a wide range of grid components to their primary risks and indicators.

8.3 Application domains

The practical application of fault detection varies significantly across grid segments, demanding adaptation of generic principles to specific operational contexts. This section explores specialized fault detection systems, beginning with microgrids, whose unique characteristics challenge classical protection paradigms, before moving to the extensive transmission and distribution networks and, finally, to the diverse landscape of power generation facilities.

8.3.1 Microgrids

Microgrids represent a fundamental departure from centralized power systems. Their high penetration of IBRs, bidirectional power flows, and dynamic topology (grid-connected vs. islanded) replace the predictable physics of synchronous machines with the complex, software-driven behavior of power electronics (Dehghanpour et al., 2021; Olivares et al. 2014). This shift invalidates traditional protection schemes, which rely on high-magnitude, predictable fault currents. The core challenge is not merely low fault current, but a fundamental loss of predictability: an inverter's response is governed by its control logic, not just immutable physics (Milano et al., 2018; Pragati et al., 2023; Zarei et al., 2019). Consequently, ensuring reliability requires a new protection philosophy built upon adaptive, data-rich, and communication-dependent systems, as summarized in Table 8.2 (Rahman Fahim et al., 2020; Zaben et al., 2024).

The dynamic state of a microgrid demands a move beyond simple current measurement toward methods that fuse a richer set of electrical signals and are aware of their operational context. However, these advanced methodologies introduce their own trade-offs between performance and complexity:

- **Adaptive and state-aware relaying:** Protection devices must dynamically alter tripping thresholds based on the microgrid's operational mode, topology, and generation profile. This is no longer optional but essential. While methods analyzing sequence components or voltage phase angles show promise (Pavankumar et al., 2023), their reliability hinges on two critical factors: the deployment of relays capable of real-time setting changes, and the

TABLE 8.1 Common deterioration risks, warning signs, and corresponding grid components.

Grid component	Deterioration risk/Fault precursor	Warning signs/Indicators
Transformers	- Insulation degradation (Oil/Solid) - Winding deformation/Shorts - Core issues (e.g., overheating) - Tap changer malfunction - Bushing deterioration - Oil leaks/Contamination	- PD activity - Dissolved gas analysis (DGA) anomalies - Abnormal temperature profiles (hot spots) - Audible noise changes (hum, crackling) - Irregular voltage/current readings - Oil level drops, discoloration
Overhead lines/Underground cables (conductors)	- Conductor corrosion/Breakage - Insulation breakdown (Cables) - Sagging/Mechanical stress - Joint/Connector failure	- Increased line impedance/resistance - Voltage drops - Intermittent faults - Visual damage (corrosion, broken strands) - Hot spots at connections (thermal imaging) - PD activity (cables)
Insulators	- Surface contamination - Cracking/Puncturing - Material degradation (e.g., UV)	- Increased leakage current - Flashovers (visible or audible) - PD activity - Visible cracks, charring, or tracking
Switchgear (Circuit breakers, disconnectors)	- Contact wear/Erosion - Mechanism failure/Sticking - Insulation degradation - Control circuit malfunctions - Low insulating medium pressure (e.g., SF6)	- Slow or incomplete operation - Increased operating times - Abnormal noise during operation - Overheating at contacts - PD activity - Gas pressure alarms
Generators (Rotating machines)	- Winding insulation failure - Bearing wear/Damage - Rotor imbalance/Damage - Stator core lamination issues - Excitation system faults	- Increased vibration levels - Abnormal temperatures (windings, bearings) - Unusual noises (grinding, whining) - Current/Voltage imbalances - PD activity - Changes in excitation current
Power electronic converters/Inverters (e.g., in Renewable Energy Sources (RES), High-Voltage Direct Current (HVDC))	- Semiconductor (IGBT, thyristor) failure - Capacitor degradation/Failure - Control system malfunction - Cooling system failure	- Abnormal output voltage/current waveforms - Increased harmonic distortion - High-frequency switching noise changes - Overheating of components - Frequent tripping or error codes
Batteries (Energy storage systems)	- Capacity fade/Cell degradation - Internal short circuits - Thermal runaway precursors - State of health (SOH) decline	- Reduced discharge capacity/duration - Increased internal resistance - Abnormal cell temperatures/voltages - Swelling of the battery casing - Off-gassing
Protection relays & control systems	- Calibration drift - Component failure - Communication link errors - Software/Firmware bugs	- Misoperation (false trips, failure to trip) - Inconsistent readings - Communication alarms - System errors or hangs

IGBT, *insulated-gate bipolar transistor;* PD, *partial discharge.*

TABLE 8.2 Summary of key challenges and detection strategies for microgrid protection.

Characteristic/Component	Key risks & fault manifestations	Detection methods & system features
IBR dominance & low inertia	Unpredictable fault response (low magnitude, high harmonics); high RoCoF; risk of control instability.	High-speed waveform analytics, voltage and frequency-based protection (e.g., RoCoF relays), adaptive settings tied to operational mode.
Bidirectional power flow	Fault current magnitude and direction vary with generation/load balance, defeating static overcurrent and distance schemes.	Communication-assisted directional and differential protection, dynamic settings based on real-time power flow data.
Mode transitions (Grid-tied/Islanded)	Drastic change in available fault current and system impedance between modes; risk of protection blinding or sympathetic tripping.	State-aware protection schemes with multiple setting groups, robust islanding detection to trigger logic changes.
Communication dependency	Reliance on data networks for adaptive and differential schemes introduces new failure modes (latency, packet loss, cyber threats).	Redundant, low-latency communication (International Electrotechnical Commission (IEC) 61850), cybersecurity protocols, and fail-safe logic in relays for communication loss events.
Protection coordination	Complex coordination required across thousands of decentralized devices; risk of cascading misoperations.	Centralized protection controllers; extensive hardware-in-the-loop validation of settings for all credible scenarios.

IBR, *inverter-based resource;* RoCoF, *rate of change of frequency.*

rigorous validation of these settings against all credible operating states via simulation and hardware-in-the-loop testing. An incorrect setting pushed in the wrong mode can be more dangerous than a static one.

- **Communication-assisted schemes:** To contend with bidirectional flows and low fault currents, differential and directional schemes are necessary. These offer superior selectivity but create a hard dependency on a communication network. The implementation of such schemes requires not just deploying the relays, but also verifying that the communication infrastructure (e.g., IEC 61850 Generic Object-Oriented Substation Event (GOOSE) messaging) meets stringent latency and reliability requirements (Belchandan & Akhtar, 2023). Furthermore, this introduces a significant cybersecurity attack surface that must be secured. A failure in the network must trigger a fail-safe, typically less-selective, protection mode.
- **High-resolution, synchronized monitoring:** Localized faults and system-wide stability threats necessitate a distributed sensor network. Time-synchronized measurements from PMUs and other intelligent electronic devices (IEDs) are critical for capturing the fast transients that characterize IBR-driven faults (Dehghanpour et al., 2021). Fusing this high-resolution data is essential for discriminating between faults and normal transients (Pinto & Moreto, 2021), but it requires a robust data acquisition and processing architecture capable of handling high-velocity data streams.

In summary, protecting a microgrid is a system integration challenge. It requires replacing static, independent devices with a coordinated, adaptive, and communication-dependent system. While this enables resilience in a complex environment, it also introduces new failure modes at the control and communication level that must be rigorously engineered and managed.

8.3.2 Transmission networks

The primary mandate of a transmission network is the stable, reliable transfer of bulk power. Protection schemes are therefore designed to mitigate high-energy faults that can trigger wide-area instability and cascading failures (Muzzammel et al., 2019). The main vulnerabilities are environmental stressors on long overhead lines (lightning, wind) and the risk of system-wide oscillations in a grid with decreasing rotational inertia (Nagpal & Henville, 2017). While DERs connect at the distribution level, their aggregated effect reduces the predictability and inertia of the entire system, making transmission-level disturbances more critical.

The modern protection philosophy for transmission has shifted from localized fault clearance to wide-area situational awareness. The objective is not just to isolate the faulted line, but to do so while actively managing the stability of the entire interconnected system.

TABLE 8.3 Transmission network: Key risks and advanced protection strategies.

Asset/System aspect	Key risks/Fault manifestations	Primary detection methodologies
High-voltage lines	Line-to-ground/multi-line faults, conductor failure.	Traveling-wave fault location; Line current differential (87L); Distance protection (21).
System stability	Power oscillations, voltage collapse, and loss of synchronism due to low inertia.	Synchrophasor data (PMUs) for real-time state estimation and out-of-step protection schemes.
Substation equipment	Transformer faults, Busbar faults, breaker failure.	Transformer differential (87T), bus differential (87B), breaker failure protection logic.

PMUs, *phasor measurement units.*

Key advancements in transmission protection reflect a paradigm shift from localized fault response to holistic, system-wide situational awareness. The cornerstone of this evolution is the deployment of synchrophasor-based monitoring. PMUs provide high-resolution, time-synchronized data from across the grid, enabling the real-time detection of angular and voltage instabilities that often precede major disturbances.

However, this data is only useful if it can be communicated and acted upon reliably and at high-speed. This highlights the critical role of communication-assisted schemes. The reliability of the protection system is now inextricably linked to the performance of its underlying high-speed, low-latency communication backbone. A classic example is line current differential (87L) protection, which compares currents at both ends of a line segment to provide fast and highly selective fault clearing, but it is entirely dependent on its communication channel. More advanced wide-area protection schemes elevate this concept further, using PMU data from across the network to take coordinated control actions and prevent system collapse (Table 8.3).

Alongside these wide-area monitoring capabilities, new methods are enhancing the precision of post-fault restoration. For long transmission lines where impedance-based methods can be imprecise, traveling-wave fault location offers superior accuracy (Gogula & Edward, 2023). By detecting the high-frequency transient waves propagating from a fault's origin at nearly the speed of light, these systems can pinpoint fault locations within meters (El-Tawab et al., 2022; Kuwa-lek & Wiczyn´ki, 2024). This drastically reduces patrol times and outage durations, improving grid resilience (Maurya et al., 2024; Xu et al., 2023). Together, these technologies create a system that can not only anticipate and prevent widespread outages but also recover from localized faults with unprecedented speed and precision.

8.3.3 Distribution networks

Distribution networks face a fundamentally different challenge: managing extreme complexity and uncertainty driven by the mass integration of DERs. The legacy model of a passive, radial network with unidirectional power flow is obsolete. Today's distribution grid is an active system with bidirectional flows, where fault currents are low magnitude, non-sinusoidal, and highly variable, rendering legacy overcurrent protection schemes unreliable and often "blind" to faults (Albarracın et al., 2018; Khavari et al., 2020). The primary risks are no longer just equipment failure but also protection mis-coordination, safety hazards (islanding, back-feed), and poor customer reliability (Table 8.4).

The response is a shift toward a hierarchical and automated protection architecture:

- **Adaptive protection:** IEDs with adaptive logic are required. These devices can dynamically change their settings based on real-time measurements of power flow direction, connected generation, and system topology. This prevents the "protection blinding" that occurs when an inverter-dominated fault fails to draw enough current to trip a legacy relay.
- **High-resolution sensing and data fusion:** Accurate fault detection requires fusing data from multiple sources (Zhang et al., 2022). Fault passage indicators confirm the path of fault current, while smart meter "last gasp" notifications can help triangulate outage locations. This granular data is essential for distinguishing true faults from transient power quality events (Mirshekali et al., 2021).
- **Automated restoration:** The complexity of the modern distribution grid makes manual service restoration too slow. FLISR systems use telemetry from the field to automatically locate a faulted section, operate automated switches to isolate it, and reconfigure the network to restore power to as many customers as possible, often in under a

TABLE 8.4 Key risks and advanced protection strategies for distribution network.

Asset/System aspect	Key risks/Fault manifestations	Primary detection methodologies
Feeders with DERs	Protection blinding from low fault current, nuisance tripping, and sympathetic tripping of DERs.	Adaptive overcurrent protection, high-resolution data from IEDs, waveform analytics.
Underground cables (Jena et al., 2023)	Insulation degradation, cable joint failure, and dielectric breakdown.	Online PD monitoring Time-domain reflectometry (TDR) for fault location.
Network operations	Sustained outages following permanent faults, mis-coordination of reclosers, and fuses.	Fault passage indicators, smart meter outage data, automated fault location, isolation, and service restoration (FLISR) systems.

DERs, *distributed energy resource;* IEDs, *intelligent electronic devices;* PD, *partial discharge.*

minute (De La Cruz et al., 2023). This requires a robust communication network (e.g., using IEC 61850) and a highly accurate grid model.

8.3.4 Power generation facilities

Fault detection within power generation facilities is critical, but the diagnostic philosophy differs profoundly between conventional synchronous plants and modern IBRs. The former is a world governed by the physics of large rotating masses and slow, predictable mechanical degradation. The latter is a world of power electronics, dominated by high-frequency switching, complex control software, and failure modes that can manifest in microseconds (Amaya-Sanchez et al., 2024).

8.3.4.1 *Conventional generation*

In thermal and hydropower plants, the primary risks are tied to electromechanical stress on large synchronous machinery. Failures are typically the result of cumulative, long-term degradation. Winding insulation faults in generators and transformers, driven by years of thermal cycling, and mechanical wear in bearings and turbines are the dominant failure modes (Ali et al., 2024).

The diagnostic approach is therefore one of proactive condition monitoring based on relatively slow-moving data. The goal is to detect incipient failures by tracking deviations from a stable baseline over weeks or months. Key methodologies include:

- **Vibration analysis:** Detects rotor-stator misalignment, bearing wear, and other mechanical imbalances.
- **Thermal monitoring:** Uses thermal imaging and embedded sensors to identify overheating in windings and connections, indicative of insulation breakdown or increased resistance.
- **PD analysis:** Online PD measurement detects small electrical discharges in insulation systems, providing an early warning of impending dielectric failure.

This is a mature field of predictive maintenance, where the asset itself telegraphs its deteriorating health over an extended period (Sarathkumar et al., 2021).

8.3.4.2 *Renewable generation*

Renewable energy systems introduce a paradigm of distributed, electronically-interfaced assets. While some failure modes are gradual, such as the performance degradation of a solar panel, the most critical and challenging faults stem from the power electronic converter—the core of every solar and wind generation unit.

Here, the diagnostic philosophy must be built for high-speed anomaly detection within vast streams of volatile data. Unlike a generator, an inverter's health is determined by the state of its semiconductor switches (IGBTs), capacitors, and control software. These can fail abruptly and catastrophically with little warning. The key challenges are:

- **Semiconductor failure:** The primary failure point in inverters. Monitoring requires analyzing high-frequency voltage and current waveforms to detect signatures of incipient failure, such as subtle changes in switching times

TABLE 8.5 Comparison of fault profiles in power generation systems.

Generation type	Dominant failure modes/Risks	Primary diagnostic approach
Conventional synchronous (thermal, hydro)	Winding insulation degradation Bearing wear/Rotor imbalance Thermal stress	Slow-data condition monitoring (vibration, thermal, PD) Predictive trend analysis
Solar PV systems	Inverter semiconductor failure Module degradation/Mismatch Bypass diode/Interconnect faults	High-frequency waveform analysis High-granularity string-level data analytics Aerial thermal imaging
Wind energy systems	Power converter (IGBT) failure Gearbox/Drivetrain wear Blade integrity damage	- Fusion of electrical and mechanical sensor data (vibration, SCADA, waveforms)

IGBT, *insulated-gate bipolar transistor;* PD, *partial discharge*

or harmonic distortion. Advanced methods like multi-scale information fusion are needed to process this complex data (Liang et al., 2023).

- **Performance degradation:** In solar Photovoltaic (PV) systems, faults like bypass diode failures or module mismatch cause subtle voltage and current imbalances that are only detectable with granular, string-level, or module-level monitoring, often combined with thermal imaging to find hot spots (Al-Shetwi et al., 2019).
- **Mechanical-electrical interaction:** In wind turbines, mechanical stress in the drivetrain (gearbox, bearings) is a major issue (Yang & Chai, 2016), but the resulting vibrations and torque ripples can propagate into the electrical system, creating complex signatures that require fusing mechanical (vibration) and electrical (waveform) data for accurate diagnosis.

A structured framework for a modern, hybrid generation portfolio must be inherently dual-natured, moving beyond siloed monitoring to create an integrated system of insight and response. The foundation of this approach is a unified, dual-regime monitoring platform engineered to ingest and analyze two fundamentally different types of information: the slow-moving condition data from conventional plants and the high-velocity, high-frequency waveform data from IBRs (Table 8.5).

Upon this platform, the framework demands the deployment of asset-specific anomaly detection, using bespoke algorithms tailored to the unique physics of each asset class. This means applying predictive models to anticipate mechanical wear in legacy generators while simultaneously using real-time pattern recognition to identify incipient electronic faults in inverters. The effectiveness of these algorithms is contingent upon their ability to fuse multimodal data streams. By instrumenting critical assets like turbines and inverters with sensor suites that capture both mechanical—such as vibration and acoustics—and electrical signatures, the system can diagnose complex, cross-domain faults that would otherwise be invisible.

Ultimately, this entire sensing and analytics architecture serves a critical operational purpose: to integrate plant protection with grid stability. This requires high-speed communication between plant-level controllers and grid operators, creating a protection scheme that not only isolates internal faults but also coordinates the plant's response to external grid disturbances, such as fault ride-through events, to maintain system-wide stability.

8.4 Data-driven methods and feature engineering

The transition from static, physics-based protection to adaptive, data-driven frameworks represents one of the most significant shifts in power system operations. While ML and DL offer the potential to detect complex fault signatures that elude traditional methods, their application in a high-reliability, safety-critical domain is fraught with challenges. The core of this challenge lies not in the algorithms themselves, but in the data they depend on and the transparency they often lack (Chudnovsky, 2017), which can lead to biased training of the DL (Alhanaf et al., 2023). This transition introduces new paradigms, each with distinct principles and limitations, as summarized in Table 8.6.

8.4.1 The data reality gap

The efficacy of any supervised ML model is contingent on a large, diverse, and accurately labeled training dataset—an ideal rarely met in power systems. The scarcity of real-world fault data forces a reliance on simulated events,

TABLE 8.6 A taxonomy of fault detection paradigms and their evidentiary basis.

Paradigm	Latency	Core principle	Primary limitation
Signal-based	Milliseconds	Compares a measured signal against a predefined, static physical threshold.	Brittle, cannot adapt to changing system conditions or complex fault types.
Model-based	ms to seconds	Compares real-time signals against the output of a simplified physics model.	Model inaccuracy, struggles with non-linearities, and unmodeled dynamics.
Data-driven (ML/DL)	ms to minutes	Learns statistical correlations and fault signatures from data.	Data dependency, opaque decision-making, potential for unforeseen failure modes.
Hybrid (DT/PIML)	seconds+	Constrains a data-driven model with a high-fidelity, physics-based simulation.	Computational complexity, latency, and dependency on accurate DT models.

DT, *digital twin;* DL, *deep learning;* ML, *machine learning;* PIML, *physics-informed machine learning.*

creating a dangerous "reality gap" where models that achieve near-perfect accuracy on simulated waveforms can degrade unpredictably when confronted with noisy, imperfect sensor data (Rafique et al., 2021).

This reality necessitates a two-pronged strategic shift away from sole reliance on supervised classifiers. One branch of this strategy circumvents the need for labeled fault data entirely through unsupervised anomaly detection. Models like autoencoders are a strategic necessity in this context, as they are trained exclusively on data from normal operations to learn a compressed representation of a "healthy" system state. Any event that the model cannot accurately reconstruct is flagged as an anomaly, enabling the detection of novel fault types and the gradual degradation of assets like solar panels (Boubaker et al., 2023).

The complementary branch of this strategy does not abandon supervised learning but instead addresses the data scarcity head-on through data augmentation. To correct the pronounced class imbalance between normal and fault conditions, synthetic fault instances are generated. While simpler methods like the Synthetic Minority Over-sampling Technique (SMOTE) are effective, the field is increasingly turning to more sophisticated, physics-informed models like generative adversarial networks (GANs), which can learn the underlying patterns of fault physics to create novel and highly realistic fault waveforms for training more robust classifiers.

8.4.2 The design spectrum

The choice of model architecture is governed by an application-driven trade-off between inference speed and inherent transparency, a tension constantly being reshaped by advances in specialized hardware. For applications demanding the highest velocity, such as primary protection requiring sub-cycle response, the paradigm of end-to-end learning is a leading option. Architectures like convolutional neural networks (CNNs) treat raw waveform data as an "image," learning feature hierarchies automatically and obviating the need for manual feature engineering. While this approach yields exceptional performance, it comes at the cost of creating an opaque "black-box," a significant barrier to adoption in a sector where every decision must be auditable.

Conversely, in applications where latency is less critical, such as post-event diagnostics or wide-area stability monitoring, a more inherently transparent philosophy of feature-based learning is preferred. In this method, raw data is first transformed into a curated feature set using established signal processing techniques like the Discrete wavelet transform or S-Transform. These human-engineered features then provide a robust and understandable input for established classifiers like support vector machine (SVM) or for time-series models like long short-term memory (LSTM) used to analyze PMU data for instabilities (Belagoune et al., 2021; Yoon & Yoon, 2022).

8.4.3 Operationalizing trust

For any data-driven model to be deployed in a protection-class application, it must transcend standard accuracy metrics and achieve verifiable trust. This is not accomplished by a single algorithm, but by an operational framework built on three reinforcing pillars: explainability, designed robustness, and physics-constrained reasoning.

1. **Explainability for auditing and forensics:** A "fault/no fault" output is insufficient. Trust demands the ability to ask *why* a model made a decision. Explainable Artificial Intelligence (XAI) techniques are primarily a tool for post-mortem analysis and model debugging (Ajayi et al., 2024). Methods like SHapley Additive exPlanations (SHAP) or local interpretable model-agnostic explanations – LIME – are applied offline to determine which features or

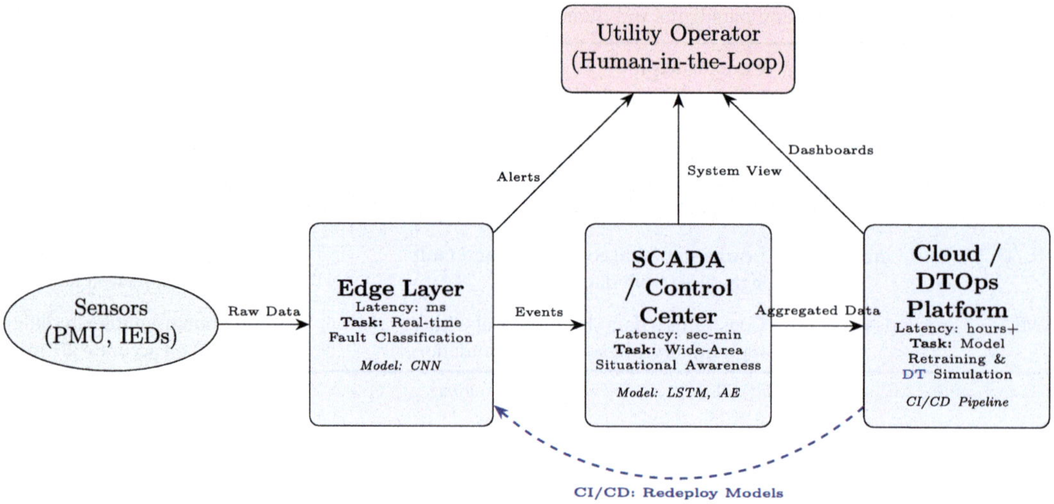

FIGURE 8.1 A reference architecture for a DTOps-enabled fault management system. The solid lines represent the real-time operational loop (sensor-to-operator) where data is processed for immediate action. The dashed blue line represents the automated governance loop, where the DTOps platform continuously retrains and redeploys improved models from the cloud to the edge, ensuring system adaptability. *DTOps*, Digital Twin operations.

waveform segments were most influential in a decision. While too slow for real-time control, this capability is essential for building operator confidence and providing the audit trail required for regulatory acceptance.

2. **Designed robustness for unforeseen events:** Trust requires resilience against both known and unknown threats – fault detection and fault tolerance (Nikolíc et al., 2021). This is achieved through two mechanisms:
   (a) **Adversarial testing:** The model is rigorously stress-tested with subtly perturbed inputs designed to find vulnerabilities an attacker might exploit. This goes beyond standard validation to probe for worst-case failure modes.
   (b) **Uncertainty quantification (UQ)** The model must know what it does not know. Bayesian methods can be used to make the model output not just a prediction, but also a confidence score. When faced with a novel or out-of-distribution (OoD) event, the model can flag its own uncertainty, escalating the decision to a human operator or a fail-safe system rather than making a confident but potentially wrong choice (Zhang et al., 2023b).
3. **Physics-constrained reasoning for plausibility:** Finally, trust is anchored by ensuring the model's outputs do not violate fundamental physical laws. PIML provides these guardrails. By embedding physical constraints (e.g., Kirchhoff's laws) directly into the model's learning process, it is penalized for producing physically implausible results. This fusion ensures that even when learning complex patterns from data, the model's solution space is bounded by the known principles of electrical engineering, dramatically improving its robustness and trustworthiness (Baker et al., 2022; Lopez et al., 2022).

8.4.4 A governed implementation architecture

The physical deployment locus of a data-driven algorithm is as consequential as its design. A resilient and trustworthy fault management system requires a hybrid edge-cloud architecture that intentionally distributes computational tasks across distinct operational tiers. This is not merely a technical configuration but an organizational one, designed to balance real-time response with deep analytical oversight while embedding governance throughout the model lifecycle (Vinothkumar et al., 2023). The structure, depicted in Fig. 8.1, allocates computation to the most appropriate layer, ensuring that decision confidence and analytical complexity increase as data moves from the edge to the cloud.

The lowest-latency tier consists of hardened edge devices executing models (e.g., CNNs on raw waveforms) to achieve sub-cycle fault isolation. The primary design principle at this layer is safe autonomy. For a model to be trusted, it must not only be fast but also recognize the limits of its own knowledge. Therefore, edge models must have built-in robustness, specifically the capacity for OoD detection. Using techniques like UQ, often via Bayesian neural networks (BNNs), the model can self-assess its confidence. When faced with a novel input, it can assign a high uncertainty value, flagging the event for human review or triggering a fail-safe rather than issuing a confident but potentially erroneous prediction (Ghahramani, 2015).

The control center functions as the hub for human-in-the-loop oversight. It aggregates alerts from the edge and wide-area data from PMUs to build system-wide situational awareness. This is the layer where explainability becomes paramount. When the edge flags an uncertain event or takes autonomous action, the operator must be able to ask why. Post-hoc XAI tools like SHAP (Lundberg & Lee, 2017) are applied here to analyze the edge model's decision. By revealing which input features most influenced the outcome, these tools provide the rationale needed for operator trust, post-event forensics, and regulatory auditing.

The apex of the architecture is the centralized cloud or on-premise data center, which serves as the system's analytical backbone and governance engine. This layer operationalizes the long-term reliability of the entire model fleet through a Digital Twin operations (DTOps) framework. Its functions are threefold:

1. **Lifecycle management:** It continuously monitors deployed models for performance degradation or concept drift (Gama et al., 2014). Upon detection of significant decay, it automatically triggers a secure continuous integration/continuous deployment (CI/CD) pipeline to manage the model's lifecycle (Djebali et al., 2024).
2. **High-fidelity validation:** The pipeline uses a high-fidelity DT as a sandbox to retrain, validate, and stress-test new model candidates. This is where computationally intensive tasks like assessing *adversarial robustness* against simulated cyber-attacks are performed (Chen et al., 2020).
3. **Secure redeployment:** Once a new model is certified, the DTOps platform securely pushes the update to the edge device fleet, closing the governance loop and ensuring the fault management system continuously adapts to the evolving physical grid.

This integrated architecture ensures that speed, safety, interpretability, and adaptability are not treated as isolated features but as designed-in properties of a holistic, defense-in-depth system.

8.5 Digital twins

The DT represents a paradigm shift from conventional state estimation, functioning not as a static model but as a high-fidelity, dynamic replica of a physical power system. Its power lies in its ability to fuse real-time telemetry with physics-based simulation, creating a uniquely powerful—and uniquely vulnerable—tool for diagnostics, prediction, and validation; especially in a world that increasingly demands resilience and sustainability (Maillard et al., 2024). The viability of the DT hinges on managing the inherent trade-offs between its capabilities and the risks they introduce.

8.5.1 Cyber-physical state replica

A DT's primary function is to assimilate high-frequency telemetry from sources like PMU and IEDs into a physics-informed model, enabling continuous cross-validation of the system state. This fusion is highly effective for identifying incipient anomalies—such as insulation degradation or emergent instabilities—that are often masked by noise in conventional estimators (Numair et al., 2023). In power electronics, for instance, DT frameworks have demonstrated accurate fault diagnosis even with sparse failure data (Xia et al., 2023).

This deep integration with physical sensors makes the DT's accuracy entirely contingent on the quality, granularity, and temporal synchronization of its input data. More critically, it creates a direct vector for cyber-physical attacks (Chen et al., 2021). An adversary can target the DT's data streams to orchestrate high-impact events like denial of service attacks (Abdelrahman et al., 2024). A false fault injection attack, for example, could involve spoofing PMU timestamps or compromising IEC 61850 GOOSE messages to trigger spurious protective actions, leading to load shedding. A more insidious threat is data poisoning, where subtly corrupted data is injected over time to gradually degrade ML-based classifiers, potentially blinding the DT to real fault signatures. Mitigation thus requires a defense-in-depth architecture incorporating secure data validation protocols and anomaly detection, with blockchain being explored as a means to create a tamper-resistant state log (Hong et al., 2024).

8.5.2 Predictive simulation environment

Beyond diagnostics, the DT provides a sophisticated environment for conducting "what-if" analyses in a risk-free virtual setting. High-fidelity simulations of fault scenarios, electromagnetic transients, and extreme contingencies allow operators to assess grid resilience and validate protection schemes. This is indispensable for grids with high penetrations of IBRs, where atypical fault currents challenge conventional protection. For example, DT-driven

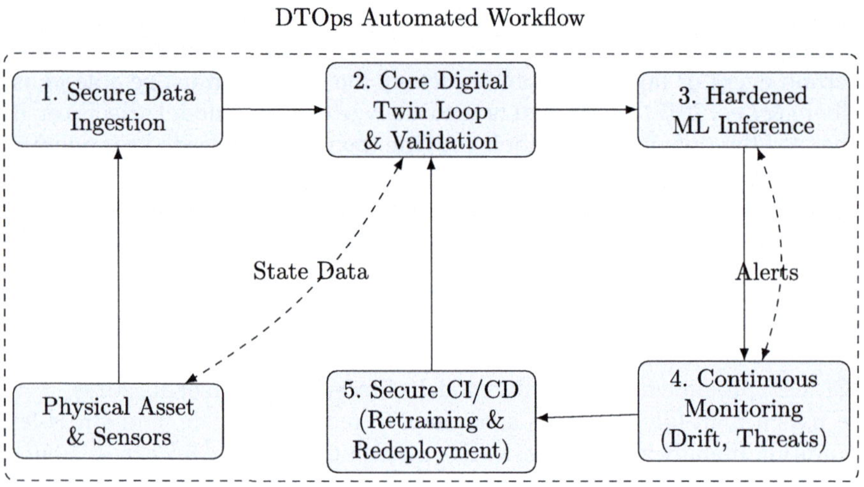

FIGURE 8.2 The DTOps pipeline for digital twin lifecycle management. *DTOps*, Digital Twin operations.

microthermal modeling can enhance real-time hotspot tracking in transformers, improving the accuracy of interturn fault detection (Feng et al., 2024).

The trustworthiness of these simulations is directly proportional to the model's fidelity. This creates a fundamental trade-off between model fidelity and computational tractability. A high-fidelity finite-element model may be too slow for real-time use, while a computationally efficient lumped-parameter model may abstract away the very dynamics it is meant to analyze. Furthermore, a DT is not a static artifact. It requires continuous calibration to counteract concept drift as physical components age and system topologies evolve. An uncalibrated or oversimplified DT is a liability; it can produce a dangerously false sense of security, leading operators to trust invalid contingency plans. Automated, artificial intelligence (AI)-driven model tuning and Bayesian inference for managing sensor uncertainty are therefore critical for ensuring the DT's long-term fidelity and reliability (Ge et al., 2023; Jain et al., 2019).

8.5.3 Secure model governance

Ultimately, the DT serves as the central engine for the DTOps governance framework discussed in Section 8.4.4. It is the high-fidelity sandbox where new ML models are rigorously trained, validated against corner cases, and stress-tested for adversarial robustness before being deployed to the edge. This role connects its two primary capabilities—state replication and simulation—into a single, operational loop. However, this also means that a compromised DT corrupts the entire governance pipeline. An attack that successfully poisons the DT will, in turn, produce compromised models that are then certified and deployed, creating a systemic, self-inflicted vulnerability. Securing the DT is therefore synonymous with securing the integrity of the entire adaptive protection scheme.

8.6 Security and trust

The transition of a DT from a research concept to a mission-critical operational asset is not a technical challenge alone, but a governance one. It requires a structured, automated, and verifiable methodology for lifecycle management. The DTOps framework provides this by extending DevSecOps principles to the cyber-physical context (Djebali et al., 2024). DTOps is not merely a workflow; it is the engine that enforces the answers to the critical questions of practical deployment. It transforms a checklist of requirements into a continuous, automated feedback loop that ensures the DT remains a verifiable, secure, and resilient tool for grid management. This operational pipeline, illustrated in Fig. 8.2, formalizes the governance process into distinct, auditable stages.

8.6.1 Governance controls

Stage 1: Verifiable data ingestion: This initial stage addresses the foundational requirements of the sensor network and data integrity. It defines the minimal viable sensor footprint (e.g., strategic PMUs, IEDs) and specifies data reporting rates based on the monitored phenomena. Crucially, it enforces cybersecurity hygiene at the source, mandating end-to-end encryption, cryptographically signed telemetry, and secure time synchronization to ensure timestamp integrity and prevent data spoofing. *Stage 2: Computationally sound synchronization:* This stage governs the computational architecture and model fidelity. It delineates the execution loci, ensuring low-latency protection models run on edge devices while intensive DT synchronization occurs centrally. It incorporates fail-safe logic to maintain local protection if central connectivity is lost. The core state estimation algorithms are subject to rigorous CI testing to ensure numerical stability and physical plausibility, preventing the DT from diverging from reality.

Stage 3: Bounded and auditable inference: Here, the framework governs how the DT's intelligence is translated into action. It establishes conservative initial thresholds and alarms, coupled with an auto-tuning methodology to refine them based on operational data. It distinguishes between critical alerts requiring automated response and advisory warnings for operators. All outputs from the hardened inference engine are logged to an immutable audit trail, providing the forensic data required for compliance and post-event analysis.

Stage 4 & 5: Continuous monitoring and automated adaptation: These tightly coupled stages form the heart of the governance loop and directly address model management and concept drift. Monitoring systems continuously track a predefined set of KPIs—such as fault detection latency, false trip rate, and localization error—against established baselines. When a KPI breach or statistical drift is detected, the secure CI/CD pipeline is automatically invoked. This pipeline manages the complete model lifecycle, including versioning, retraining, validation in the DT sandbox, and documented rollback procedures, ensuring the system adapts to the evolving grid in a controlled and verifiable manner.

By structuring the operational framework in this way, DTOps moves beyond a simple workflow to become an auditable, resilient, and adaptive governance system. It ensures that as the power system increases in complexity, the DT remains an indispensable and, above all, trustworthy tool for ensuring grid resilience.

8.6.2 Adversarial environment

The operational reliability of data-driven fault detection is not a given; it must be engineered. Trust is an emergent property built upon a defense-in-depth architecture that systematically addresses vulnerabilities at every layer of the system, from the raw data to the final decision. This requires a clear understanding of not just the inherent technical challenges, but also how malicious actors can exploit them (Mahamedi & Fletcher, 2019; Wang et al., 2021).

8.6.2.1 Trust in the data

The foundation of any data-driven system is its data. A primary challenge is the "curse of rarity" and a lack of diversity: training datasets are often sourced from specific grid topologies and operational states, limiting a model's ability to generalize. More importantly, rare but high-impact events like cascading failures or high-impedance faults are, by definition, underrepresented, impeding the model's ability to learn the most critical signatures (Belagoune et al., 2021).

- **Inherent vulnerability:** This extreme sensitivity to the training distribution is not merely a performance issue; it is a primary attack surface.
- **Data poisoning:** Adversaries can exploit this vulnerability through data poisoning, the surreptitious injection of carefully crafted corrupted data into the training set. The goal is to embed a systemic bias or a hidden "backdoor" into the model, effectively blinding it to a specific fault signature or causing it to misclassify under specific conditions orchestrated by the attacker (Narang & Bag, 2024; Yu et al., 2015). It should be noted, however, that poisoning also provides training on unusual data (Khan et al., 2023).
- **Mitigation strategy:** Building trust at this layer requires a two-pronged approach.

First, advanced techniques like transfer learning and deep representation learning can improve generalization from limited data (Zeng et al., 2023). Second, a robust DTOps pipeline must enforce data provenance and integrity checks, creating an auditable trail for all training data and flagging statistical anomalies before they can corrupt the model.

8.6.2.2 Trust in the model

While DL models can achieve high accuracy, their "black-box" nature creates significant barriers to operator trust and regulatory acceptance. If an operator cannot understand *why* a model made a decision, they are unlikely to trust it in a critical scenario. Furthermore, the high computational complexity of these models can be prohibitive for low-latency edge deployment.

- **Inherent vulnerability:** The model's complex, non-linear decision boundary, while powerful, can be incredibly brittle. It may be hyper-sensitive to small input perturbations that are semantically meaningless to a human but are sufficient to push the input across a classification threshold.
- **Adversarial sensor spoofing:** This vulnerability is targeted by adversarial attacks, where subtle, malicious modifications are made to sensor data at inference time. These perturbations, often imperceptible to standard monitoring, are designed to cause a targeted misclassification, such as forcing a "no fault" reading during an actual fault event.
- **Mitigation strategy:** Defense requires moving beyond accuracy as the sole metric.

XAI techniques like attention mechanisms in transformers can provide crucial transparency for operators (Thomas et al., 2023). More importantly, models must be hardened through adversarial training, where they are explicitly trained on such malicious examples to learn more robust decision boundaries. This, combined with model compression and quantization, aims to make robust models computationally viable for real-time protection (Ajagekar & You, 2021).

8.6.2.3 *Trust in the system*

Finally, even with trusted data and a robust model, the overall system's perception of reality can be attacked. The DT, which acts as the central brain, consolidates data to form a holistic view of the grid's state.

- **Inherent vulnerability:** The system's dependence on the DT for state awareness creates a single point of failure. If the DT's state can be desynchronized from the physical grid, any decision based on that state becomes unreliable.
- **False data injection:** Unlike sensor spoofing, which targets the ML model's perception, false data injection attacks target the state estimator of the DT itself. The goal is to fundamentally corrupt the DT's understanding of the grid state, either to mask a real fault or to trigger a false alarm that could destabilize the grid.
- **Mitigation strategy:** System-level trust relies on a fusion of cybersecurity hygiene and architectural resilience. Foundational measures include cryptographically signed telemetry and secure time synchronization. This is augmented by sensor redundancy and cross-validation protocols within the DT's state estimator. Ultimately, the DTOps framework provides the highest level of defense by maintaining an immutable audit trail of all data, model predictions, and state changes, ensuring that any anomaly can be forensically investigated and the system's integrity can be verified (Wang et al., 2022).

8.7 Future directions

The evolution of fault management is not toward three separate futures, but toward a single, unified objective: a grid with governed autonomy. Achieving this requires progressing through a hierarchy of dependencies. The foundational layer is establishing robust standards to ensure interoperability and trust. Building upon this, the grid can evolve toward self-healing capabilities. Finally, speculative but powerful next-generation analytics, like quantum computing, may one day accelerate these autonomous functions.

8.7.1 Standardization for interoperability and trust

Before any advanced autonomy can be safely deployed, the ecosystem must be built on a bedrock of standardization. The challenge is not merely technical but also institutional. Seamless data exchange between sensors, DTs, and control systems is impeded by proprietary protocols and a lack of universal data formats. Without them, the grid remains a collection of siloed components rather than an integrated, intelligent system.

However, future standards must go beyond data protocols like IEC 61850. To enable trusted autonomy, they must define:

- **AI model validation benchmarks:** Standardized test scenarios and datasets to certify that an AI model from any vendor meets minimum performance and safety requirements before deployment.
- **XAI reporting formats:** Common formats for models to report their decision-making logic, ensuring that operators and regulators can audit autonomous actions.
- **Cybersecurity protocols for Machine Learning Operations (MLOps):** Standards for securing the entire data and model pipeline, from data ingestion to model retirement, to defend against systemic threats like data poisoning.

These frameworks are the essential precursor for creating a market of interoperable, competitive, and trustworthy AI-driven solutions from diverse vendors (Abdelhamid et al., 2022; Huang et al., 2020).

8.7.2 Next-generation computational paradigms

Quantum computing represents a potential paradigm shift, not a near-term requirement. While classical computing is sufficient for many current challenges, quantum-enhanced analytics could eventually offer a step-change in performance for solving the complex optimization and high-dimensional problems inherent to a fully autonomous grid (Ajagekar & You, 2021).

The true value lies in its potential to accelerate specific, computationally intractable tasks within a mature, self-healing framework. For instance, quantum-enhanced optimization could dramatically improve real-time stability assessments or allow RL agents to explore vastly larger strategy spaces for grid restoration (Eskandarpour et al., 2020; Ma et al., 2021). However, the practical realization of fault-tolerant quantum computers remains a distant prospect. The more immediate and critical impact of quantum technology is in quantum-safe cryptography, which is essential for securing today's AI-driven systems against tomorrow's threats, ensuring the long-term integrity of the entire fault management architecture.

8.8 Conclusion

The fundamental physics of the power grid have changed. The decline of synchronous inertia and the rise of IBRs have rendered the foundational assumptions of traditional fault protection obsolete. The question is no longer if the century-old paradigm of fixed-setting, electromechanical relays is sufficient, but rather how to manage its inevitable succession in a controlled, secure, and trustworthy manner.

ML, DL, and DTs represent the necessary technological response. They provide the analytical power to navigate the complex, low inertia dynamics of the modern grid, offering the speed and precision that legacy systems cannot. Yet, these data-driven tools are not a panacea. They introduce their own profound challenges: the brittleness of models trained on biased data, the operational opacity of "black-box" architectures, and a vastly expanded cyber-physical attack surface that malicious actors are poised to exploit. Simply replacing old technology with new algorithms is a recipe for a more efficient, yet more fragile, system.

The central conclusion of this analysis is therefore not about the superiority of any single algorithm, but about the urgent need for a new operational discipline. The true innovation required is the shift from merely deploying intelligent models to managing them within a rigorous, auditable lifecycle framework like DTOps. This is the bridge from a promising research concept to a mission-critical operational asset. Such a framework transforms a collection of powerful but brittle algorithms into a resilient, adaptive, and verifiable protection system—one where trust is not assumed, but continuously engineered.

Ultimately, the trajectory is toward a grid defined by governed autonomy. The pursuit of self-healing networks and the potential of quantum computing are not ends in themselves, but milestones on a path toward a system that can be trusted to manage itself. The future grid will not be defined by the cleverness of its models, but by the robustness of the governance that ensures their integrity. The task ahead is to build not just an intelligent grid, but a dependable one.

List of abbreviations

AI	Artificial Intelligence
BMS	Battery Management System
CI/CD	Continuous integration/Continuous deployment
CNN	Convolutional neural network
DER	Distributed energy resource
DL	Deep learning
DT	Digital Twin
DTOps	Digital Twin operations
FLISR	Fault location, isolation, and service restoration
IBR	Inverter-based resource
IED	Intelligent electronic device
IGBT	Insulated-gate bipolar transistor
IoT	Internet of Things

KPI Key performance indicator
ML Machine learning
PD Partial discharge
PIML Physics-informed machine learning
PMU Phasor measurement unit
RoCoF Rate of change of frequency
SHAP SHapley Additive exPlanations
UQ Uncertainty quantification
XAI Explainable artificial intelligence

References

Abdelhamid, A., Zakzouk, N., & El Safty, S. (2022). A multi-agent approach for self-healing and res-penetration in smart distribution networks. *Mathematics, 10*, 2275.

Abdelrahman, M., Kharchouf, I., Hussein, H., Esoofally, M., & Mohammed, O. (2024). Enhancing cyber-physical resiliency of microgrid control under denial-of-service attack with Digital Twins. *Energies, 17*, 3927.

Ajagekar, A., & You, F. (2021). Quantum computing based hybrid deep learning for fault diagnosis in electrical power systems. *Applied Energy, 303*, 117628.

Ajayi, O., Mirjafari, M., Idowu, P. B., & Ullah, M. H. (2024). Explainable AI for fault detection and classification in microgrids. In *IEEE Energy Conversion Congress and Exposition (ECCE)* (pp. 1835–1840). IEEE.

Albarracın, R., Robles, G., Ardila-Rey, J., Cavallini, A., & Passaglia, R. (2018). Partial discharges: Keys for condition monitoring and diagnosis of power transformers. In *Proceeding of the 6th International Advanced Research Workshop on Transformers (ARWtr)*. Spain: IEEE.

Alcala-Gonzalez, D., Toro, E., Más-López, M., & Pindado, S. (2021). Effect of distributed photovoltaic generation on short-circuit currents and fault detection in distribution networks: A practical case study. *Applied Sciences, 11*, 405.

Alhanaf, A., Balik, H., & Farsadi, M. (2023). Intelligent fault detection and classification schemes for smart grids based on deep neural networks. *Energies, 16*, 7680.

Ali, Z., Cálasan, M., Jurado, F., & Aleem, S. (2024). Complexities of power quality and harmonic-induced overheating in modern power grids studies: Challenges and solutions. *IEEE Access*. doi:10.1109/ACCESS.2024.3477729.

Al-Shetwi, A., Sujod, M., Blaabjerg, F., & Yang, Y. (2019). Fault ride-through control of grid-connected photovoltaic power plants: A review. *Solar Energy, 180*, 340–350.

Amaya-Sanchez, Q., Argumedo, M., Aguilar-Lasserre, A., Reyes Martinez, O., & Arroyo-Figueroa, G. (2024). Fault diagnosis in power generators: A comparative analysis of machine learning models. *Big Data And Cognitive Computing, 8*, 145.

Baker, M., Fard, A., Althuwaini, H., & Shadmand, M. (2022). Real-time AI-based anomaly detection and classification in power electronics dominated grids. *IEEE Journal of Emerging and Selected Topics in Industrial Electronics, 4*, 549–559.

Belagoune, S., Bali, N., Bakdi, A., Baadji, B., & Atif, K. (2021). Deep learning through LSTM classification and regression for transmission line fault detection, diagnosis and location in large-scale multi-machine power systems. *Measurement, 177*, 109330.

Belchandan, R., & Akhtar, A. (2023). Comparative analysis of DNP3 and IEC 61850 from architectural, data mapping, data modeling and data reporting view. In *North American Power Symposium (NAPS)* (pp. 1–5). USA: IEEE.

Boubaker, S., Kamel, S., Ghazouani, N., & Mellit, A. (2023). Assessment of machine and deep learning approaches for fault diagnosis in photovoltaic systems using infrared thermography. *Remote Sensing, 15*, 1686.

Chen, J., Mohamed, M., Dampage, U., Rezaei, M., Salmen, S., Obaid, S., & Annuk, A (2021). A multi-layer security scheme for mitigating smart grid vulnerability against faults and cyber-attacks. *Applied Sciences, 11*, 9972.

Chen, L., Li, J., Peng, J., Xie, T., Cao, Z., Xu, K. He, X., Zheng, Z., Wu, B., A survey of adversarial learning on graphs. ArXiv Preprint ArXiv:2003.05730. (2020)

Chudnovsky, B. (2017). *Transmission, distribution, and renewable energy generation power equipment: Aging and life extension techniques*. CRC Press.

Collacott, R. (2012). *Mechanical fault diagnosis and condition monitoring*. Springer Science & Business Media.

Dehghanpour, E., Normandeau, M., Joós, G., & Brissette, Y. (2021). A protection system for inverter interfaced microgrids. *IEEE Transactions on Power Delivery, 37*, 2314–2325.

De La Cruz, J., Gómez-Luna, E., Ali, M., Vasquez, J. C., & Guerrero, J. M. (2023). Fault location for distribution smart grids: Literature overview, challenges, solutions, and future trends. *Energies, 16*(5), 2280.

Djebali, S., Guerard, G., & Taleb, I. (2024). Survey and insights on digital twins design and smart grid's applications. *Future Generation Computer Systems, 153*, 234–248.

El-Tawab, S., Mohamed, H., Refky, A., & Abdel-Aziz, A. (2022). Self-healing of active distribution networks by accurate fault detection, classification, and location. *Journal Of Electrical And Computer Engineering, 2022*, 4593108.

Eskandarpour, R., Ghosh, K., Khodaei, A., Paaso, A., & Zhang, L. (2020). Quantum-enhanced grid of the future: A primer. *IEEE Access, 8*, 188993–189002.

Farrokhabadi, M., Canizares, C., Simpson-Porco, J., Nasr, E., Fan, L., Mendoza-Araya, P., Tonkoski, R., Tamrakar, U., Hatziargyriou, N., Lagos, D., Wies, R. W., Paolone, M., Liserre, M., Meegahapola, L., Kabalan, M., Hajimiragha, A. H., Peralta, D., Elizondo, M. A., Schneider, K., … Reilly, J. (2019). Microgrid stability definitions, analysis, and examples. *IEEE Transactions on Power Systems, 35*, 13–29.

Feng, J., Zhao, N., & Wang, B. (2024). Digital Twin-driven hot spot capture for power transformer inter-turn short circuit fault diagnosis based on microthermal field. *IEEE Transactions on Instrumentation And Measurement, 73*, 1–13.

Gama, J., Z˘liobaite˙, I., Bifet, A., Pechenizkiy, M., & Bouchachia, A. (2014). A survey on concept drift adaptation. *ACM Computing Surveys, 46*, 1–37.

Ge, W., He, R., Wu, Q., Cai, B., Yang, C., & Zhang, F. (2023). Digital twin driven fault diagnosis method for subsea control system. *IEEE Access, 11*, 116269–116276.

Ghahramani, Z. (2015). Probabilistic machine learning and artificial intelligence. *Nature, 521*, 452–459.

Gogula, V., & Edward, B. (2023). Fault detection in a distribution network using a combination of a discrete wavelet transform and a neural Network's radial basis function algorithm to detect high-impedance faults. *Frontiers In Energy Research, 11*, 1101049.

Hong, Y., Alano, F., Lee, Y., Jiang, J., & Yeh, J. (2024). Digital Twin-based blockchain for power support in networked microgrids. *IEEE Access, 12*, 86675–86689.

Huang, F., Liu, Z., Wang, T., Zhang, H., & Yip, T. (2020). The optimization study about fault self-healing restoration of power distribution network based on multi-agent technology. *Computers, Materials & Continua, 65*(1).

Huang, Z., Wen, X., Zhu, J., & Feng, Z. (2024). Research on fault location of DC line in UHV converter station based on digital twin technology. *Measurement: Sensors, 33*, 101180.

Ibrahim, A., Sadanandan, S., Ghaoud, T., Rajkumar, V., & Sharma, M. (2024). Incipient fault detection in power distribution networks: Review, analysis, challenges, and future directions. *IEEE Access, 12*, 112822–112838.

Jain, P., Poon, J., Singh, J., Spanos, C., Sanders, S., & Panda, S. (2019). A digital twin approach for fault diagnosis in distributed photovoltaic systems. *IEEE Transactions on Power Electronics, 35*, 940–956.

Jena, S., Mishra, D., & Salkuti, S. (2023). Fault detection, classification, and location in underground cables. *Power quality in microgrids: Issues, challenges and mitigation techniques* (pp. 195–215). Springer Nature.

Khan, M., Asad, B., Vaimann, T., Kallaste, A., Pomarnacki, R., & Hyunh, V. (2023). Improved fault classification and localization in power transmission networks using VAE-generated synthetic data and machine learning algorithms. *Machines, 11*, 963.

Khavari, S., Dashti, R., Shaker, H., & Santos, A. (2020). High impedance fault detection and location in combined overhead line and underground cable distribution networks equipped with data loggers. *Energies, 13*, 2331.

Kuwa-lek, P., & Wiczyn´ki, G. (2024). Multi-point method using effective demodulation and decomposition techniques allowing identification of disturbing loads in power grids. *Electric Power Systems Research, 231*, 110335.

Liang, J., Zhang, K., Al-Durra, A., & Zhou, D. (2023). A multi-information fusion algorithm to fault diagnosis of power converter in wind power generation systems. *IEEE Transactions on Industrial Informatics, 20*, 1167–1179.

Lopez, J., Jesus Camacho, J., Ponce, P., MacCleery, B., & Molina, A (2022). A real-time digital twin and neural net cluster-based framework for faults identification in power converters of microgrids, self organized map neural network. *Energies, 15*, 7306.

Lundberg, S. M., & Lee, S. I. (2017). A unified approach to interpreting model predictions. In *Advances in neural information processing systems: 30*. USA: Curran Associates Inc.

Ma, T., Ali, S., & Yue, T. (2021). Testing self-healing cyber-physical systems under uncertainty with reinforcement learning: an empirical study. *Empirical Software Engineering, 26*, 1–54.

Mahamedi, B., & Fletcher, J. (2019). Trends in the protection of inverter-based micro-grids. *IET Generation, Transmission & Distribution, 13*, 4511–4522.

Maillard, V., Wu, X., & Muenz, U. (2024). Digital Twin: An asset to address sustainability challenges. *IEEE Electrification Magazine, 12*, 31–38.

Maurya, P., Vidhate, D., Nayak, R., Madhavi, P., Gawande, P., & Howard, E. (2024). Self-healing grids: AI techniques for automatic restoration after outages. *Power System Technology*: 48.

Milano, F., D¨orfler, F., Hug, G., Hill, D., & Verbič, G. (2018). Foundations and challenges of low-inertia systems. In *Power Systems Computation Conference (PSCC)* (pp. 1–25). IEEE.

Mirshekali, H., Dashti, R., Handrup, K., & Shaker, H. R. (2021). Real fault location in a distribution network using smart feeder meter data. *Energies, 14*(11), 3242. doi:10.3390/en14113242.

Mobashsher, M., Hosseini, S., Abdoos, A., Hashemi, S., Sanaye-Pasand, M., & Azzouz, M. (2025). *Fault type classification in the presence of inverter-based resources: Review, challenges and future works. IEEE Access.*

Muzzammel, R., Raza, A., Hussain, M., Abbas, G., Ahmed, I., Qayyum, M., Rasool, M., & Khaleel, M. (2019). MT–HVdc systems fault classification and location methods based on traveling and non traveling waves—A comprehensive review. *Applied Sciences, 9*, 4760.

Nagpal, M., & Henville, C. (2017). Impact of power-electronic sources on transmission line ground fault protection. *IEEE Transactions on Power Delivery, 33*, 62–70.

Narang, J. K., & Bag, B. (2024). Physical model learning based false data injection attack on power system state estimation. *Sustainable Energy, Grids and Networks, 40*, 101524.

Nikolíc, J., Jubatyrov, N., & Pournaras, E. (2021). Self-healing dilemmas in distributed systems: Fault correction vs. fault tolerance. *IEEE Transactions on Network And Service Management, 18*, 2728–2741.

Numair, M., Aboushady, A., Arran˜o-Vargas, F., Farrag, M., & Elyan, E. (2023). Fault detection and localisation in lv distribution networks using a smart meter data-driven digital twin. *Energies, 16*, 7850.

Olivares, D., Mehrizi-Sani, A., Etemadi, A., Can˜izares, C., Iravani, R., Kazerani, M., Hajimiragha, A., Gomis-Bellmunt, O., Saeedifard, M., Palma-Behnke, R., Jiménez-Estévez, G. A., & Hatziargyriou, N. D. (2014). Trends in microgrid control. *IEEE Transactions on Smart Grid, 5*, 1905–1919.

Pavankumar, Y., Debnath, S., & Paul, S. (2023). Microgrid fault detection technique using phase change of Positive sequence current. *International Journal of Modelling And Simulation, 43*, 171–184.

Pinto, J., & Moreto, M. (2021). Protection strategy for fault detection in inverter-dominated low voltage AC microgrid. *Electric Power Systems Research, 190*, 106572.

Pragati, A., Mishra, M., Rout, P., Gadanayak, D., Hasan, S., & Prusty, B. (2023). A comprehensive survey of HVDC protection system: Fault analysis, methodology, issues, challenges, and future perspective. *Energies, 16*, 4413.

Rafique, F., Fu, L., & Mai, R. (2021). End to end machine learning for fault detection and classification in power transmission lines. *Electric Power Systems Research, 199*, 107430.

Rahman Fahim, S. , K., Sarker, S., Muyeen, S., Sheikh, M., & Das, S. (2020). Microgrid fault detection and classification: Machine learning based approach, comparison, and reviews. *Energies, 13*, 3460.

Sarathkumar, D., Srinivasan, M., Stonier, A., Samikannu, R., Dasari, N., & Raj, R. (2021). A technical review on classification of various faults in smart grid systems. *IOP Conference Series: Materials Science And Engineering, 1055*, 012152.

Thomas, J., Chaudhari, S., Shihabudheen, K., & Verma, N. (2023). CNN-based transformer model for fault detection in power system networks. *IEEE Transactions on Instrumentation And Measurement, 72*, 1–10.

Vinothkumar, T., Sivaraju, S., Thangavelu, A., & Srithar, S. (2023). An energy-efficient and reliable data gathering infrastructure using the internet of things and smart grids. *Automatika, 64*, 720–732.

Wang, Z., Wu, Z., & Chen, B. (2021). Vulnerability of deep learning-based fault diagnosis models to adversarial attacks in power systems. *IEEE Transactions on Smart Grid, 12*, 3840–3851.

Wang, Z., Zhang, Z., Zhang, X., Du, M., Zhang, H., & Liu, B. (2022). Power system fault diagnosis method based on deep reinforcement learning. *Energies, 15*, 7639.

Xia, Q., Yue, J., Chen, J., Cui, Z., & Lyu, F. (2023). Early fault diagnosis for DC/DC converters based on digital twins and transfer learning. *Measurement Science and Technology, 34*, 115008.

Xu, X., Zhou, F., Nie, Y., Xu, W., Wang, K., OuYang, J., Zhou, K., Chen, S., & Han, Y. (2023). Fault detection and location of 35 kV single-ended radial distribution network based on traveling wave detection method. *Processes, 11*, 2494.

Yang, B., Zheng, R., Han, Y., Huang, J., Li, M., Shu, H., Su, S., & Guo, Z. (2024). Recent advances in fault diagnosis techniques for photovoltaic systems: A critical review. *Protection and Control of Modern Power Systems, 9*, 39–59.

Yang, Z., & Chai, Y. (2016). A survey of fault diagnosis for onshore grid-connected converter in wind energy conversion systems. *Renewable And Sustainable Energy Reviews, 66*, 345–359.

Yoon, D., & Yoon, J. (2022). Deep learning-based method for the robust and efficient fault diagnosis in the electric power system. *IEEE Access, 10*, 44660–44668.

Yu, W., Griffith, D., Ge, L., Bhattarai, S., & Golmie, N. (2015). An integrated detection system against false data injection attacks in the smart grid. *Security and Communication Networks, 8*, 91–109.

Zaben, M., Worku, M., Hassan, M., & Abido, M. (2024). Machine learning methods for fault diagnosis in ac microgrids: A systematic review. *IEEE Access, 12*, 20260–20298.

Zarei, S., Mokhtari, H., & Blaabjerg, F. (2019). Fault detection and protection strategy for islanded inverter-based microgrids. *IEEE Journal of Emerging and Selected Topics in Power Electronics, 9*, 472–484.

Zeng, L., Zhang, X., Yu, K., Jin, Q., Wu, Y., Chen, L., & Wu, X. (2023). Incipient fault detection and process monitoring of thermal power plant pulverizing system based on deep representation learning. *Transactions of the Institute of Measurement and Control, 47*, 3111–3123.

Zhang, M., Zhang, Z., Li, Z., Wang, J., Zhang, Y., & Liu, S. (2023a). A simple and effective open-circuit-fault diagnosis method for grid-tied power converters—A new technique based on Tellegen's theorem. *IEEE Journal of Emerging and Selected Topics in Power Electronics, 11*, 2203–2213.

Zhang, X., Du, M., Wang, Y., Zhang, H., & Guo, Y. (2023b). Research on power grid fault diagnosis based on a quantitative representation of alarm information. *IEEE Transactions on Industrial Electronics, 70*, 9582–9592.

Zhang, Y., Han, B., & Han, M. (2022). A novel distributed data-driven strategy for fault detection of multi-source dynamic systems. *IEEE Transactions on Circuits and Systems II: Express Briefs, 69*, 4379–4383.

CHAPTER

9

Digital Twin testing platforms for power grids

Felipe Arraño-Vargas, Zhiwei Shen, Georgios Konstantinou

UNSW Sydney, School of Electrical Engineering and Telecommunications, Sydney, NSW, Australia

9.1 Introduction

Power systems are in a state of constant transformation, and rapid changes associated with the energy transition are making grids less predictable. To cope with these challenges, power system control centers are introducing advanced technologies and functionalities to enable the testing of new control strategies and operational tools in a safe environment, while also facilitating the operation of modern grids.

Digital Twins (DTs) have emerged as a technology that can provide a comprehensive representation of a physical system, asset, and/or process. Enabled by software, hardware, and data integration, DTs allow real-time monitoring, controlling, prediction, optimization, and improved decision-making. In the energy sector, a power system digital twin (PSDT) can uplift how data from power grids and their equipment is processed, providing operators with new ways to visualize and understand the information and ultimately enhancing situational awareness, enabling faster, more informed, and optimized decisions, and improving the controllability and resilience of the whole system. Nevertheless, to fully harness the potential of a PSDT, further research and testing are required.

In this chapter, DTs for power systems are defined in Section 9.2. Furthermore, the need for testbed platforms to leverage the technology is examined in Section 9.3. In Sections 9.4 and 9.5, a DT testbed, based on the concept of virtual physical twin (VPT), is introduced and demonstrated.

9.2 Basic concepts

9.2.1 Definition and classification of power system digital twins

In the context of a power system and its assets, DT technology remains an evolving concept. In order to include all key elements proposed in previous definitions, the term PSDT, referred to in this chapter, is defined as *a self-adapting digital model of a power system, based on diverse information from data historians (historical datasets), real-time operations, and forecasts*. A PSDT allows for models with wider application, longer lifespans, and is not linked to specific snapshots of a system or to one particular asset. Unlike asset-level twins optimized for equipment operations and maintenance, PSDTs integrate multiasset physics and market/weather signals to optimize system-wide operations (refer to Table 9.1).

The architecture of a PSDT includes the following fundamental parts: (1) the physical twin (i.e., power system/grid), (2) the virtual representation (i.e., DT), (3) data, (4) bidirectional communication infrastructure, and (5) services and applications. Cybersecurity, as a foundational service that underpins the functions of a PSDT, and the end-users, as the ultimate beneficiaries of its capabilities, are both critical for the functional and effective implementation of the DT.

A PSDT can be further classified into the following categories (Metcalfe & Tohani, 2019):

- **Descriptive DT**: This kind of twin provides a detailed visualization of the power system in order to monitor its current state.

Digital Twin Technology and Smart Grid.
DOI: https://doi.org/10.1016/B978-0-443-36664-2.00019-8

TABLE 9.1 Key differences between power system and asset-level twins.

Characteristics	Power system digital twin (PSDT)	Asset-level digital twin
Purpose and design	To optimize the planning, control, and operation of the power grid as a whole.	To monitor health, enable predictive maintenance, and enhance performance for efficient operation of associated power equipment.
Implementation	Use of advanced information and communication technology infrastructure and sophisticated communication protocols. It involves complex systems with multiple layers of data governance and analytics.	Use of localized methods, sensors, and data.
Modeling approach	Integration of multiple physical system models to capture interactions. A PSDT requires scalability and adaptability to evolving grid conditions.	Focuses on the physical and operational characteristics of individual power equipment.
Complexity and data requirements	Extensive and diverse data sources are required. It depends on large volumes of multidimensional, diverse, and multisource data.	Simpler design with focused data inputs.
Main users	Suitable for distribution and transmission operators, market operators, regulators, and policymakers.	Suited for original equipment manufacturers, developers, and grid asset operators.

- **Informative DT**: By aggregating and analyzing historical and real-time data, an informative twin generates valuable insights into the past and current operation of the power system.
- **Predictive DT**: This type of twin, enabled by advanced analytics and machine learning algorithms, allows for predicting future states and behaviors of the power system, providing significant information on how to ensure a secure and reliable operation of the grid.
- **Comprehensive DT**: Combining descriptive, informative, and predictive twin capabilities, a holistic view of the power system is provided. A comprehensive twin enables users to fully monitor, control, and operate the grid, enabling better decision-making and strategic planning.
- **Transformative DT**: The most advanced type of twin. A transformative twin leverages data, simulations, and machine learning algorithms to self-adapt and take autonomous actions to optimize the operation of the power system. User intervention is minimized.

It is worth highlighting that a PSDT enables the integration of data across power system assets, enabling the analysis of their interactions and the prediction of overall grid behavior. In contrast, an asset-level DT is primarily used for planning, designing, maintaining, and operating individual equipment. As such, the scope of a PSDT extends beyond and can include multiple asset-level twins aimed to capture system-wide dynamics, interactions, and interdependencies. A comprehensive summary of key differences between PSDTs and asset-level twins is provided in Table 9.1.

9.2.2 Benefits and applications of power system digital twins

The deployment and operation of a PSDT unlock a range of benefits across power system planning and operations. These value-added features, derived from the modeling and use of historical and real-time data, can be categorized as:

- **Data interpretation and understanding**: The massive, continuous, real-time data measurements unlock value across the life cycle of the power system. The data provides an in-depth view of the state of the grid, that processed and analyzed is transformed into essential insights that may lead to controlled actions to optimize the operation of the system.
- **Smarter predictions:** Data processing architectures and advanced artificial intelligence (AI) algorithms allow for predicting and managing outcomes related to, for instance, scheduled maintenance, and grid operation.
- **Greater efficiency:** Monitoring the power system and using the outcomes provided by PSDT services, faster, and better decisions are made, optimizing performance and efficiency, ensuring the continuity of service.
- **Improved and cost-effective research and development (R&D):** Leveraging available data, more effective research, planning, and design can lead to the customization and refinement of algorithms and services.

A PSDT provides a set of services that enable efficient management, control, and operation of the grid. These services leverage data from not only the grid but also other relevant sources like regulators, energy market commissions, and national weather services to fulfill their purpose. A PSDT service can be supervised or self-regulated, it can run in real time, automatically or manually triggered, or executed in offline mode when required by the user. Some of the functionalities a PSDT can provide are:

- **Advanced monitoring and control** of power systems. A PSDT can function as an embedded SCADA system, incorporating state estimation and energy management systems. Most importantly, a PSDT can detect inaccurate and/or missing measurements while replacing and/or compensating for such data. As a virtual replica of the grid, a PSDT can also provide measurements at any specified resolution, thereby providing a better resolution of the grid's state, unlocking, for instance, condition monitoring with longer time horizons. In addition, a PSDT enables safe and controlled testing of new operational strategies, control schemes, and grid technologies before deploying them.
- **Forecasting and prediction** of demand, generation, and scenarios. System operators can feed the PSDT with scheduled changes in demand behavior (e.g., strikes, lockdowns, and sporting events) and thus better plan the day-to-day operation. Weather conditions can be used as additional inputs to the PSDT to forecast how the generation of renewable energy will affect the current scheduling and provide an optimal dispatch. A PSDT is a powerful technology for generating additional data and conducting *what-if* scenario testing, without risking the grid. Conducting these tests enhances situational awareness while also supporting the decision-making process of system operators (or done autonomously).
- **Integrated system planning** to meet security of supply, grid reliability and network resilience targets. A PSDT allows testing and validation of planning strategies and investment options under diverse future scenarios, ensuring robust decision-making. The PSDT could optimize generation and transmission investments based on market benefits, balancing risks and costs, considering highly controllable and flexible technologies. This service becomes even more critical as the amount of synchronous generation decreases and inverter-based resources increase.
- **Secured grid integration of renewable energy, energy storage systems, and advanced technologies** by equipping engineers with new tools to screen, analytically evaluate and assess the design and implementation of new connections. Closely linked with the integrated system planning service mentioned earlier, a PSDT provides a platform to test, for instance, the integration of inverter-based resources under a wide range of scenarios and conditions, allowing engineers to identify reliability constraints and evaluate mitigation measures.
- **Training** of system operators. A PSDT can provide a safe environment for power plant operators, transmission system operators, distributors, and dispatchers to practice and learn about the operation of the grid. An extensive hands-on training covering normal scenarios, potential faults, extreme events, and black starts can be delivered without risking the continuity of the service.
- **R&D** in order to develop new services and applications. A PSDT can serve as a platform for testing, validating, and implementing new services and solutions before their deployment.

9.2.3 Challenges and requirements for developing power system digital twins

The successful implementation of a PSDT depends, in part, on designing one or more two-way communication channels that enable the exchange of information between the power system and the DT, and their corresponding integration. As the grid evolves over time, the PSDT also needs to adapt to accurately reflect the changes. Furthermore, the service(s) provided by the PSDT should be presented in a transparent and intuitive manner to the end user. Finally, a PSDT requires a high level of security measures for purposes of data management and interoperability. All these challenges are also subjected and predisposed to the complexity, breadth, and depth of the desired purpose(s) of the DT.

The accelerated development of enabling technologies is facilitating the implementation of high-fidelity DTs. Big data, cryptography, and blockchain algorithms are enabling the secure integration, management, analysis, and storage of data. Physics-based to data-driven models are being developed to represent the behavior of the physical twin, while dedicated hardware and multicloud computing simulators are utilized to execute the models. Platforms and data transfer, real-time communication technologies such as 5G, and the Internet of Things are needed to enable the required infrastructure. Furthermore, AI algorithms have the potential to leverage DTs by introducing enhanced system operation techniques, data management, and reducing human intervention.

Despite advances in DT technology, several challenges must be addressed to leverage the viability of PSDTs. Some future directions that can unlock the full potential of a PSDT include: (1) the development of a testbed environment where the integration of PSDT components and services can be proposed, tested, and developed; (2) a framework for the seamless integration of existing systems (e.g., SCADA, EMS, and asset management systems); and (3) the development of adequate interfaces that allow users to comprehend and visualize large amounts of data and outcomes from the PSDT.

9.3 Testbeds: Foundational platforms for advancing power system applications

A significant portion of power system analyses and research relies on testbed systems. These test systems provide a standardized platform for the systematic benchmarking, evaluation, and validation of methods, algorithms, emerging applications, and technologies for modern power systems, without risking them. Furthermore, testbeds help deepen the understanding of grids' behavior, component interactions, identify potential vulnerabilities, and develop corresponding countermeasures and mitigation strategies.

9.3.1 Benchmark and synthetic testbed systems

Several benchmark models have been developed to test specific problems and/or analyze system behavior under different operational conditions. Steady-state, reliability, voltage stability, small-signal and control studies, as well as renewable energy and energy storage integration, are some of the use cases for benchmark models. Nevertheless, these testbeds no longer represent the complexities, topologies, or sizes of modern power systems, thereby limiting their practicality. Therefore there is an increasing need to move toward the use of larger, more realistic testbeds that can represent real grids.

Synthetic grids have become the new standard for benchmark models, capable of reflecting the scale and complexity of today's power systems (Gegner et al., 2016). These testbeds are developed using statistical techniques that replicate the characteristics of real grids, without revealing any sensitive or confidential information. Synthetic grids representing power systems in the United State and Australia has been developed and used for transient stability analysis and grid planning.

Benchmark and synthetic grid testbeds are commonly used for R&D, and even though they can generate and provide valuable knowledge for the industry sector, these models are fictitious representations of actual power systems, and the generated data is subjected to biases and unrealistic conditions/events. Real grids are in a state of constant change with new hardware, software, and applications affecting their planning, operation, and maintenance. Utilities and transmission system operators require models that are not only closer representations of real grids but also are constantly being updated and adapted based on system measurements, grid expansions, power plant upgrades, generator decommissions, etc.

The next step toward better understanding, learning, and gaining insights from a power system is to create an intelligent, digital replica—a DT for power systems (Arraño-Vargas & Konstantinou, 2023). DT testbeds leverage existing platforms to provide a realistic environment for generating additional insights and high-resolution data. The continuous self-adaptation of DTs allows them to mimic, in real time, the behavior of the grid, helping operators to accurately assess its state.

9.3.2 Digital Twin testbed systems

Access to DT testbeds provides a crucial platform for developing, evaluating, and refining advanced solutions across a diverse range of applications. These testbeds integrate real physical hardware with advanced simulation technologies, enabling analysis of complex scenarios or application of innovative strategies without impacting actual operational systems. Some notable examples from the available literature with large-scale support from research organizations, governments, and grid operators include:

- National Renewable Energy Laboratory (NREL)'s Advanced Research on Integrated Energy Systems testbed supports grid integration of renewable energy sources, grid modernization efforts, and cybersecurity analysis (NREL, 2025). It combines real-time simulation capabilities, power hardware-in-the-loop setups, along with extensive high-performance computing resources, allowing for validation of grid configurations and control strategies.

- Pacific Northwest National Laboratory (PNNL)'s (Pacific Northwest National Laboratory, 2025) powerNET platform merges real electrical grid hardware with dynamic simulations, facilitating cyber-physical systems research, grid resilience, cybersecurity, and performance of control algorithms.
- TwinEU (TwinEU, 2025), supported by the European Union, aims to create a federated DT representing Europe's entire electricity grid, emphasizing system resilience, renewable energy integration, and operational efficiency improvements through a sophisticated federation architecture.
- China Southern Power Grid's DT platform, designed for power grid operation and maintenance (Guo et al., 2023), integrates equipment and production data into a comprehensive 3D model, which allows monitoring and management of power equipment in real time by linking a virtual model with the actual system. This link helps simplify real-time monitoring, assist operator decisions, and proactive staff interventions.

DT testbeds have also found use cases in projects led by industry or universities. The DT of the Xianning Chibi 500-kV Substation (Bentley, 2024), developed by POWERCHINA Hubei Electric Engineering, assists with grid reliability. It is also used to streamline design, construction, and operational processes within the substation. The support for full life-cycle management through 3D and 4D modeling and the integration of real-time data assists continuous monitoring and simulation. Such an example demonstrates mature industrial application of DT technology, but with a focus on grid infrastructure and not directly on grid operations. The EPICTWIN platform (Kandasamy et al., 2022) from Nanyang Technological University focuses on cybersecurity challenges in the power sector, serving as a university-led testbed to study and simulate cyberattacks on power systems and trial effective countermeasures. ETAP, Schneider Electric, and NVIDIA have partnered to provide a virtual replica of an electrical infrastructure combined with real-time power system data and advanced analytics, helping to optimize energy use, and predict equipment faults, supporting the growth of the data center sector. Such examples highlight how leveraging DT testbeds helps address critical challenges in infrastructure, cybersecurity, and operational performance.

Despite these significant advantages, DT testbeds face multiple inherent challenges. High costs associated with the development and deployment at scale of the required infrastructure, together with the required expertise, can limit widespread adoption. The specialized skills needed for simulation and management of the DT platform also require continuous training. Flexibility can be constrained by complex initial setups and configurations, making rapid adaptation to new requirements or direct transfer to new projects challenging and slow. Additionally, discrepancies between simulation models and actual physical system behavior pose accuracy challenges, particularly when scaling simulations to large systems. The added layer of data security and data sharing complexities among different stakeholders of the DT testbed or of the different systems that form the DT testbed further complicate their uptake. Cybersecurity vulnerabilities are greater due to extensive digital connectivity within a DT testbed. Finally, ongoing maintenance and recalibration efforts are necessary to ensure that a DT testbed remains an accurate reflection of their corresponding physical systems. To that end, developing a dedicated virtual DT platform can help overcome many of these limitations without requiring substantial investment.

9.4 Virtual testbed for power system digital twins

9.4.1 Virtual physical twin

A VPT is a simplified representation of an actual physical system, in this case a power system, that draws upon data from established models (i.e., physics-based and data-driven) and measurements (e.g., SCADA, PMU) to replicate the behavior of the grid (Shen et al., 2024). A VPT is crucial in developing a PSDT as it provides a reference system that simulates the physical counterpart without exposing it to risks. This approach enables thorough experimentation and validation of DT technologies and applications in a controlled environment.

The establishment and operation of a VPT is supported by a suite of advanced information and communication technologies (ICT) that facilitate the reproduction of representative scenarios and conditions while seamlessly generating datasets. By continuously emulating known operational changes and disturbances, a VPT generates dynamic data flows that the PSDT is expected to track, process, integrate, analyze, and adapt to. This dynamic interaction enables the validation of core PSDT functionalities, including initialization, synchronization, and adaptation. The use of a VPT also facilitates the implementation and integration of PSDT applications and services. Therefore, the principal benefit of using a VPT lies in enabling the benchmarking and cross-validation of PSDT performance, ensuring its fidelity before advancing to the deployment phase.

9.4.2 Advantages of using a VPT over the grid for developing power system digital twins

When implementing a PSDT, the use of a VPT presents significant advantages when compared to the use of an actual grid. These advantages can be categorized into the configuration of the testing environment and its properties.

9.4.2.1 Configuration of the testbed

The deployment of PSDTs relies on the integration of either existing ICT systems or new infrastructure to allow the integration of data flows from the grid. These data streams are subjected to noise, delays, errors, and (cyber-) attacks as well as to different protocols and formats. Even though the use of available resources from the grid takes into account all the complexities for the deployment of a PSDT, a VPT is capable of providing tailored setups and conditions not only to cover existing scenarios, but also future ones with different technologies, protocols, noise, delays, errors, and attacks.

9.4.2.2 Properties of the testbed

Power systems are critical infrastructure in which reliability and resilience are of paramount importance as they ensure the continuous delivery of electricity. Thus their specifications and operational information are often sensitive, confidential, and protected from disclosure. This represents a constraint for the comprehensive development of PSDTs and their applications that heavily rely on data. Real-time measurements and data historians from the grid do not necessarily cover all scenarios and events to ensure that the PSDT is suitable and a fit representation of the grid. A VPT offers a safe and flexible environment to configure a wide range of customized scenarios and conditions to generate data flows for the use of the PSDT as well as its applications. In particular,

- Controllability: The VPT enables full control over simulated power system conditions, including fault scenarios, load variations, and topology changes. This controllability allows for systematic evaluation of PSDT responses under diverse and often extreme conditions that would be difficult or risky to replicate in a live grid environment.
- Observability: Through the VPT, internal states and interactions across components can be monitored with high resolution and precision. This enhanced observability provides rich datasets for validating PSDT models and algorithms, including those related to forecasting, diagnostics, and control, without relying on limited field measurements. This is a particularly valuable property in the context of AI-based applications, where the black-box nature of many models can obscure causal relationships.
- Repeatability: Events in a real power system are rare. Testing via a VPT ensures that experiments can be repeated under exactly the same conditions, which is crucial for comparing different PSDT configurations, tuning parameters, or adaptation methods. Repeatability supports the development of rigorous performance benchmarks and the identification of subtle implementation issues that may otherwise go undetected.
- Nonintrusiveness: The VPT allows for PSDT testing and development without any interference with real-world power system operations. This ensures safety and avoids the risk of unintended consequences on physical infrastructure during validation activities.
- Cost-effectiveness: By removing the need for extensive physical testing or on-site trials, the VPT offers a low-cost environment for iterative development and evaluation of PSDTs. This is particularly advantageous during early-stage prototyping or when scaling up test scenarios.

Therefore, a VPT serves as a critical enabler for the deployment of robust, scalable, and resilient PSDT solutions for real-world applications. Fig. 9.1 further illustrates how these key properties collectively support the continuous process of designing, developing, testing, implementing, and deploying PSDTs and their associated applications.

9.4.3 Unlocking enhanced AI-based applications

AI and DT technologies improve one another. AI combines computer science to create expert systems to enable problem-solving, make predictions, or classifications based on data points. At the same time, DTs can produce simulated data, providing an infinite number of scenarios. The simulated data can then be used to train AI algorithms, making them more robust and reliable. The use of a VPT can further streamline the process of creating AI-based applications before the deployment of the PSDT. The VPT is used to generate datasets, support offline training and testing, and online performance evaluation.

As mentioned in the previous section, the VPT can overcome the barriers of data acquisition from actual power systems by replicating a wide range of scenarios and generating all necessary datasets. Generating datasets in a controlled environment also allows the creation of balanced datasets, thereby reducing the risk of data imbalance,

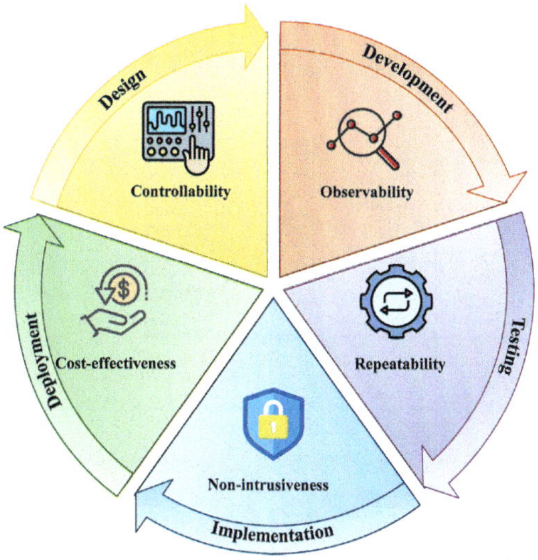

FIGURE 9.1 VPT properties. Key properties of a digital twin testbed platform and how they support the continuous deployment of power system digital twins and their applications.

which helps ensure fairness, accuracy, and generalization of the AI algorithm. The reliability of an AI algorithm is essential to ensure the practical implementation of the application. When training and testing an AI algorithm, its accuracy can be improved by generating additional scenarios not previously considered, preprocessing the datasets, and performing feature engineering, among others. Moreover, the controlled nature of a VPT-driven data generation allows users to systematically introduce variability, reducing the likelihood of overfitting.

The VPT can also be used to evaluate the performance of AI-based applications. By configuring different scenarios and conditions, the execution time and output of the solution are assessed. By taking the same steps but with different AI algorithms, a comprehensive evaluation can be performed, ensuring that the most suitable solution is selected and transferred to the field. This also provides an avenue to address the challenge of explainability. By making internal system states observable and reproducible, the detailed performance testing of black-box AI models provides opportunities for their interpretability and transparency.

9.4.4 Interfacing and integrating the virtual physical twin into power systems

In practice, the VPT should be interfaced with SCADA/EMS systems through common standards and frameworks such as CIM for network models, IEC 104, IEC 61850, and IEEE C37.118 for data streams and telecontrol, and gRCP/REST for user interfaces and services. The integration of software, hardware, and data flows within a unified architecture underpins VPT's functionalities, ensuring the interoperability, scalability, and consistency between the digital and physical domains.

Operational hardening measures, including data lineage, authenticated data streaming, and role-based access control, must be implemented throughout the ingestion, storage, and utilization of data within the VPT. These measures ensure the integrity, confidentiality, and reliability of grid operations, protecting against data corruption, unauthorized access, and cyber incidents.

Finally, a practical consideration when adopting VPTs in industry is the platform to consider. The VPT implementation should align with performance requirements (driven by services), budget constraints, and long-term deployment strategies. For instance, dedicated off-the-shelf hardware and software can offer high-fidelity modeling and simulations but come with significant capital and operational costs. Hybrid approaches can balance fidelity and cost by combining multiple flexible solutions tailored to the particular problem to solve. Alternatively, cloud-based solutions provide virtually unlimited scalability but may introduce latency and dependency on network reliability.

9.5 Virtual physical twin demonstration

9.5.1 Selection of hardware platform

Data-driven modeling is one of the main enabling technologies for the implementation of DTs. However, they are difficult to interpret. These black boxes may represent the proper behavior of a system and predict what may happen

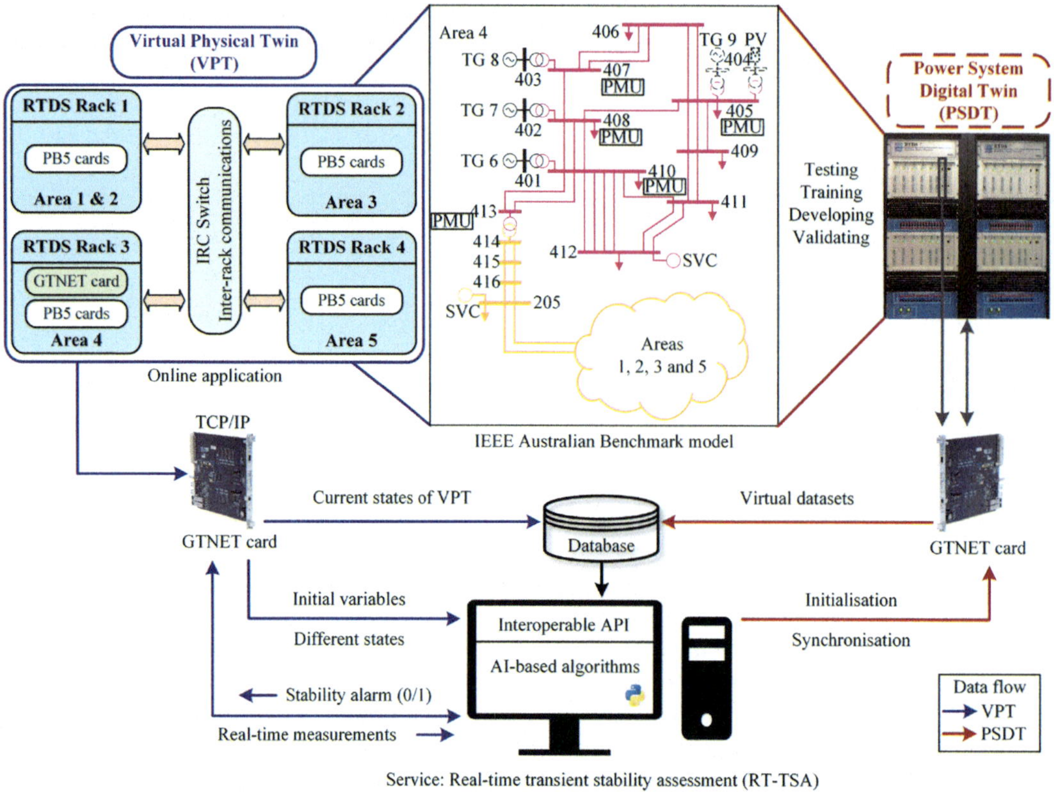

FIGURE 9.2 VPT testbed. Power system digital twin testbed system based on VPT. A real-time transient stability assessment (RT-TSA) service is illustrated. *VPT*, virtual physical twin. *From Shen, Z., Arraño-Vargas, F., & Konstantinou, G. (2024). Virtual testbed for development and evaluation of power system digital twins and their applications.* Sustainable Energy, Grids and Networks, 38, *101331. https://doi.org/10.1016/j.segan.2024.101331.*

in the future, but they make it difficult for users to understand the underlying mechanisms and rationale behind their outputs. Moreover, this is particularly challenging when developing DTs, as long-established physics-based models already exist. Consequently, physics-based models are still essential when mirroring a physical system.

Real-time digital simulators (RTSs) have proven to be the computing platform of choice for the reliable and accurate model-based simulation of modern power systems and their assets, also allowing control-, hardware-, and power HiL testing. The capability of solving model equations at a real-world clock pace while interfacing external devices and integrating data makes RTSs suitable for the design, development, and operation of high-fidelity PSDTs and, thus, a VPT testbed platform.

9.5.2 Setup

A PSDT testbed has been developed at the Real-Time Simulations Laboratory at UNSW Sydney (RTS@UNSW) (Arraño-Vargas et al., 2024). While the testbed is developed considering certain use cases, the underlying principles, architecture, and methodologies presented in the following sections are broadly applicable and can be adapted to other power system configurations and scenarios. This demonstration serves as a proof-of-concept, illustrating how key functionalities, such as data integration, model validation, and real-time interaction, can be realized in practice.

A PB5-based RTDS simulator is used to represent both the VPT (representing the actual power system) and the DT. The IEEE Australian Benchmark model is used as the network model for the VPT and DT. Three different sampling rates are acquired from the VPT: (1) high-resolution data captured from oscilloscopes (50 μs), (2) PMU datasets at a resolution of 10–50 Hz, and (3) SCADA-based measurements at a resolution of 0.25–0.5 Hz. The real-time simulators are interfaced to a workstation containing a database where all the data generated by the VPT is stored and where AI algorithms are trained and implemented using Python. The setup of the complete testbed is shown in Fig. 9.2.

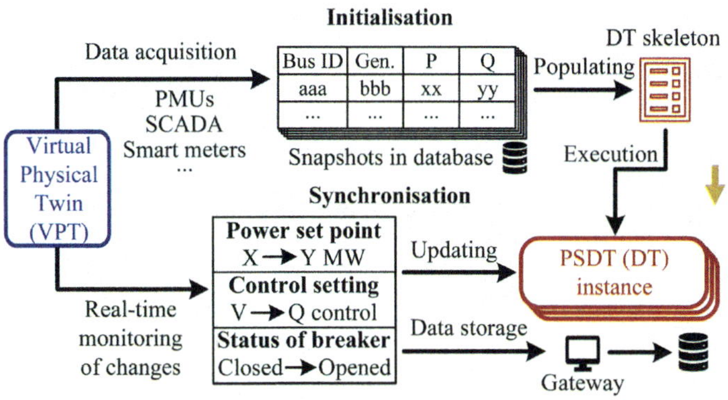

FIGURE 9.3 VPT Init & Sync. Initialization and synchronization of PSDTs using the VPT testbed. *PSDTs*, power system digital twins; *VPT*, virtual physical twin. *From Shen, Z., Arraño-Vargas, F., & Konstantinou, G. (2025). Development of adaptive model-based digital twins for evolving power systems.* Computers and Electrical Engineering, 125, *110418. https://doi.org/10.1016/j.compeleceng.2025.110418.*

9.5.3 Initialization and synchronization of a power system digital twins

A physics-based simulation model of a power system provides a clear starting point for the development of a PSDT. In this case study, the VPT testbed is used to test the initialization and synchronization of a PSDT.

Initialization and synchronization are critical functions for the operation of any DT (Palensky, Cvetkovic, Gusain, & Joseph, 2022). Initialization refers to the function that launches the operation of the DT or one of its instances (e.g., when multiple DTs are running in parallel), considering the current states of the physical twin as the starting point. Synchronization, on the other hand, is the function that reflects state changes of the physical twin into the DT, allowing for tracking dynamic changes.

In our implementation, initializations are performed using a reusable "skeleton" of the grid model and a Python script that maps all unique model identifiers with their measured values from the VPT. The skeleton works as a blueprint allowing to initialize multiple PSDT instances (i.e., simulations). Variables defined in the skeleton include the status of breakers, the operation mode of controllers, generator set points, and loads. It should be highlighted at this stage that maintaining the skeleton with up-to-date status of all components and configurations and new additions in the power system is a nontrivial challenge in itself. It involves model versioning, validation, and coordination that are aligned with system changes. However, addressing this challenge lies beyond the scope of the present implementation and this chapter.

Once the PSDT is initialized, it is crucial to keep it synchronized with the VPT. This is done by a combination of time- and event-based approaches. To reduce model-to-field skew under steady-state conditions, a periodic update, every 0.5 s, of the PSDT is performed. In each of these updates, the state of the VPT is evaluated and included in the PSDT. Event-driven updates are triggered by sudden changes in the VPT happening in the order of 30–100 ms. These changes include faults and breaker operation. Once detected, the same change is applied to the DT.

Fig. 9.3 illustrates initialization and synchronization processes for PSDTs based on physics-based models. The VPT is used as a replacement for the physical twin, generating system snapshots. The performance of the initialization and synchronization functions is carried out by benchmarking VPT measurements with PSDT outcomes. For instance, a recommended way to ensure that initialization is correct is by performing a load flow calculation.

9.5.4 Adaptation of a power system digital twins

Adaptation refers to the continuous evolution of a DT (Bazmohammadi et al., 2022). It goes beyond initialization and synchronization as adaptation allows tuning/adjusting model parameters in real-time based on measurements from the physical twin. This adaptive process can be time-driven, event-driven, intelligent, or hybrid (Bai et al., 2022). In particular, a PSDT is expected to constantly update itself using grid data flows to maintain its accurate representation.

In this case, a VPT testbed can be used to develop and validate the algorithms that identify when a deviation exists between physical and DTs, as well as approaches of changing the DT to align the two twins as quickly and as accurately as possible (Shen et al., 2025). This process, shown in Fig. 9.4, consists of two main functions: component identification and parameter estimation.

Component identification is the function capable of detecting which component in the PSDT has deviated from the VPT. In this example, an ML-based algorithm with five layers is provided with the residual errors (i.e., deviations) to then create a probability distribution (Softmax function) of the potential components that are deviating. From a practical point of view, the computational burden incurred by the component identification stage can be reduced by

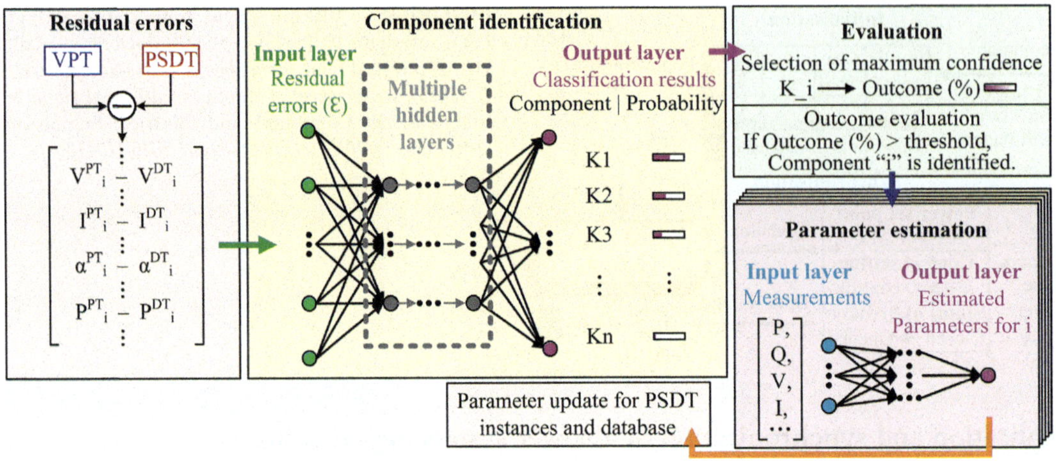

FIGURE 9.4 VPT Adapt. Adaptation of component parameters using the VPT testbed. *VPT*, virtual physical twin. *From Shen, Z., Arraño-Vargas, F., & Konstantinou, G. (2025). Development of adaptive model-based digital twins for evolving power systems.* Computers and Electrical Engineering, 125, *110418. https://doi.org/10.1016/j.compeleceng.2025.110418.*

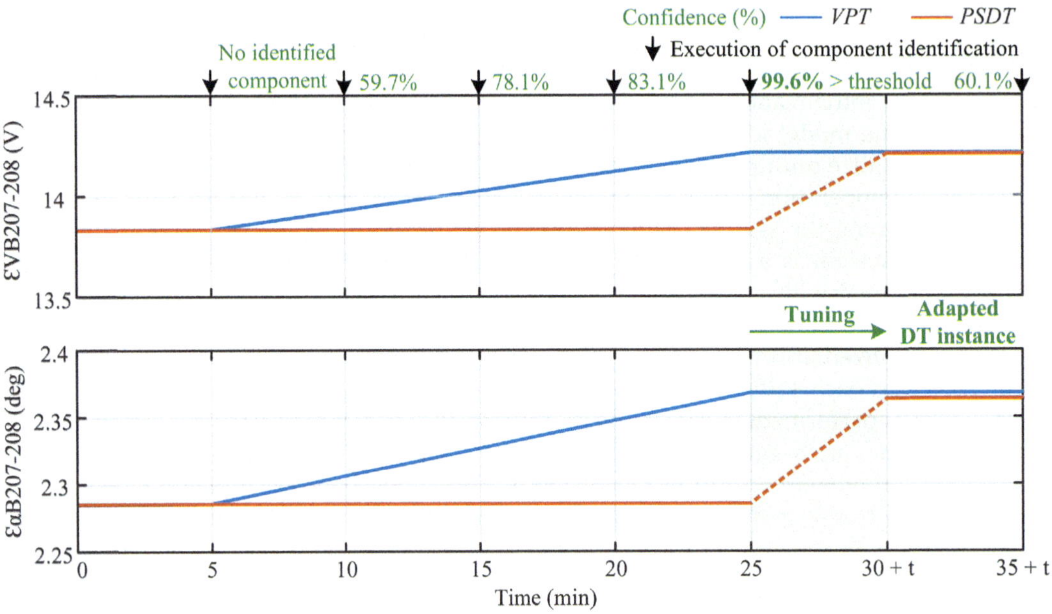

FIGURE 9.5 VPT Adapt Ex. Example of adaptation. Measured residual errors between VPT and PSDT. *PSDTs*, power system digital twins; *VPT*, virtual physical twin. *From Shen, Z., Arraño-Vargas, F. & Konstantinou, G., (2025). Development of adaptive model-based digital twins for evolving power systems.* Computers and Electrical Engineering, 125, *110418. https://doi.org/10.1016/j.compeleceng.2025.110418.*

adding additional guardrails. For instance, it can be triggered if residual errors are over a threshold for a period of time defined by the user.

The outcome of the component identification is later evaluated against a predefined threshold (in this case, 90%) to trigger the parameter estimation function. This last function uses real-time VPT measurements to estimate the component parameters based on an ML-based algorithm (i.e., a regression problem). The new values are recorded in the database and used to adapt the PSDT. This two-step process, running online, is repeated in order to reassess any remaining deviations and re-estimate the corresponding parameter, minimizing errors between PSDT and VPT until all deviations are below their predefined threshold.

Fig. 9.5 illustrates the process of adapting a transmission line whose parameters have deviated. The impedance in the VPT was subjected to a 1% increase every 5 minutes. The component identification block is executed to evaluate the residual errors against these changes. The parameter of the PSDT is not tuned until the 90% threshold is reached at minute 25. After updating the parameters of the transmission line, a new PSDT instance is initialized and executed,

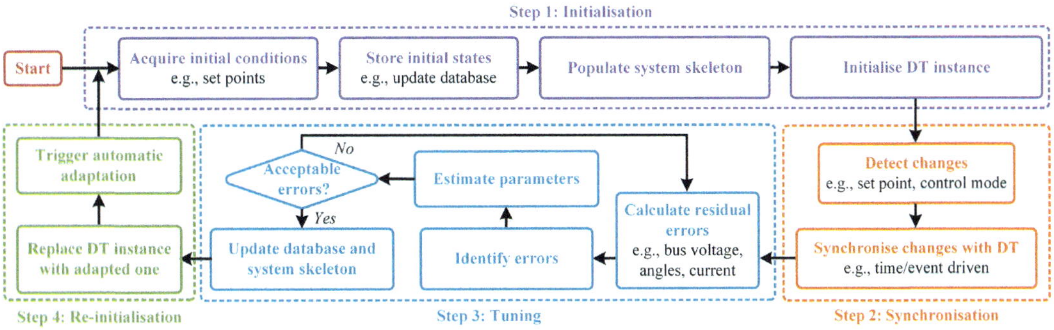

FIGURE 9.6 Adaptive workflow. Adaptive workflow: Initialization, synchronization, tuning, and reinitialization. *Modified Shen, Z., Arraño-Vargas, F. & Konstantinou, G. (2025). Development of adaptive model-based digital twins for evolving power systems.* Computers and Electrical Engineering, *125, 110418. https://doi.org/10.1016/j.compeleceng.2025.110418.*

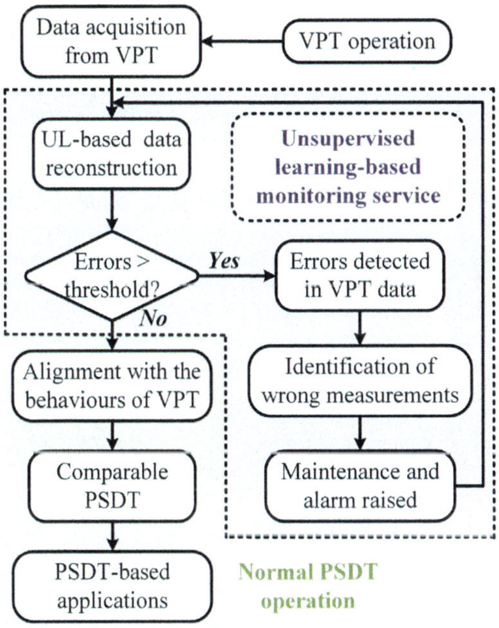

FIGURE 9.7 VPT monitoring. Monitoring services for updating PSDTs based on the VPT testbed. *VPT, virtual physical twin. Modified Shen, Z., Sun, P., Arraño-Vargas, F., & Konstantinou, G. (2024). Monitoring Service for Updating Power System Digital Twins.* 2024 IEEE PES 16th Asia-Pacific Power and Energy Engineering Conference (APPEEC), *1–5. https://doi.org/10.1109/APPEEC61255.2024.10922425.*

replacing the outdated instance while reducing downtime. The residual errors are now minimized, and the PSDT has been adapted.

The reinitialization step is usually necessary due to real-time implementation requirements, where changes to network parameters often require recompilation or conversion of the network model to meet the constraints of real-time simulation. If additional instances, as illustrated before, cannot be executed from a practical feasibility perspective, reinitialization may introduce non-negligible computational overhead, delays in PSDT services, and/or even the halting of the PSDT. Mitigation strategies include the design of symbolic or parametric networks or the use of external APIs to change internal simulation parameters. Fig. 9.6 illustrates a complete adaptive workflow including the reinitialization step.

It is worth noting that the aforementioned adaptation process is valid if the measurements provided, from the VPT to the PSDT, present a high level of fidelity and can be used directly to adapt the PSDT. Since measurements are subjected to errors and/or attacks, a wrong set of these can lead to a degradation in the adaptation process of the PSDT and, consequently, the operation of its applications. To overcome this issue, anomaly detection algorithms can be implemented when receiving measurements from the power system (Tabassum et al., 2024). Fig. 9.7 shows a generic workflow based on unsupervised learning algorithms to identify anomalies and errors in data streams. The VPT testbed is used to provide such data streams while also injecting anomalies.

TABLE 9.2 Summary of RT-TSA results for one specific operating condition.

Algorithm	ANN	DeT	SVM	DNN_1	DNN_2	LSTM
Online accuracy*	100%	100%	100%	100%	100%	100%
Execution time of stable cases (ms)						
Average	382.4	**83.8**	84.1	323.4	233.3	649.1
Maximum	752.4	**89.5**	94.4	507.1	244.8	990.2
Minimum	245.9	81.9	**81.8**	255.6	224.1	587.0
Std dev	126.4	**1.96**	3.45	87.6	6.29	107.7
Execution time of unstable cases (ms)						
Average	271.0	**83.1**	83.4	295.4	258.5	656.4
Maximum	410.1	**89.8**	97.7	487.4	432.1	1147
Minimum	247.2	**81.91**	**81.9**	271.6	230.0	596.8
Std dev	32.9	**1.85**	3.31	45.2	46.7	116.6

From Shen, Z., Arraño-Vargas, F., & Konstantinou, G. (2025). Enhancing artificial intelligence-enabled power system stability assessment through digital twin-based testbeds. 2025 IEEE 16th International Symposium on Power Electronics for Distributed Generation Systems, 385–390. https://doi.org/10.1109/pedg62294.2025.11060335.

* *30 random new cases are tested through the VPT, beyond the existing datasets. While the reported accuracy is 100% in this table, it may decrease when the algorithms are applied to other cases.*

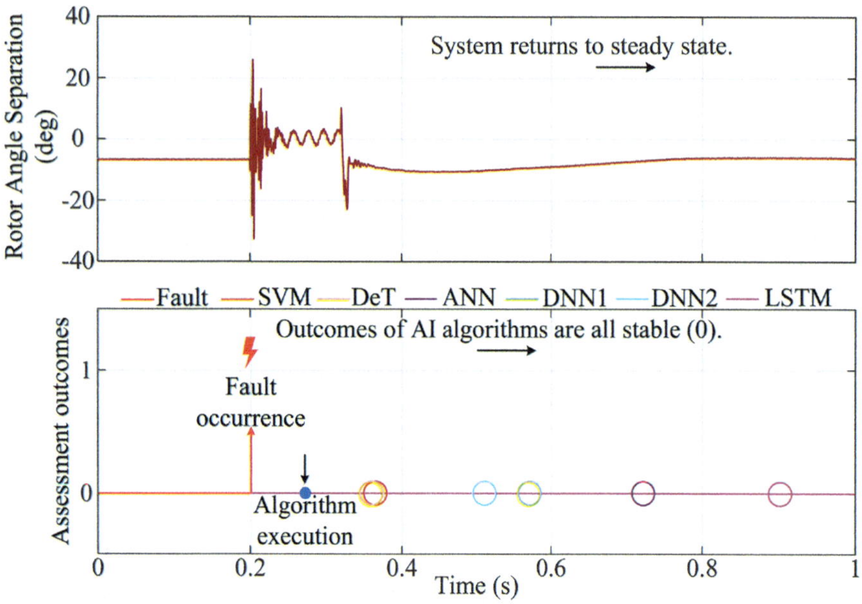

FIGURE 9.8 RT-TSA stable. Different AI-based RT-TSA algorithms for a stable case. *PSDTs*, power system digital twins; *RT-TSA*, real-time transient stability assessment. *From Shen, Z., Arraño-Vargas, F., & Konstantinou, G. (2025). Enhancing artificial intelligence-enabled power system stability assessment through digital twin-based testbeds.* 2025 IEEE 16th International Symposium on Power Electronics for Distributed Generation Systems, *385–390. https://doi.org/10.1109/pedg62294.2025.11060335.*

9.5.5 Power system digital twins application 1: Real-time transient stability assessment

Modern power systems are facing new operational and planning challenges. Reasons can be attributed to several factors, including the reduced investment in transmission infrastructure, increased system complexity, cybersecurity threats (Sridhar et al., 2012), among others. In particular, the dynamic behavior of power systems has changed due to the integration of new technologies, which has led to incorporating the effects of fast-response power electronics devices as part of the new power system stability classification (Hatziargyriou et al., 2021). Consequently, system operators require new, more advanced solutions that can support them to improve operations in a timely and effective manner. One such application is the real-time (online) transient stability assessment (RT-TSA), which has the potential to be implemented in PSDTs.

In this section, the VPT testbed is used to analyze the performance of AI-based RT-TSAs, including an artificial neural network (ANN), a decision tree (DeT), a support vector machine, two deep neural networks, and a long short-term memory (LSTM). These algorithms are tested for three-phase-to-ground faults at various locations under

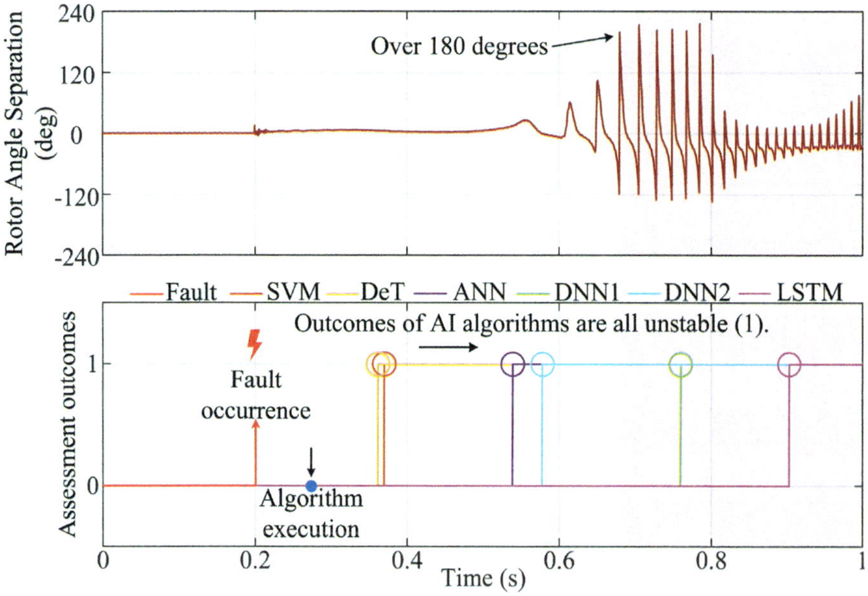

FIGURE 9.9 **RT-TSA unstable** Different AI-based RT-TSA algorithms for an unstable case. *PSDTs*, power system digital twins; *RT-TSA*, real-time transient stability assessment. *From Shen, Z., Arraño-Vargas, F. & Konstantinou, G. (2025). Enhancing artificial intelligence-enabled power system stability assessment through digital twin-based testbeds.* 2025 IEEE 16th International Symposium on Power Electronics for Distributed Generation Systems, *385–390. https://doi.org/10.1109/pedg62294.2025.11060335.*

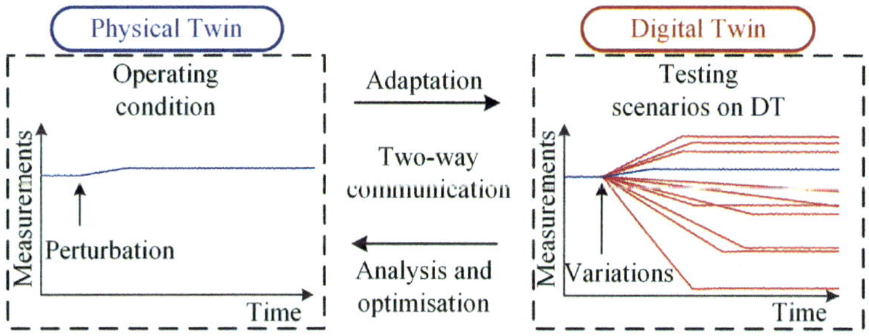

FIGURE 9.10 VPT what-if. A VPT testbed can generate multiple *what-if* scenarios to assess the evolution of the power system. *VPT*, virtual physical twin. *From Shen, Z., Arraño-Vargas, F., & Konstantinou, G. (2025). Enhancing artificial intelligence-enabled power system stability assessment through digital twin-based testbeds.* 2025 IEEE 16th International Symposium on Power Electronics for Distributed Generation Systems, *385–390. https://doi.org/10.1109/pedg62294.2025.11060335.*

different load conditions. Rotor angle separations are calculated between any two synchronous generators to assess the stability of the system after applying the disturbance. For separations larger than 180 degrees, the system is considered unstable.

The VPT provides measurements with a sampled frequency of 50 Hz in order to generate 4100 datasets, where 67% and 33% correspond to stable and unstable cases, respectively. A ratio of 8:2 is used for the training and testing of the algorithms. In the offline evaluation, all algorithms show an accuracy of over 98.5%, illustrating their potential for online implementations in a PSDT.

To evaluate the execution time of each algorithm, online tests for unknown scenarios are conducted. In this case, the VPT works as the PSDT while the algorithms run on a workstation equipped with a 2.9 GHz Intel Core i7-10700 CPU and 16 GB of RAM. Table 9.2 summarizes the execution times across 30 operating conditions and fault locations. While all algorithms successfully assess the system's stability, their average online execution times differ significantly, allowing system operators to evaluate the approach that best meets operational constraints and response-time demands under real-time conditions. Figs. 9.8 and 9.9 further illustrate the performance of each RT-TSA algorithm for a stable and an unstable case, respectively.

This RT-TSA case demonstrates the value of the VPT testbed in generating datasets, training different algorithms, and assessing their performance, which can be used by operators to select the most suitable solution to be deployed in a PSDT or an actual grid.

9.5.6 Power system digital twins application 2: *What-if* scenario testing

As discussed in Section 2.2, DTs are powerful tools for forecasting and predicting potential states of a system (Joseph et al., 2018), as shown in Fig. 9.10.

In this example, the VPT testbed is used to demonstrate the applicability of *what-if* scenario testing. To achieve this, multiple PSDT instances are run in parallel. All instances are initialized and synchronized to the VPT (as described in

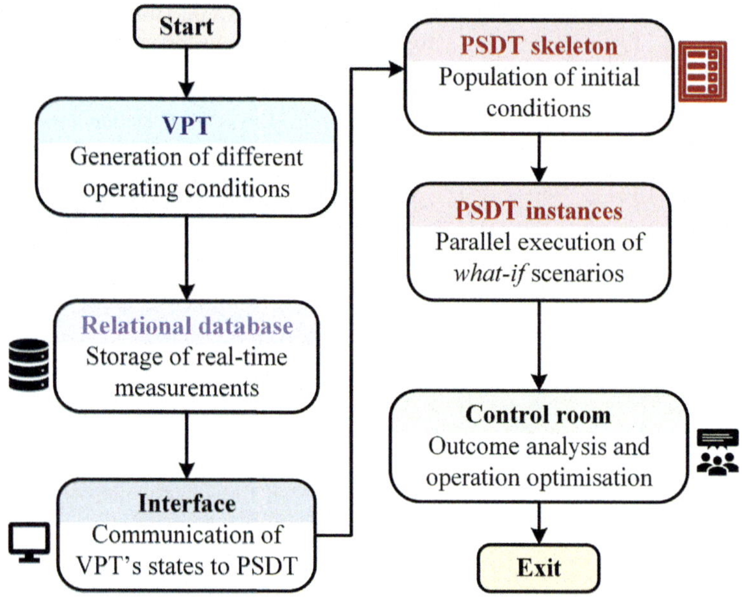

FIGURE 9.11 VPT what-if flow. Workflow for the development of *what-if* scenario testing based on the VPT testbed. *VPT*, virtual physical twin. *Modified from Shen, Z., Arraño-Vargas, F., & Konstantinou, G. (2024). What-if scenario testing through power system digital twins.* 2024 IEEE 22nd International Conference on Industrial Informatics (INDIN), *1–6. https://doi.org/10.1109/INDIN58382.2024.10774259.*

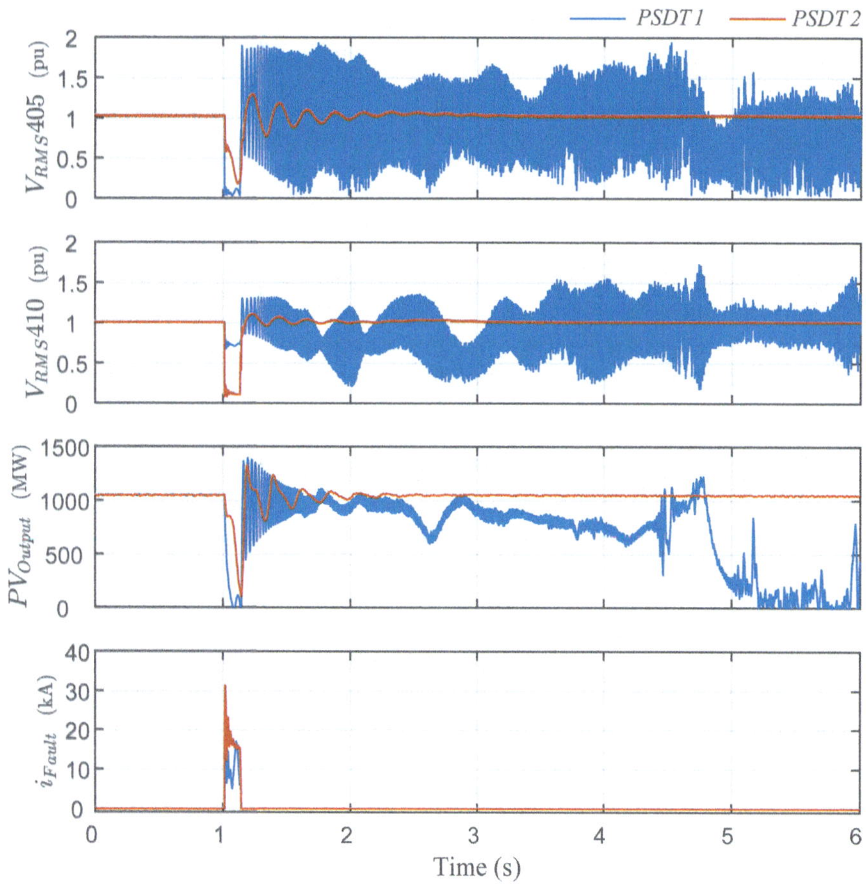

FIGURE 9.12 VPT What-if Ex. Two parallel executions of *what-if* scenario testing. *VPT*, virtual physical twin. *From Shen, Z., Arraño-Vargas, F., & Konstantinou, G. (2024). What-if scenario testing through power system digital twins.* 2024 IEEE 22nd International Conference on Industrial Informatics (INDIN), *1–6. https://doi.org/10.1109/INDIN58382.2024.10774259.*

Section 5.3), but each instance executes a predefined scenario. In this case, faults in critical buses of the system. Once the executions are done, the operators can take actions to ensure the continuity of service. A complete workflow illustrating these steps is shown in Fig. 9.11.

Fig. 9.12 shows the case in which two PSDT instances are executed in parallel for the same operating condition. As can be seen, one of the scenarios (PSDT1 in blue) results in an unstable system. This information provides system operators with invaluable insights about the system in real time. Mitigation actions can be taken manually or automatically to reduce the risk of this scenario from happening. Furthermore, this kind of testing supports the training of operators, where they can access a dedicated library of scenarios while also replaying some of the critical cases

9.6 Conclusion

PSDTs are expected to offer an unprecedented ability to replicate power grids and their behaviors. However, *how can we ensure that PSDTs and their applications are properly setup and validated in realistic, evolving environments?*

This chapter introduced the concept of a VPT testbed as a foundational enabler for the development and deployment of PSDTs. Through the VPT, it becomes possible to initialize, adapt, and evaluate DT instances in a controlled yet interactive real-time environment. This approach bridges the gap between conceptual PSDT frameworks and their practical integration into power system operations. It facilitates testing of DT algorithms and the examination of their interaction with data, system states, and real-world disturbances.

Overall, this chapter highlights how VPT-enabled testbeds can serve as an essential intermediary layer between offline development and real-world deployment, offering a robust foundation for trustworthy, adaptive, and high-impact PSDT applications. Even though a VPT testbed offers a wide range of transformative opportunities to shape the future deployment of PSDTs, several open research questions and challenges remain. These include data management and governance, representing the complex multiphysics nature of modern grids, interconnection and interoperability among twins (e.g., from different domains and entities), alongside technical readiness and changes in workforce skills, regulatory acceptance, and integration with legacy systems.

References

(NREL) National Renewable Energy Laboratory (2025). Advanced Research on Integrated Energy Systems (ARIES) https://www.nrel.gov/aries

Arraño-Vargas, F., Agelidis, V. G., & Konstantinou, G. (2024). UNSW Sydney's Real-Time Simulations Laboratory (RTS@UNSW): Supporting power electronics defined power systems, down under. *IEEE Open Journal of Power Electronics, 5*, 565–574. doi:10.1109/OJPEL.2024.3386198.

Arraño-Vargas, F., & Konstantinou, G. (2023). Modular design and real-time simulators toward power system digital twins implementation. *IEEE Transactions on Industrial Informatics, 19*(1), 52–61. doi:10.1109/TII.2022.3178713.

Bai, H., Yuan, Z., Tang, X., Liu, J., Yang, W., Pan, S., Xue, Y., & Liu, W. (2022). Automatic modeling and optimization for the digital twin of a regional multi-energy system. In *Proceedings - 2022 Power System and Green Energy Conference, PSGEC: 2022* (pp. 214–219). doi:10.1109/PSGEC54663.2022.9881075.

Bazmohammadi, N., Madary, A., Vasquez, J. C., Mohammadi, H. B., Khan, B., Wu, Y., & Guerrero, J. M. (2022). Microgrid digital twins: Concepts, applications, and future trends. *IEEE Access, 10*, 2284–2302. doi:10.1109/ACCESS.2021.3138990.

(PNNL) Pacific Northwest National Laboratory 2025 (2025). powerNET Testbed https://www.pnnl.gov/projects/center-collaborative-cyber-physical-research/powernet-testbed

Bentley 2024 (2024). POWERCHINA Hubei Optimizes city power grid with digital delivery of Xianning Chibi Substation https://www.bentley.com/wp-content/uploads/cs-powerchina-hubei-ltr-en-lr.pdf

Gegner, K. M., Birchfield, A. B., Xu, T., Shetye, K. S., & Overbye, T. J. (2016). A methodology for the creation of geographically realistic synthetic power flow models. *2016 IEEE Power and Energy Conference at Illinois, PECI 2016*. doi:10.1109/PECI.2016.7459256.

Guo, J., Zhang, J., Zheng, N., Zhang, H., & Jian, S. (2023). Research on the application of digital twin technology in power grid engineering. *Southern Energy Construction, 10*(2), 136–142. doi:10.16516/j.gedi.issn2095-8676.2023.02.018.

Hatziargyriou, N., Milanovic, J., Rahmann, C., Ajjarapu, V., Canizares, C., Erlich, I., Hill, D., Hiskens, I., Kamwa, I., Pal, B., Pourbeik, P., Sanchez-Gasca, J., Stankovic, A., Van Cutsem, T., Vittal, V., & Vournas, C. (2021). Definition and classification of power system stability—Revisited & extended. *IEEE Transactions on Power Systems, 36*(4), 3271–3281. doi:10.1109/TPWRS.2020.3041774.

Joseph, A., Cvetkovic, M., & Palensky, P. (2018). Predictive mitigation of short term voltage instability using a faster than real-time digital replica. In *Proceedings—2018 IEEE PES Innovative Smart Grid Technologies Conference Europe*. ISGT-Europe 2018. doi:10.1109/ISGTEurope.2018.8571803.

Kandasamy, N. K., Venugopalan, S., Wong, T. K., & Leu, N. J. (2022). An electric power digital twin for cyber security testing, research and education. *Computers and Electrical Engineering, 101*, 108061. doi:10.1016/j.compeleceng.2022.108061.

Metcalfe, D., & Tohani, M. (2019). *Five digital twin strategies for industrial facilities. Verdantix* https://www.verdantix.com/venture/report/five-digital-twin-strategies-for-industrial-facilities.

Palensky, P., Cvetkovic, M., Gusain, D., & Joseph, A. (2022). Digital twins and their use in future power systems. *Digital Twin, 1*(3). doi:10.12688/digitaltwin.17435.2.

Shen, Z., Arraño-Vargas, F., & Konstantinou, G. (2024). Virtual testbed for development and evaluation of power system digital twins and their applications. *Sustainable Energy, Grids and Networks, 38*, 101331. doi:10.1016/j.segan.2024.101331.

Shen, Z., Arraño-Vargas, F., & Konstantinou, G. (2025). Development of adaptive model-based digital twins for evolving power systems. *Computers and Electrical Engineering, 125*, 110418. doi:10.1016/j.compeleceng.2025.110418.

Sridhar, S., Hahn, A., & Govindarasu, M. (2012). Cyber-physical system security for the electric power grid. *Proceedings of the IEEE, 100*(1), 210–224. doi:10.1109/JPROC.2011.2165269.

Tabassum, T., Toker, O., & Khalghani, M. R. (2024). Cyber–physical anomaly detection for inverter-based microgrid using autoencoder neural network. *Applied Energy, 355*. doi:10.1016/j.apenergy.2023.122283.

TwinEU (2025). TwinEU: Developing a concept of Pan-European Digital Twin of the electricity system https://twineu.net/

CHAPTER

10

Automatic control and reinforcement learning techniques for microgrids

Davide Salaorni[a], *Federico Bianchi*[b], *Marcello Restelli*[a], *Francesco Trovò*[a]

[a]Department of Electronic, Information, and Bioengineering, Politecnico di Milano, Milano, Italy
[b]Ricerca sul Sistema Energetico, Milano, Italy

10.1 Introduction

The increasing global demand for energy and the urgent need to reduce environmental impact are significantly transforming the energy system. Central to this transition is the widespread adoption of renewable energy sources (RESs) and distributed energy resources (DERs), which are significantly reshaping traditional centralized power generation models. RESs, such as solar, wind, hydroelectric, biomass, and geothermal energy, offer a sustainable alternative to fossil fuels (Ellabban et al., 2014). Their ability to generate electricity with minimal environmental impact has positioned them as a cornerstone of modern energy strategies. However, their inherent intermittent nature and low inertia, or lack of inertial response, pose challenges for maintaining a stable, reliable, and high-quality power supply (Basit et al., 2020). To address these challenges, DERs, consisting of localized energy generation and storage systems, including rooftop solar panels, wind turbines, battery storage, and combined heat and power (CHP) units, are getting more and more prominent in the energy distribution environment. Unlike conventional power plants, DERs enable a decentralized energy model that enhances grid flexibility, reduces transmission losses, and improves overall energy resilience.

An essential component of this decentralized model is the development of microgrids (MGs), which are local power grids that can operate synchronously with the main electrical network and seamlessly switch to a standalone mode (a.k.a. islanded) without service disruptions. By integrating RESs and DERs, conventional generation, and load clusters, MGs optimize local energy management, balance supply and demand more efficiently, and contribute to grid stability (Mariam et al., 2016). Their deployment is particularly beneficial in remote areas, industrial zones, and regions prone to natural disasters, where reliable energy access is crucial. From a main grid perspective, an MG acts as a single controlled load within the power system, and having generation situated close to the load significantly reduces transmission losses. However, uncertainties in RES availability and load fluctuations can pose challenges to the secure and economical operation of MGs. The intermittent and stochastic nature of RES, such as solar and wind power, complicates real-time supply-demand balancing and often requires additional reserves, storage, or backup generation, increasing both operational complexity and costs. This calls for a suitable optimal MG operation scheme, as well as coordination and interaction mechanisms with the main grid. This scheme needs to account also for the DERs' contribution to ancillary services. Indeed, if DERs merely displace energy from a central generation without providing equivalent flexibility and capacity, the overall cost of operating the entire system will rise (Lopes et al., 2007).

In this context, the concept of a Digital Twin (DT) emerges as a promising tool for enhancing the design, validation, and operation of MGs. A DT, serving as a high-fidelity virtual replica of the physical MG, provides a safe and accurate environment for studying its inherent behavior and enabling the integration of advanced control strategies without risking instability in real-world operations. Moreover, DTs facilitate seamless coordination between the MG and the main grid, allowing predictive and adaptive strategies to be validated under diverse operating conditions before

Digital Twin Technology and Smart Grid.
DOI: https://doi.org/10.1016/B978-0-443-36664-2.00017-4

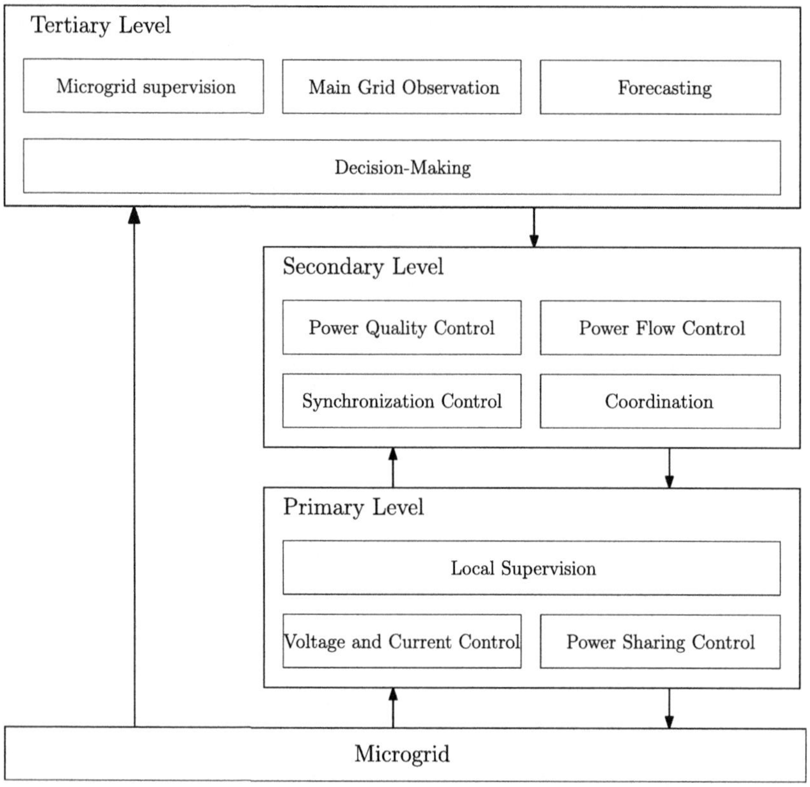

FIGURE 10.1 Hierarchical multi-layered MG framework comprising three different control levels, firstly presented in Meng et al. (2017).

deployment. In the following sections, the structure and formalization of the control problem are presented, along with the modeling of the entities that compose the MG network and are replicated by the DT.

10.1.1 Hierarchical control scheme of microgrids

MGs can be categorized into three main types: alternating current (AC) MGs, direct current (DC) MGs, and hybrid AC/DC MGs. Every MG requires an effective control strategy to meet its design goals, regardless of its type. It is widely recognized that MG control and management involve multiple objectives, spanning different technical areas, time scales, and physical levels (Bidram & Davoudi, 2012; Meng et al., 2017). According to the literature, a widely adopted MG control scheme is the hierarchical, multilayered framework, which effectively manages operations by dividing tasks across three levels, as shown in Fig. 10.1. In other works, such as Olivares et al. (2014), the tertiary control level is primarily responsible for coordinating multiple MGs or subsystems within the hosting grid, effectively serving as a hosting-grid control layer. Meanwhile, the secondary level is tasked with the operational and economic optimization of power and energy flows within the MG. In this work, we adopt the hierarchical scheme described by Meng et al. (2017), where the tertiary control level is more focused on optimizing and managing the MG internally. More specifically:

1. **Primary control** focuses on regulating local power, voltage, and current. It typically follows the setpoints provided by higher-level controllers and executes control actions through interface power converters.
2. **Secondary control** operates above the primary level and addresses system-wide challenges, including power quality regulation, synchronization with the external grid for seamless reconnection, and coordination among the MG units.
3. **Tertiary control** is responsible for overall operational optimization, cost-effective performance, system-wide management, and regulatory compliance.

Control levels can be implemented using a centralized, decentralized, distributed, or hierarchical approach, as illustrated in Fig. 10.2. These structures differ in their reliance on a central controller and the extent of communication required between the MG components. Centralized control (Fig. 10.2a) features a central control unit that collects data from all essential MG components to execute control and management procedures for secure and efficient operation. The advantages of centralized control include strong observability and controllability of the entire system, as well as

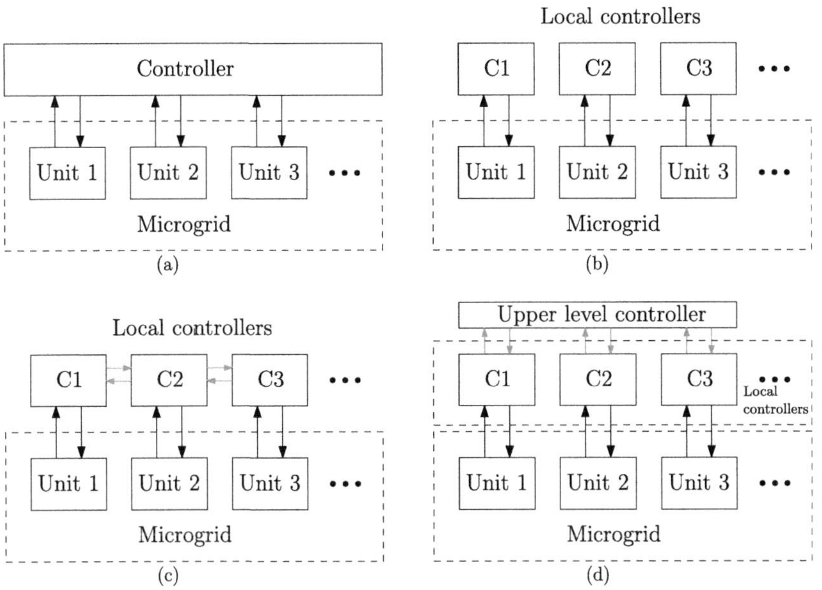

FIGURE 10.2 Typical design patterns, elaborated in Meng et al. (2017), that can be adopted to implement each control layer: centralized control (a), decentralized control (b), distributed control (c), and hierarchical control (d). *Green arrows* denote communication flows.

straightforward implementation. However, a significant drawback is the single point of failure issue, as the breakdown of the central controller leads to the loss of all functions. Furthermore, it can suffer from reduced flexibility and expandability and necessitates considerable computational resources. Consequently, centralized control is generally more suitable for small, localized MGs. In decentralized control (Fig. 10.2b), each unit regulates itself using only local information, without requiring data from other parts of the system. The advantage of this approach is the absence of real-time communication requirements. However, the lack of coordination between local regulators limits the possibility of achieving globally coordinated behaviors. Unlike centralized control, distributed control (Fig. 10.2c) does not rely on a central controller.

Instead, local controllers communicate with their neighbors to coordinate their actions based on both local measurements and exchanged information. This structure enables functionalities similar to centralized control while offering greater flexibility and resilience. However, achieving effective coordination necessitates well-designed communication protocols and consensus algorithms to enable agents to reach a unified understanding of shared information in a distributed manner.

As energy systems grow more complex and involve intricate decision-making processes, not all functions can be effectively executed in a distributed or decentralized manner. Consequently, a hierarchical control structure (Fig. 10.2d) is becoming a standard configuration in MGs. Primary control, responsible for basic functions such as voltage and current regulation and power sharing, is typically managed by local controllers. In contrast, secondary and tertiary control functions are traditionally centralized, as they rely on global information from key system components.

10.1.2 Problem formulation

In this section, we present a model for a tertiary-level controller to manage power exchanges between the MG and the main grid. Such a model can account for operational limits and costs of MG components, market conditions, grid constraints, and long-term objectives. It serves as a general framework that can be adapted based on specific operational contexts, such as grid-connected or islanded modes.

The MG's operation is modeled over a finite time horizon (typically 1 day), divided into time slots of duration Δ_t. The MG is connected to the main grid (or other MGs) and typically includes various dispatchable resources (such as batteries, thermostatically controlled loads, or generators like diesel internal combustion engines or microturbines) and nondispatchable sources (such as wind turbines and solar photovoltaic [PV] systems). The dispatchable and nondispatchable units are represented by the sets $\mathcal{D}$ and $\mathcal{N}$, respectively. Accordingly, the (average) power $p_{\text{net}}(t)$ exchanged by the MG with the grid in the time slot t is given by:

$$p_{net}(t) = \sum_{i\in\mathcal{D}} p_i^{\mathcal{D}}(t) + \sum_{i\in\mathcal{N}} p_i^{\mathcal{N}}(t), \tag{10.1}$$

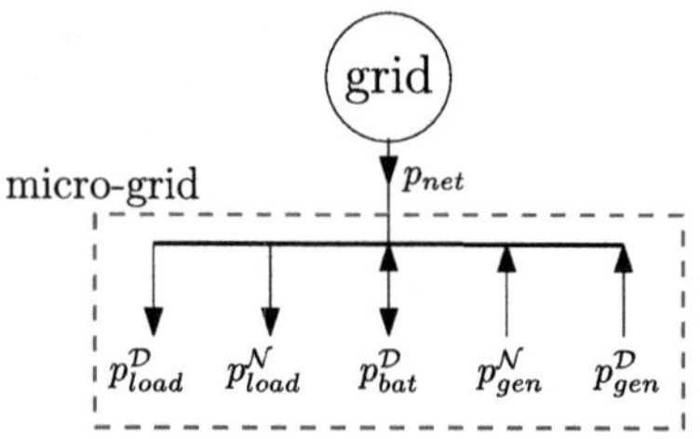

FIGURE 10.3 Schematic view of the exchange flows between an MG and the main grid. As illustrated, the net power depends on the aggregated exchanges of dispatchable and non-dispatchable units.

where $p_i^{\mathcal{D}}(t) \in \mathbb{R}$ and $p_i^{\mathcal{N}}(t) \in \mathbb{R}$ denote the power output of the i-th dispatchable and nondispatchable units, respectively. For instance, an example is provided in Fig. 10.3, where a dispatchable load, a nondispatchable load, a battery, a nondispatchable generator, and a dispatchable generator (from left to right) contribute to the overall power of the MG.

Typically, the MG and the main grid coordinate in advance to establish a power exchange profile $[\bar{p}(1), \dots, \bar{p}(T)]$, where $T \in \mathbb{N}$ represents the number of time slots in the considered time horizon. Indeed, defining in advance $\bar{p}(t)$ for each time slot t ensures that the MG neither injects surplus energy into the network nor experiences a shortage that could lead to unmet demand. This helps minimize energy losses and maintain a stable, reliable exchange with the main grid. Hence, the goal of the MG controller is to follow the predefined power exchange profile, aiming to maintain:

$$p_{\text{net}}(t) = \bar{p}(t), \; t \in \{1, \dots, T\}. \tag{10.2}$$

To do so, the controller must adjust for any discrepancies between the actual and forecasted power outputs of nondispatchable units in $\mathcal{N}$, as well as account for possible variations in energy prices. In Section 10.1.3, we provide examples of models for various dispatchable units. For forecasting models of nondispatchable units, we refer the interested reader to Palma-Behnke et al. (2013). This involves determining the optimal power allocation for the flexible units in $\mathcal{D}$ while ensuring compliance with their respective operational limits. Some remarks are needed. First, note that when $\bar{p}(t) = 0$ for all $t \in \{1, \dots, T\}$, the control objective effectively promotes full self-consumption within the MG. This aligns with the typical requirements of islanded operation, where local power balance must be maintained without external grid support. Second, depending on the nature of the source, the power values $p_i^{\mathcal{D}}(t)$ and $p_i^{\mathcal{N}}(t)$ may need to satisfy specific constraints.

The optimality of power dispatch is assessed based on unit-specific operational costs, such as fuel consumption for generators and degradation costs for batteries, alongside the economic value of energy exchanges with the main grid. Additionally, imbalance penalties resulting from deviations from the scheduled power profile are considered and quantified by:

$$\Delta_p(t) = \bar{p}(t) - p_{\text{net}}(t). \tag{10.3}$$

Other relevant objectives may also be included, such as those discussed in Sections 10.2 and 10.3, which detail additional cost components of interest.

10.1.3 Resource modeling

This section presents a possible modeling of flexible dispatchable ($\mathcal{D}$) resources within the MG and their corresponding operational costs. We adopt the following sign convention: $p_i^{\mathcal{D}}(t) > 0$ when power is absorbed by the units, and $p_i^{\mathcal{D}}(t) < 0$ when it is injected into the grid. A concise and clear summary of MG components presented in this section, along with their operational limits and cost structures, is provided in Table 10.1.

Dispatchable generator. The power output for a dispatchable generator $p_i^{\mathcal{D}}(t)$ is commonly subject to a constraint of the following kind:

$$\ell_i^p \le p_i^{\mathcal{D}}(t) \le u_i^p, \; t \in \{1, \dots, T\}, \tag{10.4}$$

ensuring it stays within the predefined minimum $\ell_i^p \in \mathbb{R}$ and maximum $u_i^p \in \mathbb{R}$ power limits, respectively. The values ℓ_i^p and u_i^p can correspond to either safe operational boundaries or preferred performance ranges. For instance, generators are typically not operated exactly at their extreme safety thresholds; instead, conservative operating margins are established to allow flexibility to increase or decrease power output in response to grid demands.

Ramp-rate constraints can also be imposed when there are limits on how quickly the generator's power output can change. Formally, we require that:

$$\ell_i^r \le p_i^{\mathcal{D}}(t) - p_i^{\mathcal{D}}(t-1) \le u_i^r, \; t \in \{1, \dots, T\}, \tag{10.5}$$

TABLE 10.1 Summary of MG components, operational limits, and cost functions.

Component	Applicable limits	Cost function
Generator	Power, ramp-rate, min up/down times Eqs. (10.4), (10.5), (10.8)	$c_i^g\left[\alpha_i^g\left(p_i^{\mathcal{D}}(t)\right)^2+\beta_i^g p_i^{\mathcal{D}}(t)+\gamma_i^g\right]$Eq. (10.6)
Battery	Power, ramp-rate, energy Eqs. (10.4), (10.5), (10.11)	$c_i^b\left(p_i^{\mathcal{D}}(t)\right)^2$ or $c_i^b\left\|p_i^{\mathcal{D}}(t)\right\|$Eqs. (10.13), (10.14)
TCL	Power, ramp-rate, temperature Eqs. (10.4), (10.5), (10.16)	$c_i^{\mathrm{tcl}}\left\|p_i^{\mathcal{D}}(t)-P_0(t)\right\|^2$ Eqs. (10.17), (10.18)
Main grid	Power Eqs. (10.1), (10.4)	Eq. (10.20)

where $\ell_i^r \in \mathbb{R}$ and $u_i^r \in \mathbb{R}$ are the lower and upper bounds for the ramp-rate, respectively. The initial power value $p_i^{\mathcal{D}}(0)$ is assumed to be known.

In the case of a small-scale fuel-based generator operating within an MG, the main operating cost arises from fuel usage, which is commonly represented by the following quadratic cost of production $-p_i^{\mathcal{D}}(t)$:

$$\mathcal{J}_i^{\mathcal{D}}\left(p_i^{\mathcal{D}}(t)\right)=c_i^g\mathcal{F}_i^g\left(p_i^{\mathcal{D}}(t)\right)=c_i^g\left[\alpha_i^g\left(p_i^{\mathcal{D}}(t)\right)^2+\beta_i^g p_i^{\mathcal{D}}(t)+\gamma_i^g\right], \tag{10.6}$$

where c_i^g represents the fuel price, $\mathcal{F}_i^g\left(p_i^{\mathcal{D}}(t)\right)$ denotes the fuel consumption rate as a function of the production, and the coefficients α_i^g, β_i^g, and γ_i^g characterize the cost of the generator output (Nguyen & Crow, 2015). Specifically, $\gamma_i^g=0$ ensures zero cost in the absence of production; $\alpha_i^g>0$ guarantees that the marginal cost increases with production; and $\beta_i^g\geq 0$ ensures the cost is nondecreasing in $-p_i^{\mathcal{D}}$, so that each additional kW of production increases the unit cost. The parameter c_i^g encompasses fuel expenses as well as additional costs such as transportation and on-site storage. Furthermore, the cost model in Eq. (10.6) can be applied to fuel cells and microturbines, though with different parameter values. For instance, in the work by Parisio et al. (2014), the quadratic cost in Eq. (10.6) is approximated using the maximum of a set of affine functions, achieving this simplification without the need to introduce binary variables. A similar approach is adopted by Palma-Behnke et al. (2013), where it is used to approximate a nonconvex cost function. Higher-order polynomial cost functions have also been proposed, e.g., by Sachs and Sawodny (2016), in which a third-degree polynomial function of power was introduced.

In addition to the fuel consumption term in Eq. (10.6), the total cost of operating a fuel-based generator may also include operation and maintenance (O&M) expenses, as well as startup and shutdown costs (Hu et al., 2023; Parisio et al., 2014). The overall cost can thus be expressed as:

$$\begin{aligned}\mathcal{J}_i^{\mathcal{D}}\left(p_i^{\mathcal{D}}(t)\right)=c_i^g&\left[\alpha_i^g\left(p_i^{\mathcal{D}}(t)\right)^2+\beta_i^g p_i^{\mathcal{D}}(t)+\gamma_i^g\right]+\\&+c_i^{\mathrm{om}}(t)\cdot\delta_i(t)+c_i^{\mathrm{su}}(t)[\delta_i(t)-\delta_i(t-1)]+\\&+c_i^{\mathrm{sd}}(t)[\delta_i(t-1)-\delta_i(t)],\end{aligned} \tag{10.7}$$

where $\delta_i(t)\in\{0,1\}$ denotes the generator's ON/OFF status, and $c_i^{\mathrm{om}}(t)$, $c_i^{\mathrm{su}}(t)$, and $c_i^{\mathrm{sd}}(t)$ represent the O&M, time-dependent startup and shutdown costs, respectively.

Finally, constraints on minimum up and down times can also be incorporated to prevent frequent switching and enforce minimum periods for which the unit must remain ON or OFF once a transition occurs. Formally:

$$\begin{aligned}&\delta_i(t)-\delta_i(t-1)\leq\delta_i(\tau), &&\text{(OFF/ON transition)}\\&\delta_i(t-1)-\delta_i(t)\leq 1-\delta_i(\tau), &&\text{(ON/OFF transition)}\end{aligned} \tag{10.8}$$

where $\tau\in\left\{t+1,\ldots,\min\left(t+T_i^{\mathrm{up}}-1,T\right)\right\}$ for enforcing the minimum uptime constraint, and $\tau\in\left\{t+1,\ldots,\min\left(t+T_i^{\mathrm{down}}-1,T\right)\right\}$ for the minimum downtime constraint, and T_i^{up} and T_i^{down} represent the minimum up and down times for the unit i (Parisio et al., 2014).

10.1.3.1 Battery

A battery unit i is primarily characterized by its power $p_i^{\mathcal{D}}$, where $p_i^{\mathcal{D}}>0$ if power is injected into the device, and $p_i^{\mathcal{D}}<0$ if power is withdrawn from it. Moreover, it may require satisfying power and ramp-rate constraints as given in Eqs. (10.4) and (10.5). Finally, it may also be necessary to impose limits on the minimum and maximum amount of stored energy. Let us first define the energy dynamics that follow the recursive formula:

$$e_i(t+1)=\zeta_i e_i(t)+\Delta_t\eta_i(t)p_i^{\mathcal{D}}(t), \tag{10.9}$$

where $\zeta_i \in (0, 1]$ represents the self-discharge coefficient, Δ_t denotes the duration of the discrete-time slot, and

$$\eta_i(t) = \begin{cases} \eta_i^+, & p_i^{\mathcal{D}}(t) \geq 0 \quad \text{(charging mode)} \\ \frac{1}{\eta_i^-}, & p_i^{\mathcal{D}}(t) < 0 \quad \text{(discharging mode)} \end{cases} \tag{10.10}$$

describes the efficiency, where $\eta_i^+, \eta_i^- \in (0, 1]$ are the charging/discharging efficiencies, respectively (Parisio et al., 2014; Hu et al., 2023).

Typically, batteries are not operated at their extreme charge levels due to degradation effects that become more pronounced at very high or very low states of charge. Moreover, operating near these limits can lead to issues such as overvoltage, undervoltage, or stability concerns, particularly at low charge levels (Zhu et al., 2021). To mitigate these risks, constraints are imposed:

$$\ell_i^e \leq e_i(t) \leq u_i^e, \; t \in \{1, \ldots, T\}, \tag{10.11}$$

ensuring the energy content stays within the predefined minimum $\ell_i^e \in \mathbb{R}$ and maximum $u_i^e \in \mathbb{R}$ energy limits, respectively. Note that by unrolling the dynamics of $e_i(t)$ in Eq. (10.9), the constraint in Eq. (10.11) can be equivalently expressed in terms of power.

The energy dynamics described in Eq. (10.9), together with the mode-dependent efficiencies defined in Eq. (10.10), are typically handled by separating charging and discharging power flows and introducing auxiliary binary variables to indicate the active operational mode at time t (Parisio et al., 2014). An alternative formulation, as proposed in Vignali et al. (2022), expresses the same dynamics without the need for binary mode indicators. Specifically, the model in Eq. (10.9), with $\eta_i(t)$ defined as in Eq. (10.10), can be equivalently written as:

$$e_i(t+1) = \zeta_i e_i(t) + \Delta_t \min\left\{\eta_i^+(t) p_i^{\mathcal{D}}(t), \frac{1}{\eta_i^-(t)} p_i^{\mathcal{D}}(t)\right\}. \tag{10.12}$$

This formulation eliminates the need for binary variables by inherently capturing the transition between charging and discharging modes via a piecewise expression.

For batteries as well, the operational cost is typically modeled as a quadratic function (He et al., 2023):

$$\mathcal{J}_i^{\mathcal{D}}\left(p_i^{\mathcal{D}}(t)\right) = c_i^b \left(p_i^{\mathcal{D}}(t)\right)^2, \tag{10.13}$$

where the coefficient c_i^b, as discussed in the work by Nguyen and Crow (2015), may explicitly incorporate the cost per kWh for charging, heat dissipation losses during both charging and discharging, and maintenance-related expenses, depending on the specific battery design. The latter can be interpreted as the cost associated with making 1 kWh of storage capacity available, which is influenced by the battery's replacement cost and its total lifetime cycling capacity, determined by factors such as the rated depth of discharge, nominal capacity, and expected lifespan. The quadratic term in Eq. (10.13) accurately represents the instantaneous stress associated with high currents, both during charging and discharging. Alternative modeling of the cost has been used by Hu et al. (2023), in which the cost was linearly dependent on the exchanged power, formally:

$$\mathcal{J}_i^{\mathcal{D}}\left(p_i^{\mathcal{D}}(t)\right) = c_i^b \left|p_i^{\mathcal{D}}(t)\right|. \tag{10.14}$$

In this case, considering the cumulative cost over a specified time horizon, the cost function penalizes the total amount of energy exchanged over time. A positive weighted combination of the cost terms in Eqs. (10.13) and (10.14) preserves the convexity of the resulting cost function.

10.1.3.2 Thermostatically controlled load

A thermostatically controlled load (TCL), such as a heat pump, air conditioner, or refrigerator, can be represented by its thermal dynamics as

$$\vartheta_i(t+1) = \xi \vartheta_i(t) + (1-\xi)(\vartheta_a(t) - \chi p_i(t)), \tag{10.15}$$

constrained to

$$\ell_i^{\vartheta}(t) \leq \vartheta_i(t) \leq u_i^{\vartheta}(t), \; t \in \{1, \ldots, T\}, \tag{10.16}$$

where $\ell_i^{\vartheta} \in \mathbb{R}$ and $u_i^{\vartheta} \in \mathbb{R}$ represent the minimum and maximum temperature limits, respectively. In Eq. (10.15), $p_i(t) \geq 0$ denotes the average power of the TCL over the interval t, obtained by switching the TCL between off (zero power) and on (nominal power P_n); ϑ_a is the ambient temperature, and ξ and χ are parameters characterizing the thermal inertia and efficiency of the controlled zone, with $\chi > 0$ for a cooling unit and $\chi < 0$ for a heating unit. By multiplying all terms by a factor C_{th}/η, where C_{th} is the thermal capacitance, modeling the amount of thermal energy that can be stored in the zone, and η is the coefficient of performance, defined as the ratio between the thermal energy supplied to the zone and the corresponding electrical consumption of the TCL— Eq. (10.15) can be expressed in terms of the (thermal) energy:

$$e_i(t) = \frac{C_{th}}{\eta}\vartheta_i(t).$$

Similarly, the temperature limits in Eq. (10.16) can be equivalently expressed as energy constraints, as shown in Eq. (10.11). Therefore, these loads can be regarded as generalized energy storage systems and thus represented by the model in Eq. (10.9), with power and capacity constraints given by Eqs. (10.4) and (10.11).

Given that the nominal power required to keep a TCL at its set-point ϑ_s is

$$P_0(t) = \frac{\vartheta_a(t) - \vartheta_s}{\chi}, \tag{10.17}$$

the operating cost of a TCL can be defined as the discomfort caused by deviating from the desired set-point:

$$\jmath_i^{\mathcal{D}}\left(p_i^{\mathcal{D}}(t)\right) = c_i^{tcl}\left|p_i^{\mathcal{D}}(t) - P_0(t)\right|^2, \tag{10.18}$$

with $c_i^{\mathrm{tcl}} > 0$. The parameter c_i^{tcl} has a natural interpretation in terms of thermal comfort. A power deviation $|p_i(t) - P_0(t)|$ leads to an approximate temperature change $\Delta\vartheta \approx \chi|p_i(t) - P_0(t)|$. By choosing $c_i^{tcl} = \alpha\chi^2, \alpha > 0$, the cost function can be normalized so that a specified temperature deviation results in a proportional increase in cost.

Bidirectional exchanges with the main grid. When operating in grid-connected mode, the MG can both export energy to and import energy from the main grid. This bidirectional energy exchange can be viewed as an additional source of flexibility available to the controller. It enables cost optimization of MG operations by, for instance, taking advantage of favorable market prices to sell excess energy or to charge storage systems in preparation for future energy deficits.

Power constraints in the form of Eq. (10.4) can model the minimum and maximum allowable power flow through the transmission line. Electricity purchase costs and sale revenues can be modeled as follows:

$$\jmath^{\mathrm{grid}}(p_{\mathrm{net}}(t)) = c_{\mathrm{buy}}(t)\cdot\max\{p_{\mathrm{net}}(t), 0\} - c_{\mathrm{sell}}(t)\cdot\max\{-p_{\mathrm{net}}(t), 0\}, \tag{10.19}$$

where the coefficients $c_{\mathrm{buy}}(t)$ and $c_{\mathrm{sell}}(t)$ represent the time-varying unitary buying and selling prices, respectively. Finally, by observing that $c_{\mathrm{buy}}(t) \geq c_{\mathrm{sell}}(t), \forall t \in \{1, \ldots, T\}$, Eq. (10.19) can be expressed in the more compact form as follows:

$$\jmath^{\mathrm{grid}}(p_{\mathrm{net}}(t)) = \max\{c_{\mathrm{buy}}(t)p_{\mathrm{net}}(t), c_{\mathrm{sell}}(t)p_{\mathrm{net}}(t)\}, \tag{10.20}$$

where the max operator automatically accounts for whether the MG is importing or exporting power, correctly assigning costs or revenues. Alternatively, to explicitly distinguish between purchasing and selling modes, an auxiliary binary variable can be introduced, as proposed by Parisio et al. (2014).

10.2 Automatic control techniques

In this section, we provide an overview of how the control problem introduced in Section 10.1.2 has been addressed in the automatic control and systems literature, with a particular focus on model predictive control (MPC) approaches (Mayne, 2014).

10.2.1 Model predictive control: A brief overview

MPC is an advanced control strategy widely adopted across various application domains to optimize complex dynamic systems while rigorously satisfying operational constraints. Initially developed in the 1970s, MPC has since evolved into one of the most powerful and widely used control techniques, thanks to its ability to manage multivariable systems and nonlinearities effectively.

The key idea behind MPC is to use a predictive model to capture the relationship between control inputs, system states, and outputs. By forecasting future system behavior, MPC determines a sequence of control actions that optimally satisfies the desired performance objectives over a finite prediction horizon. Typically, only the first control

TABLE 10.2 Comparison of MPC approaches in grid-connected MGs.

Reference	Objectives	Approach	Key Features
Parisio et al. (2014)	Economic penalty for load curtailment	MILP-MPC	Unit commitment, battery scheduling, load shedding
Clarke et al. (2016, 2018)	Economic, terminal cost of battery energy	MILP-MPC	Tightened SOC, no load curtailment
Ju et al. (2017)	Economic, ESS degradation	MILP-MPC	ESS degradation modeling, asset longevity
Parisio et al. (2016)	Economic,	Stochastic MPC	2-Stage stochastic programming, environmental impacts, expected cost min., emissions reduction
Kou et al. (2019)	Economic, power reference tracking	Stochastic MPC	2-Layer hierarchical, mixed uncertainty handling, QP approximation
Pippia (2019)	Economic	Rule-based MPC (LP)	Fast/slow model separation, low computational burden

MG, *microgrid;* MILP, *mixed-integer linear programming;* MPC, *model predictive control;* SOC, *state of charge.*

action (or a few initial steps) of this sequence is implemented before the optimization is repeated at the next time step using the newly measured system state. Interestingly, this concept mirrors the intuitive decision-making process humans often follow. For example, a skilled chess player evaluates future scenarios by simulating several possible moves and counter-moves before selecting the one that seems most advantageous. The formulation of an MPC problem typically includes the following components:

- **Objective function:** A cost function is minimized to balance competing objectives.
- **System dynamics:** A predictive model, which may be linear or nonlinear, describes how the system responds to control inputs and disturbances over time.
- **Constraints:** MPC naturally accommodates both equality and inequality constraints on system inputs, outputs, and states. These constraints ensure safe, efficient, and reliable operation within physical and regulatory limits.

Designing an MPC controller involves selecting key parameters, such as the prediction and control horizons, tuning weighting factors in the cost function, and specifying the relevant constraints. The prediction horizon determines how far into the future the controller looks when forecasting system behavior, while the control horizon defines the number of steps of the control sequence that are optimized and applied at each iteration. Although MPC requires solving a computationally demanding optimization problem at each control interval, it offers several distinct advantages over traditional control strategies. More specifically:

- **Multivariable coordination:** MPC handles multiple inputs and outputs in a unified framework, enabling optimal coordination among interconnected subsystems.
- **Constraint enforcement:** constraints are embedded directly into the control formulation, providing systematic handling of operational limits.
- **Robustness and adaptability:** the predictive nature of MPC allows it to anticipate disturbances and adapt to changes in system dynamics, improving overall performance and resilience.

10.2.2 Model predictive control–based microgrid control problem formulation

At each time slot t, the controller needs to compute the power of the dispatchable units over some prediction horizon $H_t \in \{t, t+1, \ldots, t+T-1\}$ of length $T > 0$, aiming to minimize a cost function. This function accounts for the operational costs defined in Section 10.1.3 as well as potential additional terms (see, for instance, the Objectives column in Tables 10.2, 10.3, and 10.4). The sampling interval Δ_t of a tertiary-level MPC typically ranges from several seconds to several hours and is generally larger than that of lower-level controllers. The predictive model relies on system states and available forecasts to estimate future states, typically using information derived from current and historical data. The optimization is subject to all operational constraints associated with each dispatchable unit $i \in \mathcal{D}$, which may be time-dependent. For example, battery capacity constraints depend on the initial state of charge (SOC), while generator ramp-rate limits are influenced by the unit's power output in the previous time slot $t-1$.

TABLE 10.3 Comparison of MPC approaches in islanded MGs.

Reference	Objective	Uncertainty handling	Approach
Nahata et al. (2020)	Min. generation costs, reduce gen. units mode switching	Forecast-based	Hierarchical MPC
Nair and CostaCastelló (2020)	Max. efficiency, max. asset longevity	Forecast-based	Hierarchical MPC, soft constraints on SOC
Cai et al. (2020)	Max. restored critical loads, min. renewable curtail	Worst-case scenarios	2-Layer RMPC with network reconfig
Hans et al. (2015)	Max. reliability	Scenario-based SMPC	SMPC, scenario-tree optimization
Polimeni et al. (2021)	Min. corrective actions under forecast errors	2-stage SMPC with real-time corrective rules	Piecewise-affine policies, two-layer EMS

MG, *microgrid;* MPC, *model predictive control;* RMPC, *Robust MPC;* SMPC, *stochastic MPC;* SOC, *state of charge.*

TABLE 10.4 Comparison of MPC approaches in networked/multi-MG systems.

Reference	Objective	Coordination	Uncertainty handling	Approach
Bonassi et al. (2021)	Scheduled power tracking reduces strain on units	Intercluster via supervisory layer	Deterministic	Two-layer DMPC
Baker et al. (2016)	Optimize storage use, increase RES integration	Distributed across control areas	Deterministic	DMPC
Zhao et al. (2022)	Ensure robustness, increase market participation	Cooperative energy exchange via the internal market	Worst-case scenarios	Robust DMPC
Parisio et al. (2017)	Min. total energy cost, min. inter-MG power exchange	Distributed with fairness guaranteed	Stochastic/robust variants	Cooperative DMPC

DMPC, *distributed model predictive control;* MG, *microgrid;* MPC, *model predictive control;*

An additional constraint is the power balance within the MG, which becomes particularly critical during islanded operation. The cost function can be flexibly designed to account for a range of operational objectives, including the operating costs of dispatchable units, bidirectional power exchanges with the main grid, imbalance charges due to unscheduled energy flows, load, and renewable generation curtailment, penalties for changes in the operational states of dispatchable units, user discomfort, and environmental factors such as emissions of atmospheric pollutants.

10.2.2.1 Model predictive control for grid-connected microgrids

Economic optimization is a critical component of energy management in grid-connected MGs (Hu et al., 2023). The primary objective of power flow optimization controllers is to minimize total operating costs, which encompass electricity expenses and revenues from energy trading with the utility grid, fuel costs of local thermal generators (including both running and startup costs), and costs or benefits related to energy stored in energy storage systems (ESS). The presence of both continuous variables (e.g., power flows) and binary decisions (e.g., generator ON/OFF states) naturally leads to a mixed-integer linear programming (MILP) formulation within an MPC framework.

Parisio et al. (2014) present a MILP-MPC approach aimed at minimizing total economic operating costs in MGs. Their formulation integrates fuel consumption, generator startup, and shutdown costs, O&M expenses for both generators and storage units, as well as revenues from energy sales. Additionally, a penalty term for load curtailment is included to model user discomfort. The controller determines unit commitment, economic dispatch, battery scheduling, energy trading with the grid, and load shedding at each time step. Embedded in a hierarchical control architecture, the MILP-MPC layer functions centrally for long-term optimization, while subordinate controllers address fast electrical dynamics such as voltage and frequency regulation. Building upon this foundation, Clarke et al. (2016, 2018) emphasize robustness and feasibility under disturbances and model uncertainties. Their approach introduces an artificially tightened SOC constraint to ensure battery operational limits are never violated. Unlike Parisio et al. (2014), Clarke et al. (2016) assume all loads are critical, thus prohibiting load curtailment. They further incorporate a terminal cost in the objective function to preserve battery energy at the end of the prediction horizon, supporting long-term system sustainability. To explicitly address asset longevity, Ju et al. (2017) incorporate detailed degradation models for ESSs. For batteries, degradation costs are expressed as nonlinear functions of full capacity and depth of discharge, capturing the wear from deep and frequent cycling. This approach effectively converts long-term capital degradation into short-term operational costs. In contrast, supercapacitor degradation is modeled using a simpler linear time-based function, reflecting their higher durability and suitability for high-frequency power balancing.

To better manage operations under uncertainties, Stochastic MPC solutions have been proposed (Kou et al., 2019; Parisio et al., 2016). Kou et al. (2019) focus on matching uncertain wind generation with electric vehicle (EV) charging demand, aiming to improve power balance, ensure EV user satisfaction, and bring power flow to a predefined trajectory with the utility grid. This is achieved using a two-layer hierarchical control and approximating the stochastic problem as a quadratic programming (QP) for computational efficiency, particularly handling coexisting non-Gaussian wind and Gaussian EV uncertainties. Parisio et al. (2016) address economic and environmental operation management, minimizing expected costs of correction actions and/or emissions of atmospheric pollutants. The problem is formulated as an MILP problem solved via two-stage stochastic programming. To address the computational complexity inherent in MILP-MPC formulations, Pippia et al. (2019) propose a Rule-Based MPC (RBMPC) method. This technique pre-assigns binary decision variables via if-then-else rules informed by economic signals such as electricity prices, renewable generation, and load profiles, reducing the MILP problem to a linear program and substantially decreasing computational burden without significant loss of performance. Additionally, they introduce a single-level control scheme that utilizes two MG models operating at different sampling rates: a fast model capturing rapid dynamics (e.g., ultracapacitors) for short-term predictions, and a slow model for long-term forecasting without fast dynamics. This streamlined architecture reduces control complexity and latency, enabling more frequent updates and responsive adjustments to varying conditions.

Together, these studies highlight ongoing efforts to develop MPC solutions that balance economic efficiency, system reliability, asset longevity, and computational tractability. A summary of the economic objectives and MPC structure is provided in Box 10.1

BOX 10.1

MILP-MPC structure.

Tertiary-level MPC problem: Economic objectives

$$\min_{\{p_i^{\mathcal{D}}(\tau),\delta_i(\tau)\}_{t:t+T-1,i\in\mathcal{D}}} \sum_{\tau=t}^{t+T-1} \left(\mathcal{J}^{\text{grid}}(p_{\text{net}}(\tau)) + \sum_{i\in\mathcal{D}} \mathcal{J}_i^{\mathcal{D}}(p_i^{\mathcal{D}}(\tau), \delta_i(\tau))\right)$$

$$\text{s.t.} \quad p_{\text{net}}(\tau) = \sum_{i\in\mathcal{D}} p_i^{\mathcal{D}}(\tau) + \sum_{i\in\mathcal{N}} p_i^{\mathcal{N}}(\tau), \forall\tau$$

power balance and device models;

energy/ramp limits;

on/off logic ($\delta_i(\tau) \in \{0,1\}$);

market/export bounds.

10.2.2.2 Model predictive control for Islanded microgrids

When an MG operates in islanded mode, disconnected from the main power grid, the control task becomes considerably more challenging. Without external grid support, the MG must independently balance supply and demand, maintain voltage and frequency stability, and manage the uncertainties associated with the RES variability and load fluctuations (Lopes et al., 2006). These conditions place significantly higher demands on MPC in terms of robustness, responsiveness, and operational reliability. In response to these challenges, the literature has explored a variety of strategies to adapt and improve MPC for islanded operations. These include hierarchical control architectures, which layer decision-making across timescales; robust control strategies that guard against worst-case uncertainties; and stochastic approaches that exploit probabilistic forecasts.

In the work by Nahata et al. (2020), an energy management system (EMS) based on MPC is proposed that governs power setpoints and operational modes for distributed generation units (DGUs), such as batteries, PV panels, and dispatchable generators. This supervisory layer seeks to minimize generation costs and improve DGU efficiency by reducing unnecessary mode switching, such as between charging/discharging or ON/OFF states. The EMS considers DGU saturation constraints, power models specific to each DGU type, and forecasts of both load and generation. To perform the power-to-voltage conversion, the intermediate secondary control layer solves an optimization problem based on power flow equations. The resulting voltage setpoints are then forwarded to the primary voltage controller for implementation. A hierarchical control architecture is also proposed by Nair and Costa-Castelló (2020), where MPC is employed at the highest level for energy management within the MG, while power control is handled at the lower level. This work considers an isolated MG featuring a hybrid ESS composed of batteries, supercapacitors, and fuel cells, integrated with PV generation and a dispatchable generator. The MPC aims to maximize operational efficiency and renewable energy utilization while minimizing the degradation of the ESSs. It generates power references for the ESS, the dispatchable generator, and PV curtailment, explicitly modeling the hybrid dynamics of the ESS, including charge/discharge efficiency. The cost function is designed to enhance ESS longevity by penalizing variations in SOC, as well as fuel cell power output and ramping rates. Additionally, soft constraints on ESS capacity limits are included to ensure feasibility.

While hierarchical MPC structures help manage system complexity and different time scales, they often require additional mechanisms to address uncertainty. To this end, robust MPC formulations have been proposed. The method described by Cai et al. (2020) addresses the optimal management of MGs during disconnection from the main grid, with a focus on restoring as many critical loads as possible. In such circumstances, the distribution system is reconfigured into multiple autonomous MGs, whose effective management becomes crucial to ensure continuity of service to priority loads. One of the main operational challenges lies in the uncertainty associated with the output of renewable distributed generators, whose variability makes accurate forecasting difficult. To address this issue, the proposed approach adopts a robust MPC (RMPC) strategy that enables dynamic planning of both the network structure through network reconfiguration and the output of distributed generators, including ESS and switching devices. At the core of the method is a two-layer optimization model designed to ensure the robustness of the solution even in the worst-case scenario of renewable generation. The adopted cost function aims to maximize the weighted sum of restored loads across all nodes and MGs over the optimization horizon, giving higher priority to critical loads, while minimizing the curtailment of renewable energy generation, formulated as a weighted objective.

Although robust MPC effectively handles worst-case scenarios, it may yield conservative solutions. Stochastic MPC approaches offer a probabilistic alternative that balances risk and performance more flexibly. Hans et al. (2015) propose a scenario-based stochastic MPC (SMPC) approach that explicitly incorporates the probabilistic nature of forecasts. This is achieved by representing uncertainty via a computationally tractable scenario tree, generated and reduced via Monte Carlo simulations. By solving a stochastic optimization problem over this tree, the approach enables decision-making that reduces conservatism without compromising reliability. An SMPC solution is also proposed by Polimeni et al. (2021), within a two-layer architecture. The objective is to define correction rules for the first-stage variables, which represent the optimal dispatch computed deterministically based on expected demand and renewable generation forecasts. These rules address short-term uncertainties, modeled as net demand forecast errors, by providing corrective actions. Specifically, the resulting control strategy is formulated as piecewise-affine policies that provide corrective actions based on real-time feedback from observed disturbances. This formulation enables the EMS to adapt proactively and efficiently to unforeseen deviations, thereby enhancing the system's responsiveness and operational robustness.

Together, these contributions underscore the pivotal role of MPC in managing the complex, uncertainty-laden dynamics of islanded MGs, highlighting how advancements in hierarchical, robust, and stochastic control frameworks are essential to ensure resilient, efficient, and autonomous energy management.

10.2.2.3 Model predictive control for networked microgrids

The concept of networked MGs, i.e., systems composed of multiple electrically interconnected AC and DC MGs, has attracted significant attention due to its potential to enhance RES integration and accommodate higher load demands. By enabling reciprocal power exchange among MGs, this interconnected architecture improves the overall utilization of renewable resources. However, compared to islanded MGs, the coordination and optimization of power flow in networked systems introduce greater complexity (Chen et al., 2020). To address this, decentralized/distributed control architectures and cooperative schemes have emerged as prevalent solutions, offering scalability and resilience.

One such approach is a two-layer control architecture proposed by Bonassi et al. (2021), which divides the distribution network into distinct, nonoverlapping clusters. At the lower layer, each cluster operates a local decentralized MPC (DMPC) that autonomously manages active power fluctuations. These DMPCs operate independently, without intercluster communication, and are responsible for maintaining local balance by requesting services from internal MGs. A centralized supervisory layer is activated only when a local controller requires external support. This layer coordinates power transfers between clusters via controllable DC links, ensuring adequate operational margins while preserving the autonomy of local DMPCs. The approach maintains power balance locally while enabling intercluster cooperation when necessary, all while respecting a predefined power exchange schedule with the system operator. The cost function in each DMPC penalizes deviations from scheduled power, minimizes strain on generators and storage, and treats intercluster DC power requests as a last resort. A DMPC-based framework is also explored by Baker et al. (2016), where a distributed MPC strategy coordinates storage systems and variable renewable resources across multiple control areas. The focus is on scalability and efficient resource sharing in large-scale power systems.

Building on the decentralized architecture, Zhao et al. (2022) integrate robust optimization into the MPC framework, forming a distributed robust MPC (DRMPC) framework. Targeting isolated multi-MG (MMG) systems, this approach ensures feasibility under worst-case uncertainty scenarios while maintaining real-time adaptability. Notably, the system features an internal energy market grounded in game theory, encouraging MGs to participate actively in energy exchange based on their individual contributions.

A cooperative MPC-based energy management strategy is presented by Parisio et al. (2017). This framework addresses the coordinated operation of networked MGs that may share DERs. The main goal is to optimize flexibility and minimize overall energy costs and power exchange with the main grid. A key component is the MPC-based distributed coordination algorithm, designed to ensure constraint satisfaction, promote cooperation, and guarantee fairness in the use of shared resources. The algorithm operates in three phases: a local, parallel initialization; a power distribution step coordinated by an aggregator and solved in parallel at the MG level; and a sequential redistribution step to address deviations. A distinct advantage of the framework is its ability to guarantee cost savings for each participating MG, thereby incentivizing collaboration. The study also examines robustness and the application of stochastic approaches to enhance uncertainty handling.

Collectively, these works demonstrate the growing trend toward scalable, distributed MPC architectures capable of coordinating energy resources across complex MMG systems, while addressing uncertainty and promoting cooperation. In Box 10.2, we provide some practical tips summarizing the considerations presented in the review of the existing literature to help practitioners employ MPCs in the MG field.

BOX 10.2

Deployment tips for MPC practitioners.

MPC's deployment tips

- Exploit receding-horizon structure: Cache and reuse the optimal variables and active-constraint information from the previous solve to accelerate warm starts and improve solver robustness. For example, when the MG's forecasted PV generation and load profiles remain similar across consecutive optimization cycles, initializing the solver with the prior solution can significantly reduce computation time.
- Adaptive horizon tuning: Dynamically adjust the prediction horizon based on both forecast changes and uncertainty, keeping it short under stable, low-uncertainty conditions to reduce computation, and extending it when high uncertainty or rapid changes require anticipatory control, such as managing battery scheduling during potential PV drops.
- Binary pre-fixing via sensitivity checks: Before each optimization, pre-fix binary variables (e.g., unit on/off states) that are insensitive to current price or demand variations, thereby reducing the MILP search space. For example, if a diesel generator in the MG is consistently off when PV output is high and prices are low, its state can be pre-set to OFF during those periods, decreasing the number of binary variables and speeding up the optimization.
- Hierarchical coordination fallback: If tertiary MPC fails or communication delays rise, revert to local rule-based control (e.g., fixed battery charge/discharge rules) for affected MG units, while logging infeasibilities for later analysis.

10.3 Machine learning control techniques

In this section, we review how the control problem introduced in Section 10.1.2 has been addressed in the machine learning (ML) literature, with a particular focus on reinforcement learning (RL) approaches (Sutton et al., 1998).

10.3.1 Reinforcement learning: A brief overview

RL refers to a class of techniques within the broader field of ML that learn from experience the optimal policy to follow in dynamic environments. The term RL is derived from psychology, where it describes a method of teaching humans and animals through explicit feedback, known as rewards. These rewards serve as evaluations of the subject's actions and guide behavior toward a desired outcome. For example, to train a dog to sit, one might offer a treat each time the dog performs the correct action. In this way, the dog learns to associate correct behavior with a positive outcome (i.e., the reward), while incorrect behavior yields a null or negative outcome.

RL aims to train intelligent agents by positively reinforcing desirable actions and discouraging undesirable ones. The primary objective of RL is to maximize the cumulative reward over time, favoring agents that make long-term, strategic decisions. The behavior of an agent reflects a strategy, called a policy, that determines the action to perform based on the agent's current configuration and the surrounding environment.

To formalize the problem of learning in a decision-making framework, RL approaches rely on the concept of the Markov Decision Process (MDP). This mathematical framework describes the problem $\mathcal{M}$ as a tuple:

$$M = (\mathcal{S}, \mathcal{A}, P, R, \gamma, \mu_0), \tag{10.21}$$

where $\mathcal{S}$ is the set of vectors encoding the state of the environment, $\mathcal{A}$ is the set of actions the controller can perform in each state $s \in \mathcal{S}$, P is the transition matrix, specifying the probability of moving to a state s' starting from the state s and performing an action $a \in \mathcal{A}$ with $s, s' \in \mathcal{S}$, $R \in \mathbb{R}$ is the instantaneous reward provided by performing the action a in state s, $\gamma \in [0, 1]$ is a discount factor that allows designing the problem, fostering myopic or farsighted strategies, and μ_0 is a distribution over the states $\mathcal{S}$ stating where the decision-making process starts.

More specifically, the observation space, denoted by $\mathcal{S}$, contains the information about the variables of the system, i.e., the representations of the environment as perceived by the agent. Based on this information, the agent selects actions to fulfill its objective. The action space, indicated with $\mathcal{A}$, defines the set of all possible actions the agent can execute to interact with the environment. Such actions may lead to a modification (complete or partial) of the current environment state and are evaluated with the direct scalar feedback signal from the environment, referred to as the reward. Finally, the reward function R: $=R(s, a)$ is a mapping from state-action pairs to a real-valued outcome. This function evaluates the immediate consequence of executing an action $a \in \mathcal{A}$ in a specific state $s \in \mathcal{S}$ and plays a pivotal role in guiding the agent's learning process. It should be carefully designed to reflect the objectives of the control task, as it is essentially derived from the problem's objective function.

The dynamics of the environment are expressed by the matrix P, which describes the probability distribution over successor states $s' \in \mathcal{S}$ given a current state $s \in \mathcal{S}$ and action $a \in \mathcal{A}$, i.e., $P(s'|s, a)$. In many practical scenarios, such as real-world MG applications, the transition model is unknown or only partially known.

In this setting, a policy $\pi : \mathcal{S} \rightarrow \mathcal{A}$ is a mapping from states to actions, i.e., a rule that selects an action (or defines a distribution over the available actions) in each state. The objective of RL is to find the policy that maximizes the expected cumulative reward over time. This is typically expressed in terms of the expected return, which is the discounted sum of future rewards. Formally, the optimal policy $\pi^* \in \Pi$, where Π is the space containing all the possible policies can be defined as

$$\pi^* = \arg\max_{\pi} \mathbb{E}_{s_0 \sim \mu_0, a_t \sim \pi(\cdot|s_t)} \left[\sum_{t=0}^{\infty} \gamma^t r(s_t, a_t) \right], \tag{10.22}$$

where $\gamma \in [0, 1)$ is the discount factor, which determines the relative importance of immediate versus future rewards, and the expected value is taken w.r.t. the initial state distribution μ_0 and the policy π. The choice of γ should reflect the problem's time horizon, which can be either finite or infinite.

To evaluate and improve the available candidate policies, RL relies on the Bellman equations, which provide recursive formulations of value functions. Two central functions in this framework are:

- The state-value function $V_\pi(s)$: The expected return starting from state s and following the policy π thereafter.
- The action-value function $Q_\pi(s,a)$: The expected return starting from state s, taking action a, and then following the policy π thereafter.

These functions are defined as follows:

$$V_\pi(s) = \sum_{a \in \mathcal{A}} \pi(a \mid s) \sum_{s' \in \mathcal{S}} P(s' \mid s, a)\left[r(s, a) + \gamma V_\pi(s')\right], \tag{10.23}$$

$$Q_\pi(s,a) = \sum_{s' \in \mathcal{S}} P(s' \mid s,a) \left[r(s,a) + \gamma \sum_{a' \in \mathcal{A}} \pi(a' \mid s') Q_\pi(s',a') \right]. \tag{10.24}$$

It is possible to show that the optimal policy $\pi^* = \arg\max_{\pi} E_{\mu_0}[V_\pi(s)]$ is the one that maximizes the state value function across all states s, so that $V_{\pi^*}(s) \geq V_\pi(s)$ for all $s \in \mathcal{S}$ (Sutton et al., 1998). Similar definitions can be provided using the action-value function $Q_\pi(s,a)$.

RL algorithms are typically classified into three main families: value-based, policy-based, and actor–critic methods. Value-based RL is attractive for discrete actuation (e.g., mode switching), whereas policy-gradient and actor–critic methods better suit continuous MG setpoints (generator, ESS) and can represent stochastic policies needed under uncertainty. More specifically, the first class of algorithms aims to learn a value function that estimates the expected return for each action-state pair, and derives the policy indirectly from the value function, i.e., the policy is greedy with respect to the value function. Such methods are generally well-suited for discrete action spaces but struggle in continuous ones due to the need to enumerate or discretize actions. Notable examples include Q-learning (Watkins & Dayan, 1992) and deep Q-Networks (DQN) (Mnih et al., 2013). In contrast, policy-based methods are naturally suited for continuous or high-dimensional action spaces and can represent stochastic policies. These approaches can learn the policy directly by optimizing the expected return through gradient ascent. However, they typically suffer from higher variance and are less sample-efficient. Common examples include REINFORCE (Sutton et al., 1999) and PGPE (Sehnke et al., 2008). A third category, actor–critic methods, combines direct policy learning (actor) with value function estimation (critic). Notable examples are proximal policy optimization (PPO) (Schulman et al., 2017), deep deterministic policy gradient (DDPG) (Lillicrap et al., 2020), and soft actor critic (SAC) (Haarnoja et al., 2018). For further details on RL basics, see the work of Sutton et al. (1998).

MPC and RL offer complementary strengths for managing MG energy. MPC leverages an explicit predictive model and embeds operational constraints directly into the optimization, yielding strong feasibility guarantees but requiring the solver to compute a solution at each interval and to produce accurate forecasts. RL, by contrast, can learn policies directly from interaction data and handle high-dimensional observations; however, it typically needs careful reward shaping, data collection, and safety mechanisms to respect grid constraints at learning and execution time. In practice, MPC is preferred when accurate models and tight constraints dominate, while RL shines in partially modeled, stochastic environments with rich telemetry and the opportunity to learn from historical or simulated experience.

10.3.2 Reinforcement learning–based control problem formulation

Energy management within an MG network can naturally be formulated as a control problem with continuous state and action spaces, since the system variables such as power flows, energy storage levels, and demand are represented by continuous signals. Depending on the nature of the objective, the control horizon may be either finite or infinite. Although time-dependent variables might suggest an infinite-horizon formulation, the cyclo-stationary behavior of key signals, such as energy consumption, generation from DERs, and electricity prices, allows the problem to be approximated within a limited time window. This assumption enables the design of policies over representative intervals, which can be generalized to other time periods due to the periodic characteristics of the system.

10.3.2.1 Observation space

The observation space typically includes both endogenous and exogenous variables. Endogenous (or controllable) variables generally refer to the internal state of the BESS, including indicators such as SOC and battery temperature, whose evolution can be deterministically determined by the agent's actions. On the other hand, exogenous (or uncontrollable) variables encompass both deterministic inputs, such as market price signals known in advance, and stochastic forecasts, such as solar or wind generation, load demand profiles, and weather conditions. Unlike endogenous signals, exogenous signals evolve independently of the agent's action and are generated by the environment external to the agent. Temporal variables (e.g., time of day, day of the week) are also often included to capture temporal patterns in consumption and generation behavior. An observation space including the element defined earlier is formalized as:

$$s_t = \left(\sigma(t), K(t), \hat{p}^{\mathcal{N}}(t), c_{\text{buy}}(t), c_{\text{sell}}(t), t_{\text{day}}, t_{\text{year}}\right), \tag{10.25}$$

where $\sigma(t) \in \mathbb{R}^+$ and $K(t) \in \mathbb{R}^+$ are the current internal battery status, namely its SOC and temperature, $\hat{p}^{\mathcal{N}}(t) \in \mathbb{R}^{\mathcal{N}}$ is the vector of the estimated power generated or demanded at time t by each nondispatchable energy source, e.g., household energy consumption and renewable energy generators, $c_{buy}(t) \in \mathbb{R}$ and $c_{sell}(t) \in \mathbb{R}$ are the buying and selling energy prices, and t_{day} and t_{year} are the variables addressing the daily and yearly time cycles, respectively. In a grid-connected MG, a typical observation vector may include forecasts of PV/wind generation, load, and prices over a short horizon; present SOC and battery temperature; and transformer loading indices. Actions commonly span dispatchable generation setpoints, ESS charge and discharge power, and limited shed/shift signals. To ensure feasibility, actions can be postprocessed via constraint projection (e.g., enforcing SOC bounds, ramp limits, and power balance) or trained with penalties that reflect violations observed by the environment during rollouts.

10.3.2.2 Action space

The action space defines the agent's decision-making capabilities, typically involving the dispatch strategy of energy within the MG. Formally, an action at time t can be described as a vector of commands such that:

$$a(t) = [a_i(t)]_1^{|\mathcal{D}|} := \left[p_i^{\mathcal{D}}(t)\right]_1^{|\mathcal{D}|} = \boldsymbol{p}^{\mathcal{D}}(t), \tag{10.26}$$

given to each dispatchable unit $i \in \mathcal{D}$. An element $a_i(t)$ of the vector represents the amount of power that has to be provided to or withdrawn from the considered dispatchable unit, i.e., $p_i^{D}(t)$. Notice that the chosen action is based on an estimate of $\boldsymbol{p}^{\mathcal{N}}(t)$, thus the value of net power will be $p_{net}(t) \simeq \bar{p}(t)$ and the efficacy of the action will strictly depend on the accuracy of the forecasting within the observation space.

We remark that the underlying control variables are inherently continuous, for instance, representing the amount of power charged into or discharged from the battery. However, many practical implementations discretize the action space. This is done to simplify the learning process by allowing the agent to select from a finite set of predefined energy dispatch levels.

In problems involving the control of a distribution system operator (DSO), the control action typically involves determining the retail energy price within the MG. The DSO, as an external entity, is responsible for ensuring the reliable and secure operation of the electrical grid and often serves as the market maker, i.e., the entity that sets energy prices to influence market participants' behavior. In such scenarios, the action at time t is defined as $a(t) = c_{buy}(t)$, where $c_{buy}(t)$ represents the price at which energy is purchased from prosumers or other entities within the MG.

10.3.2.3 Reward function

The reward function in MG energy management is typically designed to capture the operational costs occurring within the MG. It commonly consists of two main components:

- **Operational costs from dispatchable units** ($\mathcal{J}^{\mathcal{D}}$): This includes costs related to the usage of local generators and the degradation of battery storage systems. Refer to Section 10.1.3 for the cost functions of the different dispatchable units.
- **Costs and revenues from grid interactions** ($\mathcal{J}^{grid}$): This reflects the economic exchanges associated with buying from or selling energy to the main grid, based on market prices, following Eq. (10.20).

The objective of the controller is to choose the optimal strategy π^* that maximizes the expected cumulative reward over time, which, in this context, is equivalent to minimizing the total operational costs. Formally, the reward function at each timestep t of the episode is defined as:

$$r_t = -w_1 \mathcal{J}^{grid}(t) + w_2 \sum_{i \in \mathcal{D}} \mathcal{J}_i^{\mathcal{D}}(t), \ t \in \{0, \dots, T\}, \tag{10.27}$$

where the weights $w_1 \in \mathbb{R}^+$ and $w_2 \in \mathbb{R}^+$ are used to reflect the relative importance of the two components in the reward function. Notice that, according to Eq. (10.20), exchanges with the main grid are positive when energy is purchased and negative when it is sold. Therefore, when formulating the reward, it becomes necessary to consider the negative of this quantity. These weights must be carefully designed to ensure that the learned policy exhibits behavior consistent with the desired operational goals.

10.3.3 Reinforcement learning methods for microgrid control

The literature offers a wide array of RL techniques and perspectives for MG energy management. The predominant formulation models the problem as an optimal power-scheduling task, typically approached from an MG-centric perspective. Most works adopting this paradigm rely on centralized control and apply either traditional tabular RL methods or more advanced deep reinforcement learning (DRL) algorithms to control the amount of energy to store or retrieve from the BESS. Few other works address the problem with a grid-centric perspective by controlling the

DSO to decide the energy retail price. More recently, hybrid approaches integrating MPC capabilities have started to gain traction. A brief list of deployment tips is provided in Box 10.3.

BOX 10.3

Deployment tips for RL practitioners.

RL's deployment tips

- Offline pre-training with high-fidelity physics and domain randomization: Build the policy on large batches of historical and/or realistic data where physical and economic parameters (loads, prices, and renewable profiles) are stochastically perturbed to promote robustness to unseen conditions.
- Sim-to-real calibration and synchronization: Periodically realign the simulator's dynamics with measured system responses (e.g., inverter ramp rates, SOC efficiency drift), synchronizing the RL loop with the physical system readings that could operate at different frequencies. This reduces policy mismatch and stabilizes early real deployments.
- Safe action filtering and fallback logic: Wrap the learned policy with constraint-enforcing filters (e.g., projection layers, Lyapunov-based safety shields, or rule-based overrides) to guarantee feasibility before actuation in the live system.
- Reward function validation and goal misalignment: During deployment is necessary to rigorously check for "reward hacking" or "goal misalignment," where the agent maximizes the reward signal in an unintended way that violates the true operational goal and leads the system to evolve in an undesirable manner.

10.3.3.1 Discrete state and action spaces

The first set of methods relies on the tabular representation of the MDP, considering the states and actions as a finite set. This allows the use of value-based RL techniques to solve them.

Within this framework, Kuznetsova et al. (2013) implemented a two-step-ahead tabular Q-learning algorithm to achieve optimal power scheduling within an MG system. Their approach modeled a simplified environment that included electricity prices, demand, battery SOC, and forecasts of wind turbine generation. The action space included decisions for two time instants ahead, consisting of Boolean signals indicating whether to use the battery. Instead, the work by Foruzan et al. (2018) proposed a multiagent framework in which multiple independent Q-learning agents represent buyer and seller entities participating in an auction-based energy market within an MG. Each agent independently learns its bidding strategy based on local observations and expected outcomes. More specifically, the system was able to learn the amount of energy to trade and the associated bid price. The agents are designed to maximize individual utility functions that reflect the profit derived from successful market participation or the cost incurred through fallback interactions with the main grid when bids are rejected. Later on, Zhou et al. (2022) extended the Q-learning framework to a more sophisticated setting that incorporated multiple dispatchable and nondispatchable energy resources. In their setup, the agent observed estimated resource values generated by a deep learning–based forecasting module.

10.3.3.2 Continuous state space

Given the inherently continuous nature of the MG control problem, more recent works predominantly employ DRL techniques, as discretizing continuous observation or action spaces often leads to suboptimal approximations. A frequently adopted method is the DQN algorithm (Mnih et al., 2013), which, while effective, still requires a discretized action space. For example, Chen and Bu (2019) investigated the use of DQN in an MMG trading scenario, where each MG agent independently optimized its energy trading strategy. Their model assumed a peer-to-peer (P2P) electricity market. The agents sought to maximize a utility function encompassing trading profits and dependency on a centralized power supply. In contrast, Domínguez-Barbero et al. (2020) applied DQN to manage energy resources in an islanded MG. Their setup included hydrogen and battery storage systems, a diesel generator, and PV generation. The objective was to minimize operational costs associated with nondispatchable generation and unmet energy demand. Further advancing real-world applicability, Kolodziejczyk et al. (2021) applied DQN to real consumption and generation data, using two representative simulation periods—a winter week and a summer week. Their

TABLE 10.5 Summary of RL methods for MG control.

Reference	State/Action	Approach	Key features
Kuznetsova et al. (2013)	D/D	Q-Learning	Two-step ahead scheduling
Foruzan et al. (2018)	D/D	Multiagent Q-learning	Auction-based energy market
Zhou et al. (2022)	D/D	Q-Learning	Multi-resource setup forecasted using deep learning
Chen and Bu (2019)	C/D	Multiagent DQN	Peer-to-peer energy trading
Domínguez-Barbero et al. (2020)	C/D	DQN	Islanded MG setup with multiple generation modes
Kolodziejczyk et al. (2021)	C/D	DQN	Employment of real-world data-driven simulation
Nakabi and Toivanen (2021)	C/D	Multiple DRLs	Extended MG scenario that includes TCLs
Liu et al. (2021)	C/C	DDPG	Operational cost minimization
Guo et al. (2022)	C/C	PPO	Operational cost minimization and grid interactions

C, *Continuous;* D, *Discrete;* MG, *microgrid;* RL, *reinforcement learning.*

observation space comprised 24-hour window vectors of net demand, battery SOC, electricity price, and time-of-day features. The action space was discretized to represent various percentages of power purchased from the grid, scaled by the maximum charging capacity of the storage system.

Nakabi and Toivanen (2021) conducted a comparative study of various DRL algorithms within an extended MG control scenario. Interestingly, their experimental setting incorporates TCLs, which introduce additional flexibility into the energy management process. The RL agent is tasked with its conventional responsibilities, such as managing battery dispatch and, in addition, the direct control of TCLs. For this reason, the agent's observation space is augmented with additional environmental information, such as external temperature, which influences thermal demand, and the discrete action space is expanded to include both the battery control actions and set-point commands for TCL energy allocation.

10.3.3.3 *Continuous state and action spaces*

Several recent studies move beyond traditional value-based methods to address scenarios involving continuous action spaces, which are more representative of real-world MG operations. For instance, Liu et al. (2021) and Guo et al. (2022) apply the DDPG and PPO algorithms, respectively, to the MG energy dispatching problem. Their primary objective is to minimize operational costs, including those associated with generator usage, BESS, and power exchanges with the main grid. A summary of the described different approaches has been provided in Table 10.5.

10.3.4 Dealing with the degradation phenomenon

A common limitation in earlier works is the omission of battery aging and degradation effects, which entail high economic costs when considering battery replacement and are critical for realistic long-term planning. A partial solution to these shortcomings is offered by Sui and Song (2020), which addresses the problem of scheduling charge, discharge, and resting periods with multiple batteries through a multiagent approach. Since the study aims to minimize system degradation, the authors adopt an aging model influenced solely by temperature, while ignoring other degradation factors related to usage, such as depth of discharge, SOC, and current rate. More recent studies have begun incorporating these aspects into the problem formulation. For example, Mussi et al. (2024) employ Fitted Q-Iteration (FQI) to train a policy that maximizes revenues from energy trading while explicitly accounting for battery degradation. However, their model assumes a static energy market with fixed buying and selling prices, ignores ambient temperature variations (which significantly influence battery aging), and restricts control to discrete action spaces. These limitations are addressed in a more advanced framework proposed by Salaorni et al. (2025), which adopts a PPO-based solution that incorporates both realistic energy market dynamics and ambient temperature fluctuations. Their model operates over a continuous action space, allowing for more granular and accurate control. The goal remains the optimal energy dispatch within the MG, now under constraints and dynamics that better reflect real-world operational complexity.

10.3.5 Grid-centric reinforcement learning methods

An alternative problem formulation shifts the focus from an MG-centric perspective to a grid-centric one, aiming to optimize the behavior of the DSO, i.e., an external entity responsible for ensuring the reliable and secure operation of the electrical grid. In these scenarios, the DSO often acts as a market maker, dynamically setting energy prices to influence the behavior of connected MGs in a desirable way. Du and Li (2020) propose a Monte Carlo–based RL algorithm to train a DSO to optimize retail pricing strategies within a MMG setting. In this framework, the RL agent is solely responsible for price determination, while individual MGs are modeled using deep learning architectures, yielding a hierarchical control structure. A similar multi-level approach is presented by Shojaeighadikolaei et al. (2021), who adopt a multiagent framework based on DQN. In their model, a central agent represents the DSO, which is responsible for setting dynamic energy prices, and additional agents correspond to network prosumers, which must decide whether to charge, discharge, or idle their battery systems in response to both local conditions and the DSO's pricing signals.

10.3.6 Hybrid approaches

Hybrid frameworks combining RL and MPC have recently gained significant traction, aiming to leverage the complementary strengths of each method. MPC can supervise RL by providing expert trajectories that are then used as expert demonstrations to warm-start an RL policy via imitation learning (e.g., behavior cloning or policy distillation). The RL agent can then fine-tune this policy online, adapting to non-modeled dynamics or disturbances that the original MPC model failed to capture. Another approach integrates safety mechanisms inspired by MPC's constraint-handling capabilities (Patra & Pradhan, 2025). These safe-RL methods augment the policy with action filters, safety shields, or specialized critics designed to penalize or prevent predicted constraint violations. This provides feasibility safeguards akin to MPC while retaining the model-free adaptability of RL. A third strategy uses RL to reduce the computational complexity of the MPC's online optimization. For instance, in the context of MG systems, Da Silva et al. (2025) demonstrate using an RL agent to determine the discrete variables within a mixed-integer linear program. This simplifies the MPC's task, leaving it to solve a less complex (e.g., purely continuous) optimization problem in real-time.

10.4 Challenges and open issues

MPC and RL frameworks have proven valuable for addressing the optimal control of MG systems. The main findings regarding these methods can be summarized as follows. On one hand, MPC has proven to be a mature, interpretable, and industry-ready solution for MG tertiary control. It excels in explicitly incorporating system dynamics, physical and operational constraints, and multiple objectives, such as economic efficiency, system reliability, and asset longevity into a single optimization framework. MPC-based methods are particularly suited for scenarios where models are accurate and computational resources are sufficient, and where safety and constraint satisfaction are crucial. On the other hand, RL offers a flexible, data-driven alternative that can learn adaptive policies directly from interaction data, without requiring an explicit model of the system dynamics, demonstrating adaptability to non-stationary environments and the potential to exploit non-obvious correlations in operational data. Consequently, RL-based methods are mainly suited to complex, uncertain environments where accurate models of the system dynamics are unavailable or expensive to obtain, and where adaptability and learning from operational data are prioritized.

While MPC and RL have emerged as promising approaches for the optimal control of MGs, their practical deployment still faces challenges and open issues arising from the inherent complexity, uncertainty, and dynamic nature of decentralized energy systems. However, integrating these techniques into DTs may accelerate deployment, enabling the design of safe, explainable, and adaptive controllers for MGs.

10.4.1 Model predictive control challenges

MPC is a highly attractive solution for MG tertiary control due to its flexibility in handling multiple objectives and constraints, making it particularly effective for power flow management and economically optimal scheduling (Hu et al., 2021).

A primary concern lies in the accuracy of predictive models used to forecast the behavior of system components. While rolling-horizon strategies are commonly used to mitigate the impact of forecasting errors in input variables (e.g., load demand or renewable energy availability), many MPC implementations still treat these uncertainties implicitly or overlook them entirely. This limits the robustness of the control strategy under real-world conditions. In response, ongoing research is exploring more resilient approaches, such as RMPC and SMPC techniques, that explicitly model uncertainty and ensure constraint satisfaction across a broader range of operational scenarios.

Another significant challenge is the computational complexity of MPC algorithms, particularly when long prediction horizons are used or when switching-mode resources such as distributed generators and batteries are incorporated. These cases often lead to complex mixed-integer programming problems, which are difficult to solve in real time. Closely related is the issue of designing and tuning the weighting coefficients in multiobjective cost functions. Heuristic approaches are often adopted, but they can result in suboptimal performance and limited generalizability across different configurations or operating conditions.

Communication delays represent another source of complexity, as they can degrade data quality and disrupt the timely updating of control signals. This is especially problematic in distributed control architectures that rely on consistent, reliable communication between agents. As MG systems scale, scalability itself becomes a pressing concern. Centralized control schemes may become impractical due to increasing computational demands and heightened cybersecurity risks. Consequently, there is growing interest in distributed control methods, which offer improved scalability and fault tolerance by leveraging local data and peer-to-peer communication. However, these approaches are also more vulnerable to the aforementioned communication delays and synchronization issues.

Finally, there is a recognized need for integrated MPC frameworks that seamlessly coordinate both device-level control (e.g., power electronics) and system-level objectives, such as power flow optimization, economic dispatch, and ancillary services. This integration must also support smooth transitions between islanded and grid-connected operating modes—an area that remains an open and technically challenging problem requiring further investigation.

10.4.2 Reinforcement learning challenges

As reported by Dinata et al. (2024), adopting an RL framework can significantly enhance network efficiency by optimizing the management of energy exchanges between DERs. The data-driven nature of RL algorithms enables the discovery of non-obvious correlations within operational data and facilitates the design of adaptive policies that respond in real time to environmental changes. This improves grid stability and promotes efficient load balancing across distributed resources.

However, several challenges must be addressed to ensure the practical deployment of RL-based control systems. First, the limited availability of high-quality data often impedes learning, making it challenging to obtain generalizable policies that perform robustly across diverse, unforeseen scenarios. Second, RL algorithms must be developed with strong guarantees of safety and stability, especially when deployed in critical MG environments. Moreover, these algorithms must also ensure data privacy and protection, particularly in collaborative settings where agents may need to exchange information that could include sensitive operational or user data. Another significant limitation lies in the gap between simulation and real-world implementation. Since most RL algorithms are trained and tested in simulated environments, rigorous validation protocols and hardware compatibility assessments are essential before deployment. Finally, the issue of interpretability is often overlooked. Enhancing the transparency and explainability of data-driven methods is crucial to building trustworthy, reliable control systems.

References

Baker, K., Guo, J., Hug, G., & Li, X. (2016). Distributed MPC for efficient coordination of storage and renewable energy sources across control areas. *IEEE Transactions on Smart Grid, 7*, 992–1001.

Basit, M. A., Dilshad, S., Badar, R., & Sami ur Rehman, S. M. (2020). Limitations, challenges, and solution approaches in grid-connected renewable energy systems. *International Journal of Energy Research, 44*, 4132–4162.

Bidram, A., & Davoudi, A. (2012). Hierarchical structure of microgrids control system. *IEEE Transactions on Smart Grid, 3*, 1963–1976.

Bonassi, F., La Bella, A., Lazzari, R., Sandroni, C., & Scattolini, R. (2021). Supervised control of hybrid AC-DC grids for power balance restoration. *Electric Power Systems Research, 196*, 107107.

Cai, S., Xie, Y., Wu, Q., & Xiang, Z. (2020). Robust MPC-based microgrid scheduling for resilience enhancement of distribution system. *International Journal of Electrical Power & Energy Systems, 121*, 106068.

Chen, B., Wang, J., Lu, X., Chen, C., & Zhao, S. (2020). Networked microgrids for grid resilience, robustness, and efficiency: A review. *IEEE Transactions on Smart Grid, 12*, 18–32.

Chen, T., & Bu, S. (2019). Realistic peer-to-peer energy trading model for microgrids using deep reinforcement learning. In *2019 IEEE PES innovative smart grid technologies Europe (ISGT-Europe)* (pp. 1–5). doi:10.1109/ISGTEurope.2019.8905731.

Clarke, W. C., Manzie, C., & Brear, M. J. (2016). An economic mpc approach to microgrid control. In *2016 Australian Control Conference (AuCC)* (pp. 276–281). IEEE.

Clarke, W. C., Manzie, C., & Brear, M. J. (2018). Hierarchical economic mpc for systems with storage states. *Automatica, 94*, 138–150.

Da Silva, O., Caio, F., Dabiri, A., & De Schutter, B. (2025). Integrating reinforcement learning and model predictive control for mixed-logical dynamical systems. *IEEE Open Journal of Control Systems, 4*, 316–331. doi:10.1109/OJCSYS.2025.3601435.

Dinata, N. F. P., Ramli, M. A. M., Jambak, M. I., Sidik, M. A. B., & Alqahtani, M. M. (2024). Designing an optimal microgrid control system using deep reinforcement learning: A systematic review. *Engineering Science and Technology: An International Journal, 51*, 101651.

Domínguez-Barbero, D., García-González, J., Sanz-Bobi, M. A., & Sánchez-Úbeda, E. F. (2020). Optimising a microgrid system by deep reinforcement learning techniques. *Energies 13*. https://www.mdpi.com/1996-1073/13/11/2830. doi:10.3390/en13112830.

Du, Y., & Li, F. (2020). Intelligent multi-microgrid energy management based on deep neural network and model-free reinforcement learning. *IEEE Transactions on Smart Grid, 11*, 1066–1076. doi:10.1109/TSG.2019.2930299.

Ellabban, O., Abu-Rub, H., & Blaabjerg, F. (2014). Renewable energy resources: Current status, future prospects and their enabling technology. *Renewable and Sustainable Energy Reviews, 39*, 748–764.

Foruzan, E., Soh, L. K., & Asgarpoor, S. (2018). Reinforcement learning approach for optimal distributed energy management in a microgrid. *IEEE Transactions on Power Systems, 33*, 5749–5758. doi:10.1109/TPWRS.2018.2823641.

Guo, C., Wang, X., Zheng, Y., & Zhang, F. (2022). Real-time optimal energy management of microgrid with uncertainties based on deep reinforcement learning. *Energy, 238*, 121873. doi:10.1016/j.energy.2021.121873.

Haarnoja, T., Zhou, A., Abbeel, P., & Levine, S. (2018). Soft actor-critic: Offpolicy maximum entropy deep reinforcement learning with a stochastic actor. *arXiv*, 1801.01290. https://arxiv.org/abs/1801.01290.

Hans, C. A., Sopasakis, P., Bemporad, A., Raisch, J., & Reincke-Collon, C. (2015). Scenario-based model predictive operation control of islanded microgrids. In *2015 54th IEEE conference on decision and control (CDC)* (pp. 3272–3277). Osaka, Japan: IEEE.

He, Y., Li, Z., Zhang, J., Shi, G., & Cao, W. (2023). Day-ahead and intraday multi-time scale microgrid scheduling based on light robustness and mpc. *International Journal of Electrical Power & Energy Systems, 144*, 108546.

Hu, J., Shan, Y., Guerrero, J. M., Ioinovici, A., Chan, K. W., & Rodriguez, J. (2021). Model predictive control of microgrids-an overview. *Renewable and Sustainable Energy Reviews, 136*, 110422.

Hu, J., Shan, Y., Yang, Y., Parisio, A., Li, Y., Amjady, N., Islam, S., Cheng, K. W., Guerrero, J. M., & Rodríguez, J. (2023). Economic model predictive control for microgrid optimization: A review. *IEEE Transactions on Smart Grid, 15*, 472–484.

Ju, C., Wang, P., Goel, L., & Xu, Y. (2017). A two-layer energy management system for microgrids with hybrid energy storage considering degradation costs. *IEEE Transactions on Smart Grid, 9*, 6047–6057.

Kolodziejczyk, W., Zoltowska, I., & Cichosz, P. (2021). Real-time energy purchase optimization for a storage-integrated photovoltaic system by deep reinforcement learning. *Control Engineering Practice, 106*, 104598. doi:10.1016/j.conengprac.2020.104598.

Kou, P., Feng, Y., Liang, D., & Gao, L. (2019). A model predictive control approach for matching uncertain wind generation with pev charging demand in a microgrid. *International Journal of Electrical Power & Energy Systems, 105*, 488–499.

Kuznetsova, E., Li, Y. F., Ruiz, C., Zio, E., Ault, G., & Bell, K. (2013). Reinforcement learning for microgrid energy management. *Energy, 59*, 133–146. doi:10.1016/j.energy.2013.05.060.

Lillicrap, T. P., Hunt, J. J., Pritzel, A., Heess, N. M. O., Erez, T., Tassa, Y., Silver, D., & Wierstra, D. P. (2020). Continuous control with deep reinforcement learning. U.S. Patent No. 10,776,692. U.S. Patent and Trademark Office. https://patents.google.com/patent/US10776692B2/en.

Liu, L., Zhu, J., Chen, J., & Ye, H. (2021). Deep reinforcement learning for stochastic dynamic microgrid energy management. *2021 IEEE 4th international electrical and energy conference (CIEEC)* (pp. 1–6). doi:10.1109/CIEEC50170.2021.9511049.

Lopes, J. P., Hatziargyriou, N., Mutale, J., Djapic, P., & Jenkins, N. (2007). Integrating distributed generation into electric power systems: A review of drivers, challenges and opportunities. *Electric Power Systems Research, 77*, 1189–1203.

Lopes, J. P., Moreira, C. L., & Madureira, A. (2006). Defining control strategies for microgrids islanded operation. *IEEE Transactions on Power Systems, 21*, 916–924.

Mariam, L., Basu, M., & Conlon, M. F. (2016). Microgrid: Architecture, policy and future trends. *Renewable and Sustainable Energy Reviews, 64*, 477489.

Mayne, D. Q. (2014). Model predictive control: Recent developments and future promise. *Automatica, 50*, 2967–2986.

Meng, L., Shafiee, Q., Trecate, G. F., Karimi, H., Fulwani, D., Lu, X., & Guerrero, J. M. (2017). Review on control of dc microgrids and multiple microgrid clusters. *IEEE Journal of Emerging and Selected Topics in Power Electronics, 5*, 928–948.

Mnih, V., Kavukcuoglu, K., Silver, D., Graves, A., Antonoglou, I., Wierstra, D., & Riedmiller, M. (2013). Playing Atari with deep reinforcement learning. arXiv preprint arXiv:1312.5602. https://arxiv.org/abs/1312.5602

Mussi, M., Pellegrino, L., Pindaro, O. F., Restelli, M., & Trovò, F. (2024). A reinforcement learning controller optimizing costs and battery state of health in smart grids. *Journal of Energy Storage, 82*, 110572. doi:10.1016/j.est.2024.110572.

Nahata, P., La Bella, A., Scattolini, R., & Ferrari-Trecate, G. (2020). Hierarchical control in islanded dc microgrids with flexible structures. *IEEE Transactions on Control Systems Technology, 29*, 2379–2392.

Nair, U. R., & Costa-Castelló, R. (2020). A model predictive control-based energy management scheme for hybrid storage system in islanded microgrids. *IEEE Access, 8*, 97809–97822.

Nakabi, T. A., & Toivanen, P. (2021). Deep reinforcement learning for energy management in a microgrid with flexible demand. *Sustainable Energy, Grids and Networks, 25*, 100413. doi:10.1016/j.segan.2020.100413.

Nguyen, T. A., & Crow, M. L. (2015). Stochastic optimization of renewable-based microgrid operation incorporating battery operating cost. *IEEE Transactions on Power Systems, 31*, 2289–2296.

Olivares, D. E., Mehrizi-Sani, A., Etemadi, A. H., Cañizares, C. A., Iravani, R., Kazerani, M., Hajimiragha, A. H., Gomis-Bellmunt, O., Saeedifard, M., Palma-Behnke, R., Jiménez-Estévez, G. A., & Hatziargyriou, N. D. (2014). Trends in microgrid control. *IEEE Transactions on Smart Grid, 5*, 1905–1919.

Palma-Behnke, R., Benavides, C., Lanas, F., Severino, B., Reyes, L., Llanos, J., & Sáez, D. (2013). A microgrid energy management system based on the rolling horizon strategy. *IEEE Transactions on Smart Grid, 4*, 996–1006.

Parisio, A., Rikos, E., & Glielmo, L. (2014). A model predictive control approach to microgrid operation optimization. *IEEE Transactions on Control Systems Technology, 22*, 1813–1827.

Parisio, A., Rikos, E., & Glielmo, L. (2016). Stochastic model predictive control for economic/environmental operation management of microgrids: An experimental case study. *Journal of Process Control, 43*, 24–37.

Parisio, A., Wiezorek, C., Kyntäjä, T., Elo, J., Strunz, K., & Johansson, K. H. (2017). Cooperative mpc-based energy management for networked microgrids. *IEEE Transactions on Smart Grid, 8*, 3066–3074.

Patra, D., & Pradhan, S. (2025). Adaptive control strategies for enhancing the integration and stability of renewable energy sources in smart microgrids. *International Journal OF Electrical Engineering and Technology, 16*, 1–23.

Pippia, T., Sijs, J., & De Schutter, B. (2019). A single-level rule-based model predictive control approach for energy management of grid-connected microgrids. *IEEE Transactions on Control Systems Technology, 28*, 2364–2376.

Polimeni, S., Meraldi, L., Moretti, L., Leva, S., & Manzolini, G. (2021). Development and experimental validation of hierarchical energy management system based on stochastic model predictive control for off-grid microgrids. *Advances in Applied Energy, 2*, 100028.

Sachs, J., & Sawodny, O. (2016). A two-stage model predictive control strategy for economic diesel-pv-battery island microgrid operation in rural areas. *IEEE Transactions on Sustainable Energy, 7*, 903–913.

Salaorni, D., Bianchi, F., Trovò, F., & Restelli, M. (2025). A reinforcement learning approach for optimal control in microgrids. In *2025 International Joint Conference on Neural Networks (IJCNN)* (pp. 1–9). Rome, Italy: IEEE.

Schulman, J., Wolski, F., Dhariwal, P., Radford, A., & Klimov, O. (2017). Proximal policy optimization algorithms. *arXiv preprint arXiv:1707.06347*.

Sehnke, F., Osendorfer, C., Rückstieß, T., Graves, A., Peters, J., & Schmidhuber, J. (2008). Policy gradients with parameter-based exploration for control. In V. Kůrková, R. Neruda, & J. Koutník (Eds.), *Artificial neural networks - ICANN 2008*. Springer.

Shojaeighadikolaei, A., Ghasemi, A., Bardas, A. G., Ahmadi, R., & Hashemi, M. (2021). Weather-aware data-driven microgrid energy management using deep reinforcement learning. In *2021 North American power symposium (NAPS)* (pp. 1–6). doi:10.1109/NAPS52732.2021.9654550.

Sui, Y., & Song, S. (2020). A multi-agent reinforcement learning framework for lithium-ion battery scheduling problems. *Energies, 13*. doi:10.3390/en13081982.

Sutton, R. S., & Barto, A. G. (1998). *Introduction to reinforcement learning*: 135. Cambridge: MIT Press.

Sutton, R. S., McAllester, D., Singh, S., & Mansour, Y. (1999). Policy gradient methods for reinforcement learning with function approximation. In S. Solla, T. Leen, & K. Müller (Eds.), *Advances in neural information processing systems* (pp. 1057–1063). MIT Press.

Vignali, R., Falsone, A., Ruiz, F., & Gruosso, G. (2022). Towards a comprehensive framework for V2G optimal operation in presence of uncertainty. *Sustainable Energy, Grids and Networks, 31*, 100740.

Watkins, C. J. C. H., & Dayan, P. (1992). Q-learning. *Mach Learn, 8*, 279–292. https://doi.org/10.1007/BF00992698.

Zhao, Z., Guo, J., Luo, X., Lai, C. S., Yang, P., Lai, L. L., Li, P., Guerrero, J. M., & Shahidehpour, M. (2022). Distributed robust model predictive control-based energy management strategy for islanded multi-microgrids considering uncertainty. *IEEE Transactions on Smart Grid, 13*(3), 2107–2120. doi:10.1109/TSG.2022.3147370.

Zhou, K., Zhou, K., & Yang, S. (2022). Reinforcement learning-based scheduling strategy for energy storage in microgrid. *Journal of Energy Storage, 51*, 104379. doi:10.1016/j.est.2022.104379.

Zhu, J., Knapp, M., Sørensen, D. R., Heere, M., Darma, M. S., Müller, M., Mereacre, L., Dai, H., Senyshyn, A., Wei, X., & Ehrenberg, H. (2021). Investigation of capacity fade for 18650-type lithium-ion batteries cycled in different state of charge (soc) ranges. *Journal of Power Sources, 489*, 229422.

PART III

Digital Twin computing for power grids

CHAPTER

11

Towards digital and smart substations: A Digital Twin approach

Gabriel Rodrigues Santos, Eduardo Zancul

Production Engineering Department, Polytechnic School of the University of São Paulo, São Paulo, Brazil

11.1 Introduction

Power substations are nodes in the power grid, responsible for the timely, safe, and continuous management of electricity flow. Power substations transform voltage levels, distribute electricity, and protect and control the power flow within the grid, housing key-power systems' infrastructure to perform those tasks. From the first numeric relays until the development of Intelligent Electronic Devices (IEDs), substation automation systems have been increasingly impacted by digital technologies (Santos et al., 2024), leading to the development of smart substations (Huang et al., 2017). These facilities have been identified as a critical component of the smart grid (Li et al., 2010), instrumental to the development of more sustainable, reliable, connected, and efficient grids; potentially existing in the intersection of Industry 4.0 and power systems (Manassero et al., 2024).

Although Digital Twins (DTs) have been increasingly researched by Industry 4.0 scholars in recent years, including the development of multiple frameworks (Tao et al., 2019, 2022), their practical application remains a topic to be further explored (Zancul, Oliveira, Wang, Schützer, & Schleich, 2025). Some applications of DTs in the power industry have also been initially identified by surveying the extant literature. Reviews of DTs in the energy industry (do Amaral et al., 2023) and their application to electrical power systems (Mansour et al., 2023) have been recently published. As examples, Sleiti et al., (2022) propose a DT for power plants; Yan et al. (2024) claim to use a DT-enabled solution to localize partial discharge failures in switchgears; Li et al., (2024) report a monitoring system for substations; and the DT impacts to substation management also seem to have been investigated (Qian et al., 2023). However, none of these applications addresses a systemic DT development for power substations, especially coupled with a thorough assessment framework.

Substations are logical and power system nodes, which makes them an especially critical asset for deploying key technologies to enhance the smart grid (Li et al., 2010). However, the DT instantiation in substations has been comparatively less developed (Mansour et al., 2023). This indicates a potential opportunity. It could also indicate that the terminology "Digital Twin" has not been fully adopted by electrical substation engineers and researchers, but that the concept beyond the wording has also been developed for substations. The latter case may be partially true: power systems have had extremely advanced simulation and control schemas for several decades, in order to manage the loads, protection, and control of the power grid. For example, a fault event such as a short circuit should activate the substation's protection and control systems, triggering the circuit breaker to safely isolate the faulty load and thereby support the reliable operation of the grid. A significant part of the grid's management is performed within substations, which further emphasizes the need to understand the potential and limitations of applying new technologies to these facilities.

DTs may provide a solution to better leverage the benefits of smart substations. Santos and Zancul (2023) have deepened the understanding of the technologies and impacts that smart substation installations are expected to bring forth. Beyond calling for research on a methodology to aid investment decision-making for transmission system operators (TSO), the authors also highlighted the need for the sector to be better equipped to effectively manage the

Digital Twin Technology and Smart Grid.
DOI: https://doi.org/10.1016/B978-0-443-36664-2.00005-8

digital transition and actually reap the benefits promised by these technologies. In other words, not only knowing *what* the expected benefits of digitalization are, but actually *how*, and *by which means or tools* these benefits could be realized in a complex and evolving power grid. There is, therefore, an opportunity to bridge this gap and develop solutions based on DTs for smart substations as components of the smart grid.

This chapter argues that employing DTs can be a pathway for such a digital transition for improved substation operation and maintenance. First, a background on what has been reported in DTs for substations is presented. The Background section also argues why it is important to consider the topic and how substation DT modeling can be different from other developments of DTs across sectors. Then, drawing from an analysis of both the DT and the smart substations literature, both topics are combined in a proposed architecture. The architecture instantiates a DT for smart substations as components of the smart grid with a service-oriented approach. The characteristics and potential benefits of this application are discussed. Finally, further research recommendations are offered.

11.2 Background

In the last decade, there has been intense academic and later industrial development to conceptualize DTs as a promising concept to support digitalization in many ways. The conceptual definition of DT has been extensively discussed and progressively consolidated in the scholarly literature. Several works have addressed this through literature review, initial technological development, and analysis, especially after the late 2010s, such as Semeraro et al., 2021, Stark et al., 2019, and Tao et al., 2018).

Despite these conceptual advancements, there are still gaps in the understanding and deployment of the DT concept in several sectors. The usage of the term "Digital Twin" in the electrical systems is not standardized, and it has also been used as an "umbrella term"; there seems to be no agreement about what precisely constitutes a DT, and what, rather, constitutes, for example, a simulation model based on sensor data. There seems to be no single, agreed working definition that works. Hence, in this chapter, we broadly conceptualize what is meant by a DT, provide a few architectures developed for its realization, and detail how it has been applied to electrical substations thus far. We also point out the differences between smart substations and other contexts for which DTs have been developed, which make such applications unique. This is used to support the proposition of an architecture for DT realization in power substations, considering an informed discussion of their specific characteristics.

11.2.1 Digital Twin architectures

A DT can be conceptualized as a "set of adaptive models that emulate the behavior of a physical system in a virtual system, getting real-time data to update itself along its life cycle. The DT replicates the physical system to predict failures and opportunities for changing, to prescribe real-time actions for optimizing and/or mitigating unexpected events, observing and evaluating the operating profile system" (Semeraro et al., 2021). Other frameworks to understand and implement DTs have also been developed, with several specifications including hierarchy level, dimensions, and functions (Tao et al., 2022).

Many works structure the DT modeling in layers; prominent examples include the five-dimensional architecture by Tao et al. (2018); the six-layer architecture by Redelinghuys et al., (2020); or the seven-layer architecture by Singh et al., (2021). The latter proposition includes the following layers: (1) Physical asset (or physical twin); (2) Communication; (3) Storage; (4) Virtual model (Digital Twin); (5) Security; (6) Service; and (7) Access (user interface). Fig. 11.1 illustrates the seven-layer model by Singh et al., (2021).

Singh et al., (2021) contribution includes the explicit definition of a service layer as a layer that encompasses the services related to data-driven applications. Thinking about DTs with a service-based approach, or to support product-service systems, has been an emerging research topic (Bertoni & Bertoni, 2022; Durão, Zancul, & Schützer, 2024), and has led to the development of solutions in multiple fields and devices; for example, for a DT for a mobile electric vehicle charger (Romeral et al., 2024). Few studies have been found to demonstrate DT applications with a clear service component. Existing case studies predominantly focus on frameworks, tools, or value propositions tied to either the exchange of tangible goods or services, without fully addressing customer-oriented value offerings (Bertoni & Bertoni, 2022). By explicitly highlighting the service layer, it becomes easier to identify and design the value-adding functions, which ultimately constitute the primary objectives of DT development.

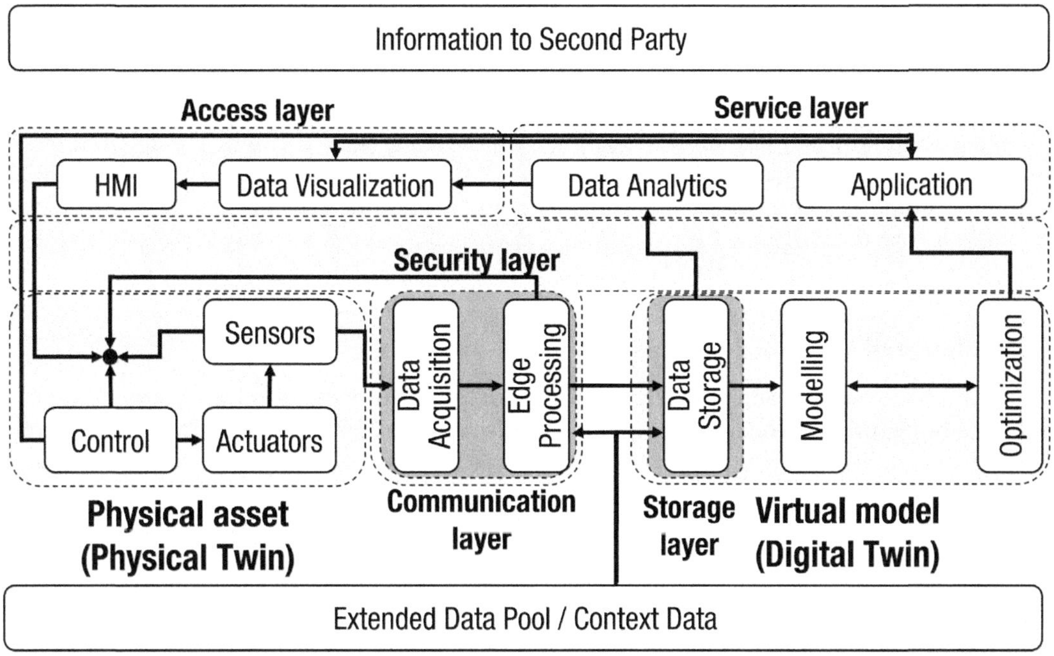

FIGURE 11.1 Seven-layer Digital Twin architecture. Seven-layer Digital Twin architecture. *Modified Singh, Soumya, Weeber, Max & Birke, Kai-Peter. (2021). Advancing digital twin implementation: a toolbox for modelling and simulation. Procedia CIRP, 99, 567–572. https://doi.org/10.1016/j.procir.2021.03.078.*

11.2.2 Digital Twins for smart grid applications

A smart grid DT can be modeled in different levels of analysis: from component or unit-level, up to system-level, or even extended to system-of-systems (SoS) level. Since the application of the concept to the smart grid is relatively new, most reported solutions are of unit-level DTs, such as individual power equipment, while system-level and SoS-level application cases are scarce (Jiang et al., 2022). Furthermore, few studies offer an architecture approach to DT modeling for the smart grid and substations. Jiang et al. (2022) describe a four-element-based model for developing Smart Grid DTs, with a case application presented in a substation circuit breaker. Udren et al. (2024) include DTs as a component of a wide-area system configuration architecture, focusing on their potential to use virtual models to mirror power systems topology, perform simulations, and improve management and maintenance. This conceptualization, however, does not fully utilize the potential applications of DTs as reported in the case of manufacturing environments.

Considering the smart grid as the ultimate SoS level, a substation DT could be conceptualized as a system-level DT. However, it is worth noting that substations are themselves made up of multiple systems: protection, control, automation, communications, etc. In this sense, the substation DT itself is also comprised of several subsystems, which should effectively interface with each other. This added complexity may partly explain why most works focus on component-level DTs, rather than developing broad and comprehensive architectures for substation-wide DTs.

11.2.3 Digital Twins within substations

Electrical substations have been appointed as key components of the smart grid (Li et al., 2010). Their technological evolution accompanied the digital transformation, with increasing degrees of automation systems digitalization (Aftab et al., 2020), and changes to technical and nontechnical characteristics (Santos et al., 2024). There have also been developments toward the future smart substation (Huang et al., 2017), which include the implementation of advanced technologies to enhance operational excellence, with increased data integration, flexibility, and cybersecurity (Manassero et al., 2024). Some of those technologies include: data analytics for enhanced predictive or condition-based maintenance, cloud computing, standardized and interoperable communications, advanced fault location, drone-based inspection, virtual and augmented reality for improved maintenance interventions, among others. An in-depth look at modern substation automation systems architecture and technologies can be found in Leonardi et al. (2014), with significant development centered around the standard IEC 61850: Communications networks and systems for power utility automation (Aftab et al., 2020).

A large proportion of publications focus on using DTs for specific functions within substations, such as investigation power system events (Tiferes et al., 2025), 3D visualization of the physical buildings or assets in the control room, an approach related to Building Information Modeling (Loggia & Flamini, 2023), or the creation of DTs of specific substation components such as transformers (Delong et al., 2021) or circuit breakers (Jiang et al., 2022). Improving substations' condition monitoring has been identified as a possible application of DTs, but with comparatively few studies reported (Mansour et al., 2023).

11.2.4 The challenge of smart substations

Electrical substations exhibit certain characteristics that distinguish them from other potential DT applications, such as those in manufacturing or healthcare, which are discussed in this section. These specific conditions drive the potential need for a tailored solution on DT architectures for electrical substations. This is especially the case with smart substations (Huang et al., 2017; Santos et al., 2024), whose application of advanced technologies leads to new capabilities, needs, and requirements. Therefore, it is worth exploring: What makes electrical substations different from other facilities, to the extent that this could impact the implementation of a DT? Some key reasons are summarized in the five points, which are discussed further.

1. **Operation and reliability**: Substations are fundamental for the continuous operation of the power grid, which is itself of infrastructural importance to society. Therefore their reliable and continuous operation must be guaranteed. The levels of availability of power equipment typically exceed those of industrial or manufacturing equipment. This reliability-centered approach drives several studies and applications in the power grid and must be thoroughly considered in any DT proposition. Furthermore, concerns over cybersecurity in digital environments should also be addressed; the impact severity of a power grid disruption caused by a cyberattack to substations cannot be understated. DTs have also been appointed as a technology that may help test and mitigate the impacts of cybersecurity threats to power systems (Srivastava et al., 2024). Security threats should be mitigated with measures such as compliance with the cybersecurity Standard IEC 62351, including its profiles, architectures, and management.
2. **Time-critical operation**: Related to the previous topic, substation protection and control systems operate with extremely tight timing constraints, with protection functions acting in a few milliseconds. Substations operate in mission-critical environments where latency, reliability, and performance are paramount. Delays in critical communication exchange could severely impact the protection ability of the system and should be avoided. Precise time synchronization is also a topic of concern (Moradi & Jahangir, 2021), with protocols and standards developed for this end. A DT architecture must reflect these temporal and reliability constraints through robust data pipelines and edge computing capabilities, without interfering with its timely operation. DT for substations must support low-latency real-time monitoring, predictive diagnostics, and control loop integration, which are not typical requirements in every generic DT use case.
3. **Legacy, capital-intensive systems integration:** Substation assets are not only capital-intensive, but also have long operational lifetimes. Regulatory lifetimes are typically a few decades, but actual assets are often operated beyond those times[a]. This naturally leads to a power grid in which new and legacy assets coexist, with challenging integration (Chaaban et al., 2015). Unlike generic DTs that may model stand-alone products or isolated processes, substation DTs must interface with legacy IEDs, SCADA systems, and IEC 61850-compliant components; for example, with architectural mechanisms for interoperability and protocol translation.
4. **Complex, layered cyber-physical systems (CPS):** Substations are prototypical CPS that integrate physical electrical infrastructure with digital control, protection, and automation, especially in more modern smart substations. Substations also feature multiple technology-specific subsystems, including electrical, electronics, communications, IT, and emerging complex data systems (Santos et al., 2024), which must be locally and regionally coordinated within the power grid. This requires multidomain modeling capabilities, integrating electrical, communication, cybersecurity, and asset management domains. Generic DTs rarely need this level of domain-spanning simulation

[a] Actually, regulatory frameworks often incentivize asset lifetime extension, which has been appointed as a potential benefit of substation digitalization (Santos & Zancul, 2023).

TABLE 11.1 Comparative summary of substations and manufacturing Digital Twins.

Aspect	Substation Digital Twin	Manufacturing Digital Twin
Timing requirements and dynamics	Real-time, very short latency (ms-level for protection functions), event-driven (e.g., fault, protection trip)	Varied (from real-time to batch), process-centric (e.g., throughput, cycle time)
Reliability and safety criticality	Very high	Varies significantly depending on the operation, from medium to high (or very high)
Standards and interoperability	Strongly governed (IEC 61850, 62351, 61970/61968)	Often proprietary or platform-dependent
Cybersecurity requirements	Very strict (compliance-focused, grid-wide impacts)	Increasing, but often factory-contained, including intellectual property protection
Domain knowledge	Power systems, protection engineering, grid codes, data engineering	Process and production engineering, operations management, data engineering
Legacy integration	Integration with decades-old hardware and firmware	Often, newer systems or greenfield digitalization
Sectoral dynamics	Highly regulated, public service with private and public providers	Typically, the private sector is governed by market dynamics and internal decision-making

fidelity. Furthermore, substations are typically conceptualized in a layered structure, from process to bay to station-level hardware and control systems; thus, developing an effective and thorough DT model for substations also requires significant attention to its vertical integration ability.

5. **Regulatory and Standards Alignment**: Substations are constrained by the electrical sector's standards and regulations, much more so than other industries[b]. The specific conditions of the electrical sector, including remuneration mechanisms, as well as its traditionally conservative approach to innovation (Biancardi et al., 2021), should also be considered when conceptualizing a substation DT. Such DT should, for example, embed compliance mechanisms and traceability features to a complex set of stakeholders and interfaces, as well as consider the local regulations.

A summary of some key differences between substation and manufacturing DTs is presented in Table 11.1.

Considering all these factors, it should be clear that developing an architecture for a substation DT is not a trivial instantiation of a generic DT framework developed for other contexts. Although specific DTs have been reported in the literature—be it for substation components, be it to aid in some substation operations—a comprehensive DT architecture for substations has not, to the best of the authors' knowledge, been reported. A proposition for this architecture is presented in the following section.

11.3 Conceptual architecture proposition and discussion

Drawing from the reference DT architecture models by Singh et al., (2021), and considering the factors detailed in the previous section, we propose a conceptual architecture for a substation DT. This architecture is presented graphically in Fig. 11.2, and its elements are detailed. Afterwards, the services enabled by the proposed architecture and potential improvements with the Substations DT are discussed. A discussion on the relationship between the substation DT and the benefits of smart substations closes this section.

11.3.1 Architecture for Digital Twins in smart substations

A brief description of the layers is as follows:

1. **Physical asset layer**: includes the physical assets contained in the substation. The core is the substation automation system, composed of its protection, control, and communications systems, arranged in the process, bay, and station levels. Physical assets include the primary power equipment (such as high-voltage transformers or circuit breakers,

[b] Some industrial sectors are also subject to extensive regulation. Aerospace and automotive industries are notable examples. Nonetheless, despite their product and quality regulations, these industries typically operate in a market-based model, unlike the power sector.

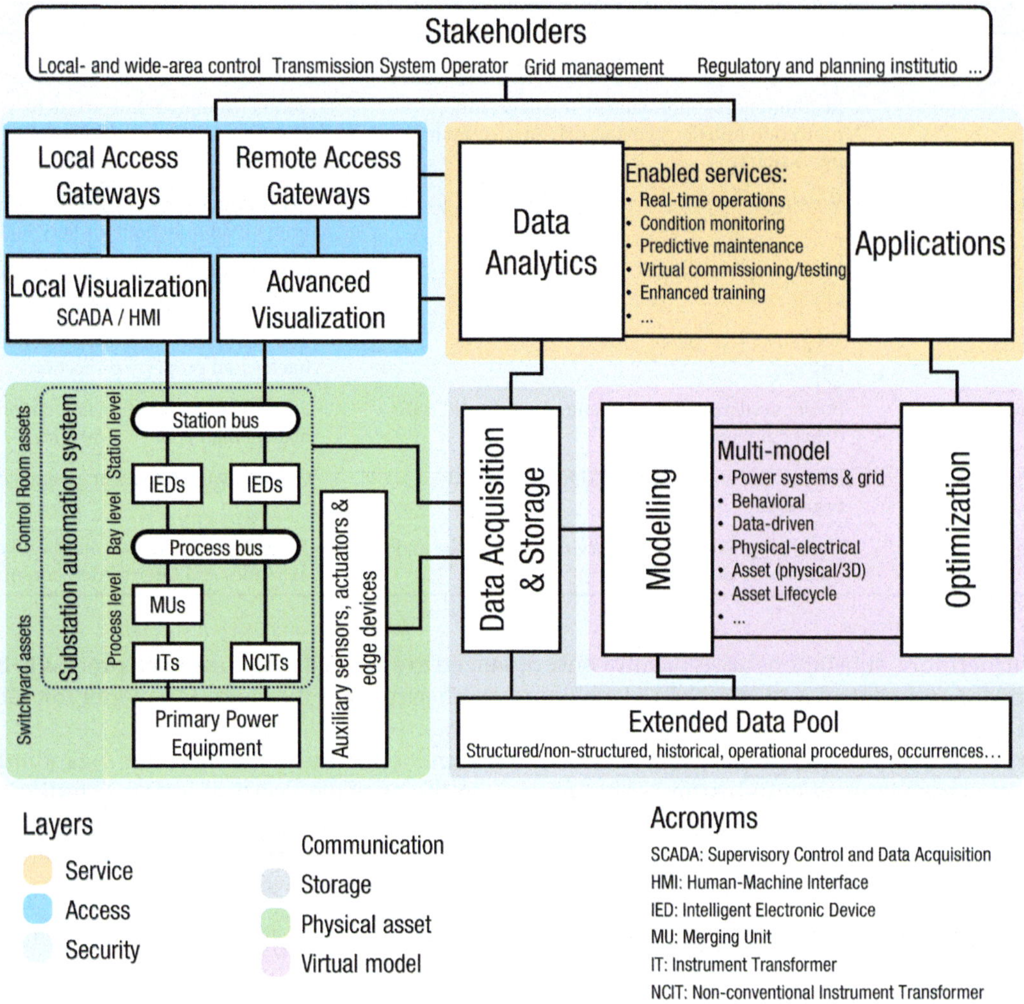

FIGURE 11.2 Visual representation of the proposed substation digital twin architecture. Proposed substation Digital Twin architecture based on seven layers. Layers are identified by colored blocks.

which interface with the high-voltage bays), instrument transformers (used to measure electrical current and voltage) and merging units (intermediate devices between conventional instrument transformers and digitized process bus), or the more advanced nonconventional instrument transformers (which already deliver digital measurements), all located in the substation switchyard; as well as the automation system IEDs, typically located in the control room (used for the implementation of the protection, control and automation functions). Furthermore, toward a full substation DT, the substation automation system is complemented with a variety of auxiliary sensors, actuators, and edge devices. These include environmental monitoring sensors, equipment monitoring sensors, cameras, and all other components that are not part of the "core automation system."

2. **Access layer**: Includes the visualization and access components, considering multiple stakeholders. In the architecture, two cases are highlighted: more typical SCADA systems and Human-Machine Interfaces used for local access at the substation, and also the remote access for more advanced visualization, for instance, in dashboards and interfaces for performing advanced analytics and services. Multiple stakeholders are considered: from the substation technicians and personnel, the TSO, local and wide-area grid management, as well as the sector's regulatory and planning institutions.
3. **Storage layer**: includes the data acquisition, preprocessing, and storage systems. The architecture should collect data both from the main substation automation system (such as process- and station-level data), and from the auxiliary devices. It is worth noting that this data can be considered *big data,* including great variety, velocity, structured and nonstructured, historical and real-time data. More broadly, the knowledge base used for the substation operations and maintenance can also be included in the data pool, such as operational work procedures

TABLE 11.2 Mapping of key standards and characteristics per digital twin layer.

Layer	Main standards or protocols	Characteristics
Physical asset	IEC 61850	Time-critical events and messages (SV/GOOSE/MMS), and millisecond-level actuation
Access	Common Information Model (CIM)/IEC 61970	HMI and SCADA interfaces, managerial interfaces, that should consider multiple stakeholders and levels
Storage	Common Information Model (CIM)/IEC 61970	Comprehensive data pool to feed advanced applications
Virtual model	Common Information Model (CIM)/IEC 61970/IEC 30173	A combination of different digital twin models (power systems, data-driven, behavior, physical assets etc.)
Communications	IEC 61850 / IEC 62541 (OPC UA)/MQTT/AMQP	Both power and auxiliary data should be integrated; multiple protocols can be used
Security	IEC 62351/IEC 27568	Must ensure the integrity, confidentiality, and availability of substation physical and virtual/data assets

and guidelines. The data is stored in an extended data pool, capable of handling and safely storing all the data, with access tailored to different use cases, ranging from high-priority, low-latency access to less mission-critical queries.

4. **Virtual model layer**: the core of the DT modeling (Tao et al., 2022), this layer encompasses all virtual modeling and optimization frameworks and systems used. The extensive data assets are an input for the construction of this layer. Substation DTs can combine multiple models to build the DTs: power systems and grid models (both physical and theoretical); behavioral and data-driven models (for electric power systems and beyond); physical substation asset 3D models; and also models of the assets across their lifecycles (a key for substations, as already discussed). All the different models can be combined to construct the DT, with multiple domains and capabilities.

Considering the criticality of operations and reliability in electric substations, the two layers present in the original architecture (Singh et al., 2021) have been represented across all other layers. Communications and security layers can draw from existing standards IEC 61850 and IEC 62351 to develop a robust and secure communications network for the entire DT:

1. **Communications layer**, includes all communication across devices and stakeholders. Especially for the substation automation systems, communication is extensively standardized, thanks to standards such as IEC 61850, as discussed in the previous section. Other standards can also be incorporated, for example, in the communication of data between the auxiliary sensors and control devices.
2. **Security layer:** represented as an all-encompassing layer, security should be present in the design of all components and interfaces of the substation DT. It must ensure the integrity, confidentiality, and availability of both operational and informational assets. It includes not only personal and work safety, but especially the cybersecurity of all virtual components of the DT. Managing and operating the power grid is a critical and sensitive activity––external attacks that could make the grid unreliable, unstable, or unsafe must not happen. Extensive research and development have been made in the cybersecurity of substation networks; it should integrate robust access control mechanisms, secure communication protocols (e.g., TLS, IEC 62351), intrusion detection systems, and real-time threat monitoring.

A summary of the main standards and characteristics mapped for these layers is presented in Table 11.2. We note that the standard and protocol list is not exhaustive, but rather a starting point for the DT development.

Finally, the service layer explicitly models the value-adding services of the DT. It includes the data analytics and the specific applications that enable advanced services for the smart substation. The enabled services are discussed in more detail in the following section.

11.3.2 Enabled services and improvements with Digital Twins in substations

Some of the services enabled by a substation DT are highlighted further:

1. **Real-time operational integration, control, and optimization:** Optimize switching operations, protection settings, or integration of distributed energy resources. This supports grid stability under increasing complexity

and variability of power flows, while improving situational awareness and decision-making at both local and system-wide levels.

2. **Condition monitoring and health diagnostics:** Real-time monitoring of IEDs, transformers, circuit breakers, and protection systems to assess performance and degradation is possible with a substation DT. IEC 61850 data models can be used to assess current states and detect anomalies; data can also be enriched with additional sensors and devices (such as environmental or equipment condition temperature, vibration, or noise sensors). Developing advanced data-driven techniques can help with early fault detection, minimize unplanned outages, and increase asset availability.
3. **Predictive maintenance with continuous updates:** By correlating historical performance records, real-time sensor data, and environmental conditions, the DT enables the forecasting of substation asset failures and the optimization of maintenance schedules. This proactive maintenance model can significantly reduce lifecycle costs and downtime, but requires alignment with the regulatory framework and the operator guidelines to ensure safe and compliant implementation.
4. **Virtual commissioning and system testing:** Simulation has been a key topic for substation automation system development. The virtual representation of substation assets and automation logic allows extensive system testing prior to physical deployment. Enabled by IEC 61850 test-mode, and with increased virtualization of substation assets and systems, an increased number of tests can be performed virtually. This can relate to faster commissioning and improved understanding of the substation automation system behavior.
5. **Enhanced training and operator support:** The DT can serve as an immersive and interactive environment for operator training and procedural rehearsal. When integrated with advanced visualization tools, such as Virtual Reality and Augmented Reality, it enables scenario-based learning, intuitive fault response exercises, and deeper comprehension of complex automation functions.

It is worth noting that the list of enabled services presented here is not exhaustive; the contribution is rather the conceptualization of the DT architecture as a service-based one. Additional services can be developed and deployed, considering the operational needs of each substation, as well as the technological developments and evolution. Indeed, the DT technicalities are not goals in themselves, but rather means to achieve value-adding benefits conceptualized in the form of enabled services. The service layer can also be monitored and managed with clear performance indicators. For instance, improvements brought by virtual commissioning can be measured through commissioning hours saved or mean time to deployment; maintenance can lead to increased asset availability and decreased time to solve fault events. Specifically for smart substations, Santos and Zancul (2023) list performance indicators that can be used to assess the digitalization impacts, which can serve as a starting point for the DT service layer performance assessment.

Smart substations promise benefits related to engineering, construction, operations, maintenance, and other more diffuse soft benefits (Santos & Zancul, 2023). Considering such a service-oriented approach, a DT can be instrumental to better realize several of the benefits of substation digitalization.

11.3.3 Roadmap for substation Digital Twin realization

Once the architecture has been described, this section provides a practical roadmap for the implementation of such a substation DT. The roadmap is presented in five steps as follows:

1 **Digitalize substation automation system and process bus:** As a baseline for further development, first the core substation automation system should be digitalized (if not natively deployed so). This includes having digital processes and station buses, and may include advanced protection and control schemes (such as centralized or virtualized protection and control). Considering the architecture presented here, it consists of preparing the Physical Asset layer for the building of the DT, alongside parts of the Communication and Access layers.

2 **Define goals for DT implementation:** This step highlights the importance of setting clear goals for the DT instantiation. For example, the primary objective could be to improve operations and maintenance routines, or improve the responsiveness of auxiliary systems through data analytics. The goals should drive the definition of the services implemented, as well as include concerns and requirements from the various substation stakeholders.

3 **Develop component-level DTs**: This phase consists on the implementation of DTs for some substation components, such as a power transformer or circuit breaker DT, with real-time data acquisition and analysis. By having a component DT, some of the goals of the DT implementation may already be achieved through a comparatively simpler process than developing a system-level DT from the outset. Auxiliary sensors, actuators and edge devices may be deployed. Furthermore, this phased approach supports gradual, step-by-step experimentation, while maintaining high reliability. Within the architecture, it includes establishing or completing the Storage,

Virtual Model, Communication, Access, and Service layers, with a focus on individual substation components or equipment.

4 **Develop substation-wide (system-level) DT:** The next step consists of modeling and implementing a system-level substation DT. More than the sum of many component-level DTs, this level should be able to capture the relationships between components across the entire substation, including primary and secondary systems. Considering the architecture described in this chapter, it also requires defining the services and infrastructure needed for their realization, encompassing all the necessary layers.

5 **Integrate DT with operations and interactively improve it:** A significant challenge for DT implementation is effectively integrating it with existing or new operational systems. This step should not be overlooked, especially considering the criticality of substation operations for the power grid (as discussed in Section 2.4). Integration should not open up security breaches or negatively impact substation reliability. An iterative approach is also recommended, that is, adjust necessary parameters, requirements, or models according the DT response under operation conditions. It should also evolve during the long lifetime of substation assets: continuous review is also recommended. Considering the regulated environments substations operate in, it is also worth noting the importance of version governance and change control. Future improvements and changes to the DT should be traceable, which must be enabled by active structures (physical and digital) for governance and control.

A pilot substation project developed in Brazil toward a smart substation has been presented by Sylvestre et al. (2021) and Manassero et al., 2024. The same pilot substation has also been assessed in terms of its expected benefits and impacts (Santos & Zancul, 2023). Even though the development did not reach the proposed substation DT degree, considering the architecture proposed in this chapter, it can be classified as a project between steps 2 and 3 of the proposed roadmap. By implementing a digitalized process bus using centralized protection and control schemes (Sylvestre et al., 2021), the first step can be completed. Furthermore, the pilot project featured the deployment of edge devices for its battery system management, an explicit element of the additional sensors, actuators, and devices present within the architecture layers. A service-based approach for the DT was not explicitly developed, but advanced analytics using real-world data for predictive capabilities was performed. Deep learning models were developed for the prediction of high-voltage transformer oil temperature, potentially contributing to improved asset monitoring (Bezerra et al., 2024). More advanced data analytics integration is planned with process bus data.

It is worth noting that the proposed roadmap is not meant to be a strict single solution for substation DT development, but rather a starting point for its instantiation. Each substation DT may feature adaptations in the steps, depending on the specific local, operational, and DT goals.

11.4 Conclusion and future research opportunities

The implementation of a DT architecture for substations represents a promising yet complex undertaking, which may help realize the full potential benefits of smart substations. This chapter proposed a service-oriented architecture, tailored to the specific requirements of substations. By enabling real-time monitoring, predictive analytics, and lifecycle asset management, DTs can enhance operational efficiency, support proactive maintenance strategies, and contribute to grid stability in increasingly dynamic energy systems.

Future research should advance substation DTs toward their realization in practice and in context; beyond theoretical developments, comprehensive substation DTs still need further development in professional practice. This is not without its challenges, though. The definition of performance metrics and assessment frameworks is essential to quantify the tangible benefits of DT adoption, particularly regarding reliability improvement, maintenance cost reduction, and response time to anomalies. Robust cybersecurity strategies can also be investigated with DTs, ensuring data integrity, secure access control, and compliance with relevant standards. Special attention must also be given to the development of interfacing layers with legacy systems and devices, allowing for incremental deployment, reduced implementation risk, and preservation of prior infrastructure investments. Ultimately, DTs should not be viewed merely as technological artifacts but as potential strategic enablers of advanced substation digitalization.

AI disclosure

During the preparation of this work the author(s) used ChatGPT, under human supervision, in some sections of this chapter, in order to assist with minor text revisions (improve sentence readability and fix typing errors). After using this tool/service, the author(s) reviewed and edited the content as needed and take(s) full responsibility for the content of the publication.

References

Aftab, M. A., Hussain, S. M. S., Ali, I., & Ustun, T. S. (2020). IEC 61850 based substation automation system: A survey. *International Journal of Electrical Power & Energy Systems, 120*, 106008. doi:10.1016/j.ijepes.2020.106008.

Bertoni, M., & Bertoni, A. (2022). Designing solutions with the product-service systems Digital Twin: What is now and what is next? *Computers in Industry, 138*, 103629. doi:10.1016/j.compind.2022.103629.

Bezerra, F. E., Di Santo, S. G., Manassero Junior, G., Tiferes, R. R, Albertini, A. R, & Santos, G. R (2024). A forecasting methodology based on second-generation wavelets and neural networks: Application to power transformer oil temperature. *Electrical Engineering, 106*(3), 3017–3029. https://doi.org/10.1007/s00202-023-02134-5. doi:10.1007/s00202-023-02134-5.

Biancardi, A., Di Castelnuovo, M., & Staffell, I. (2021). A framework to evaluate how European Transmission System Operators approach innovation. *Energy Policy, 158*, 112555. doi:10.1016/j.enpol.2021.112555.

Chaaban, F., Akkawi, M., & Nassar, I. (2015). Transferring power sectors from aging utilities into smart grids-The case of Lebanon. *Electric Power Components and Systems, 43*(16), 1793–1802. www.tandf.co.uk/journals/titles/15325008.asp. doi:10.1080/15325008.2015.1057881.

Delong, Z., Zhijun, Y., Huipeng, C., Peng, Z., & Jiliang, L. (2021). Research on Digital Twin model and visualization of power transformer. *18th IEEE International Conference on Networking, Sensing and Control: Industry 4.0 and AI*. China: Institute of Electrical and Electronics Engineers Inc http://ieeexplore.ieee.org/xpl/mostRecentIssue.jsp?punumber=9702148. doi:10.1109/ICNSC52481.2021.9702244.

do Amaral, J. V. S., dos Santos, C. H., Montevechi, J. A. B., & de Queiroz, A. R. (2023). Energy Digital Twin applications: A review. *Renewable and Sustainable Energy Reviews, 188*, 113891. doi:10.1016/j.rser.2023.113891.

Durão, L. F. C. S., Zancul, E., & Schützer, K. (2024). Digital Twin data architecture for Product-Service Systems. *Procedia CIRP, 121*, 79–84. doi:10.1016/j.procir.2023.09.232.

Huang, Q., Jing, S., Li, J., Cai, D., Wu, J., & Zhen, W. (2017). Smart substation: State of the art and future development. *IEEE Transactions on Power Delivery, 32*(2), 1098–1105. doi:10.1109/TPWRD.2016.2598572.

Jiang, Z., Lv, H., Li, Y., & Guo, Y. (2022). A novel application architecture of digital twin in smart grid. *Journal of Ambient Intelligence and Humanized Computing, 13*(8), 3819–3835. doi:10.1007/s12652-021-03329-z.

Leonardi, A., Mathioudakis, K., Wiesmaier, A., & Zeiger, F. (2014). Towards the smart grid: Substation automation architecture and technologies. *Advances in Electrical Engineering, 2014*, 1–13. doi:10.1155/2014/896296.

Li, F., Qiao, W., Sun, H., Wan, H., Wang, J., Xia, Y., Xu, Z., & Zhang, P. (2010). Smart transmission grid: Vision and framework. *IEEE Transactions on Smart Grid, 1*(2), 168–177. doi:10.1109/tsg.2010.2053726.

Li, M., Lang, Q., & Wu, Q. (2024). deepening application research on substation auxiliary equipment monitoring system enabled by advanced digital technology. *3rd International Conference on Energy, Power and Electrical Technology, ICEPET 2024*. China: Institute of Electrical and Electronics Engineers Inc. http://ieeexplore.ieee.org/xpl/mostRecentIssue.jsp?punumber=10625771. doi:10.1109/ICEPET61938.2024.10627451.

Loggia, R., & Flamini, A. (2023). Electrical safety enhanced with BIM, SCADA and Digital Twin integration: A case study of a MV-LV substation. *IEEE Transactions on Industry Applications, 60*(2), 2725–2731. https://ieeexplore.ieee.org/servlet/opac?punumber=28. doi:10.1109/TIA.2023.3332311.

Manassero, G., Di Santo, S. G., Pellini, E. L., Dos Santos, M. L., Andrade Syrio, P. M., Tiferes, R. R., … Santos, G. R. (2024). Bridging industry 4.0 and power systems: A conceptual framework. *IEEE Power and Energy Magazine, 22*(6), 112–117. http://www.ieee.org. doi:10.1109/MPE.2024.3422919.

Mansour, D. E. A., Numair, M., Zalhaf, A. S., Ramadan, R., Darwish, M. M. F., Huang, Q., Hussien, M. G., & Abdel-Rahim, O. (2023). Applications of IoT and digital twin in electrical power systems: A comprehensive survey. *IET Generation, Transmission and Distribution, 17*(20), 4457–4479. https://ietresearch.onlinelibrary.wiley.com/journal/17518695. doi:10.1049/gtd2.12940.

Moradi, M., & Jahangir, A. H (2021). A new delay attack detection algorithm for PTP network in power substation. *International Journal of Electrical Power & Energy Systems, 133*, 107226. doi:10.1016/j.ijepes.2021.107226.

Qian, P., Yu, M., Zhang, S., Jiang, S., Tang, J., & Dong, S. (2023). The application of Digital Twin technology in substation management. In *Proceedings - 2023 International Conference on Advances in Electrical Engineering and Computer Applications, AEECA*. China: Institute of Electrical and Electronics Engineers Inc http://ieeexplore.ieee.org/xpl/mostRecentIssue.jsp?punumber=10511267. doi:10.1109/AEECA59734.2023.00032.

Redelinghuys, A. J. H., Basson, A. H., & Kruger, K. (2020). A six-layer architecture for the digital twin: A manufacturing case study implementation. *Journal of Intelligent Manufacturing, 31*(6), 1383–1402. www.kluweronline.com/issn/0956-5515/. doi:10.1007/s10845-019-01516-6.

Romeral, P. A. A. F., Santos, G. R, Zancul, E., Gerz, J., Kehrer, M., Heimes, H., & Kampker, A. (2024). Digital Twin of a battery-based EV charging station to support services. *Procedia CIRP, 128*, 698–703. doi:10.1016/j.procir.2024.03.044.

Santos, G. R., & Zancul, E. (2023). A framework to assess the impacts of digital electrical substations. *Smart Energy, 12*, 100125. doi:10.1016/j.segy.2023.100125.

Santos, G. R, Zancul, E., Manassero, G., & Spinola, M. (2024). From conventional to smart substations: A classification model. *Electric Power Systems Research, 226*, 109887. doi:10.1016/j.epsr.2023.109887.

Semeraro, C., Lezoche, M., Panetto, H. é, & Dassisti, M. (2021). Digital twin paradigm: A systematic literature review. *Computers in Industry, 130*, 103469. doi:10.1016/j.compind.2021.103469.

Singh, S., Weeber, M., & Birke, K.-P. (2021). Advancing digital twin implementation: A toolbox for modeling and simulation. *Procedia CIRP, 99*, 567–572. doi:10.1016/j.procir.2021.03.078.

Sleiti, A. K., Kapat, J. S., & Vesely, L. (2022). Digital twin in energy industry: Proposed robust digital twin for power plant and other complex capital-intensive large engineering systems. *Energy Reports, 8*, 3704–3726. doi:10.1016/j.egyr.2022.02.305.

Srivastava, A., Liu, C. C., Stefanov, A., Basumallik, S., Hussain, M. M., Somda, B., & Rajkumar, V. S. (2024). Digital Twins serving cybersecurity: More than a model: Cybersecurity as a future benefit of Digital Twins 2. *IEEE Power and Energy Magazine, 22*(1), 61–71. http://www.ieee.org. doi:10.1109/MPE.2023.3325196.

Stark, R., Fresemann, C., & Lindow, K. (2019). Development and operation of Digital Twins for technical systems and services. *CIRP Annals, 68*(1), 129–132. doi:10.1016/j.cirp.2019.04.024.

Sylvestre, M. H., Nascimento, V. T., Morales Udaeta, M. E., & Junior, G. M. (2021). The energy transition, effects with the implementation of electrical substation 4.0, its technologies and sustainability in Brazil. *IEEE CHILEAN Conference on Electrical, Electronics Engineering, Information and Communication Technologies (CHILECON)*. doi:10.1109/CHILECON54041.2021.9703019.

Tao, F., Xiao, B., Qi, Q., Cheng, J., & Ping, J. (2022). Digital Twin modeling. *Journal of Manufacturing Systems, 64*, 372–389. doi:10.1016/j.jmsy.2022.06.015.

Tao, F., Zhang, H., Liu, A., & Nee, A. Y. C. (2019). Digital Twin in industry: State-of-the-art. *IEEE Transactions on Industrial Informatics, 15*(4), 2405–2415. http://ieeexplore.ieee.org/xpl/RecentIssue.jsp?punumber=9424. doi:10.1109/TII.2018.2873186.

Tao, F., Zhang, M., Liu, Y., & Nee, A. Y. C. (2018). Digital twin driven prognostics and health management for complex equipment. *CIRP Annals, 67*(1), 169–172. doi:10.1016/j.cirp.2018.04.055.

Tiferes, R. R., Lima, T. R. K., & Manassero, G. (2025). A methodology for investigating power transmission field-observed events based on Digital Twins combining physical and recorded data. *IEEE Transactions on Industry Applications, 61*(2), 2813–2824. https://ieeexplore.ieee.org/servlet/opac?punumber=28. doi:10.1109/TIA.2024.3462659.

Udren, E. A., Hunt, R., Myrda, P., & Falk, H. (2024). Introducing the unified grid control platform: A holistic road map. *IEEE Power and Energy Magazine, 22*(3), 79–89. http://www.ieee.org. doi:10.1109/MPE.2023.3344058.

Yan, J., Wang, Y., Zhou, Y., Wang, J., & Geng, Y. (2024). A novel meta-learning network for partial discharge source localization in gas-insulated switchgear via digital twin. *IET Generation, Transmission & Distribution, 18*(9), 1785–1794. doi:10.1049/gtd2.13156.

Zancul, E., Oliveira, A., Wang, Y., Schützer, K., Schleich, B., da Rocha Albertini, A., & Bezerra, F. E. (2025). A Digital Twin design and implementation approach for industrial application leveraging programmable logic controllers. *International Journal on Interactive Design and Manufacturing (IJIDeM), 19*(5), 3479–3487. doi:10.1007/s12008-024-01969-x.

CHAPTER

12

Machine learning for power system restoration

Iraklis Varlamis

Department of Informatics and Telematics, Harokopio University of Athens, Athens, Greece

12.1 Introduction

Electric power systems worldwide are facing a great transformation that is affected by the integration of distributed energy resources (DERs), renewable resources that are often variable and intermittent, and the use of smart-metering and distributed storage facilities. All the above are reshaping the topology, dynamics, and operational complexity of transmission and distribution networks. Given the growing demand for energy, the dynamic load patterns, and the exposure to extreme natural (weather, natural disasters) or cyber-physical threats, the resilience and reliability of the grid are more critical than ever. The ability to restore power swiftly and reliably after outages is quickly becoming a key pillar of grid resilience and continuity.

Modern power system restoration (PSR) is fundamentally more complex than in traditional, centrally controlled grids. The increasing penetration of renewable energy sources, inverter-based resources, and distributed generation has introduced strong variability, reduced system inertia, and tighter operating margins. At the same time, restoration actions must be coordinated across large-scale networks with incomplete observability, uncertain resource availability, and strict safety constraints. Under these conditions, static, rule-based restoration procedures-largely designed for predictable system behavior and limited control options-struggle to scale and adapt to the wide range of disturbance scenarios encountered in modern power systems.

As a result, restoration planning and execution can no longer rely solely on predefined sequences and operator heuristics (Liu et al., 2016). The restoration problem increasingly requires adaptive decision-making that can reason over complex system states, evaluate multiple competing objectives, and respond to rapidly changing conditions in near real time. These requirements naturally motivate the use of machine learning (ML) methods, which are capable of extracting patterns from high-dimensional data, supporting predictive situational awareness, and generating flexible restoration strategies beyond fixed rule sets.

A combination of ML and Digital Twin (DT) technologies emerges as a promising paradigm for PSR. The DT can provide a real-time, high-fidelity virtual replica of the physical grid (or parts of it), in which we can then simulate diverse conditions, including faults, cascading failures, and extreme events and provide a sandbox for testing restoration strategies without risking the real grid. ML algorithms can provide a data-driven, adaptive, and intelligent approach for exploiting the wealth of sensor data (supervisory control and data acquisition (SCADA), smart meters, phasor measurement units (PMUs)), historical outage logs, system telemetry, and synthetic data generated by DT-based simulations and support or even automate decision-making. This combination enables learning of complex, nonlinear relationships between system states, disturbances, and effective restoration actions relationships that may not be tractable or known in rule-based systems. ML and especially deep learning (DL) or reinforcement learning (RL)–based methods have already demonstrated value in related tasks such as fault detection, anomaly detection, load forecasting, demand response, and stability control in smart grids (Wang et al., 2024).

Moreover, recent studies highlight how DT–ML integration enhances grid resilience beyond routine operations. For instance, when the grid is dominated by power electronics (due to DERs, storage, inverters), a DT combined with ML-based anomaly detection can identify deviations from "normal" behavior in real-time, flagging faults or cyber-physical threats before they propagate (Baker et al., 2022; Idrisov et al., 2025). Thus an ML-augmented DT offers a two-fold advantage for restoration: (1) Predictive and proactive planning through the simulation of outage or disturbance scenarios and the evaluation of alternative restoration strategies under varying conditions. (2) Fast, almost real-time,

Digital Twin Technology and Smart Grid.
DOI: https://doi.org/10.1016/B978-0-443-36664-2.00009-5

decision-making for complex restoration actions (e.g., which zones to energize first, which generators to start, how to shed or reconnect the load), which would normally need more time using conventional optimization. As a result, it is possible to minimize restoration time, prevent cascading failures, and forestall generation/load imbalances.

The integration of the two worlds: ML and DT—is not trivial and involves several challenges. First of all, although DTs can generate synthetic data, capturing the full diversity of real-world disturbances and grid configurations (especially for rare large-scale blackouts) remains difficult. Another risk is to "overfit" the models to the simulated scenarios, resulting in underperformance in real-world conditions. For this reason, restoration decisions must respect physical constraints (voltage limits, power flow equations, synchronization, and transient stability) and the proposed actions must be constrained to avoid inadmissible or unsafe decisions. Consequently, as with any artificial intelligence (AI)-driven system in critical infrastructure, interpretability, explainability, and trustworthiness of the ML-based restoration decisions are essential, especially to support operator oversight and regulatory compliance. Finally, one has to consider real-time performance, latency, scalability (especially for large transmission grids), and communication delays, which can pose practical barriers.

Given these opportunities and challenges, the goal of this chapter is to explore how ML—in synergy with DT—can be leveraged for effective PSR. Specifically, the chapter will:

- Outline the fundamentals and objectives of PSR (black-start, islanding, load reconnection, load balancing, and resynchronization) in modern grids.
- Survey ML techniques particularly suited for restoration tasks (supervised, unsupervised, DL, RL) and discuss how they can be trained using real and DT-generated data.
- Present the role of DT in enabling realistic simulation, synthetic scenario generation, and predeployment testing of restoration strategies.
- Propose ML-based restoration workflows (fault detection → decision-making → action) and discuss performance metrics relevant for restoration (restoration time, load recovery ratio, reliability, and resilience).
- Examine practical considerations and challenges (data quality, physical constraints, scalability, cybersecurity, and interpretability), and highlight future research directions (e.g., hybrid ML-optimization methods, multiagent RL, physics-informed ML, DT-cloud or edge-based deployment).

In this context, ML should not be viewed as a replacement for physical models or operator expertise, but as a complementary tool that augments traditional restoration frameworks with data-driven intelligence. This introduction sets the foundation for the technical and methodological discussion in the rest of the chapter, where we dive into ML algorithms, and design and implementation considerations for ML-driven restoration in DT-enabled smart grids.

12.2 Theoretical Background: Restoration and artificial intelligence

12.2.1 The anatomy of power system restoration

PSR is the process of restoring a grid back to normal operation after a blackout. This process can be broken down to several distinct phases, namely black-start, network reconfiguration, load restoration, and synchronization/reconnection (Liu et al., 2016).

The black-start phase begins with the activation of generation units that can "self-start," without an external supply from the grid. These units comprise hydroelectric plants, gas-turbine units, or pumped-storage units that can operate under low-load conditions and are used to provide an initial energizing power that will serve as the seed for rebuilding the network backbone.

In the network reconfiguration phase that follows the black-start, a reenergized "skeleton" of the grid is configured and gradually expanded, comprising transmission lines, key substations, and backbone network paths. The reconfiguration strategy typically involves optimization of which lines/substations to energize, in what order, taking into account constraints such as line loading and reactive power, and aims to avoid overvoltages, instabilities, or operational constraint violations, which may harm the restoration stability and safety.

The load pickup or load restoration phase builds on the stable backbone and progressively reconnects consumer loads until all requests are served. Load pickup must be managed carefully since a quick restoration may cause frequency dip or voltage instability, and a slow strategy may prolong the outage for customers. Another dynamic factor that must be considered is the "cold load pickup" phenomenon, in which the loads that have been offline for a long time may have a higher peak demand on reconnection than originally expected.

During restoration, the grid may initially be restored in several isolated subgrids powered by black-start units. In the synchronization phase, these "islands" must be synchronized and merged carefully, matching voltage magnitude, frequency, phase angle, and ensuring stability before closing interconnection breakers. This merging must respect stability constraints to avoid transient faults or oscillations.

Overall, PSR is inherently a multistage, multiobjective, multiconstraint, nonlinear, and uncertain problem. It combines elements of combinatorial optimization (which units/lines/loads to restore, in what sequence), physical constraints (power flow, stability, voltage/frequency limits), and uncertainty (load demand recovery, availability of resources, system conditions). Because of this complexity—especially in large-scale transmission systems, or modern smart grids with distributed generation and renewables—manual or predefined rule-based restoration schemes become increasingly inadequate.

12.2.2 Why conventional AI failed (and why machine learning succeeds)

Over the past decades, various methods have been proposed to assist or automate parts of the grid restoration process. These include expert systems (rule-based), case-based reasoning, heuristic optimization (genetic algorithms, fuzzy logic, and neural networks), multiagent systems, and decision-support tools. These conventional approaches had some success, but often fell short for full automation of PSR.

The main reason behind the failure of expert systems and rule-based schemes is their rigidity and limited generalization. Given the enormous variability of grid states—different topologies, fault types, component statuses, load levels, renewable output, demand patterns, etc., it quickly became impractical to enumerate and maintain comprehensive rule bases for all possible scenarios. So this explicit codification of "if-then" rules in a comprehensive knowledge base of restorations remains a bottleneck. The case-based reasoning approaches depend on previous incidents and thus suffer from the scarcity of historical data about major blackouts or cascading failures. Even heuristic optimization techniques (e.g., genetic algorithms, classical neural networks), that work well in simple scenarios, can hardly assist in complex grids that require real-time constraint handling of high-dimensional state spaces. Another limitation for conventional approaches is their lack of scalability. When the grid size grows, the combinatorial explosion of restoration sequences, network reconfiguration options, load pickup patterns, synchronization constraints, and dynamic operating conditions makes classical approaches computationally expensive, slow, or infeasible for real-time decision-making.

In contrast, modern ML techniques, such as DL and deep reinforcement learning (DRL), are far better suited for modern PSR—especially when embedded within a simulation enabled environment or DT: More specifically:

- They learn policies rather than rules. DRL does not require manually coded rules; instead, it learns a control policy through trial and error (or simulation). Given a sufficiently rich training environment, which is available through the DT, a DRL agent can learn to take optimal (or near-optimal) restoration actions in diverse states—without asking for an enumeration of all possible scenarios as a prerequisite. This is a key advantage over static rule-based systems.
- They generalize to unseen conditions. Once trained with a variety of data, the DL/DRL-based approaches can potentially generalize to previously unseen scenarios. This helps in grids where rare or extreme events for which no historical data exists may occur.
- They can handle high-dimensional and nonlinear state spaces. Due to their nature, deep neural networks (DNNs) can model complex, nonlinear relationships (e.g., between grid state, load demand, generation capacities, and stability margins) in high-dimensional spaces. Thus they can decide about restoration parameters by considering many variables in tandem, such as topology, load, generation, transient stability, voltage/frequency constraints, and potential failures.
- They offer fast inference. After offline training, a DL/DRL policy can infer actions quickly - enabling real-time or near real-time restoration decisions. For a blackout, where fast restoration is critical to minimize outage impact, such speed is a major asset compared to slower optimization or manual schemes. Because the learned policy does not need labeled data per se (unlike supervised learning) but rather interaction data concerning the state transitions for different actions and the rewards, DRL is well-suited for being trained within a simulation environment (DT) in which many "what-if" scenarios, including extreme and rare ones, can be generated. This makes modern ML solutions suitable for integration with simulation/DT environments (Addo et al., 2025).

Thus DL, especially DRL, provides a promising path to overcome the limitations of conventional AI/heuristic or rule-based restoration, making feasible an automated, adaptive, resilient, and real-time PSR—particularly in modern, complex smart grids.

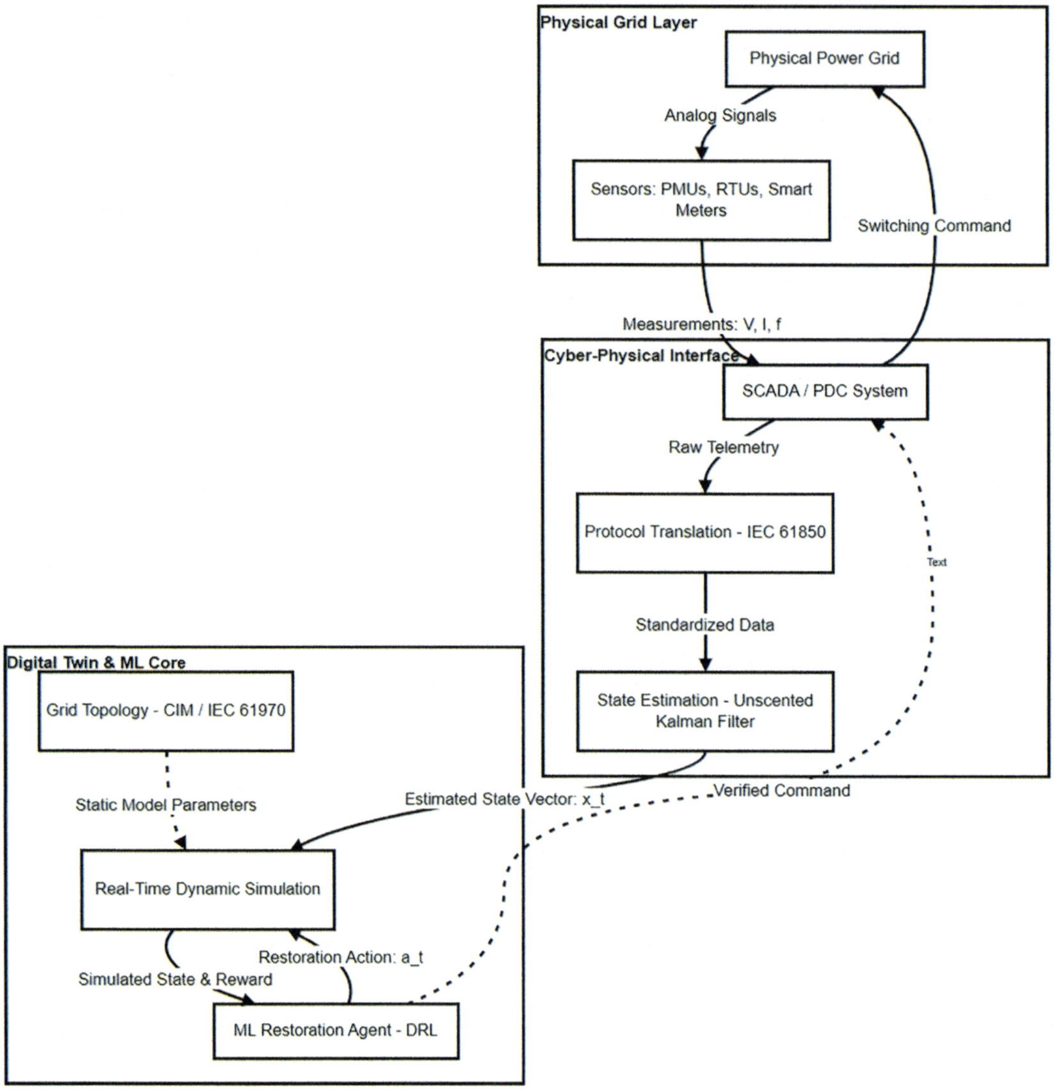

FIGURE 12.1 Digital Twin architecture for power system restoration. The diagram depicts the closed-loop data flow from physical sensors through state estimation (cyber layer) to the physics-based simulation and ML restoration agent.

12.2.3 The Digital Twin paradigm

A key enabler for modern ML/AI-based PSR is the concept of a DT applied to power systems. But in this context, DT is not merely a static database or a simple simulation model—rather, it must be a real-time dynamic simulation tightly synchronized with the physical grid. The architecture of a restoration-oriented DT relies on a layered data flow designed to bridge the gap between noisy field measurements and intelligent control agents, as illustrated in Fig. 12.1.

The synchronization process begins with real-time data ingestion from diverse grid sensors (SCADA, PMUs, smart meters) to maintain an accurate mirror of the physical system. To ensure interoperability across heterogeneous devices, data is typically aggregated using standard protocols such as IEC 61850 for substation automation. Because raw sensor data often contains noise, missing values, or time-skew errors, it passes through a state estimation layer. Here, advanced filtering techniques, such as Unscented Kalman Filters are employed to reconstruct the dynamic state of the system (voltages, phase angles). This filtered state vector provides the high-fidelity input required by the simulation core.

The simulation core itself is often built upon the Common Information Model (CIM/IEC 61970), which defines the static topology and electrical parameters of the network. This ensures that the digital replica structurally matches the physical grid. Within this environment, the DT enables closed-loop integration with AI/ML modules. The ML agent interacts with the DT by receiving the estimated system states, executing restoration actions (e.g., breaker switching) within the simulation to test their safety, and observing the resulting rewards or penalties. This "sandbox" capability allows for extensive "what-if" contingency simulations, enabling the agent to learn optimal strategies without risking the stability of the actual grid.

A DT for power systems aims at "situation awareness and virtual test" capabilities to assist decision-making under both normal and emergency conditions (e.g., faults, blackouts) (He et al., 2019). When combined with ML, the DTs significantly boost the capabilities of next-generation energy systems, enabling real-time data management, forecasting, anomaly detection, fault recovery, and operational optimization (Das et al., 2024).

A power-system DT for restoration must function as an integrated, intelligent platform characterized by several critical capabilities. It relies on real-time data ingestion from diverse grid sensors to maintain an accurate mirror of the physical system. This data feeds a high-fidelity simulation core that models physical laws and dynamic behaviors, enabling realistic prediction of restoration actions. Crucially, the DT enables closed-loop integration with AI/ML modules, allowing algorithms like RL to train and operate by interacting with the simulated environment. To overcome the scarcity of real outage data, the platform must support extensive "what-if" contingency simulation, generating synthetic scenarios for robust ML training. Finally, the entire architecture requires scalability and low-latency computation through high-performance or edge resources to ensure timely decision support for real-world, large-scale grid recovery. As indicated in (Jørgensen & Ma, 2025), the push toward real-time virtual replicas––federated across regions—reflects the strategic vision that DTs will underpin future grid planning, operation, resilience, and security.

Recent works validate this paradigm. For example, the hybrid AI–DT co-simulation framework by (Addo et al., 2025) has been shown to manage renewable-rich distribution networks under climate-driven stress, using RL to coordinate DERs and storage, improving stability and resilience. Also, a cloud-based DT combined with multiagent DRL (Pei et al., 2024) has demonstrated scalable, real-time task scheduling and system management—pointing to the practical feasibility of DT+AI frameworks for modern grids.

In conclusion, the DT paradigm transforms the way restoration can be approached: from static, offline plans and heuristics, to dynamic, data-driven, adaptive, and continuously validated strategies—enabling realistic, safe, and flexible restoration policies via ML/AI.

12.3 Machine learning techniques for power system restoration

PSR is a fundamentally complex, nonlinear, and dynamic control problem that spans topology optimization, sequential decision-making, stability management, and uncertainty handling. The integration of ML into PSR is emerging as a core enabler for next-generation resilient, intelligent, and autonomous smart grids. Unlike conventional restoration strategies-typically based on rule-based logic, human expertise, or deterministic optimization-ML approaches offer adaptability, scalability, and the ability to learn directly from operational or simulated data. This section provides a comprehensive overview of ML techniques applicable to PSR, organized into supervised, unsupervised, reinforcement, DL, and hybrid approaches, followed by key challenges observed across the literature.

12.3.1 Supervised learning for restoration

Supervised learning plays a supportive yet vital role in PSR by providing tools for rapid situational awareness and prediction, despite the inherent scarcity of labeled historical event data. Classification methods, such as decision trees and neural networks, are primarily used to enhance operator awareness by identifying fault types, locating component failures, and detecting islanded sections of the grid (Baghaee et al., 2019). These models help map a grid's state to an appropriate category of response but are limited by their need for balanced, high-quality labeled data and steady-state assumptions, making them best suited as auxiliary tools within a broader workflow.

For predictive tasks, regression methods offer valuable estimates, such as forecasting restoration time, load pickup trajectories, and postswitching stability margins. A particularly important application is the creation of surrogate models within a DT, where they emulate complex system dynamics much faster than traditional solvers (Chakraborty et al., 2021). However, because their accuracy is confined to scenarios represented in their training data, supervised regression struggles with rare or extreme grid states. Consequently, while essential for prediction and state estimation, supervised learning is typically not deployed for autonomous end-to-end restoration control but instead serves as a critical component informing larger optimization or RL systems.

12.3.2 Unsupervised learning

Unsupervised learning offers valuable capabilities for PSR by extracting actionable insights from the vast quantities of unlabeled grid data, such as SCADA and PMU measurements, where labeled historical events are lacking. A primary application is clustering, where algorithms like k-means, spectral clustering, and density-based spatial clustering of applications with noise (DBSCAN) analyze grid topology and electrical characteristics to identify useful patterns (Miraftabzadeh et al., 2023; Wang et al., 2022). This is instrumental for tasks like strategically dividing the network into stable, self-sufficient islands during a blackout, grouping loads for prioritized recovery, and defining optimal black-start zones, thereby providing a data-driven foundation for reconfiguration and resynchronization plans (Ganganath et al., 2017).

Furthermore, unsupervised anomaly detection is critical for maintaining reliability during the fragile restoration process. Techniques such as autoencoders and principal component analysis model normal system behavior, allowing them to flag deviations without prior examples of failures. This enables the early detection of abnormal conditions, such as sensor faults, unexpected voltage or frequency trajectories, or emerging instability, as the grid is being reconfigured. By operating without labeled fault data, these methods provide a vital safety net, enhancing situational awareness and resilience throughout the restoration sequence.

12.3.3 Reinforcement learning for sequential restoration

The application of RL to sequential PSR has evolved into a sophisticated field, leveraging advanced algorithmic families to address the complex Markov Decision Process (MDP) or Partially Observable MDP nature of the problem. The core formulation defines system states based on topology and operating conditions, with actions for switching or dispatch, and rewards for restored load with penalty constraints (Zhao & Wang, 2021). While single-agent approaches are common (Hosseini & Parvania, 2021), multiagent frameworks (MARL) effectively model decentralized control among microgrids or DERs (Si et al., 2023; Vu et al., 2023). Value-based methods like Deep QNetworks (DQN) and its variants are foundational for discrete action problems such as microgrid formation (Huang et al., 2022), with advanced versions incorporating action masking for improved safety and scalability in large networks (Ji et al., 2024).

For more complex control involving continuous variables like generator dispatch, policy-gradient, and actor-critic methods are preferred. Deep deterministic policy gradient (DDPG) and Twin Delayed DDPG (TD3) handle mixed discrete-continuous actions for coordinating resources like mobile storage (Yao et al., 2020), while Soft Actor-Critic (SAC) is favored for its stable, entropy-regularized exploration in uncertain environments (Dan et al., 2025). Proximal policy optimization (PPO) is prominent in large-scale, graph-based applications due to its training stability for long-horizon tasks (Jacob et al., 2024). A critical innovation is the integration of graph neural networks (GNNs), which explicitly model grid topology for superior generalization across different network sizes. Techniques like Capsule-GNN policies with PPO (Jacob et al., 2024), Graph Attention Networks with SAC, and spatiotemporal meta-GCNs allow policies to reason about connectivity and electrical relationships directly, forming a major methodological pillar for scalable restoration agents.

Given the safety-critical nature of the grid, significant research focuses on safe and robust learning paradigms. Hierarchical and hybrid methods combine the exploratory power of RL with the guaranteed feasibility of optimization, using RL for high-level sequencing and quadratic or linear programming for dispatch and constraint satisfaction (Hosseini et al., 2023). To combat the scarcity of real outage data and ensure robustness, techniques like curriculum learning, distributionally robust adversarial RL, and meta-RL for fast adaptation to new scenarios are employed. Furthermore, safety is engineered directly into the learning process through action masking to prevent invalid topologies, constrained policy optimization, and offline learning from expert demonstrations or optimization solutions (Bose & Zhang, 2023).

While most RL applications concentrate on distribution networks, transmission-level restoration presents greater challenges due to scale and dynamic complexity. Early approaches at this level often use simplified RL for tasks like repair sequencing paired with optimization for dispatch, or employ DRL to guide discrete actions while a lower layer solves power flow. For practical deployment, the choice of technique depends on the problem characteristics: DQN variants suffice for discrete switching; SAC or hybrid methods are needed for combined switch-and-dispatch; Graph-RL is essential for large, topological networks; and MARL fits decentralized multi-microgrid systems (Jacob et al., 2024). Ultimately, the trend leans toward hybrid, safe, and interpretable frameworks that combine learning with

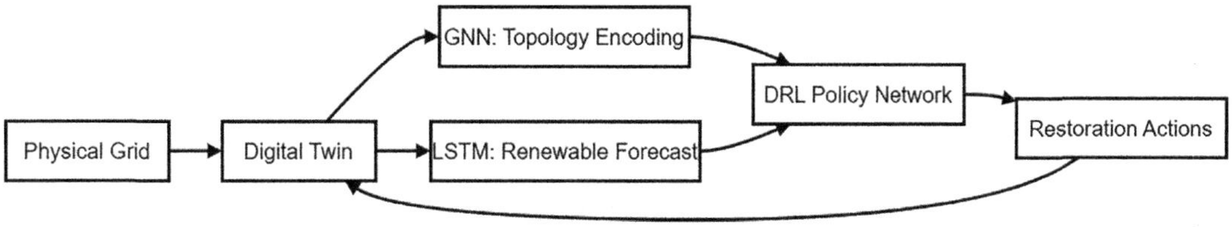

FIGURE 12.2 Integrated neural network architecture for power system restoration. The Digital Twin provides system states to topology-aware GNNs and time-series LSTM predictors, which jointly inform a DRL-based restoration policy operating in closed loop. *DRL*, deep reinforcement learning; *GNNs*, graph neural networks; *LSTM*, long short-term memory.

optimization, use robust training curricula, and provide decision transparency to build operator trust for real-world implementation.

12.3.4 Other deep learning approaches

DL provides powerful nonlinear approximation capabilities, making it well-suited for modeling the complex, high-dimensional, and dynamic relationships inherent in PSR operations. Artificial neural networks serve as a foundational tool, employed for tasks such as predicting restoration times, performing quick stability evaluations, forecasting black-start resource availability, and estimating voltage stability margins during the re-energization process (Zhao et al., 2024). Their strength lies in fast inference, making them suitable for real-time decision support. However, their performance is often limited by the requirement for large, diverse datasets-a significant challenge given the rarity of real-world restoration events from which to gather labeled examples.

For capturing the crucial temporal evolution of the grid during restoration, Recurrent neural networks, long short-term memory networks (LSTMs), and gated recurrent units are particularly valuable. Restoration is a time-dependent process where frequency, voltage, and load profiles continuously evolve as components are switched. These architectures excel at modeling such sequences, enabling dynamic state forecasting during reconfiguration, predicting transient responses to switching actions, and performing multistep load pickup forecasts (Fan et al., 2024; Liu et al., 2025). Their ability to mitigate vanishing gradients and handle long sequences is essential for restoration episodes that may last hours. Advanced frameworks integrate these temporal models with other structures, such as in a recurrent graph SAC architecture that manages uncertainties over time (Fan et al., 2024), or within spatiotemporal graph convolutional networks (GCNs) that adaptively model fault and restoration dynamics (Liu et al., 2025).

Inherently, a power grid is a graph defined by buses (nodes) and lines (edges). GNNs are, therefore a transformative DL approach for PSR, as they explicitly learn from and reason over this topological structure. GNNs can dynamically encode topology changes during restoration, handle the processes of islanding and merging, and model power flow relationships more naturally than convolutional or recurrent networks operating on non-graph data (Fan et al., 2024; Liu et al., 2025). This topology-aware learning leads to superior performance in tasks like dynamic stability prediction, line energization risk assessment, and distributed coordination under partial observability. For PSR, GNNs enable agents and predictors to understand and adapt to the grid's dynamically changing structure postfault, forming the backbone of modern, generalizable restoration policies that can scale across different network configurations.

12.4 Methodology: Neural network architectures

This section presents the core neural network architectures used for machine learning-based PSR. The restoration problem combines topology reasoning, sequential decision-making, and time-dependent uncertainty, which motivates the use of GNNs, DRL, and sequence prediction models such as LSTM networks. Each architecture addresses a distinct aspect of the restoration process and can be integrated within a Digital Twin-enabled framework.

Fig. 12.2 illustrates how GNNs, DRL, and LSTMs interact within a DT-enabled restoration framework.

12.4.1 Graph neural networks for topology learning

A power system can be naturally modeled as a graph

$$G = (V, E),$$

where each node $v \in V$ represents a bus (load, generator, or substation), and each edge $e \in E$ represents a transmission or distribution line. Node features may include voltage magnitude, frequency, load demand, or generation capacity, while edge features may represent line status, impedance, or thermal limits.

Unlike standard convolutional neural networks, which operate on fixed grid-like data, GNNs directly exploit the relational structure of the power grid. This property is crucial during restoration, where topology changes dynamically due to line outages, switching actions, islanding, and resynchronization.

A common formulation is the GCN, where node embeddings are updated through message passing:

$$\mathbf{h}_v^{(k+1)} = \sigma\left(\sum_{u \in \mathcal{N}(v)} \frac{1}{\sqrt{d_v d_u}} \mathbf{W}^{(k)} \mathbf{h}_u^{(k)} \right) \tag{12.1}$$

where $\mathbf{h}_v^{(k)}$ is the feature vector of the node v at layer k, d_v is the node degree, $\mathbf{W}^{(k)}$ is a learnable weight matrix, and $\sigma(\bullet)$ is a nonlinear activation function.

In the context of restoration, the input feature vector $\mathbf{h}_v^{(0)}$ explicitly encodes the physical state of the bus. It typically takes the form

$$\mathbf{h}_v^{(0)} = \left[|V|, \theta, P_{\text{gen}}, P_{\text{load}}, \mathbf{I}_{\text{status}} \right]^T \tag{12.2}$$

representing voltage magnitude, phase angle, active power generation, load demand, and energization status, respectively. Furthermore, the neighborhood set $\mathcal{N}(v)$ is dynamic; as the restoration agent toggles breakers or isolates faults, edges are effectively removed from the graph. This ensures that the message-passing mechanism strictly follows the currently energized topology, allowing the network to adapt to islanding conditions.

Through multiple layers, GNNs learn topology-aware representations that capture both local electrical relationships and global network structure. In PSR, this enables the model to reason about connectivity, power flow feasibility, and stability margins under changing configurations.

A key application of GNNs is grid partitioning into parallel restoration islands. By learning embeddings that reflect electrical coupling and operational constraints, GNN-based models can identify groups of buses that can be safely energized together, supporting black-start planning, island formation, and coordinated restoration across multiple subgrids.

12.4.2 Deep reinforcement learning for sequential restoration control

PSR is inherently a sequential decision-making problem, where actions taken at one stage affect future system states and feasibility. This process can be formulated as an MDP, defined by the tuple (S, A, R, P).

The system state S_t includes measurable and estimated variables such as:

$$S_t = \{\mathbf{V}_t, \mathbf{f}_t, \mathbf{b}_t, \mathbf{g}_t\} \tag{12.3}$$

where $\mathbf{V}_t$ represents bus voltages, $\mathbf{f}_t$ system frequency, $\mathbf{b}_t$ breaker and line status, and $\mathbf{g}_t$ generator availability.

The action space A_t consists of discrete and continuous control decisions, such as:

$$A_t \in \{\text{close breaker } i, \ \text{energize line } j, \ \text{ramp generator } k\}.$$

The reward function R_t is designed to encourage fast and safe restoration:

$$R_t = \alpha \cdot P_{\text{restored}} - \beta \cdot \mathbf{I}_{\text{constraint violation}} \tag{12.4}$$

where positive rewards are assigned for restored load or energized buses, and large penalties are imposed for violations such as instability, line overloads, or island collapse (e.g., +100 for load restoration and −1000 for unsafe actions).

DRL methods approximate the optimal policy $\pi(a \mid s)$ using DNNs. Value-based methods such as DQN are suitable for problems with discrete action spaces, such as breaker switching. However, restoration often involves continuous control variables (e.g., generator ramp rates and voltage setpoints), where policy-gradient methods are more appropriate.

PPO is particularly well-suited for PSR because it offers stable and sample-efficient learning while handling continuous actions. PPO constrains policy updates using a clipped objective:

$$L^{\text{PPO}}(\theta) = \mathbf{E}\left[\min\left(r_t(\theta)A_t, \text{clip}(r_t(\theta), 1-\epsilon, 1+\epsilon)A_t\right)\right] \tag{12.5}$$

where $r_t(\theta)$ is the probability ratio between new and old policies, and A_t is the advantage estimate. This prevents overly aggressive policy updates that could lead to unsafe restoration behavior.

12.4.3 Sequence prediction with long short-term memory networks

Uncertainty in renewable generation and load recovery plays a critical role during restoration. Solar and wind output, in particular, can change rapidly and affect the feasibility of restoration paths. To address this, sequence prediction models based on LSTM networks are used to forecast short-term system evolution.

Given a time series of renewable generation measurements

$$\{\mathbf{x}_{t-T}, \ldots, \mathbf{x}_t\}$$

An LSTM predicts future availability over a short horizon (e.g., the next 5 minutes):

$$\{\hat{\mathbf{x}}_{t+1}, \ldots, \hat{\mathbf{x}}_{t+H}\}$$

LSTMs maintain internal memory through gated mechanisms:

$$\begin{aligned} \mathbf{f}_t &= \sigma(\mathbf{W}_f[\mathbf{h}_{t-1}, \mathbf{x}_t]) \\ \mathbf{i}_t &= \sigma(\mathbf{W}_i[\mathbf{h}_{t-1}, \mathbf{x}_t]) \\ \mathbf{c}_t &= \mathbf{f}_t \odot \mathbf{c}_{t-1} + \mathbf{i}_t \odot \tanh(\mathbf{W}_c[\mathbf{h}_{t-1}, \mathbf{x}_t]) \\ \mathbf{h}_t &= \mathbf{o}_t \odot \tanh(\mathbf{c}_t) \end{aligned} \tag{12.6}$$

allowing the model to capture temporal dependencies and smooth transient behavior.

In restoration workflows, LSTM forecasts can be used to filter or adjust DRL actions, ensuring that restoration decisions remain sustainable given near-term renewable availability and avoiding unstable trajectories.

12.5 Key challenges in ML-based restoration

While the integration of ML and DT technologies offers a transformative path for PSR, it introduces significant challenges that must be addressed before wide-scale deployment is possible. This section explores critical issues and the future trajectory of the use of ML in grid restoration.

12.5.1 Data scarcity and imbalanced data

Major blackouts are rare events, and capturing the full diversity of real-world disturbances, particularly large-scale cascading failures, remains inherently difficult. As a result, historical data on major blackouts and restoration processes are extremely limited, which poses a significant challenge for data-driven modeling approaches.

This scarcity of data leads to several compounding issues. Only a very small number of labeled datasets exist for system restoration, restricting the ability to train and validate learning-based models effectively. Consequently, models are prone to overfitting to a narrow set of known historical events or simulated scenarios, which can cause significant degradation in performance when deployed under real-world conditions. The problem is further exacerbated by the lack of extreme-state data, such as scenarios involving multiple simultaneous line outages or complex cascading failures, which are precisely the situations where reliable restoration strategies are most critical.

To address these challenges, several complementary solutions have been proposed. One approach is the use of synthetic data generation through Digital Twins, enabling the creation of diverse "what-if" contingency scenarios that would be infeasible to observe in practice. In addition, domain randomization and robust training curricula can help expose models to a wider range of operating conditions, improving generalization. Physics-informed ML and hybrid optimization methods further constrain learning with physical laws and system models, reducing reliance on large datasets. Finally, transfer learning and meta-RL techniques allow models to adapt more effectively to previously unseen scenarios by leveraging knowledge acquired from related tasks and environments.

Restoration decisions are subject to strict time constraints. Some actions must be taken within seconds, such as protection and switching, while others occur over minutes, such as load pickup, or hours, such as full system restoration. These different time scales place strong practical limits on how decision-support tools can be used in real operations.

Because of this, ML models for restoration must meet several clear requirements. They must be fast to compute so that decisions do not delay the restoration process. They also need to handle noisy or missing data and cope with communication delays, which are common during large disturbances. In practice, these models should be able to run on edge devices or scale through cloud-based systems to manage large transmission grids. In addition, they must work reliably even when only part of the system state is known and when the availability of resources is uncertain.

More complex approaches, such as full DRL or large DNNs, can provide strong decision-making performance but often require significant computing power. To be useful in real time, these methods typically depend on hardware acceleration or cloud-edge setups to deliver results quickly enough for operators.

12.5.2 Cybersecurity and adversarial robustness

The reliance on data-driven models and real-time digital replicas expands the attack surface of the power grid. A primary concern is the susceptibility of ML models to adversarial attacks. In the context of a Digital Twin-guided restoration, a sophisticated attacker could inject subtle, nonrandom noise into the sensor data (e.g., SCADA or PMU measurements) feeding the DT.

This form of "data poisoning" or false data injection attack aims to deceive the DT into perceiving the grid state as healthy when, in reality, critical limits are being breached. For example, during a black-start sequence, an adversary might trick the neural network into believing a transmission line has sufficient thermal headroom, prompting an energization command that trips a relay and recollapses the island. Furthermore, if the ML model is trained on synthetic data generated by a compromised DT, the restoration policy itself may contain "backdoors" that trigger failures under specific, attacker-chosen conditions.

Addressing these threats requires robust adversarial training, where restoration agents are trained against active attackers in the simulation loop, and rigorous input validation protocols.

12.5.3 Explainable artificial intelligence and operator trust

Perhaps the most significant barrier to the adoption of ML in control centers is the "Black Box" problem. DNNs and DRL agents often function as opaque nonlinear function approximators, mapping complex grid states to actions without providing an intelligible rationale.

In a critical restoration scenario, human system operators carry the ultimate liability for grid safety and must maintain oversight. Consequently, the industry favors a "Human-in-the-Loop" architecture where ML/DT frameworks function as advanced decision support systems (DSS) rather than fully autonomous autopilots. Operators cannot be expected to validate a black box recommendation to "Close Breaker 16-17" without understanding why that action is safe and necessary. If an ML-based DSS recommends a counter-intuitive maneuver, such as shedding load in a stable island to prepare for a large block of generation pickup, it is imperative that the system provides a verifiable justification to the human operator.

Explainable AI (XAI) techniques are therefore essential to bridge this gap. Methods such as SHAP (SHapley Additive exPlanations) or attention mechanism visualization can highlight which specific grid variables (e.g., "Bus 12 Voltage" or "Line 4 Loading") drove the model's recommendation. By visualizing the decision logic of the restoration agent, XAI fosters the necessary trust and collaboration, ensuring that AI acts as a transparent, intelligent adviser that enhances, rather than replaces, human judgment.

12.6 Conclusion

PSR is becoming increasingly challenging as modern grids grow more complex due to renewable variability, DERs, power electronics, and evolving operating conditions. Traditional rule-based and manually driven restoration procedures are no longer sufficient to cope with this complexity, especially under large disturbances or extreme events. In this context, ML has emerged as a key enabler for intelligent, adaptive, and faster restoration decision-making.

Different classes of ML techniques contribute to restoration in complementary ways. Supervised learning supports prediction and detection tasks such as fault identification, state estimation, and restoration time forecasting. Unsupervised learning enhances situational awareness through anomaly detection, clustering, and island identification when labeled data is scarce. DL models improve the modeling of dynamic system behavior and high-dimensional relationships, while DRL enables the automated generation of sequential restoration strategies that can adapt to changing grid conditions. When combined, these approaches allow restoration decisions to move beyond static rules toward data-driven, flexible control policies.

The integration of ML with DTs represents a particularly powerful paradigm. In this framework, the DT acts as a safety engine by providing a physics-based, high-fidelity environment where restoration actions can be tested, validated, and constrained by physical laws. At the same time, ML functions as a speed engine, rapidly exploring complex decision spaces and producing near-optimal actions within the tight time limits required during restoration. DTs also address the scarcity of real blackout data by enabling large-scale synthetic scenario generation, which is essential for training and stress-testing ML models under rare and extreme conditions.

Despite these advances, important challenges remain. Limited availability of real-world restoration data, strict real-time constraints, cybersecurity risks, and the need for interpretability continue to limit large-scale deployment. However, ongoing progress in physics-informed learning, (Si et al., 2023), multiagent RL, robust training methods, and XAI is steadily reducing these barriers. Together, these developments are pushing PSR toward a future where intelligent systems can support operators with transparent, reliable, and adaptive decision-making.

Overall, the convergence of physics-based simulation and data-driven intelligence is reshaping how restoration is planned and executed. As DTs and ML mature and are increasingly integrated into control room workflows, they will play a central role in enabling self-healing, resilient, and trustworthy smart grids capable of recovering quickly and safely from major disruptions.

Reference

Addo, K., Kabeya, M., & Ojo, E. E. (2025). AI-powered digital twin co-simulation framework for climate-adaptive renewable energy grids. *Energies, 18*(21), 5593.

Baghaee, H. R., Mlakić, D., Nikolovski, S., & Dragicević, T. (2019). Support vector machine-based islanding and grid fault detection in active distribution networks. *IEEE Journal of Emerging and Selected Topics in Power Electronics, 8*(3), 2385–2403.

Baker, M., Fard, A. Y., Althuwaini, H., & Shadmand, M. B. (2022). Real-time ai-based anomaly detection and classification in power electronics dominated grids. *IEEE Journal of Emerging and Selected Topics in Industrial Electronics, 4*(2), 549–559.

Bose, S., & Zhang, Y. (2023). Load restoration in islanded microgrids: Formulation and solution strategies. *IEEE Transactions on Control of Network Systems, 11*(3), 1345–1357.

Chakraborty, S., Adhikari, S., & Ganguli, R. (2021). The role of surrogate models in the development of digital twins of dynamic systems. *Applied Mathematical Modelling, 90*, 662–681.

Dan, Y., Zhong, H., Wang, C., Wang, J., Fei, Y., & Yu, L. (2025). A graph deep reinforcement learning-based fault restoration method for active distribution networks. *Energies, 18*(16), 4420.

Das, O., Zafar, M. H., Sanfilippo, F., Rudra, S., & Kolhe, M. L. (2024). Advancements in digital twin technology and machine learning for energy systems: A comprehensive review of applications in smart grids, renewable energy, and electric vehicle optimisation. *Energy Conversion and Management: X, 24*, 100715.

Fan, B., Liu, X., Xiao, G., Xu, Y., Yang, X., & Wang, P. (2024). A memory-based graph reinforcement learning method for critical load restoration with uncertainties of distributed energy resource. *IEEE Transactions on Smart Grid*, (2), 1706–1718.

Ganganath, N., Wang, J. V., Xu, X., Cheng, C.-T., & Tse, C. K. (2017). Agglomerative clustering-based network partitioning for parallel power system restoration. *IEEE Transactions on Industrial Informatics, 14*(8), 3325–3333.

He, X., Ai, Q., Qiu, R. C., & Zhang, D. (2019). Preliminary exploration on digital twin for power systems: Challenges, framework, and applications. *arXiv preprint arXiv*, 1909.06977.

Hosseini, M. M., & Parvania, M. (2021). Resilient operation of distribution grids using deep reinforcement learning. *IEEE Transactions on Industrial Informatics, 18*(3), 2100–2109.

Hosseini, M. M., Rodriguez-Garcia, L., & Parvania, M. (2023). Hierarchical combination of deep reinforcement learning and quadratic programming for distribution system restoration. *IEEE Transactions on Sustainable Energy, 14*(2), 1088–1098.

Huang, Y., Li, G., Chen, C., Bian, Y., Qian, T., & Bie, Z. (2022). Resilient distribution networks by microgrid formation using deep reinforcement learning. *IEEE Transactions on Smart Grid, 13*(6), 4918–4930.

Idrisov, I., Veretennikov, I., & Ibanez, F. (2025). Dynamic state estimation in power electronics-dominated grids using a digital twin approach. *IEEE Access, 13*, 53938–53948.

Jacob, R. A., Paul, S., Chowdhury, S., Gel, Y. R., & Zhang, J. (2024). Real-time outage management in active distribution networks using reinforcement learning over graphs. *Nature Communications, 15*(1), 4766.

Ji, K., Qiao, J., Chen, Y., Yang, F., Zhao, Z., & Shi, M. (2024). A load transfer decision method for distribution networks based on mask-constrained deep reinforcement learning. In *2024 9th Asia Conference on Power and Electrical Engineering (ACPEE), Shanghai, China* (pp. 1526–1532). IEEE.

Jørgensen, B. N., & Ma, Z. G. (2025). Digital twin of the European electricity grid: A review of regulatory barriers, technological challenges, and economic opportunities. *Applied Sciences, 15*(12), 6475.

Liu, R., Li, C., Wang, J., Ni, J., Qin, M., & Jiang, C. (2025). *Emergency restoration strategy for the distribution network based on adaptive meta spatiotemporal graph convolutional networks* (pp. 18–22). Hohhot, China: IEEE.

Liu, Y., Fan, R., & Terzija, V. (2016). Power system restoration: A literature review from 2006 to 2016. *Journal of Modern Power Systems and Clean Energy, 4*(3), 332–341.

Miraftabzadeh, S. M., Colombo, C. G., Longo, M., & Foiadelli, F. (2023). K-means and alternative clustering methods in modern power systems. *IEEE Access, 11*, 119596–119633.

Pei, L., Xu, C., Yin, X., & Zhang, J. (2024). Multi-agent deep reinforcement learning for cloud-based digital twins in power grid management. *Journal of Cloud Computing, 13*(1), 152.

Si, R., Qiao, J., Wang, X., Ji, K., Wang, Z., Jun, Z., … Zhang, Z. (2023). A transferable multi-agent reinforcement learning method for distribution service restoration. In *2023 IEEE International Conference on Systems, Man, and Cybernetics (SMC), Oahu, Hawaii.* (pp. 1866–1871). IEEE.

Vu, L., Vu, T., Vu, T. L., & Srivastava, A. (2023). Multi-agent deep reinforcement learning for distributed load restoration. *IEEE Transactions on Smart Grid, 15*(2), 1749–1760.

Wang, K., Yang, R., Liu, C., Samarasinghalage, T., & Zang, Y. (2022). Extracting electricity patterns from high-dimensional data: A comparison of k-means and DBSCAN algorithms. *IOP Conference Series: Earth and Environmental Science, 1101*, 022007.

Wang, X., Wang, H., Bhandari, B., & Cheng, L. (2024). AI-empowered methods for smart energy consumption: A review of load forecasting, anomaly detection and demand response. *International Journal of Precision Engineering and Manufacturing-Green Technology, 11*(3), 963–993.

Yao, S., Gu, J., Zhang, H., Wang, P., Liu, X., & Zhao, T. (2020). Resilient load restoration in microgrids considering mobile energy storage fleets: A deep reinforcement learning approach. In *2020 IEEE Power & Energy Society General Meeting (PESGM)* (pp. 1–5). IEEE.

Zhao, H., Wei-Kocsis, J., Akhijahani, A. H., & Butler-Purry, K. L. (2024). Exploring a physics-informed decision transformer for distribution system restoration: Methodology and performance analysis. *arXiv preprint arXiv, 37*(2), 2407.00808.

Zhao, T., & Wang, J. (2021). Learning sequential distribution system restoration via graph-reinforcement learning. *IEEE Transactions on Power Systems, 37*(2), 1601–1611.

CHAPTER

13

Algorithm and simulation for a Digital Twin–based protection system

Kandan Mariadoss[a], *Jeni Gracia R C*[b], *Kaladi Govindaraj*[c], *V.C. Bharathi*[d]

[a]Department of Computing Technologies, School of Computing, Faculty of Engineering and Technology, SRM Institute of Science and Technology, Kattankulathur, Tamil Nadu, India [b]Saveetha School of Engineering, Saveetha Institute of Medical and Technical Sciences(SIMATS), Saveetha University, Chennai, Tamil Nadu, India [c]Department of Computer Science and Engineering, Aditya University, Surampalem, Andhra Pradesh, India [d]School of Computer Science and Engineering, VIT-AP University, Amaravati, Andhra Pradesh, India

13.1 Introduction

The evolution of modern power systems is marked by increasing complexity, primarily driven by the expansion of distributed energy resources (DERs), the integration of renewable energy sources, and the widespread adoption of advanced grid techniques. These progresses have also introduced new challenges in maintaining the stability, security, and elasticity of the electrical infrastructure while improving the efficiency and durability of the energy distribution (Shen et al., 2023). Traditional defense systems, which are primarily based on stable thresholds and predefined relay logic, struggle to cope with dynamic behavior and the decentralized nature of the contemporary power network. As such, there is an essential requirement for more intelligent, adaptive, and data-driven defense methods that can work effectively under a wide range of operating conditions and fault scenarios.

To overcome these challenges, this chapter proposes a Digital Twin (DT)based defense system designed to enhance the response, reliability, and effectiveness of the grid protection strategy. A DT is a high-fidelity, real-time virtual representation of the physical grid that is continuously updated using operational data. Recent developments in advanced grid technologies, such as phasor measurement units (PMUs) for real-time synchrophasor monitoring, AMI for fine-grained energy consumption analysis, supervisory control and data acquisition (SCADA) systems for centralized grid supervision, and Distributed Energy Resource Management System (DERMS) for coordinating renewable energy integration—have significantly enhanced the monitoring, control, and reliability of modern power systems (Abo-Khalil, 2023; Bachelor et al., 2020; Dong et al., 2022; Guo et al., 2023; Lak et al., 2023; Liu et al., 2020; Lv et al., 2022; Magargle et al., 2017; Shen et al., 2023; Wang et al., 2022; Xu et al., 2021; Yao et al., 2023; Zhang et al., 2023). This real-time virtual replica provides a continuous mirror of the behavior, status, and operational parameters of its physical equivalent. By integrating real-time sensor data, advanced analytics, and machine learning algorithms, DT forecasts facilitate fault detection, system diagnostics, and automated decision-making. Unlike traditional systems that react to events one after another, the DT-enabled system can anticipate and mitigate potential disruptions, enabling an active and self-healing approach to grid protection.

The central part of the proposed structure is its intelligent algorithm, which initiates corrective actions such as utilizing Internet of Things (IoT)-enabled sensors, analyzing system topology, and examining historical data to predict potential failures and isolate faults, redistribute loads, or reinstate the system. Dynamic system modeling ensures that the virtual presentation accurately reflects changes in load patterns, generation profiles, and grid topology. This

Digital Twin Technology and Smart Grid.
DOI: https://doi.org/10.1016/B978-0-443-36664-2.00020-4

modeling capacity significantly enhances the accuracy and adaptability of the defense mechanism, particularly in a highly dynamic environment with high penetration of renewable energy and bilateral power flow. To validate the effectiveness of a DT-based defense plan, a comprehensive co-simulation environment was developed in conjunction with MATLAB/Simulink, utilizing real-time communication interfaces. This setup mimics the actual grid's status and enables synchronous data exchange between the physical system emulator and the virtual DT framework. Through this simulation, the proposed system was tested under various defect scenarios, including short circuits, phase imbalances, and DER-related disturbances, to evaluate the accuracy of localization, response time, and system restoration and re-acquisition efforts.

Simulation results indicate that the DT-based protection system outperforms traditional security plans significantly. It demonstrates a fast response time, accurate detection and localization of high-severity faults, and enhanced adaptability to evolving grid operating conditions. The ability to separate defective segments with minimal service disruption displays the possibility of a system that enables more resilient and intelligent power networks. In conclusion, this task presents a novel framework for smart grid protection that combines DTs, real-time monitoring, and AI-powered analytics in a sophisticated system for restoration and recovery. As smart grids continue to evolve, such DT-based systems are well-positioned to serve as a foundation for future-proof power infrastructure, supporting not only advanced fault protection but also continuous optimization and autonomous grid operation. This research contributes to the growing body of knowledge on cyber-physical systems in modern power networks and paves the way toward a comprehensive, self-healing, resilient, and secure grid architecture.

13.2 Need for intelligent and adaptive protection mechanisms

The evolution of modern power systems is characterized by high penetration of DERs, bidirectional power flow, and dynamic grid configurations, rendering traditional protection methods increasingly inadequate. Conventional protection systems, which mainly rely on fixed settings and static thresholds, were originally designed for centralized grid architectures with predictable power flows. However, with the growing integration of renewable energy sources such as solar photovoltaics (PVs) and wind power, along with energy storage systems, electric vehicles, and smart loads, power grids have become highly dynamic and decentralized. These rapid changes introduce significant variability and uncertainty in grid behavior, making it difficult to anticipate all possible fault scenarios using predefined parameters. Consequently, traditional protection schemes often struggle to detect, distinguish, and mitigate faults in real time, leading to delayed responses, unnecessary tripping, or even protection failure. Moreover, modern power grids are increasingly expected to support self-healing, resilient, and real-time decision-making, which necessitates a shift from rule-based approaches to intelligent and adaptive protection methods.

Intelligent protection systems leverage advanced data analytics, artificial intelligence (AI), and machine learning (ML) techniques to enhance situational awareness and improve protection performance. These systems can learn from historical SCADA data, analyze real-time measurements, and dynamically adjust protection settings based on the current operating conditions. For instance, ML algorithms can identify patterns and anomalies in voltage, current, frequency, and power flow, enabling early fault detection and more accurate fault classification. This proactive approach not only reduces response time but also minimizes false tripping and improves fault localization accuracy. Furthermore, intelligent systems can coordinate protection actions across multiple grid levels, improving selectivity and sensitivity without compromising reliability. This capability is particularly important in active distribution networks where multiple DERs and microgrids operate concurrently, and where traditional coordination methods may lead to conflicts or blind spots.

Adaptive protection methods incorporate feedback control loops and real-time communication infrastructures to continuously update their operational parameters. These systems can automatically reconfigure in response to changes in network topology, load demand, or generation profiles, thereby ensuring continuous and reliable protection coverage. The IoT facilitates the integration of intelligent devices and sensors, enabling fine-grained monitoring and rapid decision-making. In parallel, DT models provide virtual testing environments for evaluating protection strategies under various fault scenarios without affecting the physical system. This level of adaptability ensures that protection schemes remain robust under highly variable operating conditions, fault propagation, or cyber-physical disturbances. Ultimately, the need for intelligent and adaptive protection methods in legacy power systems is evident, given the demands of a complex, dynamic, and decentralized smart grid environment. Such advanced protection approaches are essential to achieve reliable, secure, and resilient grid operation, supporting broader objectives related to energy transition, sustainability, and digital transformation in the power sector.

13.3 Role of Digital Twin in smart grid protection

The role of DT technology in smart grid protection has become increasingly pivotal as modern power systems grow in complexity due to the integration of DERs, variable load patterns, and decentralized control mechanisms. A DT serves as a real-time, dynamic virtual representation of the physical power grid and its components, continuously synchronized through streams of data collected from sensors, smart meters, intelligent electronic devices (IEDs), and SCADA systems. This virtual model provides a comprehensive and holistic view of grid operations, enabling the monitoring, simulation, analysis, and control of grid behavior under both normal and fault conditions (Bachelor et al., 2020).

In the context of power system protection, DTs represent a paradigm shift by enabling proactive fault detection, localization, and system restoration through predictive analytics and continuous feedback loops. Conventional protection mechanisms, such as static relays or fixed logic-based schemes, often lack adaptability and therefore struggle to respond effectively to the nonlinear and dynamic characteristics of modern power grids. For instance, in distribution networks with high penetration of solar PV generation, reverse power flow during periods of high generation and low demand can cause voltage rise along distribution feeders. Traditional overvoltage relays may misinterpret this operating condition as a fault, resulting in unnecessary tripping and the disconnection of healthy network sections.

DTs overcome this limitation by leveraging real-time operational data and integrating intelligent algorithms, including ML and AI-based decision-making, to distinguish between normal PV-induced voltage variations and genuine fault conditions. This predictive capability enables protection systems to respond pre-emptively, rapidly isolating faulted sections while minimizing disruption to unaffected areas.

Furthermore, DTs support what-if scenario analysis and co-simulation environments, allowing operators to evaluate a wide range of fault conditions, virtually test protection strategies, and optimize system settings without jeopardizing the physical infrastructure. By maintaining a high-fidelity, continuously synchronized representation of the grid, the DT dynamically updates its models based on operating conditions, environmental factors, and asset performance, thereby enhancing operator situational awareness. This enhanced awareness is essential for informed real-time decision-making, particularly during contingencies such as short circuits, voltage sags, or cascading failures. The application of DTs in smart grid protection also improves coordination across multiple protection layers—primary, backup, and zone-based—ensuring selective fault isolation while maintaining supply continuity.

In addition, DTs enable adaptive protection schemes that automatically reconfigure in response to topology changes, variations in DER output, or system disturbances, thereby supporting self-healing grid capabilities. The interoperability of DT platforms with SCADA, IoT, and edge computing infrastructures further enhances their effectiveness by enabling distributed intelligence and reducing latency in protective actions. Moreover, DTs provide valuable insights through historical data analysis and digital forensics, supporting post-fault diagnosis, root cause analysis, and continuous refinement of protection strategies. As cyber-physical systems continue to evolve, the integration of robust cybersecurity measures within the DT framework is essential to ensure grid integrity and resilience against both physical and cyber threats.

In summary, the DT serves as a foundational element in the development of intelligent, adaptive, and resilient innovative grid protection systems. It transforms the conventional reactive protection paradigm into a data-driven, predictive, and context-aware process, ultimately contributing to the reliability, efficiency, and sustainability of future power delivery infrastructures. Table 13.1 outlines the critical roles played by DT technology in modern innovative grid protection systems, including real-time monitoring, fault detection, predictive analysis, and dynamic system response, highlighting its advantages over conventional protection methods.

13.4 System architecture of Digital Twin–based protection

The system architecture of a DT-based protection system is a multilayered and integrated framework that mirrors the physical power grid within a virtual environment, enabling intelligent, real-time monitoring, analysis, and response to operational anomalies and fault conditions (Yao et al., 2023). At its core, the architecture consists of three primary layers: the physical layer, the DT layer, and the intelligence and control layer. These layers operate in close synchronization through robust data acquisition and communication frameworks.

The physical layer comprises the actual components of the power system, including generation units, transmission lines, distribution networks, transformers, circuit breakers, sensors, and IEDs. These components are equipped with

TABLE 13.1 Key functions and benefits of Digital Twin technology in enhancing smart grid protection mechanisms.

Aspect	Role of DT
Definition	Real-time, dynamic virtual model of the physical power grid and its components.
Data synchronization	Continuously updates using real-time data from sensors, IEDs, smart meters, and SCADA systems.
Monitoring and visualization	Enables full visibility of grid status and component behavior under normal and fault conditions.
Fault detection and localization	Uses intelligent algorithms to detect, localize, and assess faults in real time.
Predictive analytics	Predicts equipment failures or system disturbances before they occur using historical and live data.
Adaptive protection	Adjusts protection settings dynamically based on load, topology changes, and DER variability.
Simulation and testing	Allows what-if analysis, virtual testing of protection strategies, and optimization without physical risks.
Decision support	Enhances operator situational awareness and supports data-driven, real-time decision-making.
Coordination and selectivity	Improves coordination between primary, backup, and zone-based protection mechanisms.
Self-healing capabilities	Supports automatic reconfiguration and fault isolation for minimal service disruption.
Integration with other technologies	Works with IoT, edge computing, and SCADA for distributed intelligence and low-latency response.
Post-fault analysis	Aids in root cause identification and refinement of protection strategies through forensic data review.
Cybersecurity	Can include embedded security features to enhance resilience against cyber-physical threats.
Impact on grid protection	Shifts from reactive to predictive and adaptive protection, increasing grid reliability and operational efficiency.

DT, *Digital Twin;* IEDs, *intelligent electronic devices;* IoT, *Internet of Things;* SCADA, *supervisory control and data acquisition.*

advanced sensing devices such as PMUs, current transformers, and voltage transformers, which continuously collect high-resolution operational data, including current, voltage, frequency, power flow, and harmonic distortion. This data is transmitted in near real time to the DT layer using standardized and secure communication protocols such as IEC 61850, DNP3, and MQTT.

The DT layer functions as a real-time virtual replica of the physical grid and is implemented using advanced simulation tools such as MATLAB/Simulink, OpenDSS, or DIgSILENT PowerFactory. This layer hosts detailed component-level models calibrated to capture dynamic behavior under varying operating conditions. The DT models are continuously synchronized with the physical system through bidirectional data exchange, ensuring that the virtual state accurately reflects the physical system at any given time. This synchronization enables predictive simulation, what-if scenario analysis, and anomaly detection without disrupting real-world operations. In addition, the DT integrates a fault diagnosis engine that employs dynamic state estimation and signal-processing techniques to detect and classify abnormalities such as short circuits, line-to-ground faults, voltage sags, and equipment failures.

The intelligence and control layer incorporates data analytics, ML models, and protection algorithms to support real-time decision-making based on insights generated by the DT. Algorithms such as Decision Trees, support vector machines, and neural networks are applied for fault localization, severity assessment, and prognosis. When a fault or anomaly is detected, the DT simulates multiple control strategies and selects the most effective mitigation action. Once validated within the virtual environment, control commands are issued to the physical system to trigger protective actions such as circuit breaker tripping, load shedding, or islanding. This closed-loop operation ensures low latency, high reliability, and adaptive protection performance.

Furthermore, the architecture is designed to be scalable and modular, enabling seamless integration of DERs, energy storage systems, and renewable generation into the protection framework. Cybersecurity mechanisms—including encrypted communication, authentication, and blockchain-based event logging—are incorporated to preserve system integrity and trustworthiness. Overall, the DT-based protection architecture represents a paradigm shift from reactive, fixed-rule protection schemes toward proactive, context-aware, and self-adaptive mechanisms capable of evolving alongside the changing dynamics of modern smart grids. As shown in Fig. 13.1, the DT framework also integrates advanced optimization and analytics techniques, with emerging research exploring the use of quantum-inspired algorithms for addressing large-scale power system optimization challenges.

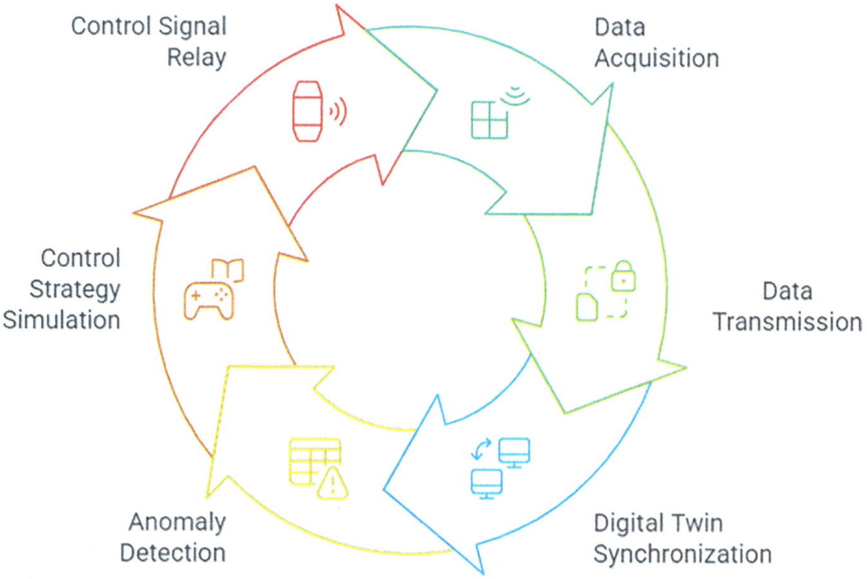

FIGURE 13.1 Digital Twin–based closed-loop control and protection process for smart grids.

13.5 Modeling of power system components in Digital Twin

Modeling power system components within a DT framework is a critical step toward achieving a high-fidelity, real-time virtual representation of the physical grid infrastructure. The DT serves as a dynamic simulation environment that continuously mirrors the operational state of the physical system by integrating real-time data streams from sensors and IEDs (Magargle et al., 2017). In the context of power system protection, accurate modeling of core components—such as transmission lines, transformers, generators, circuit breakers, relays, and DERs—is essential for replicating both electrical and mechanical behavior under normal and fault conditions.

Each component is represented using mathematical models and system equations that define key electrical parameters, including impedance, reactance, admittance, load characteristics, power flow relationships, and switching behavior. Transmission lines are commonly modeled using lumped π- or T-equivalent circuits to capture resistance, inductance, and capacitance effects. Transformers are typically represented using equivalent circuit models that account for leakage impedance, tap-changing operations, and magnetic core saturation. Generators are modeled using swing equations and excitation system models to accurately describe their dynamic response, particularly during transient conditions. DERs, such as solar PV systems and wind turbines, are represented using inverter-based dynamic models that capture their variable power output and interactions with grid voltage and frequency.

A fundamental requirement of DT modeling is synchronization with real-time operational data to ensure that the virtual environment continuously reflects the actual state of the grid. This synchronization is achieved through real-time data acquisition and communication protocols such as IEC 61850, DNP3, and MQTT, which transmit measurements from sensors and IEDs to the digital platform. The acquired data—including voltage, current, frequency, power factor, breaker status, and environmental conditions—are mapped directly to their virtual counterparts for monitoring and analysis. The DT framework employs a hybrid modeling approach that combines physics-based simulations with data-driven techniques. Physics-based models capture deterministic system behavior, while data-driven methods, such as ML and time-series forecasting, enhance state estimation, load prediction, and predictive maintenance.

Another critical aspect of DT modeling is the ability to dynamically update and adapt models in response to system changes, including topology reconfiguration, load variation, and the integration of new DERs (Guo et al., 2023). This adaptability is supported through modular modeling practices and reusable model libraries that enable plug-and-play functionality during grid expansion. In addition, co-simulation environments—such as MATLAB/Simulink coupled with real-time simulators like OPAL-RT or RTDS—enable comprehensive testing and validation of DT models under realistic operating conditions. Hardware-in-the-loop (HIL) implementations further enhance model fidelity by allowing direct interaction between physical protection devices and the virtual environment.

Overall, modeling power system components within a DT framework creates a comprehensive platform for real-time monitoring, simulation, diagnostics, and decision-making. This approach transforms traditional static protection systems into intelligent, predictive, and adaptive frameworks capable of addressing the complexity of modern smart grids. A key practical challenge in DT implementation is balancing model fidelity with computational overhead. While simplified or reduced-order models are typically employed to ensure real-time performance, higher-fidelity models are often reserved for offline analysis, validation, and optimization tasks where latency constraints are less critical.

13.6 Algorithm design for fault detection and localization

The design of an effective fault detection and localization algorithm within a DT-based protection system is essential for ensuring the stability, safety, and resilience of modern power grids. As power systems become increasingly decentralized and dynamic due to the growing penetration of DERs, traditional fault detection techniques often struggle to adapt in real time. The proposed algorithm exploits the combined capabilities of real-time data acquisition, advanced signal processing, and intelligent analytics within the DT framework to improve fault detection accuracy and localization precision.

The process begins with continuous monitoring of key electrical parameters—including voltage, current, frequency, and impedance—at multiple locations throughout the grid. These measurements are streamed from physical sensors and IEDs into the DT, ensuring continuous synchronization between the virtual and physical systems (Liu et al., 2020). The DT continuously simulates expected system behavior under normal operating conditions, and deviations beyond predefined thresholds are identified as anomalies. Signal preprocessing techniques such as filtering, wavelet transforms, and Fast Fourier Transform (FFT) analysis are applied to remove noise and extract meaningful features associated with transient disturbances. These features are then analyzed using rule-based logic, pattern recognition methods, or trained ML models to identify fault types, including line-to-ground, line-to-line, and three-phase faults.

Once a fault is detected, the localization stage is initiated. Fault location is determined using methods such as impedance-based estimation, traveling wave analysis, and time-synchronized measurements. The DT compares real-time observations with simulated fault scenarios to estimate fault location with high accuracy. ML models, including Decision Trees and support vector machines (SVMs), are incorporated to further enhance localization performance by learning from historical fault data. Decision Trees are selected for their interpretability and transparent classification logic, which supports operator trust and diagnostic clarity. SVMs are employed for their robustness in high-dimensional feature spaces and their ability to separate overlapping fault signatures, particularly in noisy measurement environments.

Within the DT framework, impedance-based fault location is implemented by continuously estimating line impedance using synchronized voltage and current measurements from both the physical system and its virtual replica. The estimated impedance trajectory is compared against baseline models to infer fault distance and type in real time. In parallel, traveling wave analysis is implemented by detecting high-frequency transients generated at fault inception using wavelet-based feature extraction. These transient signals are time-aligned and analyzed within the DT to estimate fault location based on wave arrival times, while allowing more detailed offline analysis using high-fidelity electromagnetic transient models when required.

The algorithm dynamically adapts its detection thresholds and model parameters based on real-time feedback from the DT, accounting for DER variability, load fluctuations, and switching actions. The use of synchronized PMU measurements and high-resolution, time-stamped data further improves localization speed and accuracy. Once a fault is accurately identified and located, the algorithm interfaces with the control layer to initiate corrective actions such as isolating the affected section or rerouting power flow, thereby minimizing service disruption. The modular design of the algorithm supports scalability and deployment across multiple protection zones within a smart grid.

In summary, the proposed fault detection and localization algorithm integrates advanced signal processing, data-driven analytics, and real-time simulation within the DT framework to provide a robust and adaptive protection solution. Its seamless integration into the DT enhances situational awareness and enables predictive, autonomous responses, making it a key enabler of intelligent and resilient smart grid protection systems.

13.7 Predictive analytics and failure forecasting

Predictive analytics and failure forecasting play a pivotal role in enhancing the reliability and responsiveness of DT-based protection systems in modern power grids. As power systems become more decentralized and dynamic with the integration of DERs, the need to anticipate potential failures before they occur has become paramount [10]. Predictive analytics involves the application of statistical models, ML algorithms, and data mining techniques to analyze historical and real-time data from various sensors and IEDs embedded within the grid infrastructure. These data streams include voltage, current, frequency, harmonic distortion, and thermal signatures that are continuously monitored and fed into the DT environment. Within this virtual model, algorithms are trained to recognize patterns that precede faults or component degradation. Failure forecasting, a subset of predictive analytics, focuses specifically on identifying the likelihood and timing of equipment failures or system anomalies based on current operational conditions and historical behavior.

The implementation of predictive analytics begins with data preprocessing, where noisy and incomplete data are filtered and normalized to ensure accurate modeling and prediction. Feature extraction techniques such as principal component analysis, FFT, and wavelet transform are used to identify critical parameters that influence system health. These features are then input into supervised or unsupervised learning models, such as SVM, Random Forests, artificial neural networks, and, more recently, long short-term memory networks, which are well-suited for time-series forecasting. The trained models are deployed within the DT framework, where they operate in parallel with the physical grid, continuously assessing real-time data and updating forecasts. For example, suppose a transformer exhibits an unusual rise in winding temperature and deviations in the load profile. In that case, the predictive model may flag a potential insulation failure within a specific time window, enabling proactive maintenance.

One of the significant advantages of integrating predictive analytics within a DT-based protection system is its ability to perform dynamic risk assessment. Unlike older protection systems that rely on fixed limits, the DT-enabled system can assess how assets are changing and take action before the protective relays need to kick in. This preemptive decision-making allows the system to reconfigure grid topology, isolate vulnerable components, or issue alerts to operators well in advance. Additionally, by using reinforcement learning or adaptive learning methods, the predictive models can improve their performance by learning from new faults or changes in how things operate. This results in a self-evolving protection system that adapts to the unique behavior of each asset and the overall grid configuration. If the optimal isolation plan—derived from high-fidelity models and global optimization—is unavailable due to incomplete data, communication delays, or computational limitations, the system will prioritize actions using a hierarchical decision strategy. At the highest priority level, fast, locally validated protection rules based on simplified DT models and predefined protection settings are executed to ensure immediate fault isolation. If sufficient time and data are available, the system refines these actions using more detailed DT simulations and optimization routines.

In conclusion, predictive analytics and failure forecasting form the backbone of intelligent, proactive protection in DT-based innovative grid systems. By leveraging real-time monitoring, advanced data analytics, and machine learning, these techniques enable early fault detection, reduce downtime, and minimize the impact of failures. The seamless integration of these capabilities within the DT framework not only enhances operational reliability but also supports the evolution toward autonomous, self-healing power systems capable of sustaining the growing demands of future energy networks.

13.8 Control logic for protection and isolation

The control logic for protection and isolation in a DT-based power system plays a crucial role in ensuring timely and intelligent responses to abnormal grid conditions, such as faults, equipment failures, or instability caused by fluctuations in DERs [12]. This logic is built on a hierarchical decision-making framework that integrates real-time data acquisition, fault analysis, and automated control execution. At its core, the control logic continuously receives synchronized data from sensors, IEDs, and PMUs, which is mirrored in the DT to maintain an up-to-date virtual representation of the power system. When anomalies such as sudden voltage drops, current surges, or frequency deviations are detected, the DT executes a predefined protection algorithm that determines the type, location, and severity of the fault. This involves comparing real-time parameters with protection thresholds and utilizing signal-processing techniques, such as wavelet transforms, harmonic analysis, or ML classifiers, to filter transient noise and accurately identify fault signatures.

Once a fault is confirmed, the control logic evaluates possible isolation strategies by simulating their outcomes in the DT. For example, it may virtually disconnect a transmission line or a busbar to test if the fault is cleared without affecting system stability. These simulations, executed within milliseconds, allow the system to choose the optimal isolation plan. Based on this analysis, the DT triggers the real-world actuation commands through communication with breakers, relays, or reclosers, ensuring the faulty section is isolated swiftly while keeping the healthy parts of the network operational. In cases of DER-based microgrids, the control logic is further enhanced with local intelligence to handle bidirectional power flows and prioritize critical loads during islanding scenarios. Moreover, adaptive protection schemes are integrated, which adjust protection settings in real-time based on the network topology, load conditions, and variability of renewable generation.

The control logic also includes coordination mechanisms to avoid false tripping or protection overlap, particularly in meshed networks or interconnected zones. This is achieved through the coordination of inverse-time relays, time grading, or centralized decision-making in a protection coordination center. Importantly, the DT logs all actions and system states before, during, and after the fault event, allowing for postevent analysis and future improvement of protection schemes. Additionally, ML models can be embedded within the control logic to learn from past events and improve predictive accuracy over time. In this setup, both rule-based and AI-enhanced decision engines can operate in parallel, offering redundancy and flexibility. The use of high-speed communication protocols, such as IEC 61850, ensures the low-latency execution of control actions, while cybersecurity measures ensure data integrity and authentication during command execution.

In summary, the control logic in a DT-based protection and isolation framework serves as the system's brain, merging real-time monitoring, intelligent fault detection, predictive simulation, and autonomous control to achieve rapid, reliable, and adaptive protection. This logic transforms traditional passive protection mechanisms into an active, context-aware, and self-healing network infrastructure, which is crucial for managing the increasing complexity and dynamism of modern smart grids.

13.9 Communication framework and data acquisition

Communication framework and data acquisition play a pivotal role in the effectiveness of a DT-based protection system, enabling seamless interaction between the physical power grid and its virtual counterpart. The communication framework serves as the backbone that supports real-time data flow, synchronization, and control commands between grid components and the DT environment. This continuous exchange of information ensures that the digital replica reflects the current state of the physical grid with high fidelity, which is essential for accurate fault detection, system diagnostics, and predictive maintenance. The data acquisition process begins at the field level, where sensors, IEDs, PMUs, and smart meters are installed across various parts of the grid. These devices capture a wide range of electrical parameters, including voltage, current, frequency, harmonics, power factor, and fault signals. High-resolution, time-synchronized data acquisition is achieved through GPS-based timestamping, ensuring consistency across distributed grid assets.

The communication framework incorporates both wired and wireless technologies to transport this sensor data to the control center or edge computing nodes, where the DT resides. Standard communication protocols employed in this setup include IEC 61850 for substation automation, DNP3 (Distributed Network Protocol), Modbus for legacy systems, and MQTT or OPC UA for lightweight, event-driven communication in IoT-based setups. The selection of a communication protocol depends on the latency requirements, bandwidth availability, and reliability constraints of the specific grid environment. In high-speed protection schemes, low-latency and deterministic communication is crucial; thus, protocols like Generic Object Oriented Substation Event under IEC 61850 are often preferred for transmitting time-critical data such as protection commands. For broader system monitoring and condition-based analysis, less time-sensitive data can be handled through protocols like MQTT or HTTP.

An essential feature of this framework is its support for bidirectional communication. Not only does the physical grid transmit operational data to the DT, but the DT can also send control signals, optimization commands, and reconfiguration instructions back to the physical grid in real-time. To maintain reliability and security, communication channels are often encrypted and utilize virtual private networks, firewalls, and authentication mechanisms to protect against cyber threats and unauthorized access. Furthermore, edge computing is frequently integrated to reduce data transmission delays and processing loads on central servers. In such setups, local processing units at substations or microgrid nodes pre-process raw data and send only relevant, filtered, or aggregated information to the DT system.

In summary, the communication framework and data acquisition layer are foundational components of the DT-based protection system. It enables high-speed, reliable, and secure exchange of real-time data, ensuring that the DT operates with up-to-date information to perform accurate simulations, predictions, and protection functions. As smart grids continue to evolve with increased decentralization and digitalization, robust and flexible communication architectures will become indispensable for enabling adaptive, self-healing, and intelligent protection systems. Fig. 13.2 illustrates the integration of physical power grid elements, real-time data acquisition, virtual twin modeling,

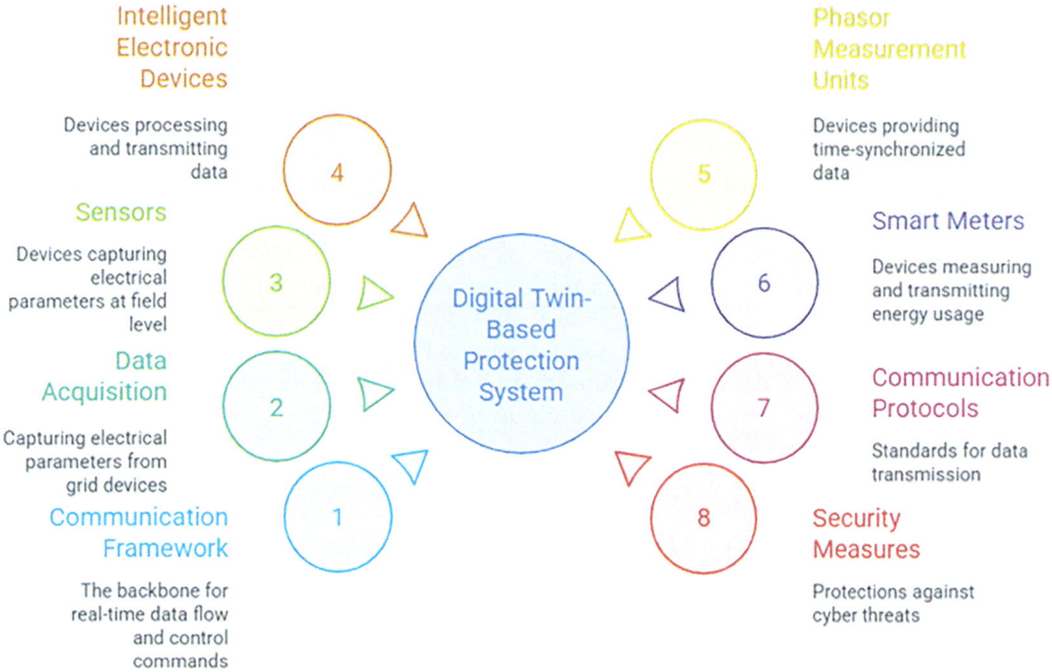

FIGURE 13.2 Components of a Digital Twin–based protection system.

predictive analytics, control logic, and communication interfaces for enhanced fault detection, system monitoring, and adaptive response.

13.10 Co-simulation environment setup

The co-simulation environment setup is a critical aspect of implementing and validating a DT-based protection system, as it enables synchronized interaction between real-time physical grid data and virtual simulations of system behavior. In this study, a co-simulation environment is designed to integrate MATLAB/Simulink with a real-time communication framework, enabling the modeling, analysis, and testing of protection mechanisms under dynamic conditions. MATLAB/Simulink is used as the core simulation platform due to its powerful capabilities in system modeling, control design, and simulation of electrical networks. The Simscape Power Systems toolbox facilitates the accurate modeling of grid components such as transmission lines, transformers, circuit breakers, and DERs. These components are mirrored virtually in the DT, enabling the real-time replication of physical system states.

To ensure the DT receives timely and accurate data, sensor readings, and control commands are exchanged using communication protocols like IEC 61850, OPC UA, or MQTT, depending on the system's requirements. These protocols provide a standardized and efficient mechanism for interfacing between physical devices, such as IEDs and PMUs, and the simulation model. A key feature of the co-simulation is the real-time data acquisition module that collects voltage, current, frequency, and breaker status signals and forwards them to the DT environment with minimal latency. This allows the system to perform condition monitoring, detect abnormalities, and trigger predictive diagnostics through the DT algorithm.

In parallel, the simulation environment incorporates event-based triggers and decision logic that emulate fault conditions such as short circuits, line-to-ground faults, and load imbalances. These simulated faults are injected at various points in the network to evaluate the accuracy and speed of the protection algorithm. The decision-making algorithm in the DT utilizes incoming real-time data to identify fault types and locations. Then it simulates control actions such as isolating the faulty section, rerouting power, or sending signals to protective relays. The co-simulation also allows for seamless switching between fault scenarios and operating conditions, making it a robust platform for stress-testing the system.

To achieve synchronization between the real-time data stream and the simulation environment, a time-step control mechanism is employed to align the sampling intervals of the physical devices with those of the digital model.

The architecture can be extended with middleware platforms such as Functional Mock-up Interface or LabVIEW, depending on the desired level of hardware integration and real-time performance requirements. HIL testing is optionally incorporated to interface physical protection relays or controllers with the simulation, enhancing realism and validation. Overall, this co-simulation environment provides a highly dynamic and responsive testbed that effectively demonstrates the functionality of the DT-based protection system, ensuring that it operates with improved accuracy, faster response, and higher adaptability than conventional protection schemes.

13.11 Comparative analysis with conventional protection schemes

The transition from traditional grid protection mechanisms to DT-based systems represents a fundamental shift in how faults and disturbances are detected, diagnosed, and mitigated in modern power networks. Conventional protection schemes—primarily based on static relay configurations—rely on predefined settings and fixed thresholds to identify abnormal operating conditions. These schemes typically employ overcurrent, differential, or distance protection techniques derived from electrical parameters such as current, voltage, and impedance. While effective for well-structured radial networks with predictable power flows, conventional approaches often struggle to accommodate the dynamic and decentralized characteristics of modern smart grids, which include high penetration of renewable energy sources, bidirectional power flow, and rapidly varying load profiles.

In contrast, DT-based protection systems introduce a real-time virtual replica of the physical grid that is continuously synchronized using live sensor measurements and communication networks. This capability enables not only real-time monitoring and visualization but also predictive analytics and enhanced situational awareness—features largely absent from conventional schemes. The DT dynamically updates its internal models in response to changes in operating conditions, allowing for rapid detection and assessment of emerging anomalies. This adaptability results in a more flexible and accurate protection response, particularly in environments characterized by DER integration, fluctuating grid topologies, and complex fault dynamics.

A key differentiator between the two approaches lies in the level of intelligence and adaptability embedded within the protection logic. Traditional systems depend on static coordination and fixed protection settings, making them vulnerable to false tripping or delayed operation when unexpected grid behavior occurs. By contrast, DT-based protection systems integrate ML algorithms that learn from both historical and real-time data, enabling reliable discrimination between transient disturbances and genuine faults. This capability significantly reduces nuisance tripping and improves the selectivity and dependability of fault isolation.

Fault localization and isolation capabilities further highlight the advantages of DT-based systems. Conventional protection often relies on predefined protection zones and manual relay coordination, which can result in unnecessarily large portions of the network being disconnected during fault events. In comparison, DT-based systems provide granular, component-level visibility, allowing only the affected sections to be isolated while maintaining service continuity elsewhere. The incorporation of predictive failure models also enables proactive mitigation actions—such as controlled load shedding or network reconfiguration—before severe system degradation occurs, a functionality largely unattainable in reactive conventional schemes.

Moreover, DT-based protection offers a superior simulation and testing environment. While traditional systems require offline relay coordination studies and physical testing, DT frameworks support continuous simulation and co-simulation under real-time operating conditions. This capability accelerates validation, reduces commissioning time, and enables rapid refinement of protection strategies without hardware modifications.

Overall, this comparative analysis demonstrates that although conventional protection schemes remain effective for static and predictable networks, they are increasingly inadequate for modern, dynamic grid environments. DT-based protection systems offer superior adaptability, responsiveness, fault prediction, and resilience—attributes that are essential for ensuring secure and reliable operation in the face of growing grid complexity. As smart grids continue to evolve, DT-based protection is poised to become a cornerstone of next-generation, self-healing, and intelligent power infrastructure.

13.12 Implementation challenges and mitigation techniques

The implementation of DT-based protection systems in modern power grids introduces a range of technical, computational, and operational challenges that must be addressed to ensure reliable and efficient performance.

One of the primary challenges is real-time data acquisition and synchronization between physical grid assets and their virtual counterparts. DTs rely on high-frequency, time-aligned data from PMUs, IEDs, and various sensors. However, inconsistencies in sampling rates, communication latency, and packet loss can degrade DT fidelity, leading to inaccurate fault detection or delayed protective actions. To mitigate these issues, high-speed communication standards such as IEC 61850 and IEEE C37.118 are employed, along with time-synchronization mechanisms such as GPS clocks and Precision Time Protocol, ensuring accurate temporal alignment of measurement streams.

Another major challenge is the computational burden associated with real-time DT modeling, updating, and analytics. The complex dynamic behavior of grid components—combined with DER variability and fault dynamics—requires substantial processing resources, particularly when advanced ML or deep learning techniques are used for prediction and decision-making. Edge and fog computing architectures provide an effective mitigation strategy by decentralizing computation and reducing latency. Additionally, algorithm optimization through parallel processing and GPU acceleration enables the timely execution of fault detection and control routines.

Data security and integrity pose further challenges due to the continuous exchange of information between physical and digital layers. Cyber threats such as data spoofing, unauthorized access, and man-in-the-middle attacks can compromise both system accuracy and operational safety. These risks can be mitigated through strong encryption, secure authentication protocols, blockchain-based data verification, and the integration of cybersecurity-aware AI models capable of detecting anomalous communication patterns.

Scalability and interoperability also remain critical concerns, particularly in large-scale smart grids comprising heterogeneous devices from multiple vendors. Seamless integration of legacy infrastructure with modern DT platforms requires adherence to open standards and interoperable communication protocols. Middleware solutions that abstract device-specific interfaces and enable plug-and-play functionality can significantly reduce integration complexity and enhance system scalability.

Model accuracy and calibration represent another key challenge in DT deployment. Since DT-based diagnostics and decision-making depend directly on the fidelity of the virtual model, inaccuracies in parameters or assumptions can result in incorrect responses, especially during transient events. Initial calibration is typically performed using historical data, system specifications, and commissioning test results. However, power system characteristics evolve due to equipment aging, environmental effects, topology changes, and sensor drift. Consequently, periodic recalibration is required, with lightweight updates performed continuously or weekly, and more comprehensive recalibration conducted following major network changes or fault events. Automated calibration techniques, adaptive parameter estimation, and data-driven correction layers are therefore essential for maintaining long-term DT accuracy while reducing operational overhead.

Finally, economic and organizational barriers—such as high deployment costs, workforce skill requirements, and integration with existing protection frameworks—may limit widespread adoption. Pilot projects, cost–benefit analyses, and targeted training programs can help utilities overcome these barriers and transition toward full-scale deployment. In summary, while DT-based protection systems offer transformative potential, their successful implementation depends on a holistic strategy that addresses technical, cybersecurity, scalability, and operational challenges.

13.13 Future enhancements in Digital Twin–based protection

As power systems evolve toward increasingly dynamic, decentralized, and data-driven architectures, future enhancements in DT-based protection systems are expected to reshape grid operations into intelligent, self-adaptive, and resilient infrastructures. One of the most prominent directions is the deeper integration of advanced AI and ML techniques within the DT framework (Abo-Khalil, 2023; Bachelor et al., 2020; Dong et al., 2022; Guo et al., 2023; Lak et al., 2023; Liu et al., 2020; Lv et al., 2022; Magargle et al., 2017; Shen et al., 2023; Wang et al., 2022; Xu et al., 2021; Yao et al., 2023; Zhang et al., 2023). These techniques enable predictive and prescriptive analytics, allowing protection systems not only to detect and localize faults but also to anticipate disturbances and recommend optimal corrective actions.

Deep learning models trained on historical fault records, sensor data trends, and operational conditions can enhance fault classification, prognosis, and risk assessment. The integration of DTs with edge and fog computing architectures further reduces latency by processing data closer to its source, enabling faster protective responses—particularly in remote or resource-constrained environments.

Blockchain technology offers another promising enhancement by ensuring data integrity, traceability, and trust across the DT ecosystem, especially in systems involving multiple stakeholders and distributed assets. Hybrid

DT models that combine physics-based simulation with data-driven learning approaches are also gaining traction, offering improved adaptability under uncertainty and evolving grid conditions. Cybersecurity remains a critical focus, necessitating multilayered defense mechanisms, intrusion detection systems, and secure communication frameworks to protect against emerging cyber-physical threats. Interoperability and standardization will also play a key role in enabling seamless integration across diverse platforms and legacy systems. Enhanced interoperability through standardized communication protocols, such as IEC 61850 and Common Information Model, ensures that DTs communicate effectively with existing substation automation systems, SCADA, and EMS platforms.

Additionally, future enhancements may include the deployment of self-learning agents and autonomous DTs that can evolve their behavior based on continual learning from operational feedback, thereby creating a self-healing and self-optimizing grid. The integration of immersive technologies, such as augmented reality and virtual reality, with DTs could further support operator training, remote diagnostics, and interactive system visualization, thereby enhancing situational awareness. Lastly, scaling DT-based protection systems to operate across large-scale, interconnected intelligent grid networks with multi-agent coordination and hierarchical decision-making structures will be pivotal for nationwide or transnational grid operations. These advancements collectively represent a paradigm shift in how protection systems are conceived and deployed—moving from static, rule-based mechanisms to adaptive, intelligent, and resilient systems that form the foundation of next-generation energy infrastructure.

13.14 Conclusion

The increasing complexity of modern power systems demands intelligent and adaptive defense methods, governed by the integration of DERs and advanced smart grid techniques. This study successfully depicts the development and simulation of a DT-based defense system capable of enhancing grid reliability, response, and conditions. By leveraging real-time sensor data, ML algorithms, and dynamic system modeling, the proposed approach enables early fault detection, failure forecasting, and timely control actions. The co-simulation environment integrates MATLAB/Simulink and a real-time communication framework, effectively validating system defects and reducing disruptions. Comparative evaluations under various fault views confirmed significant improvements in accuracy, response time, and adaptability compared to traditional security plans. Overall, this research paves the way for elastic, self-healing, and future-ready smart grid infrastructures, demonstrating the transformative potential of DT technology in power system defense.

References

Abo-Khalil, A. G. (2023). Digital twin real-time hybrid simulation platform for power system stability. *Case Studies in Thermal Engineering, 49*, 103237. doi:10.1016/j.csite.2023.103237.

Bachelor, G., Brusa, E., Ferretto, D., & Mitschke, A. (2020). Model-based design of complex aeronautical systems through Digital Twin and thread concepts. *IEEE Systems Journal, 14*(2), 1568–1579. https://ieeexplore.ieee.org/xpl/mostRecentIssue.jsp?punumber=4267003. doi:10.1109/JSYST.2019.2925627.

Dong, Y., Chen, Q., Ding, W., Shao, N., Chen, G., & Li, G. (2022). State evaluation and fault prediction of protection system equipment based on Digital Twin technology. *Applied Sciences, 12*(15), 7539. doi:10.3390/app12157539.

Guo, M., Fang, X., Hu, Z., & Li, Q. (2023). Design and research of digital twin machine tool simulation and monitoring system. *The International Journal of Advanced Manufacturing Technology, 124*(11-12), 4253–4268. doi:10.1007/s00170-022-09613-2.

Lak, P. Y., Key, S., Yoon, S. M., & Nam, S. R. (2023). Digital Twin application for the evaluation of protection performance in IEC-61850-based digital substations. In *International Conference on Advanced Power System Automation and Protection, APAP 2023* (pp. 68–72). South Korea: Institute of Electrical and Electronics Engineers Inc. http://ieeexplore.ieee.org/xpl/mostRecentIssue.jsp?punumber=10347561. doi:10.1109/APAP59666.2023.10348444.

Liu, Z., Zhang, A., & Wang, W. (2020). A framework for an indoor safety management system based on Digital Twin. *Sensors, 20*(20), 5771. doi:10.3390/s20205771.

Lv, Z., Guo, J., Singh, A. K., & Lv, H. (2022). Digital Twins based VR simulation for accident prevention of intelligent vehicle. *IEEE Transactions on Vehicular Technology, 71*(4), 3414–3428. http://ieeexplore.ieee.org/xpl/tocresult.jsp?isnumber=8039128&punumber=25. doi:10.1109/TVT.2022.3152597.

Magargle, R., Johnson, M., Davoudabadi, P., Kesarkar, O., Krishnaswamy, S., & Pitchaikani, A. (2017). A simulation-based Digital Twin for model-driven health monitoring and predictive maintenance of an automotive braking system. *Modelica, 132*, 35.

Shen, Z., Xu, W., Li, W., Shi, Y., & Gao, F. (2023). Digital twin application for attach detection and mitigation of PV-based smart systems using fast and accurate hybrid machine learning algorithm. *Solar Energy, 250*, 377–387. doi:10.1016/j.solener.2023.01.007.

Wang, B. o, Zhang, C., Zhang, M., Liu, C., Xie, Z., & Zhang, H. (2022). Digital Twin analysis for driving risks based on virtual physical simulation technology. *IEEE Journal of Radio Frequency Identification, 6*, 938–942. doi:10.1109/jrfid.2022.3203694.

Xu, Q., Ali, S., & Yue, T. (2021). Digital Twin-based anomaly detection in cyber-physical systems. In *Proceedings - 2021 IEEE 14th International Conference on Software Testing, Verification and Validation, ICST 2021* (pp. 205–216). Norway: Institute of Electrical and Electronics Engineers Inc. http://ieeexplore.ieee.org/xpl/mostRecentIssue.jsp?punumber=9438541. doi:10.1109/ICST49551.2021.00031.

Yao, Z., Li, D., Li, Z., Zhou, P., & Li, L. (2023). Relay protection mirror operation technology based on digital twin. *Protection and Control of Modern Power Systems, 8*(1). https://www.springer.com/journal/41601. doi:10.1186/s41601-023-00328-4.

Zhang, C., Zhou, Q., Shuai, B., Williams, H., Li, Y., Hua, L., & Xu, H. (2023). Dedicated adaptive particle swarm optimization algorithm for Digital Twin based control optimization of the plug-in hybrid vehicle. *IEEE Transactions on Transportation Electrification, 9*(2), 3137–3148. https://www.ieee.org/membership-catalog/productdetail/showProductDetailPage.html?product=PER473-ELE&utm_source=Mainsite_CSE&utm_medium=CSE_Promotion&utm_campaign=Catalog_Promotion-PER473. doi:10.1109/TTE.2022.3219290.

CHAPTER

14

Cloud computing for Digital Twin power grids

Sana Saleh, Zohaib Saadat, Muhammad Tariq

Smart Technologies and Applied Intelligence Research (STAIR) Lab, Department of Computer Engineering, School of Engineering, National University of Computer and Emerging Sciences (NUCES), Islamabad, Pakistan

14.1 Overview of Digital Twins and power sector smart grids

The pressing worldwide demand for carbon neutrality and the quest for resilient, intelligent grid architectures are redefining modern power systems. Essential to this development are two complementary technology concepts: smart grids and Digital Twins (DTs). A DT is a dynamic, virtual representation of real-world power system assets and processes that enables advanced monitoring, analysis, prediction, and optimization. Simultaneously, smart grids transform power networks through digital communication, control, and sensor technologies. Together, these technologies, especially when enhanced by the robust capabilities of cloud computing, are poised to transform the design, operation, and maintenance of power systems.

Beyond synthesizing existing literature, this chapter contributes a unique perspective by proposing a structured integration model for cloud-enabled DTs in smart grids. Specifically, it highlights: (1) A multilayered framework linking cloud architecture, scalability, and interoperability; (2) a categorization of challenges and opportunities in cloud-based DT deployment; and (3) evaluation criteria for cloud readiness in power systems. These contributions distinguish this work from prior reviews, positioning the chapter not only as a synthesis but also as a reference framework for researchers and practitioners.

Fueled by developments in data analytics and the connectivity enabled by the Internet of Things (IoT), DTs stand at the forefront of the Industry 4.0 transformation. Devices like sensors, smart meters, and phasor measurement units (PMUs) built into the power grid are necessary to gather and transmit massive amounts of real-time operational and environmental data from the physical system to its digital equivalent. (Shi et al., 2018).

Once collected, this enormous volume of data necessitates sophisticated technologies for efficient processing, administration, and storage. Machine learning, predictive analytics, and optimization approaches are used in the DT environment to gather information, identify trends, predict future occurrences, and improve grid functions, including load balancing, demand response, and predictive maintenance. Big data, IoT, and most importantly, cloud computing, are propelling the use of DTs and greatly enhancing their potential. Technologies like blockchain, advanced data analytics,and other high-speed communication networks are crucial for creating reliable solutions, especially in the security space. Therefore, the development and smooth integration of several supporting technologies are directly related to the efficiency of a power grid DT. It is anticipated that power grid DTs will become increasingly complicated, autonomous, and intelligent as these technologies advance.

14.1.1 The five-dimensional Digital Twins model

Researchers have proposed a 5G model to help better understand the components and operational dynamics of DT (Tao et al., 2019). This model typically consists of:

1. A physically existent system or object set up in the real-world (Ciotta et al., 2021).
2. The digital framework or paradigm that is present in virtual worlds (Patacas et al., 2020).

Digital Twin Technology and Smart Grid.
DOI: https://doi.org/10.1016/B978-0-443-36664-2.00015-0

3. An updating mechanism or connection that permits bidirectional data exchange and synchronization between virtual and physical components (Gandrup et al., 2020).
4. An engine for forecasting that uses both historical and current data to predict events and actions in the future (West & Blackburn, 2018).
5. A system that analyzes and investigates different occurrences to offer recommendations for enhancements or preventive measures (Zhu et al., 2022).

Certain models specifically identify "data" as a fundamental element and "services" as an enhanced capacity for digital transformation.

This 5G model highlights its active and dynamic features, setting a DT apart from passive simulation models. It highlights how the DT can not only replicate and evaluate, but also forecast and suggest actions. This capability shows that DTs can do more than just monitor power grids; they can also be involved in operational decision-making, real-time control systems, and long-term strategic planning. They provide asset management, load balancing, and problem identification.

14.1.2 The smart grid paradigm

The traditional electrical grid is characterized by one-way electricity flow and centralized control; it is progressively evolving into a "smart grid." This shift is necessary to address the issues facing modern energy systems.

14.1.2.1 Definition, key features, and technologies

A smart grid uses sensors, software, digital technology, and two-way communication to manage and control the distribution of power from several producing sources, meeting end users' changing needs in real-time. Among the objectives are cost control, improved grid stability and dependability, efficient integration of renewable energy sources, lower energy losses, and encouragement of customer and utility active involvement.

Smart grids integrate sophisticated distributed energy resources (DERs), such as rooftop solar panels and battery storage devices, under astute management (Adnan et al., 2025; Cicioni et al., 2023). They implement demand response programs and real-time pricing systems that empower customers, in addition to making considerable use of advanced metering infrastructure (AMI). AMI enables consumers and stakeholders to make better decisions by providing real-time insights on electricity usage, fluctuating prices, and associated carbon emissions through a two-way information flow (Zheng et al., 2022). Among the several smart grid technologies are IoT devices that improve sensing and control, blockchain for safe energy transactions, artificial intelligence and machine learning for predictive analytics and operational optimization, advanced energy storage systems that give grid flexibility, and high-voltage direct current transmission for effective long-distance power transfer (Amin & Molinas, 2017). For DTs to be used correctly, a data-rich environment is created by these smart grid technologies. Thus, DTs offer the advanced modeling and analytical skills required to manage the increasing complexity brought on by smart grid operations. Especially, DTs are being used in intelligent resource allocation and scheduling in 5G-enabled distribution networks with latency sensitivity, hence improving system responsiveness and performance in real-time energy contexts (Zhou et al., 2022).

14.1.2.2 Role in modern energy systems and decarbonization

The global shift to greener, more sustainable energy is a result of smart grids. They guarantee the ability to meet the increasing electricity demand brought on by the electrification of sectors like transportation, as well as the expansion of the usage of alternative energy sources like solar and wind. By classifying these sporadic resources based on national and global carbon reduction targets, smart grids may better control fluctuations in usage, boost the use of renewable energy, and lower carbon emissions.

As a result, electric vehicles (EVs) can be added to the grid through intelligent charging and effective load control. Vehicle-to-grid (V2G) features, a large part of this infrastructure, make it easier for EVs to send electricity back to the grid at the right times, lowering the risk of overloading the network (Saba et al., 2023; Saleem et al., 2019). Local energy systems expand as a result of this support, enabling people to closely monitor the energy supply in their communities and generate more creative green solutions. Smart grid technology is growing as a result of the need to lower emissions and increase the amount of renewable resources in the grid, both of which require improved management tools. These interactions typically result in the need for cloud computing to manage the pertinent activities and data. The combination of cloud computing, smart grids, and DTs is currently altering how we approach appropriate energy use and manage their intricacy.

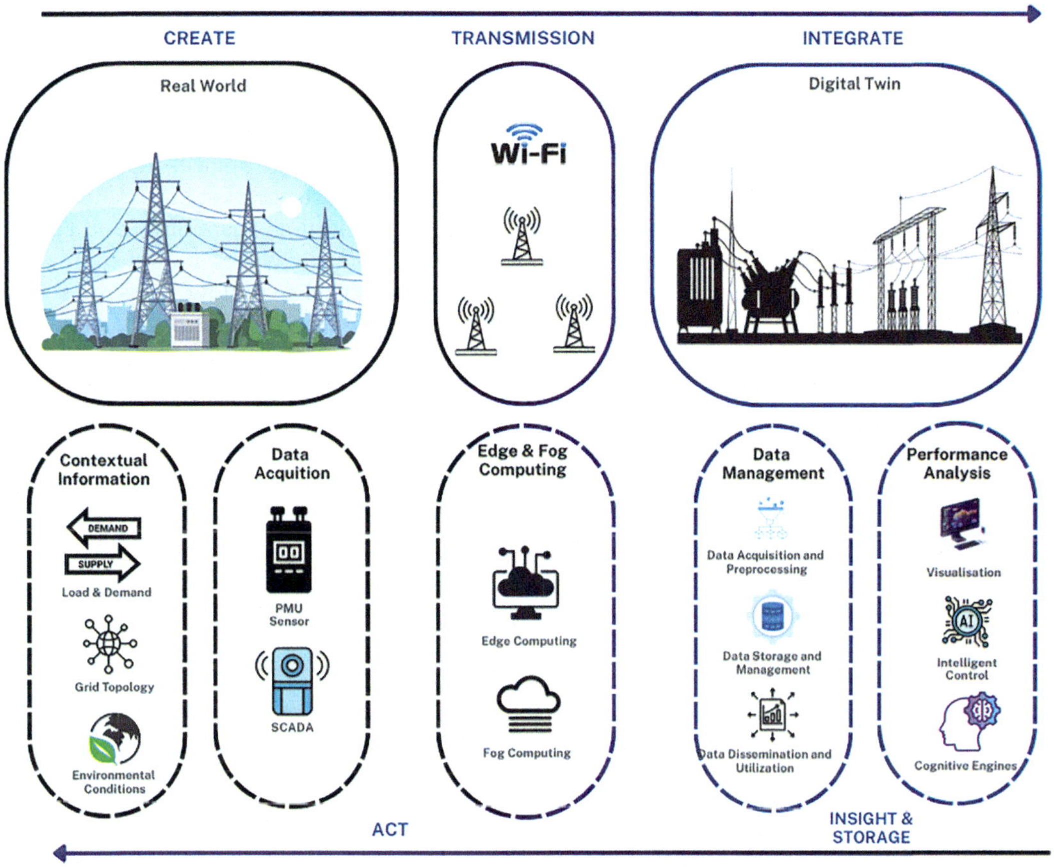

FIGURE 14.1 Enhancing smart grid capabilities with cloud-based DTs: This DT architecture leverages cloud computing to integrate real-time data and simulations, enabling predictive control and proactive grid management. *DT*: Digital Twin.

14.1.3 The convergence: Digital Twins for enhanced smart grid capabilities

Cloud computing can be used to integrate DT technology with the smart grid architecture, opening up incredible, game-changing possibilities for the energy industry. Smart grids present significant operational challenges due to their distributed resources, bidirectional energy flows, and active user interaction. By enabling greater interaction between network components and providing grid operators with intelligent decision-making support, DT technology is emerging as a great tool to reduce complexity, especially in dynamic events like cyberattacks or breakdowns.

DTs allow operators to combine real-time operational data, interactive smart grid capabilities, predictive analytics, and sophisticated simulation skills by creating a dynamic virtual reality of the actual grid. Significant gains in grid flexibility, dependability, and efficiency may come from this collaboration. Improved demand response management, increased load balancing, and the smooth integration of variable renewable energy sources through impact simulation are crucial applications in this networked environment that allow operators to anticipate and reduce anomalies. From reactive to proactive and predictive grid management, the transformation in the operation, planning, and optimization of power systems promises increased performance and resilience in a progressively complex energy environment. One finds a conceptual model illustrating this integration in Fig. 14.1. It shows how physical grid infrastructure flows to cloud-based DT systems, therefore enabling closed-loop input for real-time monitoring, analysis, and control.

14.2 The imperative of cloud computing for power grid Digital Twins

The creation and functioning of extensive DTs for power grids impose significant computing and data management requirements. Cloud computing has become an essential facilitator, offering the requisite infrastructure, services, and scalability to actualize the complete potential of these advanced virtual copies.

14.2.1 Addressing the demands of power grid Digital Twins

DTs of power grids are fundamentally data-intensive and computationally exigent. Their operations necessitate the integration and analysis of extensive data from various sources, including real-time sensor devices and intelligent electronic equipment deployed across geographically dispersed assets. This includes PMUs, smart meters, supervisory control and data acquisition (SCADA) systems, and IoT-enabled edge devices that continuously monitor grid conditions. Alongside these, historical operational data, weather forecasts, market information, geospatial data, and comprehensive asset specifications enrich the DT environment. Developing, sustaining, and employing a high-fidelity DT that accurately represents the dynamic condition of a complex system requires substantial computational resources. These include ongoing model updates, intricate simulations (e.g., power flow, transient stability, and fault analysis), artificial intelligence (AI)/machine learning (ML) model training (e.g., predictive maintenance or load forecasting), and advanced analytics.

The structural complexity of contemporary power networks, which escalates dramatically with the incorporation of new technologies such as DERs and intelligent sensors, intensifies these demands. Conventional on-premises IT infrastructure frequently fails to provide these requirements in a cost-efficient and flexible way (Saeed et al., 2021). The capabilities of most individual utility companies to develop and sustain the vast volume of data and computation required for a thorough power grid DT generally fall short. Offering practically infinite resources on demand, cloud computing directly overcomes these problems and makes it not just a necessary tool but also a major facilitator for big-scale power grid DTs.

14.2.2 Core benefits of cloud adoption: Scalability, flexibility, and advanced analytics

Cloud computing for electrical grid DTs offers many basic benefits that are difficult to achieve with traditional on-site solutions. Among these, the most important are scalability, adaptability, and access to advanced analytical capabilities. Excellent flexibility provided by cloud platforms helps businesses dynamically adjust computing and storage capacity based on the evolving needs of their digital transformation projects without requiring large initial physical infrastructure capital investments. This elasticity is crucial since a DT may need major computational resources for certain simulations or analytical activities, integrate other data sources over time, or grow in complexity. One major benefit is the ability to allocate resources on demand, ensuring that DTs remain responsive even during high-demand events such as peak load periods or system disturbances (Adnan et al., 2019; Ali et al., 2019).

Furthermore, cloud systems offer a whole range of preconfigured, managed services for complex analytics, AI/ML, and big data processing. Advanced features for linking and monitoring vast sensor networks, data stream management, and predictive algorithm execution abound on platforms such as Microsoft Azure IoT, Amazon Web Services (AWS) IoT, and Google Cloud IoT. By abstracting most of the underlying infrastructure complexity, these services significantly lower the entrance barriers for utilities to use new technologies and free teams to focus on developing value-added digital transformation applications. Cloud-based simulation tools provide efficient and cost-effective solutions for comprehensive power system analysis and decision-making assistance, facilitating "what-if" scenario modeling for various contingencies, which is essential for evaluating the impacts of potential events—such as sudden demand surges, renewable intermittency, equipment failures, or cyberattacks—before they occur in the physical grid. This capability enhances operational readiness, supports risk-informed decision-making, and strengthens strategic planning within the framework of increasingly complex smart grids (Thurner et al., 2018). The democratization of access to high-performance computers and advanced AI technologies allows a greater range of utilities and researchers to design and apply complex power grid DTs, hence increasing industry innovation.

14.2.3 Limitations of traditional cloud approaches for real-time grid operations

Although centralized cloud computing provides considerable benefits, conventional cloud architectures exhibit specific limitations, such as increased latency, potential bandwidth constraints, and reduced real-time responsiveness, particularly in scenarios requiring instantaneous control or protection actions.

Numerous essential smart grid functions, including protective relaying, real-time voltage regulation, swift fault detection and isolation, necessitate decision-making and action within milliseconds. As a result, there comes a point at which typical cloud round-trip times are inadequate. Table 14.1 provides a quantitative analysis of various time scales and their appropriateness for centralized versus edge deployment.

The transmission of substantial quantities of high-frequency sensor data from extensive grid assets to a remote centralized cloud, followed by data processing and the subsequent dispatch of control signals back to the physical

TABLE 14.1 Latency requirements versus cloud constraints.

Grid application	Time scale	Cycle equivalence (60Hz)	Cloud suitability	Constraint
Protection (fault isolation)	1–16 ms	< 1 Cycle	**Unsuitable**	The total permitted reaction time is frequently exceeded by the network round-trip time (RTT)
Frequency regulation	2–4 (Sec)	~100 + Cycles	**Marginal**	Edge prefers variable WAN latency (jitter), which adds risk
Economic dispatch	5–15 (Min)	N/A	**Suitable**	When compared to the decision window, cloud RTT is insignificant
PMU data stream (Ingest)	Continuous (30–60 Hz)	Continuous	**Adequate (with aggregation)**	High-bandwidth is needed; without edge filtering, straight streaming to the cloud is expensive

grid, may result in intolerable delays caused by network latency. This intrinsic latency can make solely centralized cloud-based DTs ineffective for time-sensitive operations.

Moreover, if all data needs to be constantly sent to a central cloud, the significant volume of data generated by a highly monitored smart grid can overload network bandwidth and cause congestion and higher prices. Centralizing large volumes of critical operational grid data makes it a more attractive target for cyberattacks, necessitating robust cybersecurity measures (Li et al., 2020). In particular, privacy concerns may arise when sensitive user-level data, such as individual household consumption patterns or location-specific operational details, are exposed during transmission or centralized storage, potentially leading to misuse or unauthorized profiling. The operational constraints and vulnerabilities connected with only centralized cloud models have greatly driven the research and deployment of alternative architectures, including edge and hybrid cloud solutions, which try to localize computation and data processing near the data sources at the grid edge (Shen et al., 2020). This suggests a more general inclination toward distributed intelligence in DT models of power grids.

14.3 Cloud architectures for power grid Digital Twins

The choice of cloud architecture is a crucial decision that directly affects the performance, scalability, cost, and applicability of a power grid DT. Several architectural models have emerged, each offering distinct advantages and trade-offs.

14.3.1 Architectural models: Centralized, edge, and hybrid approaches

- *Centralized cloud architecture*: Under a standard centralized cloud architecture, a central cloud platform receives all data from the sensors, smart meters, and operational systems of the power grid. Hosted in this central cloud, the DT runs all data storage, computation, modeling, and analytical operations (Zhang et al., 2019). This architecture utilizes the economies of scale and extensive computational resources provided by massive cloud data centers. It is typically appropriate for applications that are not very sensitive to latency, including activities like long-term planning, conducting offline system analyses, training extensive AI models with historical data, and performing comprehensive asset management analytics across fleets. The principal advantages include streamlined administration and access to vast computational resources (Li et al., 2022). Nonetheless, disadvantages encompass possible data transmission latency issues for real-time applications, reliance on uninterrupted network connectivity to the central cloud, substantial bandwidth consumption for raw data transfer, and potential data sovereignty or privacy concerns when sensitive grid data is transferred off-premises.
- *Edge computing architecture*: Edge computing embodies a decentralized paradigm wherein data processing and certain DT operations are executed nearer to the data production source—at the network's "edge." Within power grids, edge nodes may be situated at substations, on intelligent electronic devices (IEDs), at local control centers, or at aggregated DER sites. These edge nodes are capable of hosting partial DT models, conducting preliminary data filtering and aggregation, executing real-time analytics, and initiating local control actions (Deng et al., 2020).

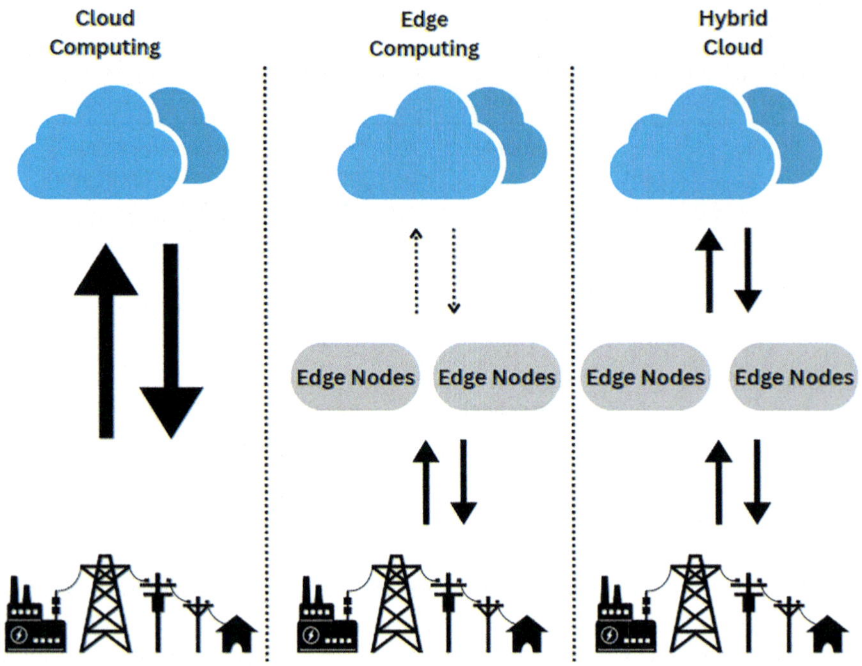

FIGURE 14.2 Architectural paradigms for deploying DTs in power systems: A comparison of computational approaches: (A) centralized cloud, (B) edge computing, and (C) a hybrid architecture leveraging both paradigms. *DT*: Digital Twin.

This design is essential for applications with strict low-latency demands, including real-time fault detection and isolation, local voltage regulation, and swift anomaly detection for cybersecurity. Fog computing serves as an intermediary layer between edge devices and the central cloud, offering supplementary processing and storage resources nearer to the edge than the primary cloud, thus diminishing network traffic to the core. DT edge networks aims to combine mobile edge computing with DTs to build and manage DT models directly at the network edge, thereby enabling real-time interactions and decision-making (Cao et al., 2020b). The primary advantages of an edge architecture encompass markedly diminished latency, optimized bandwidth utilization through local data processing, improved privacy and security for sensitive local information, and the capacity for specific functions to function independently in the event of temporary disconnection from the central cloud. Nonetheless, edge nodes generally possess more constrained computing and storage capabilities relative to the central cloud, and overseeing an extensive, distributed edge infrastructure might present additional complications.

- *Hybrid cloud architecture*: A hybrid cloud architecture aims to integrate the advantages of both centralized cloud and edge computing models, providing a balanced solution for comprehensive power grid DTs. This model allocates real-time operations, data preprocessing, and latency-sensitive analytics to edge components, while the central cloud is reserved for tasks necessitating significant computational power, extensive data storage, intricate simulations, training advanced AI models, and delivering comprehensive grid-wide analytics and coordination (Mao et al., 2017). A practical demonstration of this approach can be found in the development of a cognitive Digital Twin for a real-world microgrid in Singapore, where a client-side component located adjacent to the microgrid handles real-time control, while a cloud-based server executes computationally intensive algorithms for energy management and optimization. This hybrid architecture enables seamless real-time communication and enhanced decision-making, while also reducing operational costs and improving reliability (Sivaneasan et al., 2025). The AWS DT framework (DTF) specifically accommodates a hybrid design, allowing essential components to be deployed on-premises at the customer location while interfacing with data, models, and processing capabilities stored on AWS. This method facilitates optimum resource use, improved system resilience (since essential local functions can persist during cloud outages), and a more effective equilibrium between real-time responsiveness and comprehensive analytical capabilities. Hybrid architectures, while they have considerable benefits, necessitate meticulous design of data synchronization methods and workload distribution strategies between edge and cloud and may include more intricate integration efforts. The shift from fully centralized systems toward integrated edge intelligence suggests that hybrid architecture will likely emerge as the primary model for power grid DTs, as it can accommodate operational needs ranging from millisecond-level control to long-term strategic planning.

In Fig. 14.2A, a visual comparison of the three architectures is shown. The figure illustrates the fundamental differences in data flow and processing locations among centralized cloud computing, decentralized edge computing, and the integrated hybrid model. In the centralized model, all data is sent to the cloud for processing. The edge computing model emphasizes local processing near data sources, reducing latency. The hybrid architecture combines the strengths of both, with edge nodes handling latency-sensitive tasks and the cloud performing complex analytics and large-scale coordination.

14.3.2 Cloud service models and their relevance

Cloud computing services are often provided in three primary models: Infrastructure as a Service (IaaS), Platform as a Service (PaaS), and Software as a Service (SaaS). The selection of a service model profoundly influences the degree of control, management accountability, and development effort necessary for a power grid DT.

- *IaaS*: IaaS offers access to essential computing resources, including virtual machines (VMs), storage (both block and object), and networking components via the internet. The cloud provider oversees the foundational physical infrastructure, whereas the user is accountable for the management of the operating system, middleware, applications, and data.
- *Relevance to power grid DTs:* IaaS provides the utmost flexibility and control, rendering it appropriate for utilities with specific infrastructure needs, those requiring deep integration with legacy systems, or those aiming to deploy custom DT platforms or specialized simulation software necessitating particular OS configurations. The main benefits include the ability to scale computing and storage on demand, flexible pricing options, and the ability to replicate on-premises infrastructures in the cloud. Nonetheless, it also involves the highest managerial burden for the utility.
- *PaaS*: PaaS offers an all-encompassing platform for the development, deployment, execution, and management of applications, eliminating the intricacies associated with constructing and sustaining the foundational infrastructure. PaaS solutions generally encompass operating systems, environments for executing programming languages, databases, messaging systems, AI/ML frameworks, and IoT device management services.
- *Relevance to power grid DTs*: PaaS accelerates power grid DT development and deployment. Managed services can be used for IoT data ingestion and device management (e.g., AWS IoT Core and Azure IoT Hub), time-series databases, data analytics platforms (e.g., Azure Synapse Analytics and Google BigQuery), and AI/ML model development and deployment (Jahn et al., 2019). This significantly reduces operational overhead and allows development teams to focus on DT application logic rather than infrastructure management. Modern AI/ML and containerized application teams are adopting PaaS due to its simplified application lifecycle management, enhanced developer productivity, and built-in security and monitoring (Song et al., 2020).
- *SaaS*: SaaS provides fully functional software programs via the internet, usually on a subscription model. The cloud provider oversees all facets of the software service, encompassing the application, the foundational infrastructure, and all maintenance and upgrades. Users utilize the software via a web browser or specialized client applications.
- *Relevance to power grid DTs*: Utilities may use specific DT functions or specialized analysis tools for power systems, which are offered as SaaS solutions by external providers. Certain features of complete DT platforms, such as AWS IoT TwinMaker or Azure DTs, as well as specialized partner solutions developed on these platforms, may be perceived as SaaS-like from the end user's viewpoint. This idea appeals to utilities wanting quick application of specific capabilities with minimum internal development or management effort. Key benefits include easy accessibility, no installation or maintenance burden for users, and predictable subscription-based pricing (Schaefer et al., 2020).

The choice of a suitable service model, or a combination thereof, for a power grid DT is contingent upon factors including the utility's internal technical proficiency, current IT/operational technology (OT) infrastructure, preferred degree of customization and control, specific functional needs of the DT, and the maturity of available solutions. A discernible trend toward utilizing PaaS solutions is apparent, especially for expedited development and the utilization of specialized cloud services for IoT, data analytics, and AI/ML, which are essential to contemporary digital transformations.

14.3.3 Cloud platforms and reference architectures for power grid Digital Twins

Leading cloud service providers (CSPs), AWS, Microsoft Azure, and Google Cloud Platform (GCP), offer a wide range of services and tailored frameworks that can be used to build advanced DTs for power grids. These platforms provide tools for data ingestion, modeling, and storage, along with advanced simulation engines, analytics, and user-friendly visualization tools.

- *AWS*: AWS offers several tools specifically designed to support Digital Twin development and digital transformation. By allowing users to combine data from many sources, including AWS IoT SiteWise for industrial equipment data, Amazon Kinesis Video Streams for video feeds, and connectors for services like Amazon Timestream or Amazon S3 for data lakes, together enabling the development of DTs. It enables the creation of virtual representations of physical systems and the amalgamation of existing 3D models with real-world data. One notable aspect of IoT TwinMaker is its capacity to autonomously create a knowledge graph that represents the connections between physical assets and their data sources. It connects with Amazon-Managed Grafana and other tools for visualization. The AWS DTF offers an extensive reference architecture for developing and expanding DT solutions. It outlines three primary domains: the Spatial Data Plane (for the management of 3D assets), the Industrial Data Fabric (utilizing services such as AWS IoT Greengrass for edge data collection and AWS IoT SiteWise for the organization of industrial data), and the simulation compute environment (employing high-performance computing (HPC) services for predictive modeling). Essential AWS services for power grid DTs comprise AWS IoT Core for secure device connectivity, Amazon S3 for scalable data lake storage, Amazon SageMaker for the development and deployment of ML models, and AWS IoT Greengrass for facilitating edge computing capabilities, which enable local processing and decision-making.
- *Microsoft Azure*: Azure DT is Microsoft's primary solution for building large-scale digital representations of physical environments, including energy networks, manufacturing facilities, and urban areas. It enables users to establish custom models of their physical environments using the digital twins definition language (DTDL), an open modeling language based on JSON-LD. Essential features comprise a real-time execution environment that depicts the DT as a dynamic graph, functionalities to assimilate data from Azure IoT Hub (for IoT devices) and additional business systems, through Logic Apps and REST APIs, with the ability to stream twin change events to Azure services such as data explorer, synapse analytics, and event hubs for analysis and storage. Azure IoT Hub functions as the primary gateway for the connection, monitoring, and management of IoT devices at scale. Azure offers Azure IoT Central for expedited, low-code IoT application development; Azure Stream Analytics for real-time data processing; Azure ML for constructing AI models; and Azure Data Lake Storage for extensive data storage. Azure's solutions for the energy sector primarily target use cases such as grid asset maintenance, energy management, and load balancing.
- *GCP*: Google Cloud offers a comprehensive array of services that establish a robust framework for constructing DTs of power grids. The Google Cloud IoT Platform, which includes IoT core for device management, though its future remains uncertain as Google refines its IoT strategy, provides functionalities for data ingestion (through MQTT and HTTPS), device authentication, telemetry storage, data processing, and alerting, frequently incorporating features beneficial for DT development. Essential GCP services for data teams comprise cloud storage for constructing data lakes, BigQuery as a highly scalable data warehouse for sophisticated analytics, Vertex AI for a consolidated ML platform, Pub/Sub for resilient real-time data intake, and Dataflow for both stream and batch data processing. Google Cloud highlights partner solutions; for instance, Cognite Data Fusion, running on GCP, offers an industrial DataOps platform that uses AI-driven contextualization to generate digital replicas of industrial assets by delineating intricate data linkages.
- *Reference architectures and initiatives*: In addition to individual CSP products, other reference architectures and overarching initiatives direct the development of DTs for power systems:
  - The cognitive DT for a Microgrid in Singapore illustrates a hybrid edge-cloud design, featuring client-side components proximate to the actual microgrid for real-time control and a cloud-based server for computationally demanding energy management and optimization algorithms (Su & Shi, 2024).
  - TwinArch is a proposed three-dimensional DT reference architecture emphasizing functionality, dependability, and lifecycle, designed for cross-industry adaptability and offering a unified platform for the development and discourse of DT applications (Madni et al., 2019).
  - The notion of DT edge networks delineates a hierarchical architecture that includes a virtual plane (including DT models at both the edge and in the cloud) and a user/physical plane, aimed at mitigating communication burdens linked to exclusively centralized DT modeling (Rasheed et al., 2020).
  - A notable endeavor is the National Grid ESO's Virtual Energy System (VirtualES) in Great Britain. This program seeks to establish an integrated ecosystem of DTs that reflects the whole of the Great Britain energy system. The objectives are to enable secure data sharing across organizational boundaries, support complex scenario modeling, and promote collaborative decision-making to achieve decarbonization targets. The VirtualES initiative underscores the significance of interoperability, standardized frameworks, and cooperation among industry stakeholders (O'Dwyer et al., 2020).

14.3.3.1 *Real-world architectures: Orchestrating cloud services for complex grid use cases*

Building the suite of services presented by AWS, Azure, and GCP, their value is best understood not when viewed in isolation but when they are orchestrated to address particular, complex grid problems. The following exemplars illustrate this kind of orchestration, associating architectural decisions with real-world results.

14.3.3.1.1 Dynamic voltage control in a high-DER penetration feeder

A primary challenge in current distribution grids with high levels of DERs, such as rooftop solar, is having to deal with unstable voltage profiles. To address this, a DT architecture uses a hybrid edge-cloud model to combine rapid local response with sophisticated cloud-based forecasting.

The architecture starts at the edge, where nodes running frameworks such as AWS IoT Greengrass or Azure IoT edge are installed in substations. These nodes harvest data from smart inverters and PMUs, running lightweight algorithms to carry out real-time, local voltage control, hence minimizing latency. This real-time data stream is then ingested into the cloud via services such as AWS IoT Core or Azure IoT Hub. Here, the architecture splits into two: raw, high-bandwidth telemetry is stored in a time-series database like Amazon Timestream or Azure Data Explorer for historical investigation and anomaly detection, while a graph database like Amazon Neptune keeps a real-time digital model of the feeder's topology. This topological model can be built using Azure DTs and its DTDL, which describes the interconnections between assets such as lines, transformers, and DERs. The core predictive capability is enabled in the cloud. An ML model trained and deployed on Amazon SageMaker or Azure ML processes this aggregated data lake. It forecasts possible voltage deviations through the correlation of real-time grid data against extrinsic variables such as weather forecasts, which can be obtained from public data sets in Google BigQuery. The result of this process is coordinated, anticipatory action. The brain of the DT comes from closing the control loop. When the ML model foresees an imminent voltage surge, it will calculate optimal setpoints automatically and send these commands back to the inverter controllers through the same IoT Hub/edge framework, modifying reactive power output to stabilize the feeder in advance of a violation.

This architectural pattern, depicted in Fig. 14.3, is not just theoretical; utilities such as Singapore's SP Group have piloted such cloud-enabled DTs for real-time voltage regulation, ensuring reliability despite high and fluctuating renewable generation.

14.3.3.1.2 Real-time grid stability assessment and fault response

For transmission system operators, a paramount concern is assessing grid stability in near-real-time to avoid localized faults growing into mass blackouts. This use case demands an architecture prioritizing high-speed data processing and large-scale simulation, tightly coupling edge sensing with cloud-based analytics.

The data infrastructure is based on high-speed ingestion. Millions of data points are provided every second by PMUs from a wide-area monitoring system (WAMS). Real-time processing tools like AWS Kinesis Data Streams and Azure Stream Analytics manage this massive data firehose by carrying out crucial aggregation and filtering. An integrated, dynamic grid model is filled with this processed data. The entire transmission network can be modeled as a live graph of assets and their relationships, including crucial contextual awareness, using a service like Azure DTs. Simultaneously, the grid's actual state is saved for detailed analysis in scalable data warehouses like Azure Data Explorer or Google BigQuery. AI-driven prediction constitutes the analytic core of this DT. The stability assessment model, located in Google Vertex AI or AWS SageMaker, continuously examines the state of the grid. At the occurrence of a triggering event, e.g., a line trip, the DT can immediately spin up a sophisticated contingency analysis simulation on HPC infrastructure such as AWS ParallelCluster or Azure CycleCloud to simulate cascading risk and verify contemplated mitigation measures. Last but not least, the architecture facilitates an orchestrated response. The outputs of such simulations are automatically visualized for system operators on Amazon-Managed Grafana-based dashboards. Crucially, the DT can also recommend specific remedial actions (e.g., generation re-dispatch and topology changes) directly to energy management systems (EMS), or even initiate automated responses to stabilize the grid within milliseconds.

This approach is operational in projects like those by Siemens and TenneT in Germany, where DTs provide critical real-time stability monitoring, drastically reducing decision-making time during contingency events and enhancing overall system resilience.

These scenarios illustrate that the value of CSP platforms is not in any one service, but in their orchestration together. The decision between AWS, Azure, or GCP typically hinges on pre-existing utility partnerships and individual strengths of service (e.g., AWS industrial emphasis, Azure DT simulation, Google AI/ML, and data

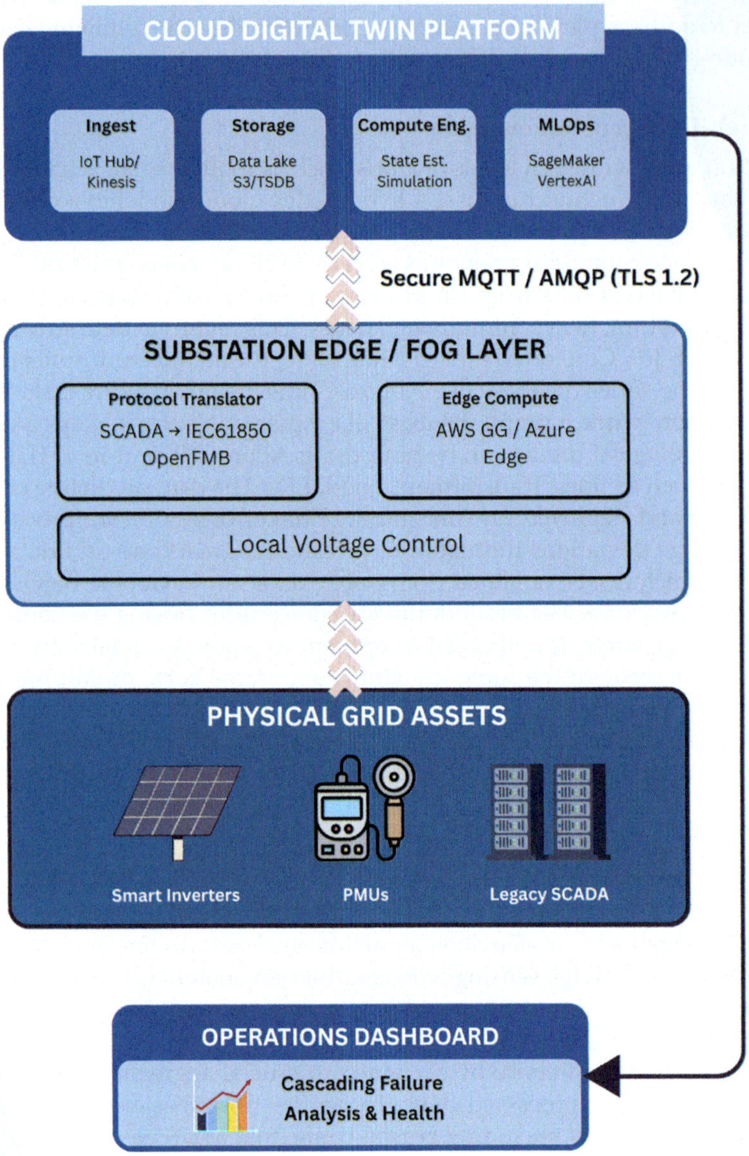

FIGURE 14.3 End-to-end hybrid cloud architecture for power grid DTs. The substation edge/fog layer, the cloud platform, and physical assets are all illustrated in the figure together with the data flow and workload distribution. It emphasizes where important services are located, such as local control and protocol translation at the edge and scalable ingestion, storage, computing, and MLOps in the cloud, coupled by secure protocols. *DT*: Digital Twin.

analysis). The resulting architectural trade-offs, such as latency versus analytic depth, need to be managed carefully, as addressed in the following section.

14.3.3.2 Critical considerations and trade-offs

While the architectures described earlier demonstrate significant potential, their deployment is not without inherent challenges and trade-offs.

Latency and physical constraints: For certain grid operations like protective relaying or fast frequency response, even sophisticated edge computing nodes are subject to inherent physical limits. The feasibility of moving all essential protection functions into cloud or hybrid environments is limited by wide-area network jitter, propagation delays, and synchronization problems. Deployment is slow, but research on FPGA-based edge devices, ultra-reliable low-latency communication with 5G/6G, and time-sensitive networking aims to fill these gaps. In practice, utilities typically leave cloud resources for supervisory and predictive analytics and maintain local or hardwired relays for ultra-high-speed response.

Legacy system integration: It is highly challenging to map proprietary SCADA and EMS protocols into open standards [like open platform communications unified architecture (OPC UA) and International Electrotechnical Commission common information model (IEC CIM)]. Semantic mismatches, vendor lock-in, and inconsistent data continue to be major obstacles in practice. In addition to generic gateways, advanced middleware design patterns such as semantic mediation layers or message bus designs (e.g., Kafka streams) improve interoperability but increase latency, complexity, and even the possibility of single points of failure. Because excessive integration may cause well-tested legacy infrastructures to become unstable, utilities must choose between interoperability and system stability.

Although cloud-based DTs are excellent at analytics, proprietary SCADA protocols make it difficult to integrate them with traditional grid infrastructure. In order to overcome this, contemporary Hybrid DT designs depend more and more on Open Field Message Bus for edge interoperability and IEC 61850 for substation-level object modeling. These standards enable edge nodes to use the common information model (CIM/IEC 61970) to standardize data from various legacy devices before sending it to the cloud. Regardless of the underlying vendor hardware, this translation layer guarantees that the cloud-based Digital Twin receives a standardized data structure.

Cybersecurity and operational flexibility: Although CSPs provide robust security primitives (Identity and access management (IAM), encryption, and zero-trust networking), the essence of the challenge is balancing security with real-time operability. For instance, multifactor authentication, high-frequency key rotation, or deep packet inspection could add latency or computational overhead that conflicts with millisecond-level grid requirements. Likewise, extensive encryption makes fast failover between edge and cloud nodes difficult. Pilot projects highlight the need for risk-based adaptive security, where protection mechanisms dynamically adjust to operational criticality rather than applying uniform, resource-intensive measures across all DT layers.

In summary, although AWS, Azure, and GCP offer comprehensive toolkits for DT development, the effectiveness of cloud-based DTs in power grids depends not only on adopting these solutions but also on acknowledging their boundaries and managing the associated trade-offs. The DT cloud platform ecosystem is active and competitive. Although all major CSPs provide essential functionalities such as data ingestion, storage, and analytics, their specific DT services (such as AWS IoT TwinMaker and Azure DTs) and partner ecosystems offer unique benefits tailored to specific use cases or aligned with a utility's current technological preferences and investments. The advent of comprehensive reference architectures and collaborative efforts such as VirtualES indicates a significant movement toward standardization and interoperability, essential for achieving the vision of interconnected, system-wide DTs for power grids. Table 14.2 presents a comprehensive assessment of the principal cloud platforms and their significance to power grid DTs.

14.4 Scalability considerations for cloud-based power grid Digital Twins

Scalability is a critical issue when deploying DTs for power grids due to the extensive nature of these systems, the substantial amount of data they produce, and the processing demands of the necessary analysis. Cloud computing has intrinsic benefits in tackling scalability challenges; however, effective strategies are still essential.

14.4.1 Scalability dimensions in power grid Digital Twins

Scalability for power grid DTs is a multidimensional requirement that encompasses various factors.

- *Data volume*: Power grids produce vast amounts of data from a growing array of sources, including potentially millions of smart meters, various sensors on transmission and distribution lines, substation equipment such as IEDs, PMUs, SCADA systems, meteorological services, GIS databases, and historical operational records (Wang et al., 2019b). The DT must have the capability to acquire, store, and process these expanding datasets, which may easily reach terabyte- or even petabyte-scale.
- *Data velocity*: A significant portion of this data, especially from operating sensors and PMUs, is received in rapid, real-time streams. The DT infrastructure must accommodate this high pace to guarantee that the virtual model accurately represents the current condition of the actual grid.
- *Complexity and granularity of the model*: As DTs advance, they increasingly exhibit complexity, integrating more intricate physics-based models, enhanced representations of individual assets, and wider system interdependencies. The system must expand to accommodate these progressively complex models.

TABLE 14.2 Comparative overview of major cloud platforms for power grid DTs.

Feature Category	AWS	Microsoft Azure	GCP
Key DT service(s)	AWS IoT TwinMaker and AWS DTF	Azure DTs	Google Cloud IoT Platform (evolving), partner solutions (e.g., Cognite data fusion)
Core data ingestion	AWS IoT Core, AWS IoT Greengrass, Amazon Kinesis, and custom connectors	Azure IoT Hub, Azure Event Hubs, Logic Apps, and REST APIs	Google Cloud IoT Core (evolving), Pub/Sub, Dataflow, and Transfer service
Data modeling	Entities, Components, knowledge graph (IoT TwinMaker), and DTDL support through partners	DTs definition language (DTDL) and live graph representation	Custom modeling, Partner solutions (e.g., Cognite for contextualization)
Data storage and analytics	Amazon S3 (data Lake), AWS IoT SiteWise (industrial data), Amazon Timestream (time-series), Amazon Redshift, and Amazon Athena	Azure data lake storage, Azure data explorer, Azure synapse analytics, Azure time-series insights	Google Cloud Storage (Data Lake), BigQuery (Data Warehouse/Analytics), and Bigtable (NoSQL)
AI/ML services	Amazon SageMaker and AWS AI/ML services	Azure machine learning and Azure AI services	Vertex AI, Google Cloud AI/ML services
Edge computing support	AWS IoT Greengrass, AWS IoT SiteWise edge	Azure IoT edge and Azure Stack edge	Google Distributed Cloud edge, IoT Greengrass on GCP (via Anthos)
Visualization	Amazon-Managed Grafana, 3D scene composer (IoT TwinMaker), and partner solutions (e.g., Nvidia Omniverse)	Power BI, Azure DTs explorer, and partner solutions (e.g., Unity and Unreal Engine)	Looker, Google Data Studio, partner solutions
Key strengths for power grids	Mature IoT ecosystem, Industrial data focus (SiteWise), Scalable simulation (HPC), and DTF reference architecture	Strong DTDL-based modeling, Live execution graph, Integration with enterprise services, and energy-specific solutions	Leading AI/ML capabilities, Powerful data analytics (Big Query), Strong partner ecosystem for industrial data
Example power sector use cases/partners	Predictive maintenance, asset optimization, and process optimization (e.g., Siemens collaboration). Schneider Electric and ETAP	Grid asset maintenance, energy optimization, load balancing, and renewable energy integration (e.g., E.ON nd Schneider Electric)	Energy optimization (Properate with Cognite), Weather-dependent energy forecasting (WeatherNext), and Smart home energy management (Carrier)

- *Number of users and applications*: A successful DT will likely accommodate numerous users (operators, planners, maintenance personnel, analysts) and support a diverse array of applications simultaneously (monitoring dashboards, simulation tools, predictive analytics services, and reporting systems). The architecture must be capable of scaling to accommodate these simultaneous needs without performance degradation.
- *Geographic scale*: Distribution technologies may first concentrate on a particular substation or distribution feeder but might subsequently extend to encompass entire regions or even national grids. The foundational framework must facilitate this geographic expansion effortlessly.
- *Computational requirements*: Complex simulations (e.g., transient stability and cascade failure analysis), training extensive AI/ML models (e.g., for anomaly detection or load forecasting), and real-time analytics necessitate substantial and frequently variable computer resources.

The dynamic characteristics of power grids, characterized by the ongoing incorporation of renewable energy sources, the expansion of EVs and their charging infrastructure, and the changing regulatory frameworks, highlight the need for DT solutions that can adapt and scale efficiently. Cloud adoption inherently provides unmatched scalability, enabling enterprises to dynamically modify their computing, storage, and networking resources according to demand without the limitations and substantial initial investments linked to fixed on-premises infrastructure. Nonetheless, merely utilizing cloud deployment does not ensure scalability; the DT application and its foundational architecture must be conceived with scalability as a fundamental premise from the beginning. Neglecting to address these multifaceted scaling requirements may result in performance constraints, increased operational expenses, and the inability of the DT to adjust to the changing demands of the smart grid.

TABLE 14.3 Digital Twin workload placement decision matrix.

DT function	Latency tolerance	Data volume	Computational load	Recommended placement	Reasoning
Protective relaying	Ultra-Low (< 4 ms)	Low (Local)	Low	Edge (IED/Local)	To avoid damaging equipment, reactions must occur within physical grid cycles (50/60Hz)
Volt/VAR enhancement	Low (Sec)	Medium	Low/Medium	Edge/fog	Requires quick feedback loops, which local smart inverters or substation nodes can handle well
State estimation (local)	Medium (Sec/Min)	Medium	Medium	Hybrid	Local topology processed at edge; aggregated findings forwarded to Cloud for system-wide perspective
Prognosis and asset health	High (Hrs/Days)	High (Historical)	High (ML Models)	Cloud	requires the use of managed AI services for training on terabytes of historical data (such as vibration and temperature)
System-wide simulation	High (Min/Hrs)	Very High	Very High	Cloud	Elastic cloud compute scaling is necessary for large-scale contingency analysis (such as cascading failures)

14.4.2 Architectural orchestration for representative use cases

To translate the broad discussion of cloud services and scalability into a tangible framework, this section revisits two complex, representative use cases for power grid DTs: dynamic voltage control in high-DER feeders and real-time grid stability assessment. While detailed architectures were outlined in Section 3.3.1, here we emphasize their scalability dimensions and orchestration requirements.

Dynamic voltage control in high-DER feeders: This use case illustrates the need to scale across thousands of edge nodes while ensuring millisecond-level responsiveness. Edge services (e.g., AWS IoT Greengrass and Azure IoT edge) handle immediate control, while cloud-hosted ML models (AWS SageMaker and Azure ML) scale elastically to retrain and deploy updated setpoints. The orchestration layer (AWS Step Functions and Airflow) must manage high-frequency data ingestion from millions of IoT devices into time series databases like Amazon Timestream, ensuring the DT remains accurate at the feeder and regional levels.

Real-time grid stability assessment: This wide-area DT highlights scalability challenges in streaming and computation. With PMU networks producing gigabytes per second, real-time ingestion (AWS Kinesis and Azure Stream Analytics) must elastically expand to avoid bottlenecks. HPC resources (AWS ParallelCluster and Azure CycleCloud) are orchestrated on demand to run stability simulations in parallel, while graph databases (Azure Digital Twins and Amazon Neptune) maintain network-scale context. The DT scales not only across compute resources but also across geographic regions, enabling operators to assess and mitigate cascading instabilities in near-real-time.

These examples underscore that scalability is not merely about adding more resources, but about coordinating edge, cloud, and workflow orchestration services so that DTs can adapt elastically to rapidly shifting data volumes, computational loads, and operational needs. Table 14.3 summarizes these placement choices based on computation intensity and latency.

14.4.3 Cloud-native scalability mechanisms and resource management

Cloud platforms offer a comprehensive array of native processes and services aimed at enhancing scalability and optimizing resource management for high-demand applications such as power grid DTs. Effectively using these is essential for constructing resilient and cost-effective digital transformation solutions.

Key cloud-native scalability mechanisms include:

- *Elasticity*: This refers to the capacity to obtain computing resources (such as VMs, containers, or serverless operations) when necessary and relinquish them when they are no longer needed. This means that the DT can allocate resources for extensive simulation executions or AI model training, then reduce them to optimize expenses.
- *Auto-scaling*: Most cloud platforms include auto-scaling functionalities that automatically modify the quantity of active computing resources according to established parameters (e.g., CPU utilization and queue length) or

predetermined timetables. During times of elevated data ingestion from the grid or when numerous users execute intricate queries against the data terminal, the system can autonomously scale out to maintain performance and scale in during less active periods. Predictive auto-scaling policies, potentially guided by the DT's predictive capabilities, can foresee load surges and proactively modify resources (Wang et al., 2019a).

- *Load balancing*: Load balancers allocate incoming network traffic (e.g., data streams from IoT devices and user requests to digital transformation applications) among various backend resources (e.g., server instances and containers). This mitigates the risk of any singular resource becoming a bottleneck, augments responsiveness, and improves fault tolerance (Municio et al., 2019).
- *Serverless computing*: Through AWS Lambda, Azure Functions, and Google Cloud Functions, serverless computing enables code to run in response to events, without requiring manual server management. The platform automatically controls the size of these functions in response to the number of incoming events (Baldini et al., 2017). It fits well with event-driven processes used in DT architecture, enabling both raw data transformation and the generation of alarm signals.

Along with the other key mechanisms, managing resources is gaining greater importance. The new deep reinforcement learning technique with Dynamic Twin assistance (Luong et al., 2019) is made to support elements of the Dynamic Twin system and allocate resources in edge computing networks efficiently. By employing decision trees, these algorithms are better able to adapt to variations in edge node workloads and network performance, which enhances choices about the placement of microservices or the distribution of bandwidth, resulting in quicker completion times and lower energy consumption (Ren et al., 2019). Additionally, by combining learned insights from data with models to optimize the network's resource allocation, DTs help numerous sensors in big IoT networks share resources for on-jdemand collaborative sensing.

Furthermore, the construction of highly adaptive, self-regulating systems is made possible by the integration of AI and ML into DT's operational framework (Madni et al., 2019). In these situations, the DT, augmented with real-time data and AI insights, can mimic the physical power grid and optimize its cloud resource usage and resource allocation within the physical grid. This trend indicates a transition from manual or basic rule-based scaling to increasingly automated and intelligent resource management, crucial for managing the dynamic and frequently unpredictable workloads linked to real-time power grid operations and intricate DT simulations.

14.4.4 Strategies for managing large-scale simulations and data throughput

The simulation functionalities of DTs are fundamental to their worth, enabling the examination of "what-if" situations, forecasting future conditions, and enhancing designs and operations. Power grid simulations, particularly at scale, can be exceedingly demanding and require the management of substantial data flows.

Managing large-scale simulations in power grid DTs presents unique challenges owing to the interdisciplinary characteristics of the models (encompassing electrical, thermal, market, and communication dimensions), the diverse time scales involved (ranging from microseconds for electromagnetic transients to years for planning analyses), and the variability of the simulation models and tools employed. Co-simulation techniques, involving the interaction of numerous specialized simulators, and co-modeling approaches are frequently essential to accurately represent the many interdependencies within the power system (Portante et al., 2017). Cloud-based HPC services, including AWS Batch, AWS ParallelCluster, and Google Cloud's HPC solutions, offer the requisite on demand computational power to execute extensive, distributed simulations without necessitating substantial initial investment in on-premises HPC clusters (Hong et al., 2017).

In parallel, real-time DTs for operational monitoring demand robust data ingestion pipelines capable of processing high-velocity and high-volume data. To accept sensor and telemetry data from IoT devices or grid monitoring systems, cloud providers provide scalable data-streaming services (such as AWS Kinesis, Azure Event Hubs, and Google Cloud Pub/Sub). DTs can accomplish millisecond-level anomaly detection, state estimation, and situational awareness by combining these pipelines with real-time analytics services like AWS Kinesis Data Analytics or Azure Stream Analytics (Cao et al., 2020a; Li et al., 2018; Vögel et al., 2018).

Furthermore, orchestration frameworks like Apache Airflow or AWS Step Functions oversee the workflow across diverse data pipelines and simulations, guaranteeing automatic job coordination, scalability, and fault tolerance (Li et al., 2023). By ensuring that the latency-sensitive stream analytics and the computationally demanding batch simulations work together harmoniously, the orchestration enables DTs to deliver actionable insights over a variety of timescales.

14.4.4.1 Real-world scenario: Nationwide cascading failure analysis

When dealing with situations that are extremely complicated and large in scope, such as modeling a nationwide cascading failure study, the necessity of these cloud-based HPC services becomes evident. Following a major unforeseen contingency (e.g., the sudden loss of a critical transmission corridor), system operators must understand the potential for a cascading failure across the entire national grid. Simulating this requires running thousands of nonlinear, differential-algebraic power flow and dynamics simulations under different initial conditions and mitigation strategies, a task that is computationally prohibitive for on-premise clusters sized for average workloads. In practice, a DT would leverage AWS ParallelCluster or Google Cloud's HPC Toolkit to automatically provision a massive, purpose-built cluster of high-performance compute instances in minutes. An orchestration tool like AWS Batch would then submit a job array of thousands of concurrent simulations. Each simulation tests a different "what-if" scenario (e.g., "What-if we shed load at substation X within 200ms?").

The cloud's elasticity allows this to be done in a fraction of the time and cost of traditional solutions. A utility could run 10,000 simulations in parallel, completing a full national risk assessment in an hour. An on-premise cluster would be overwhelmed by this task, causing critical delays during a crisis. As results stream in, they are aggregated and analyzed using cloud-native data analytics tools. The DT synthesizes the outcomes to identify the most probable failure pathways and the most effective mitigation strategies, providing operators with a data-driven playbook for preventing a large-scale blackout.

The same elastic, on demand HPC infrastructure can also be leveraged for optimizing grid resilience against extreme weather events (e.g., hurricanes or heatwaves), where operators must explore thousands of contingency scenarios involving infrastructure damage, fluctuating DER availability, and changing demand patterns within hours.

14.4.4.2 Real-world scenario: Real-time grid stability assessment in high-distributed energy resources feeders

Another critical application where cloud services become indispensable is real-time stability assessment in feeders with high penetration of DERs. Such feeders are prone to rapid voltage fluctuations, reverse power flows, and oscillatory instabilities that must be detected and mitigated within seconds. A DT designed for this purpose must efficiently manage high-velocity data streams. This is achieved by using cloud services like Google Cloud Pub/Sub for resilient data ingestion from IoT devices and smart sensors, and managed stream processing services like AWS Kinesis, Azure Stream Analytics, or Google Cloud Dataflow to build and scale real-time data pipelines for instant processing, anomaly detection, and continuous model updating.

The architecture would combine these data pipelines with edge computing for ultra-low-latency local control and cloud services for large-scale analysis. Time-series databases (e.g., Amazon Timestream) store high-frequency sensor data from PMUs and DER inverters, while a graph database (e.g., Amazon Neptune) represents the feeder topology. Cloud-based AI/ML services such as AWS SageMaker or Azure Machine Learning train and deploy predictive models to identify early signs of instability.

The orchestration of these components is crucial. The entire process, including data intake, DT state updates, ML inference, and corrective control signal triggering, is coordinated by a workflow engine like AWS Step Functions or Amazon-managed workflows for Apache Airflow. With actionable data in a matter of seconds, the DT essentially serves as an early-warning system and a decision-support tool for operators.

14.4.4.3 Synthesis: Orchestration for scale and insight

Beyond specific tasks, developing large-scale power grid DT tools requires effectively managing intricate simulations and data streams. It calls for controlling how many models and data sources interact, maintaining consistency, coordinating work, resolving issues, and making the most use of cloud resources.

The necessity for this effective orchestration is highlighted by the intricacy of co-simulation, which involves data transmission and execution synchronization among multiple separate models. Without it, even a wealth of cloud resources could lead to an ineffective, unmanaged DT system that is unable to provide timely insights. The scenarios described earlier demonstrate how cloud-native services for HPC, data-streaming, and workflow orchestration are strategically combined to overcome these challenges, enabling utilities to perform critical analyses, from national failure simulations to real-time stability assessment, that were previously impractical.

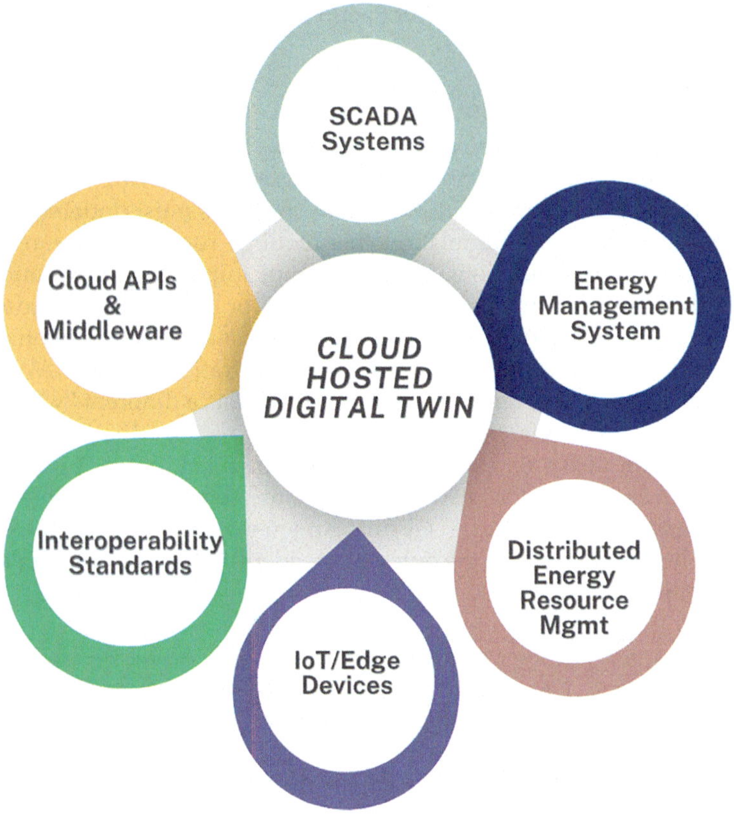

FIGURE 14.4 Incorporation of cloud-based DTs in power system operations: Integration landscape of cloud-hosted DTs within modern power grid operations. The central DT is connected with critical grid components and supported by communication standards and middleware to enable seamless data flow and control. *DT*: Digital Twin.

14.5 Incorporation of cloud-based Digital Twins in power system operations

For a cloud-based DT to provide significant value, it must smoothly integrate with the current operational procedures and technological architecture employed by the power sector. The creation of resilient interfaces with OT systems, including SCADA and EMS, is achieved by the implementation of data exchange standards and clearly defined application programming interfaces (APIs) in this technological methodology, as illustrated in Fig. 14.4.

14.5.1 Interfacing with operational technology

Contemporary power system operations depend significantly on a collection of OT systems:

- *SCADA systems*: For real-time monitoring and control of grid assets (circuit breakers and transformers). The DT ingests SCADA data for synchronization and can feed insights back to operator displays.
- *EMS*: For advanced network analysis (state estimation and optimal power flow). The DT provides enhanced forecasts and scenario modeling to inform EMS decisions.
- *WAMS*: For a dynamic, high-resolution view of grid stability. The DT consumes high-speed PMU data from WAMS for real-time stability assessment.
- *Distributed energy resource management systems (DERMS):* For managing distributed energy resources. The DT can optimize setpoints for DERs, which are executed via the DERMS.

The DT ingests real-time data and contextual information (topology and setpoints) from these systems and provides advanced analytical insights, predictions, and optimization recommendations, primarily serving as a decision-support layer that enhances, rather than replaces, existing OT. A cloud-based DT must integrate with these OT platforms for many reasons:

- *Data ingestion*: This process involves acquiring real-time operational data, including measurements, status indicators, and alerts, which are crucial for maintaining synchronization between the DT and the physical grid (Singh et al., 2021).

- *Contextual information*: To acquire network topology data, asset parameters, and operational setpoints that delineate the framework for DT models and simulations (Kutzner et al., 2020).
- *Providing insights*: To give operators the DT's advanced analytical insights, predictions (such as upcoming faults and load estimates), and optimization suggestions. These actions may occur via specialized DT dashboards or by integrating data into current SCADA/EMS operator displays (Sacks et al., 2020).
- *Closed-loop control (future)*: In more sophisticated contexts, the DT may offer recommendations or even automated control signals to the OT systems, necessitating elevated levels of trust, security, and reliability (Batty, 2018).

The architecture of a DT frequently reflects the necessity for OT integration. A cognitive DT for a microgrid can include an edge client placed side by side to the physical microgrid for instant data gathering from local controllers and for executing rapid control actions, while the cloud-based server element takes care of more computationally intensive operations like energy management optimization. Integrating DTs with enterprise systems, such as systems, applications, and products for plant maintenance, entails capturing IoT events from physical assets, potentially transmitted through SCADA, and converting them into actionable business activities, including the generation of maintenance notices or work orders (Mihai et al., 2022). Services such as AWS IoT SiteWise can be employed to gather and contextualize data from industrial control systems (PLCs, DCSs, and SCADA) before their utilization by advanced applications or the DT.

In the short to medium term, cloud-based DTs are expected to serve mainly as an intelligent advisory layer and decision-support system that enhances current OT systems, rather than directly supplanting them. The integration approach will focus on secure and reliable data exchange, providing advanced analytical information to human operators or the current control logic in the defined OT framework. The long-term vision of a "twin-centric digital control center architecture," in which the DT assumes a more pivotal role in real-time control, hinges on surmounting considerable technical and trust-related challenges. Although these interfacing needs set the stage, implementation in practice tends to run into difficulties, especially in bringing legacy systems aboard and conforming to standard protocols. These issues are explored in the following section.

14.5.2 Overcoming challenges with legacy system integration and standards

Integrating contemporary, cloud-based DT platforms with existing, frequently antiquated OT systems in power grids poses considerable problems. These OT systems, SCADA, EMS, IEDs, protection relays, and PLCs were never designed for cloud connectivity or open data exchange, and bridging them with modern DT architectures remains one of the most technically demanding aspects of deployment.

Key challenges include:

- *Proprietary protocols and data formats*: Many legacy systems use vendor-specific data models and proprietary communication protocols by SCADA systems, IEDs, and PLCs, making direct data extraction more difficult and calling for specialist drivers or protocol converters (Ma et al., 2019). Mapping these private semantics into standards models, even with middleware, frequently results in misalignment or information loss, necessitating careful engineering trade-offs between interoperability and fidelity.
- *Data silos*: OT systems (such as asset management databases, SCADA, and protection relays) usually have compartmentalized data. It takes a lot of integration work to combine this data into an extensive DT. This problem is made worse by inconsistent update schedules and disparate data ownership regulations throughout departments.
- *Absence of standardization*: The lack of widely adopted data standards in older devices results in inconsistencies in naming conventions, measurement units, and semantics. This creates persistent issues in merging datasets from equipment of different vendors and vintages, especially when attempting real-time co-simulation.
- *Security concerns*: Integrating traditional OT with external networks or cloud platforms presents considerable cybersecurity risks. Strict segmentation can mitigate exposure, but often at the expense of limiting the DT's ability to perform high-frequency data exchange for advanced analytics (Nazir et al., 2017).
- *Insufficient sensorization*: Many grid segments lack the sensors needed to feed a high-fidelity DT. Retrofitting with PMUs or smart sensors can be costly and disruptive, forcing utilities to balance between investment in instrumentation and reliance on estimation algorithms.
- *Dynamic data environment*: As assets are replaced, systems upgraded, or new DERs added, the grid's "data environment" changes continuously. Maintaining integration adapters and semantic mappings, therefore, becomes an ongoing, resource-intensive task rather than a one-time implementation.

Addressing these difficulties requires a layered, standards-driven approach.

- **OPC UA** is widely adopted for industrial interoperability, offering secure, platform-agnostic data access between OT systems and DT platforms (Nazir et al., 2017). It provides defined methods for securely transmitting control commands or setpoints, as well as retrieving historical and real-time data. However, its information model is quite rigid and typically requires additional mediation layers to match DT-native models (like DTDL in Azure DTs), which add latency.
- **MQTT** enables lightweight publish–subscribe messaging, which is highly suitable for retrofitted devices or edge hardware with limited resources. Its simplicity is a strength for low-resource environments, but also a limitation, as MQTT alone is not highly semantic, so it needs to be used in tandem with more metadata or higher-level models to facilitate meaningful DT integration.
- **IEC CIM** (standards IEC 61970, IEC 61968, and IEC 62325) provides a standardized methodology for representing grid assets and operations (Ascher & Kondzialka, 2018). It enables semantic consistency across SCADA, GIS, EMS, and asset management systems, supporting modular DT applications like state estimation, outage prediction, or asset lifecycle management. Yet, CIM's complexity makes implementation challenging: integrating models from different vendors can lead to conflicts (e.g., multiple identifiers for the same resource), and maintaining large CIM repositories requires specialized expertise (Kirpes et al., 2019). RDF and namespace management can mitigate these issues, but at the cost of additional infrastructure overhead.

Middleware and edge gateways (e.g., AWS IoT Greengrass and Azure IoT edge) are often deployed in substations or control centers to interface with legacy devices. These perform local data filtering and transformation, translate proprietary protocols into OPC UA or MQTT, and securely forward normalized data to the cloud DT (Nazir et al., 2017). While effective, this approach still requires constant upkeep, custom drivers must be updated for each firmware change, and edge deployments introduce another potential point of failure in the data pipeline.

API management adds the final layer of interoperability by exposing DT functionality securely to applications, operators, and third parties. Cloud-native tools like Amazon API Gateway, Azure API Management, and Google Apigee provide mechanisms for:

- Secure access control (authentication, authorization, and throttling).
- Monitoring and analytics for performance optimization.
- Developer portals to enable internal/external use of DT data and services.
- Lifecycle management of APIs and backward compatibility (Mathijssen et al., 2020).

In practice, APIs allow DTs to move beyond internal utility use cases, e.g., enabling mobile apps for field engineers to query transformer health, or secure third-party access for market operators. However, stricter API governance can reduce agility, and utilities must balance openness with resilience against cyberattacks.

In conclusion, it is not a case of "adopting standards," but rather a case of managing trade-offs: fidelity over interoperability, security over openness, and standardization over customization. This multilayered approach, standards (OPC UA, MQTT, and IM), edge middleware, and managed APIs, underlie scalable, interoperable DT installations in actual power grids. While these integration strategies and standards overcome most technical impediments, their performance is most effectively illustrated through actual implementations. The following sections contain case studies where these principles have been used in practice.

14.5.3 Illustrative case studies: Real-world implementations and lessons learned

Case studies are an essential link between theoretical integration strategies and their operational impact. Analysis of concrete applications of DT in the energy area and related complex sectors, together with pilot projects, offers valuable insights into their advantages, difficulties, and gained experiences.

- Cognitive DT for microgrid optimization (Singapore) is an important case study on the development of a cognitive DT for a Singaporean pragmatic microgrid. With real-time control, it employs a hybrid architecture with a client-side edge component alongside the physical microgrid and a cloud server executing computationally expensive algorithms for energy management and optimization at operational expenses within 0.24% of the theoretical optimum (Sivaneasan et al., 2025). The project demonstrated the effectiveness of the DT for real-time optimization. This example highlights the rational advantages of a hybrid edge-cloud DT structure for enhancing microgrid resilience and efficiency.
- National grid, the ESO's virtual energy system, known as VirtualES, which is focused on the UK. The aim is to build an interconnected system of DTs that mirrors the total energy system across Great Britain (Fuller et al., 2020). The goal of VirtualES is to support safe and secure sharing of data among organizations, to improve scenario creation

and to help achieve decarbonization aims through informed decision-making. Building a common structure and exploring various aspects of social and technical factors are necessary to support interoperability among DTs from numerous stakeholders. The project points out that extensive digital transformations now involve many stakeholders and that strong data governance rules are needed.

14.6 Data storage and management strategies in the cloud

The effectiveness of a power grid DT is essentially reliant on the data it utilizes and produces. An effectively crafted data storage and management approach is essential, catering to the distinctive attributes of power system data in a cloud context.

14.6.1 Data ecosystem for power grid Digital Twins: Volume, velocity, variety, veracity

The data ecosystem underpinning a power grid DT is defined by the "4 V's" of big data:

- *Volume*: Power grids produce a substantial amount of data. This encompasses ongoing data streams from potentially millions of smart meters, high-frequency measurements from thousands of sensors (PMUs, IEDs on lines and substations), operational data from SCADA and EMS systems, historical maintenance records, asset information, geospatial data (GIS), weather forecasts and historical weather data, electricity market transactions, and results from numerous simulations run. This data can rapidly grow to terabytes and petabytes, necessitating scalable storage systems.
- *Velocity*: A substantial amount of this data is received in real-time or near-real-time streams. For instance, PMUs can deliver measurements several times per second. The DT system must be able to collect, process, and respond to this high-velocity data to ensure an accurate and timely depiction of the grid's status.
- *Variety*: The data is presented in many formats. This type of information encompasses structured data (e.g., relational database records from asset management systems, SCADA point values), semistructured data (e.g., JSON or XML messages from APIs, configuration files), and unstructured data (e.g., text from maintenance logs, weather satellite imagery, and video feeds from substation cameras). The data management plan must address this diversity.
- *Veracity*: This pertains to the precision, dependability, and credibility of the data. The efficacy of analytics and ML models (e.g., decision trees, random forests, neural networks, and Bayesian models) is contingent upon the quality of the input data. Ensuring data authenticity entails methods for data cleansing, validation, managing missing values, and detecting and rectifying mistakes from defective sensors or inconsistent sources. This task represents a substantial ongoing challenge in power grid DTs, since inconsistent or erroneous data can result in defective models, inaccurate predictions, and potentially harmful operational decisions.

Fig. 14.5 represents the core data ecosystem supporting power grid DTs. The 4 V's characterize the complexities of big data handling in DT systems, highlighting the need for scalable storage, real-time analytics, heterogeneous data integration, and rigorous data validation.

DTs necessitate stringent data management during their complete lifecycle. This includes storage, data governance, designing data architecture, data modeling (with a focus on semantic consistency), data integration from diverse sources, assuring interoperability, managing data quality and uncertainty, effective data visualization, and strong data security. The data is crucial for constructing and perpetually refining the virtual models, establishing and sustaining the cyber-physical linkages, and facilitating the intelligent operations and services offered by the DT. The quality of this data is essential; a broad, high-quality database is necessary for the success of any digital transformation program.

14.6.2 Cloud-based data storage solutions

A power grid DT's data strategy demands a polyglot persistence approach, employing specialized cloud storage solutions tailored to distinct data types and access patterns inherent to grid operations.

14.6.2.1 Data lakes for heterogeneous data

Data lakes are the foundational storage layer for a DT, essential for consolidating the vast and heterogeneous data streams generated by the grid. This "schema-on-read" adaptability is optimal for power grid DTs, which must

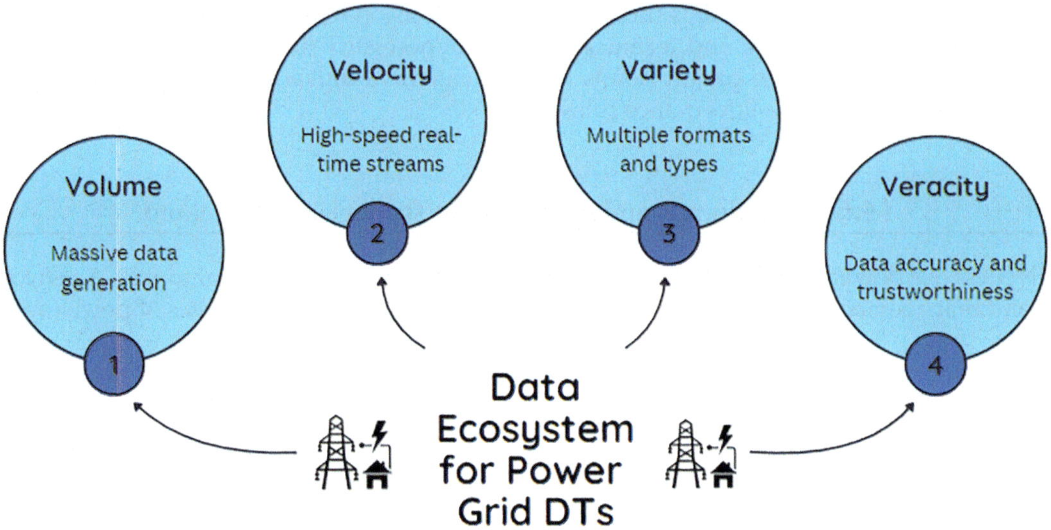

FIGURE 14.5 Data ecosystem for power grid DTs. The four V's of big data in power grid DTs. *DT*: Digital Twin.

assimilate data from diverse sources and in several formats. Cloud providers deliver scalable object storage solutions such as Amazon S3, Azure Data Lake Storage, and Google Cloud Storage, which often constitute the core of a data lake (Ehwerhemuepha et al., 2020).

Within a digital transformation framework, the data lake functions as the principal store for all unrefined data gathered from the grid, including sensor readings, SCADA logs, meteorological feeds, and GIS files, in addition to processed data, simulation results, and historical archives. This unified repository enables many analytical tasks, including exploratory data science, AI model training, batch reporting, and long-term trend analysis. Traditional cloud data lake systems consist of various services operating collaboratively. An AWS data lake may utilize Amazon S3 for the storage of raw and staged data, AWS Glue for extract, transform, load processes, and the management of a data catalog (which supplies metadata and schema details), and Amazon Athena for serverless querying of data directly in S3 employing standard SQL. A Google Cloud data lake architecture often utilizes Google Cloud Storage for data durability, services such as Cloud Dataflow or Pub/Sub for data ingestion, Cloud Dataproc or Dataflow for data processing, and BigQuery for advanced analytics (Saputra et al., 2020).

Although data lakes serve as a fundamental storage layer, they frequently do not constitute the exclusive data solution for a power grid DT. Specialized databases are frequently employed alongside the data lake for certain data types or access patterns that necessitate high-performance. The data lake serves as the singular source of "unrefined truth", with selected subsets or processed data sent to other optimized data repositories for particular digital transformation activities, such as real-time monitoring or complex relationship analysis.

14.6.2.2 Time-series databases for operational data

While data lakes store everything, time-series databases (TSDBs) are optimized for the core operational heartbeat of the grid: massive volumes of time-stamped sensor data from PMUs, smart meters, and IEDs. They are a fundamental element of the data management strategy for power grid DTs (Liu et al., 2019).

In contrast to general-purpose databases, TSDBs provide enhanced performance for standard time-series tasks, including trend identification, moving average calculations, windowed aggregations, and anomaly detection within designated durations. They frequently incorporate attributes such as optimized compression algorithms specifically designed for time-series data, automated data lifecycle management (e.g., down-sampling older, high-precision data into lower-resolution aggregates for prolonged storage and establishing data retention policies), and integrated analytical functions explicitly intended for time-series analysis.

Notable instances of cloud-based or cloud-compatible TSDBs include Amazon Timestream, a fully managed service that can ingest tens of gigabytes of time-series data per minute and execute SQL queries across terabytes of data in a few seconds. InfluxDB is a widely recognized open-source time-series database (also offered as a cloud service) noted for its exceptional performance and capability for millisecond precision, which might be vital in specific power system study contexts. Microsoft Azure provides Azure time-series insights for comprehensive IoT analytics, encompassing time-series data storage and visualization (Liu et al., 2019). A high-performance, specialized time-series database is essential for effectively replicating the dynamic state of the power grid and enabling swift analysis of operational data for monitoring, control, and diagnostics.

14.6.2.3 Graph databases for asset modeling and network topology

Graph databases are uniquely suited to model the power grid's fundamental nature as a network of interconnected assets. They are especially adept at modeling and querying these heavy interrelationships, treating connections as primary elements alongside the entities they link, which is vital for DT applications that require understanding connectivity and dependencies (Fabregat et al., 2018).

In a power grid DT, a graph database may be utilized to:

- Construct a comprehensive electrical network structure, encompassing the interconnections among all assets.
- Illustrate the interdependencies of various components (e.g., a transformer relies on its cooling system).
- Chart the information flow and control signal pathways.
- Catalog metadata pertaining to assets and their interrelations.

This facilitates efficient querying for tasks like

- Tracking the dissemination of a fault throughout the network.
- Determining all customers or essential loads impacted by a particular equipment failure.
- Assessing the effect of a component malfunction on system stability.
- Visualizing subnetworks or comprehending intricate interdependencies.

Cloud providers deliver managed graph database services, including Amazon Neptune, which accommodates widely used graph models such as Property Graph, queried via Apache TinkerPop Gremlin, and RDF, queried using SPARQL, capable of scaling to accommodate billions of relationships and processing over 100,000 queries per second, and Azure Cosmos DB, which features an API for Apache Gremlin, capitalizing on Cosmos DB's global distribution, elastic scalability, and multimodel functionalities (Angles et al., 2018). Certain DT solutions, such as AWS IoT TwinMaker, integrate the notion of a knowledge graph, which is a graph-based depiction of the system's components and their interrelations, frequently generated or enhanced automatically from interconnected data sources. A DT designed to offer a comprehensive, system-wide comprehension of the power grid and facilitate advanced analytics, such as root cause analysis or cascade failure impact assessment, can effectively use a graph database to model and navigate the grid's structure and interdependencies.

14.6.3 Real-time data processing and stream analytics frameworks

A DT needs to process and evaluate data in real-time in order to be more than just a historical record. The engines that convert raw data streams into instantaneous, useful insights for grid operators are known as stream processing frameworks.

These frameworks are designed to process data in transit with high-throughput and low-latency. They make it easier to quickly analyze incoming data, enabling operations like

- *Real-time anomaly detection*: Identifying unusual trends or deviations from normal operational conditions that could indicate a security incident or impending issue.
- *Real-time computations*: Utilizing raw sensor data to compute derived metrics or key performance indicators.
- *Event correlation*: Combining multiple facts or events from various sources to understand a more complex situation.
- *Initiating notifications and automated responses*: Triggering alerts to operators or automated control actions predicated on real-time conditions.

Various open-source and cloud-native stream processing frameworks exist:

- **Apache Kafka** is frequently employed as a distributed, high-throughput, fault-tolerant messaging system that serves as a central intake layer or buffer for data streams from several sources. It separates data producers (e.g., sensors) from data consumers (e.g., analytics apps or the DT itself) (Damjanovic-Behrendt & Behrendt, 2019).
- **Apache Flink** is a robust open-source framework designed for stateful computations on both unbounded and constrained data streams. It excels in complex event processing, windowed aggregations, and applications necessitating exactly-once processing semantics, rendering it appropriate for advanced real-time analytics within a DT.

- **Apache Spark** Streaming offers scalable, high-throughput, and fault-tolerant stream processing for live data streams, generally handling data in micro-batches.
- Cloud providers deliver managed services that facilitate the deployment and operation of stream processing pipelines (Mampage et al., 2022). This encompasses AWS Kinesis (for the collection, processing, and analysis of real-time streaming data), Azure Stream Analytics (an event processing engine capable of analyzing large-scale streaming data from IoT devices and other sources), and Google Cloud Dataflow (a unified, serverless, and fully managed service for stream and batch data processing). Dataflow pipelines often use Google Cloud Pub/Sub as their ingestion foundation (Cravero et al., 2022).

The dynamic nature of a DT requires the integration of real-time stream processing abilities. This allows the DT to maintain a current representation of the power grid and react to, and in certain cases forecast, physical system events very instantaneously. This ability to facilitate dynamic optimization, quick fault response, and improved situational awareness is essential for shifting grid management from a reactive to a predictive and proactive approach.

To migrate from theoretical AI models to practical deployment, Digital Twins must embrace a rigorous machine learning operations (MLOps) lifecycle. Data versioning in the cloud data lake is the first step in this procedure, which guarantees the traceability of the training datasets. A centralized feature store standardizes feature engineering, enabling models (such as load forecasting models) to be trained on past data. Models are released to the edge using a canary deployment technique after they have been validated, first managing a small proportion of devices to track performance. The cycle completes with drift detection, where the DT analyzes forecast accuracy against real-time grid behavior, initiating automatic retraining pipelines in the cloud if the model weakens.

14.6.4 Data governance, quality, and privacy in cloud environments

As power grid DTs get more complex and data-centric, robust data governance, meticulous data quality assurance, and extensive privacy protection measures are essential, particularly when utilizing cloud settings. Data governance is an organized strategy for controlling data across its entire lifecycle, from creation and acquisition to use, storage, and eventual destruction. Power grid DTs rely on a robust data governance framework to handle a variety of sensitive operational data, vital infrastructure data, and potentially consumer-specific data. This framework ought to specify in detail:

- *Data ownership and stewardship*: This entails designating who is in charge of managing particular datasets and allocating accountability for them.
- *Data policies*: Establishing internal guidelines for the gathering, storing, accessing, processing, sharing, and retaining of data.
- *Access controls*: Implementing strategies like role-based access control (RBAC) to ensure that, in accordance with the principle of least privilege, only authorized individuals and systems can access designated data (Mamun et al., 2022).
- *Compliance*: Observing pertinent industry norms and external rules, such as the California Consumer Privacy Act, the General Data Protection Regulation for personal data, and specific energy sector requirements (Cruz et al., 2022). As suggested in projects like National Grid ESO's VirtualES, the growing trend of linked DTs and data sharing among many stakeholders,e.g., transmission system operators, distribution system operators, market operators, consumers, and third-party service providers, makes cross-organizational data governance and customer consent mechanisms ever more important (Mittelstadt et al., 2018).
- *Data quality*: This is essential to any DT's usefulness and credibility. A decision tree built from faulty, insufficient, or inconsistent data can produce flawed models and untrustworthy forecasts, which may lead to incorrect or harmful operational decisions. This is why the adage "garbage in, garbage out" is particularly pertinent. To ensure data integrity, you must:
- *Standardization*: Establishing defined data models and formats, such as the IEC common information model (CIM), to ensure that data from different power system sources are semantically consistent and interoperable.
- *Validation and cleaning*: Implementing procedures to check incoming data for timeliness, correctness, completeness, consistency, and originality, and to correct or update inaccurate data.
- *Data provenance*: Monitoring the origin and development of data to comprehend its ancestry and guarantee traceability. A large, high-quality database is a prerequisite for the legitimacy and success of DTs.

Particularly when DT handles data pertaining to potentially individual users or exposes confidential operating aspects of critical infrastructure, data privacy becomes a significant concern. The importance of privacy protection techniques is growing.

- *Anonymization and pseudonymization*: Personally identifiable information should be removed or covered over before datasets are used for in-depth analysis or model training.
- *Differential privacy*: Introducing statistical noise into query results or datasets to conceal unique records while permitting important group analysis.
- *Data minimization*: Collecting and keeping only the information necessary to achieve the stated goal.
- *Secure data handling*: Employing robust encryption for data both at rest and in transit while guaranteeing secure storage and processing environments on the cloud.

Not only is good data governance, which comprises thorough privacy protections, tight security policies (expounded in Section 6), and rigorous quality assurance,a compliance need. A fundamental requirement is building confidence in power grid DTs among stakeholders: operators, legislators, and the public. Data sharing and cooperation needed to realize the full revolutionary possibilities of digital technology in the energy sector depend on this confidence.

14.7 Cybersecurity for cloud-hosted Digital Twins in power grids

Especially in cloud situations, the increasing complexity and interconnection of power grid DTs create a variety of cybersecurity issues. The safe and reliable operation of the physical power system depends on safeguarding these fundamental virtual assets and their data.

14.7.1 The changing threat landscape for smart grid Digital Twins

Because smart grids combine digital technologies with extensive communication networks, they naturally expand the attack surface relative to conventional power systems. Cyberattacks on smart grids can have major consequences, including operational disruptions, large-scale power outages, significant financial losses, and even public safety hazards (Tariq & Poor, 2018). Because DT is a full, data-intensive virtual grid with active connections to functional systems, it is a particularly appealing target for bad actors (Tariq et al., 2021).

Specific cybersecurity issues that are pertinent to DTs include:

- *Infrastructure vulnerabilities*: One may gain unauthorized access to the DT or disrupt its functioning by taking advantage of flaws in the core IT and OT infrastructures, such as network systems, IoT devices, edge computing nodes, or the cloud platform. Industrial assets frequently make use of outdated communication technology with possibly insufficient security safeguards.
- *Data manipulation*: The integrity of data communicated to or stored on the DT may be compromised by attackers. Altered sensor readings or operational parameters can cause the DT to falsely depict the state of the grid, either hiding sinister behavior inside the physical system or producing false and negative operating decisions.
- *Expanded attack surface*: The necessity to connect multiple sensors, actuators, and systems to supply the DT greatly increases the probable area of risk. Every linked component, data link, or API could provide a possible access point for attackers.
- *Denial of service (DoS/DDoS)*: DTs rely on computing resources and continuous data feeds. DoS, or even Distributed DoS (DDoS), attacks may flood the DT's communication channels or processing capabilities, thereby incapacitating it and therefore impeding operators' capacity to monitor the grid or make informed decisions.
- *Compromise of DT models or logic*: Advanced attackers might concentrate on the basic models, algorithms, or simulation engines built into the DT. Including backdoors or vulnerabilities in these parts could silently weaken DT's dependability or enable clandestine manipulation.

One important issue is that, while DTs are typically regarded as tools to enhance the cybersecurity posture of the physical power grid, through capabilities such as anomaly detection and attack simulation, the security of the DT itself remains underexplored. This emphasizes a major shortfall: DT is not only a defensive tool but also a necessary asset requiring a strong and complete protection plan. Because of its dual function, cybersecurity rules must address both protecting DT against breaches and utilizing its attributes to enhance the security architecture of the whole power system.

14.7.2 Securing the cloud infrastructure (identity and access management, network security, and encryption)

The foundational cloud architecture must be strictly protected to maintain the security of a cloud-hosted power grid DT. This entails using the security features and best practices offered by the selected CSP in accordance with the concept of shared responsibility.

- *IAM*: A strong IAM is essential. This includes
- *AWS*: By carefully granting users and services accessing DT resources specific permissions, AWS IAM ensures the least privilege access. It is imperative to apply multifactor authentication (MFA) to every account. IAM is used in AWS IoT TwinMaker for access control, thereby ensuring data security
- *Azure*: For identity services, Azure Active Directory (Azure AD), Azure RBAC, and MFA implementation. A basic best practice is using Zero-Trust Architecture concepts, which call for clear validation for every access request
- *Google Cloud*: Managing resource access with Google Cloud IAM. By imposing additional access limitations depending on user identification, IP address, and device trust, Virtual private clouds (VPC) service controls help to create security perimeters around sensitive DT data and services
- *Network security*: Isolating DT resources and regulating network traffic is essential
- *VPCs/virtual networks (VNets)*: Implementing DT components within segregated virtual networks (e.g., Amazon VPC, Azure VNet, and Google Cloud VPC) to regulate inbound and outbound traffic using firewalls, security groups, and network access control lists (ACLs)
- *Network segmentation*: Segmenting the network (e.g., for development, testing, production DT environments, or isolating OT-interfacing components from general analytics components) to mitigate the impact of a potential compromise. Azure enables administrators to utilize Network Segmentation Groups (subnets)
- *Secure connectivity*: This entails utilizing secure communication channels, such as virtual private networks or dedicated interconnects (e.g., AWS Direct Connect, Azure ExpressRoute, and Google Cloud Interconnect), to enable traffic between on-premises OT systems and the cloud-hosted DT, as well as among various cloud services
- *Private endpoints*: Employing private endpoints to access cloud services while safeguarding traffic from public internet exposure
- *Data encryption*: Safeguarding data both at rest and in transit
- *Encryption in transit*: All data transmitted between sensors/OT devices and the cloud, among cloud services, and between the cloud and users must be encrypted utilizing robust protocols such as transport layer security (TLS) 1.2 or above (HTTPS)
- *Encryption at rest*: Data retained in cloud storage services (e.g., object storage, such as S3 or Azure Blob Storage, databases, and data lakes) must be protected with strong encryption techniques. CSPs frequently offer default server-side encryption along with options for customer-managed encryption keys. Utilize services such as Azure Key Vault or AWS Key Management Service (KMS) for the secure management of cryptographic keys
- *Logging and monitoring*:
- *Activity logging*: Implementing extensive logging of API requests, user actions, and system events with services such as AWS CloudTrail, Azure Monitor, or Google Cloud's operations suite (formerly known as Stack driver)
- *Security monitoring*: Establishing ongoing security surveillance and threat identification utilizing CSP-native tools (e.g., Amazon Guard Duty, Microsoft Defender for Cloud, and Google Security Command Center) or external SIEM/SOAR solutions

Fundamental cybersecurity practices, including routine security evaluations, vulnerability management, prompt system patching (where appropriate under the shared responsibility model), and the maintenance of comprehensive incident response plans, are critical. The shared responsibility model is fundamental to cloud security: the CSP is accountable for the security of the cloud infrastructure and core services, while the customer is responsible for securing their use of the cloud, which includes configuring services securely, managing identities and access, appropriately encrypting data, and implementing essential network controls. A comprehensive understanding of the specific security tools and best practices provided by the selected CSP is essential for safeguarding a power grid DT.

TABLE 14.4 Security measures in the hybrid Digital Twin stack.

Layer	Security risk	Recommended control	Standard/protocol
Device / sensor	Spoofing and physical tampering	Hardware-based trust and identity certifications	**TPM 2.0, X.509 certificates**
Edge node	Execution of rogue code	Isolation of containers and signed updates	**Docker/Kubernetes benchmarks**
Communication	Man-in-the-Middle (MITM)	Encryption and mutual authentication	**mTLS (Mutual TLS 1.2+), AES-256**
Cloud data	Data leak / Unauthorized access	Role-based access control (RBAC)	**OAuth 2.0, OIDC, and KMS**
API / application	Injection attacks and DDoS	API gateway throttling	**OWASP Top 10 Standards**

Note: *Cloud provider services are perpetually advancing. Table 14.2 provides a summary based on the available data.*

14.7.3 Application and data-level security for Digital Twins

In addition to safeguarding the cloud infrastructure, targeted security measures must be implemented at the application and data layers of the DT to ensure its integrity, confidentiality, and availability.

- *Data integrity and authenticity*: As DTs are inherently data-centric, it is essential to guarantee the integrity and authenticity of the data they utilize and produce.
- *Blockchain technology*: Blockchain technology is increasingly advocated for essential data exchanges, especially between IoT devices and DT, or among many stakeholders within a distributed DT ecosystem. Its intrinsic immutability and distributed ledger attributes can furnish a verifiable and tamper-proof record of data transactions, guaranteeing data integrity and authenticity (Ferrag et al., 2019). Smart contracts on a blockchain can automate secure data exchange and authentication procedures.
- *Verification of data sources*: Establishing protocols to authenticate data sources (e.g., sensor identities and message origins) and to guarantee that the format and content of incoming data align with anticipated criteria is essential for averting the introduction of malicious or corrupted material into the DT. Digital signatures and message authentication codes may be utilized (Gupta & Jain, 2023).
- *Data confidentiality*: Safeguarding important grid operational data and any personal information handled by the DT is imperative.
- *Encryption*: Comprehensive encryption protocols should be implemented not only for data at rest and in transit (as previously addressed about infrastructure security) but also possibly for data within the DT application if it is processed in untrusted environments.
- *Lightweight cryptography*: Resource-efficient methods are essential for IoT devices at the grid edge, ensuring secure communication without overwhelming limited processing power (Turner et al., 2023).
- Access control: The DT application should implement fine-grained, attribute-based access control or RBAC techniques to ensure that users and services have access solely to the data and functions for which they are permitted (El Majdoubi et al., 2022).
- *Privacy-enhancing technologies*: Techniques such as differential privacy and homomorphic encryption can be utilized to analyze sensitive information while maintaining the secrecy of individual data points, especially during the training of ML models or the dissemination of aggregated insights (El Majdoubi et al., 2022).
- *Secure communication channels*: All communication channels associated with the DT—between sensors and edge gateways, edge gateways and the cloud, various microservices within the DT, and between the DT and user interfaces or external systems—must be fortified using robust, mutually authenticated encryption protocols such as TLS or datagram TLS (DTLS) (López et al., 2018).
- *Best practices for application security*: The software components of the DT must be created in accordance with secure coding standards, subjected to regular security assessments (including penetration testing and vulnerability scanning), and equipped with protocols for secure updates and patch management. API security is especially critical when the DT exposes functionalities to external systems (see Section 5.3).

Prioritizing data-centric security is essential, as any compromise of the data supporting the DT renders its outputs inaccurate, which may result in erroneous operational decisions with grave repercussions for the physical power system. The distinctive issues at the data source, especially the multitude of potentially susceptible IoT devices, necessitate focused attention.

A layered protection system is necessary since threats can be found at any level of the stack, from the physical sensor to the API. Table 14.4 provides an overview of various cross-layer controls.

14.8 Conclusion

Cloud computing has emerged as the ubiquitous backbone for scalable and smart DTs in contemporary power systems. With its provision of on demand scalability, high-end analytics, and strong data management, cloud-based DTs are revolutionizing grid management from reactive to proactive and sustainable. Although this chapter has presented the underpinnings of architectures, integration practices, and data factors, future development directions for DTs will be governed by a number of key dimensions.

A key direction lies in advancing autonomy through deeper AI integration, where DTs evolve from decision-support to predictive and prescriptive systems capable of self-optimization and self-healing grid operations. Equally important is the imperative of interoperability and standardization, since the vision of interconnected "DTs of DTs" depends on shared data models, open APIs, and widely accepted frameworks like CIM and OpenFMB. At the same time, cybersecurity demands will intensify, requiring real-time anomaly detection, model integrity verification, and zero-trust architectures to safeguard closed-loop decision-making.

The rise of hybrid edge-cloud architectures will also be pivotal, with latency-sensitive tasks handled at the edge while the cloud supports large-scale coordination, optimization, and long-term learning. This hybrid paradigm underscores the need for seamless orchestration of distributed intelligence. Finally, several research opportunities remain open, including lightweight AI for edge devices, federated learning for privacy-preserving collaboration, sim-to-real transfer techniques, and frameworks for assessing DT maturity and trust.

In conclusion, cloud-enabled DTs offer a flexible and future-ready pathway to managing grid complexity and advancing decarbonization. Realizing their full potential will require collaboration across utilities, vendors, policy-makers, and researchers to address challenges in autonomy, interoperability, and cybersecurity, ultimately steering power systems toward resilience, intelligence, and sustainability.

References

Adnan, M., Iqbal, M. S., Siddiqi, S. J., Ahmed, I., Shah, A., Ullah, I., & Tariq, M. (2025). Modeling the future: Mathematical insights for smart grid 3.0. *Computers and Electrical Engineering, 123*. https://www.sciencedirect.com/science/journal/00457906. doi:10.1016/j.compeleceng.2025.110283.

Adnan, M., Tariq, M., Zhou, Z., & Poor, H. V. (2019). Load flow balancing and transient stability analysis in renewable integrated power grids. *International Journal of Electrical Power and Energy Systems, 104*, 744–771. doi:10.1016/j.ijepes.2018.06.037.

Ali, M., Adnan, M., & Tariq, M. (2019). Optimum control strategies for short term load forecasting in smart grids. *International Journal of Electrical Power & Energy Systems, 113*, 792–806. doi:10.1016/j.ijepes.2019.06.010.

Al Mamun, A., Azam, S., & Gritti, C. (2022). Blockchain-based electronic health records management: A comprehensive review and future research direction. *IEEE Access, 10*, 5768–5789. doi:10.1109/access.2022.3141079.

Amin, M., & Molinas, M. (2017). Understanding the origin of oscillatory phenomena observed between wind farms and HVdc systems. *IEEE Journal of Emerging and Selected Topics in Power Electronics, 5*(1), 378–392. doi:10.1109/jestpe.2016.2620378.

Angles, R., Arenas, M., Barceló, P., Hogan, A., Reutter, J., & Vrgoč, D. (2018). Foundations of modern query languages for graph databases. *ACM Computing Surveys, 50*(5). http://dl.acm.org/citation.cfm?id=J204. doi:10.1145/3104031.

Ascher, D., & Kondzialka, C. (2018). Towards model-driven CIM-based data exchange for DSOs. *Energy Informatics, 1*(S1). doi:10.1186/s42162-018-0032-4.

Baldini, I., Castro, P., Chang, K., Cheng, P., Fink, S., Ishakian, V., Mitchell, N., Muthusamy, V., Rabbah, R., Slominski, A., & Suter, P. (2017). Serverless computing: Current trends and open problems. In *Research advances in cloud computing* (pp. 1–20). Springer. http://www.springer.com/in/book/9789811050251. doi:10.1007/978-981-10-5026-8_1.

Batty, M. (2018). Digital Twins. *Environment and Planning B: Urban Analytics and City Science, 45*(5), 817–820. doi:10.1177/2399808318796416.

Cao, D. i, Hu, W., Zhao, J., Zhang, G., Zhang, B., Liu, Z., Chen, Z., & Blaabjerg, F. (2020a). Reinforcement learning and its applications in modern power and energy systems: A review. *Journal of Modern Power Systems and Clean Energy, 8*(6), 1029–1042. doi:10.35833/mpce.2020.000552.

Cao, K., Liu, Y., Meng, G., & Sun, Q. (2020b). An overview on edge computing research. *IEEE Access, 8*, 85714–85728. doi:10.1109/access.2020.2991734.

Cicioni, G., De Angelis, A., Janeiro, F. M., Ramos, P. M., & Carbone, P. (2023). Battery impedance spectroscopy embedded measurement system. *Batteries, 9*(12). www.mdpi.com/journal/batteries. doi:10.3390/batteries9120577.

Ciotta, V., Asprone, D., Manfredi, G., & Cosenza, E. (2021). Building information modelling in structural engineering: A qualitative literature review. *CivilEng, 2*(3), 765–793. doi:10.3390/civileng2030042.

Cravero, A., Pardo, S., Sepúlveda, S., & Muñoz, L. (2022). Challenges to use machine learning in agricultural big data: A systematic literature review. *Agronomy, 12*(3). www.mdpi.com/journal/agronomy/. doi:10.3390/agronomy12030748.

Damjanovic-Behrendt, V., & Behrendt, W. (2019). An open source approach to the design and implementation of Digital Twins for smart manufacturing. *International Journal of Computer Integrated Manufacturing, 32*(4-5), 366–384. doi:10.1080/0951192x.2019.1599436.

Deng, S., Zhao, H., Fang, W., Yin, J., Dustdar, S., & Zomaya, A. Y. (2020). Edge intelligence: The confluence of edge computing and artificial intelligence. *IEEE Internet of Things Journal, 7*(8), 7457–7469. http://ieeexplore.ieee.org/servlet/opac?punumber=6488907. doi:10.1109/JIOT.2020.2984887.

Ehwerhemuepha, L., Gasperino, G., Bischoff, N., Taraman, S., Chang, A., & Feaster, W. (2020). HealtheDataLab – A cloud computing solution for data science and advanced analytics in healthcare with application to predicting multi-center pediatric readmissions. *BMC Medical Informatics and Decision Making, 20*(1). doi:10.1186/s12911-020-01153-7.

El Majdoubi, D., El Bakkali, H., Sadki, S., Maqour, Z., & Leghmid, A. (2022). The systematic literature review of privacy-preserving solutions in smart healthcare environment. *Security and Communication Networks, 2022*. https://www.hindawi.com/journals/scn/. doi:10.1155/2022/5642026.

Fabregat, A., Korninger, F., Viteri, G., Sidiropoulos, K., Marin-Garcia, P., Ping, P., Wu, G., Stein, L., D'Eustachio, P., & Hermjakob, H. (2018). Reactome graph database: Efficient access to complex pathway data. *PLoS Computational Biology, 14*(1). http://www.ploscompbiol.org/article/browseVolume.action?field=volume. doi:10.1371/journal.pcbi.1005968.

Ferrag, M. A., Derdour, M., Mukherjee, M., Derhab, A., Maglaras, L., & Janicke, H. (2019). Blockchain technologies for the internet of things: Research issues and challenges. *IEEE Internet of Things Journal, 6*(2), 2188–2204. http://ieeexplore.ieee.org/servlet/opac?punumber=6488907. doi:10.1109/JIOT.2018.2882794.

Fuller, A., Fan, Z., Day, C., & Barlow, C. (2020). Digital Twin: Enabling technologies, challenges and open research. *IEEE Access, 8*, 108952–108971. doi:10.1109/access.2020.2998358.

Gandrup, J., Ali, S. M., McBeth, J., Van Der Veer, S. N., & Dixon, W. G. (2020). Remote symptom monitoring integrated into electronic health records: A systematic review. *Journal of the American Medical Informatics Association, 27*(11), 1752–1763. http://jamia.oxfordjournals.org/content/22/e1. doi:10.1093/jamia/ocaa177.

Gupta, N., & Jain, A. K. (2023). RSA based consensus algorithm for resource-constrained distributed devices. *Internet Technology Letters, 6*(6). https://onlinelibrary.wiley.com/journal/24761508. doi:10.1002/itl2.471.

Hong, C. H., Spence, I., & Nikolopoulos, D. S. (2017). FairGV: Fair and fast GPU virtualization. *IEEE Transactions on Parallel and Distributed Systems, 28*(12), 3472–3485. http://ieeexplore.ieee.org/xpl/RecentIssue.jsp?punumber=71. doi:10.1109/TPDS.2017.2717908.

Jahn, C., Weigell, J., Borremans, A., & Lepekhin, A. (2019). Smart grid – The concept, trends and stakeholders. In *International conference on digital technologies in logistics and infrastructure (ICDTLI 2019)* (pp. 497–501). Atlantis Press. doi:10.2991/icdtli-19.2019.88.

Kirpes, B., Danner, P., Basmadjian, R., Meer, H. d., & Becker, C. (2019). E-mobility systems architecture: A model-based framework for managing complexity and interoperability. *Energy Informatics, 2*(1). https://energyinformatics.springeropen.com/about. doi:10.1186/s42162-019-0072-4.

Kutzner, T., Chaturvedi, K., & Kolbe, T. H. (2020). CityGML 3.0: New functions open up new applications. *PFG - Journal of Photogrammetry, Remote Sensing and Geoinformation Science, 88*(1), 43–61. https://rd.springer.com/journal/volumesAndIssues/41064. doi:10.1007/s41064-020-00095-z.

Li, J., Cai, J., Khan, F., Rehman, A. U., Balasubramaniam, V., Sun, J., & Venu, P. (2020). A secured framework for SDN-based edge computing in IoT-enabled healthcare system. *IEEE Access, 8*, 135479–135490. http://ieeexplore.ieee.org/xpl/RecentIssue.jsp?punumber=6287639. doi:10.1109/ACCESS.2020.3011503.

Li, J., Gu, C., Xiang, Y., & Li, F. (2022). Edge-cloud computing systems for smart grid: state-of-the-art, architecture, and applications. *Journal of Modern Power Systems and Clean Energy, 10*(4), 805–817. doi:10.35833/mpce.2021.000161.

Li, S., Gerver, P., MacMillan, J., Debrunner, D., Marshall, W., & Wu, K. L. (2018). *Challenges and experiences in building an efficient apache beam runner for IBM streams: 11* (pp. 1742–1754). http://vldb.org/pvldb/11.

Li, X., Chen, T., Yuan, D., Xu, J., & Liu, X. (2023). A novel graph-based computation offloading strategy for workflow applications in mobile edge computing. *IEEE Transactions on Services Computing, 16*(2), 845–857. http://ieeexplore.ieee.org/xpl/RecentIssue.jsp?punumber=4629386. doi:10.1109/TSC.2022.3180067.

Liu, R., Yuan, J., & Huang, X. (2019). Benchmarking time series databases with IoTDB-benchmark for IoT scenarios. *arXiv*. https://arxiv.org. doi:10.48550/arXiv.1901.08304.

López, D. D., Uribe, M. B., Cely, C. S., Murgueitio, D. T., Garcia, E. G., Nespoli, P., & Mármol, F. G. (2018). Developing secure IoT services: A security-oriented review of IoT platforms. *Symmetry, 10*(12) https://res.mdpi.com/symmetry/symmetry-10-00669/article_deploy/symmetry-10-00669.pdf?filename=&attachment=1. doi:10.3390/sym10120669.

Luong, N. C., Hoang, D. T., Gong, S., Niyato, D., Wang, P., Liang, Y. C., & Kim, D. I. (2019). Applications of deep reinforcement learning in communications and networking: A survey. *IEEE Communications Surveys and Tutorials, 21*(4), 3133–3174. http://ieeexplore.ieee.org/xpl/RecentIssue.jsp?punumber=9739. doi:10.1109/COMST.2019.2916583.

Ma, Z., Schultz, M. J., Christensen, K., Værbak, M., Demazeau, Y., & Jørgensen, B. N. (2019). The application of ontologies in multi-agent systems in the energy sector: A scoping review. *Energies, 12*(16). https://www.mdpi.com/1996-1073/12/16. doi:10.3390/en12163200.

Madni, A. M., Madni, C. C., & Lucero, S. D. (2019). Leveraging digital twin technology in model-based systems engineering. *Systems, 7*(1). https://www.mdpi.com/2079-8954/7/1/7/pdf. doi:10.3390/systems7010007.

Mampage, A., Karunasekera, S., & Buyya, R. (2022). A holistic view on resource management in serverless computing environments: Taxonomy and future directions. *ACM Computing Surveys, 54*(11s), 1–36. doi:10.1145/3510412.

Mao, Y., You, C., Zhang, J., Huang, K., & Letaief, K. B. (2017). A survey on mobile edge computing: The communication perspective. *IEEE Communications Surveys and Tutorials, 19*(4), 2322–2358. http://ieeexplore.ieee.org/xpl/RecentIssue.jsp?punumber=9739. doi:10.1109/COMST.2017.2745201.

Mathijssen, M., Overeem, M., & Jansen, S. (2020). Identification of practices and capabilities in API management: A systematic literature review. *arXiv*. https://arxiv.org. doi:10.48550/arxiv.2006.10481.

Mihai, S., Yaqoob, M., Hung, D. V., Davis, W., Towakel, P., Raza, M., Karamanoglu, M., Barn, B., Shetve, D., Prasad, R. V., Venkataraman, H., Trestian, R., & Nguyen, H. X. (2022). Digital Twins: A survey on enabling technologies, challenges, trends and future prospects. *IEEE Communications Surveys and Tutorials., 24*(4), 2255–2291. http://ieeexplore.ieee.org/xpl/RecentIssue.jsp?punumber=9739. doi:10.1109/COMST.2022.3208773.

Mittelstadt, B., Benzler, J., Engelmann, L., Prainsack, B., & Vayena, E. (2018). Is there a duty to participate in digital epidemiology? *Life Sciences, Society and Policy, 14*(1). doi:10.1186/s40504-018-0074-1.

Municio, E., Daneels, G., Vučinić, M., Latré, S., Famaey, J., Tanaka, Y., Brun, K., Muraoka, K., Vilajosana, X., & Watteyne, T. (2019). Simulating 6TiSCH networks. *Transactions on Emerging Telecommunications Technologies, 30*(3). http://onlinelibrary.wiley.com/journal/10.1002/(ISSN)2161-3915. doi:10.1002/ett.3494.

Nazir, S., Patel, S., & Patel, D. (2017). Assessing and augmenting SCADA cyber security: A survey of techniques. *Computers & Security, 70*, 436–454. doi:10.1016/j.cose.2017.06.010.

O'Dwyer, E., Pan, I., Charlesworth, R., Butler, S., & Shah, N. (2020). Integration of an energy management tool and Digital Twin for coordination and control of multi-vector smart energy systems. *Sustainable Cities and Society, 62*, 102412. doi:10.1016/j.scs.2020.102412.

Patacas, J., Dawood, N., & Kassem, M. (2020). BIM for facilities management: A framework and a common data environment using open standards. *Automation in Construction, 120*. https://www.journals.elsevier.com/automation-in-construction. doi:10.1016/j.autcon.2020.103366.

Cruz, P.-De La, Rabetino, R., & Kantola, J. (2022). Unveiling the shades of partnerships for the energy transition and sustainable development: Connecting public–private partnerships and emerging hybrid schemes. *Sustainable Development, 30*(5), 1370–1386. https://doi.org/10.1002/sd.2288.

Portante, E. C., Kavicky, J. A., Craig, B. A., Talaber, L. E., & Folga, S. M. (2017). Modeling electric power and natural gas system interdependencies. *Journal of Infrastructure Systems, 23*(4). doi:10.1061/(ASCE)IS.1943-555X.0000395.

Rasheed, A., San, O., & Kvamsdal, T. (2020). Digital Twin: Values, challenges and enablers from a modeling perspective. *IEEE Access, 8*, 21980–22012. doi:10.1109/access.2020.2970143.

Ren, J., Wang, H., Hou, T., Zheng, S., & Tang, C. (2019). Federated learning-based computation offloading optimization in edge computing-supported internet of things. *IEEE Access, 7*, 69194–69201. doi:10.1109/access.2019.2919736.

Saba, I., Bukhari, M., Ullah, M., & Tariq, M. (2023). Threats, vulnerabilities, and mitigation in V2G networks. *Green Energy and Technology, 1271*, 1–30. https://www.springer.com/series/8059. doi:10.1007/978-3-031-35911-8_1.

Sacks, R., Brilakis, I., Pikas, E., Xie, H. S., & Girolami, M. (2020). Construction with Digital Twin information systems. *Data-Centric Engineering, 1*(6). doi:10.1017/dce.2020.16.

Saeed, M. H., Fangzong, W., Kalwar, B. A., & Iqbal, S. (2021). A review on microgrids' challenges perspectives. *IEEE Access., 9*, 166502–166517. http://ieeexplore.ieee.org/xpl/RecentIssue.jsp?punumber=6287639. doi:10.1109/ACCESS.2021.3135083.

Saleem, Y., Crespi, N., Rehmani, M. H., & Copeland, R. (2019). Internet of things-aided smart grid: Technologies, architectures, applications, prototypes, and future research directions. *IEEE Access, 7*, 62962–63003. http://ieeexplore.ieee.org/xpl/RecentIssue.jsp?punumber=6287639. doi:10.1109/ACCESS.2019.2913984.

Saputra, D. I. S., Handani, S. W., Indartono, K., & Wijanarko, A. (2020). SMART-in english: Learn english using speech recognition. *Journal of Robotics and Control (JRC), 1*(4), 109–113. https://journal.umy.ac.id/index.php/jrc/article/view/7822/5112. doi:10.18196/jrc.1423.

Schaefer, J. L., Mairesse Siluk, J. C., de Carvalho, P. S., Pinheiro, J. R., & Schneider, P. S. (2020). Management challenges and opportunities for energy cloud development and diffusion. *Energies, 13*(15). https://www.mdpi.com/1996-1073/13/16/4048. doi:10.3390/en13164048.

Shen, F., Wu, Q., Huang, S., Chen, X., Liu, H., & Xu, Y. (2020). Two-tier demand response with flexible demand swap and transactive control for real-time congestion management in distribution networks. *International Journal of Electrical Power & Energy Systems, 114*, 105399. doi:10.1016/j.ijepes.2019.105399.

Shi, Y., Tuan, H. D., Nasir, A. A., Duong, T. Q., & Poor, H. V. (2018). PMU placement optimization for smart grid observability and state estimation. arXiv preprint arXiv:1806.02541. https://arxiv.org/abs/1806.02541

Singh, M., Fuenmayor, E., Hinchy, E. P., Qiao, Y., Murray, N., & Devine, D. (2021). Digital twin: Origin to future. *Applied System Innovation, 4*(2). https://www.mdpi.com/2571-5577/4/2/36/pdf. doi:10.3390/asi4020036.

Sivaneasan, B., Tan, K. T., & Zhang, W. (2025). Cognitive Digital Twins for the microgrid: A real-world study for intelligent energy management and optimization. *IEEE Internet Computing, 29*(1), 39–47. https://ieeexplore.ieee.org/servlet/opac?punumber=4236. doi:10.1109/MIC.2024.3488896.

Song, X., Jiang, T., Schlegel, S., & Westermann, D. (2020). Parameter tuning for dynamic digital twins in inverter-dominated distribution grid. *IET Renewable Power Generation, 14*(5), 811–821. http://www.theiet.org/. doi:10.1049/iet-rpg.2019.0163.

Su, W., & Shi, Y. (2024). Distributed energy sharing algorithm for micro grid energy system based on cloud computing. *IET Smart Cities, 6*(3), 225–236. https://ietresearch.onlinelibrary.wiley.com/journal/26317680. doi:10.1049/smc2.12049.

Tao, F., Liu, W., Zhang, M., Hu, T., Qi, Q., Zhang, H., Sui, F., Wang, T., Xu, H., Huang, Z., Ma, X., Zhang, L., Cheng, J., Yao, N., Yi, W., Zhu, K., Zhang, X., Meng, F., Jin, X., … Luo, Y. (2019). Five-dimension digital twin model and its ten applications. *Jisuanji Jicheng Zhizao Xitong/Computer Integrated Manufacturing Systems, CIMS., 25*(1), 1–18. http://www.cims-journal.cn/CN/volumn/home.shtml. doi:10.13196/j.cims.2019.01.001.

Tariq, M., Ali, M., Naeem, F., & Poor, H. V. (2021). Vulnerability assessment of 6g-enabled smart grid cyber-physical systems. *IEEE Internet of Things Journal, 8*(7), 5468–5475. http://ieeexplore.ieee.org/servlet/opac?punumber=6488907. doi:10.1109/JIOT.2020.3042090.

Tariq, M., & Poor, H. V. (2018). Electricity theft detection and localization in grid-tied microgrids. *IEEE Transactions on Smart Grid, 9*(3), 1920–1929. doi:10.1109/TSG.2016.2602660.

Thurner, L., Scheidler, A., Schafer, F., Menke, J.-H., Dollichon, J., Meier, F., Meinecke, S., & Braun, M. (2018). Pandapower—An open-source python tool for convenient modeling, analysis, and optimization of electric power systems. *IEEE Transactions on Power Systems, 33*(6), 6510–6521. doi:10.1109/tpwrs.2018.2829021.

Turner, S. W., Karakus, M., Guler, E., & Uludag, S. (2023). A promising integration of SDN and blockchain for IoT networks: A survey. *IEEE Access, 11*, 29800–29822. http://ieeexplore.ieee.org/xpl/RecentIssue.jsp?punumber=6287639. doi:10.1109/ACCESS.2023.3260777.

Vögel, H. J., Sü, C., Hubregtsen, T., Ghaderi, V., Chadowitz, R., André, E., Cummins, N., Schuller, B., Härri, J., Troncy, R., Huet, B., Nen, M., Ksentini, A., Conradt, J., Adi, A., Zadorojniy, A., Terken, J., Beskow, J., Morrison, A., … Müller, S. (2018). Emotion-awareness for intelligent vehicle assistants: A research agenda. In *Proceedings of the 1st International Workshop on Software Engineering for AI in Autonomous Systems* (pp. 11–15). IEEE. doi:10.1145/3194085.3194094.

Wang, N., Varghese, B., Matthaiou, M., & Nikolopoulos, D. S. (2019a). ENORM: A framework for edge node resource management. *IEEE Transactions on Services Computing, 13*(6), 1. doi:10.1109/tsc.2017.2753775.

Wang, Y., Chen, Q., Hong, T., & Kang, C. (2019b). Review of smart meter data analytics: Applications, methodologies, and challenges. *IEEE Transactions on Smart Grid, 10*(3), 3125–3148. doi:10.1109/tsg.2018.2818167.

West, T. D., & Blackburn, M. (2018). Demonstrated benefits of a nascent Digital Twin. *INSIGHT, 21*(1), 43–47. doi:10.1002/inst.12189.

Zhang, C., Patras, P., & Haddadi, H. (2019). Deep learning in mobile and wireless networking: A survey. *IEEE Communications Surveys & Tutorials, 21*(3), 2224–2287. doi:10.1109/comst.2019.2904897.

Zheng, T., Liu, M., Puthal, D., Yi, P., Wu, Y., & He, X. (2022). Smart grid: Cyber attacks, critical defense approaches, and Digital Twin. *arXiv*. https://arxiv.org. doi:10.48550/arXiv.2205.11783.

Zhou, Z., Jia, Z., Liao, H., Lu, W., Mumtaz, S., Guizani, M., & Tariq, M. (2022). Secure and latency-aware digital twin assisted resource scheduling for 5G edge computing-empowered distribution grids. *IEEE Transactions on Industrial Informatics, 18*(7), 4933–4943. doi:10.1109/tii.2021.3137349.

Zhu, L.-T., Chen, X.-Z., Ouyang, B., Yan, W.-C., Lei, H., Chen, Z., & Luo, Z.-H. (2022). Review of machine learning for hydrodynamics, transport, and reactions in multiphase flows and reactors. *Industrial & Engineering Chemistry Research., 61*(28), 9901–9949. doi:10.1021/acs.iecr.2c01036.

CHAPTER

15

Fog computing for Digital Twin power grids

Amit Kumar Tyagi[a], *Saurabh Singhal*[b], *Ajeet Kumar Sharma*[c], *Avinash Kumar Sharma*[d]

[a]School of Law Forensic Justice & Policy Studies, National Forensic Sciences University, Gandhinagar, Gujarat, India [b]Department of CSE, Greater Noida Institute of Technology (Engg. Institute), Greater Noida, Uttar Pradesh, India [c]School of Computer Science & Engineering, IILM University, Greater Noida, Uttar Pradesh, India [d]Department of Computer Science & Engineering (CSE), Sharda School of Computing Science & Engineering (SSCSE), Sharda University, Greater Noida, Uttar Pradesh, India

15.1 Introduction

Modernization of power grids, referred to as smart grids, brought a new era of automation, data gathering, and intelligent decision making in the energy sector. Pacing this change is the Digital Twin (DT)—a dynamic virtual model that simulates the physical grid's action in real time (Abo-Khalil, 2023). A DT of a power grid constantly reflects real-time conditions, predicts future events, and supports proactive decision-making for grid stability, fault diagnosis, load balancing, and energy optimization. Such capabilities, however, require enormous computational resources and latency-low communication, which conventional cloud-centric methods frequently fail to deliver. Here, fog computing comes into play.

DTs can model such patterns, but only if response speed is involved. Fog computing enables DTs to take action on events, such as a precipitous decline in solar production—within milliseconds by rerouting energy or activating backup sources locally (Biswas & Wang, 2023). In addition, the decentralization enhances the resilience of the grid by decreasing reliance on central infrastructure that can be overwhelmed or inaccessible in case of network congestion or hack attacks.

Fog computing extends the cloud by offloading computation and storage resources near the sources of data—at the edge of the network (Singh et al., 2019). In contrast to cloud computing, where data is centralized in big data centers that might be far from the end devices, fog computing adds localized processing that cuts down latency and traffic congestion. In the case of power grids, fog nodes may be placed strategically close to substations, transformers, or control units so that real-time analysis and control are provided—essential for an efficient DT setup.

For power grids, DTs may be used to represent electrical substations, transformers, transmission lines, or the whole grid (Mansour et al., 2023). They are updated in real-time with live data from Internet of Things (IoT) sensors installed throughout the grid infrastructure. The system offers real-time visualization and monitoring, which has various advantages, such as:

- Predictive maintenance and fault detection.
- Scenario simulation (e.g., blackout events or demand spikes).
- Increased control and planning with artificial intelligence (AI)/machine learning (ML) models.

But deploying such a system is not easy:

- Processing large amounts of streaming data across geographically distributed sources.
- Providing low-latency communication for real-time decisions.
- Protecting data privacy and cybersecurity.

Digital Twin Technology and Smart Grid.
DOI: https://doi.org/10.1016/B978-0-443-36664-2.00007-1

TABLE 15.1 Comparison of cloud and fog in power grid.

Feature	Cloud computing	Fog computing
Latency	High	Low
Bandwidth usage	High	Reduced via edge filtering
Real-time responsiveness	Limited	High
Scalability	High	Moderate
Security complexity	Moderate	High (distributed nature)
Suitable for	Batch processing, archival, and heavy, non-time—sensitive computation	Real-time monitoring and quick actions

Fog computing solves most of these issues by decentralizing the computation. Fog computing brings processing to the edge of data, minimizing latency and bandwidth consumption. It enables real-time analysis and decision-making at the network edge, enhancing system efficiency and responsiveness overall. The fog layer serves as an intermediary layer between IoT devices and cloud servers, with the benefit of:

- Lower latency through local processing.
- Effective utilization of bandwidth through filtering and preprocessing data prior to transmission to the cloud.
- Control mechanisms are localized when there is a loss of communication with the cloud.

In a power grid context, this can be interpreted as better fault recovery times, more precise load forecasts, and a timely response to energy dynamics. The DT architecture for the power grid based on fog includes the following levels:

1. Physical Grid Layer: The physical electrical infrastructure—substations, transformers, and transmission lines.
2. IoT and Sensor Layer: Voltage, current, temperature, frequency, etc., measuring devices spread across the grid.
3. Fog Layer: Distributed edge nodes doing real-time analytics, control operations, and filtering.
4. Cloud Layer: Centralized storage, large-scale analytics, AI model training, and archival.
5. DT Layer: Real-time simulation, visualization, and intelligent control systems emulating the physical grid virtually.

This chapter shows how fog computing improves the performance and dependability of power grid DTs.

15.1.2 Cloud vs fog in power grids

Table 15.1 shows the difference between cloud computing and fog computing, to highlight the significance of fog computing.

15.1.3 Challenges and opportunities

The convergence of fog computing and DT infrastructures for power grids has brought transformative promise. But this integration brings along a variety of technical, operational, and organizational challenges as well. Awareness about these challenges is necessary not only to help reduce the associated risks but also to seize opportunities they bring for research, innovation, and scalable deployment. The following discusses these in detail.

15.1.3.1 Key challenges

The key challenges in fog-enabled DT power grids are ensuring interoperability across diverse devices, platforms, and protocols. Heterogeneous data formats and vendor-specific standards hinder seamless integration, making real-time synchronization and control difficult.

15.1.3.1.1 Heterogeneity and interoperability

Among the leading challenges is the heterogeneity of platforms, protocols, and devices in power grid ecosystems. DTs need to bring together data from a vast range of sources: IoT sensors, phasor measurement units (PMUs), SCADA systems, and smart meters, commonly in proprietary formats. Achieving semantic interoperability between fog, edge, and cloud layers is still far from easy. Lack of common data models and standard application programming interface (APIs) results in isolated implementations, which slow down scalability and vendor-agnostic adoption.

15.1.3.1.2 Optimization of latency and bandwidth

Although fog computing decreases latency by processing data locally, near the source, dynamic workload balancing among fog and cloud layers complicates. Unpredictable traffic spikes, particularly during grid events such as faults or switching operations, can overburden fog resources. Ensuring quality of service (QoS) under such limitations needs adaptive orchestration algorithms, which are yet to mature.

15.1.3.1.3 Security and privacy

Power grids are a part of vital infrastructure, and the introduction of fog computing increases the attack surface. Authentication, secure communication, and real-time intrusion detection at the fog level are challenging because edge devices have limited computational capabilities. Data privacy under federated learning (FL) or decentralized analytics is also a challenge, particularly while complying with regulations such as the General Data Protection Regulation (GDPR). NERC CIP compliance in North America mandates strict cybersecurity for the Bulk Electric System, but its distributed nature clashes with fog computing's decentralized architecture, creating challenges in unified security, access control (like US Person Ops in cloud), auditing, and managing varied regional laws.

15.1.3.1.4 Resource management

Fog environments are usually composed of resource-scarce devices with less storage, processing capabilities, and energy. Proper scheduling of analytics operations, AI workloads, and data synchronization in the fog environment is challenging due to the stringent real-time requirements of applications. In addition, power utilization at the fog level must be reduced to meet green computing objectives.

15.1.3.1.5 Scalability and maintenance

Managing geographically distributed thousands of fog nodes in a power grid requires solid orchestration and monitoring tools. Fault tolerance, remote debugging, and constant software updates (CI/CD) are more difficult than in centralized cloud systems. Scalable DevOps support tools and platforms are still evolving.

15.1.3.1.6 Standardization and compliance

Lack of global standards on fog–DT integration, data governance, and interoperability gives rise to uncertainty. Furthermore, mapping DT deployments with cybersecurity standards, utility regulations, and operational standards (e.g., International Electrotechnical Commission [IEC] 61850, IEEE 2030.5) takes a lot of work and coordination among stakeholders.

15.1.3.2 Emerging opportunities

In spite of the challenges, a number of opportunities emerge that can reshape the design and operation of future power systems.

15.1.3.2.1 Real-time grid intelligence

Real-time simulation, diagnostics, and optimization at substation or even feeder levels are made possible by fog-enabled DTs. This creates the potential for autonomous grid control where fog nodes independently make localized decisions without depending on central approvals, significantly enhancing fault detection and load balancing capabilities.

15.1.3.2.2 Edge artificial intelligence and federated learning

The meeting of edge AI and FL is a strong prospect. Fog nodes are capable of training and deploying light models locally, learning grid behavior, and coordinating with other nodes without sending raw data. This is better for privacy and less dependent on high-bandwidth cloud communications. Resource-aware AI for energy forecasting and anomaly detection in the fog layer is an area of active research.

15.1.3.2.3 Resilient and decentralized control

Fog computing enables multiagent control systems that enable various grid elements to be semiautonomous. This is particularly useful for microgrids, which might isolate and operate independently during outages. Fog nodes enable fault resilience, rapid restoration, and in situ fault handling—vital capabilities for future smart grids.

15.1.3.2.4 Digital Twin-as-a-Service

One of the promising business models arising from this convergence is Digital Twin-as-a-Service, where third parties and utilities subscribe to DT services deployed across fog infrastructures. This model can democratize access to advanced grid simulations, analytics, and optimization tools, even for regional or small utilities. It also enables ecosystem development where various vendors add modular building blocks into common DT platforms.

15.1.3.2.5 Standardization and interoperability efforts

Efforts by IEEE, IEC, National Institute of Standards and Technology (NIST), and open-source consortia are speeding up the creation of standards for DTs, IoT, and fog computing. Adopting these standards will open up cross-platform interoperability and speed up innovation. There is also an emerging trend toward open DTs and reference architectures, facilitating collaborative development and simpler integration between systems.

15.1.3.2.6 Sustainable and green computing

Fog computing can support sustainability objectives by reducing data transfers, optimizing grid activities to avoid energy wastage, and executing workloads on nodes powered by local renewable energy. In addition, DTs are able to model the environmental consequences of grid decisions in real time and allow for greener operational decisions. Subsequent studies might design carbon-aware scheduling and energy-saving protocols specifically for fog environments.

15.1.3.2.7 Integration with 5G/6G and mobile edge computing

The deployment of 5G and subsequent 6G networks will offer low-latency, high-reliability links crucial for fog–DT integration. Mobile edge computing (MEC) architectures coupled with telecom operators can further increase the footprint of fog computing, facilitating mobile DTs, e.g., for EV fleets, mobile substations, or drones. The integration is also viable for supporting network slicing for differentiated QoS in grid applications.

15.2 Literature review

DTs of energy systems have increasingly been adopted for the optimization of operations, fault diagnosis, asset management, and security verification. An exhaustive review on these fields, presenting how DTs converge IoT, big data, ML, and cloud technologies to leverage smarter grid performance, especially in fault detection and asset modeling (Cioara et al., 2022). Equally, the authors in El-Hajj et al. (2024)) outline far-edge DT architectures that simulate physical grid conditions in real time, supporting efficient event processing at the network edge.

For grid-connected microgrids, the authors in Kumari et al. (2023) pointed out that DTs need to enable real-time data ingestion, cyber-physical coupling, and model-based anomaly detection—vital for providing system resilience and predictive control. Such studies indicate that although cloud-centric DTs provide strong analysis capabilities, latency becomes a bottleneck for real-time grid operations. Fog computing, as an extension of edge computing, puts processing nodes nearer to data sources. For smart grids, in Hussain and Beg (2019), the authors suggested a fog-assisted cloud architecture that provides ~50% reduction in latency and more than 40% energy saving over cloud-only implementations. Additionally, the study of Naha et al. (2018) examines fog computing requirements, fault tolerance, and microservice-based deployments in IoT paradigms. A critical analysis by authors in Gilbert et al. (2019) confirms that edge/fog integration reduces network congestion and enhances QoS but creates challenges in security, interoperability, and programming. The authors in Knebel et al. (2020) introduce a DT architecture in cloud–fog, and via deployment, show the ways in which response time is considerably lowered with fog nodes and the real-time requirements of DT systems are met. The authors in Badidi et al. (2020) also support fog's efficacy in smart grid analytics, especially where latency-sensitive and energy-limited applications are concerned.

Security is a key facilitator of DTs in power systems. The authors in El-Hajj et al. (2024) examine frameworks such as ANGEL and inverter-based DTs to be used for real-time anomaly detection and robustness against false-data injection and Denial-of-Service (DoS) attacks. Hybrid DT frameworks between MATLAB/SIMULINK and single-board computers are also examined for secure microgrid prototyping. Nevertheless, these systems tend to be centered around computational resources, which makes real-time responsiveness impossible under adversarial environments. Microgrid research demonstrates that DTs that are augmented with AI, IoT, big data, and fog computing facilitate predictive maintenance, integration of renewables, and scalable operation (Negi et al., 2024). In Ranawaka et al. (2024), the authors indicated toward utilizing AI and blockchain for decision-making in energy

TABLE 15.2 Identified gaps and their implications for fog-based Digital Twins (DTs).

Identified gap	Implication for fog-based DTs
High latency in cloud-based DT systems	Enables edge-level processing for real-time grid control
Limited cybersecurity resilience	Necessitates secure fog node deployment
Underutilization of AI/ML at the edge	Calls for distributed intelligence across fog nodes
Scalability challenges in heterogeneous grids	Supports hierarchical fog-cloud architectures

AI, *artificial intelligence;* ML, *machine learning.*

TABLE 15.3 Details of layers in the fog-based DT architecture.

Layer	Components	Functions
Physical Grid Layer	Transformers, substations, smart meters, transmission lines	Real-world electrical infrastructure producing operational data
IoT Sensor Layer	Current, voltage, temperature, frequency sensors	Real-time data acquisition from physical grid
Fog Layer (Tier 1)	Local controllers, data aggregators	Local filtering, preprocessing, control loop execution
Fog Layer (Tier 2)	Edge servers, AI engines, security checkpoints	Predictive analytics, anomaly detection, event response
Cloud Layer	Central servers, ML model repository, historical databases	Long-term storage, model training, high-level simulations
Digital Twin Layer	Control dashboards, simulation engines, user interfaces	Real-time representation of the power grid for operators and planners

DT, *Digital Twin.*

DTs. Most of the implementations, however, remain cloud-centric, which indicates that there is a need for distributed analytics. Table 15.2 shows the identified gaps and their implications for fog-based DTs.

15.3 Architecture of fog-based Digital Twin for power grids

The architecture of a fog-based DT in power grids is designed to handle real-time monitoring, local decision-making, and distributed intelligence. Unlike traditional cloud-dependent systems, this layered architecture ensures rapid response to grid fluctuations, decentralized analytics, and enhanced resilience. The architecture comprises several layers that are explained in detail in Table 15.3.

15.3.1 Data flow in the architecture

Sensing data is gathered from sensors in the physical grid, cleaned for noise removal and threshold verification by fog nodes (Abo-Khalil, 2023). Collected data is processed at the edge with aggregated data to perform predictive maintenance and threat assessment against security threats. Model training, analysis, and storage are performed through cloud offloading. Real-time DTs are updated through fog and cloud information, which is used for visualization and control. The system provides efficient and predictable grid operations.

Data from sensors in the physical grid are collected, filtered by fog nodes for removal of noise and threshold validation (Russell et al., 2019). Collected data are aggregated and subjected to edge analytics for security threat detection and predictive maintenance. Model training, analysis, and archival are done via cloud offloading. A real-time digital copy is updated with fog and cloud data and used for visualization and control.

TABLE 15.4 Latency comparison.

Architecture type	Computation location	Comment
Cloud-only DT	Remote Data Center	Not ideal for fast-response scenarios
Edge-only DT	Device-level	Limited compute resources
Fog-enabled DT	Distributed Fog Nodes	Ideal for real-time anomaly detection

DT, *Digital Twin*.

15.3.2 Interlayer communication

A system of interlayer communication allows for an easy and dependable method of data interchange between the physical level, fog level, edge level, and cloud level of the architecture. Data from sensors on the physical level travels up to the fog and edge levels through low-latency communication for filtering, aggregating, and analyzing in real-time, while insights that have been processed and decisions made regarding controls travel down to support prompt actuating and control of the power grid. Interoperable interfaces and secure communications protocols maintain the integrity, confidentiality, and synchronization of data between the various layers of the architecture. Enough information sharing has been established through the two-way communication link that updates models and synchronizing DTs can occur efficiently, allowing for improved adaptive decision-making and therefore more resilient and predictable operations based upon the experiences of the five-level architecture.

15.3.3 Advantages over traditional architectures

The traditional cloud-based systems for a power grid can derive numerous advantages through the implementation of fog architecture when compared to the traditional cloud-based system. These include decreased latency from processing data nearer the data source (edge of network), allowing for improved real-time analysis and detection of faults; greater privacy, support for localized control, and easy-to-scale integration of edge devices. In fact, fog architecture suits changing and widely dispersed energy applications.

15.4 Role of fog computing in enhancing Digital Twin capabilities

DTs live on real-time interaction with their physical counterparts. Centralized cloud-based systems, however, provide latency that can interfere with mission-critical operations such as fault isolation in power grids. Fog computing, being brought closer to the sensors and actuators, negates that. A cloud–fog DT deployment, where response times are dramatically better when computation is pushed to the edge, was illustrated in Jalasri and Lakshmanan (2023).

In power systems, PMUs produce high-frequency data streams requiring near real-time processing. In Yang et al. (2018), the authors demonstrated that PMU fog nodes with anomaly detection minimize end-to-end delay while preserving data fidelity. Likewise, Barbara's Thin Edge systems—and industrial deployments with thousands of fog nodes—emphasize millisecond-level DT synchronization and AI execution at substations. Table 15.4 shows the latency comparison of different architectures to show their impact on fog-based DTs.

15.4.1 Bandwidth optimization and data management

Sensor networks over a grid produce massive amounts of data. Offloading all raw data to the cloud is wasteful. The authors in Jaiswal et al. (2020) prioritized fog as a scalable, distributed system, important for managing smart-meter data and renewable energy, and also stated that utilities with fog platforms realized quantifiable decreases in congestion in their networks and faster outage response times.

15.4.2 Enhancing grid resilience and fault isolation

When faults happen in the grid—because of equipment malfunction or cyberattack—fog nodes provide an essential line of defense. They can localize faults, redirect power, and preserve local operation even when connectivity from the cloud is lost. Fog-based anomaly-detection processes play a central role. Yang et al. (2018) employed secure service access (SSA) and k-nearest neighbors (k-NN) in PMU fog nodes to identify and respond to grid disturbances in real time with high accuracy. Further, Hussain and Beg (2019) presented fog-supported smart grids that include outage management system (OMS), distribution management system (DMS), and geographic information system (GIS) systems to avoid cascading outages using local intelligence and control.

15.4.3 Distributed intelligence and edge-based artificial intelligence

Fog computing pushes AI and analytics out to the edge of power grids. Lightweight models on fog nodes support real-time prediction and control—characteristics at the heart of an active DT capability. Low-latency FL with blockchain, as envisioned in 6G DT networks, demonstrates distributed ML in the fog layer, enhancing reliability and data privacy. Deep reinforcement learning (GD-DRL) has also been applied to distribute the workloads of distributed transformation systems across a combination of cloud and fog computing environments, which helps to reduce latency and cost, while increasing the chances of successful execution. The fog computing environment plays an especially critical role, allowing many AI functions (such as forecasting, anomaly detection, and self-healing) that require immediate action at the local level to be executed locally and to be updated through learning-based mechanisms within the cloud computing environment.

15.4.4 Cybersecurity advantages

Fog computing enhances security within DT ecosystems by enabling distributed processing to be less vulnerable to failure at a single location. The presence of many fog nodes acts as a decentralized means of providing trust in each other, identifying any threats, enforcing any established policies, and isolating compromised areas (Rehman et al., 2023). This helps to mitigate any potential threats originating from within by providing a framework for protecting against potential data losses. Distributed Denial-of-Service (DDoS)attacks may target a number of fog processing nodes that are located within a substation to inhibit automatic failure isolation.

15.4.5 Managing renewable energy variability

Renewables such as solar and wind bring in volatility that has to be managed at the local level. Fog nodes performing forecasting and balancing algorithms can adjust in near real time. Fog's importance in demand-response and distributed energy balancing—using fog to compute data from rooftop PV or wind at the point of origin, minimize blackouts, and make dispatch decisions (Haghnegahdar et al., 2022).

15.4.6 Scalability and hierarchical control

Distributed fog layers enable large-scale DT deployment through a scalable architecture. Hierarchical fog-cloud levels enable regional analytic aggregation with local autonomy (Alli & Alam, 2020). Zig-zag topologies facilitate master–slave fog coordination in fault cases as presented in IoT-supported smart grid literature.

15.5 Communication and security framework

The incorporation of fog computing in DT systems for electric power grids not only brings improved computational abilities but also a significant necessity for a strong communication and security infrastructure (Kabir et al., 2024). The framework needs to enable high-throughput, low-latency, as well as secure data exchange among sensors, fog nodes, cloud platforms, and DT interfaces. Due to the cyber-physical nature of power grids, the security and reliability of data transport take precedence. The security and communication architecture should guarantee availability, real-time access control, data confidentiality, and data integrity, and withstand threats that span from man-in-the-middle attacks to insider tampering.

The architecture of communication for a fog-based DT system is generally organized in several tiers. The sensor or perception layer forms the bottom layer and comprises field-deployed IoT devices like PMUs, smart meters, and environmental sensors (Yang et al., 2018). These devices are tasked with aggregating high-frequency data points on voltage, current, frequency, and temperature. These data points are communicated to Tier 1 fog nodes, which are situated near the source of the data and conduct upfront preprocessing operations such as filtering, threshold detection, and rudimentary anomaly detection. Tier 2 fog nodes, which run at a higher level, are in charge of sophisticated edge analytics such as real-time forecasting, AI-driven decision-making, and threat analysis or security incidents. The cloud layer, positioned above the fog layer, provides long-term data storage, complex model training, and historical trend analysis. Finally, the DT interface offers visualization, operator control, and simulation-based decision support, drawing on both cloud-trained models and real-time fog computations.

Each layer in this hierarchy communicates both vertically (e.g., sensor-to-fog to cloud) and horizontally (e.g., fog-to-fog, or intercloud synchronization). To facilitate such distributed communication effectively and securely, several lightweight and scalable protocols are utilized. For sensor-to-fog communication, protocols like Message Queuing Telemetry Transport (MQTT) and Constrained Application Protocol (CoAP) are used more often because of their minimal overhead. They are usually secured by Transport Layer Security (TLS) or Datagram TLS (DTLS), which offer encryption and authentication specific to constrained environments. Fog-to-fog or fog-to-cloud communications frequently employ RESTful APIs over HTTPS, which are protected by mutual TLS or IPsec tunnels. Peer-to-peer communications between fog nodes, necessary for redundancy and collaborative decision-making, are facilitated using data distribution services like DDS or secure gRPC with mutual authentication to avoid impersonation or rogue node entry.

Security is implemented across all layers by a range of mechanisms. The most basic of these is authentication. Mutual authentication is required, especially for control signal-sending and -receiving devices. This usually entails the utilization of digital certificates and public-key infrastructure, in which the communicating parties authenticate one another's identities. Attribute-Based Access Control, typically achieved through Ciphertext-Policy Attribute-Based Encryption (CP-ABE), offers granular access permissions and guarantees data manipulation and decryption can only be done by approved entities. Adding a Zero-Trust Architecture improves security even more by requiring authentication at each step of communication and removing the presumption of trust based on the network location.

Data confidentiality and integrity are protected through symmetric and asymmetric encryption combined. Symmetric encryption algorithms like AES-128 or AES-256 are used in most fast and secure communication between fog nodes and sensors. Asymmetric techniques provide secure key exchange and digital signatures, allowing tamper-proof and verifiable communication. Hashing functions like SHA-256 and keyed-hash message authentication codes (HMACs) are utilized to authenticate the integrity of the data being transmitted. In applications where fog nodes need to compute data without disclosing its content—e.g., in privacy-critical applications—homomorphic encryption methods could be employed for computation over encrypted data.

Aside from cryptographic security, trust management is gaining more and more importance in distributed fog systems (Rehman et al., 2023). Blockchain and other distributed ledger technologies can be utilized to store immutable transaction and event logs, especially for auditing purposes or postincident analysis. Fog nodes may be used to support lightweight blockchain clients to make sure data updates or control commands are traceable and cannot be repudiated (Kabir et al., 2024). Reputation-based systems can also be utilized to track node behavior over time and dynamically adjust trust levels depending on how well they follow anticipated patterns.

Intrusion detection and ongoing monitoring are essential elements of an effective responsive security posture. Lightweight Intrusion Detection Systems (IDSs) in the fog layer can observe network traffic for abnormalities like spikes in data, strange patterns of commands, or protocol infringements. These IDS systems tend to utilize ML for identifying new threats in real-time. Combined with automated response systems like deactivating a suspicious node or quarantining a stream of data, fog-based IDS ensures grid stability and authenticity of data. All significant events are recorded, frequently in encrypted format or through blockchain-supported tamper-proof technologies, for accountability and to facilitate forensic investigation.

Given the nature of power grid operations being critical, the threat environment is extensive as well as severe. Threats are in the form of passive attacks like eavesdropping, where attackers intercept messages to obtain sensitive information, and active threats like man-in-the-middle attacks, whereby attackers alter data in transit. Replay attacks are especially risky in DT setups, since they can make actuators act inappropriately following incorrect instructions that were received in the past. DoS and DDoS attacks can leave fog nodes unresponsive and even freeze local control loops. Rogue nodes or Sybil attacks—where several false identities are employed to distort trust metrics—are also specific to fog and must be tackled. Effective countermeasures are mutual TLS, HMAC verification, cryptographic time stamping, rate limiting, and node reputation systems. For a clearer explanation of how these elements work together, imagine the following practical example of secure fault detection in a power grid situation. A fault has been detected by a transformer, generating abnormal current and temperature readings. The sensor digitally signs and encrypts the data with DTLS and forwards it to a Tier 1 fog node. The node checks the message's authenticity and integrity, compares it to past thresholds, and concludes that a fault condition exists. It sends an immediate notification to the Tier 2 fog node and the DT interface. At the same time, the Tier 2 fog node cross-references the fault pattern using its model of anomalies and records the event in a local blockchain ledger. The fault summary is encrypted and

TABLE 15.5 Various threat landscape and defenses.

Threat class	Description	Countermeasures
Eavesdropping	Passive interception in sensor-to-fog/fog-fog communication	TLS/DTLS encryption
Man-in-the-Middle (MitM)	Interception or data manipulation during communication	Mutual TLS, certificate pinning
Replay attacks	Replay of recorded communication to trigger false actions	Timestamping, nonce-based protocols, and HMAC verification
DDoS	Flooding fog nodes to render them unresponsive	Rate limiting, IDS alerts, automated throttling
Impersonation	Fake node injection to manipulate system behavior	PKI, node attestation, physical-layer fingerprinting
Rogue fog nodes	Insider or compromised fog nodes are distributing false data	Reputation systems, blockchain audit trails
VM attacks	Exploitation of resource usage or virtualization vulnerabilities	Secure virtualization layers

DDoS, *Distributed Denial-of-Service;* DTLS, *Datagram Transport Layer Security;* DT, *Digital Twin;* HMAC, *hash message authentication codes;* PKI, *public-key infrastructure;* TLS, *Transport Layer Security.*

sent to the cloud, where a permanent analytics engine analyzes the effect on overall grid performance. The cloud then sends back a signed model update to fog nodes and the DT interface to ensure subsequent alerts can be processed even more effectively. This entire process occurs in milliseconds to seconds, and the integrity, privacy, and traceability of data are maintained throughout. Adherence to global standards further enhances the credibility of the proposed framework. MQTT and CoAP protocols are standardized by OASIS and IETF, respectively. NIST and IEC standards, like IEC 62443, specify security guidelines for smart grid communications. These standards provide interoperability, impose robust authentication policies, and require hardware and software components to undergo security testing. Several upgrades will be planned in the future to enhance the communication and security architecture further. Postquantum cryptography algorithms will be required to shield against potential quantum-based attacks in the future. Secure boot mechanisms and tamper-resistant hardware modules will maintain device integrity even when a node is physically attacked. FL models will enable fog nodes to update AI algorithms without the exchange of raw data, thus maintaining privacy. Lightweight Security Information and Event Management solutions designed for fog environments would also be able to offer improved threat correlation and response at the edge.

In summary, the communication and security infrastructure in fog-enabled DT architectures for power grids is critical to making those systems work securely and reliably. By using stacked communication protocols, end-to-end encryption, strong authentication, trust management, intrusion detection, and alignment with international standards, the architecture not only safeguards data but also supports timely and trustworthy decision-making. As DTs are increasingly integrated into critical infrastructure, ongoing innovation in fog-level communications and security will be essential to unlocking their full capabilities.

15.5.1 Threat landscape and defenses

The threat landscape in fog-based DT power grids includes cyberattacks such as data breaches, malware, DDoS attacks, and insider threats, alongside physical vulnerabilities like sensor tampering. These threats can compromise grid stability and data integrity. Defenses include lightweight IDSs, end-to-end encryption, blockchain for data provenance, and zero-trust architectures. AI-driven anomaly detection and secure boot mechanisms further enhance system resilience against both cyber and physical security challenges. Table 15.5 shows various threat landscapes and defenses.

15.6 Future research directions

Subsequent studies on fog-enabled DT power grids will emphasize scalable AI-powered orchestration, ultra-lightweight edge analytics, and robust cybersecurity. FL, semantic interoperability, and postquantum cryptography will be emphasized to spur grid operations that are resilient and adaptive. Combining 5G/6G and sustainable mechanisms will continue to aid intelligent, real-time energy management.

15.6.1 Artificial intelligence–driven resource and task orchestration

Future complexity and volume of fog-enabled DT infrastructure necessitate smart orchestration strategies beyond traditional static policies. The use of AI, especially DRL, presents a promising avenue for task scheduling and resource allocation (Zhou et al., 2024). DRL agents can be trained to optimize the allocation of computational tasks over heterogeneous fog and cloud resources by constantly interacting with their environment, optimizing toward energy efficiency, latency, and task completion. Future studies ought to investigate multiobjective reinforcement learning models that take dynamic factors like network bandwidth, power constraints, and fault tolerance into account. Hybrid scheduling frameworks blending heuristic methods with AI predictors can also be designed to provide increased robustness in scheduling.

15.6.2 Lightweight federated learning

The traditional ML models can pose privacy concerns, such as the need for a central point of collection to store all the data collected, thus using up more bandwidth (Shokri & Shmatikov, 2015). FL offers an alternative to these limitations by decentralizing the model training process and allowing model training to take place at the edge or "fog" level. However, many of the existing FL models are not well-suited for use within the constrained environments found in most fog-based deployments. Therefore, there is a need for future design efforts aimed at creating lightweight architectures, such as TinyML, which can be deployed directly on edge-based devices. In addition, there is an opportunity to explore the use of hierarchical models, where fog nodes act as an intermediate between edge devices and cloud-based servers, reducing the latency of training and personalizing the model. Exploring adaptive aggregation approaches, based on the reliability of nodes and the quality of the data being produced by each node, should be further explored to improve accuracy when training models.

15.6.3 Semantic interoperability and Digital Twin compatibility

Interoperability among DT components from multiple vendors and systems continues to be a significant challenge (Acharya et al., 2024). Semantic technologies, such as ontologies and knowledge graphs, can offer standardized vocabularies to represent assets, processes, and events for power grid systems. Future work can involve the development of domain-specific ontologies for smart grids that will allow for uniform interpretation of the data and cross-platform integration. Semantic annotation methods should also be created to autonomously add contextual metadata to sensor readings. These semantic models can then be combined with AI reasoning engines to improve decision-making in the DT environment so that more dynamic and responsive management of the grid is possible (Ranawaka et al., 2024).

15.6.4 Blockchain-based trust and provenance

Given the decentralized operation of fog computing, data integrity and trust across nodes are paramount. Blockchain technology can provide a secure and transparent means for logging transactions, configurations, and events (Komalavalli et al., 2020). Lightweight consensus algorithms like Proof of Authority or Delegated Proof of Stake would be an area of research for future studies, which would be more appropriate for resource-limited fog nodes. Blockchain integration with DTs can provide tamper-proof audit trails to maintain accountability in control decisions and system alterations. Access control and actions can also be automated by smart contracts based on predefined conditions, increasing autonomy and minimizing the need for human intervention (Zaidi et al., 2021).

15.6.5 Cyber-physical security

The use of cyber-physical systems in today's electric grids creates new attack vectors that fall well beyond the capabilities of traditional IT security systems. For example, fog node IDSs need to be lightweight, provide real-time detection capability, and detect both network and operational anomalies (Martins et al., 2022). Future research should also include an exploration of hybrid IDS architecture designs that provide both signature-based detection and anomaly-based detection, leveraging ML. The resiliency of these systems should be strengthened by adding defensive capabilities such as Moving Target Defense and Deception Technology. Cooperative defense approaches that allow for the sharing of threat intelligence at the fog node level and addressing those threats collectively need to be explored to improve situational awareness.

15.6.6 Privacy, data ethics, and regulatory compliance

Handling personal and operational data in large quantities in DT environments requires strict privacy controls and ethics. Future work should research how to integrate privacy-preserving measures such as homomorphic encryption, differential privacy, and secure multiparty computation into the fog layer. In addition, compliance frameworks should be established to comply with data protection laws like the GDPR and domestic energy legislation. Fairness, accountability, and transparency-based ethical AI models should be integrated into decision-making to foster stakeholder trust and responsible deployment (Krook et al., 2025).

15.6.7 Edge analytics for renewable integration and virtual power plants

The growth of distributed energy resources like solar panels, wind turbines, and battery storage necessitates high-end edge analytics for real-time monitoring and control. DTs can predict and model the behavior of these assets, allowing grid operators to optimize load balancing, energy trading, and fault management. Future studies should concentrate on creating adaptive analytics models that can handle multimodal data (e.g., weather, load, and frequency) at the edge. V2G system integration with DTs is another potential area, enabling electric cars to serve as mobile energy storage devices within virtual power plants.

15.6.8 Resilient multiagent control and decentralized coordination

Conventional centralized control mechanisms for power systems are ever more unsuitable to handle distributed and dynamic settings. Multiagent systems, in which every fog node or grid entity is an intelligent agent, provide an extensible solution to decentralized control (Mahela et al., 2022). Cooperation protocols for agents to negotiate, cooperate, and agree on grid operations, even under adversarial scenarios, should be the focus of future research. Techniques such as game theory and distributed optimization can be employed to simulate agent interactions. Fault-tolerance and self-healing properties of these systems should also be emphasized so that they operate uninterrupted during failures.

15.6.9 Software toolchains: DevOps and orchestration for fog

Software engineering practices should be strong in deploying and maintaining fog-based DTs at scale. Future research should investigate extending DevOps principles to the fog environment, such as continuous integration, testing, and deployment of models and control logic. Container orchestration systems such as Kubernetes can be extended to accommodate the special needs of fog computing, including heterogeneity of resources and geo-distribution. Fog-native programming paradigms and declarative configuration languages will be researched to simplify development and minimize operational overhead. Observability frameworks must also be incorporated to observe the health and performance of DT components in real-time.

15.6.10 Quantum-safe and postquantum security

The emergence of quantum computing is a great challenge to traditional cryptographic protocols, risking the confidentiality and integrity of the DT system data. Research in the future must explore postquantum cryptographic primitives that can be applied within the resource limitations of fog nodes. These could be lattice-based, hash-based, or code-based cryptosystems. Quantum key distribution may be investigated as a method of realizing provably secure communication channels, particularly for essential grid infrastructure (Alléaume et al., 2014). The development of hybrid cryptographic protocols integrating classical and quantum-safe approaches will provide backward compatibility with future-proofing the system.

15.6.11 5G/6G Integration and network slicing

The convergence of 5G and next-generation 6G technologies provides unprecedented potential for improving DT system performance (Sheraz et al., 2024). Ultra-reliable low-latency communications through 5G can be used to facilitate time-sensitive grid operations, whereas network slicing facilitates the generation of exclusive virtual networks to support various DT applications. Future studies need to concentrate on the creation of adaptive slicing algorithms with the ability to dynamically allocate network resources in relation to workload patterns and QoS demands. Also, there is a need to explore the integration of fog computing with MEC to provide computational resources nearer to the end devices in order to support mobility and minimize latency.

15.6.12 Digital Twin co-simulation and testing

DT as-built models must be validated to guarantee the accuracy and trustworthiness of the models in real-life conditions as well (Piras et al., 2024). Co-simulation tools that combine physical and virtual entities as a real-time, comprehensive testing platform are possible. The scalability and high-fidelity future work should concentrate on developing such tools, which could include high-resolution grid component models and interfacing with hardware-in-the-loop systems. Standardized benchmarking procedures must be created to measure DT solution performance in different fault and stress scenarios. Furthermore, DTs need to be stress-tested for predictive maintenance and automated recovery protocols support.

15.6.13 Scalability, energy-efficiency, and sustainability

With DT implementations scaling up within regions and utility grids, their energy utilization and environmental footprint need to be studied. Research must concentrate on creating energy-efficient hardware and algorithm platforms for fog nodes, which can be energized with renewable energy sources. Low-power design and thermal management methodologies can minimize the carbon footprint of the deployments (Li et al., 2020). Also, sustainability indicators need to be created to measure the life-cycle cost of DT elements, from production to disposal. Incentive mechanisms can be investigated to stimulate the adoption of green computing practices among utility players and tech vendors.

15.6.14 Standardization, ecosystems, and market models

Fog-based DT systems' large-scale adoption needs established standards and cooperative ecosystems. Future research needs to concentrate on establishing open APIs, data exchange modes, and interoperability protocols based on prevailing standards like IEC 61850 and IEEE 2030.5. Collaboration with standardization organizations and industry associations will be necessary for ensuring compatibility and certification. Further, studies should examine the potential market models for DT offerings, such as subscription, pay-per-use, and DT-as-a-Service offerings. Business models that enable value co-creation among utilities, technology companies, and regulators will ensure a sustainable and scalable implementation of DTs. Table 15.6 summarizes the key research areas along with their key challenges and opportunities.

15.7 Case study

A simulation-based case study was carried out to evaluate the practical effectiveness of the proposed fog-enhanced DT architecture for smart grid operations. Due to the complexity and safety constraints associated with real-world power grid experimentation, the study relies on simulation using iFogSim with network behavior modeled to reflect realistic latency and communication overheads. The simulated scenario represents a smart grid environment comprising distributed smart meters and PMUs deployed across multiple substations. Two system configurations were analyzed: a Cloud-only DT model, in which all sensor-generated data are transmitted directly to a centralized cloud for processing and DT updates, and a fog-enhanced DT model, where data are first processed at fog nodes located near substations before selective offloading to the cloud. The simulation involves a total of 1000 sensors. Each generates one data sample every second and sends that data to 10 Fog nodes. The fog nodes are responsible for filtering out noise, validating threshold conditions, and performing preliminary analyses (e.g., identifying potential security threats and inferring predictive maintenance). The cloud layer performs computationally intensive functions:

TABLE 15.6 Summary of key research areas.

Research area	Key challenges	Opportunities
AI resource orchestration	Dynamic task allocation across fog/cloud	Lower latency, improved QoS
Federated learning & edge AI	Lightweight models, privacy, bandwidth limits	Local intelligence, reduced cloud dependency
Semantic interoperability	Multi-vendor alignment, ontology integration	Seamless integration, cross-domain DT development
Blockchain & provenance	Consensus overhead, privacy	Immutable audit logs, enhanced trust
Cyber-physical security	Real-time detection, zero-trust scalability	Resilience, automated defense
Privacy	Data ethics, legal compliance	Regulatory alignment, user trust
Multiagent control	Decentralized coordination, fault recovery	Scalable islanding, autonomous subgrids
DevOps & fog orchestration	Heterogeneity, CI/CD complexity	Versioned DT deployment, agile modeling workflows
Energy-efficiency & sustainability	Nonrenewable energy use, life-cycle concerns	Green computing, circular economy in DT deployments
Standardization & market models	API alignment, investment viability	DT ecosystems, services monetization, vendor-neutral platforms

DT, *Digital Twin;* QoS, *quality of service.*

training ML models, conducting long-term analytics, and archiving data. During the hour-long simulation period, multiple instances of both node and communication failures will occur at regular intervals to test the robustness of the system. Key metrics will be used to evaluate the system's performance: End-to-End latency, Cloud Bandwidth Utilization, and Fault Recovery Time (defined as the time required to reacquire sync between the DT and the Cloud after a failure occurs). The outcome of the study supports that fog solutions will provide more benefits than Cloud-Dedicated Solutions over time. If localized processing is applied at a fog node, the latency of the end-to-end process will decrease and allow DTs to be updated in a timely manner. The use of fog will improve how control actions are taken. When using fog, overall data transfer from the sensors to the cloud will be considerably less than with a cloud-centric approach, as only the aggregate information will be transferred. In addition, the ability of fog to provide a quick recovery in a failure mode is superior to that of using a cloud, which will provide better continuity. These data point to a quantitative conclusion of the fog integration advantages to DTs in smart grid environments.

15.8 Conclusion

This chapter provides evidence that in order to develop the energy networks of the future, there will need to be an evolution away from the traditional cloud-centric infrastructures toward an intelligent, distributed approach, combining fog computing with DT technologies. These advances in technology allow for improved operational performance of the smart power grid by allowing for lower latency times of processing large amounts of data from grid resources and monitoring their health remotely, due to continued increases in the complexity of the grids, increasing data workloads from each grid, and the continued increase in physical and cyber threats to the grid. The framework and architecture outlined in this paper present an all-encompassing framework for a multilevel approach to building the smart grid through integration of all resources, including sensors, fog nodes, and cloud services, with a goal of moving computational intelligence closer to the data source, providing for a secure and efficient operation of the smart power grid. A primary point of this research is the undeniable advantage of fog-enhanced DT systems over traditional cloud-only systems in terms of performance. The simulation demonstrates a clear performance increase, having measured reductions in end-to-end latency, as well as cloud bandwidth consumption. These quantifiable performance improvements will improve such functions as real-time decision-making, localized fault detection, dynamic load balancing, and effective predictive maintenance—all critical functions to maintain grid stability and resilience for current and future power systems.

References

Abo-Khalil, A. G. (2023). Digital twin real-time hybrid simulation platform for power system stability. *Case Studies in Thermal Engineering, 49*, 103237. doi:10.1016/j.csite.2023.103237.

Acharya, S., Ali Khan, A., & Päivärinta, T. (2024). Interoperability levels and challenges of digital twins in cyber–physical systems. *Journal of Industrial Information Integration, 42*, 100714. doi:10.1016/j.jii.2024.100714.

Alléaume, R., Branciard, C., Bouda, J., Debuisschert, T., Dianati, M., Gisin, N., Godfrey, M., Grangier, P., Länger, T., Lütkenhaus, N., Monyk, C., Painchault, P., Peev, M., Poppe, A., Pornin, T., Rarity, J., Renner, R., Ribordy, G., Riguidel, M., Salvail, L., Shields, A., Weinfurter, H., & Zeilinger, A. (2014). Using quantum key distribution for cryptographic purposes: A survey. *Theoretical Computer Science, 560*(1), 62–81. doi:10.1016/j.tcs.2014.09.018.

Alli, A. A., & Alam, M. M. (2020). The fog cloud of things: A survey on concepts, architecture, standards, tools, and applications. *Internet of Things, 9*, 100177. doi:10.1016/j.iot.2020.100177.

Badidi, E., Mahrez, Z., & Sabir, E. (2020). Fog computing for smart cities' big data management and analytics: A review. *Future Internet, 12*(11), 190. doi:10.3390/fi12110190.

Biswas, A., & Wang, H.-C. (2023). Autonomous vehicles enabled by the integration of IoT, edge intelligence, 5G, and blockchain. *Sensors, 23*(4), 1963. doi:10.3390/s23041963.

Cioara, T., Anghel, I., Antal, M., Antal, C., Arcas, G. I, & Croce, V. (2022). An overview of digital twins application in smart energy grids. In *Proceedings - 2022 IEEE 18th International Conference on Intelligent Computer Communication and Processing Conference, ICCP: 2022* (pp. 25–30). doi:10.1109/ICCP56966.2022.10053945.

El-Hajj, M., Itäpelto, T., & Gebremariam, T. (2024). Systematic literature review: Digital twins' role in enhancing security for Industry 4.0 applications. *Security and Privacy, 7*(5). doi:10.1002/spy2.396.

Gilbert, G. M., Naiman, S., Kimaro, H., & Bagile, B. (2019). A critical review of edge and fog computing for smart grid applications. *IFIP Advances in Information and Communication Technology, 551*, 763–775. doi:10.1007/978-3-030-18400-1_62.

Haghnegahdar, L., Chen, Y. u, & Wang, Y. (2022). Enhancing dynamic energy network management using a multiagent cloud-fog structure. *Renewable and Sustainable Energy Reviews, 162*, 112439. doi:10.1016/j.rser.2022.112439.

Hussain, M. d. M., & Sufyan Beg, M. M. (2019). Fog computing for Internet of Things (IoT)-aided smart grid architectures. *Big Data and Cognitive Computing, 3*(1), 8. doi:10.3390/bdcc3010008.

Jaiswal, R., Davidrajuh, R., & Rong, C. (2020). Fog computing for realizing smart neighborhoods in smart grids. *Computers, 9*(3), 76. doi:10.3390/computers9030076.

Jalasri, M., & Lakshmanan, L. (2023). Managing data security in fog computing in IoT devices using noise framework encryption with power probabilistic clustering algorithm. *Cluster Computing, 26*(1), 823–836. doi:10.1007/s10586-022-03606-2.

Kabir, M. dR., Halder, D., & Ray, S. (2024). Digital Twins for IoT-driven energy systems: A survey. *IEEE Access, 12*, 177123–177143. doi:10.1109/access.2024.3506660.

Knebel, F. P., Wickboldt, J. A., & De Freitas, E. P. (2020). A cloud-fog computing architecture for real-time digital twins. *arXiv*. https://arxiv.org. doi:10.48550/arxiv.2012.06118.

Komalavalli, C., Saxena, D., & Laroiya, C. (2020). Overview of blockchain technology concepts. In *Handbook of research on blockchain technology* (pp. 349–371). Elsevier. https://www.sciencedirect.com/book/9780128198162. doi:10.1016/B978-0-12-819816-2.00014-9.

Krook, J., Winter, P., Downer, J., & Blockx, J. (2025). A systematic literature review of artificial intelligence (AI) transparency laws in the European Union (EU) and United Kingdom (UK): A socio-legal approach to AI transparency governance. *AI and Ethics, 5*(4), 4069–4090. doi:10.1007/s43681-025-00674-z.

Kumari, N., Sharma, A., Tran, B., Chilamkurti, N., & Alahakoon, D. (2023). A comprehensive review of digital twin technology for grid-connected microgrid systems: State of the art, potential and challenges faced. *Energies, 16*(14), 5525. doi:10.3390/en16145525.

Li, B. o, Kuo, H., Wang, X., Chen, Y., Wang, Y., Gerada, D., Worall, S., Stone, I., & Yan, Y. (2020). Thermal management of electrified propulsion system for low-carbon vehicles. *Automotive Innovation, 3*(4), 299–316. doi:10.1007/s42154-020-00124-y.

Mahela, O. P., Khosravy, M., Gupta, N., Khan, B., Alhelou, H. H., Mahla, R., Patel, N., & Siano, P. (2022). Comprehensive overview of multi-agent systems for controlling smart grids. *CSEE Journal of Power and Energy Systems, 8*(1), 115–131. https://ieeexplore.ieee.org/xpl/RecentIssue.jsp?punumber=7054730. doi:10.17775/CSEEJPES.2020.03390.

Mansour, D.-E. A., Numair, M., Zalhaf, A. S., Ramadan, R., Darwish, M. M. F., Huang, Q. i, Hussien, M. G., & Abdel-Rahim, O. (2023). Applications of IoT and digital twin in electrical power systems: A comprehensive survey. *IET Generation, Transmission & Distribution, 17*(20), 4457–4479. doi:10.1049/gtd2.12940.

Martins, I., Resende, J. S., Sousa, P. R., Silva, S., Antunes, L., & Gama, J. (2022). Host-based IDS: A review and open issues of an anomaly detection system in IoT. *Future Generation Computer Systems, 133*, 95–113. https://www.journals.elsevier.com/future-generation-computer-systems. doi:10.1016/j.future.2022.03.001.

Naha, R. K, Garg, S., Georgakopoulos, D., Prakash Jayaraman, P., Gao, L., Xiang, Y., & Ranjan, R. (2018). Fog computing: Survey of trends, architectures, requirements, and research directions. *IEEE Access, 6*, 47980–48009. doi:10.1109/access.2018.2866491.

Negi, G. S, Singh, R., Gehlot, A., Malik, P., Sharma, R., Obaid, A. J., Alferaidi, A., Alharbi, Y., & Kumar, S. (2024). Microgrids 4.0: Digitalization of microgrid with IoT and recent technology interventions. *IET Communications*. doi:10.1049/cmu2.12864.

Piras, G., Agostinelli, S., & Muzi, F. (2024). Digital Twin framework for built environment: A review of key enablers. *Energies, 17*(2), 436. doi:10.3390/en17020436.

Ranawaka, A., Alahakoon, D., Sun, Y., & Hewapathirana, K. (2024). Leveraging the synergy of Digital Twins and artificial intelligence for sustainable power grids: A scoping review. *Energies, 17*(21), 5342. doi:10.3390/en17215342.

Rehman, A., Ahmad Awan, K., Din, I. Ud, Almogren, A., & Alabdulkareem, M. (2023). FogTrust: Fog-integrated multi-leveled trust management mechanism for Internet of Things. *Technologies, 11*(1), 27. doi:10.3390/technologies11010027.

Russell, L., Kwamena, F., & Goubran, R. (2019). Towards reliable IoT: Fog-based AI sensor validation. In *Proceedings - 2019 3rd IEEE International Conference on Cloud and Fog Computing Technologies and Applications, Cloud Summit 2019* (pp. 37–44). Canada: Institute of Electrical and Electronics Engineers Inc. http://ieeexplore.ieee.org/xpl/mostRecentIssue.jsp?punumber=9035980. doi:10.1109/CloudSummit47114.2019.00013.

Sheraz, M., Chuah, T. C., Lee, Y. L., Alam, M. M., Al-Habashna, A., & Han, Z. (2024). A comprehensive survey on revolutionizing connectivity through artificial intelligence-enabled Digital Twin network in 6G. *IEEE Access, 12*, 49184–49215. http://ieeexplore.ieee.org/xpl/RecentIssue.jsp?punumber=6287639. doi:10.1109/ACCESS.2024.3384272.

Shokri, R., & Shmatikov, V. (2015). Privacy-preserving deep learning. In *Proceedings of the ACM Conference on Computer and Communications Security* (pp. 1310–1321). IEEE. doi:10.1145/2810103.2813687.

Singh, S. P, Nayyar, A., Kumar, R., & Sharma, A. (2019). Fog computing: From architecture to edge computing and big data processing. *The Journal of Supercomputing, 75*(4), 2070–2105. doi:10.1007/s11227-018-2701-2.

Yang, Z., Chen, N., Chen, Y., & Zhou, N. (2018). A novel PMU fog-based early anomaly detection for an efficient wide area PMU network. In *IEEE 2nd International Conference on Fog and Edge Computing, ICFEC 2018 - In conjunction with 18th IEEE/ACM International Symposium on Cluster, Cloud and Grid Computing, IEEE/ACM CCGrid 2018* (pp. 1–10). United States: Institute of Electrical and Electronics Engineers Inc. http://ieeexplore.ieee.org/xpl/mostRecentIssue.jsp?punumber=8358330. doi:10.1109/CFEC.2018.8358730.

Zaidi, S. Y. A., Shah, M. A., Khattak, H. A, Maple, C., Rauf, H. T, El-Sherbeeny, A. M., & El-Meligy, M. A. (2021). An attribute-based access control for IoT using blockchain and smart contracts. *Sustainability, 13*(19), 10556. doi:10.3390/su131910556.

Zhou, G., Tian, W., Buyya, R., Xue, R., & Song, L. (2024). Deep reinforcement learning-based methods for resource scheduling in cloud computing: A review and future directions. *Artificial Intelligence Review, 57*(5). doi:10.1007/s10462-024-10756-9.

CHAPTER

16

Edge computing for Digital Twin-enabled future electrical grid systems

Abubakar Ahmad Musa[a], *Hengshuo Liang*[b], *Qingqing Li*[a], *Milon Biswas*[a], *Cheng Qian*[c], *Siping Li*[a], *Wei Yu*[a]

[a]Department of Computer and Information Sciences, Towson University, Towson, MD, United States [b]School of Engineering & Technology, University of Washington Tacoma, WA, United States [c]Department of Computer Science and Information Technology, Hood College, MD, United States

16.1 Introduction

The traditional power grid is the critical safety infrastructure essential for delivering electrical energy to the public (e.g., organizations and consumers). Recent advances in information and communication technology (ICT), particularly in communication, computation, and storage, combined with successive industrial revolutions, have driven the evolution from traditional grid systems to modern smart grids (Fang et al., 2012; Xu et al., 2017). Unlike the traditional grid, which primarily focuses on one-way power transmission as a foundational component of national critical infrastructures, the smart grid enables bidirectional power and information flows between power plants and consumers. This capability promotes more efficient, reliable, and resilient operations while supporting the integration of diverse renewable energy resources, including solar and wind (Moulema et al., 2015). Consequently, the smart grid serves as a continuous point of attraction for national and international attention, generating ongoing interest from industry players, the research community, and government agencies worldwide.

As a key representative of the smart-world cyber-physical system (CPS) in the energy domain, the smart grid leverages ICT to enable two-way communication between its physical infrastructure and corresponding cyber components. It generates operational data that enables real-time monitoring and control of safety-critical processes within the energy CPS. However, the large-scale data processing, generated by two-way communication, calls for cloud computing solutions that have high computing capability but fall short in critical grid operations due to latency and bandwidth constraints. Edge computing addresses these limitations by enabling localized computation, giving rise to smart grid edge computing (Feng et al., 2021; Liang et al., 2024; Minh et al., 2022; Shi et al., 2016; Yu et al., 2018). The integration of smart grid with edge computing allows transmission and distribution networks to coexist and share actionable data. This capability enhances decision-making for both power plants and consumers, ultimately supporting more efficient energy generation, distribution, and consumption. To effectively manage these decentralized edge devices in the smart grid, robust and efficient communication technologies are critical (Navarro-Ortiz et al., 2020; Sutton et al., 2019; Yu et al., 2016a, 2016b). Future networking techniques aim to support low-latency, high-reliability, and high bandwidth communication and computation. These capabilities facilitate the seamless integration of DER, including renewables and electric vehicles (EVs), into the smart grid's transmission and distribution networks, thereby improving the robustness, reliability, and sustainability of the smart grid. In parallel with advances in networking techniques, the concept of Digital Twins (DTs) aims to generate virtual replicas of physical assets that are synchronized with their real-world counterparts (Qian et al., 2022). Integrating DT into a smart grid can enable real-time synchronization and state tracking of these distributed edges and processes through their corresponding virtual components (Qian et al., 2024a). Specifically, the DT will act as the dynamic mirror of the distributed edge resources, enabling real-time monitoring and interaction across components by leveraging

Digital Twin Technology and Smart Grid.
DOI: https://doi.org/10.1016/B978-0-443-36664-2.00023-X

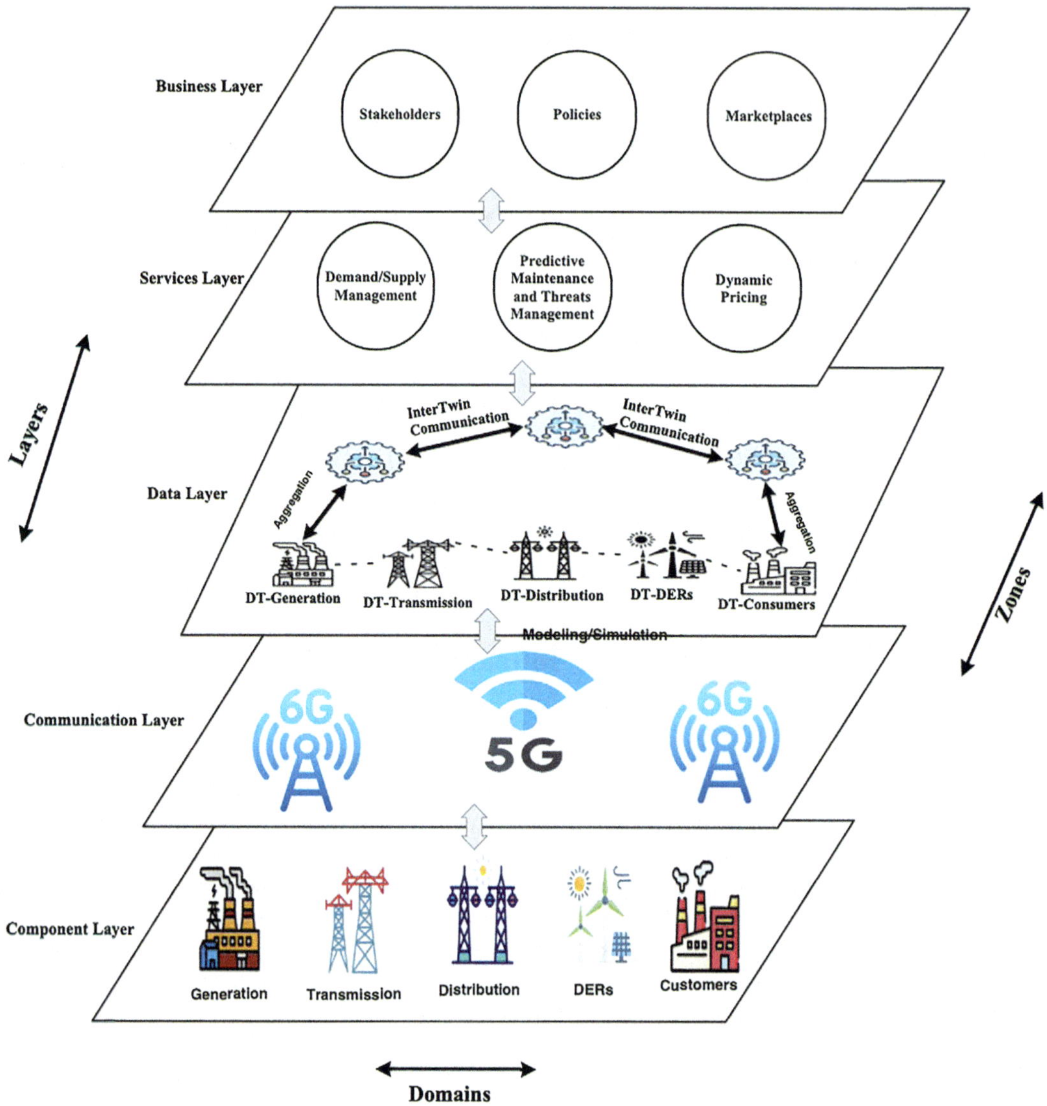

FIGURE 16.1 DT-enabled smart grid architecture. *DT*, Digital Twins.

operational data. Furthermore, this integration will also promote collaboration among distributed grid edges for efficient resource sharing and management while minimizing communication and computation overhead.

While DT supports the effective management of highly distributed operations at the edge of the smart grid, it also generates massive amounts of data. To fully realize the benefits of DT integration, the data must be managed efficiently. A promising approach to handling data is embedding DT capabilities as a dedicated data layer within the hierarchical structure of CPS (Qian et al., 2024a), in which the smart grid is a subset. This smart grid metaverse concept enables data-driven computation and communication at the edge of future smart grids (Qian et al., 2024a; Tightiz et al., 2024). Such an architecture allows for the real-time tracking of safety-critical events, thereby generating high-quality data that can power advanced, distributed data-driven services at the distributed edges of the smart grid.

As Fig. 16.1 illustrates, in the DT-enabled smart grid framework, the data layer is a host for digital replicas of all operational domains of the grid (Grid, 2013; Gopstein et al., 2021; Qian et al., 2024a; Tightiz et al., 2024). This enables real-time status tracking of the entire safety-critical infrastructure and supports efficient data-driven computation by leveraging the speed, bandwidth, and flexibility of the communication layer.

This chapter makes the following contributions:

TABLE 16.1 Essential grid monitoring subsystems.

Subsystems	Role	Application domain
Supervisory control and data acquisition (SCADA)	Real-time visibility and control	Transmission and distribution
Phasor measurement unit (PMU)	High-resolution system awareness	Transmission
Energy management systems (EMS)	Forecasting, optimization, and dispatch	Transmission and distribution
Advanced metering infrastructure (AMI)	Consumer data, demand-side management	Distribution

- We review the existing research efforts on DT and edge computing in power grid systems.
- We propose a DT-enabled smart grid edge computing framework that incorporates DT capabilities as a dedicated layer within the hierarchical grid architecture to address emerging challenges in the grid.
- We conduct a case study that demonstrates the proposed DT-enabled smart grid framework and its capability of managing the dynamic energy demand and supply of the smart grid.
- We also outline some aspects of future research directions and potential innovations in this domain.

The remainder of the chapter is organized as follows: Section 16.2 presents the technological background of edge computing, DT, and smart grids. In Section 16.3, we define a three-dimensional (3D) problem space that covers the various smart grid domains, potential research objectives, and intelligent techniques that guide data handling and analysis. Section 16.4 presents the integration of DT and edge computing in the smart grid. Section 16.5 presents the details of the case study, along with the evaluation methodology and results analysis. We outline promising research avenues for DT-enabled smart grids in Section 16.6. Finally, Section 16.7 concludes the chapter.

16.2 Preliminaries

To contextualize the integration of DT and edge computing technologies in smart grid systems, it is essential to first establish a foundational understanding of each component. This section provides an overview of the smart grid systems, followed by key concepts in edge computing and DT technology.

16.2.1 Smart grid as an energy cyber-physical system

Generally speaking, CPS drives the system design of integrating computation (cyber) into physical processes, leading to one system. In CPS, embedded sensors, computing devices, and communication networks are utilized to enable the monitoring and control of physical processes (things or objects). In many CPS, real-time, reliable computing and communication are desired to support interaction between the digital and physical worlds, combining sensing, communication, computation, and actuation to support autonomous or semi-autonomous decision-making. In CPS, state monitoring and control of physical components are driven by real-time operational data, which often varies across spatial and temporal dimensions (Musa et al., 2023a). Fortunately, advancements in the ICT field have extended CPS concepts to various smart-world critical infrastructure systems, including energy, transportation, manufacturing, healthcare, agriculture, and cities, among others (Liang et al., 2019, 2020; Liu et al., 2019a, 2023; Mallapuram et al., 2017; Mulo et al., 2025; Musa et al., 2023b; Qian et al., 2024a, 2024d; Rajkumar et al., 2010; Stankovic, 2014).

The smart grid is a representative example of an energy CPS, designed to optimize energy generation and distribution across the grid's transmission and distribution networks, thereby realizing a reliable and sustainable energy ecosystem (Esenogho et al., 2022). The grid consists of two main domains (illustrated in Fig. 16.2): the transmission network, which encompasses generation sources and high-voltage transmission lines, and the distribution network, which delivers electricity from the transmission lines to end-users. Across both domains, vast amounts of operational data is continuously generated during safety-critical processes. This generated data must be processed and acted upon in real-time to ensure system reliability and resilience.

Consequently, there is a growing need for advanced techniques to support efficient data handling and robust mechanisms for tracking and managing critical grid operations. These capabilities are enabled through the integration of intelligent monitoring subsystems, summarized in Table 16.1 and conceptually embedded within the grid architecture as illustrated in Fig. 16.2. These subsystems play a critical role in guiding utility decision-making

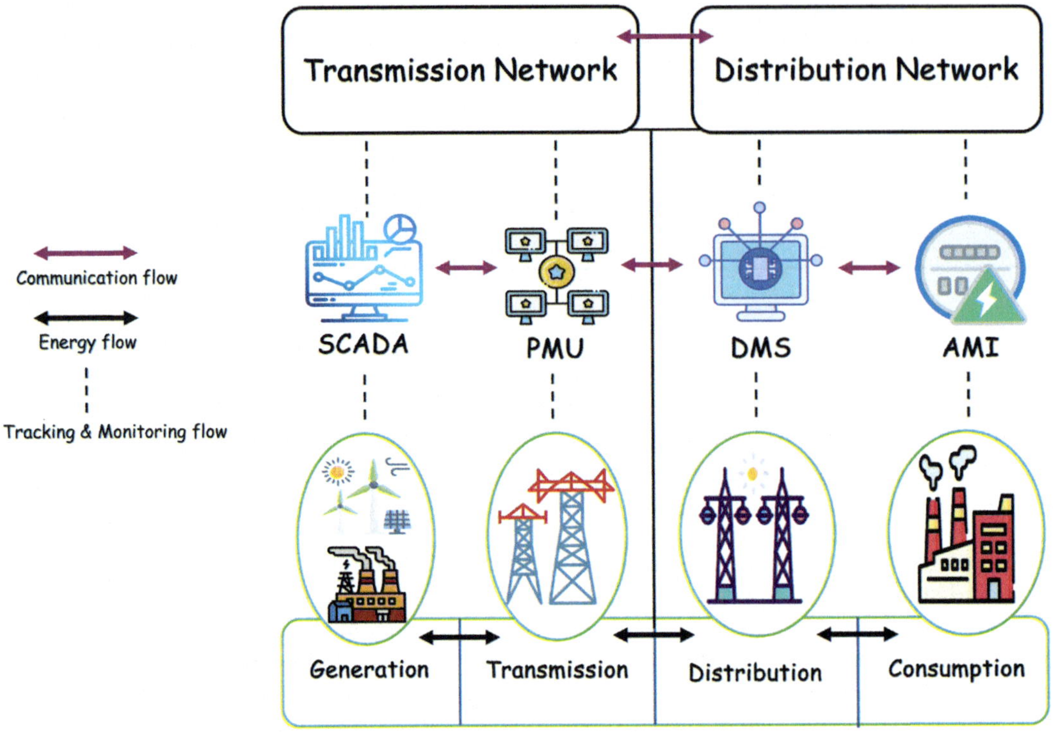

FIGURE 16.2 Smart grid with intelligent monitoring subsystems.

by generating and analyzing operational data. While some adopt centralized architectures, others are increasingly adopting decentralized approaches, aligning with modern grid transformation efforts.

Supervisory control and data acquisition (SCADA): First and foremost, the SCADA subsystem is the core backbone that offers real-time visibility and control over power grid operations. The subsystem integrates sensing and actuation devices, remote terminal units, and a central monitoring and control room to collect and transmit data on power flows, voltages, breaker status, and equipment health conditions. The subsystem is fundamental across transmission and distribution networks, enabling and guiding utilities' decisions based on the provided visualizations and alarms. The SCADA subsystem is widely considered the traditional backbone of power grid monitoring, providing the essential operational data for situational awareness and tracking (Boyer, 2009).

Phasor measurement unit (PMU): The PMU serves as the "Grid Stability Watchdog" that provides high-fidelity and time-synchronized data, i.e., Global Positioning System (GPS)-timestamped measurements of voltage and current phasors at microsecond intervals. These observations are aggregated into wide-area monitoring systems (WAMS), which span large geographical areas and offer near-instantaneous insights into the dynamic state of the grid. Unlike SCADA, which samples data every few seconds, PMUs operate at 30–120 samples per second, enabling early detection of oscillations and instabilities. PMUs are especially critical within the transmission network for maintaining system reliability and detecting blackouts before they escalate (Terzija et al., 2010).

Energy management systems (EMS): The EMS serves as the central intelligence for grid balance and stability, dispatch planning, and reliability across regions. It provides centralized intelligence that manages generation, consumption, economic dispatch, and grid frequency. The EMS ensures optimal scheduling of generation resources while maintaining system frequency and reliability. The management subsystem often incorporates SCADA data, PMU insights, and market signals to balance operational efficiency and economic performance (Rathor & Saxena, 2020). In decentralized grid architectures, such as microgrids and DER management systems, EMS features are incorporated into the distribution management system (DMS) to provide distribution domain functionalities (Kudzin et al., 2025).

Advanced metering infrastructure (AMI): The AMI provides real-time, detailed visibility into consumer-side energy usage and power quality management. This granular visibility empowers utilities to implement demand response strategies and allows customers to track their consumption patterns. AMI is foundational in smart grids by enabling more dynamic pricing models and decentralized energy management (Mohassel et al., 2014).

16.2.2 Edge computing

Generally speaking, edge computing emerged in response to the limitations of centralized cloud computing, particularly in latency-sensitive, safety-critical systems, in which handling data processing in real-time is essential (Yu et al., 2018). Initially popularized through content delivery networks aimed at reducing latency by caching web content closer to users, edge computing has since evolved to address the challenges posed by the proliferation of Internet of Things (IoT) devices, which are often resource-constrained. These devices generate large volumes of real-time operational data across various domains, which must be processed locally to ensure timely responses and system reliability (Cao et al., 2020; Shi et al., 2016). Edge computing addresses these issues as it deploys computation much closer to data sources directly so that low-latency analytics, control, and decision-making can be promptly conducted within the operational environment. Recent advances in wireless and mobile networks (e.g., 4G, . . . , 6G), along with the development of edge-specific hardware and software (e.g., Raspberry Pi, NVIDIA Jetson, AWS Greengrass, Azure IoT Edge), have made edge computing more accessible and scalable (Yu et al., 2024). These innovations empower edge nodes to perform complex tasks such as artificial intelligence (AI) inference, anomaly detection, and data analytics locally without relying on distant cloud servers.

Edge computing is typically organized into several layers: the terminal layer (front-end) that comprises resource-constrained, battery-powered sensing and actuating devices; the edge layer (near-end) that consists of edge servers to handle computational tasks beyond the capabilities of terminal devices; and the cloud layer (far-end) that handles resource-intensive tasks that cannot be executed at the edge (Yu et al., 2018). The terminology used for edge computing varies based on the deployment context. For instance, it is referred to as "fog computing" or "cloudlets" when involving distributed nodes (Babar et al., 2021; Yi et al., 2015), "mobile edge computing" when mobility is a key factor (Mach & Becvar, 2017), or "in-device computing" when computation occurs entirely within resource-constrained nodes (Lu & Shi, 2023; Yu et al., 2024). The core research topics in edge computing include computation offloading, resource allocation, and scheduling, all of which aim to ensure efficient resource management within edge environments. Additional focus areas include data caching and storage for timely data access, as well as security and privacy protection to mitigate emerging threats. Techniques for energy efficiency and computation cost optimization are also explored. These topics represent some actively explored research directions in the edge computing domain (Cao et al., 2020; Shi et al., 2016).

16.2.3 Digital Twin

The concept of creating digital replicas of physical systems originated with NASA's space missions in the late 1960s. During the Apollo 13 mission, engineers responsible for ground-based operations created a virtual replica of the spacecraft to simulate real-time operations and resolve synchronization-related issues. The term "Digital Twin" was later formalized by Michael Grieves in 2002 under the "Mirrored Spaces Model" and popularized in NASA's digital engineering initiatives by 2010. These early developments laid the theoretical and foundational groundwork for the broader application of DT across multiple domains (Grieves, 2023, 2005; Hussaini et al., 2022; Musa et al., 2023c; Qian et al., 2024b, 2024c; Singh et al., 2021).

With the advent of IoT-enabled smart CPS, advances in high-performance computing, big data analytics, and cloud platforms have transformed DT from static simulations into dynamic, data-driven systems capable of real-time monitoring, prediction, and control. Initially prominent in aerospace, manufacturing, and product lifecycle management, DT is now increasingly deployed across diverse CPS domains, including healthcare, transportation, and energy systems. From the CPS perspective, a DT represents a family of techniques and concepts used to track the status of physical objects in real-time via their digital counterparts. The cyber-physical synchronization is achieved through modeling and simulation, often powered by machine learning or deep learning (DL) techniques (Qian et al., 2022). DTs thus serve as the foundation for the emerging CPS metaverse, enabling interconnected, data-driven collaboration across safety-critical infrastructures such as smart grid and intelligent manufacturing (Qian et al., 2024a).

Within the energy CPS domain, the growing demand for clean energy, the increasing complexity of DERs, and the limitations of traditional fossil fuel-based infrastlructure have driven the evolution of grid architectures from centralized systems to decentralized smart grids, microgrids, and even virtual power plants, each of them introducing a considerable level of automation, decentralization, and digital control (Kabir et al., 2024). In this context, the electric digital twin grid (EDTG) (Sifat et al., 2023) has emerged as a key innovation. The EDTG generates a virtual mirror of the physical grid, combining historical and real-time data to support monitoring, forecasting, fault detection,

TABLE 16.2 Comparison of existing smart grid monitoring subsystems with the proposed DT-edge architecture.

Reference	Subsystem/Architecture	Monitoring type	Scope/Domain	Capabilities
Boyer (2009)	SCADA (supervisory control and data acquisition)	Centralized	Transmission and distribution	Real-time data collection, remote control, and alarm generation
Terzija et al. (2010)	PMU (phasor measurement unit) with WAMS	Centralized (high-fidelity)	Transmission (wide-area)	Microsecond-level time-data, synchronized phasor instability detection
Rathor and Saxena (2020), Kudzin et al. (2025)	EMS (energy management system)	Centralized (planning + control)	Generation to transmission	Frequency balancing, load dispatch, forecasting, grid stability
Mohassel et al. (2014)	AMI (advanced metering infrastructure)	Decentralized (consumer-side)	Distribution and end-user	Granular energy usage tracking, demand response, quality monitoring
This work	**DT-enabled edge framework**	Hybrid (central + edge)	Generation to end-user (entire networks and domains)	Real-time tracking, local decision-making, distributed analytics, proactive and reactive threat detection, DT-based modeling and coordination

DT, *Digital Twins;* WAMS, *wide-area monitoring systems.*

and adaptive control. It enables capabilities such as online analytics, self-healing, and intelligent decision-making, supported by simulation and AI-based learning.

The modern grid's DT operates through multi-layered architectures that include real-time data acquisition, cloud-based platforms, and bidirectional communication. Key communication protocols, such as MQTT, IEC 61850 MMS (Jha et al., 2025), open platform communications (OPC), and generic object oriented substation event (GOOSE), enable fast and secure data exchange between physical sensors, SCADA systems, and control centers. These capabilities make DT essential for advancing grid safety, reliability, demand-side management, and operational optimization.

A comparative overview of the smart grid subsystems (SCADA, PMU, EMS, and AMI) is provided in Table 16.2, along with the proposed DT-enabled distributed edge architecture. The table contrasts their monitoring approaches, scopes of operation, and key functionalities. While traditional systems focus on centralized control and domain-specific tasks, the DT-enabled architecture introduces decentralized, predictive, and real-time intelligence across the grid. This paradigm shift is essential for addressing the increasing complexity and emerging challenges in modern energy systems.

16.3 Problem space for edge-enabled Digital Twin in smart grid

The smart grid is inherently complex and multidimensional, involving various objectives, operational layers, and dynamic interactions across the CPS. To better understand the integration of edge-enabled DT, we introduce a 3D problem space, as shown in Fig. 16.3, which provides a holistic structural view for analyzing the integration of DT-enabled edge computing into the smart grid. The 3D problem space comprises the following dimensions: (1) *X* dimension: It represents the core operational domains of the grid, each with unique data and control requirements.

(2) *Y* dimension: It captures the functional objectives that represent the primary goals of grid operations. (3) *Z* dimension: It encompasses the intelligent techniques used to handle real-time data. By mapping challenges and opportunities across these three axes, the 3D problem space helps identify high-impact intersections where DTs and edge computing can enhance grid performance, resilience, and efficiency.

16.3.1 Grid operational domains (*X* dimension)

The *X* dimension of the 3D problem space represents the operational domains of the smart grid, such as power generation, transmission, and distribution, and the end-user (consumption). Each domain encompasses distinct

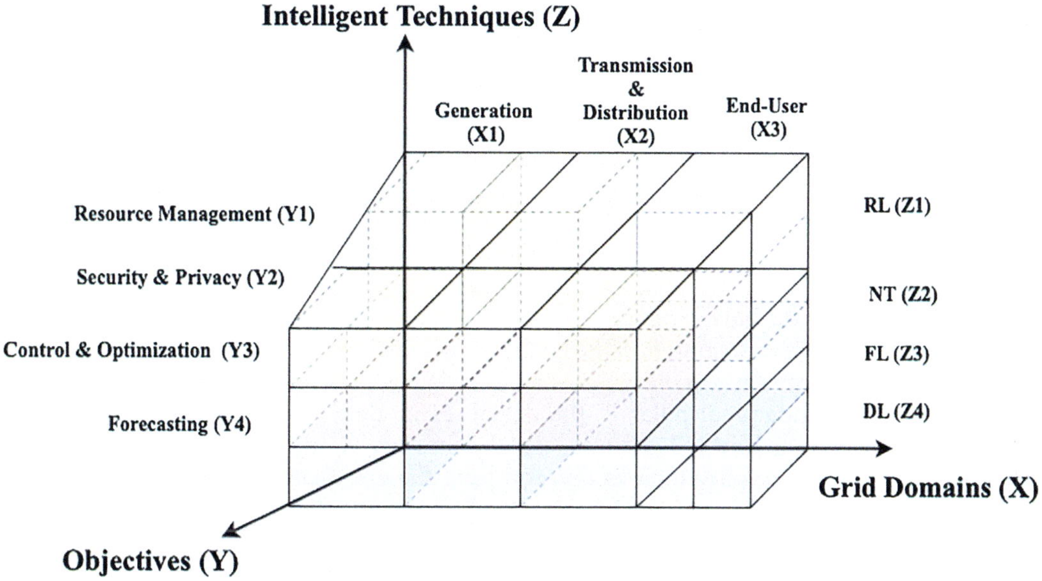

FIGURE 16.3 A problem space for DT-enabled edge intelligence in power grid. *DT*, Digital Twins.

processes and data flows, all of which are critical to the overall functionality of the grid. The integration of the edge-enabled DT enhances observability, control, and responsiveness across these domains.

Generation (X_1): The generation domain of the grid encompasses both centralized (e.g., thermal, hydroelectric, nuclear) and DER, including solar panels, biomass units, and wind farms, among others. With the increasing adoption of renewable sources, this domain is shifting toward a more decentralized architecture. In DT-enabled edge computing environments, generation systems benefit from enhanced capabilities such as real-time forecasting of renewable energy output using environmental data (e.g., solar irradiance, wind speed); predictive maintenance through condition-based monitoring of critical components like turbines and inverters; and deployment of edge-resident control logic for dynamic dispatch and fault response. Additionally, the integration of DT enables seamless cyber-physical synchronization, allowing for continuous tracking and simulation of the physical state of generation assets, thereby improving reliability and efficiency.

Transmission and distribution (X_2): The transmission and distribution domain segments serve as the critical infrastructure responsible for transporting electricity from generation facilities to end-users, spanning high-voltage transmission lines and medium-to-low-voltage distribution networks. It ensures the stability, reliability, and security of energy flow across vast geographical areas. With the integration of DT and edge computing, these highly safety-critical domains gain advanced operational intelligence. For instance, wide-area situational awareness is achieved through a PMU integrated with edge-enhanced SCADA systems, offering real-time visibility into voltage, frequency, and load dynamics. Localized feedback loops enable precise voltage and frequency regulation, while intelligent routing algorithms anticipate congestion and dynamically balance loads to optimize system performance. Moreover, edge-based anomaly detection at substations and feeders supports proactive fault detection and isolation, minimizing downtime. Through edge-fog coordination, regional hubs can aggregate multi-source data, enabling distributed control decisions and enhancing grid resilience.

End-user (X_3): The end-user domain segment encompasses the final point of electricity consumption, including residential households, commercial buildings, industrial facilities, and prosumers—entities that both consume and produce electricity (e.g., via solar photovoltaic system (PVs) and battery systems). At this edge of the grid, edge computing is pivotal for enabling localized intelligence and user-centric services. Demand-side management systems dynamically adjust loads in real-time in response to grid conditions or price signals, thereby improving grid balance and cost efficiency. Home energy management systems (HEMS) leverage real-time local data to optimize appliance usage, reducing energy waste while enhancing comfort. EVs charging control at the edge allows for adaptive scheduling based on grid load and dynamic tariffs. Moreover, blockchain-integrated peer-to-peer energy trading platforms enable prosumers to buy and sell energy locally, facilitated by DT agents that ensure transparency and trust. Finally, edge computing supports privacy-preserving analytics by keeping sensitive user data local, mitigating risk exposure while still allowing actionable insights for both users and grid operators.

16.3.2 Functional objectives of the grid (Y dimension)

The Y dimension of the 3D problem space encapsulates the key functional objectives that drive the design and deployment of smart grid systems. These functional objectives reflect the purpose of deploying intelligent technologies and emphasize how DT and edge computing integration are applied to improve reliability, efficiency, responsiveness, and intelligence in future energy infrastructures. Each objective corresponds to a distinct application area essential for grid performance and sustainability.

Resource management (Y_1): Resource management focuses on the efficient allocation, scheduling, and utilization of both physical and virtual grid assets. This includes managing DER, storage systems, and transmission capacity to meet dynamic energy demands. In DT-enabled edge systems, real-time data from sensors and asset-level DT enables localized decisions for load balancing, energy dispatch, and fault-tolerant scheduling. These mechanisms ensure operational continuity and contribute to sustainability goals by minimizing energy losses and maximizing the integration of renewable energy sources.

Security (Y_2): Safety is a foundational pillar in the design of modern power systems, particularly due to the increasing attack surface introduced by IoT devices, cloud connectivity, and distributed grid structures. The Y_2 dimension of the proposed problem space focuses on protecting the grid from cyber-physical threats, such as false data injection attacks (FDIAs), denial-of-service (DoS) attacks, or malicious control signal manipulation. Edge computation fused with DT-assisted systems enables distributed anomaly detection, real-time threat analysis, and intrusion detection systems (IDS) that act autonomously at local nodes. Secure communication protocols and blockchain-based authentication can further protect data integrity and ensure trust across the grid network.

Control and optimization (Y_3): This layer addresses the dynamic control of electrical parameters, including voltage, frequency, and power flow, to make sure that the grid operates within safe and optimal conditions. Automated optimization techniques are deployed at the edge to respond to fluctuations in demand and generator output, as well as unforeseen disturbances (e.g., faults, congestion). DT play a critical role in simulating grid scenarios, predicting the impact of control decisions, and orchestrating real-time adjustments across distributed control zones. The combined DT-enabled edge approach enables decentralized yet coordinated control actions, improving resilience and reducing operational overhead.

Forecasting (Y_4): Forecasting involves the use of predictive analytics to anticipate grid conditions, including load demand, generation availability (especially from renewables), market prices, and environmental factors like weather. Accurate forecasts are essential for proactive grid planning, energy trading, and contingency management. Edge-based forecasting solutions enable localized prediction capabilities, reducing latency and improving responsiveness. DT enhance forecasting accuracy by incorporating real-time feedback from physical components, historical trends, and machine learning-based scenario modeling. This foresight empowers operators to make informed decisions, optimize energy flow, and avoid disruptions.

16.3.3 Intelligent techniques utilized (Z dimension)

The Z dimension of the 3D problem space encompasses the intelligent techniques and algorithms utilized to process operational data at the edge, enabling real-time tracking and monitoring of safety-critical operations through DT (Hatcher & Yu, 2018; Liang et al., 2022; Mao et al., 2018; Mohammadi et al., 2018; Olowononi et al., 2021; Xu et al., 2020). Key strategies are explained in the following section.

Nature-inspired techniques (NT) (Z_1): This entails using algorithms that mimic the natural phenomena (processes, behaviors) observed in biological, physical, or ecological systems to model and solve nonlinear, multi-objective, and high-dimensional optimization problems, which are highly common for smart grid operations. Relevant to this, genetic algorithms and particle swarm optimization are representative examples of these algorithms (Pop et al., 2022). **Reinforced learning (RL) (Z_2):** This is trial-and-error learning with memory, where behavior improves through feedback in the form of rewards or penalties. The goal here is to learn a strategy that maximizes cumulative reward over a specified period. RL explores decision-making in uncertain and dynamic environments (Liu et al., 2019b). This makes it applicable to various smart-world applications, such as autonomous driving and energy optimization, especially when paired with simulation environments or DT.

Federated learning (FL) (Z_3): This is a distributed collaborative learning approach in which multiple edge devices train without sharing their raw data (Chen et al., 2023). The approach preserves privacy by keeping data within its origin, especially in privacy-sensitive application domains such as healthcare or smart grids, where data is geographically dispersed across multiple locations. Despite these developments, issues related to data heterogeneity

TABLE 16.3 Summary of Digital Twins in power grids.

Reference, Year	Objective	Approach	Future research
Zhou et al. (2021)	X2, Y1/Y2, Z3	DT-enabled safe and threat-free collaboration among distributed grid edge devices.	DT-enabled task offloading among collaborating distributed grid edge devices.
Liao et al. (2023)	X2, Y1, Z2	Age of information (AoI) metric for the optimum synchronization among collaborative edge node's resource allocation.	AoI for managing information timeliness in renewables integrated energy grids.
Fan et al. (2023)	X1, Y1, Z1	Nature-inspired AI techniques assisted DT for efficient, scalable, and cost-effective renewable grid operations.	Decentralized monitoring and control for scalable and resilient DT-enabled renewable grid operations.
Onile et al. (2024)	X3, Y4, Z4	A hybrid model that combines physics-based and machine learning-based modeling to represent/mimic/predict end-users' energy demand behavior.	Hybrid framework for secure, resilient, self-healing, and intelligent grids.

AI, *artificial intelligence;* DT, *Digital Twins.*

challenges (heterogeneous data distributions) and communication constraints remain open for further exploration within the research community.

Deep learning (DL) (Z_4): This mimics how humans learn from perception, using deep structures to interpret complex data. It is a learning technique in which computational models, such as neural networks with multiple layers, learn to extract features and represent patterns from data in a hierarchical manner. Deep neural architectures (e.g., convolutional neural networks (CNNs), recurrent neural networks (RNNs), and transformers) have achieved impressive performance across various fields and disciplines, including computer vision and spatial-temporal CPS data handling (Musa et al., 2023a).

16.4 Integrating Digital Twin and edge computing in smart grid

16.4.1 Digital Twin in power grid

DT has emerged as a transformative innovation in modernizing power grids, offering synchronization between physical assets and their virtual replicas in real-time. By continuously reflecting the state and behavior of grid components, DTs enable enhanced situational awareness, predictive analytics, and autonomous control capabilities that are increasingly vital in decentralized, data-intensive energy systems.

Some research has demonstrated the application of DT concepts across various aspects of grid operations. Notable use cases include the prevention of malicious updates across collaborating edge nodes (Zhou et al., 2021), tracking radio resource usage to prevent network congestion (Othman et al., 2023), and monitoring and forecasting end-user power demand behavior (Onile et al., 2024). Other applications address critical concerns, such as avoiding data tampering in energy generation forecasting (Boi et al., 2024), optimizing response times for multi-scheduled distribution grid resources (Liao et al., 2023), and efficiently managing energy in renewable-based systems (Fan et al., 2023). These efforts illustrate the expanding role of DTs in building intelligent, secure, and responsive grid infrastructures (Saad et al., 2020).

As summarized in Table 16.3, most DT-enabled grid efforts to date have focused on enabling reliable and safe monitoring and state tracking across various domains and network layers. These foundational capabilities serve as the basis for more advanced functionalities, including real-time decision-making, self-healing, and coordinated control in future power systems.

16.4.2 Edge computing in power grid

Edge computing in the power grid involves real-time processing of operational data at or near their sources, primarily generated by resource-constrained sensing and actuation devices. To overcome limitations in computation, storage, and energy, these devices often collaborate with more capable edge nodes, enabling efficient, data-driven operations and proactive responses to maintenance needs and cyber threats. Edge computing has emerged as

TABLE 16.4 Summary of edge computing in power grids.

Reference, Year	Objective	Approach	Future Research
Islam et al. (2023)	X2, Y1/Y2, Z1	Deep reinforcement learning (DRL) halts the false state information injection from the collaborating edge nodes during offloading decisions	Short-learning-based offloading in energy demand management systems.
Abdelmoula et al. (2023)	X1, Y3, Z4	Real-time, distributed monitoring of PV systems on campuses	Demonstration of the embedded hardware-based PV testbed.
Lu et al. (2024)	X2, Y1, Z1	An effective and optimum resource management strategy among collaborative grid inspection robots	DRL for the effective management of collaborative inspection robots.
Li et al. (2024)	X2, Y1, Z2	Nature-inspired AI for load balancing and optimization of energy consumption among MEC-based grid networks	DRL for the energy efficient, MEC-assisted grid networks.
Raihan Uddin et al. (2024)	X3, Y2, Z3	Edge-based false data injection detection for smart metering network	Encryption methods assist FL for detecting false data within the edge.
Liu et al. (2025)	X2, Y1, Z1	DRL for the optimum offloading policy generation among the grid-distributed computation entities	Optimum policies and migration strategies among mobile distributed grid entities.

AI, *artificial intelligence;* FL, *federated learning;* MEC, Maximum Export Capacity; PV, photovoltaic.

a key direction in the development of intelligent, responsive smart grid infrastructure. A major advantage of edge computing in the smart grid is its ability to support end-to-end, real-time monitoring and control across distributed network domains. This facilitates the seamless integration of DERs, including energy storage systems and renewable generation sources (e.g., solar photovoltaics, wind turbines) into both centralized and decentralized grid architectures, including microgrids (Abdelmoula et al., 2023; Sun et al., 2024). It also enables real-time operation and optimization of smart distribution networks (Ge, 2024), allowing for continuous visibility from generation to consumption (Alaguraj & Kathirvel, 2024).

Edge computing also supports collaborative offloading and scheduling strategies between heterogeneous grid components. For instance, Liu et al. (2025) proposed a collaborative offloading policy that enables coordination between end devices, edge nodes, and the cloud. In contrast, Lu et al. (2024) introduced a scheduling scheme for collaborative computation among robotic agents inspecting transmission lines. These examples demonstrate the importance of adaptive policies that balance workloads between resource-constrained and resource-rich nodes. Furthermore, effective management at the edge must encompass both physical and virtual resources, as highlighted in Zou et al. (2024), enabling efficient energy consumption and optimal utilization of grid resources (Li et al., 2024).

Edge nodes have finite computational, memory, and network resources. A typical Raspberry Pi 4 (R. P. Foundation, 2023) or Jetson Nano (NVIDIA Corporation, 2023) provides a quad-core advanced RISC machine (ARM) processor at 1.5 GHz, 4 GB RAM, and a GPU with roughly 128 compute unified device architecture (CUDA) cores. Such devices can maintain on the order of tens of DT replicas, perform inference at 5–10 queries per second (QPS), and store 24–48 hours of sensor data. More demanding applications (e.g., real-time fault detection across dozens of DERs) may require multi-node clusters or offload to nearby microservers. Therefore, the DT layer must support adaptive placement and load balancing to respect these resource budgets while preserving low-latency processing.

As summarized in Table 16.4, edge computing in smart grid systems is primarily focused on enabling efficient, secure, and low-latency data handling across all layers and domains of the smart grid. This local intelligence is crucial for addressing the real-time requirements and operational complexities of modern power systems.

16.4.3 Combining edge computing with Digital Twin in power grid

Integrating edge computing with DT technologies creates a powerful synergy for modernizing the power grid. While DTs provide real-time virtual replicas of physical components for monitoring, simulation, and predictive analytics, edge computing brings the required computational power closer to these components, enabling rapid decision-making and control without relying on centralized systems. In the following, we present a smart grid architecture structured into five interoperable layers that combine DT and edge computing to enhance computational capabilities, as shown in Fig. 16.1.

Component layer: This layer accommodates the physical grid components such as smart sensors, actuators, meters, transformers, and switches. They record and transmit energy-related information, such as voltage, current, temperature, and equipment status. Each component generates domain-specific operational data that supports monitoring, control, and decision-making across the grid infrastructure.

Communication layer: This layer facilitates the data exchange across different domains of the grid and enables the real-time state tracking of the component layer by overlying the DT layer. It ensures the necessary communication performance, such as low-latency, high bandwidth, and reliability, which is required for effective monitoring, synchronization, and control of grid operations.

Data layer: This layer hosts the virtual replicas of physical components from the grid's component layer. The layer is built on the concept of federated DT (Vergara et al., 2023), which enables independent virtual replicas to interact and collaborate across various grid domains. The federation involves regular distributed learning and its corresponding aggregation mechanism, leveraging hybrid-adaptive consensus mechanisms (Alghamdi & Albassam, 2024; Thwe et al., 2025) for effective state synchronization between the various twins and their corresponding physical objects. The layer also supports inter-twin communication (Othman et al., 2023) to enable cross-domain data exchange and cooperation between the virtual replicas.

Services layer: Having the data layer accommodate the entire power grid component layer's twin replicas makes service layer operations easier. This is due to the whole abstraction of the grid into a layer. From a service layer's perspective, the entire abstraction facilitates effective energy demand-supply management, adequate protection of the grid against data-driven threats and faults, and dynamic pricing mechanisms that further empower grid stakeholders to drive dynamic demand response.

Business layer: This layer accommodates the various stakeholders within the energy ecosystem (i.e., consumers, utilities, etc.). It defines the guiding operating policies and supports various energy marketplaces responsible for managing energy trading, pricing mechanisms, and contractual agreements, thereby facilitating efficient, transparent, and policy-compliant energy transactions.

Besides the five interoperability layers, several operating domains and zones span the *X*-axis and *Z*-axis of the grid edge framework, aiding in the management and safe operation of the grid.

On the interoperability details, it is true that distributed monitoring components operate at different sampling rates: for example, PMUs report phasor measurements at 50–120 Hz, whereas AMI meters typically record aggregate consumption every 15 60 minutes. Our data layer is designed to absorb these heterogeneities and provide real-time state tracking of all edge operations by normalizing message formats, units, and timestamps. We adopt a federated DT paradigm in which raw sensor data remains on the edge device, and only the global (guiding) model parameters are exchanged between the data layer and edge nodes. This approach reduces heterogeneity because the data layer orchestrates synchronization and aggregation, while edge devices update or download the global model without exposing local measurements.

Overall, it is evident that the introduced DT layer within the hierarchical structure of the smart grid serves as an *"abstraction"* reference that tracks the various operations within the components layer, as well as hasten or enable the ideal operation of the services and business layer.

16.4.4 Security and privacy controls

The decentralized nature of a DT-enabled edge grid introduces a broad threat surface, including FDIAs (Liang et al., 2016), DoS attacks (Gu & Liu, 2007), and model poisoning in FL (Fang et al., 2020). To systematically address these risks, we propose a threat control matrix (Table 16.5) that maps each interface in our architecture to concrete security controls. At the *component-DT interface*, hardware-rooted trust mechanisms such as secure enclaves (e.g., ARM TrustZone (Pinto & Santos, 2019) or trusted platform module (TPM)) provide tamper-resistant storage for keys and enable remote attestation; telemetry streams should be authenticated and timestamped to mitigate spoofing. At the *DT-edge interface*, the primary risks are model poisoning and tampering. These are mitigated by robust aggregation in FL, differential privacy to obscure client updates, gradient clipping, and signed DT model updates. At the *edge-cloud interface*, mutual authentication, encrypted channels, and rate limiting help counter DoS and man-in-the-middle (MITM) attacks (Conti et al., 2016). This mapping clarifies how each layer in our architecture can be secured without compromising low-latency processing.

TABLE 16.5 Threat control matrix for DT-enabled edge architecture.

Interface	Main threats	Controls and mechanisms
Component ↔ DT	Sensor spoofing (Park et al., 2016), FDIA	TEE/TPM attestation (Shang et al., 2024), telemetry signing, freshness checks (Zhong et al., 2024)
DT ↔ Edge	Model poisoning, tampering (Ananthakrishnan et al., 2025)	Robust FL aggregation, differential privacy, update signing (Zhang et al., 2025)
Edge ↔ Cloud	DoS, MITM	Mutual auth (Otway & Rees, 1987), TLS/DTLS (Kumar & Gandhi, 2020)

DoS, *denial-of-service;* DT, *Digital Twins;* DTLS, Datagram Transport Layer Security; FDIA, *false data injection attacks;* FL, *federated learning;* MITM, *man-in-the-middle;* TEE, Trusted Execution Environment;TPM, *trusted platform module;* TLS, Transport Layer Security.

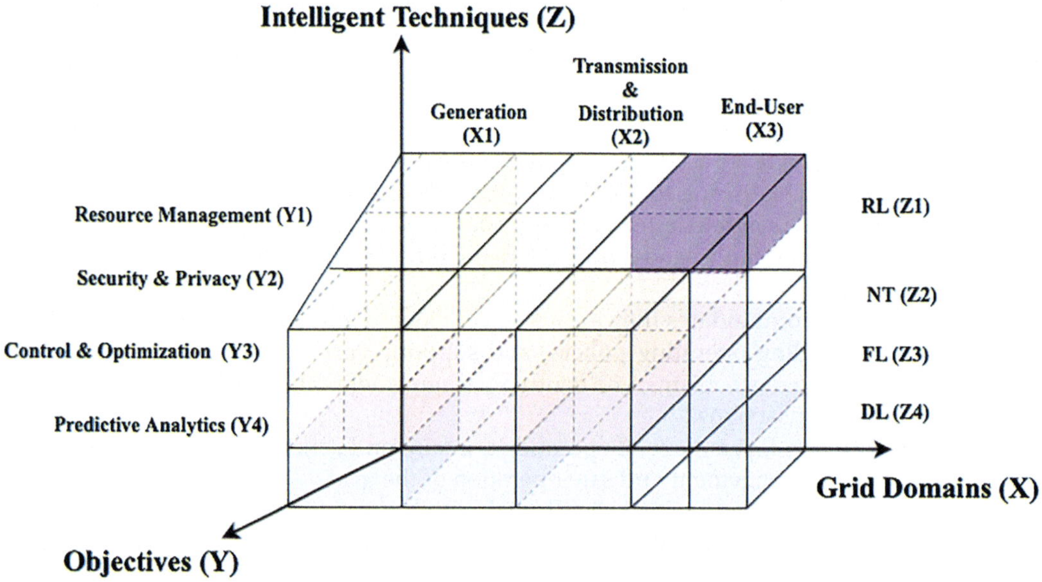

FIGURE 16.4 Case study in problem space.

16.5 Case study: Edge controlled for demand and supply in smart grid

This section presents a case study that consolidates the concepts introduced in previous sections, specifically, the framework presented in Fig. 16.1 of Section 16.4.3. The study demonstrates one of the several objectives envisioned to be achieved by DT-enabled edge computation within emerging power grids. The study fulfills the subspaces < X_3, Y_1, Z_1> as indicated by the proposed roadmap (Fig. 16.4), which is highlighted in purple and covers energy resource management by balancing demand and supply within the context of a microgrid, leveraging deep reinforcement learning (RL). Similar case studies can be considered as well, for example, to fulfill the subspaces < X_3, Y_2, Z_3> to address security issues, i.e., false data injection, anomaly detection at the smart meter leveraging FL, etc. In the following, we provide a detailed design and evaluation of the conducted case study.

16.5.1 Overview

The smart grid undergoes a fundamental transformation due to the pervasive integration of DERs that continue to grow in popularity (Akorede et al., 2010; Kar et al., 2024; Mohamad et al., 2021). The decentralized components substantially complicate grid management operations, particularly within residential areas. Utility companies must manage demand variability and intermittent solar generation while optimizing costs in real-time.

DT technology provides a practical answer through its ability to simulate virtual duplicates of physical grid assets. The twin-duplicate system models physical system behavior to enable predictive analytics, providing real-time monitoring of energy flow and consumption, as well as device health status. The combination of DT with edge computing enables fast, localized decision-making through operations near data sources, including residential homes and microgrid nodes, without relying on centralized servers.

The success of these systems depends on the intelligence that they contain. The DT receives adaptive control policies via RL, particularly the model-free soft actor-critic (SAC) (Haarnoja et al., 2018), enabling it to learn optimal energy storage and dispatch strategies through interaction with its environment.

Based on the proposed framework, as depicted in Fig. 16.1, the case study conducted here utilizes simulation to model a residential energy system as a digital replica, incorporating solar power generation and dynamic pricing. The evaluation assesses an RL-based edge-hosted agent against a traditional rule-based system in terms of cost reduction, system stability, and supply-demand management.

16.5.2 Scenario

We aim to design a scenario that illustrates how DT-enabled edge computing concepts enable a smart, decentralized residential microgrid. In this scenario, the resident does not rely solely on the local electric company's generation but also utilizes renewable energy sources (e.g., solar power). The decentralized microgrid has solar photovoltaic generation, a battery energy storage system (BESS), and a dynamic energy pricing scheme (Parida et al., 2011). The goal is to minimize supply-demand imbalance and reduce energy costs by deploying an RL model against a traditional rule-based system while operating within the physical constraints of the battery and generation system. To be precise, this scenario involves optimizing the timing of charging or discharging the battery using RL to achieve a lower cost bill and minimize the imbalance between demand and supply compared to a rule-based approach.

To yield a realistic scenario, a few premises are set as follows:

- **Scenario assumptions:** We use a simulation model to mirror a decentralized residential microgrid using real photovoltaic data and synthesized load patterns. A three-day sample of measured solar generation is repeated to construct a 30-day (720 h) profile, while household load is modeled as a daily sinusoidal curve with added Gaussian noise. All models run on edge computing devices with limited computational resources, so control decisions must be lightweight and executed between an edge server and its DT.
- **Solar generation and charging:** Instead of fixing sunrise and sunset hours, we use the real PV profile to capture natural daylight variation (U. A. C. S. Department, 2012). The battery is permitted to charge whenever solar output is available, and electricity prices are low, and to discharge when household demand exceeds the combined solar and battery supply or when electricity prices are high. This emulates realistic charging behavior driven by both generation and economic signals.
- **Dynamic electricity price:** To reflect real-world energy markets, time-of-use pricing is employed. We consider a peak mid-off-peak tariff (e.g., \$0.20/kWh during 5–10 PM, \$0.15/kWh during normal hours, and \$0.10/kWh overnight) as well as a flat-rate tariff (\$0.15/kWh). These price signals influence both the baseline and RL controllers when scheduling battery charging and discharging.

Having these premises in place, we designed two strategies (baseline and proposed methods) as follows: (1) *Baseline strategy:* The baseline uses a simple, price-aware rule: the battery charges when photovoltaic generation exceeds household usage or electricity prices are in the off-peak range and discharges when household demand exceeds photovoltaic output or prices are in the peak range. Battery actions are constrained to keep the state of charge (SOC) between 20% and 80%. (2) *Proposed method strategy:* The proposed method trains an SAC RL agent (Haarnoja et al., 2018) to derive a more adaptive and cost-sensitive control policy. SAC is chosen for its sample efficiency and ability to cope with stochastic environments such as variable solar generation and load noise. Unlike the baseline, the SAC agent observes the current load, solar generation, the battery's SOC, and the dynamic electricity price and learns to adjust the battery's charging or discharging power to balance energy flows and minimize operating costs.

16.5.3 Methodology

In the following, we introduce the profile setup for the environment, including house power usage configuration, solar system generation configuration, and battery configuration. We also present the SAC model configuration and evaluation metrics.

Environment profile: (1) Battery storage: We evaluate two capacities (10 and 20 kWh). In all cases, the battery's maximum charge/discharge rate is limited to 5 kW. (2) Solar generation: To capture realistic diurnal variation, we construct a 30-day (720-hour) generation profile by repeating a three-day sample of real photovoltaic data. Analytically, the PV output at hour t is approximated by the noisy sinusoid generation $= \max\left(0, 6\sin\left(\frac{2\pi t}{24} - \frac{\pi}{2}\right)\right) + \mathcal{N}(0, 0.3)$.

(3) Household load: Household demand fluctuates daily and is modeled as $\text{Load}_t = 10 + 3\sin\left(\frac{2\pi t}{24}\right) + \mathcal{N}\left(0, \sigma^2\right)$, where the noise standard deviation σ governs variability (e.g., $\sigma \in \{0.1, 0.5\}$). (4) Dynamic pricing: We adopt a time-of-use tariff: \$0.10/kWh during off-peak hours (midnight–7 AM), \$0.15/kWh during normal hours (7 AM–5 PM and 10 PM–midnight), and \$0.20/kWh during peak demand hours (5 PM–10 PM). A flat-rate scheme of \$0.15/kWh is also considered for comparison.

Baseline strategy: The baseline controller employs a price-aware rule: the battery charges whenever photovoltaic generation exceeds household demand *or* the real-time electricity price is in the off-peak range and discharges whenever demand exceeds generation *or* the price is in the peak range. The controller keeps the SOC between 20% and 80% to prolong battery life.

SAC RL model configuration: The SAC agent is trained for 200,000-time steps in each scenario. The agent observes the current load, solar output, SOC, and electricity price and outputs a continuous charging or discharging power. We use a neural network with two hidden layers of 256 units and an adaptive entropy coefficient (ent_coef = auto); the discount factor is $\gamma = 0.99$. The reward at time t is defined as $r_t = -\left|\text{Load}_t - (\text{Generation}_t + \text{BatteryAction}_t)\right| \times \text{Price}_t - \lambda_{\text{SOC}} \forall_{\{\text{SOC}_t \notin [0.2, 0.8]\}}$, where λ_{SOC} is a small penalty applied when SOC leaves the safe band.

Evaluation metrics: We report and compare the following metrics for both baseline and SAC policies: (1) Total cost: for each time t, define $\text{Cost}_t = |\text{Load}_t - (\text{Generation}_t + \text{BatteryAction}_t)| \times \text{Price}_t$; the total cost is $\sum_{t=1}^{T} \text{Cost}_t$. We record BaselineTotalCost and RLTotalCost using the baseline and RL battery actions, respectively. (2) Average absolute imbalance (AAI): at each time t define $\text{imbalance}_t = \text{Load}_t - (\text{Generation}_t + \text{BatteryAction}_t)$; then AAI $\text{AAI} = \frac{1}{T} \sum_{t=1}^{T} |\text{Imbalance}_t|$ Imbalance$_t$|. We record BaselineAvgImb and RLAvgImb computed under baseline and RL actions, respectively. (3) Peak imbalance: PeakImb $\text{PeakImb} = \max_t |\text{Imbalance}_t|$. We report BaselinePeakImb and RLPeakImb. (4) Peak imbalance reduction (%): quantifies the reduction in peak imbalance achieved by the RL agent relative to the baseline:

$$\text{PeakReduction}(\%) = 100 \times \frac{\max\limits_t |\text{Imbalance}_t|_{\text{baseline}} - \max\limits_t |\text{Imbalance}_t|_{\text{RL}}}{\max\limits_t |\text{Imbalance}|_t \Big|_{\text{baseline}}}$$

These metrics—*Total cost, AAI, peak imbalance,* and *peak reduction*—are the basis for comparing baseline and RL performance across all scenarios in our experiments.

In addition to constraining the SOC to the 20%–80% band, we impose hard limits on the battery's ramp rate, restricting the change in charging/discharging power to ±2 kW per hour, and maintain a 10% reserve margin to handle unforeseen load spikes. During training and deployment, actions that violate SOC or ramp rate constraints are projected back onto the admissible set. Furthermore, suppose the communication channel between the edge and the DT is disrupted, or the RL model produces an infeasible action. In that case, the controller reverts to the price-aware baseline strategy until normal operation resumes. Future work may explore safe reinforcement learning or model-predictive-control wrappers that explicitly enforce these constraints and provide formal stability guarantees.

16.5.4 Evaluation results

In this experiment, we first describe the time-series patterns used to construct the synthetic dataset. A three-day real photovoltaic sample shows clear daily cycles: output is near zero in the early morning, rises sharply in the morning, peaks around noon, and then declines in the afternoon with random fluctuations (see Fig. 16.5). The corresponding three-day synthetic load pattern oscillates around a 10-kW baseline with a 3 kW sine wave and added noise (Fig. 16.6). To build a 30-day (720-hour) dataset, we repeat these three-day solar and load sequences.

The resulting 30-day solar series preserves daily peaks and random noise (Fig. 16.7), while the 30-day baseline load series exhibits long-period oscillations plus noise (Fig. 16.8). These underlying patterns set the stage for comparing the price-aware baseline controller and the RL controller.

Table 16.6 summarizes the performance metrics across eight scenarios (two battery capacities, two pricing schemes, and two noise levels). Under the peak time-of-use tariff, the SAC-based RL controller markedly outperforms the baseline: total cost drops by roughly 50%, AAI decreases by about half, and peak imbalance is nearly halved. For example, with a 20 kWh battery and noise 0.1, the baseline total cost is \$1105.86, whereas the RL cost is \$531.34; the average imbalance falls from 11.37 to 5.47 kW, and the peak imbalance from 18.82 to 9.48 kW, yielding a peak reduction of nearly 50%. Conversely, under the flat-rate tariff, the RL controller performs worse than the baseline: total costs increase by 9%–12%, and both average and peak imbalances rise slightly, leading to negative peak reduction values. This outcome suggests that price signals are critical for the RL agent to optimize battery dispatch. Battery capacity and

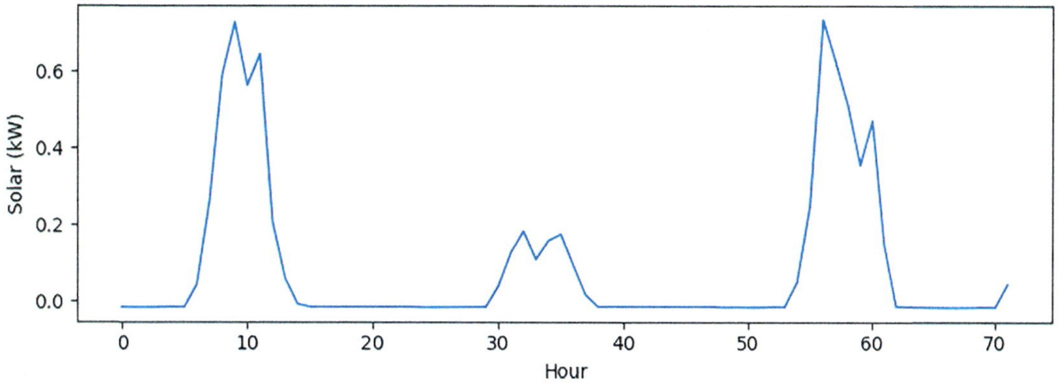

FIGURE 16.5 Real solar generation sample (3 days).

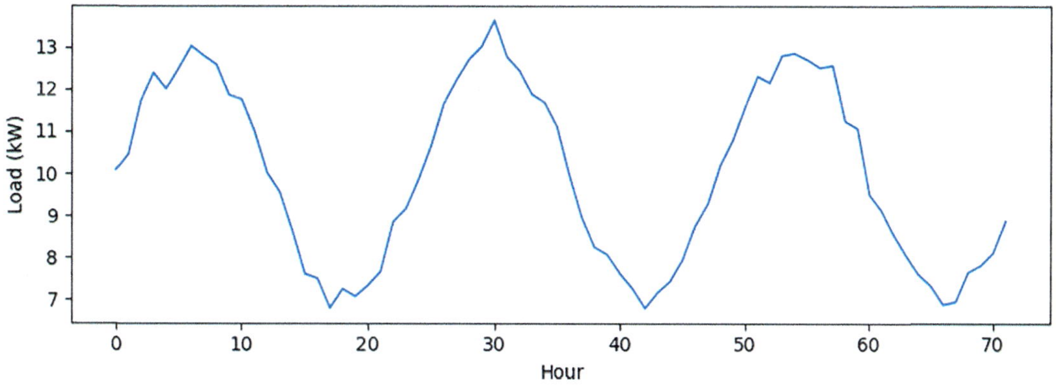

FIGURE 16.6 Synthetic load sample (3 days).

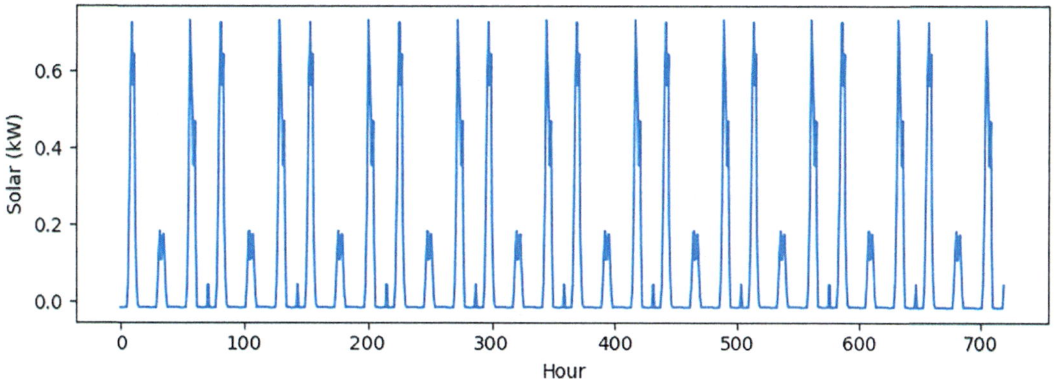

FIGURE 16.7 Solar generation (30 days).

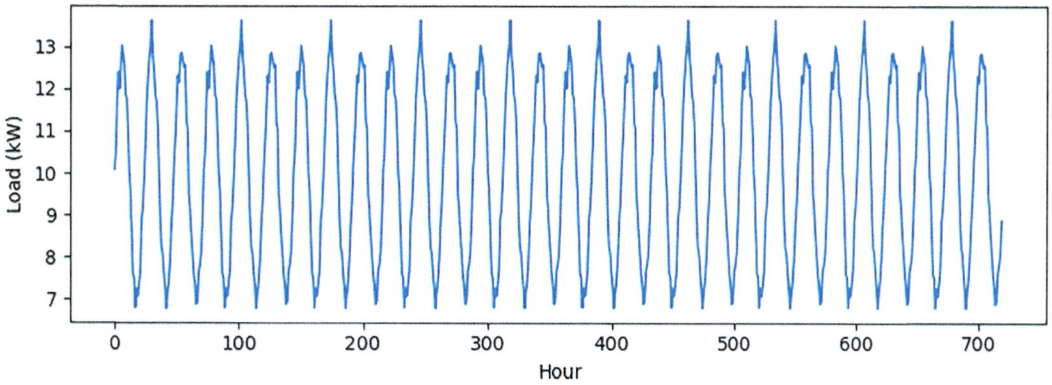

FIGURE 16.8 Synthetic base load (30 days).

TABLE 16.6 Smart grid control: Baseline vs. RL performance (30-day simulation).

Battery (kWh)	Scheme	NoiseStd	Baseline-Cost ($)	Baseline-AvgImb	Baseline-PeakImb	RL Cost($)	RL-AvgImb	RL-PeakImb	PeakReduction (%)
10.0	peak	0.1	1105.655029	11.366220	18.815012	544.367188	5.601904	9.497070	49.523972
10.0	peak	0.5	1104.244385	11.349007	19.671671	541.189880	5.555809	9.929382	49.524460
10.0	flat	0.1	530.651855	4.913443	8.815012	580.155151	5.371808	9.261112	−5.060688
10.0	flat	0.5	528.792725	4.896229	9.671671	572.486694	5.300802	10.083452	−4.257603
20.0	peak	0.1	1105.855103	11.368999	18.815012	531.340271	5.472689	9.478748	49.621353
20.0	peak	0.5	1104.444336	11.351785	19.671671	529.723755	5.447462	10.135525	48.476543
20.0	flat	0.1	530.351807	4.910665	8.815012	577.878601	5.350728	9.260576	−5.054607
20.0	flat	0.5	528.492676	4.893451	9.671671	592.148743	5.482859	10.607779	−9.678862

RL, *reinforced learning*.

noise levels have relatively minor effects on the relative performance; the price scheme is the dominant factor. Overall, the RL controller excels in dynamic pricing environments but offers no benefit under flat pricing, underscoring the importance of price-aware control policies.

16.6 Challenges and future research directions

As shown by the proposed framework in Fig. 16.1 and the case study in Section 16.5, Fig. 16.4 provides a roadmap for integrating DT-enabled edge computing into power grids. The case study fulfills the subspaces $< X_3, Y_1, Z_1>$ to minimize imbalance between demand and supply within the case of microgrids. Similar case studies can be considered in the future. For example, the subspaces $< X_3, Y_2, Z_3>$ can be considered to deal with the security and privacy of the consumption (end-user) domain within the distribution network. The same applies to the transmission network and many other systems. Yet, the smooth integration of DT-enabled edge computing to address emerging power grid issues is not free of challenges, some of which will be discussed.

Scalable deployment issues: Evident from our framework (as illustrated in Fig. 16.1), the DT supports sensing and actuating devices by tracking and monitoring their computational state in real-time. This necessitates deploying DT-based models on resource-constrained, battery-powered devices. While the integration of the duo techniques (edge and DT) is seen to be beneficial for addressing future grid challenges, during the deployment phases, it is critical to find a means of balancing computational efficiency, performance, and scalable resource management trade-offs across the pervasive, heterogeneous grid ecosystem (Aghazadeh Ardebili et al., 2024).

Emerging security and privacy issues: The pervasive distributed structure of the emerging grids widens the grid interfaces for privacy and safety threats, while the data layer is envisioned to provide a means of tracking the safety-critical operations in real-time, giving room for preventing or detecting threats. It is important to ensure that both the data and component layers remain synchronized all the time, i.e., the two layers must be ensured to be free from latency issues that may lead to misalignment between the two layers, thereby disrupting the tracking, which malicious adversaries might leverage to launch several attacks like replay, spoofing, MITM, etc. Regarding privacy-related issues, the data layer is built on the federated DT concept, in which multiple systems and virtual replicas are envisioned to coordinate, collaborate, and function as a unified system. At the same time, the raw data remains local, and a considerable level of privacy requirements can be preserved. However, considerable effort is required to transition this vision into reality.

Data-driven collaborative services: As illustrated in our framework (Fig. 16.1), the data layer enables the services provided within the services layer. From the business layer's perspective, it is envisioned that future electrical grids could incorporate marketplaces (Qian et al., 2025) that facilitate energy trading among various stakeholders in the energy ecosystem (utilities, consumers, service providers, etc.), leveraging the several enacted policies that guide interactions and transactions within the marketplaces. The marketplaces could pave the way for real-time, data-driven demand response; for example, consumers or other energy ecosystem members could trade in surplus energy they may have during peak hours through an auction mechanism, where the highest bidder wins the trade. The same applies to anyone in need of energy to meet their needs. All these activities are envisioned to occur in real-time.

Security and resource management in distributed DT architectures: As intelligence becomes decentralized, ensuring cybersecurity and safety across all facets of the edge is critical. Future research could focus on the development of lightweight, distributed intrusion detection and response systems that can operate autonomously at the edge while being reinforced by cloud-level analytics for threat correlation and policy enforcement. DT can enhance resource scheduling and energy efficiency by combining edge-local autonomy with cloud-wide coordination. Future work could explore hybrid architectures where local edge nodes manage short-term decisions while cloud models optimize long-term planning, especially in DER-dense and EV-integrated grids.

Distributed coordination of edge intelligence with DT: Future smart grids will require seamless coordination across generation, transmission, and consumption domains. Research can focus on how edge, fog, and cloud systems collaboratively manage grid stability through the synchronized use of DT. This includes designing control frameworks that enable real-time load balancing across layers while minimizing latency. Another research area is building interoperable DT that can communicate and collaborate across different grid layers and platforms. This includes developing standardized data models, APIs, and semantic frameworks to ensure DT at the edge, fog, and cloud can work together without fragmentation.

16.7 Conclusion

The global energy ecosystem is experiencing a fundamental shift from centralized infrastructure to decentralized, intelligent, and resilient systems driven by automation, reliability, and sustainability. During this transition, DT technologies emerge as a key enabler, providing real-time state tracking, predictive insights, and coordinated control across distributed grid edges. This chapter introduced a framework that integrates DT paradigms and concepts with distributed edge computation, supported by a case study that validates its practical relevance. Our results demonstrated a nearly 19% reduction in energy costs due to dynamic pricing that neutralizes the demand-supply imbalance. This paves the way for more efforts to address the identified future research directions and realize the full potential of the proposed layer. By leveraging data-centric intelligence and advancing AI-driven modeling and simulation, the proposed framework lays the groundwork for scalable and adaptive smart grid operations. Looking ahead, the envisioned data layer, capable of fusing contextual awareness with predictive analytics, opens new frontiers for innovation across diverse and pervasive energy domains. Ultimately, the chapter concludes by outlining clear pathways for future research and development, inspiring the design of next-generation smart grid systems that are not only intelligent and secure but also sustainable and future-ready.

References

Abdelmoula, I. A., Kaitouni, S. I., Lamrini, N., Jbene, M., Ghennioui, A., Mehdary, A., & El Aroussi, M. (2023). Towards a sustainable edge computing framework for condition monitoring in decentralized photovoltaic systems. *Heliyon, 9*(11), e21475.

Aghazadeh Ardebili, A., Zappatore, M., Ramadan, A. I. H. A., Longo, A., & Ficarella, A. (2024). Digital twins of smart energy systems: A systematic literature review on enablers, design, management and computational challenges. *Energy Informatics, 7*(1), 94.

Akorede, M. F., Hizam, H., & Pouresmaeil, E. (2010). Distributed energy resources and benefits to the environment. *Renewable and Sustainable Energy Reviews, 14*(2), 724–734.

Alaguraj, R., & Kathirvel, C. (2024). Edge computing and IoT-driven renewable energy integration for decentralized smart grid optimization. In *7th international conference on circuit power and computing technologies (ICCPCT): 1* (pp. 1775–1780). IEEE.

Alghamdi, W., & Albassam, E. (2024). Synchronization patterns for digital twin systems. *Journal of Applied Data Sciences, 5*(3), 1026–1037.

Ananthakrishnan, G., Sengan, S., Mohanraj, E., Palanisamy, T., Veeramallu, B., & Srinivasarao, B. (2025). Mitigating data tampering in smart grids through community blockchain driven traceability frameworks. *Journal of Machine and Computing, 5*(3), 1745–1762.

Babar, M., Khan, M. S., Ali, F., Imran, M., & Shoaib, M. (2021). Cloudlet computing: Recent advances, taxonomy, and challenges. *IEEE Access, 9*, 29609–29622.

Boi, A., Rossi, M., & Kim, H. (2024). Preventing data tampering in renewable energy forecasting using adversarial digital twins. *IEEE Transactions on Sustainable Computing, 9*(2), 789–798.

Boyer, S. A. (2009). *SCADA: Supervisory control and data acquisition*. International Society of Automation.

Cao, K., Liu, Y., Meng, G., & Sun, Q. (2020). An overview on edge computing research. *IEEE Access, 8*, 85714–85728.

Chen, Z., Tian, P., Liao, W., Chen, X., Xu, G., & Yu, W. (2023). Resource-aware knowledge distillation for federated learning. *IEEE Transactions on Emerging Topics in Computing, 11*(3), 706–719.

Conti, M., Dragoni, N., & Lesyk, V. (2016). A survey of man in the middle attacks. *IEEE Communications Surveys & Tutorials, 18*(3), 2027–2051.

Esenogho, E., Djouani, K., & Kurien, A. M. (2022). Integrating artificial intelligence internet of things and 5G for next-generation smartgrid: A survey of trends challenges and prospect. *IEEE Access, 10*, 4794–4831.

Fan, W., Zhang, L., & Ahmad, I. (2023). Energy management of renewable microgrids using edge-enabled digital twins. *Applied Energy, 333*, 120551.

Fang, M., Cao, X., Jia, J., & Gong, N. (2020). Local model poisoning attacks to {Byzantine-Robust} federated learning. In *29th USENIX security symposium (USENIX Security 20)* (pp. 1605–1622). USENIX Association.

Fang, X., Misra, S., Xue, G., & Yang, D. (2012). Smart grid the new and improved power grid: A survey. *IEEE Communications Surveys Tutorials, 14*(4), 944–980.

Feng, C., Wang, Y., Chen, Q., Ding, Y., Strbac, G., & Kang, C. (2021). Smart grid encounters edge computing: Opportunities and applications. *Advances in Applied Energy, 1*, 100006.

Ge, C. (2024). Strategies and practices of edge computing in real-time optimization and control of smart distribution networks. In *International conference on new power system and power electronics (NPSPE)* (pp. 224–229). IEEE.

Gopstein, A., Nguyen, C., O'Fallon, C., Hastings, N., & Wollman, D. (2021). NIST framework and roadmap for smart grid interoperability standards, release 4.0. *NIST framework and roadmap for smart grid interoperability standards, release 4.0*: 10. Department of Commerce, National Institute of Standards and Technology.

Grid, S. (2013). *IEEE smart grid vision for computing: 2030 and beyond* (pp. 1–133). *IEEE.*

Grieves, M. W. (2005). Product lifecycle management: The new paradigm for enterprises. *International Journal of Product Development, 2*(1-2), 71–84.

Grieves, M. W. (2023). Digital twins: Past, present, and future. In N. Crespi, A. T. Drobot, & R. Minerva (Eds.), *The digital twin* (pp. 97–121). Springer.

Gu, Q., & Liu, P. (2007). Denial of service attacks. In H. Bidgoli (Ed.), *Handbook of computer networks: Distributed networks, network planning, control, management, and new trends and applications, 3* (pp. 454–468). Wiley.

Haarnoja, T., Zhou, A., Abbeel, P., & Levine, S. (2018). Soft actor-critic: Off-policy maximum entropy deep reinforcement learning with a stochastic actor. In *International conference on machine learning* (pp. 1861–1870). PMLR.

Hatcher, W. G., & Yu, W. (2018). A survey of deep learning: Platforms, applications and emerging research trends. *IEEE Access, 6*, 24411–24432.

Hussaini, A., Qian, C., Liao, W., & Yu, W. (2022). A taxonomy of security and defense mechanisms in digital twins-based cyber-physical systems. In *IEEE international conferences on Internet of Things (iThings) and IEEE green computing communications (GreenCom) and IEEE cyber, physical social computing (CPSCom) and IEEE smart data (SmartData) and IEEE congress on cybermatics (Cybermatics)* (pp. 597–604). IEEE.

Islam, S., Zografopoulos, I., Hossain, M. T., Badsha, S., & Konstantinou, C. (2023). A resource allocation scheme for energy demand management in 6G-enabled smart grid. In *IEEE power & energy society innovative smart grid technologies conference (ISGT)* (pp. 1–5). IEEE.

Jha, S. K., Samantaray, S. R., & Kumar, D. (2025). Design and demonstration of an IEC 61850-based distribution substation using digital twin technology. In *IEEE 1st international conference on smart and sustainable developments in electrical engineering (SSDEE)* (pp. 1–6). IEEE.

Kabir, M. R., Halder, D., & Ray, S. (2024). Digital twins for IoT-driven energy systems: A survey. *IEEE Access, 12*, 177123–177143.

Kar, M. K., Kanungo, S., Dash, S., & Parida, R. R. (2024). Grid connected solar panel with battery energy storage system. *International Journal of Applied, 13*(1), 223–233.

Kudzin, A., Takayama, S., & Ishigame, A. (2025). Energy management systems (EMS) for a decentralized grid: A review and analysis of the generation and control methods impact on EMS type and topology. *IET Renewable Power Generation, 19*(1), e70008.

Kumar, P. M., & Gandhi, U. D. (2020). Enhanced DTLS with CoAP-based authentication scheme for the internet of things in healthcare application. *The Journal of Supercomputing, 76*(6), 3963–3983.

Li, S., Xie, Y., Shi, M., Zheng, X., & Lu, Y. (2024). Hybrid intelligent algorithm aided energy consumption optimization in smart grid systems with edge computing. *Intelligent Systems with Applications, 24*, 200444.

Liang, F., Hatcher, W. G., Liao, W., Gao, W., & Yu, W. (2019). Machine learning for security and the internet of things: The good, the bad, and the ugly. *IEEE Access, 7*, 158126–158147.

Liang, F., Qian, C., Yu, W., Griffith, D., & Golmie, N. (2022). Survey of graph neural networks and applications. *Wireless Communications and Mobile Computing, 2022*(1), 9261537.

Liang, F., Yu, W., Liu, X., Griffith, D., & Golmie, N. (2020). Toward edge-based deep learning in industrial internet of things. *IEEE Internet of Things Journal, 7*(5), 4329–4341.

Liang, G., Zhao, J., Luo, F., Weller, S. R., & Dong, Z. Y. (2016). A review of false data injection attacks against modern power systems. *IEEE Transactions on Smart Grid, 8*(4), 1630–1638.

Liang, S., Jin, S., & Chen, Y. (2024). A review of edge computing technology and its applications in power systems. *Energies, 17*(13), 3230.

Liao, J., Wang, R., & Liu, H. (2023). Ultra-reliable scheduling for distribution grids via digital twin synchronization. *IEEE Transactions on Smart Grid, 14*(3), 2340–2351.

Liu, W., Wang, Y., Xu, C., & Zheng, M. (2025). A smart grid computational offloading policy generation method for end-edge-cloud environments. *Journal of Reliable Intelligent Environments, 11*(1), 1–13.

Liu, X., Qian, C., Hatcher, W. G., Xu, H., Liao, W., & Yu, W. (2019a). Secure internet of things (IoT)-based smart-world critical infrastructures: Survey, case study and research opportunities. *IEEE Access, 7*, 79523–79544.

Liu, X., Qian, C., Yu, W., Griffith, D., Gopstein, A., & Golmie, N. (2023). Using deep reinforcement learning to automate network configurations for internet of vehicles. *IEEE Transactions on Intelligent Transportation Systems, 24*(12), 15948–15958.

Liu, X., Xu, H., Liao, W., & Yu, W. (2019b). Reinforcement learning for cyber-physical systems. In *IEEE international conference on industrial internet (ICII)* (pp. 318–327). IEEE.

Lu, S., & Shi, W. (2023). Vehicle computing: Vision and challenges. *Journal of Information and Intelligence, 1*(1), 23–35.

Lu, X., Yuan, S., Nian, Z., Mu, C., & Li, X. (2024). Optimizing task offloading for power line inspection in smart grid networks with edge computing: A game theory approach. *Information, 15*(8), 441.

Mach, P., & Becvar, Z. (2017). Mobile edge computing: A survey on architecture and computation offloading. *IEEE Communications Surveys & Tutorials, 19*(3), 1628–1656.

Mallapuram, S., Ngwum, N., Yuan, F., Lu, C., & Yu, W. (2017). Smart city: The state of the art, datasets, and evaluation platforms. In *IEEE/ACIS 16th international conference on computer and information science (ICIS)* (pp. 447–452). IEEE.

Mao, Q., Hu, F., & Hao, Q. (2018). Deep learning for intelligent wireless networks: A comprehensive survey. *IEEE Communications Surveys Tutorials, 20*(4), 2595–2621.

Minh, Q. N., Nguyen, V.-H., Quy, V. K., Ngoc, L. A., Chehri, A., & Jeon, G. (2022). Edge computing for IoT-enabled smart grid: The future of energy. *Energies, 15*(17), 6140.

Mohamad, F., Teh, J., & Lai, C.-M. (2021). Optimum allocation of battery energy storage systems for power grid enhanced with solar energy. *Energy, 223*, 120105.

Mohammadi, M., Al-Fuqaha, A., Sorour, S., & Guizani, M. (2018). Deep learning for IoT big data and streaming analytics: A survey. *IEEE Communications Surveys Tutorials, 20*(4), 2923–2960.

Mohassel, R. R., Fung, A., Mohammadi, F., & Raahemifar, K. (2014). A survey on advanced metering infrastructure. *International Journal of Electrical Power & Energy Systems, 63*, 473–484.

Moulema, P., Yu, W., Griffith, D., & Golmie, N. (2015). On effectiveness of smart grid applications using co-simulation. In *24th international conference on computer communication and networks (ICCCN)* (pp. 1–8). IEEE.

Mulo, J., Liang, H., Qian, M., Biswas, M., Rawal, B., Guo, Y., & Yu, W. (2025). Navigating challenges and harnessing opportunities: Deep learning applications in internet of medical things. *Future Internet, 17*(3). https://www.mdpi.com/1999-5903/17/3/107.

Musa, A. A., Hussaini, A., Liao, W., Liang, F., & Yu, W. (2023a). Deep neural networks for spatial-temporal cyber-physical systems: A survey. *Future Internet, 15*(6), 199.

Musa, A. A., Hussaini, A., Qian, C., Guo, Y., & Yu, W. (2023b). Open radio access networks for smart IoT systems: State of art and future directions. *Future Internet, 15*(12), 380.

Musa, U. S., Guo, Y., Qian, C., & Yu, W. (2023c). Optimal sampling for moving object trajectory tracking in smart transportation systems: A transformer-based approach. In *IEEE international conference on big data (BigData)* (pp. 1230–1235). IEEE.

Navarro-Ortiz, J., Romero-Diaz, P., Sendra, S., Ameigeiras, P., Ramos-Munoz, J. J., & Lopez-Soler, J. M. (2020). A survey on 5g usage scenarios and traffic models. *IEEE Communications Surveys Tutorials, 22*(2), 905–929.

Olowononi, F. O., Rawat, D. B., & Liu, C. (2021). Resilient machine learning for networked cyber physical systems: A survey for machine learning security to securing machine learning for CPS. *IEEE Communications Surveys Tutorials, 23*(1), 524–552.

Onile, E., Li, K., & Wu, H. (2024). Smartgrid-twin: A digital twin framework for user demand behavior monitoring in power systems. *IEEE Access, 12*, 12345–12356.

Othman, M., Ali, H., & Zikria, Y. B. (2023). Digital twin-enabled radio resource management for edge-intelligent smart grid. *IEEE Transactions on Industrial Informatics, 19*(5), 5674–5683.

Otway, D., & Rees, O. (1987). Efficient and timely mutual authentication. *ACM SIGOPS Operating Systems Review, 21*(1), 8–10.

Parida, B., Iniyan, S., & Goic, R. (2011). A review of solar photovoltaic technologies. *Renewable and Sustainable Energy Reviews, 15*(3), 1625–1636.

Park, Y., Son, Y., Shin, H., Kim, D., & Kim, Y. (2016). This ain't your dose: Sensor spoofing attack on medical infusion pump. In *10th USENIX workshop on offensive technologies (WOOT 16)*. USENIX Association.

Pinto, S., & Santos, N. (2019). Demystifying arm trustzone: A comprehensive survey. *ACM Computing Surveys (CSUR), 51*(6), 1–36.

Pop, C. B., Cioara, T., Anghel, I., Antal, M., Chifu, V. R., Antal, C., & Salomie, I. (2022). Review of bio-inspired optimization applications in renewable-powered smart grids: Emerging population-based metaheuristics. *Energy Reports, 8*, 11769–11798.

Qian, C., Liu, X., Ripley, C., Qian, M., Liang, F., & Yu, W. (2022). Digital twin cyber replica of physical things: Architecture, applications and future research directions. *Future Internet, 14*(2). https://www.mdpi.com/1999-5903/14/2/64.

Qian, C., Guo, Y., Hussaini, A., Musa, A. A., Sai, A. M. V. V., & Yu, W. (2024a). A new layer structure of cyber-physical systems under the era of digital twin. *ACM Transactions on Internet Technology*. doi:10.1145/3674974.

Qian, C., Qian, M., Hua, K., Liang, H., Xu, G., & Yu, W. (2024b). Digital twin based internet of vehicles. In *33rd international conference on computer communications and networks (ICCCN)* (pp. 1–6). IEEE.

Qian, M., Qian, C., Hussaini, A., Xu, G., Liao, W., & Yu, W. (2024c). Secured digital-twin data service for the internet of smart things. In Z. Lyu (Ed.), Smart spaces. *Intelligent data-centric systems* (pp. 71–102). Academic Press. https://www.sciencedirect.com/science/article/pii/B9780443134623000194.

Qian, M., Qian, C., Xu, G., Tian, P., & Yu, W. (2024d). Smart irrigation systems from cyberphysical perspective: State of art and future directions. *Future Internet, 16*(7), 234. https://www.mdpi.com/1999-5903/16/7/234.

Qian, M., Musa, A. A., Biswas, M., Guo, Y., Liao, W., & Yu, W. (2025). Survey of artificial intelligence model marketplace. *Future Internet, 17*(1), 35.

NVIDIA Corporation. (2023). *Jetson Nano Developer Kit*. NVIDIA Developer Documentation. https://developer.nvidia.com/embedded/jetson-nano-developer-kit.

R. P. Foundation. (2023). Raspberry Pi 4 model B specifications. https://www.raspberrypi.com/products/raspberry-pi-4-model-b/specifications/.

Rajkumar, R. R., Lee, I., Sha, L., & Stankovic, J. (2010). Cyber-physical systems: The next computing revolution. *Proceedings of the 47th design automation conference, ser. DAC '10. New York, NY, USA*. Association for Computing Machinery https://doi.org/10.1145/1837274.1837461.

Rathor, S. K., & Saxena, D. (2020). Energy management system for smart grid: An overview and key issues. *International Journal of Energy Research, 44*(6), 4067–4109.

Raihan Uddin, M., Rahman, R., & Nguyen, D. C. (2024). False data injection attack detection in edge-based smart metering networks with federated learning. arXiv preprint arXiv:2411.01313. https://doi.org/10.48550/arXiv.2411.01313.

Saad, A., Faddel, S., Youssef, T., & Mohammed, O. A. (2020). On the implementation of IoT-based digital twin for networked microgrids resiliency against cyber attacks. *IEEE Transactions on Smart Grid, 11*(6), 5138–5150.

Shi, W., Cao, J., Zhang, Q., Li, Y., & Xu, L. (2016). Edge computing: Vision and challenges. *IEEE Internet of Things Journal, 3*(5), 637–646.

Sifat, M. M. H., Choudhury, S. M., Das, S. K., Ahamed, M. H., Muyeen, S., Hasan, M. M., Ali, M. F., Tasneem, Z., Islam, M. M., Islam, M. R., Badal, M. F. R., Abhi, S. H., Sarker, S. K., & Das, P. (2023). Towards electric digital twin grid: Technology and framework review. *Energy and AI, 11*, 100213.

Singh, M., Fuenmayor, E., Hinchy, E. P., Qiao, Y., Murray, N., & Devine, D. (2021). Digital twin: Origin to future. *Applied System Innovation, 4*(2), 36.

Shang, K., Lin, J., Qin, Y., Shen, M., Ma, H., Feng, W., & Feng, D. (2024). CCxTrust: Confidential computing platform based on TEE and TPM collaborative trust. arXiv preprint arXiv:2412.03842. https://doi.org/10.48550/arXiv.2412.03842.

Stankovic, J. A. (2014). Research directions for the internet of things. *IEEE Internet of Things Journal, 1*(1), 3–9.

Sun, L., Bai, S., Shang, Y., Zhou, L., Xie, F., & Song, C. (2024). Collaborative control of photovoltaic-storage-charging integrated microgrid based on edge computing. In *3rd international conference on energy and electrical power systems (ICEEPS)* (pp. 825–831). IEEE.

Sutton, G. J., Zeng, J., Liu, R. P., Ni, W., Nguyen, D. N., Jayawickrama, B. A., Huang, X., Abolhasan, M., Zhang, Z., Dutkiewicz, E., & Lv, T. (2019). Enabling technologies for ultra-reliable and low latency communications: From PHY and mac layer perspectives. *IEEE Communications Surveys Tutorials, 21*(3), 2488–2524.

Terzija, V., Valverde, G., Cai, D., Regulski, P., Madani, V., Fitch, J., Skok, S., Begovic, M. M., & Phadke, A. (2010). Wide-area monitoring, protection, and control of future electric power networks. *Proceedings of the IEEE, 99*(1), 80–93.

Thwe, M. M., Stefanov, A., Rajkumar, V. S., & Palensky, P. (2025). Digital twins for power systems: Review of current practices, requirements, enabling technologies, data federation and challenges. *IEEE Access, 13*, 105517–105540.

Tightiz, L., Dang, L. M., Padmanaban, S., & Hur, K. (2024). Metaverse-driven smart grid architecture. *Energy Reports, 12*, 2014–2025.

U. A. C. S. Department, "UMass Smart* Solar Panel Dataset," https://traces.cs.umass.edu/docs/traces/smartstar/, 2012, contains 1-minute resolution power output data from 50 rooftop photovoltaic systems.

Vergara, C., Bahsoon, R., Theodoropoulos, G., Yanez, W., & Tziritas, N. (2023). Federated digital twin. In *IEEE/ACM 27th international symposium on distributed simulation and real time applications (DS-RT)* (pp. 115–116). IEEE.

Xu, G., Yu, W., Griffith, D., Golmie, N., & Moulema, P. (2017). Toward integrating distributed energy resources and storage devices in smart grid. *IEEE Internet of Things Journal, 4*(1), 192–204.

Xu, H., Liu, X., Yu, W., Griffith, D., & Golmie, N. (2020). Reinforcement learning-based control and networking co-design for industrial internet of things. *IEEE Journal on Selected Areas in Communications, 38*(5), 885–898.

Yi, S., Li, C., & Li, Q. (2015). A survey of fog computing: Concepts, applications and issues. In *Proceedings of the 2015 workshop on mobile big data* (pp. 37–42). ACM.

Yu, A., Qian, C., & Guo, Y. (2024). A real-time hand gesture recognition system on raspberry pi: A deep learning-based approach. In *IEEE 21st consumer communications & networking conference (CCNC)* (pp. 499–506). IEEE.

Yu, W., Liang, F., He, X., Hatcher, W. G., Lu, C., Lin, J., & Yang, X. (2018). A survey on the edge computing for the internet of things. *IEEE Access, 6*, 6900–6919.

Yu, W., Xu, H., Hematian, A., Griffith, D., & Golmie, N. (2016a). Towards energy efficiency in ultra dense networks. In *IEEE 35th international performance computing and communications conference (IPCCC)* (pp. 1–8). IEEE.

Yu, W., Xu, H., Zhang, H., Griffith, D., & Golmie, N. (2016b). Ultra-dense networks: Survey of state of the art and future directions. In *25th international conference on computer communication and networks (ICCCN)* (pp. 1–10). IEEE.

Zhang, C., Weng, J., Weng, J., Zhong, Y., Liu, J.-N., & Deng, C. (2025). Robust and secure federated learning with verifiable differential privacy. *IEEE Transactions on Dependable and Secure Computing, 22*, 5713–5729.

Zhong, P., Zhou, F., Feng, L., & Li, W. (2024). FINT: Freshness-based in-band network-wide telemetry in resource-constrained environments. *IEEE Transactions on Network and Service Management, 22*, 1314–1329.

Zhou, Z., Jia, Z., Liao, H., Lu, W., Mumtaz, S., Guizani, M., & Tariq, M. (2021). Secure and latency-aware digital twin assisted resource scheduling for 5G edge computing-empowered distribution grids. *IEEE Transactions on Industrial Informatics, 18*(7), 4933–4943.

Zou, J., Xin, P., Wang, C., Zhang, H., Wei, L., & Wang, Y. (2024). AI services-oriented dynamic computing resource scheduling algorithm based on distributed data parallelism in edge computing network of smart grid. *Future Internet, 16*(9), 312.

CHAPTER

17

Quantum computing with Digital Twins and power grids

S. Chidambaranathan

Department of Computer Applications, St. Xavier's College, Palayamkottai, Tamil Nadu, India

17.1 Introduction

The rising penetration of distributed energy resources (DERs), other forms of renewable energy, and an expanding demand for real-time control, optimization, and protection are all factors influencing the evolution of modern power grids. Although these developments can enhance sustainability and make energy more accessible to a broader audience, they also introduce significant complexity into grid operations. When it comes to dealing with the high-dimensional, nonlinear, and dynamic characteristics of modern systems, the traditional computational and analytical methods used for monitoring, fault detection, load forecasting, and control are frequently insufficient (Saini et al., 2024). This has resulted in an immediate demand for calculations that are more powerful and sophisticated, enabling them to address the challenges of scalability, adaptability, and uncertainty faced by the power network in future revenue generation. This research examines conversion as a transformative solution for the operation of modern power grid systems, leveraging quantum computing (QC) and DT technologies. The purpose of this research is to consider the demands placed on the industry. Digital Twin (DT) technology provides a dynamic and virtual representation of components and systems that form a physical grid, which is constantly synchronized with data collected in real-time. With DTs, which mimic the status and behavior of the physical grid in a digital environment, it is possible to conduct continuous supervision, forecast maintenance, predict faults, and make more informed decisions. Conversely, the costs of high-fidelity DTs are rapidly rising due to increases in load, grid size, and the complexity of maintenance and analysis. This phenomenon presents obstacles to processing, storage, and real-time analytics.

The promising answer to these restrictions is given by QC, which can carry out calculations in parallel and solve complex problems that classical computers cannot. When it comes to solving problems that involve large joint search spaces and nonlinear dependencies, quantum systems utilize the principles of superposition, entanglement, and quantum tunneling. The integration of algorithms such as quantum annealers and variable quantum encoders (VQE) with DT-based power grid models enable the optimization of load balancing and energy forecasts, allowing for dynamic fulfillment, fault testing, and communication. Quantum algorithms, for example, can be used to solve computationally intractable real-time issues that are otherwise beyond the capabilities of classical solvers. These problems include unit commitment and the best power flow.

A simulation environment is constructed within the proposed framework. This environment couples' traditional DT models with QC backends. The utilization of this hybrid technique makes it possible to test and validate quantum-enhanced algorithms on power grid situations that are representative of real conditions. When compared to standard DT models on their own, the results of these simulations indicate that computational performance is enhanced, response times are shortened, and the accuracy of system diagnostics and control decision-making is improved. Furthermore, the quantum-enhanced DT system demonstrates a higher level of resilience when dealing with conditions characterized by uncertainty and fluctuation. These variables are intrinsic to environments dominated by renewable technology and distributed grids.

In brief, the study's findings underscore the significant potential of integrating QC with DT technology to develop intelligent, adaptive, and scalable safety and control systems for future power grids. The combination of these two

Digital Twin Technology and Smart Grid.
DOI: https://doi.org/10.1016/B978-0-443-36664-2.00003-4

emerging technologies provides a solid foundation for constructing next-generation smart grids that are both efficient and robust in the face of growing energy demands and operational complexity. This is achieved by pushing the boundaries of real-time analytics, fault resilience, and energy optimization. This chapter, therefore, proposes and explores a framework for integrating QC with DT technology to overcome these computational limitations and enable a new paradigm of real-time, proactive, and resilient smart grid management.

17.2 Digital Twin architecture for power grids

The DT architecture for the power grid represents a shift toward an instance that is monitored, analyzed, and controlled by a contemporary energy routing system. A DT is a dynamic, virtual representation of the physical grid system that constantly reflects the status, behavior, and performance of real-world components through real-time data exchange and powerful analytics. At its core, DTs represent the physical grid system. Numerous levels, tightly interconnected, make up the DT architecture for the power grid. Each layer is responsible for a distinct set of features that, when combined, enable intelligent grid management. Physical assets from the real-world form the foundation. These assets include substations, transformers, transmission lines, DERs, and smart meters. These components, equipped with sensors and intelligent electronic devices (IEDs), collect various operating parameters, including voltage, current, frequency, and temperature. To ensure reliability, low delay, and interoperability between heterogeneous devices, this data is transferred using a communication network that is both secure and high-speed. This process often utilizes protocols such as International Electrotechnical Commission (IEC) 61850, Distributed Network Protocol version 3 (DNP3), or Message Queuing Telemetry Transport (MQTT). This is controlled by a high-volume, high-velocity data management layer, which is responsible for consolidating, preprocessing, and storing data. To ensure consistency and accuracy, real-time data streams are synchronized with historical databases, and robust filtering, feature extraction, and temporal alignment are carried out.

The modeling and simulation layer, which is the analytical heart of the DT, is situated on top of this foundation. In this case, the physical behavior of the grid components is modeled by utilizing a combination of physics-based modeling techniques and data-driven methods, such as machine learning. For example, long short-term memory (LSTM) neural networks are commonly employed for short- and medium-term load and generation forecasting, as they can effectively capture temporal dependencies and seasonal patterns in historical time-series data. In addition, convolutional neural networks (CNNs) can be applied to high-resolution sensor measurements or transformed signal representations (e.g., time–frequency maps) to detect anomalies, incipient faults, or abnormal operating conditions within grid components. Simulation of operational dynamics, prediction of system behavior, and the ability to perform what-if analyses under various load, fault, or generation scenarios are all capabilities of these models. Simulating transient responses, power flows, and fault propagation is accomplished through the utilization of sophisticated simulation engines, which are frequently coupled with platforms such as MATLAB/Simulink, OpenDSS, or DIgSILENT PowerFactory. When coupled with real-time data, these simulations enable the DT to accurately and adaptively reflect the current status of the grid. In addition, the control and decision-making layer utilizes intelligent algorithms to analyze anomalies, anticipate failures, and either recommend or execute control actions autonomously. These control actions include islanding, load shedding, and reconfiguration. In critical, time-sensitive scenarios, this layer can handle both human-in-the-loop decision-making systems and entirely automated responses. The creation of a closed-loop system that enables continual optimization and learning is made possible by a feedback mechanism, which ensures that the results of actions are communicated back to the physical layer.

In conclusion, the user interface and visualization layer provide grid operators, engineers, and decision-makers with tools for situational awareness, diagnostics, and strategic planning. It offers intuitive dashboards, 3D visualizations, and augmented reality and virtual reality capabilities. To protect critical grid data and enable secure interactions across the entire DT ecosystem, the design incorporates measures for data privacy, cybersecurity, and access control. These mechanisms are interwoven throughout the architecture. The architecture of the DT acts as a crucial enabler of real-time observability, adaptive control, and predictive maintenance. This is because the power grid is becoming increasingly decentralized, bidirectional, and variable due to the integration of renewable energy sources and DERs. This architecture, which is layered, modular, and scalable, not only improves the resilience and efficiency of the grid but also relies on the integration of emerging techniques such as QC, blockchain, and artificial intelligence, which in turn helps support more autonomous and intelligent systems.

FIGURE 17.1 Quantum computing's role in power systems.

17.3 Quantum computing in power systems

Increasing complexity, non-linearity, and distributed energy resources (DERs), as well as renewable power generation and data generated by smart grid infrastructures, are emerging as transformative techniques in QC, which is capable of calculating modern power systems. These challenges are because QC is capable of eliminating them. When dealing with high-dimensional and dynamic problems, traditional power system analysis, optimization, and control are often hindered by the limitations of classical computing (Saini et al., 2024). This is especially true when facing challenges that include many objectives. Superposition, phase, and quantum tunneling are some basic new methods that can be applied to information processing with QC, based on the principles of quantum physics. With the help of these quantum principles, quantum computers can perform a large number of simultaneous calculations, which significantly enhances both their processing and efficiency. QC enables the implementation of advanced applications in power systems. Some examples of these applications include large-scale unit commitment, real-time optimal power flow (OPF), rapid fault diagnostics, and predictive maintenance. Quantum annealing, for instance, is an approach suitable for solving combined optimization problems, such as power distribution networks, fault isolation, and resource allocation.

Similarly, quantum approximate optimization algorithms (QAOA) and VQE can be used for approximate non-linear and stochastic power flow problems, complementing classical solvers. Quantum machine learning (QML) utilizes the validity of high dimensional patterns in quantum-uplift facility spaces to enable rapid load forecasts, discrepancy investigations, and cyber-physical threat analysis. This is possible through the use of potential quantum improvements. When it comes to handling the unpredictable and constantly changing behavior brought about by renewable energy sources and electric vehicles, these capabilities are especially significant. QC, when combined with DT technology, is likely to accelerate the process of real-time analysis of grid data. This, in turn, enables faster simulations, evaluations of what-if scenarios, and more responsive decision-making. Through the implementation of DT simulations on quantum backends, operators can analyze various grid scenarios simultaneously, thereby delivering grid management techniques that are more robust and proactive.

Additionally, hybrid quantum–classical architectures enable their application in realistic settings. In this architecture, quantum subroutines are utilized for activities that need a significant amount of processing, while classical processors are responsible for handling conventional control logic and data management. The current quantum hardware is still in the noisy intermediate-scale quantum (NISQ) period, which means that it has a limited number of qubits and is susceptible to noise. However, rapid improvements in error correction, qubit fidelity, and quantum cloud platforms are making quantum-enhanced power system applications increasingly feasible. It has already been established that several utility firms and academic organizations have initiated experiments with quantum algorithms for grid optimization and resilience augmentation.

On the other hand, to fully capitalize on the advantages of QC, ongoing research and development are necessary in the areas of algorithm design, quantum software frameworks, and domain-specific quantum–classical integration techniques (Ren et al., 2024). In the not-too-distant future, as quantum technology continues to develop and mature, it is anticipated to become a crucial enabler in the construction of intelligent, adaptable, and self-healing power systems capable of meeting the energy demands and sustainability goals envisioned for the 21st century. For the sake of this discussion, QC is not only an incremental enhancement; rather, it represents a paradigm shift in the manner in which power systems might be optimized, protected, and otherwise managed. Fig. 17.1 is the illustration of QC's role in power systems, highlighting its application in high-dimensional optimization, fault prediction, load forecasting, and real-time decision-making within DT-enhanced innovative grid environments.

17.4 Integration of quantum algorithms with Digital Twins

When it comes to modeling, simulating, and controlling large power grid systems, the incorporation of quantum algorithms with DT technology represents a revolutionary leap (Ullah et al., 2022). DTs act as dynamic virtual representations of physical assets, providing real-time insights through the continuous synchronization of operational data, historical records, and simulation models. However, traditional computing methods employed within DT frameworks increasingly struggle with computational limitations as power grids become more decentralized and data-intensive. This challenge is driven by the proliferation of distributed energy resources (DERs), renewable generation, and stringent real-time control requirements. These limitations are particularly evident when addressing high-dimensional, nonlinear, and combinatorially complex problems.

QC presents a significant advantage in this context. Quantum algorithms—particularly quantum annealing and the variational quantum eigensolver (VQE)—are capable of exploring large solution spaces more efficiently than classical optimization and numerical solvers. When integrated with DT platforms, these algorithms enhance predictive analytics, optimization, and fault detection capabilities by accelerating the solution of problems such as load flow optimization, unit commitment, grid stability analysis, and real-time energy dispatch.

Quantum annealing is especially well-suited for combinatorial optimization tasks in grid operations, including minimization of transmission losses and optimal scheduling of DERs. Within a DT framework, quantum annealers can evaluate a vast number of operational scenarios in near real-time, enabling the identification of optimal grid configurations that maintain balance and reliability. Similarly, the VQE algorithm, which employs a hybrid quantum–classical approach, can be used to solve nonlinear system equations describing power flow, fault propagation, and state estimation. By learning complex system dynamics that are difficult to capture using purely physics-based or conventional data-driven models, VQE improves modeling accuracy in data-rich DT environments.

In this integrated architecture, the DT functions as a real-time data interface and simulation platform, supplying the quantum processor with relevant operational parameters while receiving optimized outputs for actuation, visualization, and further simulation. The resulting closed-loop feedback between the DT and the quantum backend enables not only responsive behavior to faults or instability, but also proactive prediction and prevention of adverse grid events.

Furthermore, the integration of quantum algorithms with DTs improves the modeling of uncertainty and probabilistic phenomena, such as stochastic equipment failures, fluctuating demand, and intermittent renewable generation (Din et al., 2025). By encoding these uncertainties into quantum states, the system can perform probabilistic simulations that capture a wide range of possible outcomes, which is essential for designing robust and adaptive control strategies. In practice, this integration is realized through hybrid quantum–classical computational architectures that explicitly divide computational responsibilities between classical DT platforms and quantum processors. Classical processors embedded within DT environments—such as MATLAB/Simulink or Python-based DT engines—handle continuous data acquisition, routine state estimation, forecasting, and rule-based control actions, while quantum co-processors are selectively invoked to address computationally intensive tasks, including large-scale reoptimization, combinatorial decision-making, or nonlinear system analysis. For example, under normal operating conditions, the DT performs real-time state estimation and load forecasting using classical solvers; however, when a fault prediction or severe renewable fluctuation is detected, a quantum co-processor is triggered to solve a complex reoptimization problem, such as optimal re-dispatch or network reconfiguration, within strict time constraints. Quantum software development kits, including IBM Qiskit, D-Wave Ocean, and Xanadu PennyLane, facilitate this interaction by translating DT-generated problem formulations into quantum-compatible representations and converting quantum measurement outcomes into actionable control decisions. Despite current limitations in quantum hardware related to noise and qubit fidelity, the coupling of near-term quantum algorithms with DTs offers a scalable and forward-compatible pathway for advancing smart grid protection, control, and optimization.

17.4.1 Data flow and integration mechanics

This subsection details the technical process through which data flows between the DT and the QC layer, focusing on data encoding, quantum execution, and result feedback.

Real-time measurements from the physical power grid—such as voltage levels, power flows, frequency deviations, and equipment status—are collected via sensors, SCADA systems, and phasor measurement units (PMUs). These data streams are ingested by the DT's data management layer, where they undergo preprocessing steps including filtering, normalization, dimensionality reduction, and problem formulation. At this stage, grid optimization or analysis tasks are expressed in forms compatible with quantum algorithms, such as quadratic unconstrained binary optimization (QUBO) models for quantum annealing or Hamiltonian formulations for VQE-based approaches.

The pre-processed DT data are encoded into quantum states by mapping system variables onto qubits. Depending on the algorithm and hardware constraints, encoding schemes such as basis encoding, angle encoding, or amplitude encoding are employed. For example, discrete decision variables related to unit commitment or switching configurations can be represented using binary qubit states, while continuous parameters such as power magnitudes may be embedded into parametrized quantum circuits.

Once encoded, the quantum circuit or annealing process is executed on the quantum processing unit (QPU). Quantum annealers search for low-energy states corresponding to optimal grid configurations, while VQE employs iterative hybrid quantum–classical optimization to approximate system eigenstates. Classical optimizers update circuit parameters based on measurement results, forming a feedback loop that continues until convergence criteria are met.

The outcomes of quantum computations are obtained through repeated qubit measurements, yielding probabilistic solution distributions. These results are decoded into classical variables representing actionable grid decisions, such as optimal dispatch levels, control setpoints, or stability indicators. The decoded outputs are then fed back into the DT's control and decision layer, where they update the digital model, inform operator decisions, or trigger automated control actions. This closed-loop process enables near real-time adaptation of the DT-based on quantum-enhanced computations.

The overall integration between the physical power grid, the DT, and the QC backend follows a closed-loop data flow designed to support near real-time analysis and control. Operational data are continuously collected from the physical grid through sensors, Supervisory Control and Data Acquisition (SCADA) systems, and PMUs and are ingested into the DT's data management layer, where they are filtered, normalized, and structured into physics- and data-driven models. These processed data are then translated into quantum-compatible representations, such as QUBO formulations or Hamiltonian models, and encoded into quantum states for execution on the QPU using algorithms such as quantum annealing or VQE. The quantum processor generates probabilistic solution outputs, which are decoded and post-processed into classical variables representing optimal control actions, system states, or stability indicators. These results are subsequently reintegrated into the DT's control and decision layer, where they update the virtual model, inform operator decisions, and trigger automated control responses. The feedback provided by the DT to the physical grid completes the loop, enabling adaptive, predictive, and quantum-enhanced operation of large-scale power systems.

17.5 Predictive analytics and fault forecasting

Predictive analytics and fault forecasting have emerged as essential components for ensuring grid stability, reliability, and resilience in the context of modern power grids, where complexity, decentralization, and dynamic behavior are continuously increasing due to the integration of DER, renewable energy sources, and real-time control mechanisms. Therefore, these components are essential for ensuring grid resilience, reliability, and stability. The goal of predictive analytics is to foresee possible failures or abnormalities before they occur by utilizing historical and real-time data, system behavior modeling, and advanced computing tools. In a system that is based on DT, predictive analytics becomes even more effective. This is because the virtual reproduction of physical power grid components enables continuous monitoring, simulation, and performance evaluation in a risk-free digital environment. Because of this, operators can proactively identify degradation trends, anomalous patterns, or precursors to problems, and then carry out preventive maintenance or remedial measures with minimal interruption to the system (Lv et al., 2022). The application of QC in DT models yields a substantial enhancement in the accuracy of predictive skills.

In comparison to conventional algorithms, quantum algorithms can handle high-dimensional data spaces and nonlinear system behaviors with greater efficiency. For example, quantum annealing can be utilized to optimize system settings and simultaneously evaluate a large number of fault scenarios, which ultimately results in an improvement in the granularity and accuracy of forecasts. Similarly, VQEs, also known as VQE, can be utilized to simulate dynamic responses and identify the circumstances under which a component is most likely to fail. These techniques have the potential to identify intricate interdependencies among grid variables, such as changes in voltage, harmonic distortions, load imbalances, and switching transients, which conventional analytics may fail to evaluate in real-time or may overlook entirely. By incorporating these capabilities into the DT environment, it is now possible to

TABLE 17.1 Overview of predictive analytics techniques and fault forecasting methods used for power grid protection.

Aspect	Description
Purpose	Anticipate faults and anomalies in power grid systems before they occur
Role of DT	Provides real-time, virtual replicas of grid components for continuous monitoring, simulation, and performance analysis
Use of historical & real-time data	Analyzes trends, patterns, and behaviors to detect early signs of failure
Quantum computing contribution	Enables faster, more accurate modeling of complex, nonlinear grid behaviors using high-dimensional computation
Quantum algorithms used	- **Quantum annealing:** Optimizes parameters and simulates multiple fault scenarios - **VQE :** Models dynamic responses
Fault types detected	There can be transient disturbances, short circuits, equipment degradation, cascading failures, etc.
Advantages over traditional methods	- Greater accuracy and granularity in fault prediction - Faster simulation and decision-making - Better understanding of variable interdependencies
Real-time forecasting capability	Allows near-instantaneous analysis and response using live data from sensors and IEDs
Maintenance optimization	Identifies high-risk components for proactive maintenance, reducing downtime and operational cost
Support for grid resilience	Enables self-healing, risk-aware, and adaptive protection strategies for smart grids
Outcome	A system for proactive, probabilistic, and autonomous forecasting and protection against grid faults

DT, *Digital Twin;* IEDs, *intelligent electronic devices;* VQE, *variable quantum encoders.*

model and anticipate a wide range of fault events, including transient disturbances, short circuits, cascading failures, and equipment deterioration, under various operational settings and weather conditions. It is possible to continually train and update predictive failure detection models by utilizing live data from sensors and IEDs, which further refine the models' accuracy and relevance. Furthermore, the capability of running quantum-enhanced predictive simulations in parallel significantly reduces the amount of processing time necessary to evaluate complex what-if scenarios, which in turn enables near-instantaneous decision-making in critical situations. Not only does this real-time forecasting help to improve operational efficiency, but it also plays an integral part in risk management, contingency planning, and adaptive protection tactics.

Additionally, predictive analytics optimizes resource allocation by determining which components are most at risk and require inspection or maintenance (Otgonbaatar et al., 2023). This helps reduce operating expenses and downtime, which in turn optimizes resource allocation. In short, the combination of QC with DT technologies makes it possible to create a paradigm shift in fault forecasting. This shift brings about a transition from reactive to proactive fault forecasting, as well as from deterministic to probabilistic approaches. This shift enables systems for managing power grids that are more durable, intelligent, and autonomous. Self-healing grids are capable of identifying, forecasting, isolating, and recovering from problems with minimal human intervention. This skill is essential for the realization of self-healing grids, which ensure high availability and stability in next-generation energy infrastructures. Table 17.1 showcases the *overview of predictive analytics techniques.*

17.6 Resilience and active control in quantum-Digital Twin systems

Both resilience and active control are essential characteristics of modern power grid systems, particularly as these systems continue to evolve into highly distributed, data-driven, and complex infrastructures. Combining QC with DT technology provides an innovative and potentially fruitful strategy for significantly enhancing these characteristics (Nouma & Yavuz, 2024). The term "resilience" refers to the capability of a quantum-DT-enabled power grid to predict, withstand, adapt to, and recover from disturbances, including faults, cyberattacks, equipment failures, and variability in renewable energy generation. DTs act as continuously updated virtual replicas of physical grid components by ingesting real-time sensor and operational data, enabling immediate situational awareness and predictive analysis.

These DTs become substantially more powerful when integrated with QC, particularly for the complex, nonlinear, and high-dimensional optimization problems inherent in power systems, as quantum algorithms offer the ability to explore large state spaces far more efficiently than conventional computing approaches.

Quantum algorithms, such as quantum annealing, can rapidly explore numerous system states and optimization pathways in parallel, enabling the identification of effective control strategies under abnormal or fault-induced conditions. For example, when a grid experiences instability due to an overload or short circuit, a quantum-enhanced DT can simulate multiple corrective actions in near real-time, evaluate their effectiveness, and trigger the most suitable control response. In addition, VQEs and hybrid quantum–classical models can monitor system eigenstates and stability margins, providing early warning indicators of oscillatory behavior or abnormal frequency deviations before the system reaches a critical state.

In the context of active control, the quantum-DT system enables dynamic coordination of grid assets—such as transformers, circuit breakers, and DERs—based on continuous assessment of grid health. Rather than relying on static protection settings or delayed operator intervention, quantum-DT systems can autonomously generate adaptive control setpoints, protection thresholds, and dispatch strategies. Quantum-enhanced predictive models allow the evaluation of future operating scenarios and enable proactive reconfiguration of grid topology, which is particularly valuable for managing uncertainty introduced by renewable energy sources such as wind and solar. Moreover, QC facilitates real-time reconfiguration of microgrids and islanded networks, helping to maintain power supply continuity during large-scale disturbances.

A self-healing grid refers to a power system capable of autonomously detecting faults, diagnosing their impact, and executing corrective actions with minimal or no human intervention (Jaschke et al., 2024). The real-time feedback loop between the physical grid and its DT—augmented by quantum computation—forms the core mechanism enabling self-healing behavior. Compared with conventional systems, quantum-DT-enabled grids can isolate faults, reroute power, and restore service more rapidly, thereby reducing downtime and associated economic losses.

Furthermore, the integration of quantum-enhanced detection techniques improves cyber-physical resilience by enabling advanced anomaly detection across communication networks, device behavior, and power usage patterns. Quantum-assisted machine learning and optimization methods can identify subtle deviations indicative of cyber intrusions or coordinated attacks, strengthening grid security. As smart grids become increasingly reliant on interconnected and intelligent devices, cyber-resilience becomes as critical as physical resilience. Together, QC and DT technologies provide a robust foundation for next-generation power systems that are resilient, adaptive, autonomous, and capable of evolving alongside increasing system complexity.

17.6.1 Case study: Self-healing in a microgrid

This section presents a sequential example illustrating how a quantum-DT-enabled system enables self-healing in a microgrid environment. First, a fault is detected when the DT identifies abnormal voltage and current patterns from real-time sensor data, indicating a feeder-level short circuit within the microgrid. The DT immediately updates its virtual model to reflect the fault location and its potential impact on connected loads and DERs. Second, the DT triggers a quantum optimization task, invoking a quantum co-processor to evaluate approximately 1000 possible isolation and power re-routing configurations in milliseconds using a quantum annealing algorithm. These scenarios consider constraints such as load priorities, line capacities, DER availability, and stability margins. Third, the optimal configuration—minimizing load shedding and maintaining system stability—is selected based on the quantum algorithm's output and is automatically executed through control signals sent to circuit breakers, switches, and inverter-based resources. Finally, feedback from the physical system confirms successful fault isolation and power restoration, and the DT updates its state to reflect the new steady operating condition. This closed-loop process demonstrates how quantum-enhanced computation enables rapid, autonomous decision-making, allowing the microgrid to recover from faults with minimal disruption and without human intervention.

17.7 Scalability and deployment considerations

The integration of QC and DT systems into modern power grids holds both enormous promise and significant technological challenges. Scalability and deployment of these systems are also important considerations. Scalable solutions are essential to ensure that real-time monitoring, control, and optimization functions can function effectively across large, heterogeneous networks. This is because power systems are becoming increasingly complex due to the widespread integration of DER, renewable generation, and dynamic load behavior. Within the context of an asynchronous and unified virtual environment, the DT framework can simulate thousands of physical components, such as transformers, lines, substations, and DERs. This is a crucial factor to consider when evaluating scalability. Maintaining precise state synchronization, data fidelity, and communication that are free of latency between the physical grid and its digital copy becomes increasingly complex as the number of components increases. This necessitates the establishment of robust computing and communication infrastructures (Wang et al., 2024). Additionally, the deployment of such a large-scale DT system necessitates the use of effective edge and fog computing layers. These layers can preprocess data close to its source, which lowers bandwidth usage and speeds up decision-making.

Due to the constraints of the currently available hardware, scalability remains a significant obstacle in the field of QC. Due to their limited qubit counts, short coherence times, and susceptibility to noise, the current quantum processors are not suitable for addressing large-scale problems. On the other hand, hybrid quantum–classical computing architectures present a significant opportunity for an intermediate answer. The system can strike a compromise between performance and feasibility by offloading computationally heavy optimization tasks to quantum processors. These activities include load forecasting, fault localization, and energy management. At the same time, the system maintains real-time control loops within classical DT settings. In addition, the deployment of such hybrid systems on a large-scale would necessitate the use of strong middleware platforms that can orchestrate the movement of data between classical and quantum environments, maintain synchronization, and gracefully handle faults.

Interoperability with the existing grid infrastructure is another significant development to consider. Because power utilities often use legacy systems and proprietary protocols, incorporating a contemporary quantum-DT platform requires adherence to communication standards such as IEC 61850, Common Information Model (CIM), and DNP3 to ensure the smooth flow of data and control operations (Ajagekar & You, 2024). It is also possible to scale computation workloads by utilizing cloud-based deployment methodologies, which can provide elastic resources for applications such as simulation, training, and optimization. On the other hand, this raises concerns about data privacy and cybersecurity, particularly when sensitive grid data is transmitted over networks or processed on platforms provided by third parties. When it comes to securing the deployment environment, strong encryption, secure application programming interfaces (APIs), and access control mechanisms based on blockchain technology are likely necessary.

The deployment lifecycle must also consider system calibration, model training, and the continual validation of the DT against real-time sensor data. This is because the deployment lifecycle addresses practical concerns. To accomplish this, a significant amount of automation, the capacity for self-learning, and adaptive feedback mechanisms are required (Yoon et al., 2025). Furthermore, to deploy such systems at the national or regional grid level, it is necessary for numerous stakeholders, including utility operators, quantum hardware providers, software developers, and regulators, to collaborate. To ensure the grid's dependability and resilience, it is necessary to establish policies that include governance, fail-safe measures, and data-sharing protocols. While the scalability and deployment of quantum-DT systems are technically demanding, advancements in cloud computing, quantum algorithms, distributed systems, and AI-driven orchestration will make such systems viable, intelligent, and adaptive solutions for the next-generation of resilient energy infrastructure. In general, this is the case.

17.8 Conclusion

The fusion of QC and DT techniques represents transformative progress in the management and defense of modern power grids. DERs and renewable payment generation continue to increase the complexity of the system; traditional calculation approaches are no longer sufficient to meet the demand for real-time monitoring, dynamic control, and predictive defect management. This study demonstrates that integrating QC, utilizing algorithms such as quantum algorithms and VQE, into DT models provides a powerful solution to overcome these limitations. By enabling the high-speed processing of large-scale, nonlinear, and uncertain data sets, the accuracy of the quantum-uplift DT framework's forecasts, fault detection accuracy, and energy will significantly improve the influence of optimization. The co-simulation environment further validates the feasibility of this hybrid architecture, demonstrating superior computational efficiency and response compared to classical systems. Ultimately, this research highlights a new direction for smart grid innovation, where intelligent, scalable, and self-contained structural quantum techniques, along with real-time digital replicas, can be achieved. The proposed approach lays a solid foundation for the construction of the next-generation energy system, which is not only elastic and flexible but also capable of meeting the evolving demand of the global energy zone network.

References

Ajagekar, A., & You, F. (2024). Variational quantum circuit learning-enabled robust optimization for AI data center energy control and decarbonization. *Advances in Applied Energy, 14*, 100179. doi:10.1016/j.adapen.2024.100179.
Din, I. U., Taj, I., Awan, K. A., Almogren, A., & Altameem, A. (2025). Quantum and GAN-driven Digital Twin approach for IoT-based consumer electronics manufacturing. *IEEE Internet of Things Journal, 12*(4), 3734–3741. doi:10.1109/JIOT.2024.3499737.
Jaschke, D., Pagano, A., Weber, S., & Montangero, S. (2024). Ab-initio tree-tensor-network Digital Twin for quantum computer benchmarking in 2D. *Quantum Science and Technology, 9*(3), 035055. doi:10.1088/2058-9565/ad5585.
Lv, Z., Cheng, C., & Song, H. (2022). Digital Twins based on quantum networking. *IEEE Network, 36*(5), 88–93. doi:10.1109/mnet.001.2200131.
Nouma, S. E., & Yavuz, A. A. (2024). Trustworthy and efficient Digital Twins in post-quantum era with hybrid hardware-assisted signatures. *ACM Transactions on Multimedia Computing, Communications and Applications, 20*(6). doi:10.1145/3638250.
Otgonbaatar, S., Nurmi, O., Johansson, M., Mäkelä, J., Gawron, P., Puchała, Z., & Dumitru, C. O. (2023). Quantum computing for climate change detection, climate modeling, and climate Digital Twin. *TechRxiv*. doi:10.36227/techrxiv.24478663.v1.
Ren, C., Dong, Z. Y., Yu, H., Xu, M., Xiong, Z., & Niyato, D. (2024). ESQFL: Digital Twin-driven explainable and secured quantum federated learning for voltage stability assessment in smart grids. *IEEE Journal on Selected Topics in Signal Processing, 18*(5), 964–978. doi:10.1109/JSTSP.2024.3485878.
Saini, K., Singh, A., Ahuja, A., Arora, N., & Saini, R. (2024). Research advancements in quantum computing Digital Twins. In *Digital Twins for smart cities and villages* (pp. 37–53). Elsevier. doi:10.1016/B978-0-443-28884-5.00002-6.
Ullah, M. H., Eskandarpour, R., Zheng, H., & Khodaei, A. (2022). Quantum computing for smart grid applications. *IET Generation, Transmission and Distribution, 16*(21), 4239–4257. doi:10.1049/gtd2.12602.
Wang, P., Zhang, D., Gan, L., & Zhang, Y. (2024). Key technologies and applications of collaboration between digital power grid and Internet of Things. *Digital Twins and Applications, 1*(1), 26–37. doi:10.1049/dgt2.12013.
Yoon, J., Alkhudary, R., Talluri, S., & Féniès, P. (2025). Risk management and macroeconomic disruptions in supply chains: The role of blockchain, Digital Twins, generative AI, and quantum computing. *IEEE Transactions on Engineering Management, 72*, 2995–3009. doi:10.1109/tem.2025.3585433.

CHAPTER

18

Integrating generative AI and LLMs with digital twins in smart grids

*Narges Fatemi, Siavash Hosseini, Javad Fattahi**
School of Electrical Engineering and Computer Science, Faculty of Engineering, University of Ottawa, Ottawa, ON, Canada
*Corresponding author: javad.fattahi@uottawa.ca.

18.1 Introduction

The evolution of modern power systems has brought about the introduction of digital technology to enhance grid stability, resiliency, reliability, and energy planning effectively. This requires smart grid assets, including distributed energy resources (DERs), smart meters, electric vehicle supply equipment (EVSE), supervisory control and data acquisition (SCADA), phasor measurement units (PMUs), and a DER management system (DERMS), to make the grid sustainable and resilient (Nguyen et al., 2024). In parallel with the development of power systems, the sequence of industrial revolutions has deeply altered industry operations to ultimately lead to the establishment of Industry 5.0. This new phase signifies a shift toward intelligent systems that integrate advanced technologies such as big data, high-performance computing (HPC), and deep learning to execute complex tasks with high efficiency. These innovations drive automation, improve decision-making, and enable the handling of large datasets, forming the foundation for contemporary industrial functions. However, the integration of these heterogeneous technologies presents challenges in communication and coordination.

Tackling the coordination challenges, DERMS have become essential tools for unlocking the full potential of DERs in smart grids. These systems offer sophisticated features for coordinating DERs, forecasting energy needs, and optimizing performance (Bazmohammadi et al., 2022). With advanced DERMS, we can manage power flows in real-time, enhance demand response (DR) strategies, and cut costs by adapting controls based on load patterns and market signals (Strezoski, 2023; Shayanfar, 2016). Control strategies, energy scheduling methods, and AI-driven analytics further extend DERMS capabilities for efficient DER management (Ahmad, Md, Ahmad, & Alowaifeer, 2023; Yuanyuan, 2022). Expanding beyond DERMS, the concept of software-defined microgrids (SDMG) introduces additional flexibility and resilience. The scholarly work presented in (Ndiaye et al., 2021) introduces the innovative concept of microgrid software definition, a paradigm aimed at optimizing the malleability of microgrid management. To ensure that communications and control systems for microgrids are cost-effective and reliable. It allows for the rapid sharing of energy resources between communities (Rehmani et al., 2019). A software-defined smart grid testbed was proposed in You et al. (2020) to integrate real-world wireless communication systems with AI algorithms. Furthermore, the integration of SDMG for cyber-physical protection within microgrids has been elucidated in Vu et al. (2020), offering a comprehensive exploration into the enhancement of microgrid resiliency against cyber threats.

Consequently, this orchestrated consolidation of SDMG technologies contributes directly to Digital Twins and scalability within smart grid ecosystems. The integration of Digital Twins (Bazmohammadi et al., 2022), alongside software decoupling techniques (Kabbara et al., 2022; Sun et al., 2020), containerization (Abuabdo & Al-Sharif, 2019), and serverless computing (Li et al., 2022), emerges as techniques that leverage the versatility of SDMG to harmonize real-time data with Digital Twins. The outcome is expected to be an enriched decision-making process and an elevated level of resilience against disruptions.

Digital Twin Technology and Smart Grid.
DOI: https://doi.org/10.1016/B978-0-443-36664-2.00021-6

To further enhance the state prediction, situational awareness (SA), and situational intelligence (SI) empower operators with proactive decision-making toward grid reliability and resilience (Arrano-Vargas & Konstantinou, 2023; Renold & Kathayat, 2024). Real-time cyber-physical visualization gained attention, using technologies like Digital Twins to improve grid security, flexibility, and resilience through advanced visualization tools (Danilczyk et al., 2019; Kumar et al., 2023). Furthermore, supervised learning techniques are increasingly applied for energy storage system optimization, enabling predictive control strategies that maximize storage utilization and operational efficiency (Li et al., 2021; Park et al., 2020). In addition, multiagent and self-organizing systems offer promising approaches for self-adaptive energy management, facilitating decentralized control and autonomous coordination of DERs (Alramlawi et al., 2022; Glass & Di Marzo Serugendo, 2023). Lastly, ancillary services and demand prediction are actively studied to enhance energy efficiency and dynamic load management (Bazmohammadi et al., 2022; Francisco et al., 2020; Srinivasan et al., 2020). Among these transformative digital technologies, Digital Twins have emerged as a promising solution for real-time system monitoring, predictive maintenance, and grid optimization. A more comprehensive review of Digital Twins in smart grids is presented in Cioara et al. (2022), highlighting their transformative role in modern power systems. Coupled with machine learning and AI, Digital Twins offer predictive insights and enable dynamic simulations such as load flow analysis and contingency planning. All together, it provides a cost-effective alternative to traditional hardware-in-the-loop (HIL) validation without the need for extensive physical infrastructure.

Taking this further, Digital Twins in the Loop (DTiL) have shown significant potential to enhance smart grid applications. DTiL integrates real-time sensor data with simulation environments, enabling continuous system evaluation, predictive analytics, and fault tolerance strategies. The DTiL framework also addresses the high costs of traditional enterprise platforms, allowing for cost-effective electrification in remote and underserved areas, and supports enhanced operational intelligence and fault resilience.

Simultaneously, the rise of generative artificial intelligence (generative AI) opens new avenues for innovation in Digital Twins and SDMG. Generative AI is a branch of AI that leverages advanced algorithms to process and create new content by learning from large datasets. It mimics human creativity, making it a transformative tool across sectors such as entertainment, marketing, and education. By identifying and learning data patterns, generative AI can produce content that is often indistinguishable from human-generated outputs, such as realistic dialogues or imaginative visuals. Within generative AI, transformers and reinforcement learning (RL) have notably accelerated advancements in AI technology. Transformers play a foundational role by enabling the development of Large Language Models (LLMs), which are capable of understanding and generating complex textual content. LLMs have advanced the field of natural language processing (NLP), reshaping how machines interpret and produce human language (Devlin, Chang, Lee, & Toutanova, 2019).

Despite these promising advancements, several challenges persist in grid digitization and advanced smart grids. These include data integration, model accuracy, ethical concerns, and the need for human-machine collaboration. Future advancements, such as Digital Twins, federated learning, blockchain, and AI-powered collaborative robots, will continue to enhance the capabilities of predictive maintenance, making industrial systems more autonomous and resilient (Ucar et al., 2024).

The integration of advanced digital technologies such as DERMS, SDMG, Digital Twins, and large Digital Twins, generative AI, and LLMs, into modern smart grid infrastructure is an exciting space for innovation in energy management, system resilience, and operational intelligence. In the following sections, we break down each of these components, starting with the SDMG architecture, which shows how layered modular systems from field-level sensors to cloud-based controllers enable secure, scalable, and real-time energy coordination. Then we dive into the transformative power of Digital Twins, highlighting how they can mirror physical assets through reduced-order modeling, scenario simulation, and adaptive control for predictive maintenance, grid fault analysis, and operational forecasting. Building on this foundation, it extends to the integration of LLMs and generative-AI, which are used not only for semantic interoperability but also for load and generation forecasting, prescriptive maintenance, and intelligent decision support through natural language interfaces. This chapter also covers advanced applications such as edge-based LLM deployment, parameter-efficient fine-tuning (PEFT), and physics-informed learning for contingency planning. Throughout, it provides practical frameworks, implementation strategies, and future research directions.

18.2 Software-defined microgrid architecture

Real-time monitoring and control capabilities in modern DERMS have opened new avenues for optimizing DER dispatch and power-flow management. These capabilities underpin efficient DR initiatives, load-shedding

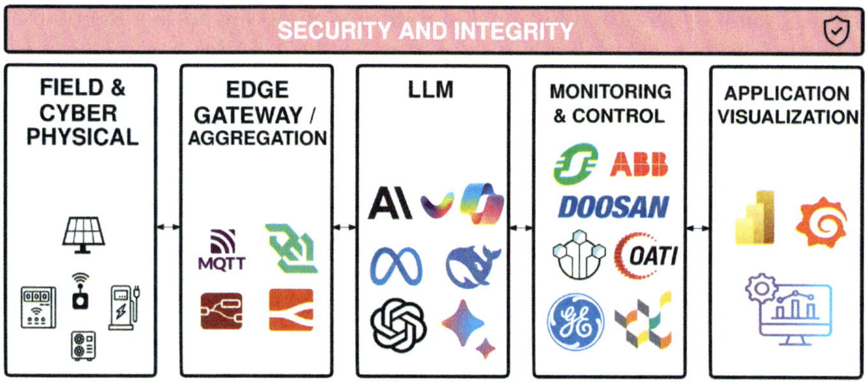

FIGURE 18.1 Layered architecture of a software-defined microgrid (SDMG), integrating field assets, edge aggregation, LLM-based semantic processing, monitoring and control, and application visualization with cross-layer security and integrity.

strategies, and the potential to lower energy costs by responding to real-time market signals. Building on this foundation, SDMG can enable agile coordination and communal sharing of energy resources while reinforcing cyber-physical protection. Moreover, the integration of Digital Twins, software decoupling techniques, containerization, and serverless computing paves the way for enriching decision-making processes against diverse operational challenges.

Fig. 18.1 shows a general structure of an SDMG, which is a modular system that enables flexible and effective management of DERs in small-scale or virtual utility environments. The architecture begins at the physical level, where hardware components such as energy resources, sensors, and meters interface with the electrical infrastructure to generate real-time data. This information is collected at the edge gateway (i.e., located at the houses or distributed transformer level) before being transmitted to the central system, where it is harmonized by LLM and used to construct Digital Twins assets. In monitoring an advanced control layer, intelligent forecasting and scheduling logic then optimize energy flows based on real-time conditions, economic signals, and operational goals. At the interface level, it provides visibility, insights, and assistance for both operations and troubleshooting. Security and identity mechanisms are embedded throughout the architecture to ensure secure data handling and regulatory compliance. Next, we describe each layer in detail.

18.2.1 Physical and field layer

This covers hardware components (e.g., DERs, sensors, meters, and local control equipment) that directly interact with the electrical infrastructure by producing or consuming power. Real-time operational data, including voltage, current, and power flow, is gathered by energy meters and sensors. In more modern smart grids, meters often transmit data via Wi-SUN to enable wide-area, self-healing mesh networks, making them ideal for rural or campus-scale systems. This layer aligns with the vision of AMI 2.0, emphasizing resilient, interoperable, and data-rich edge infrastructure.

18.2.2 Edge gateway/aggregation

This is an interface between low-level cyber-physical hardware and higher-level subsystems (e.g., Digital Twins, DERMS, visualization interfaces). Field device data are preprocessed, secured, aggregated, and delivered, as well as local automation and control logic simplified. Low-cost computing platforms and commodity industrial-grade hardware are often employed, selected on the basis of cost, reliability, and ease of integration. In this layer, diverse communication protocols (Wi-Fi, Zigbee, Low-Rank Adaptation [LoRA], or Modbus/TCP) will be grouped into standardized data payloads, usually in JSON or XML, that support uniform processing at higher layers. Minimal local processing is done to conserve bandwidth and provide reliable operation even with data validation, filtering, and aggregated usage calculation.

18.2.3 Large Language Model layer

This layer manages data heterogeneity, advanced predictive and prescriptive planning, which commonly arise when integrating DERs and sensors from diverse manufacturers. Instead of manually coding numerous custom adapters, data streams from different devices are interpreted and normalized in real time, allowing uniform representation of generation, consumption, and other operational metrics. Using LLM, new DER types and data formats can be incorporated more efficiently via natural language instructions or template-based prompts. This enables the system to adapt quickly when equipment evolves or additional monitoring points are added. Regulatory reporting requirements can be simplified through semantic tagging, which automatically classifies parameters such as consumption, generation, or peak demand periods. From an implementation perspective, the LLM semantic layer can be deployed on cloud infrastructure or on local hardware with resource-optimized language models, based on factors like latency, connectivity, or privacy concerns. Cloud deployment provides more compute power and easier updates, but adds latency and privacy risks. Edge deployment gives faster response and keeps data local, though with limited resources.

This approach increases flexibility and scalability, enables easier integration of new standards and devices, and ensures that subsequent layers, such as Digital Twins and DER management systems, receive clean, standardized data for more accurate modeling, optimization, and analysis. A lightweight real-time message broker (such as Message Queuing Telemetry Transport (MQTT) or WebSocket) could typically be used to establish reliable telemetry channels for sensor data, power generation metrics, and environmental observations. By maintaining a chronological archive of parameters such as solar output, battery state-of-charge, and voltage profiles, the ingestion layer supplies essential input to Digital Twin models and forecasting engines, facilitating scenario simulations and predictive analytics. In Section 18.4, we will further discuss the implementation details of this layer.

In practice, the placement of the LLM semantic layer could introduce a trade-off between cloud and edge deployment. Cloud-based deployment provides access to large-scale computational resources, enabling the use of higher-parameter LLMs and more sophisticated fine-tuning methods. However, cloud deployment adds latency due to round-trip communication. These factors can become critical in applications requiring real-time control or where regulatory frameworks restrict the movement of sensitive energy data. On the other hand, edge deployment of the LLM layer could reduce latency and enhance data privacy by keeping sensitive operational details local. At the same time, edge hardware is constrained in both memory and compute capacity, limiting the size of LLMs that can be executed and potentially reducing model accuracy. Resource-efficient approaches, such as quantization or LoRA, are often required to make deployment feasible in these constrained environments. A hybrid approach is also possible, in which lightweight inference runs locally while more computationally demanding tasks are offloaded to the cloud. This allows routine monitoring to be executed at the edge with minimal latency, while large-scale forecasting or retraining is performed in the cloud.

18.2.4 Digital Twin layer

This technology serves as a simplified software abstraction in which physical assets are virtually represented to enhance system observability, support informed decision-making, and facilitate operational control. Each DER and associated sensing or control devices are modeled through a virtual representation that captures key operational characteristics. Parameters such as generation capacity, power output, state-of-charge, and environmental measurements are continuously updated using real-time data received from the data ingestion layer. This ensures synchronization between the physical and virtual states, thereby maintaining an accurate and up-to-date system representation. To facilitate communication with higher-level control layers, standardized application programming interfaces can be provided to support data queries and control updates. In addition, the Digital Twin layer supports scenario simulation capabilities that allow operational behavior to be evaluated under a range of planning and control strategies. Simulations such as load shifting, battery dispatch scheduling, DR events, and islanding operations can be executed using simplified, resource-efficient analytical tools. In the next section, we will elaborate more on the Digital Twin structure and its integration.

18.2.5 Monitoring and control

At the core, forecasting and control logic can be applied to predict energy demand and renewable generation, particularly from photovoltaic sources. These forecasts can be derived using statistical averaging, rule-based heuristics, compact machine learning models, or LLM models trained on historical consumption and weather data. Forecast outputs can then be used to determine dispatch schedules, including battery charging and discharging, surplus energy allocation, and local consumption versus grid export decisions. These control strategies are continuously updated based on real-time operational data provided by the Digital Twin layer. Optimization functions are focused on objectives such as minimizing operational costs, maximizing local energy utilization, or enhancing participation in net-metering schemes. Local electricity tariffs or pricing signals can be incorporated into scheduling routines using linear programming approaches or simple rule-based methods. These routines consider the system constraints such as frequency ride-through (FRT) and voltage ride-through (VRT), battery capacity, state-of-charge limits, and generation forecasts. Once control decisions are computed, setpoints can be communicated to field devices, secured with encryption and basic authentication to ensure data integrity.

Analytics functionality in this layer is supported by historical and real-time data acquired from the Digital Twin. Predictive maintenance capabilities can be enabled by applying machine learning models to tempo-spatial key performance indicators from DERs. Deviations from expected operating behavior can be detected, and alerts can be generated to notify operators of potential faults or degradation, thereby reducing unplanned downtime and extending asset lifespan. Cost optimization analytics may also be implemented to identify opportunities for improving economic efficiency. Energy usage, generation profiles, and pricing signals can be visualized to inform strategies such as battery scheduling or load shifting. In addition, intelligent assistance can be provided through the integration of an LLM-powered chatbot.

18.2.6 Security and identity (cross-cutting)

The security and identity layer should be applied as a cross-cutting layer to ensure data integrity, confidentiality, and secure access across all system components. Device-level authentication should be implemented using lightweight certificates or one-time-use tokens, enabling only authorized sensors, controllers, and gateways to connect to the network. Communication between devices and gateways should be secured using Transport Layer Security (TLS 1.2 and after). User access to the application layer can be managed through standard authentication protocols like OAuth 2.0 or JWT, with role-based access control restricting privileges based on user roles. Data at rest should be encrypted using standard methods such as AES-256 to safeguard operational and billing records. For regulatory compliance, lightweight logging and auditing mechanisms can be used to track user actions and system events.

18.3 Digital Twins model

Digital Twins are increasingly recognized for improving reliability, sustainability, and efficiency in industrial applications, particularly in power systems. Their roles range from asset management and predictive maintenance to grid optimization and resilience enhancement. Unlike traditional validation methods such as HIL testing, Digital Twins offer a more flexible and cost-effective way to simulate and analyze systems without relying on extensive hardware. Integration with AI and optimization tools enables adaptive control, efficient fault diagnosis, and restoration planning. As power grids transition to cleaner and more electrified systems, Digital Twins can bridge digital intelligence with physical operations. Digital Twin architectures vary by precision and response time, from high-fidelity physics-based models to data-driven and hybrid approaches. Techniques such as data-driven modeling, physics-informed neural networks, and reduced-order models are used for system abstraction. Reduced-order models are especially valuable for their speed and low computational demand, offering a practical alternative to HIL for prototyping and real-time control.

The flowchart in Fig. 18.2 describes an adaptive Digital Twin construction process. Initially, a reduced-order model of the cyber-physical system (CPS) serves as a baseline to efficiently approximate real-world system dynamics. Subsequently, the initial policy, system state, and status indicators will be added to tune this abstracted model. To this end, a policy explicitly refers to a decision-making rule set that governs control actions under specific operating conditions. It could be mathematically defined as a mapping from observed states to actions. For example, a policy may specify how active and reactive power should be dispatched based on measured voltage, or how storage units should be charged and discharged given load forecasts and state-of-charge constraints. In RL terms, the policy corresponds to ($\pi(a \mid s)$), where actions (a) such as load shedding, inverter setpoint adjustment, or topology switching are chosen given a system state (s). Simulation policies can be tested offline to explore candidate strategies, while operational policies are applied directly to the live system once validated. Following initialization, the process enters an iterative loop, starting with the continuous observation of system states and status data. Each new data observation leads to fine-tuning the model parameters using data-driven methods to ensure model accuracy and alignment with real-world conditions. After parameter adjustment, the adequacy of the current control policy is evaluated. If the predicted outputs remain acceptable, the same policy continues to guide system actions. If the predictions fall outside acceptable ranges, a structured policy updating process begins. This involves three steps: (1) identifying the aspects of control that are no longer valid, such as thresholds for voltage deviations or timing of dispatch, (2) generating a candidate revised policy by adapting decision rules or retraining simplified controllers, and (3) validating the

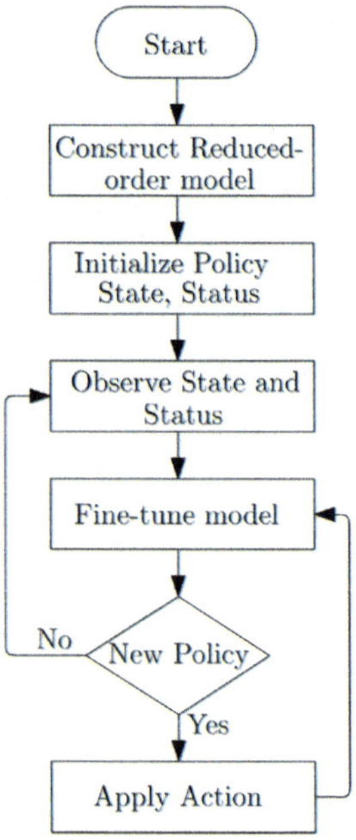

FIGURE 18.2 Adaptive Digital Twin workflow showing reduced-order model construction, initialization of policy and system states, continuous observation of operating status, model fine-tuning, policy evaluation, and action execution.

revised policy against predicted outcomes. If the new policy produces acceptable results, it replaces the old policy and becomes active. Otherwise, the revision cycle repeats until an acceptable policy is achieved. At each iteration, the necessity for a new policy is assessed. If a new policy is required, the system predicts future outputs using the updated model. If the predicted outputs are acceptable, actions are applied according to the current policy. Otherwise, the policy is updated, and outputs are re-predicted until acceptable outcomes are achieved.

Box 18.1 describes the procedure for continuously adapting a Digital Twin model and its policies. Using Markov Decision Processes (MDPs) with a set of states (s_t), actions (a_t), a transition function that describes the probability of moving from one state to another given an action, a reward function (R_t) that assigns a numerical value to each action taken in a state, and a policy (π) that maps states to actions to maximize long-term reward, we can model sequential decision-making under uncertainty. It starts with creating a computationally efficient reduced-order model of the CPS, initializing essential parameters like the initial policy, state, and status. The algorithm then repeatedly collects real-time system data, fine-tunes the Digital Twin using observed information, and decides whether adjustments to the control policy are necessary. Upon determining that a policy update is required, it predicts system outputs and evaluates their acceptability. Acceptable predictions trigger immediate action, while unacceptable results lead to policy revisions and re-evaluations, thus ensuring consistent and robust model performance and effective real-time decision-making.

BOX 18.1

Adaptive Digital Twin algorithm

1: Construct a reduced-order CPS model
2: Initialize policy, state, status
3: While True do
4: Observe and acquire current state and status (s_t,σ_t)
5: Fine-tune model parameters $\hat{M}_t \leftarrow \left(\hat{M}_{t-1}, s_t, \sigma_t\right)$
6: If a new policy is required, **then**
7: Predict system output $\hat{y}_t \leftarrow \hat{M}_t(s_t)$
8: If$\hat{y}_t$ is acceptable **then**
9: $a_t \leftarrow \pi_t(s_t)$
10: Apply actiona_t
11: else
12: Update Policy$\pi_t \leftarrow (\pi_t, s_t, \hat{y}_t)$
13: Re-predict output using updated policy
14: End if
15: End if
16: Go to step **4**
17: End while

To make this loop concrete, the short pseudocode in Box 18.2 shows the integration of an inverter DTiL of an MDP for Volt/Var control. Candidate reactive setpoints are first tested on the Digital Twin under operating limits, the best safe setpoint is selected, and only then is the action applied to the physical inverter and rechecked for violation.

BOX 18.2

Pseudo code for inverter DTiL for Volt/Var control

Inputs:
- BESS Digital Twin
- Policy that proposes reactive power candidates
- Operating limits: voltage bounds, inverter capability, rate of change, temperature
- Voltage deadband and control step

Initialization:

- Initialize policy and Digital Twin
- Set the previous reactive setpoint to zero

Loop each control step:

1: Read field measurements: voltage, active power, reactive power, state of charge, temperature.
2: If the voltage is inside the deadband, set the reactive setpoint to zero and go to **Step 7**.
3: Get a list of candidates reactive setpoints from the policy. Include a no-change option.
4: For each candidate:
4.1: Enforce capability and rate limits.
4.2: Use the Digital Twin to predict the next state for one step.
4.3: Check predicted violations: voltage, thermal, capability, etc.
4.4: Score the candidate. Prefer a smaller voltage error. Penalize reactive usage. Reject any predicted violation.
5: Select the highest-scoring candidate. If all candidates violate limits, select the fallback (i.e., the smallest magnitude setpoint).
6: Apply the selected setpoint to the real inverter.
7: Wait one control step. Read new measurements.
8: Check actual limit violations. If a violation is found, apply a protective fallback (i.e., return to the zero setpoint).
9: Compute a reward from the new measurements. Penalize voltage error, reactive usage, and any violation.
10: Update the policy using the step data and the reward.
11: Update the Digital Twin so predictions stay aligned with the new measurements.
12: Store the applied setpoint as the previous setpoint.
13: Go to 1 and repeat the loop.

Reliable execution of the MDP loop requires explicit definitions of state-variable ranges, the temporal cadence of state transitions, and the mechanisms that keep the Digital Twin aligned with the physical grid. In distributed energy environments, fast-changing electrical variables such as voltage magnitudes (typically 0.95–1.05 per unit), voltage angles (approximately plus or minus 30–40 degrees in distribution feeders), reactive power injections, and line currents must be refreshed every 1–5 seconds to support Volt/Var control, stability monitoring, and protection-aware behavior. Several other slower operational variables, including storage state-of-charge (SoC) (generally 10%–90%), feeder loading trajectories, and market-based commitments, can be updated on 5–15 minutes intervals. The Digital Twins shall thus maintain a multirate state vector in which each component has (1) a defined operating band and (2) a maximum permissible age before it is considered stale. Policy evaluation and policy updates are only carried out when the relevant components of the state satisfy these currency requirements. To prevent the Digital Twins from drifting away from real system behavior, they can incorporate an observer design that reconstructs the true system state from noisy, partial, and asynchronous telemetry. This synchronization can be achieved using an Extended Kalman Filter (EKF), Unscented Kalman Filter (UKF), or a moving-box estimator, depending on deployment constraints. The observer reconciles high-rate measurements from SCADA and PMU units with slower smart meter and DERMS telemetry. It applies the power-flow equations as consistency checks. It then estimates internal variables that are not directly measured, including inverter states, latent load components, and battery internal states. The resulting estimated state ($\hat{s}_t$) will be used as the reference input for both the Digital Twins and the MDP policy loop, ensuring that control decisions are not based on unfiltered or time-misaligned measurements.

To ensure reliability over time, the Digital Twins and their associated control policies must follow a formal model-management lifecycle. Rather than overwriting the Digital Twins each time a parameter or structure is updated, every change is stored as a versioned Digital Twins schema with a unique identifier. Each schema is linked to a calibration dataset, which explicitly documents the historical operating intervals used to fit model parameters,

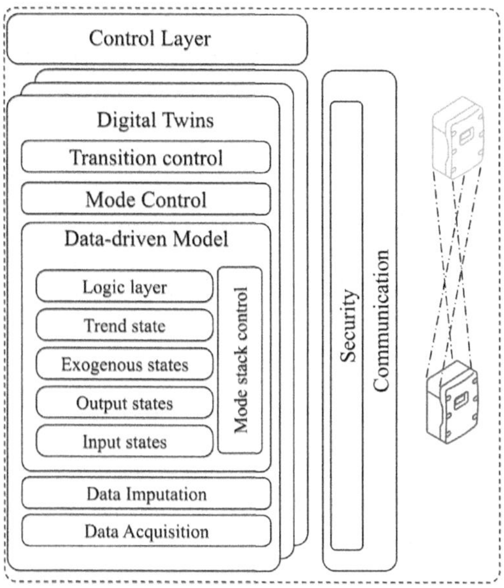

FIGURE 18.3 Data-driven Digital Twin architecture for smart grid assets, integrating data acquisition, state modeling, control layers, and secure communication with physical devices.

including disturbance events, representative seasonal load profiles, and high-variability renewable periods. Before a new version of the Twin or policy π is allowed to influence the physical system, it must undergo a back-testing phase. During back-testing, the candidate schema is evaluated over a window of archived measurements, and its predicted trajectories are compared against ground truth using defined metrics such as trajectory error, constraint-violation counts, and policy-induced stress indicators. Deployment of updated Digital Twins and policies should use a sectional rollout rather than a full system-wide replacement. Therefore the updated instance operates in parallel with the existing production Digital Twins on a limited and controlled subset of assets, such as a single feeder, a microgrid segment, or a representative group of DERs. The system compares the two Digital Twin versions by evaluating their predictions, recommended actions, and violation patterns against each other and against real measurements. The updated Digital Twin is promoted to broader deployment only after it demonstrates stable and consistent behavior over a defined evaluation interval, typically ranging from several hours to several days, depending on the control domain. If instability or divergence appears at any stage, the sectional rollout is halted, and the prior version remains active.

The increasing complexity and dynamic nature of modern power systems make it difficult for a single centralized Digital Twin to represent the entire system effectively. Therefore a better approach is to integrate multiple Digital Twins into a unified, software-defined platform to support a fully connected power system. One proposed hierarchical architecture for the energy internet uses Digital Twins to process data from generators, storage units, meters, and sensors. A central approach handles monitoring, forecasting, and energy analysis under uncertainty, acting like a co-simulation framework. While effective, this centralized model faces limitations in scalability and resilience as systems grow and decentralize. To address these limitations, Fig. 18.3 illustrates a scalable, multilayered Digital Twins architecture designed for reliable smart grid management. The software stack includes (1) data processing to aggregate and normalize telemetry; (2) AI-driven models for prediction, anomaly detection, and optimization; (3) mode control for real-time adjustments; and (4) transition control to manage operational shifts. Supporting layers handle integration across systems and enforce security through authentication and encryption. Cyber-physical components like inverters, meters, storage units, and EV chargers act as clients, publishing data to a broker. Each Digital Twin functions as a micro-service proxy, bridging these assets with upper-layer systems like DERMS. WebSocket or MQTT protocols support stable, secure, bidirectional communication with the broker, allowing real-time control commands in response to system updates while maintaining low latency. The data processing layer collects real-time and historical data (voltage, current, power, etc.) and device status, and applies advanced analytics and filtering to detect anomalies and predict system behavior. It also merges insights from neighboring Digital Twins for a broader system view. These insights inform control actions like energy dispatch, fault mitigation, and parameter tuning. A data-driven method is also used to compute optimal control gains for inverters.

To ensure data continuity, advanced imputation methods are used, including statistical, clustering-based, and neural network-based techniques. One can utilize Generative Adversarial Imputation Nets (GAIN), which use deep

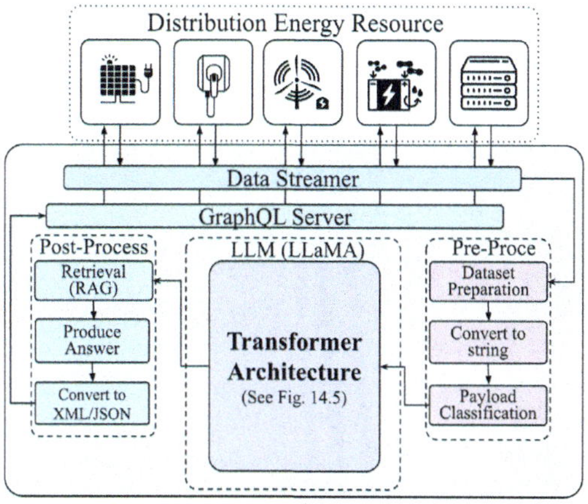

FIGURE 18.4 Architecture of an LLM-integrated SDMG, where DER data streams are pre-processed, interpreted through a transformer-based LLM, enhanced using retrieval-augmented generation, and converted into standardized control-ready formats.

learning to estimate missing values by learning data distributions (Read Section 18.5, for further details on generative-AI models). GAIN can effectively improve the reliability of real-time predictions and control decisions by minimizing bias and maintaining the accuracy of the Digital Twins.

The mode control layer in Digital Twins identifies key system sequential and operational states such as ON, OFF, connected, fault, and islanded. To model system behavior effectively, system dynamic input (e.g., active/reactive power), output (e.g., voltage, frequency, current), and external factors (e.g., temperature) should be tracked through the entire Digital Twin stack. As mentioned before, the reduced-order models, like state-space or machine-learning-based surrogates, will be used to capture dynamics efficiently with minimal computational load. These models integrate with tertiary control strategies and advanced Generative-AI planning to support real-time monitoring, predictive maintenance, and energy optimization.

On the state-transition layer, instead of the conventional MDP method proposed earlier, one can employ RL (Vamvakas et al., 2023), Partially Observable Markov Decision Processes (POMDPs) (Lauri et al., 2023), and finite state machines (FSMs) (Merlin Sajini et al., 2023) to model and optimize state transitions and constraints, ensuring optimal operation and control of components in smart grids. In the context of a DER in a microgrid, the set of events capable of triggering state transitions encompasses various electrical and operational events such as voltage fluctuations, frequency deviations, faults, load demands, and system initialization or shutdown. The set of states for constructing Digital Twins can include:

The logic layer contains the mathematical model of the system, which can be as complex as a full model encompassing the nonlinear model, primary controls, and small-signal model, or it may employ a reduced-order model with lower computational demands. Furthermore, to maintain the consistency of Digital Twins operation, we ensure the coherence of the model by employing a stack control layer that timestamps and tags the system states according to the operation mode. The set of state-transition events

This structured approach enhances the Digital Twins' ability to predict system behavior, optimize control actions, and improve the resilience of smart grid operations.

18.4 Integrating Large Language Models in smart grid

LLMs can be utilized in SDMG by enhancing interoperability across various payload scenarios. Their architecture is crafted to function at a high level of abstraction, prioritizing NLP tasks instead of being limited to specific hardware or operating systems. As shown in Fig. 18.4, an LLM-based architecture can be designed to streamline message exchange and support predictive contingency planning within the SDMG framework. The LLM acts like a service brokerage, offering solutions for predictive and prescriptive maintenance, troubleshooting, interoperability, energy planning, and more. The brokerage stack is made up of three key components: preprocessing, LLM inference, and postprocessing.

In the preprocessing stage, the dataset is constructed through three explicit steps. First, all relevant artifacts (device payloads, protocol messages, troubleshooting instructions, configuration manuals, and vendor-specific data dictionaries) are collected from in-field assets. Each artifact is normalized into a plain-text string so that it can be ingested by the LLM tokenizer without retaining binary or protocol-specific encoding. Second, each sample is reorganized into a structured Question–Answer (QA) template with three fields: *Input*, *Output*, and *Assistant*.

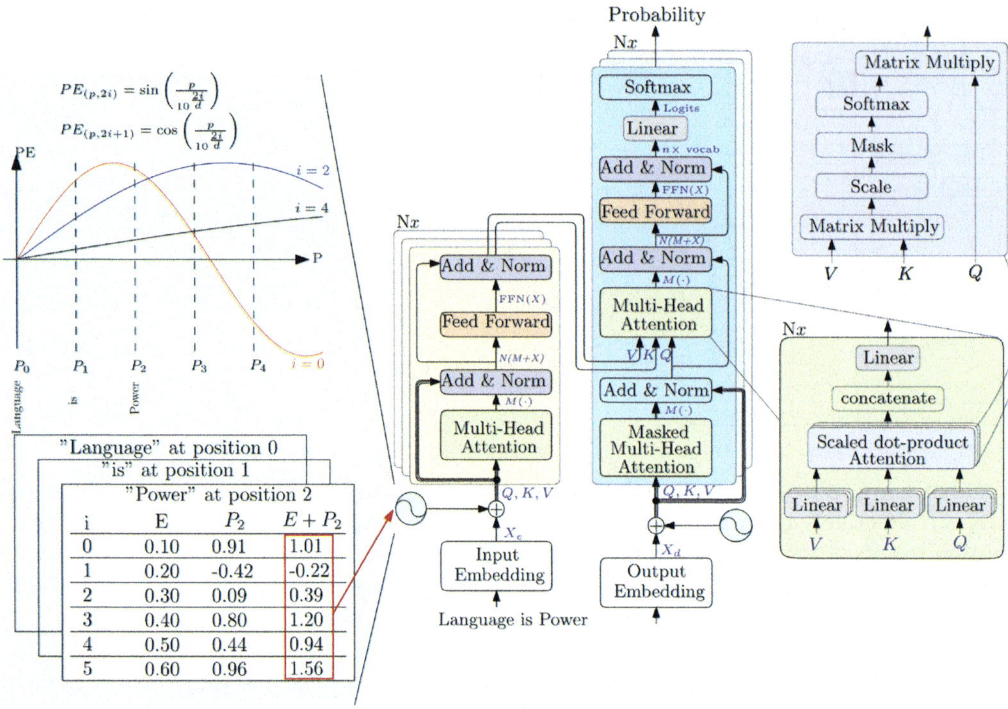

FIGURE 18.5 Transformer architecture overview showing token embedding with positional encoding, encoder and decoder blocks, multi-head self-attention, feed-forward layers, normalization, masked attention, and output probability generation.

The *Input* field contains the raw message or payload exactly as it would be observed on the device interface. The *Output* field specifies the desired structured transformation, such as converting the payload into a different protocol, generating the equivalent data model, or producing a corrected version with missing fields restored. The *Assistant* field provides explicit contextual guidance, including protocol semantics, allowable tag ranges, field definitions, and constraints that the model must observe during inference. Then, the dataset can be augmented to ensure broad generalization across operational conditions. Samples will be labeled according to four task types: (1) identification, where the model must determine the device type, protocol, or message category; (2) completion, where missing fields, attributes, or enumerations are restored; (3) prediction, where the model infers the correct payload type or expected fields based solely on a device or message name; and (4) conversion, where a message is translated between two protocols while preserving semantic meaning and field constraints. The final preprocessed dataset provides a consistent, fully annotated corpus for supervised fine-tuning, allowing the LLM to handle protocol translation, payload correction, message normalization, and device-specific reasoning reliably within smart energy systems.

18.4.1 Large Language Models architecture and fundamentals

LLM employs a transformer architecture, which consists of an encoder and a decoder. The encoder processes the input and passes only the Key (K) and Value (V) matrices to the decoder, discarding the Query (Q) matrix. The decoder receives the KVmatrices and, after passing through the masked multihead attention mechanism, generates its own Query (Q) matrix. This Q matrix is then used to compute attention over the $K$and Vmatrices received from the encoder. This mechanism ensures that the model produces sequential outputs while maintaining attention over the relevant encoder representations.

As graphically shown in Fig. 18.5, each word in both the input and output sequences is first converted into a token through a process called *tokenization* (the left pale blue box in Fig. 18.5). These tokens are then mapped to continuous numerical vectors known as *embeddings*. Since Transformer models do not inherently preserve the order of tokens after embedding, positional encodings (PE) are added to the embedding vectors (red box in Fig. 18.5).

The PE provides information about the token order by assigning each position pa unique encoding vector. These vectors are computed using sine and cosine functions to ensure that each token's position is represented relative to others in the sequence. For a sequence length $n$and model dimensiond_m, the positional encoding matrix $PE \in \mathbb{R}^{n \times d_m}$

is constructed such that:

$$PE_{p,i} = \begin{cases} sin\left(\frac{p}{\lambda_i^{d_m}}\right) & if\ i\ is\ even \\ cos\left(\frac{p}{\lambda_i^{d_m}}\right) & if\ i\ is\ odd \end{cases} \tag{18.1}$$

where, the $p$is the position index in the sequence, i is the dimension index of the encoding, $d_m$represents the dimensionality of the model, and $\lambda_i^{d_m}$creates a geometric progression of frequencies across the embedding dimensions (i.e., $\lambda_i^{d_m}$=10, $000^{2i/d_m}$).

As shown in Fig. 18.5, the multihead attention mechanism operates on three: Query (Q), Key (K), and Value (V). These matrices are generated from the embedded input and share the same dimensions, $Q, K, V \in \mathbb{R}^{n \times d_m}$where $n$is the sequence length. The embeddings are linearly projected into three different spaces, creating the Q, K, and Vmatrices, each containing the same input information but serving different roles within the attention mechanism. Each matrix is computed by multiplying the input embedding $X$with a corresponding learnable weight matrix W^Q, W^K, and W^V, respectively:

$$Q' = XW^Q \quad K' = XW^K \quad V' = XW^V \tag{18.2}$$

Once Q', K', and $V'$are computed, they are split into multiple heads, allowing the model to attend to different aspects of the input. As depicted in Fig. 18.5, each matrix is divided into $h$different subspaces, each with a reduced dimensionality d_k, with

$$d_k = \frac{d_m}{h}, \tag{18.3}$$

Each head processes a different part of the sequence, enabling the model to capture multiple dependencies simultaneously. For each head, the self-attention can be calculated by

$$A = S\left(\frac{QK^T}{\sqrt{d_k}}\right)V, \tag{18.4}$$

where $S(\bullet)$denotes the softmax function applied for normalization of the attention scores, and the division by $\sqrt{d_k}$stabilizes gradients by scaling. After computing the attention outputs for all heads, the outputs are concatenated and linearly transformed. The multihead attention mechanism is then defined as

$$M(Q, K, V) = [A_1 \| A_2 \| \cdots \| A_h]W^O, \tag{18.5}$$

where $A_i$represents the attention output from the ith head, $\|$denotes concatenation along the feature dimension, and $W^O \in \mathbb{R}^{hd_k \times d_m}$is a learnable weight matrix that projects the concatenated outputs back to the required dimensionality d_m. After computing the multihead attention, the output $M(Q, K, V)$is passed through a feed-forward network, denoted as $F(\bullet)$. Unlike the attention mechanism, the feed-forward layers operate independently at each sequence position, enabling the model to perform nonlinear transformations of the representations produced by attention. Given an input matrix $X \in \mathbb{R}^{n \times d_m}$, representing a batch of $n$vectors, the feed-forward network is applied as

$$F(X) = \Phi(XW_1 + b_1)W_2 + b_2, \tag{18.6}$$

where $W_1 \in \mathbb{R}^{d_m \times h_m}$projects the input into a higher-dimensional space of size h_m, $b_1 \in \mathbb{R}^{h_m}$is the associated bias, $\Phi(\bullet)$denotes a nonlinear activation function such as ReLU, $W_2 \in \mathbb{R}^{h_m \times d_m}$projects back into the original dimensionality d_m, and $b_2 \in \mathbb{R}^{d_m}$is the second bias term. Since $F(X)$operates position-wise, it enhances the model's capacity to learn complex feature transformations that are not restricted to weighted combinations from the attention outputs.

After passing through the feed-forward network, the output is normalized using layer normalization, denoted $N(\bullet)$ yielding to;

$$N(F(M(Q, K, V))), \tag{18.7}$$

This normalization step helps ensure stability during training and maintains consistent feature scaling across batches. After these transformations, the updated Key (K) and Value (V) matrices are passed to the decoder, providing necessary contextual information for subsequent decoding stages. The feasibility of deploying the LLM semantic layer depends not only on architecture design but also on the available compute platform.

This is already a complex iterative computation, but it does not quite cut it for achieving semantic interoperability in smart energy systems. To really make it work, we need to take the model a step further than its initial pre-training.

This is where fine-tuning comes in, allowing us to tailor the pretrained model to specific tasks and datasets relevant to the domain.

The purpose of fine-tuning is to adjust the model's parameters based on a targeted dataset, allowing the LLM to better capture the nuances and contextual patterns required for a specific application. Fine-tuning significantly improves the ability to perform domain-relevant tasks with higher accuracy and reliability. There are three main approaches to fine-tuning LLMs:

- **Self-supervised learning**: The model learns patterns directly from the structure of the input data, without needing manually labeled annotations.
- **Supervised learning**: The model is trained on labeled datasets, where clear input-output pairs guide the learning toward specific tasks.
- **Reinforcement learning**: The model is optimized based on feedback or reward signals, such as in Reinforcement Learning from Human Feedback (RLHF), where human evaluators refine model outputs to improve quality.

Given the structured message exchange payloads (JSON or XML) in a smart grid, the supervised learning approach is a better option for fine-tuning the LLM. The number of trainable weights in LLMs, such as parameters in multihead self-attention mechanisms, feed-forward subnetworks, layer normalization, and embedding layers, can range from 1 billion to over 70 billion. Fine-tuning all these parameters requires significant computational resources, resulting in high memory usage, prolonged training times, and substantial energy costs, even when using optimized training techniques. To address this, PEFT can be employed to reduce the computational and memory overhead of full fine-tuning. Rather than updating all weights, PEFT freezes the original model parameters and introduces a small set of trainable parameters, drastically reducing computational demands. A common PEFT method is *LoRA*, which injects lightweight trainable matrices into specific layers while keeping the base model frozen. This enables efficient task adaptation with minimal memory and compute overhead, making PEFT a practical solution for fine-tuning large models. In traditional fine-tuning, the hidden layer output is computed as:

$$h(x) = W_0 x \tag{18.8}$$

where W_0represents the pretrained model weights and $x$is the input batch. In *LoRA*, instead of modifying the full weight matrix W_0, two small trainable matrices $A$and $B$are introduced, yielding:

$$h(x) = W_0 x + BAx \tag{18.9}$$

where, $W_0 \in \mathbb{R}^{d \times k}$denotes the frozen pretrained weights, and $A, B \in \mathbb{R}^{d \times k}$are low-rank decomposition matrices with $r \ll \min(d, k)$.

Fine-tuning efficiency also depends critically on selecting appropriate hyperparameters, including the *optimization algorithm, learning rate, and number of epochs*:

- **Optimization algorithm:** Choosing effective optimizers, such as Adam or AdamW, helps the model converge efficiently without deviating excessively from its pretrained knowledge.
- **Learning rate:** This controls the update speed of model parameters. Smaller models (1B–3B parameters) typically require learning rates between $1 \times 10^{-4}$and 5×10^{-4}, while larger models (30B–70B) require smaller rates, usually between $2 \times 10^{-6}$and 5×10^{-5}.
- **Number of epochs:** This refers to how many complete passes through the training dataset are performed. Too few epochs can lead to underfitting, while too many causes overfitting. The optimal number depends on the dataset size and complexity.

Careful tuning of fine-tuning hyperparameters, such as learning rate, number of epochs, optimizer choice, and LoRA rank and scaling enables efficient and high-quality adaptation of large-scale LLMs to smart grid payloads. However, a fine-tuned model remains bounded by the data it has seen during training. To address this limitation, retrieval augmented generation (RAG) can be integrated into the LLM semantic layer so that, at inference time, the model is grounded in external artifacts such as protocol specifications, schema definitions, and payloads. Instead of relying solely on parametric memory, the LLM queries a retrieval index built from representative message samples and documentation from multiple smart grid communication protocols and data models, and uses the retrieved context to reconstruct incomplete payloads, normalize message fields, and identify protocol or profile types.

Operationally, the RAG pipeline proceeds in three steps. First, the incoming payload or error message is embedded into a vector representation using the same or a compatible encoder as the LLM. Second, a similarity search is performed against an index of embedded reference payloads or snippets from technical documentation in order to retrieve the top k most relevant items. Third, these retrieved segments are concatenated with the prompt and passed to the LLM, which attends jointly to the user payload and the retrieved context. This changes the generation behavior, and instead of synthesizing structure purely from memory, the model is biased to copy, adapt, and complete fields according to concrete, standard-compliant patterns. This directly improves tasks such as reconstructing missing tags, correcting enumeration values, mapping vendor-specific fields to canonical attributes, and disambiguating protocol families that share similar naming.

The effectiveness of this RAG layer depends on the structure and scale of the retrieval corpus, which here refers to the collection of documents, payload instances, and protocol snippets that are stored for retrieval. A sufficiently large and diverse repository allows similarity search to return items that closely match both the syntax and semantics of the incoming payload, providing strong contextual cues and reducing hallucinated or structurally invalid outputs. When the corpus covers multiple protocols and message variants, the model can exploit differences between retrieved items to infer the most likely target schema for an ambiguous payload. If the corpus is too small or poorly curated, the system tends to retrieve weak or incomplete matches that offer limited support for semantic normalization. In contrast, a very large repository increases memory consumption and may require more sophisticated indexing strategies to preserve acceptable query latency.

Document size and chunking together determine both the number of entries in the retrieval index and the number of tokens sent to the LLM. When large documents are split into very small chunks, the repository grows quickly, and each query must search over many more vectors, which increases retrieval latency. Small chunks also tend to fragment the structured context hierarchies, breaking the relationship between headers, metadata, and nested elements, and leading to the retrieval of partial fragments that are difficult for the LLM to use as a template. Larger, schema-aware chunks, for example, a full message instance or a major subtree per chunk, preserve structural consistency and reduce the total number of index entries, but each retrieved chunk carries more tokens, which increases transfer time and LLM inference time. Selecting an appropriate chunk size, therefore requires balancing semantic completeness, token efficiency, and both retrieval and inference latency, especially when operating under tight memory and bandwidth constraints at the edge.

The retrieval process usually utilizes an indexing method and is implemented as approximate nearest neighbor search over embedding vectors. Libraries such as Facebook AI Similarity Search (FAISS), which implement highly optimized approximate indexes using techniques like inverted files and product quantization, enable fast and low-latency similarity search when querying thousands to millions of payload embeddings, particularly when Graphics Processing Unit (GPU) acceleration is available. For very large collections, graph-based structures such as Hierarchical Navigable Small World graphs (HNSW), maintain a multilayer proximity graph in which queries traverse from coarse to fine neighborhoods, which improves recall at the cost of higher memory usage and more complex search operations. Better recall at a fixed latency budget translates into higher quality retrieved content, which in turn improves the LLM's ability to generate syntactically valid and protocol-compliant messages. The choice of index, therefore determines the trade-off between retrieval speed, memory footprint, and the quality of retrieved samples, particularly when dealing with payloads that vary in length and structural complexity.

As these design choices, corpus size, document size, index type, and chunking granularity, take effect, latency becomes the next constraint. Retrieval delay increases with larger repositories, more complex index structures, or chunks that are so large that they push the model close to its context limit. In SDMG settings, the LLM semantic layer typically supports supervisory functions such as protocol translation, payload repair, and tagging, rather than millisecond-level protection tasks, so subsecond end-to-end latency is generally sufficient. Experiments show that RAG-enabled inference can remain below 0.6 second per payload, even for messages exceeding 1000 tokens on cloud GPUs, which indicates that semantic interoperability services can be supported without violating operational constraints at the control center level. For a fixed base model, adding RAG also reduces structural errors and improves protocol-level accuracy compared to pure fine-tuning, which confirms that retrieval provides a more scalable path to covering new or rarely seen message formats (Fatemi & Fattahi, 2025).

Hardware constraints also influence deployment options, where on local gateways without GPUs, compact configurations allow models in the 1B–3B range, combined with quantization and PEFT, to run on commodity platforms with 4–16 GB of RAM, providing basic semantic normalization and limited RAG capabilities, for instance, a small local retrieval index built from payloads observed at a single site. GPU acceleration for larger models, such as 7B–8B, and for high-throughput RAG search over large, multiprotocol repositories is needed due to the higher computational load of embedding, vector search, and decoding. In one configuration, an 8B model saturated approximately 98% of GPU utilization while CPU load remained below 3.5%, which confirms that GPU resources are the primary bottleneck.

TABLE 18.1 Comparison of different parameters of LoRA for fine-tuned unsloth-Llama-3-3B.

Rank (r)	Alpha (α)	Loss function	BLEU score
16	32	0.0306	92%
32	64	0.0183	94%
64	128	0.0124	94%
128	256	0.0040	95%

BLEU, *Bilingual Evaluation Understudy;* LoRA, *Low-Rank Adaptation.*

Cloud-hosted deployments can therefore offload heavy retrieval and inference workloads, while edge nodes execute smaller models for latency-sensitive or privacy-constrained tasks. Detailed platform specifications, GPU sizing, and performance measurements are reported in Fatemi & Fattahi (2025).

18.4.2 Semantic interoperability

To make the most of LLMs for a specific job, like ensuring messages can work together seamlessly, you need to tweak the language models by adjusting their settings based on a particular dataset. This fine-tuning process is essential because general-purpose LLMs often do not grasp the intricacies of structured data formats or specific tasks. By going through this adaptation, the LLM becomes much better at noticing the subtle details and context of the data, which boosts its performance for the task at hand.

Numerous parameters affect the quality and effectiveness of fine-tuning LLMs. This spans from the LLM model to the parameter size, learning rate, epoch, and LoRA method. Therefore evaluation of the model across a range of hyperparameters is required to identify the optimal configuration for fine-tuning the model. To assess performance, the Bilingual Evaluation Understudy (BLEU) score is commonly used to measure the correspondence between the generated output and a ground truth reference. A high BLEU score illustrates better alignment with reference outputs. Therefore it indicates that the fine-tuned model is reliable for the task.

Several open-source LLMs are available with fine-tuning features. These models allow access to the weights and architecture, enabling complete customization and adaptation to specific tasks. LLaMA, BERT, and DeepSeek are examples of open-source models that have different parameters and architectures. LLaMA is a popular choice for fine-tuning, which is a transformer-based architecture. It is available in a range of parameter sizes such as 1B, 3B, and 8B. To determine the optimal configuration, it is essential to conduct several experiments using different model sizes and training settings. Models with higher parameters are more capable of handling complex tasks and more effectively adapt to the structure and style of the dataset during fine-tuning.

The LoRA allows efficient fine-tuning by introducing a small set of trainable parameters into each layer, while keeping the original model weights frozen. Two key parameters define the LoRA method of Rank (r) and Alpha (α). The r controls the size of the low-rank matrices added to the model's weights, and α adjusts the impact of these additions during fine-tuning. Increasing the ratio of r to αa certain point led to a rise in the loss function, due to overfitting, where the model becomes complex and starts learning noise instead of meaningful patterns. Table 18.1 summarizes the impact of increasing LoRA rank (r) alongside proportionally larger α on model performance. As the $r - \alpha$pair increases, the loss consistently decreases, showing that larger LoRA configurations improve training stability and convergence. The BLEU score, however, rises from 92 at rank 16–94 at rank 32, remains unchanged at rank 64, and increases only slightly to 95 at rank 128. This indicates that while higher r are effective at reducing training loss, and improvements in translation quality plateau beyond rank 64. Overall, larger LoRA settings enhance optimization efficiency but yield only modest accuracy gains after a certain threshold, highlighting the trade-off between computational cost and performance.

Another effective hyperparameter in the fine-tuning process is the number of epochs and the learning rate. Learning rate controls the speed at which model weights are updated during training. If the learning rate is set too low, the learning process becomes excessively slow, requiring a much longer time for the model to reach an optimal state. On the other hand, if the learning rate is high, the model may fail to converge. Additionally, the number of epochs refers to a complete process through the dataset. Therefore depending on the number of epochs, each sample may be seen multiple times during fine-tuning. If the number of epochs is too low, the model does not learn sufficiently, leading to underfitting. The appropriate number of epochs should be determined based on the size of the dataset. In Fig. 18.6A, which shows the loss function, higher loss values appear when the learning rate is set too high or

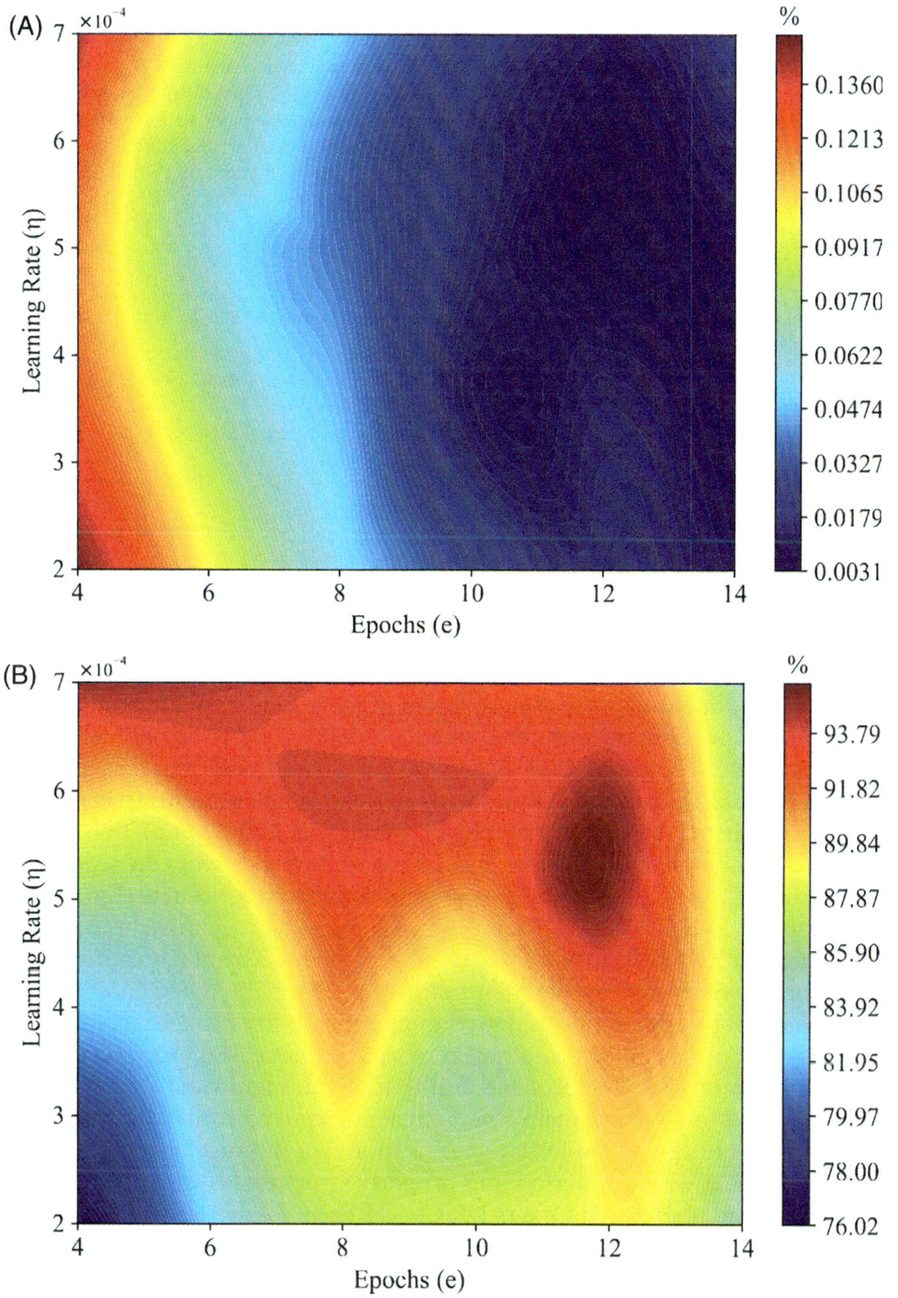

FIGURE 18.6 Effect of learning rate and epoch on loss function using fine-tuned LLaMA-3-8B using different ranking and rating parameters (A) LoRA r = 128, α = 256, (B) LoRA r = 128, α = 256.

when training is limited to very few epochs, while the lowest loss (dark blue regions) occurs at moderate learning rates (around 3–5 × 10^{-4}) with sufficient epochs (8–12), indicating stable convergence. In Fig. 18.6B, which shows the BLEU score, the highest performance (red regions) is also observed in the midrange of learning rates with enough epochs, while excessively high learning rates reduce BLEU despite longer training. Together, the two plots highlight the trade-off between training stability and final model quality, showing that the optimal region lies where loss is minimized and BLEU score is maximized.

After fine-tuning the model with appropriate values for hyperparameters, RAG can be utilized to improve accuracy. Fine-tuning enables the model to better understand the structure and semantics of the dataset. However, it does not access external data or information that may not have been included in the dataset, which is a fine-tuning model. RAG can address this limitation by integrating a retrieval mechanism with the generative model. During generative output, the model searches for relevant data in an external dataset, which RAG used to support generated output. In Fig. 18.7, the Unsloth/LLaMA-3-8B was fine-tuned over part of the IEEE SEP 2.1 message dataset with a

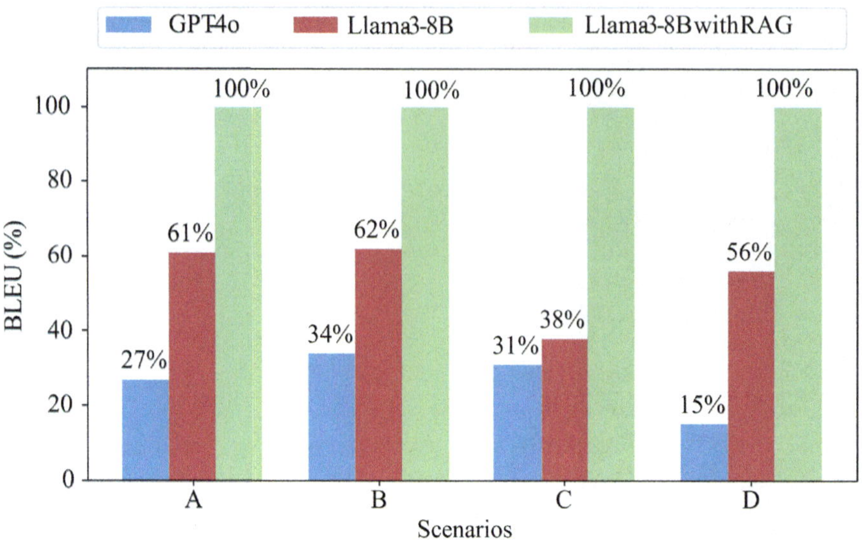

FIGURE 18.7 BLEU score comparison for four scenarios: (A) Identifying the protocol/payload name, (B) completing the missing data/attributes/tags, (C) reconstructing the payload by name, and (D) converting payloads from different protocols. *BLEU*, Bilingual Evaluation Understudy.

learning rate of 5×10^{-4}, 12 training epochs, and with LoRA parameters of $r = 128$ and $\alpha = 256$. The performance of the fine-tuned model with and without RAG was computed with GPT-4 on 500 test payloads randomly selected from different protocols such as IEEE 2030.5, IEC 61968/61970, and OpenADR.

Fig. 18.7 shows the BLEU scores achieved across four scenarios when different models were evaluated on protocol and payload-related tasks. In each scenario, higher performance was consistently obtained by Llama3-8B with RAG, with perfect scores being recorded in all cases. Lower scores were observed for GPT-4o and standard Llama3-8B, with GPT-4o being outperformed in every task. In scenario A (protocol/payload name identification), a 27% score was produced by GPT-4o, while 61% was achieved by Llama3-8B. In scenario B (data completion), slight improvements were noted, with 34% and 62% generated by GPT-4o and Llama3-8B, respectively. In scenario C (payload reconstruction by name), only 31% and 38% were produced. The most challenging task, scenario D (protocol conversion), led to the lowest GPT-4o score of 15%, while 56% was recorded by Llama3-8B. In contrast, a consistent 100% BLEU was delivered by the RAG-enhanced model in every case, highlighting the advantage conferred by retrieval augmentation. RAG compensates for potential gaps in the fine-tuning data, especially when encountering unseen inputs, so it reduces incorrect responses. Table 18.2 outlines how the interoperability broker reconstructed incomplete SEP 2.1 and OpenADR payloads into valid, fully compliant message formats. The examples shown represent only a subset of the tested cases but demonstrate the broker's ability to repair significant structural and semantic errors. In each example, missing fields were restored, malformed elements were corrected, and protocol-specific tags were reintroduced or fixed to conform fully to the relevant specification. For instance, incomplete DER Program and Historical Reading objects from SEP 2.1 were converted into properly structured XML with corrected `<description>`, `<mRID>`, and link elements, while the malformed OpenADR Cancel Party Registration payload was transformed into a syntactically valid, schema-compliant message with the appropriate namespace declarations, registration identifiers, and object wrappers reinstated. These examples highlight the system's capacity to reconstruct complex, incomplete payloads into standards-aligned outputs.

18.4.3 Preventive and prescriptive maintenance

Preventive maintenance is a proactive strategy involving routine inspections, servicing, or replacements at scheduled intervals (based on time or usage) to minimize the risk of failure, regardless of current equipment condition (Zeng et al., 2024). Smart grids comprise multiple components linked by a common bus for sharing surplus capacity (Xiao et al., 2022). Each component has unique performance and demand profiles, with health affected by external shocks and maintenance actions. Random events, categorized as invalid, valid, or extreme, can degrade performance or cause failures (Zhou et al., 2023). To manage this, preventive maintenance restores components to "good-as-new" at regular intervals, while minimal repairs restore pre-failure performance after faults. Together, they support system availability and reliability. As outlined in Algorithm 18.2, integrating generative AI and LLMs can enable an interactive operator interface. The LLM interprets conversation context, predicts failure likelihood and timing, and recommends optimal actions such as adjusting settings, scheduling repairs, or replacing parts

TABLE 18.2 Examples of IEEE 2030.5 (SEP 2.1) and OpenADR command payloads shown in their incomplete form and after being completed and made compliant using an LLM-based interoperability broker.

Payload function	Incomplete input	Complete output
DER controller	<?xml version="1.0"?> <DERProgramList subscribable="1" all="1" results="1" pollRate="600" xmlns="urn:ieee:std:2030.5:ns"> <DERProgram href="/derp/1" subscribable="1"> <mRID>10000000000000341000000000000034</mRID> test for DERProgram</description> <version>0</version> <ActiveDERControlListLink href=\"/derp/1/actderc\" all="1"/> <DefaultDERControlLink href="/derp/1/dderc"/> <DERControlListLink href="/derp/1/derc" all="1"/> <DERCurveListLink href="/derp/1/derc/1" all="1"/> <primacy>0 </DERProgram> </DERProgramList>	<?xml version="1.0"?> <DERProgramList href="/derp" subscribable="1" all="1" results="1" pollRate="600" xmlns="urn:ieee:std:2030.5:ns"> <DERProgram href="/derp/1" subscribable="1"> <mRID>10000000000000341000000000000034</mRID> <description>test for DERProgram</description> <version>0</version> <ActiveDERControlListLink href=\"/derp/1/actderc\" all="1"/> <DefaultDERControlLink href="/derp/1/dderc"/> <DERControlListLink href="/derp/1/derc" all="1"/> <DERCurveListLink href="/derp/1/derc/1" all="1"/> <primacy>0</primacy> </DERProgram> </DERProgramList>
Historical reading	<?xml version="1.0"> <HistoricalReading href="/bill/1/ca/1/ver/1" xmlns="urn:ieee:std:2030.5:ns"> <mRID>8000000000000037100000000000mRID> <description>Te> <version>56</version> <BillingReadingSetListLink href="//1/rs" all="1"/> <ReadingTypeLink href="/upt/1/mr/1/rt"/> </HistoricalReading>	<?xml version="1.0"> <HistoricalReading href="/bill/1/ca/1/ver/1" xmlns="urn:ieee:std:2030.5:ns"> <mRID>80000000000000371000000000000025</mRID> <description>Test</description> <version>56</version> <BillingReadingSetListLink href="/upt/1/mr/1/rs" all="1"/> <ReadingTypeLink href="/upt/1/mr/1/rt"/> </HistoricalReading>
OpenADR created event	<?xml version="1.0" ?> <oadr:oadrPayload xmlns:oadr="http://openadr.org/oadr-2.0b/2012/07"> <oadr:oadrSignedObject xmlns:oadr="http://openadr.org/oadr-energyinterop/20111"ei:schemaVersion="2.0b"> <requestID xmlns="http://docs.oasis-open.org/ns/energyinterop/201110/payloads">4fb73f76-16ab-4c3c-8a04-bbb77bc637b7</requestID> <ei:rc1113c-9512-4f42-a858-9a97b98f5597 <ei:venID>123ABC</ei:venID> </oadr:oadrCancelPartyRegistration> </oadr:oadrSignedObject>	<?xml version="1.0" ?> <oadr:oadrPayload xmlns:oadr="http://openadr.org/oadr-2.0b/2012/07"> <oadr:oadrSignedObject xmlns:oadr="http://openadr.org/oadr-2.0b/2012/07oadr:id="oadrSignedObject"> <oadr:oadrCancelPartyRegistration xmlns:ei="http://docs.oasis-open.org/ns/energyinterop/20111"ei:schemaVersion="2.0b"> <requestID xmlns="http://docs.oasis-open.org/ns/energyinterop/201110/payloads">4fb73f76-16ab-4c3c-8a04-bbb77bc637b7</requestID> <ei:registrationID>31c1113c-9512-4f42-a858 9a97b98f5597</ei:registrationID> <ei:venID>123ABC</ei:venID> </oadr:oadrCancelPartyRegistration> </oadr:oadrSignedObject> </oadr:oadrPayload>

LLM, *Large Language Models.*

within cost, time, and resource limits (Santiago et al., 2024). Generative-AI, leveraging architectures like conditional generative adversarial networks (cGANs) and variational autoencoders (VAEs), simulates realistic degradation scenarios, allowing maintenance models to learn dynamically. This improves prescriptive maintenance by optimizing schedules and predicting long-term system health impacts.

In addition to strategic planning, LLMs can support engineers with real-time guidance and decision-making in operational scenarios. RL is transforming prescriptive maintenance by making it dynamic and adaptive. Unlike traditional fixed-interval strategies, deep RL learns from real-time system behavior and environmental changes. It continuously updates degradation models, ensuring predictions stay accurate as systems evolve and helping operators make better maintenance decisions.

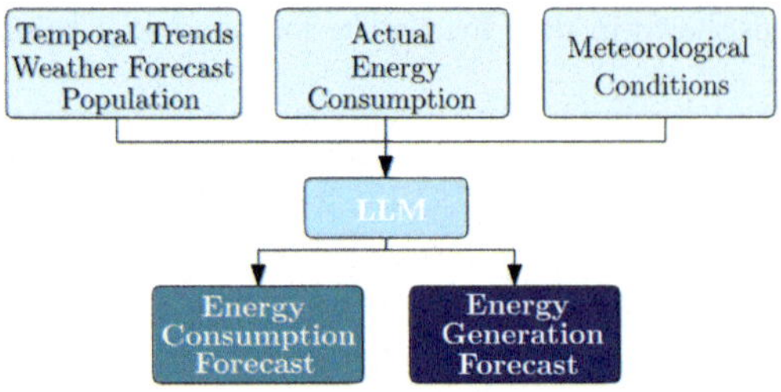

FIGURE 18.8 LLM framework for energy planning, where historical, demographic, consumption, and meteorological data are processed to forecast energy demand and generation.

In complex smart grids, multiagent RL allows decentralized components to optimize maintenance locally while still supporting overall system performance (Giacotto et al., 2025). Meanwhile, generative AI pushes this further by solving optimization problems where explicit system models are too complex or unavailable. Models like GANs explore maintenance actions and are guided by RL feedback, finding strategies that balance short-term fixes with long-term system health. Together, generative-AI and RL create adaptive, data-driven maintenance systems that improve reliability and extend operational life (Yuan, 2023).

However, real-world microgrids pose unique challenges, where engineers often struggle with overwhelming amounts of error data, scattered across different device models and brands, each with its own troubleshooting manual. Finding the right solution is time-consuming and impractical, especially without centralized, intuitive tools.

To address this, LLMs can deliver real-time, brand-agnostic troubleshooting. A structured dataset was created by extracting hundreds of troubleshooting scenarios from various device models. Each data entry is designed with an input representing a device error, and an output specifying the corresponding solution. Additionally, an instruction field provides contextual guidance, explaining under which conditions the error occurs and what corrective action should be applied. Using this dataset, a 7-billion-parameter LLM is fine-tuned with LoRA and quantization techniques (learning rate $r = 5 \times 10^{-4}$ with 12 epochs).

The result of this fine-tuning will be a highly capable model that functions as an assistant for engineers. Upon encountering an error from a device, the system will rapidly analyze the issue and provide the correct solution in real-time, regardless of the brand or model of the device. This solution will be instantly available to the engineer, eliminating the need for manual reference to separate troubleshooting manuals and dramatically reducing the time required to diagnose and resolve system issues.

18.4.4 Energy planning

LLM possesses great potential to improve energy planning in smart grid applications. Effective energy planning requires accurate forecasting, optimal resource utilization, and adaptive decision-making based on real-time data analysis. Applying the capability of LLMs to process vast amounts of data and perform complex analytical operations can greatly improve these core areas.

18.4.4.1 Demand and generation forecasting

Another significant application field of LLMs in energy planning is load forecasting, as shown in Fig. 18.8. With accurate predictions of patterns of energy usage, effective functioning and grid stability are guaranteed. LLMs, using past trends, weather forecasts, population, and actual consumption patterns, can provide precise estimates of future energy demand. With this capability, grid managers can predict and modify supply plans and utilize renewable resources more effectively, guaranteeing the best use of available resources. Similarly, LLMs facilitate renewable energy generation forecasting by analyzing meteorological conditions, historical generation trends, and other related data. Improved predictions for intermittent resources such as solar and wind power allow grid operators to predict variability, reducing reliance on backup generation sources and optimizing storage management.

To practically apply the LLM architecture and its foundational equations to load and generation forecasting, input data such as historical load profiles, weather forecasts (including temperature, wind speed, and irradiance), timestamps, and demographic factors should be structured into time-indexed sequences. Each element in the sequence should be tokenized and embedded into high-dimensional vectors using the approach described in Eq. (18.2). *PE*, defined in Eq. (18.1), should be added to the embeddings so that temporal relationships are preserved. This ensures that time-dependent patterns, such as daily demand peaks or seasonal variation, are recognized by the model.

Once embedded, the input sequence should be passed through the multihead attention mechanism described in Eqs. (18.4)–(18.5). In this step, the model is enabled to identify and emphasize dependencies across time steps.

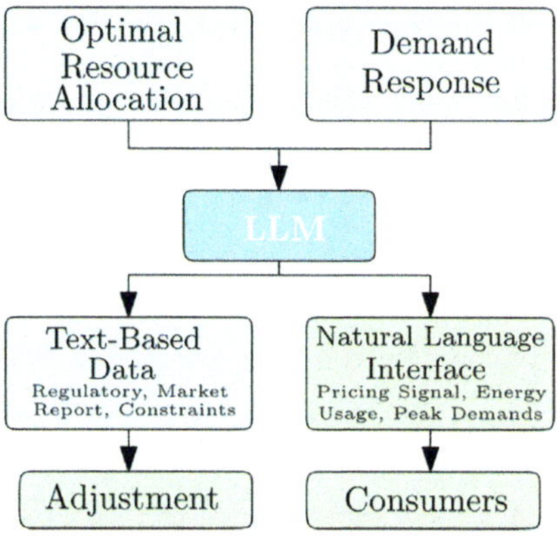

FIGURE 18.9 Resource planning and ancillary services framework using operational, regulatory, market, pricing, and consumer energy-use data to support control adjustments and demand-response actions.

For instance, it can be inferred that a weekday evening in winter may resemble previous evenings with similar environmental conditions. Attention weights are assigned accordingly, allowing the model to focus on historically relevant situations. This mechanism supports the learning of long-term dependencies and interactions, such as the impact of temperature changes on heating demand or the delayed effect of cloud cover on solar output. The resulting attention outputs should then be processed through Feed-Forward Networks, as formulated in Eq. (18.6). These networks apply nonlinear transformations independently to each position in the sequence. This allows local fluctuations in the data, such as an unexpected surge in demand or a drop in wind generation, to be captured without interference from unrelated time steps. The independence across positions supports the modeling of granular, localized features that are critical in operational forecasting. For predictive tasks such as generating future load or renewable generation profiles, the decoder module should be used. It incorporates masked attention derived from the same query, key, and value mechanisms outlined earlier in Eqs. (18.2)–(18.3). This masking ensures that only past and current data influence each prediction step, thereby maintaining the causal structure required for time-series forecasting.

To adapt a general pretrained model to specific energy datasets, fine-tuning should be applied using supervised learning. This process involves training the model on structured input-output pairs, such as past consumption data paired with actual future loads. To reduce the computational burden, PEFT methods such as LoRA can be used, as described in Eq. (18.9). In this approach, the original model weights remain frozen, and only small trainable matrices are introduced. This allows the model to be efficiently adapted to domain-specific tasks using modest computational resources. To enable deployment in constrained environments, such as edge devices in distributed energy systems, post-training quantization techniques should be applied. By reducing the numerical precision of the model parameters, for example, from 32-bit to 8-bit, both memory usage and inference time can be reduced while maintaining acceptable levels of accuracy. Through this series of steps, from input preparation and attention-based modeling to efficient fine-tuning and deployment optimization, LLMs can be effectively applied to forecast energy load and generation. These forecasts can then be used by grid operators to adjust supply strategies, manage storage resources, and better integrate intermittent renewable sources, improving the stability and efficiency of power systems.

18.4.4.2 Resource allocation and ancillary services

As shown in Fig. 18.9, LLMs can be used to enhance resource allocation procedures by contextually comprehending natural language instructions into practical suggestions for energy distribution. By processing text-based data from diverse sources, such as regulatory texts, market reports, and system constraints, LLMs recommend adjustments that maximize grid performance while keeping operational costs to a minimum. Besides, LLM-driven platforms can amplify ancillary service and DR measures. Through communication with prosumers directly through natural language interfaces, the models facilitate effective sharing of energy usage, pricing signals, and peak demand curtailment offers. The interactive capability allows consumers to participate actively in demand-side management, reducing aggregate grid strain during peak demand conditions.

In this case, model development starts by preparing input data from various and often unstructured text sources such as market forecasts, regulatory directives, technical manuals, and operational logs. These inputs need to be tokenized and embedded as described in Eq. (18.2) so the model can process domain-specific terminology and instruction patterns. To ensure the semantic ordering of constraints and resource prioritization rules is preserved, *PE* from Eq. (18.1) should be applied. For example, a regulatory instruction that prioritizes renewable dispatch before

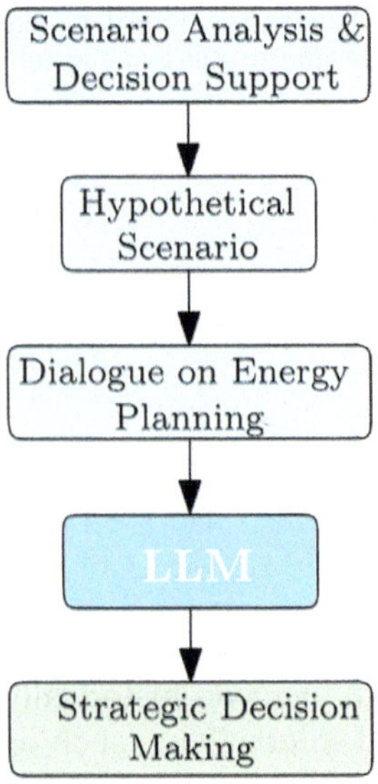

FIGURE 18.10 Scenario analysis workflow for energy planning, showing how hypothetical conditions and operator dialogue support strategic decision-making.

conventional sources must be kept in the same order, so downstream layers process it with the correct precedence. Through this step, the model gains the ability to distinguish context-dependent roles, such as whether a generation source is dispatchable or has curtailment limits.

The embedded inputs should then be passed into the attention layers, where contextual associations across multiple operational variables and decision criteria can be identified. Here the model learns to associate specific clauses in policy documents with cost impacts or match consumer energy profiles with incentive-based DR offers. This cross-referencing enables the generation of flexible resource coordination strategies while observing operational and regulatory boundaries.

Unlike forecasting, where attention is temporal, attention in this domain needs to be configured to prioritize relevance across heterogeneous features. For instance, priority may need to be given to cost coefficients, grid congestion data or carbon intensity signals depending on operational goals. Multihead attention allows this selective focus to be achieved across parallel subspaces, each modeling a different optimization vector.

The outputs from attention are then passed through the feed-forward layers Eq. (18.6) where interdependencies among pricing signals, resource types, and spatial availability are refined. This is especially important when decisions require integrating distributed resource availability across multiple sites or reconciling load shedding with customer preferences. The model at this stage forms a latent representation that can balance multiple competing objectives.

In DR applications the decoder is used to translate system-level decisions into human-readable offers or control signals. This can include generating messages to residential or commercial customers proposing load reduction in exchange for dynamic pricing benefits. By using masked attention and contextual memory, the decoder can preserve continuity across time or multiturn conversations and ensure customers receive consistent and understandable engagement. Fine-tuning should be done using curated examples that capture energy-specific semantics such as structured response templates, incentive thresholds, and site-level operating constraints. To make this parameter efficient, fine-tuning using LoRA Eq. (18.9) should be used. This allows operational knowledge and linguistic nuances to be injected into the model with minimal memory overhead and can be deployed in edge environments or cloud-integrated systems. Post-training quantization can then be applied to ensure real-time performance in grid operation centers or local energy management systems, and enable rapid synthesis of resource schedules or user-facing communication.

18.4.4.3 Scenario analysis and decision support

LLMs possess the ability to perform advanced scenario analysis by conceptualizing various scenarios of energy planning through dialogue. Operators can input hypothetical scenarios or potential future results, and the LLM is then able to rapidly process implications and produce strategic guidance. Such an ability for rapid creation and interpretation of sophisticated scenarios facilitates more informed, fast decision-making to react to emergent events or changes in demand-supply balance. As illustrated in Fig. 18.10, the LLM must have special abilities to systematically interpret hypothetical situations, explore alternative outcomes, and give strategic guidance in real-time. This means

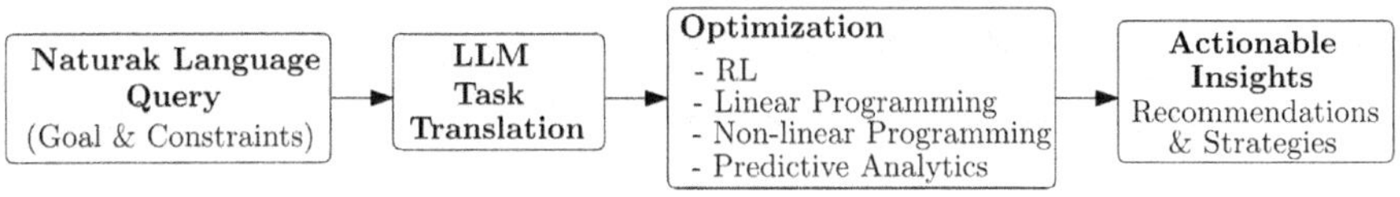

FIGURE 18.11 Smart grid analytics workflow that converts natural-language goals and constraints into optimization tasks for generating actionable recommendations and operational strategies.

functionalities different from general-purpose language modeling, scenario generation, consequence evaluation, and iterative interaction with operators.

PE as per Eq. (18.1) are essential to maintain logical sequencing so the model can evaluate scenarios where sequence-dependent outcomes and chained conditional events are common. The model must be trained to understand and differentiate conditional language and scenario descriptors that indicate hypothetical or speculative contexts. Through the embedding procedures defined in Eq. (18.2), the LLM must encode textual markers (such as what if, in case of, or assuming) with enough distinction to trigger scenario branching internally. Attention mechanisms as per Eqs. (18.4)–(18.5) must be explicitly tuned to identify causal, conditional, and comparative relationships within scenario descriptions. LLM should be able to see how changing one parameter (e.g., increased EV penetration) affects multiple related parameters (e.g., peak load, transformer capacity, and grid stability). Attention heads should be specifically configured to track dependencies across multiple hypothetical dimensions simultaneously so the model can efficiently explore multidimensional scenario spaces. Feed-forward layers as per Eq. (18.6) must transform these attention-driven insights into robust internal representations suitable for consequence prediction. Here the focus shifts to encoding potential impacts rather than just descriptive data, so the LLM can infer downstream effects of scenario assumptions, evaluating economic consequences of renewable curtailment or resilience impacts of infrastructure disruptions.

Most importantly, the decoder must support iterative and interactive generation of actionable scenario outcomes. Masked attention enables step-by-step elaboration of future conditions or strategic options. Explicit design consideration must be given to the model's ability to produce coherent scenario narratives, outcome summaries, or sequential decision paths that operators can understand and query. This includes, for example, clearly enumerating grid vulnerabilities under specific hypothetical conditions or presenting decision trade-offs in response to sudden changes in renewable generation.

PEFT techniques in this case ensure the scenario-based reasoning capability. Fine-tuning emphasizes the model's skill in conditional reasoning, risk assessment, and strategic prioritization. Rapid scenario elaboration is key in real-time or near-real-time decision-making environments where delays can kill decision quality. Quantization reduces the compute required to evaluate scenarios on the fly so operators can iterate through scenarios, change parameters or assumptions, and get instant strategic feedback.

18.4.4.4 Integration with advanced analytics and optimization techniques

The integration of LLMs with other techniques, such as RL, linear and nonlinear programming, and predictive analytics, further enhances their ability in energy planning. As shown in the block diagram in Fig. 18.11, by interpreting natural language requests and providing actionable insights, LLMs effectively bridge the gap between human operators and sophisticated computational tools.

The LLM must be capable of parsing ambiguous planning inputs, recognizing mathematical intent, resolving intervariable dependencies, and emitting structured outputs that satisfy the syntactic and logical requirements of advanced optimization frameworks. This enables a seamless integration between natural language interfaces and robust decision-making engines, forming a unified pipeline from human query to algorithmic resolution. Tokenization and embedding must preserve the syntactic and semantic distinctions between operational terms (e.g., "maximize utilization") and symbolic expressions. Positional encoding using Eq. (18.1) maintains the order of expressions in logical blocks, which is needed for encoding solvable models.

The model must attend across variable definitions, interrelated constraints, and objective functions, learning how each component influences the feasible solution space. For example, attention should enable the model to detect when a constraint modifies a decision boundary or when a time-indexed variable requires multistep consistency. To perform the translation from descriptive text to formal optimization structure, feed-forward layers through Eq. (18.6) consolidate abstract representations into output-ready encodings. These layers must be configured to capture nonlinear mappings from plain language into structured forms that adhere to the syntactic and semantic requirements of solvers. For instance, this includes converting phrases like "limit battery discharge between 6 and 9 p.m." into time-indexed inequality constraints with bound logic. During generation, masked attention helps to enforce token dependency, ensuring that variables introduced earlier are properly referenced later, and that the resulting output remains parseable and executable without manual correction. For an advanced analytics application, the fine-tuning should be performed using paired samples of text-based planning instructions and their corresponding analytical formulations. Through PEFT using methods such as LoRA using Eq. (18.9), the model can be adjusted to align with the structure and expectations of specific optimization techniques without requiring retraining of the full parameter set. This is especially important for integrating with proprietary systems where memory and compute constraints must be respected. Quantization may also be applied to respond promptly with solver inputs, scenario definitions, or analytic summaries, even when deployed on hardware with limited processing capabilities.

LLM-assisted optimization: To support data-driven decision-making in smart grids, energy planning workflows should integrate LLMs with deterministic optimization solvers and operator oversight in a structured and systematic manner to translate high-level planning objectives into solver-ready formulations and actionable control policies. This integration enables the automation of model specification, the enforcement of physical and operational constraints, and the interpretation of optimization results within a single, modular architecture. Within this architecture, the LLM can function as an interface that standardizes the transition from a conceptual engineering description to a solver-ready optimization model and can remain involved throughout the planning cycle. When an operator specifies an energy planning task in natural language, the LLM can parse the request and map key phrases to a predefined library of system states and decision variables. These states may include nodal active and reactive demands/supplies and bus voltages and angles, while decision variables can represent controllable actions such as generator dispatch levels, storage charging and discharging schedules, DR setpoints, and network reconfiguration decisions. In parallel, the LLM can identify candidate performance criteria and planning objectives, such as operating cost, renewable utilization, emissions, security margins, and resilience metrics, and relate them to key inputs like load forecasts, renewable output, market prices, and network availability to support subsequent sensitivity and uncertainty analysis. The LLM can identify the appropriate planning problem type, such as day-ahead economic dispatch with network constraints, storage co-optimization over multiple time periods, or resilience-aware scheduling that includes emergency operating modes. As part of this step, it can help classify constraints into hard and soft categories and into functional and non-functional groups. Hard constraints capture physical feasibility, safety limits, protection requirements, and essential operational conditions, including power-flow equations, thermal limits of lines and transformers, admissible voltage and frequency ranges, and ride-through capabilities. Soft constraints represent preferences and service quality standards that may be relaxed at a penalty in the objective function, such as comfort ranges, voluntary demand reductions, and emission limits. Functional constraints are directly linked to physics, protection, or regulation, whereas non-functional constraints represent economic requirements, contractual obligations, or policy goals. This classification explicitly separates non-negotiable conditions from those that can be adjusted based on cost or other criteria, and the LLM can encode that separation in the evolving model schema.

Once variables, objectives, and constraints are structured, the LLM can determine the optimization problem class by analyzing whether the relevant states and decision variables are continuous, integer, or mixed, and by identifying how uncertainty will be represented. Uncertainty may be modeled using deterministic scenarios from forecast ensembles, probabilistic distributions associated with forecast errors, or bounded sets for worst-case analysis. Depending on these choices and the structure of the constraints and objective, the LLM can map the planning problem to an appropriate formulation class, linear programming (LP), quadratic programming (QP), mixed-integer linear programming (MILP), mixed-integer quadratic programming (MIQP), robust optimization, or stochastic optimization, and select a corresponding optimization template. This step can be supported by prior fine-tuning on paired text–schema examples and by RAG that retrieves stored model fragments, constraint blocks, and solver configurations from a technical knowledge base. The LLM can then instantiate the selected template with the specific assets, time horizons, limits, objective weights, and uncertainty representations provided by the operator or retrieved from the Digital Twin. The result can be a structured, machine-readable schema that explicitly lists state variables, decision variables, hard constraints, soft constraints with associated penalty coefficients, the form of the objective function, and the chosen representation of uncertainty (scenario sets, probability distributions, or worst-case sets). The schema can also record the declared problem class and intended solver family. Units, numerical limits, and references to buses, lines, and devices in the underlying network model can be attached to enable consistency checks against the Digital Twin representation.

This schema can be forwarded to a validation and safety-checking component rather than directly to the solver. The validation step can verify that variables have consistent physical units and realistic bounds, that asset and topology references are valid, and that hard constraints satisfy safety margins and regulatory requirements. It can also check that soft constraints and penalty weights are well-scaled and that the selected problem class is compatible with the chosen solver. If inconsistencies are detected, such as missing bounds, incompatible units, ambiguous constraints, or an infeasible mix of uncertainty modeling and solver capabilities, the schema can be rejected or returned to the LLM with structured feedback for revision under explicit guidelines. In this way, the model specification can be refined iteratively rather than issued as a single one-shot description. Only schemas that pass validation are translated into the specific input format required by the deterministic solver and submitted via an application programming interface or an Open Platform Communications (OPC) interface to the optimization platform. The solver then computes a solution and returns optimal decision variables along with diagnostic information, including feasibility indicators, statistics on constraint violations, objective values, and, when available, dual variables and optimality measures. Dual variables and reduced costs can be used to support classical sensitivity analysis by quantifying how marginal changes in constraints, costs, or limits affect the optimal solution. The LLM can interpret these diagnostics, highlighting binding constraints, stressed resources, and influential parameters for the operator.

Sensitivity analysis, uncertainty modeling, and LLM-based assistance can then be combined to support systematic planning. Deterministic optimization provides a baseline solution for a specific scenario or parameter set. The LLM can derive and interpret sensitivity measures, using dual variables, reduced costs, and parametric variations, to identify critical constraints and locations where small changes in demand, renewable output, or network capacity significantly affect the objective. Using these indicators, the LLM can assist the operator in specifying structured what-if analyses, such as modifying line limits, changing reserve requirements, or adjusting penalty weights on curtailment, and in configuring parametric studies in which selected parameters vary over prescribed ranges. The LLM can automatically generate and schedule these optimization runs, organize the resulting solutions, and summarize their impacts on costs, emissions, and constraint violations, effectively automating large portions of the planning loop while keeping the engineer in control. The same architectural pattern can be extended to probabilistic, stochastic, and robust optimization. When uncertainty is represented by discrete scenarios, the LLM can formulate a scenario-based stochastic program that introduces scenario indices, anticipativity constraints, and scenario weights, and then orchestrate solver runs that compute a policy optimizing expected performance across all scenarios. When uncertainty is described probabilistically, the LLM can incorporate chance constraints so that line flows or voltages satisfy limits with specified probabilities and convert these chance constraints into tractable approximations for the selected solver. For worst-case analysis, the LLM can formulate robust optimization models by defining uncertainty sets around loads, renewable output, or outages and generating constraint sets that remain feasible for all realizations within these sets. In all these cases, the LLM can configure, parameterize, and iterate on the optimization model rather than replace the underlying mathematical methodology. It can organize scenario data, define uncertainty sets, link them to Digital Twin data streams, and generate formulations that satisfy the required mathematical structure. After the solver returns solutions and statistics, the LLM can interpret the results, including risk metrics such as expected cost, cost variance, violation probabilities, and worst-case performance measures. It can present these metrics in a form that supports comparison of alternative plans, such as a cost-minimizing deterministic schedule, a more conservative robust schedule, or a probabilistic schedule that trades cost against risk, and can propose revised formulations or updated parameterizations in response to operator feedback.

Throughout this integration workflow, the operator remains the final decision authority. Candidate schedules, setpoints, uncertainty models, and scenario families can be presented for review. The operator can accept, adjust, or reject them in accordance with operational policies, SA, market conditions, and contingency forecasts. Only after explicit approval are the selected setpoints transmitted to the Digital Twin layer and then to field devices. In this sequence, the LLM serves as a planning and optimization assistant that supports problem formulation, schema generation, constraint classification, sensitivity exploration, uncertainty modeling, and result interpretation; it can translate solver outputs into plain-language summaries and visualizations that highlight consequences, associated risks, and anticipated benefits for the operator. As an integration technology framework, this approach can accelerate model specification, automate repetitive planning analyses, support interoperability with existing platforms and protocols, and enable systematic analysis of planning trade-offs under explicit human oversight.

18.4.4.5 Human-aligned Large Language Model operations

LLMs can progress from passive analytics to directly influencing operational decisions, but doing so requires their outputs to be embedded in explicit and auditable workflows rather than informal assistant-style interactions. In SDMG and Digital Twins solutions, LLM-generated recommendations may affect schedules, setpoints, contingency responses, and switching actions that carry physical, economic, and safety implications. To manage this effectively, an active LLM interaction layer must operate under HIL governance, coordinating with operators, Digital Twins, and control systems to ensure that outputs are integrated into grid operations in a safe, transparent, and accountable manner. A practical LLM integration framework should distinguish three mutually exclusive output classes of informational, assisted, and autoapplied. Informational outputs consist of advisory interpretations or analyses that remain confined to the interface and are not sent to field assets. Assisted actions consist of concrete operational proposals, such as adjusted dispatch schedules, modified reactive power setpoints, or revised maintenance timings, that require explicit operator approval before execution. Autoapplied actions consist of narrowly scoped, preauthorized adjustments that may be executed automatically once they satisfy predefined deterministic safety checks. This classification is determined by a policy layer that evaluates both the scope of the proposed action and its corresponding risk category.

Before any action is considered for execution, the framework should apply constraint validation and DTiL simulation. The LLM should generate a machine-readable proposal based on the current system state, which includes telemetry, forecasts, Digital Twin variables, and applicable operational constraints. Each proposed action should be evaluated against static and dynamic limits such as thermal ratings, voltage and frequency bounds, SoC thresholds, ramp-rate restrictions, and interlocking rules. Any action that violates operational hard constraints should be reclassified as advisory and returned to the operator, together with an explanation. Actions that satisfy all constraints should then be evaluated using a Digital Twin forecast over a short horizon to determine whether they remain feasible under expected operating conditions. Only proposals that are both constraint-compliant and simulation-feasible should be eligible for either assisted or autoapplied execution pathways. A HiL governance should be applied according to the risk classification assigned to each action. Low-risk actions that lie within predefined and preauthorized operating envelopes may be eligible for automatic execution after passing both constraint checks and Digital Twin validation. Actions classified as medium or high risk should always be directed to an operator approval queue. Actions that affect critical assets or operational states, such as switching, topology changes, or protection-parameter adjustments, should require a two-person approval rule. Under this rule, execution is permitted only after confirmation from two distinct authenticated operators, providing independent oversight before any irreversible command is issued.

To support interpretability and reduce operator workload, each proposed action should include a structured explanation derived from the LLM and augmented by the validation process. This explanation should identify the constraints that motivated the proposal, the forecasts or state trajectories that influenced the recommendation, and the expected effects on operational metrics such as loading, stability margins, congestion, or cost. In the operator interface, these explanations should be presented through structured elements such as a justification panel, constraint indicators, and visual summaries of Digital Twin projections. This presentation approach helps maintain alignment between LLM-generated recommendations and the operator's understanding of system behavior. Box 18.3 summarizes the simplified human-aligned LLM operations workflow by which LLM-generated actions are categorized, validated, simulated, routed for approval, and logged. It explicitly distinguishes informational, assisted, and autoapplied action paths; incorporates Digital Twins validation prior to any control command; and applies the two-person rule to high-impact actions.

Key Point: Interactive Interface with NLP and generative-AI

While Operator interacts do
Read Input: Receive the user's natural language query
Parse Query: Extract entities, keywords, and intent using NLP
Intent Classification: Predict the user's intent for the query
Transform Query: Map parsed query to ontology terms
If Intent is data retrieval, **then**
Retrieve Data: Query relevant data based on the transformed query
Display Visualization: Present real-time data insights using visualizations
End if
If Intent is insight generation, **then**
Generate Context: Prepare contextual input for the generative-AI model
Query Generative-AI: Get textual insights from the generative-AI server
Process Response: Extract and format relevant parts of the generated text
Display Insight: Present generated insights to the operator
End if
Generate Response: Formulate a comprehensive response based on data/insight
Display Response: Present the response to the operator
End while

BOX 18.3

LLM-assisted human-in-the-loop operation

1: While the control-room session is active:
2: Receive operator query, event trigger, or periodic task invocation.
3: Collect system state S_t (telemetry, forecasts, Digital Twin state, constraints).
4: Query LLM with (S_t, operator context, policies) to obtain a proposal P.
5: Parse P into:
6: P_{type} in {INFO, ACTION}
7: Candidate action set $A = \{a_1, \ldots, a_k\}$
8: Explanation object E_0.
9: IfP_{type} = INFO or A is empty then
10: Display informational response with explanation E_0.
11: Log (S_t, P, E_0, nan).
12: Continue loop.
13: End if.
14: Step 1: Constraint validation
15: For each action a_i in A do:
16: Evaluate $C(a_i, S_t)$.
17: If$C(a_i, S_t)$ reports any constraint violation then
18: Mark a_i as unsafe.
19: End if.
20: End for.
21: If all a_i are unsafe then
22: Return advisory response "Rejected by constraints"; augment E_0.
23: Log; continue loop.
24: End if.
25: Step 2: Digital Twin feasibility simulation
26: For each remaining feasible a_i do
27: Simulate $\hat{S}(t : t + H) = \Phi_{\mathrm{DT}}(S_t, a_i, H)$.
28: Evaluate $G_{\mathrm{DT}}\left(\hat{S}(t : t + H)\right)$.
29: If$G_{\mathrm{DT}}\left(\hat{S}(t : t + H)\right)$ reports any feasibility violation, then
30: Mark a_i as DT unsafe.
31: End if.
32: End for.
33: If all a_i are DT unsafe then
34: Return advisory response `Rejected by Digital Twin`; augment E_0.
35: Log; continue loop.
36: End if.
37: Step 3: Risk classification and routing
38: For each remaining a_i do:
39: Compute risk class ρ_i = RISK(a_i,S_t).
40: Compute routing category γ_i in {AUTO, ASSISTED}.
41: Ifa_i affects critical assets then
42: Set γ_i = ASSISTED and ρ_i = HIGH.
43: End if.
44: End for.
45: Partition A into A_{auto} and A_{assist} based on γ_i.
46: Step 4: Autoapplied actions
47: For each a_i in A_{auto} do:
48: Ifρ_i = LOW and a_i lies within the approved envelope then
49: $U(a_i)$; display rationale; log execution.
50: Else:
51: Move a_i to assist the queue.
52: End if.
53: End for.
54: Step 5: Assisted actions
55: For each a_j in A_{assist} do
56: If ρ_j = *HIGH* then
57: $r_{\mathrm{reg},j} = 2$.
58: Else 59: $r_{\mathrm{reg},j} = 1$.
60: End if.
61: Create approval record $R_j = \{a_j, S_t, E_j, \rho_j, r_{\mathrm{reg},j}\}$.
62: Place R_j in pending operator queue.
63: End for.
64: Display pending actions and explanations in operator interface, including affected assets, constraints, Digital Twin results, and expected impacts.
65: Upon operator responses for record R_j:
66: Let Ω_j be the set of distinct operator approvals for R_j.
67: If $|\Omega_j| = r_{\mathrm{reg},j}$ then 68: Compute execution time τ_j.
69:$U(a_j)$; log provenance (S_t, a_j, E_j, ρ_j, Ω_j, τ_j).
70: Else ifR_j is rejected or expires then
71: Log rejection and rationale.
72: End if.
73: End loop

In this process, S_t represents the current system state at time t, including telemetry, forecasts, Digital Twins state variables, and active operational constraints. P denotes the output proposal generated by the LLM, and P_{type} is the proposal type flag that indicates whether P is informational (`INFO`) or contains operational actions (`ACTION`). The set $A = \{a_1, \ldots, a_k\}$ contains the candidate actions proposed by the LLM, where each a_i (or a_j when indexed in the assisted set) is a concrete operational modification such as a change in setpoints, schedules, or switching commands.

E_0 and E_j here denote structured explanation objects associated with the initial proposal and with individual actions. The function $C(a_i,S_t)$ is the constraint-checking operator that evaluates an action against physical and policy limits. The $G_{DT}\left(\hat{S}(t:t+H)\right)$ is the Digital Twins operator that generates a predicted trajectory $\hat{S}(t:t+H)$ over a horizon H when action a_i is applied to state S_t. The $G_{DT}(\bullet)$ is the feasibility evaluator that inspects a simulated trajectory and determines whether it violates any Digital Twins–based limits or stability conditions. The quantity ρ_i (and ρ_j) is the risk level assigned to an action, while γ_i is the routing label that determines whether the action is eligible for automatic execution (AUTO) or must go through operator approval (ASSISTED). The sets A_{auto} and A_{assist} contain the actions routed to auto-execution and assisted approval paths, respectively. $U(a_i)$ is the execution function that issues a validated action to the control system. For assisted actions, $r_{reg,j}$ is the number of operator approvals required under the governance policy, R_j is the approval record that bundles the action, state, explanation, risk level, and approval requirement, Ω_j is the set of distinct operators who have approved R_j, and τ_j is the timestamp recorded at the moment the action is executed.

The flow begins when the control room session receives a query, event, or periodic trigger and constructs the current state S_t. This state, combined with operator context and applicable policies, is passed to the LLM to generate a proposal P. If P_{type} indicates that the proposal is purely informational, or if the candidate set A is empty, the system presents the explanation E_0 to the operator, logs the interaction, and takes no further action. When A contains one or more candidate actions, the workflow first applies constraint validation. Each action a_i in A is examined by $C(a_i,S_t)$, which checks the action against thermal limits, voltage and frequency bounds, SoC limits, ramp-rate constraints, and any encoded interlocking or policy rules. Actions that trigger violations in this stage are marked unsafe and removed from A. If no actions remain after constraint validation, the system reports that the proposal has been rejected by constraints, augments E_0 with this information, it logs the outcome and returns to the next loop iteration.

Actions that survive constraint validation are then subjected to Digital Twins feasibility analysis. For each remaining action a_i, the Digital Twins process$\Phi_{DT}(S_t,a_i,H)$ produces a short-horizon trajectory $\mathcal{S}(t:t+H)$, reflecting how the system would evolve if that action were applied under current and forecasted conditions. The evaluator G_{DT} inspects each $\mathcal{S}(t:t+H)$ to determine whether any limit or stability criterion would be violated. Actions that lead to unacceptable behavior in the simulation are marked as Digital Twin unsafe and removed. If all actions are removed at this stage, the system reports that the proposal has been rejected by the Digital Twin, logs this result, and terminates the iteration.

The remaining actions, which have passed both constraint and Digital Twins checks, are then classified according to risk and routed to appropriate execution paths. The function RISK assigns a risk level ρ_i to each action based on its scope, affected assets, and operating context. The routing label γ_i is then set to AUTO or ASSISTED according to policy. Any action that touches critical elements, such as switching devices, topology configuration, or protection settings, is forced into the ASSISTED path with a high-risk label. The candidate set is thus divided into A_{auto}, containing actions eligible for direct execution under strict conditions, and A_{assist}, containing actions that require human approval. In the autoapplied path, actions a_i in A_{auto} are executed immediately through $U(a_i)$ only if their risk level ρ_i is low and they fall within predefined, policy-approved envelopes. When such an action is executed, the system displays the rationale to the operator and records the event in the log. Actions in A_{auto} that do not satisfy these criteria are downgraded into the assisted path and moved into A_{assist} to ensure human review. In the assisted path, each action a_j in A_{assist} is wrapped into an approval record R_j that includes the action, the associated state, an explanation E_j, the risk level ρ_j, and the required number of approvals $r_{reg,j}$. High-risk actions, especially those involving critical switching or protection changes, are assigned $r_{reg,j} = 2$ to enforce a two-person rule, whereas other actions require a single approval. These records are placed in a pending queue and presented to operators through the interface, along with clear explanations of constraints, Digital Twin outcomes, and expected impacts.

As operators review the queue, their approvals for each record R_j are collected into Ω_j. When the number of distinct approvals in Ω_j matches $r_{reg,j}$, the corresponding action a_j is executed through $U(a_j)$, and the system records the execution time τ_j along with the full provenance (state, explanation, risk level, and identity of approvers). If a record is rejected or expires without sufficient approvals, that decision and its rationale are logged, and no command is issued to the field.

18.5 Integration of generative artificial intelligence into smart grids

By leveraging real-time data and generative-AI algorithms, predictive maintenance is refined to ensure the efficiency and longevity of assets in smart grids (Hosseini et al., 2022). Recent advances in generative AI offer a set of new technologies and tools to strengthen predictive models and enhance model accuracy and reliability

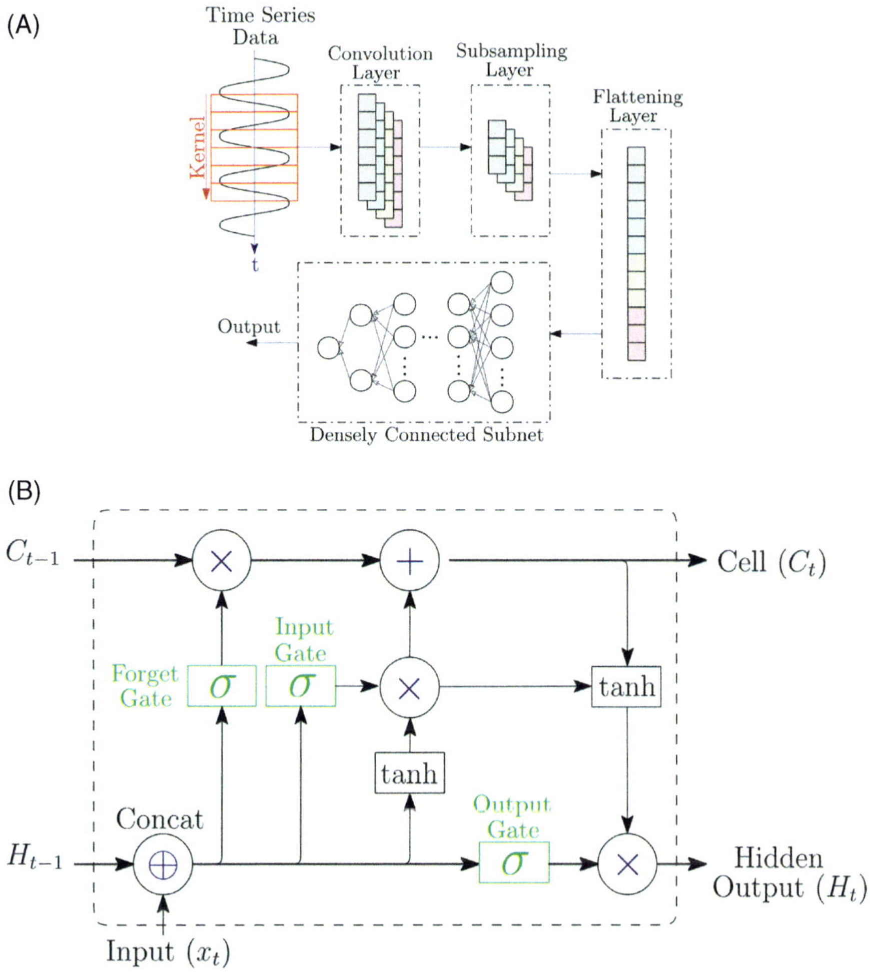

FIGURE 18.12 Deep learning architectures for time-series analysis in smart grid applications: (A) a one-dimensional Convolutional Neural Network (CNN) regressor that extracts local temporal features through convolution, subsampling, flattening, and dense layers; and (B) a standard LSTM cell that models sequential dependencies using input, forget, and output gates to update the cell and hidden states. The symbol ⊕ denotes concatenation, and ⊗ denotes element-wise multiplication.

(Bhagat, 2024). However, high development and implementation costs of generative AI may limit accessibility in SDMG for smaller organizations and communities or lower-value assets. To overcome these barriers, LLMs can enhance generative-AI driven solutions by bridging the gap between technical predictions and engineer-friendly explanations. LLMs can address the challenges limiting the adoption of generative AI in predictive maintenance by operating at the network edge, summarizing raw sensor data into compact, anonymized text reports. This ensures that sensitive operational details remain local while still enabling insights to be shared safely.

Among all, deep learning models utilize unique spatial-local connectivity features, which make them efficient for supervised learning.

Fig. 18.12A illustrates the structure of a 1-D CNN regressor, which relies on layers that perform convolution operations on the input data or feature maps using pre-configured kernels. The Long Short-Term Memory (LSTM) model, shown in Fig. 18.12B, on the other hand, leverages its memory capabilities to retain information across sequential data points, making it suitable for time-series analysis. Through recurrent connections and memory mechanisms utilizing input, forget, and output gate operations, the LSTM module learns new useful information to add to its memory and omits details that are no longer important. This contributes to the development of intelligent maintenance strategies, reinforcing the importance of data-driven methodologies in optimizing the performance and longevity of Smart Grid equipment. In the next section, we explore more sophisticated AI-driven methodologies that can further enhance predictive, prescriptive, and preventive actions.

18.5.1 Health monitoring of grid-scale battery energy storage systems

Fig. 18.13 represents generative deep learning architectures, including VAEs, RL, cGANs, and adversarial autoencoders (AAEs) for addressing the multifaceted challenges of State-of-X (SoX) estimation in grid-scale battery energy storage systems (BESS).

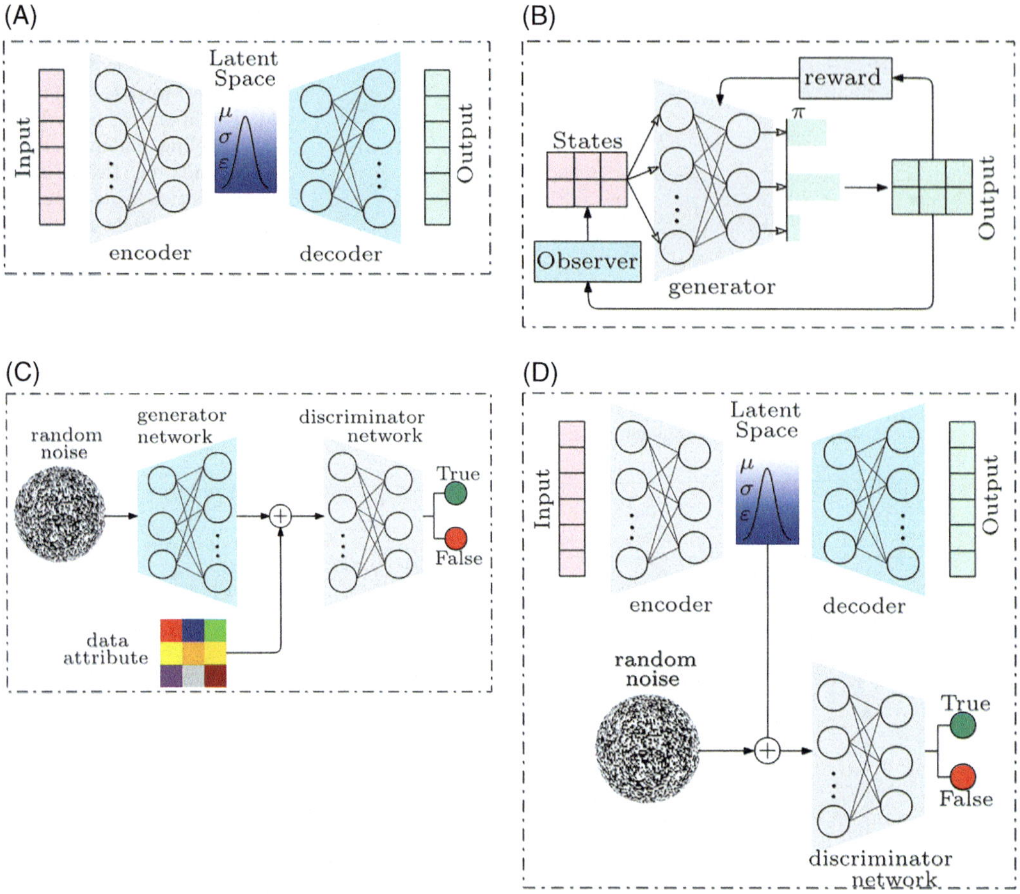

FIGURE 18.13 Overview of generative deep learning architectures (A) variational autoencoder (VAE) where an encoder maps the input data to a probabilistic latent space with mean (μ) and variance (σ) samples from this space are passed through a decoder to reconstruct the output, (B) reinforcement learning-based generator where the generator network maps observed states to outputs (actions) and learns through a reward-based feedback loop, (C) conditional generative adversarial network (cGAN) where a generator creates data from random noise and conditional attributes, and a discriminator evaluates the generated data as true or false, and (D) adversarial autoencoder (AAE) combining autoencoding with adversarial training where the encoder maps input to a latent distribution regularized using a discriminator that distinguishes between true and generated latent samples.

Fig. 18.13A represents the VAE, which models the data generation process by learning an encoder $q_\phi(z \mid x)$that approximates the posterior distribution over latent variables z, and a decoder $p_\theta(x \mid z)$that reconstructs the input $x$from z. The optimization objective includes a reconstruction term and a Kullback–Leibler (KL) divergence (D), formalized as

$$\mathcal{L} = \mathbb{E}_{q_\phi(z|x)}\left[\log p_\theta(x|z)\right] - D\left(q_\phi(z|x) \parallel p(z)\right), \tag{18.10}$$

where $p(z)$is typically a standard Gaussian prior. This enables sampling of latent representations that can generate novel electrode microstructures or degradation trajectories, which are invaluable in SoX prognosis models and inverse design workflows. In grid-scale BESS, the VAE can address the high dimensionality and temporal dependencies present in the degradation data, such as time-series voltage, current, temperature, and environmental parameters, by projecting them into a compact and structured latent space. This latent space captures the underlying degradation modes and hidden health states of the BESS system, effectively compressing operational histories into a lower-dimensional representation amenable to monitoring and forecasting. By continuously feeding in sensor data streams x_t, the VAE can be used to learn a latent trajectory z_t, which evolves and correlates with critical SoX indicators. This approach allows the inference of SoX by decoding the latent variables into expected observable behaviors and anomaly detection. Furthermore, the stochastic nature of the latent space enables quantification of the SoX uncertainties, which is needed for stable operation and dispatch strategies in grid applications. The flexibility of VAEs also enables their coupling with surrogate physical models of degradation mechanisms, wherein the latent space can be regulated by physics-informed constraints, enhancing the interpretability and reliability of the SoX estimations under varying load profiles, environmental conditions, and duty cycles typical of grid-scale BESS operations. In practice, such capabilities allow grid operators to detect hidden degradation early and act before failures compromise availability.

The RL generators, shown in Fig. 18.13B can also be used as a sequential decision-making problem where an agent interacts with an environment, generating battery usage profiles through a policy $\pi(a \mid s)$. The agent is optimized to maximize the expected cumulative reward.

$$J(\pi) = \mathbb{E}_{\tau \sim \pi}\left[\sum_t r(s_t, a_t)\right], \tag{18.11}$$

where the reward r can be linked to BESS stability and desired SoX behaviors, such as maximizing cycle life or minimizing degradation rates under given duty cycles. RL framework can be extended beyond the SoX estimation and management problem as a sequential decision-making process, where the BESS system itself constitutes the environment, and the RL agent learns to interact with it by taking actions $a_t$such as charging, discharging, or rest cycles, based on the observed system state s_t, which encapsulates sensor readings, historical usage patterns, and environmental variables. In a BESS system, the RL agent learns policies that inherently capture the trade-offs between performance (e.g., providing frequency regulation or peak shaving) and SoX degradation mechanisms, by associating rewards or penalties with the observed SoX evolution under different operating scenarios. Worth noting, the agent does not require explicit supervision of SoX data but instead leverages self-supervised learning through interaction with high-fidelity BESS simulators or Digital Twins, which can emulate electrochemical, thermal, and aging dynamics in response to various control strategies. By embedding physics-informed constraints or using hybrid model-free/model-based RL strategies, the agent can be guided to respect operational limits and safety margins while optimizing for SoX outcomes such as remaining useful life (RUL) maximization, degradation rate minimization, or state-of-health (SoH)–preserving dispatch.

The RL model can be combined with operational scenarios, where the agent is trained not only to select actions but also to generate realistic duty cycles or environmental stress patterns that challenge the BESS system. This dual generative-discriminative capability allows the RL framework to serve both as an estimator and as a data augmentation tool, enriching the SoX prediction pipelines by creating synthetic operational sequences that expose the model to diverse stress trajectories, thereby improving uncertainty quantification of SoX prognostics.

Fig. 18.13C shows the architecture of cGANs, where the data generation is a two-player adversarial game, where a generator network $G(z, c)$synthesizes candidate samples from random noise $z$conditioned on attributes c, while a discriminator $D(x, c)$attempts to distinguish between real samples from the data distribution $p_{\text{data}}(x)$and the synthetic samples generated by G. The adversarial loss governing this competition is given by

$$\min_G \max_D \mathbb{E}_{x \sim p_{\text{data}}}\left[\log D(x, c)\right] + \mathbb{E}_{z \sim p(z)}\left[\log(1 - D(G(z, c), c))\right]. \tag{18.12}$$

The cGAN offers a way for generating synthetic yet realistic datasets that reflect complex electrochemical, thermal, and degradation behaviors under diverse operating conditions. The conditioning vector c can encapsulate SoX targets or operational attributes, environmental stressors (e.g., temperature, humidity), or grid service profiles (e.g., DR events). By training the cGAN on BESS datasets, where each sample x includes time-series measurements of system variables alongside annotated SoX labels, the generator learns to produce plausible operational data streams that correspond to specific SoX conditions encoded in c.

This is particularly valuable for overcoming the data sparsity and imbalance issues, where degradation events, faults, or rare grid stress scenarios are underrepresented in the available data. The cGAN can serve as a data augmentation engine, enriching the SoX estimation models with synthetic yet high-fidelity data that reflects early failure signatures, or atypical usage cycles that may otherwise be absent from historical records. Moreover, the cGAN can be used to explore extreme regimes by simulating aggressive cycling under heatwave conditions or deep discharge under grid blackouts. For operators, this provides a way to stress-test contingency plans against rare or future failure modes, ensuring preparedness under extreme grid conditions.

Beyond data augmentation, cGANs can be employed for conditional generation in SoX-aware decision support systems, where operators can query the generator to visualize expected system behaviors given planned dispatch strategies or operational constraints. Additionally, the discriminator component of the cGAN can act as an anomaly detector, identifying whether new operational data sequences are consistent with the learned distribution of healthy states.

Therefore the cGAN offers a forward generator of synthetic, SoX-conditioned data streams for model training, scenario analysis, and risk assessment, and as a discriminator that enhances the detection of off-distribution behaviors

TABLE 18.3 Generative architectures and their BESS applications.

Architecture	Dominant feature	BESS application
VAE	Probabilistic latent representation with uncertainty quantification	State-of-charge (SoC) and state-of-health (SoH) estimation, degradation trajectory modeling, anomaly detection, and missing-data imputation
RL	Policy learning through reward-based sequential decision-making	Adaptive charge/discharge scheduling, lifecycle-aware dispatch, safety-constrained control for grid services such as frequency regulation
cGAN	Conditional data generation with realism discrimination	Synthetic generation of rare fault or extreme cycling data, augmentation for SoC/SoH training datasets, anomaly detection using the discriminator
AAE	Structured latent space enforced by adversarial regularization	Health-state clustering, predictive maintenance, inverse design of charging profiles, and early warning via off-manifold detection

AAE, *adversarial autoencoders;* BESS, *battery energy storage systems;* cGAN, *conditional generative adversarial networks;* RL, *reinforcement learning;* VAE, *variational autoencoders.*

indicative of emerging faults or anomalous degradation modes, ensuring enhanced reliability, safety, and longevity of grid-scale energy storage assets.

The AAE, as shown in Fig. 18.13D, represents a hybrid generative modeling that merges the reconstruction of autoencoders with the distribution-matching prowess of adversarial networks, addressing some of the key limitations of conventional VAEs and GANs. In the AAE, the encoder network $q_\phi(z\,|\,x)$ learns to map input multivariate time-series data (x) of operational data, including current, voltage, temperature, and grid interactions, into a latent space z. Simultaneously, a decoder $p_\theta(x\,|\,z)$reconstructs the data from z, ensuring the preservation of critical information required for SoX inference. However, unlike in VAEs, where the latent distribution is regularized via the KL divergence, the AAEs employ an adversarial mechanism wherein a discriminator $D(z)$is used to distinguish between samples drawn from the encoder $q_\phi(z\,|\,x)$and those sampled from a predefined prior $p(z)$. The encoder is trained adversarially to fool the discriminator, thereby ensuring that the learned latent representations are indistinguishable from the prior:

$$\min_{q_\phi}\max_{D}\mathbb{E}_{z\sim p(z)}[\log D(z)]+\mathbb{E}_{x\sim p_{\text{data}}}[\log(1-D(q_\phi(z|x)))]. \tag{18.13}$$

For the SoX estimation of BESS, the structured latent space enables compact, disentangled, and semantically meaningful representations of the high-dimensional, noisy, and often incomplete operational data in a heterogeneous grid. By enforcing the latent space to follow a known prior, AAEs support interpretability, where specific regions of the latent space can be associated with distinct degradation, SoX, or operational anomalies. Additionally, this structured latent enables inverse optimization for SoX management, where practitioners can navigate the latent space systematically, using gradient-based methods, Bayesian optimization, or RL. This can be used to identify cycling patterns or maintenance protocols that map to correct latent regions associated with battery life and stable SoH profiles. This latent space engineering can be used by operators to balance grid ancillary services against asset longevity. Furthermore, the AAE can provide anomaly detection, as any observed data $x$yields a latent representation deviated from the structured prior distribution. This can be utilized for predictive maintenance and early warning systems in grid-scale BESS deployments. To this end, latent clusters map naturally into interpretable health states, which can be linked to maintenance schedules or operational derating strategies.

Table 18.3 compares four generative architectures in the context of BESS. VAEs are most suitable for representing latent battery states and providing uncertainty-aware estimates of SoC and SoH. RL differs by focusing on adaptive control and policy learning, making it effective for real-time scheduling of charging and discharging actions. Conditional GANs provide high-fidelity synthetic data generation that can enrich training datasets, especially for rare or extreme operating conditions. AAEs combine the autoencoder structure with adversarial training to yield structured latent spaces, which can support predictive maintenance and early fault detection. These generative architectures not only could advance modeling accuracy but also connect directly to operator decision-making, where VAEs can support early anomaly alerts, RL could enable adaptive dispatch strategies, using cGANs to enhance the stress-testing with synthetic rare events, and AAEs can be used to translate complex states into actionable maintenance triggers.

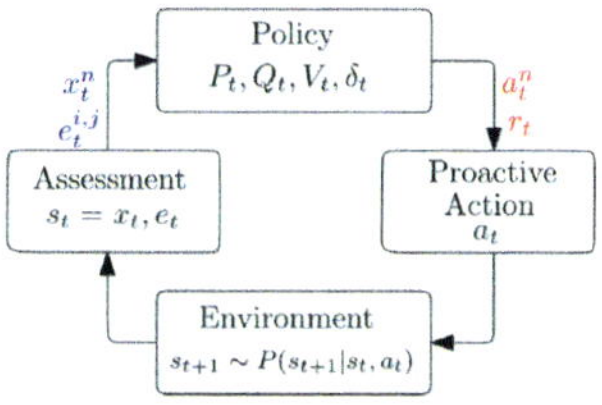

FIGURE 18.14 Reinforcement learning process for grid contingency planning, showing state assessment, policy-based action selection, and environment feedback for proactive grid control.

18.5.2 Smart grid contingency planning

Generative deep learning can be extended beyond asset-level applications to support system-wide contingency planning and resilience optimization, ensuring the maintenance of smart grid stability and the recovery from disruptive events. The transmission power grid can be modeled as a spatial-temporal graph structure $\mathcal{G} = (\mathcal{V}, \mathcal{E})$, where buses represent the nodes $\mathcal{V}$, and transmission lines constitute the edges $\mathcal{E}$. In particular, let $\mathbf{A} \in \mathbb{R}^{|\mathcal{V}| \times |\mathcal{V}|}$be the adjacency matrix of $\mathcal{G}$, and let $\mathbf{D}$ be the diagonal degree matrix with $D_{ii} = \sum_j A_{ij}$. The grid dynamic states are subject to continuous evolution driven by operational actions, exogenous disturbances, and system contingencies. To address these complexities, spatio-temporal graph neural networks (ST-GNNs), integrated with reinforcement learning (RL) algorithms, can be used for smart grid contingency planning, enabling the learning of control policies that are capable of proactively managing grid vulnerabilities and ensuring system resilience under extreme conditions. By effectively capturing the intricate spatial dependencies and temporal dynamics of grid states, ST-GNNs are positioned to serve as intelligent agents that can anticipate cascading failures, recommend corrective actions such as load shedding, redispatch, or DR, and assess the long-term impacts of these interventions on grid stability and operational efficiency.

Through this approach, the power grid can be formulated as a dynamic RL environment in which the operators are trained to make sequential decisions in response to evolving system states. Fig. 18.14 shows a four-step RL process for grid contingency planning comprised of (1) Assessment of current grid state, (2) Policy based on optimal power flow (OPF) and resilience, (3) Action via control strategies such as load shedding or redispatch maneuver, and (4) Reward measured by system metrics like load shed, voltage/frequency violations, and operational cost.

One can formulate the spatio-temporal power grid dynamics, where each bus $n \in \mathcal{V}$can be characterized by node-level state features at the timestep t:

$$x_t^n = \left\{ P_{d,t:t+T}^n, \quad Q_{d,t:t+T}^n, \quad P_{g,t}^n, \quad Q_{g,t}^n, \quad V_t^n, \quad \delta_t^n \right\}, \quad \forall n \in \mathcal{V}, \tag{18.14}$$

where $P_{d,t:t+T}^n, Q_{d,t:t+T}^n$ denote forecasted active and reactive demands over a prediction horizon T, $P_{g,t}^n, Q_{g,t}^n$ represent generation injections, and V_t^n, δ_t^nrepresent bus voltage magnitudes and angles. Similarly, transmission line features (edge-level state features) at timestep t can be defined by:

$$e_t^{i,j} = \left\{ \rho_{t:t+T}^{f,i,j}, \quad f_t^{i,j} \right\}, \quad \forall (i,j) \in \mathcal{E}, \tag{18.15}$$

where $\rho_{t:t+T}^{f,i,j}$ represents predicted failure probabilities of transmission lines, derived from fragility curves and weather forecast models, and $f_t^{i,j}$denotes current loading levels. The combined node and edge features define the RL state of the power grid at time $t$is $s_t = \{x_t, e_t\}$.

In the RL paradigm, the power grid operator acts as an agent sequentially interacting with the grid (environment). At each timestep t, the agent observes the state s_t and selects actions a_t to control and stabilize the grid:

$$a_t^n = \left\{ P_{g,t+1}^n, \quad Q_{g,t+1}^n, \quad P_{t+1}^{\text{sh},n}, \quad Q_{t+1}^{\text{sh},n} \right\}, \forall n \in \mathcal{V}, \tag{18.16}$$

where P^{sh}, Q^{sh}represent load-shedding actions, and $P_{g,t+1}$, $Q_{g,t+1}$correspond to adjustments in generator outputs.

According to RL conventions, the environment transitions probabilistically from state $s_t$to a new state s_{t+1} based on the environment's state-transition probability function $P(s_{t+1} | s_t, a_t)$ which reflects physical grid dynamics and stochastic disturbances. The reward function r_t quantifies the immediate performance of selected actions a_t. Reward here can capture both resilience and economic efficiency.

Characterizing the recursive relationship between current and future grid states, enabling systematic optimization through an RL algorithm. This model can utilize a GNN to precisely approximate this value function online, leveraging historical and real-time data to predict system vulnerabilities and recommend corrective actions. The GNN framework captures spatial dependencies between buses and transmission lines, enabling the estimation of key electrical variables such as voltage magnitude and angle for each bus. The predicted values for the next timestep can be expressed by

$$\hat{V}_i^{(t+1)} = \mathcal{F}_V\left(h_i^{(t+1)};\Theta_V\right),\quad \hat{\delta}_i^{(t+1)} = \mathcal{F}_\delta\left(h_i^{(t+1)};\Theta_\delta\right), \tag{18.17}$$

where $\mathcal{F}_V, \mathcal{F}_\delta$are trainable mapping functions, and Θ_V,Θ_δ are the learnable parameters of the model. Unlike traditional recurrent networks, which suffer from vanishing gradients and slow training times, tactical communication networks (TCNs) utilize causal convolutions to capture long-range dependencies efficiently. The temporal evolution of bus embeddings can be captured using an attention-weighted formulation:

$$h_i^{(t+1)} = \sum_{\tau=t-T+1}^{t} \alpha_\tau \mathrm{TCN}\left(h_i^{(\tau)}\right), \tag{18.18}$$

where $\alpha_\tau$is the attention weights and can be computed via:

$$\alpha_\tau = \frac{\exp\left(\mathbf{w}^\top h_i^{(\tau)}\right)}{\sum_{k=t-T+1}^{t} \exp\left(\mathbf{w}^\top h_i^{(k)}\right)}. \tag{18.19}$$

The model integrates past states for real-time predictions, helping to anticipate abrupt changes like cascading failures. By considering essential constraints, it maintains OPF during learning and inference, minimizing the risk of infeasible operating points. The objective function integrates data-fitting loss terms with physics-based constraints to jointly train state estimation and resilience prediction models:

$$\mathcal{L}_\mathrm{k} = \mathbb{E}\left[{V_i^{(t+1)} - \hat{V}_i^{(t+1)}}^2 + \lambda {\delta_i^{(t+1)} - \hat{\delta}_i^{(t+1)}}^2 + \mu V^\pi(s_t) - \hat{V}^\pi(s_t)^2 \right] + \alpha_\mathrm{k}\mathcal{L}_\mathrm{k} + \alpha_\mathrm{v}\mathcal{L}_\mathrm{v} + \alpha_\mathrm{n}\mathcal{L}_\mathrm{nd} + \alpha_\mathrm{kv}\mathcal{L}_\mathrm{kv} \tag{18.20}$$

where $\mathcal{L}_\mathrm{k}$and $\mathcal{L}_\mathrm{kv}$ are physics-based constraints and integrate Kirchhoff's Current Law (KCL) and Kirchhoff's Voltage Law (KVL), respectively. Coefficient α_k, α_v, α_n, α_kv are hyperparameters, and an additional term may be included to incorporate economic dispatch considerations.

In simple terms, the ST-GNN can be thought of as learning two kinds of "grammar" from the power grid. The first is the spatial grammar, which captures how buses in the network influence one another through their physical and electrical connections. The second is the temporal grammar, which describes how operating conditions and disturbances evolve over time. By combining these two perspectives, the ST-GNN can anticipate how local failures might spread across the grid and suggest effective interventions to prevent cascading outages.

The operational impact of the ST-GNN and RL-based contingency planning framework can be evaluated under the contingency scenarios for an IEEE 118-bus benchmark and with about 1,000 synthetic training samples that capture system behaviors under contingency scenarios. Here, the sequential decision-making was implemented over 5-minute discrete time steps, spanning a 24-hour period (totaling 288 steps). The system state was updated dynamically, reflecting variations in load demand, changes in grid topology due to faults, and the corresponding grid response to maintain stability. To assess system vulnerabilities, we simulated a 2-N outage scenario (failure in two consecutive transmission lines) on transmission lines between buses 48 and 68, and between buses 48 and 53. This event caused disturbances in the distribution of voltage angles, particularly impacting the buses near the affected lines and neighboring buses.

Fig. 18.15 shows the effectiveness of the ST-GNN and RL framework in mitigating peak stress while maintaining overall system stability. Fig. 18.15A depicts the predicted active power demand (in MW) as a spatial-temporal heatmap, overlaid with line plots for five selected PQ buses (2, 15, 21, 50, and 66). The dashed lines represent forecasted trajectories, while the solid lines show actual demand after DR actions are applied. During peak hours, several buses exhibit forecasted demand above the 180 MW threshold, triggering corrective DR interventions. This comparison highlights the framework's ability to anticipate overloads and reduce demand through targeted actions. Fig. 18.15B shows the spatial distribution of applied DR across the full network. The heatmap indicates where and when interventions occur, with darker red regions corresponding to higher levels of load reduction. The responses are not uniform across buses: those near stressed network regions require stronger interventions, while others remain largely unaffected. This spatial variation demonstrates the framework's capacity to localize vulnerabilities and apply DR selectively, avoiding unnecessary curtailments.

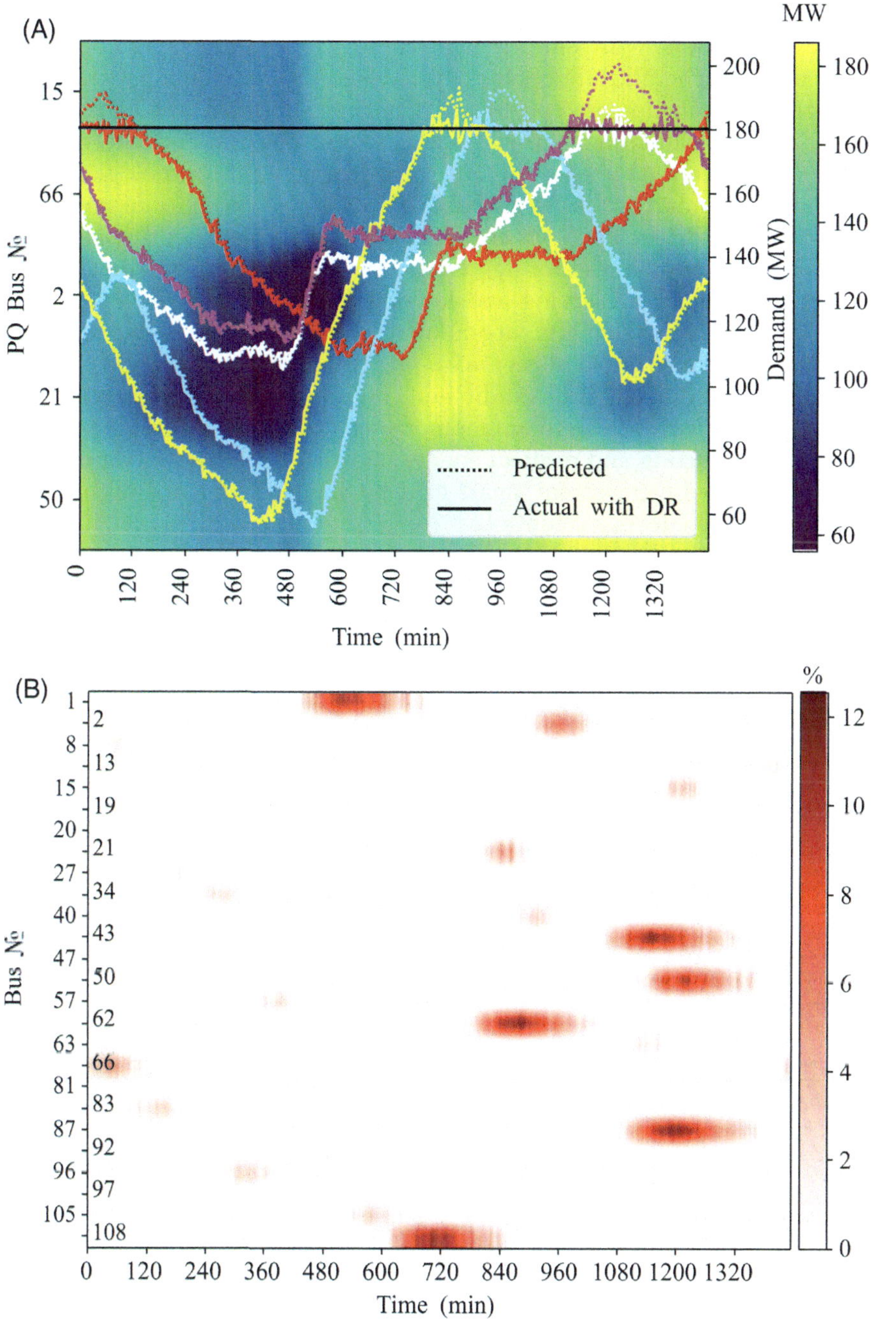

FIGURE 18.15 Spatio-temporal demand-response results for the IEEE 118-bus benchmark: (A) predicted and actual demand profiles for selected PQ buses, and (B) bus-level distribution of applied demand response over time.

18.6 Conclusion

This chapter takes a closer look at how generative AI and LLMs are being integrated with Digital Twins in smart grids. We laid out a SDMG architecture that includes field devices, edge gateways, an LLM semantic layer for managing heterogeneous data, and Digital Twin models that support forecasting, scheduling, and setpoint dispatch under operational and market constraints. We also discussed streamlined Digital Twin workflows with iterative policy updates, trade-offs between cloud and edge deployment of the LLM layer, and security and identity measures across the system. In terms of AI integration, the chapter focused on domain-specific adaptation. Parameter-efficient fine-tuning techniques (LoRA/PEFT) were applied to LLMs for protocol translation and semantic tagging, with evaluation using BLEU scores. RAG was introduced to address gaps when dealing with unfamiliar protocols or payloads. Applications included preventive and prescriptive maintenance with LLM-assisted troubleshooting, load and renewable generation forecasting, resource allocation and DR messaging, and scenario-based planning. For asset health, we compared generative approaches (VAE, cGAN, and AAE) and RL for grid-scale BESS. At the system level, we proposed and examined ST-GNNs in combination with reinforcement learning for contingency analysis and targeted actions such as DR. The key themes include semantic interoperability, reduced-order modeling, parameter-efficient adaptation, and decision support grounded in domain constraints. Challenges remain in standardizing interfaces, addressing latency and privacy trade-offs, validating learned policies within physical and safety limits, and orchestrating SDMG components at scale. As we look ahead, the integration of new AI technologies could really open up exciting research opportunities for smart grids. Explainable AI (XAI) is likely to be crucial in ensuring that recommendations and control decisions driven by LLMs are both auditable and trustworthy in safety-critical operations. Smaller language models might pave the way for lightweight deployment on edge devices like substations, feeders, and smart meters, enabling privacy-preserving and low-latency inference right at the data source. Multimodal AI has the potential to merge sensor telemetry, operational logs, and visual inspection data, enhancing our SA and improving fault detection. Spatial AI could bolster reasoning that takes topology into account, facilitating localized decision-making across distributed assets and coordinated control among feeders, microgrids, and storage systems. Additionally, agent-based approaches, where autonomous agents represent grid assets and negotiate in real time, could lead us toward self-organizing and resilient energy networks. All these advancements suggest that the next generation of AI might gradually transform smart grid intelligence from merely reactive assistance to more proactive, adaptive, and interpretable forms of control.

Inverter's Digital Twins states.

Several inverter states are used to construct their Digital Twins		
Auto/Manual mode	Coupled	Decoupled
Emergency shutdown request	External fault	Floating
FRT	Frequency deviations	Grid outage
Grid reconnection	High THD	High temperature
Internal failure	Islanding detection	Load restoration
Load shedding	Low DC link	ON/OFF
Overcurrent/overvoltage protection	Power control	Self-test
Setpoint changed	Synchronization	VRT

Inverter's Digital Twins state-transition.

Several inverter state-transition flags are used to construct their Digital Twins		
Active power control	Battery SOC low	Communication loss
DC link low	Detection of islanding	Emergency shutdown requests
fault detection	Fault-related events	Frequency deviations
Grid connection status	Grid voltage quality	Harmonics detection
Load demand	Load shedding/restoration	Maintenance triggers
Manual mode	Reactive power control commands	Synchronization events

References

Abuabdo, A., & Al-Sharif, Z. A. (2019). Virtualization vs. containerization: Towards a multithreaded performance evaluation approach. In *Proceedings of IEEE/ACS international conference on computer systems and applications, AICCSA*. IEEE Computer Society Jordan. doi:10.1109/AICCSA47632.2019.9035233.

Ahmad, S., Md, S., Ahmad, C. B., & Alowaifeer, M. (2023). A review of microgrid energy management and control strategies. In *IEEE Access: 11* (pp. 21729–21757). IEEE.

Alramlawi, M., Monsalve, C., Ruhe, S., Schafer, K., Nicolai, S., & Bretschneider, P. (2022). Working principle of self adaptive distributed digital twins in power systems. *IEEE power and energy society general meeting*. IEEE Computer Society Germany. doi:10.1109/PESGM48719.2022.9916753.

Arrano-Vargas, F., & Konstantinou, G. (2023). Modular design and real-time simulators toward power system digital twins implementation. *IEEE Transactions on Industrial Informatics, 19*(1), 52–61. doi:10.1109/TII.2022.3178713.

Bazmohammadi, N., Madary, A., Vasquez, J. C., Mohammadi, H. B., Khan, B., Wu, Y., & Guerrero, J. M. (2022). Microgrid Digital Twins: Concepts, applications, and future trends. *IEEE Access, 10*, 2284–2302. doi:10.1109/ACCESS.2021.3138990.

Bhagat, A. (2024). Applying generative AI in predictive maintenance: A new paradigm. *ESP International Journal of Advancements in Computational Technology (ESP-IJACT), 2*(4), 100–103.

Cioara, T., Anghel, I., Antal, M., Antal, C., Arcas, G. I., & Croce, V. (2022). An overview of digital twins application in smart energy grids. In *Proceedings - 2022 IEEE 18th international conference on intelligent computer communication and processing conference, ICCP 2022*. Romania: Institute of Electrical and Electronics Engineers Inc. doi:10.1109/ICCP56966.2022.10053945.

Danilczyk, W., Sun, Y., & He, H. (2019). ANGEL: An intelligent digital twin framework for microgrid security. *51st North American power symposium, NAPS 2019*. United States: Institute of Electrical and Electronics Engineers Inc. doi:10.1109/NAPS46351.2019.9000371.

Devlin, J., Chang, M.-W., Lee, K., & Toutanova, K. (2019). BERT: Pre-training of deep bidirectional transformers for language understanding, in Proc. 2019 Conf. North American Chapter of the Association for Computational Lingui. *Human Language Technologies, 1*, 4171–4186. doi:10.18653/v1/N19-1423.

Fatemi, N., & Fattahi,, J. (2025). Adaptable semantic interoperability in heterogeneous smart grids using Large Language Models. *Energy Reports, 14*, 5774–5789.

Francisco, A., Mohammadi, N., & Taylor, J. E. (2020). Smart city Digital Twin-enabled energy management: Toward real-time urban building energy benchmarking. *Journal of Management in Engineering, 36*(2). doi:10.1061/(ASCE)ME.1943-5479.0000741.

Giacotto, A., Marques, H. C., & Martinetti, A. (2025). Prescriptive maintenance: A comprehensive review of current research and future directions. *Journal of Quality in Maintenance Engineering, 31*(1), 129–173. doi:10.1108/JQME-07-2023-0064.

Glass, P., & Serugendo, G. DiM. (2023). Coordination model and Digital Twins for managing energy consumption and production in a smart grid. *Energies, 16*(22), 7629. doi:10.3390/en16227629.

Hosseini, S., Shahbandegan, A., & Akilan, T. (2022). Deep neural network modeling for accurate electric motor temperature prediction. *Canadian conference on electrical and computer engineering*. Canada: Institute of Electrical and Electronics Engineers Inc. doi:10.1109/CCECE49351.2022.9918222.

Kabbara, N., Belaid, M. ON., Gibescu, M., Camargo, L. R, Cantenot, J., Coste, T., Audebert, V., & Morais, H. (2022). Towards software-defined protection, automation, and control in power systems: Concepts, state of the art, and future challenges. *Energies, 15*(24), 9362. doi:10.3390/en15249362.

Kumar, P., Kumar, R., Aljuhani, A., Javeed, D., Jolfaei, A., & Islam, A. K. M. N. (2023). Digital twin-driven SDN for smart grid: A deep learning integrated blockchain for cybersecurity. *Solar Energy, 263*, 111921. doi:10.1016/j.solener.2023.111921.

Lauri, M., Hsu, D., & Pajarinen, J. (2023). Partially observable markov decision processes in Robotics: A survey. *IEEE Transactions on Robotics, 39*(1), 21–40. doi:10.1109/tro.2022.3200138.

Li, H., Kaleem, M. B., Chiu, I.-J., Gao, D., & Peng, J. (2021). A digital twin model for the battery management systems of electric vehicles. *Conf. High Performance Computing & Communications; 7th Int. Conf. Data Science & Systems; 19th Int. Conf. Smart City; 7th Int. Conf. Dependability in Sensor, Cloud & Big Data Systems & Application (HPCC/DSS/SmartCity/DependSys)*, 1100–1107. doi:10.1109/HPCC-DSS-SmartCity-DependSys53884.2021.00171.

Li, Z., Guo, L., Cheng, J., Chen, Q., He, B., & Guo, M. (2022). The serverless computing survey: A technical primer for design architecture. *ACM Computing Surveys, 54*(10s), 1–34. doi:10.1145/3508360.

Merlin Sajini, M. L., Suja, S., Gilbert Raj, S. M., Kowsalyadevi, S., & Maria, C. (2023). Finite state machine-based load scheduling algorithm for smart home energy management. *IETE Journal of Research, 69*(10), 7460–7475. doi:10.1080/03772063.2021.2010607.

Ndiaye, M., Hancke, G. P., Abu-Mahfouz, A. M., & Zhang, H. (2021). Software-defined power grids: A survey on opportunities and taxonomy for microgrids. *IEEE Access, 9*. http://ieeexplore.ieee.org/xpl/RecentIssue.jsp?punumber=6.

Nguyen, L. H., Nguyen, V. L., Hwang, R. H., Kuo, J. J., Chen, Y. W., Huang, C. C., & Pan, P. I. (2024). Towards secured smart grid 2.0: Exploring security threats, protection models, and challenges. *IEEE Communications Surveys and Tutorials*. doi:10.1109/COMST.2024.3493630.

Park, H.-A., Byeon, G., Son, W., Jo, H.-C., Kim, J., & Kim, S. (2020). Digital twin for operation of microgrid: Optimal scheduling in virtual space of digital twin. *Energies, 13*(20), 5504. doi:10.3390/en13205504.

Rehmani, M. H., Davy, A., Jennings, B., & Assi, C. (2019). Software defined networks-based smart grid communication: A comprehensive survey. *IEEE Communications Surveys and Tutorials, 21*(3), 2637–2670. doi:10.1109/COMST.2019.2908266.

Renold, A. P., & Kathayat, N. S. (2024). Comprehensive review of machine learning, deep learning, and digital twin data-driven approaches in battery health prediction of electric vehicles. *IEEE Access, 12*, 43984–43999. doi:10.1109/ACCESS.2024.3380452.

Santiago, R. A. F., Barbosa, N. B., Mergulhão, H. G., Carvalho, T. F., Santos, A. A. B., Medrado, R. C., de Melo Filho, J. B., Pinheiro, O. R., & Nascimento, E. G. S. (2024). Data-driven models applied to predictive and prescriptive maintenance of wind turbine: A systematic review of approaches based on failure detection, diagnosis, and prognosis. *Energies, 17*(5), 1010. doi:10.3390/en17051010.

Shayanfar, H. A. (2016). Multiobjective optimization of distribution networks with distributed generation and capacitor banks for system planning. *IEEE Transactions on Power Systems, 31*, 2204–2215. doi:10.1109/TPWRS.2015.2456865.

Srinivasan, R. S., Manohar, B., & Issa, R. R. A. (2020). *Urban building energy CPS (UBE-CPS): Real-time demand response using digital twin cyber-physical systems in the built environment* (pp. 309–322). Springer International Publishing. doi:10.1007/978-3-030-41560-0_17.

Strezoski, L. (2023). Distributed energy resource management systems—DERMS: State of the art and how to move forward. *WIREs Energy and Environment, 12*(1), e460.

Sun, G., Xu, Z., Yu, H., Chen, X., Chang, V., & Vasilakos, A. V. (2020). Low-latency and resource-efficient service function chaining orchestration in network function virtualization. *IEEE Internet of Things Journal, 7*(7), 5760–5772. doi:10.1109/JIOT.2019.2937110.

Ucar, A., Karakose, M., & Kırımça, N. (2024). Artificial intelligence for predictive maintenance applications: Key components, trustworthiness, and future trends. *Applied Sciences, 14*(2), 898. doi:10.3390/app14020898.

Vamvakas, D., Michailidis, P., Korkas, C., & Kosmatopoulos, E. (2023). Review and evaluation of reinforcement learning frameworks on smart grid applications. *Energies, 16*(14), 5326. doi:10.3390/en16145326.

Vu, T. V., Nguyen, B. L. H., Cheng, Z., Chow, M. Y., & Zhang, B. (2020). Cyber-physical microgrids: Toward future resilient communities. *IEEE Industrial Electronics Magazine, 14*(3), 4–17. doi:10.1109/MIE.2019.2958039.

Xiao, H., Yi, K., Peng, R., & Kou, G. (2022). Reliability of a distributed computing system with performance sharing. *IEEE Transactions on Reliability, 71*(4), 1555–1566. doi:10.1109/TR.2021.3111031.

You, M., Zhang, X., Zheng, G., Jiang, J., & Sun, H. (2020). A versatile software defined smart grid testbed: Artificial intelligence enhanced real-time co-evaluation of ICT systems and power systems. *IEEE Access, 8*, 88651–88663. doi:10.1109/ACCESS.2020.2992906.

Yuan, T. (2023). *Prescriptive maintenance and operation with deep reinforcement learning*. ETH Zurich. doi:10.3929/ethz-b-000616958.

Yuanyuan, L. (2022). Short-Term Prediction Method of Solar Photovoltaic Power Generation Based on Machine Learning in Smart Grid. *Mathematical Problems in Engineering* p. 8478790. Wiley Online Library.

Zeng, Y., Zhang, Z., Wu, T., & Liang, W. (2024). Integrated optimization and engineering application for disassembly line balancing problem with preventive maintenance. *Engineering Applications of Artificial Intelligence, 127*, 107416. doi:10.1016/j.engappai.2023.107416.

Zhou, X., Bai, G., Tao, J., & Xu, B. (2023). An improved method to search all minimal paths in networks. *IEEE Transactions on Reliability, 72*(4), 1420–1431. doi:10.1109/TR.2023.3234055.

CHAPTER

19

Harnessing generative AI and large language models for Digital Twin-enabled smart grids

Abiodun Emmanuel Onile[a], Hossein Nourollahi Hokmabad[a], Eduard Petlenkov[b], Juri Belikov[a]

[a]Software Science Department, Tallinn University of Technology, Tallinn, Estonia [b]Department of Computer Systems, Tallinn University of Technology, Tallinn, Estonia

19.1 Introduction

The energy sector globally grapples with pressing challenges in sustainability and resource efficiency. Rising climate change pressures, resource depletion, and decarbonization goals demand innovative solutions (Tomar & Grover, 2023). In light of this critical situation, the International Energy Agency (IEA) advises that global carbon emissions from electricity generation be reduced to net zero by 2045, with advanced economies reaching this goal by 2035, to align with the aims of the Paris Agreement (Ranawaka et al., 2024). Advanced and intelligent digital technologies can tackle these technical challenges while maintaining the energy efficiency, reliability, stability, and environmental sustainability of the grid. In fact, the IEA has highlighted several digital solutions to address the current and future challenges facing power grids, including embedded sensing, automation and control, real-time optimization, improved design and predictive monitoring, and distributed generation (DG) and demand response (Ranawaka et al., 2024). In the era of Industry 4.0 and Industry 5.0, utilizing cyber-physical system (CPS) technologies like the Internet of Things (IoT), cloud and edge computing, Big Data, machine learning (ML), artificial intelligence (AI), and Digital Twins (DTs) can greatly improve the effectiveness of these digital solutions (Eleftheriou & Anagnostopoulos, 2022; Inderwildi et al., 2020; Ranawaka et al., 2024). Specifically, DTs facilitate the simulation and modeling of the behavior of electricity grid assets for real-time and predictive analytics (Abdelrahman et al., 2025). Time series data consists of a chronologically ordered sequence of values for a particular statistical indicator, providing analytics tools with highly valuable insights for strategic planning and decision-making, especially if future data can be accurately predicted in critical sectors (Zhang et al., 2025). The profound insights gained from this DT modeling approach allow them to predict system behavior, enhance performance, and make data-driven decisions that boost efficiency and productivity (Qin et al., 2024). In the same vein, DG, along with the growing use of renewable energy sources (RESs), has become a key driver in the transformation of the power industry (Ranawaka et al., 2024). While their contribution to achieving carbon neutrality is widely acknowledged, they also enhance efficient power generation by bringing loads closer to the sources, thus reducing losses (Ranawaka et al., 2024). However, these emerging trends present technical challenges that negatively affect the stability and reliability of the power grid, arising from the uncertainty, complexity, and volatility inherent in RESs (Ranawaka et al., 2024). These developments thus have some shortcomings as their dynamic nature prevents accurate modeling of microgrid assets using solely data-driven DTs. Scholarly discussions regarding user digital twins (UDTs) focus on a key question that relates to the necessary levels of fidelity and synchronicity (Wu & Guan, 2024). The prevailing view is that attaining perfect fidelity and synchronicity is difficult in terms of practicality for the majority of applications (Wu & Guan, 2024).

Hybrid DTs embody a revolutionary technological advancement characterized by their ability to integrate multi-physics and multiscale probabilistic simulations. These digital representations utilize intricate physical models in conjunction with sensor-derived data and historical records to faithfully replicate the operational behaviors exhibited by their counterpart real-world assets (Nele et al., 2024). Physics-informed neural networks (PINNs) have emerged as

Digital Twin Technology and Smart Grid.
DOI: https://doi.org/10.1016/B978-0-443-36664-2.00022-8

a promising method that integrates physical laws into neural network training to solve partial differential equations (PDEs), yielding substantial improvements in convergence, generalization, and interpretability (Azhahudurai & Veeramanikandan, 2023). This high-fidelity DTs model promises to enhance the integration of RES.

Nonetheless, the growing complexity and scale of contemporary power systems present new challenges for the development and implementation of Digital Twin networks (DTNs). Key concerns include how to conduct dynamic real-time simulations of interconnected DTs of physical entities, manage the large amounts of data produced by the IoT (i.e., Internet of Vehicles (IoV)), optimize system performance with limited resources, and ensure the fidelity of the models (Bujari et al., 2021; Qin et al., 2024). For example, authors in Azhahudurai and Veeramanikandan (2023) noted that while hybrid DTs show promise, recent studies have identified inherent limitations, particularly when dealing with PDEs that involve high frequency or multiscale features. Furthermore, contemporary hybrid DTs prediction models like those based on RNN (i.e., long short-term memory (LSTM)) are often only capable of a short-term predictive horizon and lack context of distant information. This challenge has led to explorations from multiple perspectives, including the development of diverse model architectures and learning strategies (Azhahudurai & Veeramanikandan, 2023). Lately, there has been a significant increase in the development of transformer-based solutions (i.e., Generative AI or large language models (LLMs)), and their integration into DT-based smart grids opens the avenue for a number of benefits (Zeng et al., 2023). On the one hand, these transformer-based solutions for long-term time series forecasting (LTSF) have shown significant improvements in prediction accuracy compared to traditional methods in sequence modeling (Zeng et al., 2023; Zhou et al., 2021). Transformers are widely regarded as the most effective method for capturing precise long-range dependencies and extracting semantic correlations among elements in extended temporal sequences (long sequences). Large context training windows associated with transformers ensure that the entire dataset/context is captured by the model of interconnected DTs of physical entities. In time series modeling, where the focus shifts to identifying temporal relationships within an ordered set of continuous data points, positional encoding and tokenizing subseries in transformers help preserve information ordering, thus helping to capture temporal dynamics (Zeng et al., 2023). Transformers use their multi-head self-attention (MHSA) mechanism to extract semantic correlations among elements within long sequences while utilizing parallel computation to facilitate significant speed efficiency (Zeng et al., 2023; Zhao et al., 2023). On the other hand, generative models have the potential to revolutionize DTs by enabling realistic simulations for improved decision-making. Generative models such as LLMs have gained considerable attention in recommender systems due to their impressive capabilities in understanding the world and their potential to be useful in assisting users in reconciling conflicting information interests (Cheng et al., 2025).

Personalized demand-side recommendation services (PRSs) act as a framework for integrating users' feedback into the energy management decision-making process (Onile et al., 2023, 2024). This approach is vital, as it provides essential information to assist users in reconciling conflicting information interests through a comprehensive analysis of their energy consumption behavior, thus empowering stakeholders such as electricity consumers, prosumers, facility managers, distribution system operators (DSOs) etc. (Karnouskos, 2011), thus reducing reliance on guidance from specialized professionals (Schmid et al., 2011). By synthesizing insights from diverse sources, PRSs improve decision-making and troubleshooting in complex systems such as the smart-grid. LLM-based recommender systems leverage the robust world knowledge capabilities of LLMs to extract and enrich semantic information from recommendation data. This approach addresses key challenges such as data sparsity and stakeholder disengagement with demand-side management (DSM) solutions (Cheng et al., 2025; Onile et al., 2023).

In this light, this chapter analyzes the power sector's key challenges and explores how generative AI and DT approaches, such as novel physics-informed transformer architecture (PITA) and decision support schemes (DSS), can be explored to drive innovation toward the sustainability of the smart energy grids. The rest of the chapter is organized as follows. Section 19.2 presents the literature review. The proposed PITA-based DTs model and DSS are described in Section 19.3, while Section 19.4 contains the results. The concluding remarks are presented in Section 19.5.

19.2 Literature

From a literature perspective, the concept of the DT was first introduced by Michael Grieves in 2002. Since then, both the idea and its associated technologies have undergone significant advances (Segovia & Garcia-Alfaro, 2022). A comprehensive overview of the history of DTs architecture that leverages ML and advanced analytics for real-time monitoring and continuous prediction of machines' operational status, end-of-life, and failure likelihood was presented for early warnings and timely interventions to prevent potential failures.

A common challenge across all types of DT implementations lies in system modeling, as there is no established industry standard for constructing these models (Eleftheriou & Anagnostopoulos, 2022). For example, the work of

Qin et al. (2024) identified obsolescence in data-driven DT models caused by changes in the operating environment. This is evident in DT prediction gaps associated with the impact of the increased integration of extreme volatility renewable energy in DT prediction. Notably, integration of volatile RES (or distributed energy resources (DERs)) affects reliability and requires advanced/improved modeling. In this context, PINNs as a promising alternative have been examined (Zhao et al., 2023). The author described a DT framework that integrates advanced big data analytics by utilizing highly accurate synthetic data generated by DTs combined with physics-based features to improve predictive capabilities. While conventional PINNs have achieved significant success in various applications, recent studies have highlighted their limitations. Specifically, PINNs struggle with solutions that exhibit high frequency or multiscale features, often providing overly smooth or simplistic approximations that diverge from the true solution, even when the analytical solutions are straightforward (Zhao et al., 2023). The authors noted that minor inaccuracies in the DT model can arise due to inadequate capture of physical conditions, flawed modeling, time delays, or synchronization issues that normally occur during continuous model updates.

This challenge has spurred research efforts from multiple angles, such as designing diverse model architectures and learning frameworks (Azhahudurai & Veeramanikandan, 2023). The paper presented in Segovia and Garcia-Alfaro (2022) concentrates on the architecture planning, design, and construction of DTs, noting that DTs can be implemented across cyber physical systems, with examples spanning from individual robots to comprehensive environments like smart cities and 5G networks (Segovia & Garcia-Alfaro, 2022). The authors further presented in-depth insights on the formulation of key research questions on the implementation, which includes: What approach should we take to design the architecture of a DT? How can we achieve real-time data synchronization between the physical and virtual components of a DT? NASA defines a DT as a unified simulation that combines physics-based models across multiple scales and incorporates probabilistic elements (Abdelrahman et al., 2025). Furthermore, Segovia and Garcia-Alfaro (2022) noted that the DT architecture should be adaptable over time to accommodate system changes, such as the addition of new components, interfaces, or connections to other components, as well as adjustments to system behavior. The architecture should also support multiple models and interconnectivity with other DTs. While there are common components in all DTs, those at different scales (human, building, district, urban, city, global, and universal) vary in the volume of components. Future research should focus on developing multiscale DT frameworks that integrate data and insights across different scales seamlessly, facilitating a holistic approach to building environment (BE) management (Abdelrahman et al., 2025).

In contrast, the transformer architecture allows for parallel computation of sequential data without significantly increasing complexity, enabling it to leverage the parallel processing power of Tensor Processing Units (TPUs) and Graphics Processing Units (GPUs). Unlike RNNs and their variants, transformers employ a self-attention mechanism that facilitates parallel computation, offering significantly higher computational efficiency and speed compared to RNNs (Zhao & Xiong, 2024). Furthermore, transformers do not suffer from vanishing gradients, as their attention-based operations allow them to correlate information from all elements in the sequence in parallel (Azhahudurai & Veeramanikandan, 2023). Recently, the transformer architecture has been introduced, utilizing an attention mechanism to process sequential data. Unlike RNN-based methods, the transformer enables the model to access any part of the historical data regardless of distance, potentially making it more suitable for capturing recurring patterns with long-term dependencies, such as in long-term sequence forecasting tasks (Li et al., 2019). In addition, selecting a diverse training dataset is crucial for creating a robust foundational model, capturing the dynamic nature of real-world data where trends and patterns shift due to various influences (Eleftheriou & Anagnostopoulos, 2022; Garza et al., 2023). This training set integrates time series data from a wide variety of domains, such as demographics, weather, IoT sensor data, and energy. In Hong et al. (2024), an LLM-Twin framework, which leverages LLMs, has been proposed (Qin et al., 2024).

One crucial question raised by Zeng et al. (2023) is how effective are transformers at time series forecasting? However, there are several significant challenges with transformers that hinder their direct application to long sequence time series forecasting (LSTF). Despite being impressed by transformer performance in the previous studies examined, the authors in Anagnostopoulos et al. (2024), Ramirez et al. (2025) identified a significant weakness, such as locality-agnosticism, where the point-wise dot product self-attention in the standard is insensitive to local context, making the model vulnerable to anomalies in time series data. For example, transformers may miss subtle local patterns (e.g., sudden spikes in energy consumption) while overemphasizing global dependencies. Work (Anagnostopoulos et al., 2024) noted that physics-informed architectures can complement locality-aware mechanisms in transformer architectures while identifying hybrid designs (e.g., comprising self-attention and physics losses) to balance global physical consistency and local pattern sensitivity as effective solutions. While DTs have the potential to provide numerous benefits for the power systems, they have faced challenges that have limited their widespread adoption and use (Eleftheriou & Anagnostopoulos, 2022), with one major obstacle being the difficulty stakeholders encounter in comprehending and utilizing analytics generated by the DTs (Najafi et al., 2023). Work

(Wu & Guan, 2024) noted that LLMs can be utilized not only for pattern recognition and process optimization but also to boost the productivity of humans. Work (Sun et al., 2024) further presented a study on LLM, DT, and multi-agent systems, noting that DT systems driven by intelligent agents excel in analysis and decision-making support (Nele et al., 2024; Sun et al., 2024).

Contributions: To holistically address the previously mentioned challenges and provide answers to those aforementioned questions, this chapter proposes a novel scheme for designing DT models utilizing an ensemble of existing tools, such as transformer architecture, smart-grid asset physics, and conversational chatbot DSS for their implementation. Thus, this study presents the following contributions.

- The chapter proposes innovative hybrid physics transformer DTs, introduced as a novel design paradigm called PITA. This architectural shift aims to enhance DT modeling capabilities by integrating physics-based modeling and transformer architecture for jointly predicting the smart-grid asset profile.
- The chapter conducts and presents an experimental validation to compare state-of-the-art DT models with the proposed PITA-based DT architecture.
- The chapter introduces DT-based DSS founded on LLMs, which enhance communication and stakeholders' engagement with DSM schemes in smart-grids.

19.3 Methodology

This section outlines the proposed generative AI approach to DT modeling. It specifically covers (1) the hybrid physics-informed transformer-based DT modeling (PITA) scheme, and (2) decision support methods, which involve demand-side recommendation services and a conversational chatbot approach for extending stakeholders' engagement with the DSM scheme. The approach is detailed in the subsequent sections.

19.3.1 Physics-guided Digital Twins modeling

DTs are characterized as multi-physics, multiscale simulations of an asset. The physics-based DTs approach incorporated fundamental laws of physics, adhering to white-box principles, as noted in Onile et al. (2024). The following describes the physics of selected smart-grid assets (i.e., solar photovoltaic (PV) power, Heating, Ventilation, and Air Conditioning (HVAC) air filter).

HVAC air filter physics model: The HVAC model was based on modeling the pressure drop across air filters, which is crucial to achieving optimal cooling satisfaction, as the pressure drops due to clogging can lead to increased energy consumption due to reduced efficiency. Neglecting the maintenance of an HVAC system can have severe consequences, and a detailed model representation of an air filter has been developed based on the investigation of filter clogging. This model considered factors such as fan power consumption, filter efficiency estimation, and air quality related to the concentration of measured outdoor air, as described in Eqs. (19.1) and (19.2).

$$m = Q_v \, \nabla_t \, (C_m(e) \; - C_m(S))[\mu g] \; and \tag{19.1}$$

$$V \times \frac{dc_m}{d_t} = c_m(s) \, Q_v - c_m \, Q_v, \tag{19.2}$$

In this context, m represents the mass accumulated in the filter at each time step t. The variables C_m,$C_m(e)$, and $C_m(S)$ denote the homogeneous air concentration in the room, the outdoor air concentration, and the supply air concentration, respectively. In addition, Q_v is the outside air flow rate, and V represents the volume of the room.

Battery physics model: The battery physics equations used in this experiment were based on the fundamental principles of electrochemistry and thermodynamics. It ensures that the prediction of capacity was not based solely on historical data but also ensures that the predictions align with the known physics of batteries. The key battery physics equations that describe the behavior of a battery and are relevant for capacity prediction are described in Eqs. (19.3) and (19.4).

The state of charge (SOC) estimation techniques (see Eq. (19.3)) is used to determine the remaining capacity of a battery. The SOC estimation problem, which is about determining the remaining charge in a battery based on voltage and current measurements. It provides the percentage of remaining charge, essential for capacity prediction and management.

$$SOC = \frac{Capacity_{remaining}}{Capacity_{total}} \times 100\% \tag{19.3}$$

Here SOC is the state of charge (%), $Capacity_{remaining}$ is the current capacity, the amount of charge left in the battery, or the remaining battery capacity at time t, representing how much charge the battery can hold after degradation over time or cycles (Ah) . $Capacity_{total}$ is the maximum charge capacity of the battery when fully charged (Ah). The Li-ion battery aging model depicted in Eq. (19.4) describes the decrease in battery capacity over time due to usage and environmental factors. Incorporating degradation over time is a critical factor in the prediction of long-term capacity.

$$Capacity_{remaining} = Capacity_{initial} \times exp(-kt), \tag{19.4}$$

where k is the constant of the aging rate, and t is time. $Capacity_{initial}$ is the initial or rated capacity of the battery before degradation begins and can be measured in ampere-hours (Ah).

19.3.2 Data-driven transformer architecture

Although physics models play a crucial role in the design phase of DTs, the increasing availability of DT data and the accessibility of advanced open-source libraries like OpenAI, Torch, and TensorFlow, along with affordable computational resources (such as GPUs, TPUs, and Central Processing Units (CPUs)), data-driven models are gaining popularity. Unlike physics-based modeling, data-driven ML models rely on the idea that both known and unknown physical principles are embedded in and revealed by data. Consequently, these models can capture all the properties of a system using data (Onile et al., 2024). Transformers are data-driven AI models by design, as their architecture inherently relies on learning patterns from large datasets through attention mechanisms. The transformer-inspired model for time series forecasting uses self-attention to analyze historical data windows. Transformers employ a self-attention mechanism that facilitates parallel computation, offering significantly higher computational efficiency and speed compared to RNNs (Zhao & Xiong, 2024). The multi-head attention mechanism helped the model learn multiscale features, provide contextual awareness, and effectively capture relationships between different positions in a sequence (Zhao & Xiong, 2024). The comprehensive overview of the developed transformer scheme integrated key components such as MHSA, normalization layers (LayerNorm), and a multi-layer perceptron (MLP) (Shin et al., 2024).

19.3.2.1 *Positional encoder*

The positional encoder class transforms input positions into fixed-length vectors, allowing the transformer to capture temporal relationships within the data. The features of the encoder are fed into the weighted cross-attention mechanism.

19.3.2.2 *Multi-head self-attention*

Since global features encompass all the information about smart-grid assets, it is essential to identify the global characteristics that are relevant to the extracted energy asset features and prioritize those related global characteristics. To achieve this, correlation factors can be used to quantify the relationship between the characteristics of global and specific electricity assets. Using MHSA, a connection between the two can be established, enabling the calculation of correlation between individual assets and global features using Eq. (19.5) (Zhao & Xiong, 2024). The multi-head self-attention mechanism is composed of two components: multi-head and self-attention. The self-attention part computes the weights between the query and key, and these weights are then used to sum the values, resulting in the final attention weight. An MHSA layer employed multiple self-attention mechanisms as separate heads to extract various types of features. Local self-attention dynamically computed the relationships between tokens in a sequence, learned context-dependent patterns directly from data, and was calculated by dividing each time series input into non-overlapping windows. Within each window, the query (Q), key (K), and value (V) matrixes was computed (see Eq. (19.5)).

$$Attention\,(Q,\ K,\ V) = softmax\left(\frac{QK^T}{\sqrt{d}}\right)V, \tag{19.5}$$

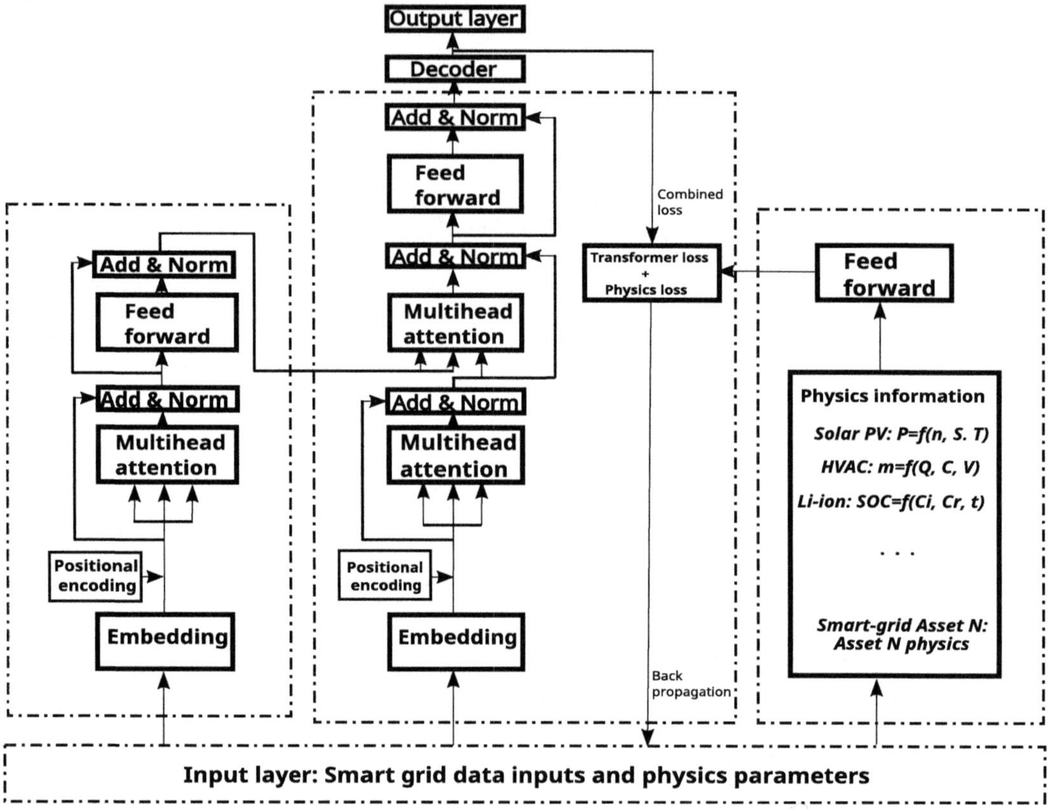

FIGURE 19.1 Schematic diagram of the proposed PITA architecture with loss-level physics fusion. The transformer model path effectively captures global patterns, while the physics model path leverages physical laws to extract fine-grained physics parameters and local variations, ensuring that predictions adhere to physical constraints. *PITA*, Physics-informed transformer architecture.

where, $\sqrt{d}$ represents a scale factor based on d, the dimension of each vector. The multi-head mechanism projects the query Q, key K, and value V into different subspaces, where it calculates the attention weights for each (profiles in) assets' subspace. The variable T denotes a specific timestep or position within the input sequence or token.

19.3.3 Proposed PITA-based Digital Twin model

In this section, the proposed PITA-based DT model was introduced. The adopted PITA model was developed based on a hybrid approach. Recreating energy assets using DTs solely based on physics can be problematic, as inaccuracies in variable definitions may lead to significant consequences affecting the DT's overall behavior. Similarly, a purely data-driven approach has limitations in capturing complex physical constraints, making it challenging to meet certain temporal and intelligent requirements of the energy system. Therefore, a hybrid approach is often considered a more effective solution (Onile et al., 2024). Hybrid DT models combine the strengths of data-driven and physics-based approaches, offering enhanced precision that depends on the model's complexity. The core features of the proposed PITA architecture are described in Fig. 19.1.

The proposed PITA architecture in Fig. 19.1 falls under the fused model architecture or structure. The data-driven (transformer) models learned directly from data, but were regularized by physics equations during training. Physics was embedded into the learning process through the loss function or network constraints (see Eq. (19.9)). Physics and data-driven models were combined in the loss function, thus guiding training but not changing the network architecture. Notably, the transformer never explicitly takes the physics model output as input (like in serial) or produces outputs separately (like in parallel). Instead, physics was fused into the optimization process, not as an independent layer but as a regularization pathway in the loss function. Based on the hybrid architecture, the model integrates multiple blocks over a given time, where a sliding window size (W) (i.e., 24 hourly input for one day) was applied to input sequences ($x_0, x_1, ..., x_t$) of smart-grid individual asset data (i.e., power, irradiance, temperature, Li-ion capacity, etc.) at time step t fed into the encoder to yield a context representation expression x_t^T which is projected to predict $\hat{y}_t$ and is physics-regularized by comparing it against the physics-based output y_t^{phy} followed by fusing both pathways in the loss function to obtain the final output y_t. The encoder integrated four layers comprising one input embedding layer(with model dimension or channel depth of 64), two transformer encoder layers with four attention heads (with effective head dimension of 16 per head), and one output projection layer for asset prediction based on the specific prediction task. In parallel, the decoder received previous timesteps' outputs or ground truths and updates its hidden state h_t to produce a prediction $\hat{y}_t$ through cross-attention with the encoder feature, which was constrained by a physics model, ensuring that predictions respect both physics constraints and data consistency.

19.3.4 Network training, validation (testing), and hyperparameter tuning

All network models were developed using PyTorch, with training conducted using a batch size of 35 and capped at 30 epochs. Training details for our models include the use of the Adam optimizer with mean squared error (MSE) loss. We employ an initial learning rate of 0.001 for all networks, which was determined empirically through trial and error. After integrating the physics equations into the model, validation and testing were crucial to ensure that the model generalizes well and adheres to the physical constraints. During model validation, the use of a validation set to tune hyperparameters like λ (lambda) was considered to balance prediction accuracy and physical consistency.

19.3.5 Consumer behavior rating and generative recommendation

A rating of DT-predicted consumers' behavior was accomplished using a graph-based scoring method, such as PageRank (see Eq. (19.6)), which provided an estimated score of their energy behavior. Typically, consumer behavior can be modeled by analyzing the frequency of transitions in the energy consumption profiles (i.e., base (b), mid (m), and peak (p)) of end-consumers.

$$\begin{bmatrix} b_{11} & m_{12} & \cdots & p_{1n} \\ b_{21} & m_{22} & \cdots & p_{2n} \\ \vdots & \vdots & \ddots & \vdots \\ b_{m1} & m_{m2} & \cdots & p_{mn} \end{bmatrix}$$

In this context, each row in the rating matrix corresponds to a specific microgrid asset, such as a user or a battery energy storage system (BESS), while the columns represent the load profiles associated with these assets.

$$PR(V_i) = (1 - d) + d \sum_{i-1}^{N} \frac{PR(V_i)}{N(V_i)}, \tag{19.6}$$

Here, d represents the damping factor, $N(V_i)$ denotes the total number of nodes, and $PR(V_i)$ signifies the current ranking of the consumption profile.

LLMs have gained considerable attention in recommender systems due to their impressive capabilities in understanding the world (Cheng et al., 2025). Notably, users occasionally face challenges comprehending recommendation tips. Dialogue-based systems like chatbots, which use conversations to filter information, can help close this gap by encouraging users to adopt energy-efficient behaviors (Onile et al., 2023). They achieved this by engaging users in extended interactions, leveraging their role as companions rather than just sources of information. In this case, an important function of conversational chatbots involves enhancing existing recommender systems by employing strategies like generating supplementary textual data capable of improving end-user understanding and extending engagement with DSM solutions (Cheng et al., 2025). To implement the proposed conversational chatbot interface, an open-source OpenAI model called generative pretrained transformer (GPT) was utilized to facilitate interaction between the developed hybrid DT system and electricity consumers while also enhancing consumer engagement on the demand side of recommender systems. In this scenario, the outputs from the recommender system served as input data or prompts that directly linked to the chatbot, enabling users to continue the conversation to further clarify the recommendations provided.

19.3.6 Evaluation metrics

Our evaluation metrics include MSE and root mean square error (RMSE), which were both used for evaluating the prediction accuracy of the proposed PITA-based DT model. This selection was motivated by the additional interpretive value these metrics provided, as they quantify performance improvements relative to a baseline, enhancing the clarity of our findings. The MSE between predicted and actual values was computed and then averaged across alltime series, evaluation time points, and forecasting horizons to provide a comprehensive assessment of predictive accuracy. Furthermore, the physics adherence metrics add terms that enforce the physics or adherence to physical laws, and helped the model generalize better. This approach was defined based on a loss function that includes both the prediction error and the violation of physical constraints (see Eq. (19.7). The overall experiment loss function is a linear combination of transformer (target) and physics loss (auxiliary) functions. The PITA model loss function consists of two main components:

1. Prediction loss: The following loss equations are used in the loss function of a transformer model. It measures the discrepancy between predicted and actual assets' profiles (i.e., battery capacity) using MSE metrics.

$$L_{pred} = \frac{1}{N}\sum_{i=1}^{N}\left(y_{pred} - y_{true}\right)^2 \tag{19.7}$$

2. Physics-informed loss: Enforces adherence to physical laws by incorporating the physics equations. For example, in Eq. (19.4), the physics loss $L_{physics}$ term was formulated to penalize deviations from the expected battery capacity (exponential decay law) and was calculated as the MSE between actual (i.e., $Capacity_{physics}$) and the model's predicted $Capacity_{remaining}$ (i.e., $Capacity_{pred}$) which was computed by $Capacity_{initial} \times exp(-kt)$ using the known value of k.

$$L_{physics} = \frac{1}{M}\sum_{j=1}^{M}\left(i_{pred} - i_{physics}\right)^2 \tag{19.8}$$

The expression for the combined loss function is depicted as follows:

$$L_{total} = L_{pred} + \lambda L_{physics} \tag{19.9}$$

Here $i_{physics}$ and y_{true} refer to the actual observed values, ground truth, or physics parameters obtained from the dataset, while i_{pred} and y_{pred} represents the values predicted by the physics and transformer models, respectively. N and M are the number of samples or data points. λ is a hyperparameter controlling the trade-off between prediction accuracy and adherence to physical constraints such as initial and boundary conditions, termed λ. It controls the impact of the physics loss function, turning the PITA-based model training into a constrained optimization problem. The value of the Lagrange multiplier λ was tuned by testing and monitoring different values that ensure efficient training convergence and accuracy using experimental adjustment, taking into account the problem's boundary condition ratio and initial neural network weights, as these factors interplay strongly with the choice of λ. To measure this effect, we trained the PITA-based model using various values of λ and the value set that yields the best validation accuracy (MSE loss) was chosen, with the result presented in Section 19.4 (see Figs. 19.6 and 19.7). While experimental adjustment or manual tuning of λ is a hyperparameter controlling the trade-off between prediction accuracy and adherence to physical constraints such as initial and boundary conditions, termed as has been considered for experimentation, adaptive Lagrangian methods dynamically adjust λ is a hyperparameter controlling the trade-off between prediction accuracy and adherence to physical constraints such as initial and boundary conditions, termed based on constraints violation, thus eliminating a tedious trial-and-error approach during the optimization process. This adaptive adjustment aims to enhance convergence speed and robustness, especially in large-scale problems.

An additional system assessment was conducted using the consumer engagement index (EI). The customer engagement score for the demand-side recommender chatbot was calculated by dividing the total number of independent consumer interactions with the top-N recommendations by the total number of recommendations received by consumers during the review period (Onile et al., 2023).

$$EI = \frac{R_T}{A_T} \times 100 \tag{19.10}$$

where R_T represents the total number of reactions of user to recommendations during the analysis period T, and A_T denotes the total number of recommendations given during the analysis period T.

TABLE 19.1 Experimental dataset table.

#	Variables	Sampling frequency	Time period [h]	Source
1	PM2.5 HVAC data	1 [h]	680	Air quality at Paris 1er Arrondissement Measurement Station (2005)
2	Solar PV	1 [min.]	1 (60 min.)	Leva et al. (2020)
3	Li-ion capacity	1 [h]	183	Li-ion battery aging datasets (2022)
4	Ambient temperature	1 [h]	183	Li-ion battery aging datasets (2022)

PV, photovoltaic.

19.4 Results

This section outlines the outcomes of comprehensive experiments conducted to validate the effectiveness of the proposed PITA-based DT methods. The section further presents the simulation results of DSS, focusing on evaluating our approach using metrics such as accuracy, MSE, RMSE, and consumer EI. Specifically, the study employs a comparative analysis of the proposed PITA DT approach in relation to vanilla transformer ML methods. The experiments were conducted on a dataset comprising PM2.5 HVAC system data obtained from (Air quality at Paris 1er Arrondissement Measurement Station, 2005; Walther, 2021), while the solar PV and Li-ion dataset were acquired from Leva et al. (2020) and Li-ion battery aging datasets (2022), respectively (details in Table 19.1). The data preprocessing steps involved handling missing values, removing columns irrelevant to remaining useful life (RUL) estimation in the Li-ion dataset, and correcting data skewness for improved statistical accuracy. An end-user study was conducted to evaluate the effects of a demand-side recommender chatbot using EI metrics on both the delivery of recommendations on the engagement of stakeholders (i.e., electricity consumers) with the demand-side recommender framework. To support this, data related to recommendation prompts linked to multiple microgrid (MG) assets were gathered. Additionally, 14 participants aged between 20 and 50, all residents of the cybernetic building at Tallinn University of Technology, were recruited to take part in the study. The findings or outcomes of the study are outlined the following.

Comparative analysis: An extensive empirical analysis of traditional transformer-based approaches and PITA based DT vis-à-vis their ability to handle long input sequences and their sensitivity to time series data was conducted. Figs. 19.2 and 19.3 depict the modeled predictive DT for PITA-based on the HVAC air filter and solar PV smart-grid asset.

Figs. 19.4 and 19.5 respectively describe the comparative analysis of key metrics (i.e., MSE and RMSE loss functions) for evaluating forecasting performance toward minimizing the loss of the developed PITA-based DT architecture compared to the vanilla transformer architecture for asset behavior prediction of HVAC PM2.5 air filter, indoor air quality, and solar PV over the training duration (epoch). Fig. 19.5 highlights the superior performance of PITA DTs' architecture (blue curve) in comparison to the transformer-based DT approach (red curve) for solar PV with about 49.41% decrease in MSE loss capable of contributing to improvement in demand reduction for key microgrid assets' energy use. Similarly, in Fig. 19.4, the experimental results indicated modest improvement in prediction performance of proposed PITA architecture for the HVAC air filter DT model (blue curve) with about 47.87% reduction in MSE prediction loss compared to the vanilla transformer DT model (red curve).

Furthermore, the study analyzes the PITA-based model MSE loss performance at various values of λ hyperparameter compared to vanilla transformer-based models toward establishing the model accuracy across different combination of physics and data-driven loss for selected assets prediction tasks, as illustrated in Figs. 19.6 and 19.7. This metric was adopted for controlling the trade-off between prediction accuracy and adherence to physical constraints using λ hyperparameter which controls the impact of the physics loss function. The coefficient λ is defined in Eq. (19.9). The physics loss function extends the MSE loss by incorporating a λ hyperparameter component. As illustrated in Fig. 19.6, the PITA-based DT for HVAC (blue curve) demonstrates overall lower MSE loss and performance improvements over the vanilla transformer baseline model (green curve). The experiments demonstrate that setting λ to 0.0008, produces a MSE loss of 0.1184, resulting in the most accurate PITA prediction results

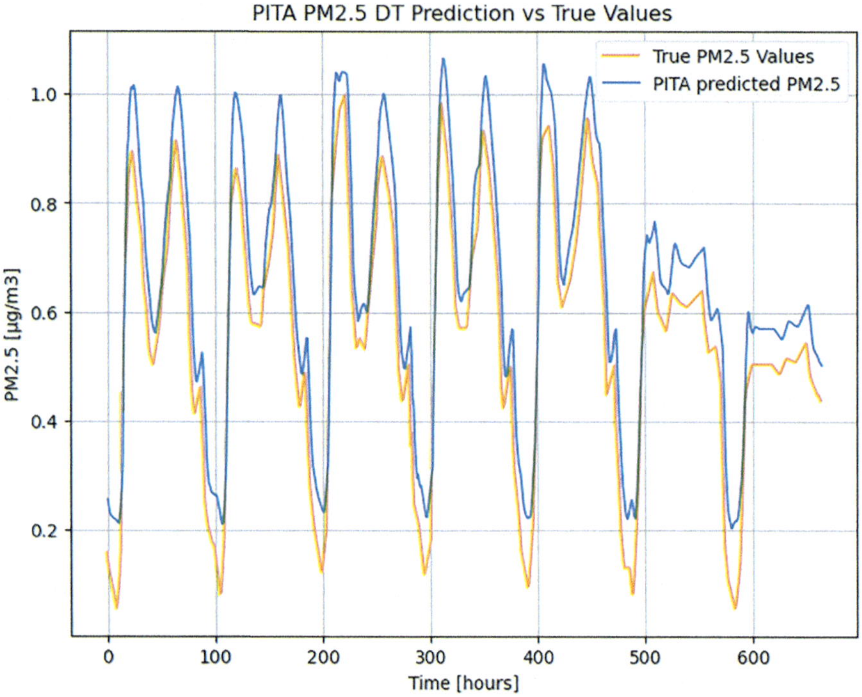

FIGURE 19.2 Experimental results of PITA Digital Twins architecture for HVAC PM2.5. *PITA*, Physics-informed transformer architecture.

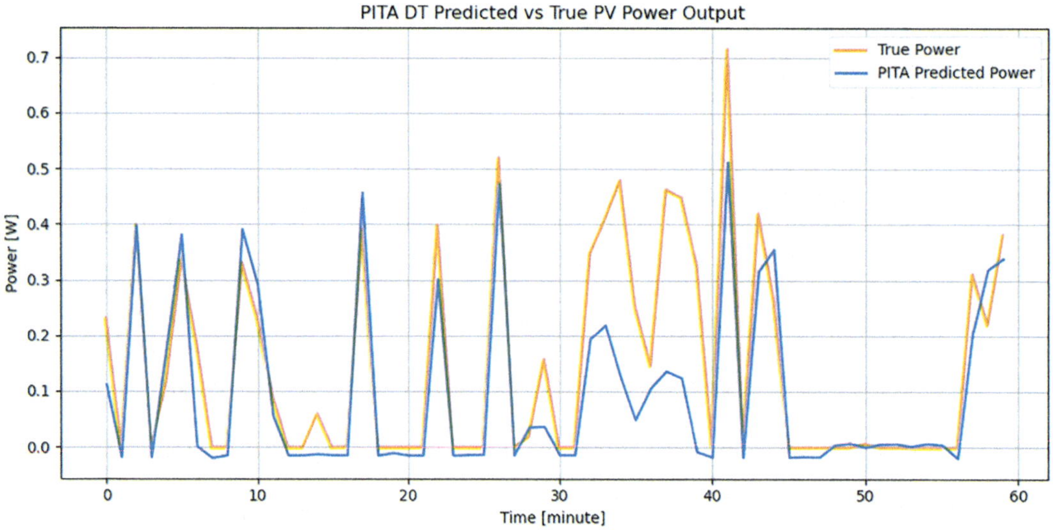

FIGURE 19.3 Experimental results of PITA Digital Twins architecture for Solar PV. *PITA*, Physics-informed transformer architecture; *PV*, photovoltaic.

(as presented in Fig. 19.6). Again, Fig. 19.7 describes the PITA-based DT for solar PV, indicating the best MSE loss of 0.0036 by setting λ to 0.9. Overall, the range of λ experimental values tested varied between 0.0008 and 0.9.

The optimal λ is a hyperparameter controlling the trade-off between prediction accuracy and adherence to physical constraints such as initial and boundary conditions. The value differed significantly between assets (0.0008 for HVAC vs. 0.9 for solar PV), indicating that the balance between data-driven and physics-based learning is highly asset-dependent.

Table 19.2 describes the result of comparative analysis of key loss valuation metrics (i.e., MSE and RMSE) for the prediction of the proposed PITA model against traditional vanilla transformer model-based DT models for selected microgrid assets. The table highlights the exceptional performance of PITA-based DT models when compared to the baseline transformer-based DT model for selected smart-grid assets. The table revealed that the developed PITA-based DT models for solar PV models achieved up to 49.41% decrease in MSE loss compared to transformer-based DT

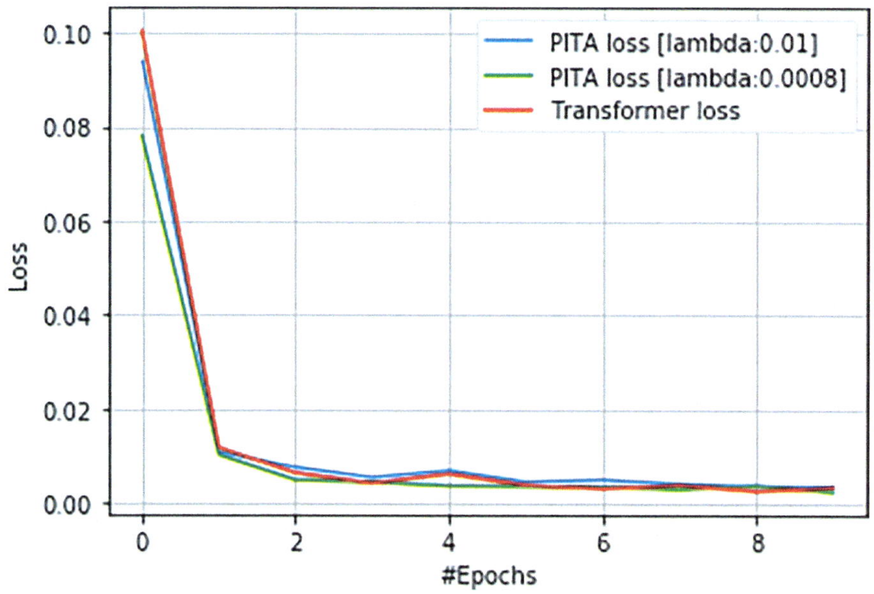

FIGURE 19.4 Experimental results of vanilla transformer vs. PITA Digital Twins architecture prediction loss for HVAC PM2.5. *PITA*, Physics-informed transformer architecture.

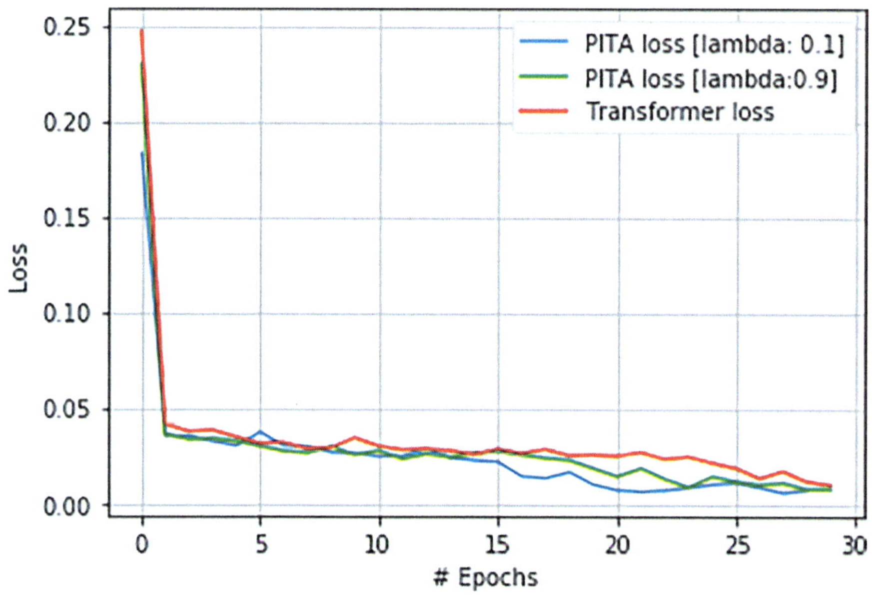

FIGURE 19.5 Experimental results of vanilla transformer vs. PITA Digital Twins architecture prediction loss for solar PV. *PITA*, Physics-informed transformer architecture.

and 54.67% decrease in MSE loss when compared to PINN-based DT model. Similarly, experimental results indicated that the developed PITA-based DT models for Li-ion achieved a 94.16% reduction in MSE prediction loss compared to the vanilla transformer DT model and a 98.28% decrease in MSE loss when compared to the PINN-based DT model.

Decision support and demand-side recommendation services: Following the DT prediction generated by PITA, individual asset profiles were ranked (see Eq. (19.6)) to generate actionable recommendations to assist users in resolving conflicts arising from competing information interests through a thorough analysis of their predicted consumption behavior and energy profile. Table 19.3 depicts the top-N actionable energy efficiency recommendations presented to consumers based on ranking scores derived from predicted DT profiles using the PageRank model. By effecting the energy efficiency recommendations, an overall decrease in load operation hours is expected to be achieved through the demand-side recommendation strategy. Notably, some studies have shown that in certain instances, consumers manually disregard or override efficiency signals and recommendations, thereby further expanding the "knowledge-action gap." In case where advice is not clear, smart-grid stakeholders are able to improve their understanding using the top-N recommendations as prompts or input token for the integrated conversational chatbot interface. Conversational chatbots are employed to enhance user interaction with demand-side recommendation services, particularly when the recommendations are not clear. Dialogue-based systems like chatbots, which use conversation

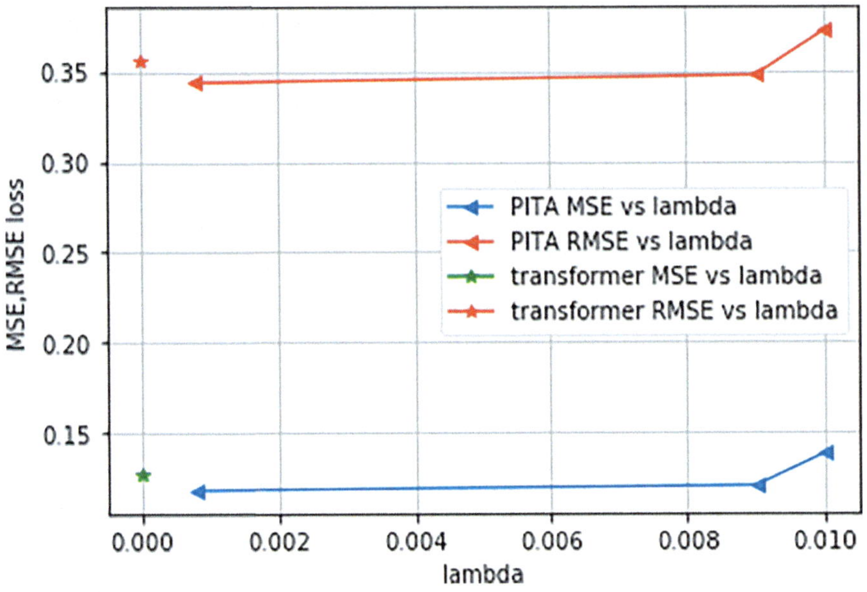

FIGURE 19.6 Experimental results of vanilla transformer vs. PITA Digital Twins architecture prediction loss for solar PV. *PITA*, Physics-informed transformer architecture; *PV*, photovoltaic.

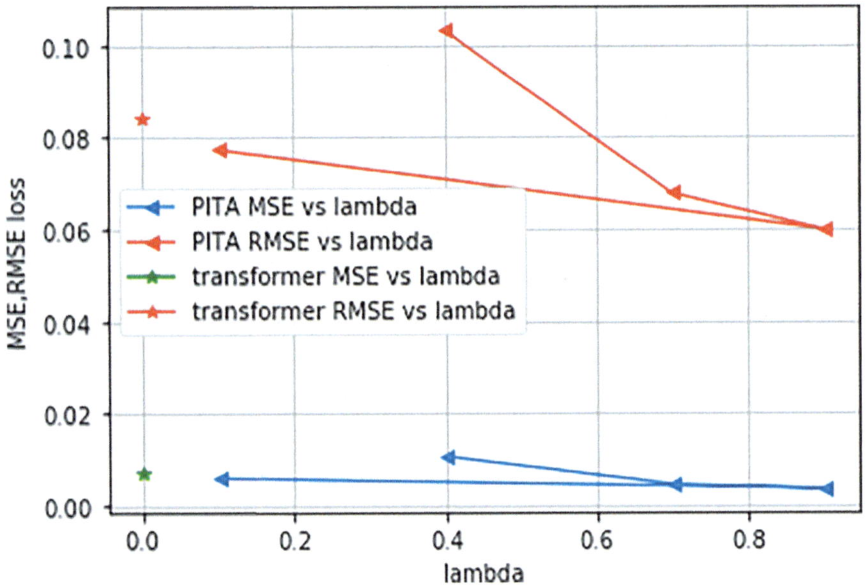

FIGURE 19.7 Experimental results of λ vs. MSE and RMSE loss for PITA Digital Twins architecture for solar PV. *MSE*, Mean squared error; *PITA*, physics-informed transformer architecture; *PV*, photovoltaic; *RMSE*, root mean square error.

TABLE 19.2 Comparative experiment results on evaluation metrics for predictions made by PITA-based Digital Twins versus vanilla transformer-based Digital Twins models for selected grid assets.

Metrics	MSE [PITA]	MSE [trans.]	MSE [PINN]	RMSE [PITA]	RMSE [trans.]	RMSE [PINN]
HVAC	**0.06620**	0.12700	0.11270	**0.25730**	0.35640	0.33570
Li-ion	**0.00051**	0.00873	0.029607	**0.02248**	0.09344	0.17207
Solar PV	**0.00359**	0.00710	0.00792	**0.05994**	0.08426	0.08902

MSE, *mean squared error;* PINN, *physics-informed neural networks;* PITA, *physics-informed transformer architecture;* PV, *photovoltaic;* RMSE, *root mean square error.*

to filter information, can help close this gap by gently encouraging users to adopt energy-efficient behaviors, as shown in Table 19.4 where a chatbot achieved up to 68.68% improvement in user engagements. These systems engage users in extended interactions, leveraging chatbots' unique ability to act as companions rather than just sources of information (Onile et al., 2023). Fig. 19.8 depicts the integrated DTs conversational chatbot interface. The interface

TABLE 19.3 Illustrative example of energy service recommendation result.

Device ID	Recommendation tips	Predicted rating
Solar PV	Install renewable energy storage on site to reduce climate change levy payments	1.0000
HVAC	Install insulation in cavity/solid walls to increase thermal mass & reduce up to 1/3 heat loss	0.4713
HVAC	Replace HVAC filters to achieve up to 30% energy savings	0.4290
Li-ion battery	Operate/Charge battery at optimal temperature 25°C to protect the battery and to reduce demand	0.4259
Solar PV	Clean off dirt on solar PV surface to improve energy output	0.4202

PV, photovoltaic

TABLE 19.4 Improved engagement index (EI) based on consumer click rate using conversational chatbot for enhancing recommender system.

	Sole recommendation	Recommendation & chatbot	Increase [%]
Engagement	34.583	58.333	68.675

enables stakeholders to click on any of the Top-N recommendation tips and seamlessly continue a bi-directional conversation with the chatbot of individual asset for further clarification of the recommendation tip.

While the potential advantages of the enhanced and more precise physics-informed transformer-based DTs architecture for energy asset modeling were emphasized, a number of challenges were also acknowledged. The specificity of the physics equations to chosen assets is critical for DT modeling, and a simplified linearized physics model may not be effective for capturing all nonlinear effects, such as shading, inverter clipping, and temperature extremes in energy assets such as solar PV systems, which can cause the physics-informed loss to regularize toward an incomplete representation. Scalability is also a concern, as attention operations scale quadratically and the computational cost of the hybrid PITA model grows significantly with input window size (Kim et al., 2025), making long-term forecasting computationally expensive. Furthermore, the transformer architecture is parameter-heavy, posing a risk of overfitting given the limited asset (i.e., solar PV) data available. Additionally, the balance between physics and data influence requires careful tuning of the regularization parameter λ, setting it too low reduces physics impact, while setting it too high compromises data fit, making training sensitive and experimental (Xiang et al., 2022). Privacy concerns associated with smart meters and IoT devices collecting fine-grained energy usage/asset data as inputs to the PITA model could result in major privacy leaks (Carmody et al., 2021). The cybersecurity risks for physics-informed transformer-based DT forecasting systems come from three layers: Model-level risks (adversarial attacks, data poisoning), user energy data risks (privacy leakage, tampering, unauthorized access), and IoT device/network risks (malware, insecure firmware, unencrypted communication, DoS attacks). These cybersecurity issues are capable of preventing possible real-world implementation/deployments, prompting the need for improved data security to promote safe modeling of distributed energy assets. Finally, the study's user evaluation involved a small sample size ($n = 14$), which may be influenced by biases inherent in the LLM-based chatbot (i.e., limited online data augmentation), potentially affecting the reliability and generalizability of the results/findings as well as the developed models (Jin et al., 2024).

19.5 Conclusion

This chapter examines the role of foundational models, such as transformer architecture-based DT for smart-grid energy efficiency. This approach combines physics and transformer-based models for the analysis of both temporal and spatial data components of the smart energy grid. In particular, the chapter focuses on careful selection of relevant asset physics equations and integrating them into the loss function of a transformer model, toward the development of PITA-based DT that not only learns from data but also respects fundamental physical principles. While the hybrid DT approach ensured reliable predictions of key smart-grid asset behavior, it was equally crucial

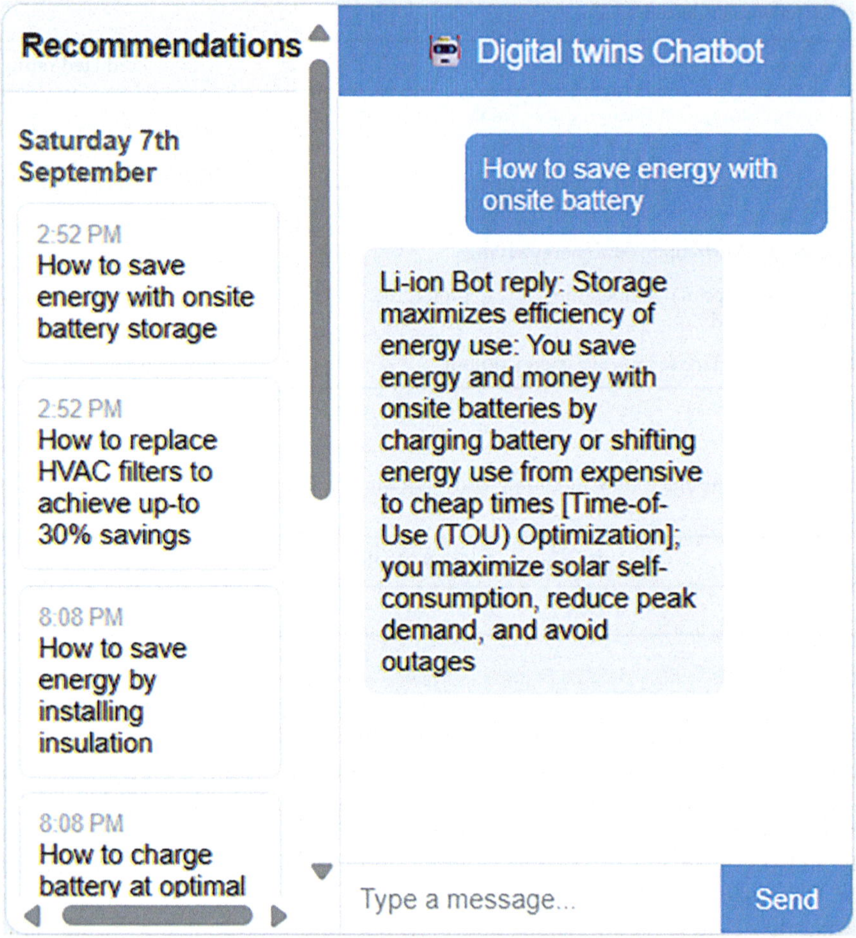

FIGURE 19.8 Illustrative example of integrated Digital Twins conversational chatbot interface.

for decision support and stakeholder engagement. The results showed that the proposed DT modeling approach achieved the highest overall accuracy with about 94.16% decrease in MSE loss. The chapter also demonstrated that the LLM-based chatbot achieved about 69% improvement in user engagement with DSM schemes. Future study will focus on increasing the range of models that can be addressed to incorporate essential electricity grid assets such as electric vehicle (EV) charging stations. Furthermore, research on cybersecurity for smart-grid interconnected DTs toward resilience, and secure communication across distributed grid assets based on implementing secure data protocols for DT communication, and an approach based on testing of extreme conditions or adversarial attacks on the PITA model will be examined.

Acknowledgment

The work was partly supported by the Estonian Research Council grant PRG1463 and by the Estonian Centre of Excellence in Energy Efficiency, ENER (grant TK230), funded by the Estonian Ministry of Education and Research. The authors have no competing interests to declare that are relevant to the content of this article.

References

Abdelrahman, M., Macatulad, E., Lei, B., Quintana, M., Miller, C., & Biljecki, F. (2025). What is a Digital Twin anyway? Deriving the definition for the built environment from over 15,000 scientific publications. *Building and Environment, 274*, 112748. doi:10.1016/j.buildenv.2025.112748.

Air quality at Paris 1er Arrondissement Measurement Station. (2005). PM2.5 dataset. In E. Walther (Ed.), Building physics: Applications in Python. https://github.com/eddes/buildingphysics/blob/master/chapter_2/2_IAQ_filtration/PM25.txt

Anagnostopoulos, S. J., Toscano, J. D., Stergiopulos, N., & Karniadakis, G. E. (2024). Residual-based attention in physics-informed neural networks. *Computer Methods in Applied Mechanics and Engineering, 421*. https://www.sciencedirect.com/science/journal/00457825. doi:10.1016/j.cma.2024.116805.

Azhahudurai, K., & Veeramanikandan, V. (2023). Time series forecasting of air pollutant PM2.5 using transformer architecture. *International Journal of Science and Research (IJSR), 12*(11), 2075–2082. doi:10.21275/sr231125192357.

Bujari, A., Calvio, A., Foschini, L., Sabbioni, A., & Corradi, A. (2021). A Digital Twin decision support system for the urban facility management process. *Sensors, 21*(24). https://www.mdpi.com/1424-8220/21/24/8460/pdf. doi:10.3390/s21248460.

Carmody, J., Shringarpure, S., & Venter, G. V. (2021). AI and privacy concerns: A smart meter case study. *Journal of Information, Communication and Ethics in Society, 19*(4), 492–505. doi:10.1108/jices-04-2021-0042.

Cheng, W., Qin, Z., Wu, Z., Zhou, P., & Huang, T. (2025). Large language models enhanced hyperbolic space recommender systems. *Proceedings of the 48th International ACM SIGIR Conference on Research and Development in Information Retrieval.* https://doi.org/10.1145/3726302.3730019.

Eleftheriou, O. T., & Anagnostopoulos, C.-N. (2022). Digital twins: A brief overview of applications, challenges and enabling technologies in the last decade. *Digital Twin, 2*, 2. doi:10.12688/digitaltwin.17581.1.

Hong, Y., Wu, J., & Morello, R. (2024). LLM-Twin: Mini-giant model-driven beyond 5G digital twin networking framework with semantic secure communication and computation. *Scientific Reports, 14*(1), 19065. doi:10.1038/s41598-024-69474-5.

Inderwildi, O., Zhang, C., Wang, X., & Kraft, M. (2020). The impact of intelligent cyber-physical systems on the decarbonization of energy. *Energy & Environmental Science, 13*(3), 744–771. doi:10.1039/c9ee01919g.

Jin, D., Mehri, S., Hazarika, D., Padmakumar, A., Lee, S., & Liu, Y. (2024). Data-efficient alignment of large language models with human feedback through natural language. *Proceedings of the AAAI Conference on Artificial Intelligence, 38*(16), 18123–18131. https://arxiv.org/abs/2311.14543.

Karnouskos, S. (2011). Demand side management via prosumer interactions in a smart city energy marketplace. *IEEE PES International Conference and Exhibition on Innovative Smart Grid Technologies (ISGT Europe)* (pp. 1–7). IEEE. doi:10.1109/ISGTEurope.2011.6162818.

Kim, J., Kim, H., Kim, H. G. i, Lee, D., & Yoon, S. (2025). A comprehensive survey of time series forecasting: Architectural diversity and open challenges. *Artificial Intelligence Review, 58*(216). https://doi.org/10.1007/s10462-025-11223-9.

Garza, A., Challu, C., & Mergenthaler-Canseco, M. (2023). TimeGPT-1. arXiv preprint arXiv:2310.03589. https://arxiv.org/abs/2310.03589

Leva, S., Nespoli, A., Pretto, S., Mussetta, M., Ogliari, E., Photovoltaic power and weather parameters , IEEE Dataport, (2020), doi: 10.21227/42v0-jz14.

Li, S., Jin, X., Xuan, Y., Zhou, X., Chen, W., Wang, Y. X., & Yan, X. (2019). Enhancing the locality and breaking the memory bottleneck of transformer on time series forecasting. *Advances in neural information processing systems 32 (NeurIPS 2019)*. ACM Digital Library https://papers.nips.cc/32.

NASA Ames Prognostics Center of Excellence. Li-ion battery aging datasets. (2022), https://www.kaggle.com/datasets/mystifoe77/nasa-battery-data-cleaned.

Najafi, P., Mohammadi, M., van Wesemael, P., & Le Blanc, P. M. (2023). A user-centred virtual city information model for inclusive community design: State-of-art. *Cities, 134*, 104203. doi:10.1016/j.cities.2023.104203.

Nele, L., Mattera, G., Yap, E. W., Vozza, M., & Vespoli, S. (2024). Towards the application of machine learning in digital twin technology: a multi-scale review. *Discover Applied Sciences, 6*(10). https://link.springer.com/journal/42452. doi:10.1007/s42452-024-06206-4.

Onile, A. E., Belikov, J., Petlenkov, E., & Levron, Y. (2023). Leveraging digital twins and demand side recommender chatbot for optimizing smart grid energy efficiency. *IEEE PES innovative smart grid technologies - Asia, ISGT Asia.* IEEE http://ieeexplore.ieee.org/xpl/mostRecentIssue.jsp?punumber=10372539. doi:10.1109/ISGTAsia54891.2023.10372761.

Onile, A. E., Petlenkov, E., Levron, Y., & Belikov, J. (2024). Smartgrid-based hybrid digital twins framework for demand side recommendation service provision in distributed power systems. *Future Generation Computer Systems, 156*, 142–156. https://www.sciencedirect.com/science/journal/0167739X. doi:10.1016/j.future.2024.03.018.

Qin, B., Pan, H., Dai, Y., Si, X., Huang, X., Yuen, C., & Zhang, Y. (2024). Machine and deep learning for digital twin networks: A survey. *IEEE Internet of Things Journal, 11*(21), 34694–34716. doi:10.1109/jiot.2024.3416733.

Ramirez, I., Pino, J., Pardo, D., Sanz, M., del Rio, L., Ortiz, A., Morozovska, K., & Aizpurua, J. I. (2025). Residual-based attention physics-informed Neural Networks for spatio-temporal ageing assessment of transformers operated in renewable power plants. *Engineering Applications of Artificial Intelligence, 139*, 109556. doi:10.1016/j.engappai.2024.109556.

Ranawaka, A., Alahakoon, D., Sun, Y., & Hewapathirana, K. (2024). Leveraging the synergy of digital twins and artificial intelligence for sustainable power grids: A scoping review. *Energies, 17*(21), 5342. doi:10.3390/en17215342.

Schmid, U., Ragni, M., Gonzalez, C., & Funke, J. (2011). The challenge of complexity for cognitive systems. *Cognitive Systems Research, 12*(3-4), 211–218. doi:10.1016/j.cogsys.2010.12.007.

Segovia, M., & Garcia-Alfaro, J. (2022). Design, modeling and implementation of digital twins. *Sensors, 22*(14), 5396. doi:10.3390/s22145396.

Shin, M., Seo, M., Yoo, S. S., & Yoon, K. (2024). TFUSFormer: Physics-guided super-resolution transformer for simulation of transcranial focused ultrasound propagation in brain stimulation. *IEEE Journal of Biomedical and Health Informatics, 28*(7), 4024–4035. http://ieeexplore.ieee.org/xpl/RecentIssue.jsp?punumber=6221020. doi:10.1109/JBHI.2024.3389708.

Sun, Y., Zhang, Q. i, Bao, J., Lu, Y., & Liu, S. (2024). Empowering digital twins with large language models for global temporal feature learning. *Journal of Manufacturing Systems, 74*, 83–99. doi:10.1016/j.jmsy.2024.02.015.

Tomar, P., & Grover, V. (2023). Transforming the energy sector: Addressing key challenges through generative AI, Digital Twins, AI, data science and analysis. *EAI Endorsed Transactions on Energy Web, 10*. doi:10.4108/ew.4825.

Walther, E. (2021). *Building physics – Applications in python.* DIY Spring https://hal.science/hal-04580928v1.

Wu, T. G., & Guan, C. (2024). Advancing intra and inter-city urban digital twins: An updated review. *Journal of Planning Education and Research.* doi:10.1177/0739456X241260887.

Xiang, Z., Peng, W., Liu, X. u, & Yao, W. (2022). Self-adaptive loss balanced physics-informed neural networks. *Neurocomputing, 496*, 11–34. doi:10.1016/j.neucom.2022.05.015.

Zeng, A., Chen, M., Zhang, L., & Xu, Q. (2023). Are transformers effective for time series forecasting? *Proceedings of the AAAI Conference on Artificial Intelligence, 37*(9), 11121–11128. doi:10.1609/aaai.v37i9.26317.

Zhang, Z., Wang, Y., Tan, S., Xia, B., & Luo, Y. (2025). Enhancing Transformer-based models for long sequence time series forecasting via structured matrix. *Neurocomputing, 625*, 129429. doi:10.1016/j.neucom.2025.129429.

Zhao, K., & Xiong, W. (2024). Region-guided transformer for remote sensing image captioning. *International Journal of Digital Earth, 17*(1), 1753–8947. doi:10.1080/17538947.2024.2400988.

Zhou, H., Zhang, S., Peng, J., Zhang, S., Li, J., Xiong, H., & Zhang, W. (2021). Informer: Beyond efficient transformer for long sequence time-series forecasting. *Proceedings of the AAAI Conference on Artificial Intelligence*, *35*(12), 11106–11115. doi:10.1609/aaai.v35i12.17325.

Zhao, Z., Ding, X., & Prakash, B.A. (2023). PiNNsFormer: A transformer based framework for physics-informed neural networks. arXiv preprint arXiv:2307.11833. https://arxiv.org/abs/2307.11833

CHAPTER

20

Integration of generative artificial intelligence and large language models into Digital Twin-based smart grids

Saverio Ieva[a,b], *Davide Loconte*[a], *Giuseppe Loseto*[b,c], *Michele Ruta*[a,b], *Floriano Scioscia*[a,b], *Davide Marche*[d], *Marianna Notarnicola*[d]

[a]Department of Electrical and Information Engineering, Polytechnic University of Bari, Bari, Italy [b]donkeyPower S.r.l., Bari, Italy [c]Department of Engineering, LUM Giuseppe Degennaro University, Casamassima, Italy [d]Lutech S.p.A., Bari, Italy

20.1 Introduction

The global energy sector has undergone a significant transformation due to concerns about climate change, sustainability initiatives, and recent technological advancements in renewable energy (Cronin et al., 2018; Joseph & Balachandra, 2020; Schleussner et al., 2016). As energy systems become more complex and need to be more environmentally friendly, an increasing number of infrastructures are incorporating information and communication technology (ICT) to move toward smart grids (Dileep, 2020). A Digital Twin (DT) is a cyber-physical system that contains a dynamic digital representation of a physical asset, which is continuously updated through real-time data streams. This new paradigm enables cost analysis and process optimization (Amaral et al., 2023), as well as simulation and "what-if" analysis on the digital replica without disrupting the physical asset. In the context of energy systems, DTs integrate data from heterogeneous sources, support model-driven analytics, and enable predictive maintenance and operational resilience. These systems enable real-time monitoring, bidirectional communication, and enhanced reliability through advanced metering and analytics. Connected devices can provide users and operators with more precise data, which enables them to save energy, perform preventive maintenance, and automatically resolve issues (Alasali et al., 2023; Alotaibi, Abido, Khalid, & Savkin, 2020). For these reasons, yearly global smart grid investments are expected to grow from $50 billion in 2022 to $130 billion by 2028 (Statista, 2023).

Nevertheless, there are still obstacles to a complete digitalization of energy infrastructures, as well as concerns regarding cognitive overload, interoperability, and standardization in this domain. In particular, smart grid DTs introduce substantial complexity for end users due to their scale and heterogeneity. To overcome these issues, the DT is a paradigm under highly active research. Traditional user interfaces (UIs) that only use 3D graphics and standard interaction models can make it harder to find resources, make decisions, and navigate effectively (Manickam et al., 2023). The integration of *generative Artificial Intelligence (generative AI)* and *Large Language Models (LLMs)* into DT architectures is emerging as a possible way to solve such usability constraints. These models offer new interaction paradigms based on natural language, support semantic reasoning, and enable innovative automation pipelines. In particular, the combination of LLMs with domain-specific Knowledge Graphs (KGs) and Retrieval-Augmented Generation (RAG) pipelines offers promising directions for improving the accessibility, responsiveness, and decision support capabilities of DT systems for energy infrastructures.

Digital Twin Technology and Smart Grid.
DOI: https://doi.org/10.1016/B978-0-443-36664-2.00024-1

What this chapter delivers.

- A DT-LLM reference architecture aligned with Smart Grid Architecture Model (SGAM), layering a KG and a RAG pipeline on top of streaming grid data and physics-based models.
- A pragmatic view of deployment options (on-prem vs. cloud LLMs), balancing capability, latency, and data residency/privacy constraints.
- A forward-looking map of multimodal/world model and collaborative/federated AI for evolving DTs from passive virtual counterparts to proactive reasoning systems.
- Section 20.2 provides a comprehensive background on Generative AI and LLMs, introducing LLM concepts and capabilities (Section 20.2.1), RAG architectures grounding LLM outputs in a core of factual knowledge (Section 20.2.2), and open-source frameworks enabling LLMs integration into existing systems (Section 20.2.3).

The remainder of the chapter is structured as follows:

- Section 20.3 describes the reference architectural model for DT-based Smart Grids for facilitating decision support through semantic enrichment (Section 20.3.1) and real-time monitoring (Section 20.3.2). The section also explores the integration of LLMs and DTs (Section 20.3.3) and the application of data-driven methodologies in energy grids (Section 20.3.4).
- Section 20.4 presents real-world applications of DTs for energy infrastructures, illustrating practical use cases and implementation scenarios.
- Section 20.5 discusses the main challenges and considerations related to incorporating Generative AI technologies into DTs, paying attention to issues such as interoperability, data governance, and system reliability (Section 20.5.1).
- Section 20.6 delineates emerging trends and potential research directions, including multimodal/world model (Section 20.6.1) and collaborative/federated AI techniques (Section 20.6.2).
- Finally, Section 20.7 summarizes the main findings, also highlighting the overall contributions of the work.

20.2 Generative artificial intelligence and large language models

Generative AI refers to a specific area of AI aiming to design models able to create new content by learning patterns from large datasets. A core component of this field is the use of LLMs, which are advanced neural networks trained on very large amounts of documents and data to perform different language-related tasks, including Question Answering (QA), summarization, and code writing. State-of-the-art models demonstrated exceptional efficacy in several tasks, promoting new perspectives and applications in various use cases.

20.2.1 Large language models: Core concepts and capabilities

From a technical perspective, LLMs are Deep Learning (DL) models comprising billions of parameters, trained on massive text corpora to generate and understand natural language (Raza et al., 2025). Main state-of-the-art LLMs are based on the *transformer* architecture (Vaswani et al., 2017), which uses self-attention mechanisms to capture long-range dependencies in natural language texts. Compared to earlier recurrent and convolutional neural networks, this architecture has consistently performed better and has been shown to be scalable in terms of both data volume and model size. Training LLMs typically follows a two-phase process: an unsupervised pretraining stage on large unlabeled datasets (using objectives such as next-word prediction or masked token infilling), followed by supervised fine-tuning or instruction tuning on task-specific data. The nature of the pretraining phase enables these models to learn linguistic and semantic patterns from large-scale corpora without manual annotation (Wolf et al., 2020). Once trained, these models show strong capabilities in both language understanding and generation. They achieve state-of-the-art performance on various Natural Language Processing (NLP) benchmarks and downstream tasks, including text classification, QA, summarization, translation, and code generation. Recent larger models exhibit novel emergent abilities, such as multi-step reasoning and instruction-following capabilities. In particular, a new prompting technique, named *Chain of Thought (CoT)* (Wei et al., 2022), has been devised to guide the model for producing intermediate reasoning steps before generating the final answer. For example, in a smart grid scenario, to predict peak energy demand, a CoT-enabled model might first reason through factors such as historical consumption patterns, weather forecasts, and scheduled industrial activity, before providing the final demand prediction combining insights from the previous steps. This strategy has frequently outperformed smaller and fine-tuned models in terms of accuracy on challenging tasks like math or logic, suggesting that LLMs could approximate contextual reasoning.

One of the key features of LLMs is the ability to adapt to specific tasks without explicit training through prompt engineering techniques, such as *few-shot* learning (Wang et al., 2020). This involves guiding the system to solve problems by providing relevant context in the input prompt, instead of updating the underlying model with task-specific training data, which may be problematic due to the computational cost or the lack of meaningful data. In this context, the user can also supply a minimal set of examples in the request, in the form of question-answer pairs, and the model infers the task format and applies it to new queries. The few-shot ability was illustrated for the first time (Patil & Gudivada, 2024) related to *the Generative Pre-trained Transformer* (*GPT*)*-3* model developed by *OpenAI*, which showed that, even without fine-tuning, it could achieve strong performance on diverse tasks by being prompted with a few demos of each task. GPT-3 demonstrated strong few-shot learning capabilities, effectively leveraging a handful of in-context examples to perform new tasks without weight updates. Few-shot prompts can include any number of examples, up to the limit of the model's context window. In some cases, however, *zero-shot* tasks (Pourpanah et al., 2022) may produce satisfactory results without any examples at all. The exploitation of prompting techniques represented a significant turning point that accelerated the widespread adoption of LLMs in real-world applications.

20.2.2 Retrieval-augmented generation

One major disadvantage of LLMs is that once they complete the training, their knowledge is frozen. An LLM pre-trained on a snapshot of the internet can encode a significant amount of knowledge, but it will not know about facts outside its training data; it has no direct mechanism to cite sources or update its knowledge base. Furthermore, LLMs are prone to hallucinations, which are plausible-sounding but incorrect statements. RAG is one of the possible solutions to these issues. This approach combines an LLM and its memory encoded in the network weights with an external nonparametric memory that can be queried on the fly (Lewis et al., 2020). In a RAG system, the model does not rely solely on what it has memorized; instead, the RAG system actively retrieves relevant information from an outside knowledge source to guide the model in generating its answers.

This approach addresses several limitations of LLMs. Firstly, it enables the model to retrieve information and specifics that were either absent from the training set or had changed since the training set. Secondly, it offers a means of grounding; that is, the output of the model can be directly derived from retrieved documents, increasing verifiability and reducing the risk of hallucinations. Third, it allows interpreting the model output, since it is possible to point to specific retrieved passages as the basis for its answer, and users can trace where a given piece of information came from (Lewis et al., 2020).

A typical RAG architecture comprises two primary components: the generator and the retriever. The retriever's responsibility is to search external knowledge bases in accordance with the context of the conversation or the user's requests. Modern implementations often use a vector space retrieval method: the query is converted into an embedding, which is used to find semantically similar documents. The top- k relevant documents or passages are then fed, along with the original query, into the generator model. The generator then produces a final answer based on both the query and the retrieved text. It is typically a standard LLM, such as GPT-4 or LlaMa 2, specifically prompted to incorporate the retrieved context from an external knowledge source into its response. By conditioning the model on this context, the generator can produce more accurate, relevant, and grounded outputs while leveraging its inherent language processing capabilities.

RAG has proven to be effective across a range of use cases. By augmenting generation with retrieval, these models have achieved state-of-the-art results on tasks that were previously challenging for parametric-only models. The accuracy is better with this method, and the outputs are more specific and factual than with a model that was based on parameters alone. This is because the language model could draw concrete details from relevant Wikipedia articles, reducing guesswork. Beyond academic benchmarks, RAG is useful whenever up-to-date or domain-specific information is required. As an example, a recent study (Ke et al., 2025) combined the GPT-4 with a retrieval system over medical guidelines to evaluate patients' surgical fitness. In that particular task, the RAG-enhanced GPT-4 model outperformed human experts with an exceptionally high accuracy rate of 96.4%), and the model's answers showed no hallucinations and were more consistent, since it always pulled directly from validated medical protocols. Similarly, as described in Gramegna et al. (2025), a RAG system can be used in the smart grid domain to retrieve real-time sensor data, maintenance information, and safety protocols to provide context-aware recommendations to engineers.

TABLE 20.1 RAG evaluation checklist for smart grids.

Metric	Description and application	How to measure
Answer supported ratio	Quantifies response fraction logically entailed by context; prevents hallucination of safety parameters.	Natural Language Inference (NLI) classification (e.g., DeBERTa He et al. (2021)) of hypothesis-premise entailment.
Citation precision and recall	Verifies claim-source mapping (precision) and retrieval completeness (recall) for auditability.	Automated regex token checks; human audit for semantic relevance.
Groundedness probes	Tests refusal behavior against adversarial queries (e.g., non-existent assets) to prevent fabrication.	Calculation of "Refusal Rate" on curated negative constraint datasets.
Latency-SLA under load	Validates response time under peak load against strict control room Service Level Agreements (SLAs).	P95/P99 latency tracking via concurrent load simulation (e.g., Apache JMeter).
Offline evaluation sets	Benchmark accuracy vs. a verified "golden set" of expert Question and Answer (Q&A) pairs to prevent regression.	Semantic similarity scoring (BERTScore Zhang et al. (2020), Ragas) vs. reference answers.

RAG, *retrieval-augmented generation.*

TABLE 20.2 Operational guardrails and protections.

Guardrail	Description and application	How to implement
Source whitelisting	Restricts retrieval to validated/signed repositories (e.g., International Electrotechnical Commission (IEC) standards) to block unverified web data.	Vector DataBase (DB) metadata filters (e.g., source_type="verified").
Top- k retrieval thresholds	Discards low-relevance data ($< \tau$) to force "Unknown" state and prevent guessing.	Post-retrieval filtering based on Cosine similarity scores.
Max context windows	Truncates input to the effective reasoning window; avoids "lost-in-the-middle" phenomena.	Tokenizer-based counting and chunk truncation (e.g., tiktoken).
Personally Identifiable Information (PII) scrubbing	Redacts sensitive entities (names, Internet Protocol addresses) from queries prior to inference.	Middleware Named Entity Recognition (NER) pipelines (e.g., Presidio) or regex pattern masking.
Strict prompt templates	Enforces persona constraints and citation formatting; prevents jailbreaking.	Hardcoded system instructions and delimiter wrappers in API calls.

API, *application programming interface.*

Integrating RAG in safety-critical domains like smart grids requires rigorous quality assurance. Table 20.1 presents an evaluation framework where specific *metrics* are correlated with their *description and application* to grid reliability, alongside standardized technical methodologies on *how to measure* them. Complementing these benchmarks, Table 20.2 details *production guardrails* necessary for operational security. This matrix maps each protective *Guardrail* (e.g., source whitelisting) to its specific risk mitigation strategy and provides engineering guidance on *how to implement* these controls to filter low-quality data and prevent model hallucinations (Ji et al., 2023).

20.2.3 Open-source frameworks for large language models integration

Since the emergence of LLMs, numerous open-source libraries and frameworks have been developed for research and integration in production solutions. These frameworks aim to serve, fine-tune, and design pipelines that leverage the capabilities of these models with relative ease. Some main frameworks are:

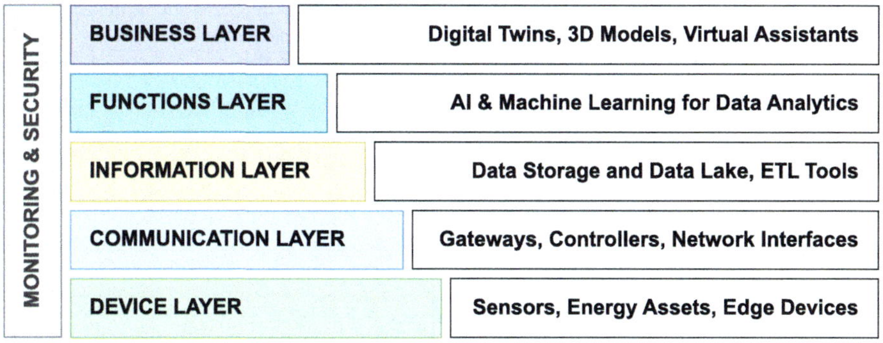

FIGURE 20.1 Architectural model for Digital Twins-based smart grids.

- **Hugging face transformers**: (https://huggingface.co/docs/transformers/v4.17.0/en/index) The *transformers* library has become a common tool for using pre-trained LLMs in Python. It offers a unified Application Programming Interface (API) to various state-of-the-art transformer models, as well as an online model repository for the download and serving of models contributed by the community. The main design principles of the library are its extensibility and research-friendliness, while also retaining robustness and code quality for production deployments. Users can load a pre-trained LLM with a few lines of code, then further fine-tune it on a custom dataset or deploy it for inference. This democratization of model access has been a driving goal of the project. The transformers library greatly lowers the barrier to entry for LLM experimentation and ensures that results can be replicated across environments. Its emphasis on open community involvement reflects how open-source infrastructure accelerates progress in Generative AI.

- **LangChain:** (https://www.langchain.com) LangChain is one of the most popular open-source Python frameworks for developing solutions integrating LLMs. It lets developers connect several LLM calls and add external data or tools (like databases, APIs, or knowledge bases) to a pipeline that can complete specific tasks. LangChain provides abstractions for building complex workflows like chatbots and agents that use an LLM alongside the retrieval of documents, calculators, or other utilities. For example, one can build a QA system that first uses a vector database to fetch relevant context and then feeds it into an LLM to generate a sensible answer—all coordinated by LangChain. The framework interfaces with various components in the LLM ecosystem; it can seamlessly incorporate a vector store (e.g., ChromaDB for embeddings) into an LLM-driven pipeline (Vidivelli et al., 2024). LangChain functions as the glue that connects LLMs with other modules, thereby simplifying the implementation of RAG workflows, tool-using agents, or multi-step reasoning chains.
- **Deepset Haystack**: (https://haystack.deepset.ai/) Haystack is a framework specifically designed for building QA and RAG pipelines. Haystack provides a tool for coupling search indexes, retriever models, and generator models into a unique system. It supports functionalities like indexing, searching document indexes, and extracting answers or passing them to a model to generate an answer. Haystack includes annotation tools for creating QA datasets and a representational state transfer (REST) API for deploying QA systems. This makes it a popular choice in industrial settings (Syed et al., 2021).

Other tools focus on specific tasks, like LlamaIndex for document indexing, or DeepSpeed and ColossalAI for large-scale model training. All these frameworks are open-source, promoting an active community of users, accelerating innovation, and facilitating scientific reproducibility (Wolf et al., 2020.

20.3 Digital Twin architecture for smart grids

The smart grid concept refers to an integrated system in which power infrastructures and Information and Communication Technology (ICT) networks are tightly interconnected to enable responsive, efficient, and resilient energy management. In general, a smart grid combines traditional power infrastructure with advanced digital technologies, including sensors, communication networks, and data-oriented analysis, to improve energy production and distribution. Such complex systems can be designed according to the SGAM (Panda & Das, 2021). As shown in Fig. 20.1, it defines five abstraction layers, ranging from the physical components up to high-level business processes, each responsible for specific functionalities and interaction patterns, with semantic technologies and AI-based services integrated across them. At the lower level, edge devices, including sensors and actuators, interact with

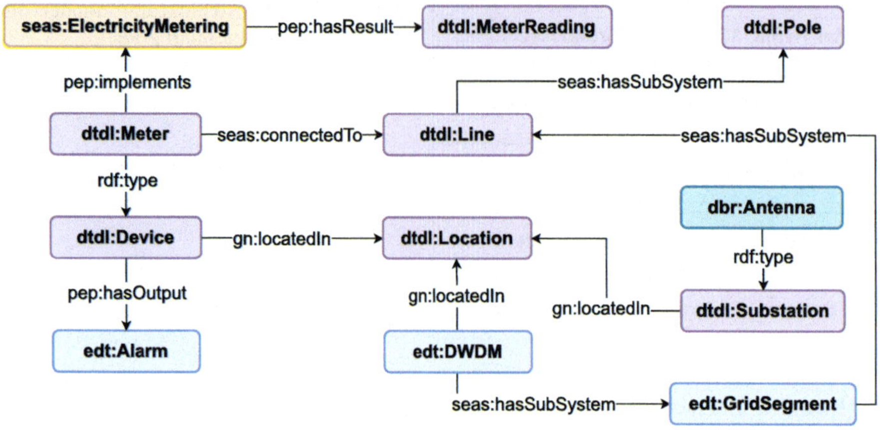

FIGURE 20.2 Semantic representation of the power grid domain within the KG. *KG*, Knowledge graph.

energy assets and field equipment. Data collected from these sources is transferred through device controllers and network gateways to ensure compatibility with various protocols. Then, Extract, Transform, Load (ETL) tools are used to facilitate the aggregation and harmonization of heterogeneous information. The information layer also orchestrates the lifecycle of collected information, overseeing tasks such as data normalization, storage, and distribution across the system. A centralized data lake is usually exploited as the backbone of the architecture, enabling scalable, schema-flexible storage of structured and unstructured data. On top of this, data analysis modules support advanced analytics and real-time insights, leveraging graph databases, in-memory storage, and AI/Machine Learning (ML) services. These can be accessed via higher-level interfaces, including dashboards, Natural Language User Interfaces (NLUIs), and virtual assistants, to support operational decision-making and human interaction. Finally, asset simulation techniques provide a digital representation of cyber-physical entities using semantic models and 3D DT, enabling enhanced monitoring and control. Following this model, both engineers and researchers can develop flexible, data-driven frameworks that support different applications based on DT as virtual models that mirror real-world energy systems.

The SGAM architecture emphasizes modularity, scalability, and semantic interoperability, ensuring that both legacy and modern components can be effectively integrated into a cohesive and intelligent smart grid ecosystem. The following subsections present the architectural components that enable the integration of LLM-based intelligence within smart grids.

20.3.1 Knowledge graphs for grid representation

The three lower layers (device, communication, and information) of the SGAM model presented in Section 20.3 provide the essential technical foundation for achieving interoperability among heterogeneous devices, communication protocols, and data representations. These layers enable a consistent abstraction of physical assets and network interactions, which is critical for the deployment of reliable automation solutions in smart grid environments (Panda & Das, 2021; Yang et al., 2020). Building upon this foundation, the proposed approach adopts knowledge-based data annotation as a core design principle. Logical data models are constructed to unify system information related to different aspects of the energy domain, leveraging standardized ontologies and linked data principles (Bizer et al., 2023) to create a reference KG. This modeling strategy serves multiple purposes: (1) facilitates automated reasoning and the discovery of implicit relationships within the infrastructure; (2) promotes uniformity in documentation and system design also reducing implementation errors in the development of new functionalities; (3) supports transparent communication between domain experts, engineers, and data analysts. Furthermore, this approach sets the basis for advanced Model Driven Engineering (MDE) techniques (Lehner et al., 2025), where high-level semantic descriptions can be systematically transformed into executable control logic, thus bridging the gap between specification and deployment (Yang et al., 2020).

Examples of semantic representations modeled in the KG are shown in Figs. 20.2 and 20.3, describing a section of the energy network and illustrating how physical components and their relationships are digitally represented. All entities, detailed in Table 20.3, are modeled as resource description framework (RDF) (Cyganiak et al., 2014) resources following the linked data best practices (Bizer et al., 2023) in order to make the proposed approach general-purpose and support semantic-based data annotation and interpretation. By adopting RDF, each entity (e.g., generators, transformers, sensors, and consumption nodes) can be uniquely identified and semantically described, facilitating interoperability across diverse systems. Linked Data principles further enable the integration of

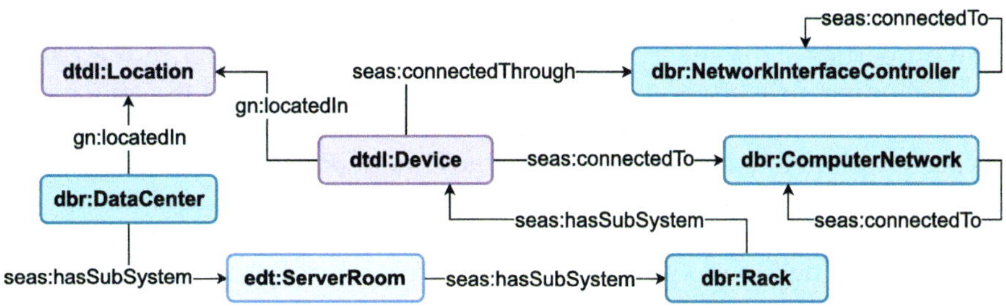

FIGURE 20.3 Semantic representation of the IT infrastructure within the KG. *KG*, Knowledge graph.

TABLE 20.3 Entities and ontologies exploited to model smart grid concepts.

Entity	Description
Digital Twins Definition Language (DTDL) ontology	
Device	A device included within the energy infrastructure
Meter	Physical asset used for measuring energy consumption and detecting events of interest
Meter reading	The amount of electricity consumption recorded through a meter
Location	Physical location of a device or energy asset
Substation	An electrical station is part of a generation, transmission, or distribution system
Line	A line composing a telecommunications or power transmission infrastructure
Pole	A utility pole used to support overhead power lines or other public utilities
Energy Digital Twin (EDT) ontology	
Alarm	Specific event related to an issue identified by a device
DWDM	Dense wavelength division multiplexing (DWDM) system supporting fiber-optic transmissions
Grid segment	Section of a power or communication network
Server room	Air-conditioned room devoted to the continuous operation of computer servers in a data center
DBpedia Resources (DBR) ontology	
NIC	Network interface controller (NIC) representing a physical connection through which the device interacts with external peripherals
Computer network	A group of interconnected devices capable of communicating with each other
Antenna	Local secondary substation installed to support critical scenarios requiring high data availability in the smart grid
DataCenter	Centralized facility designed to store and process vast amounts of data related to the energy grid
Rack	Physical container designed to house servers, networking devices, and other data center equipment
Smart Energy Aware Systems (SEAS) ontology	
Electricity metering	Process of measuring and recording electrical energy consumption

distributed data sources, supporting unified reasoning over the network and enhancing compatibility with existing semantic web tools. This approach allows for more effective knowledge representation, automated inference, and seamless data sharing, which are essential for advanced analytics and decision-making in smart grid environments. In particular, the following well-known vocabularies are used as upper ontologies modeling basic concepts and properties concerning energy infrastructures, network communication, and metering procedures: Digital Twins *Definition Language (DTDL)* ontologies for energy grid (https://github.com/Azure/opendigitaltwins-energygrid) and smart cities (https://github.com/Azure/opendigitaltwins-smartcities), specifically proposed for modeling DT solutions, including monitoring of grid and urban assets (concepts in blue); *Smart Energy Aware Systems (SEAS)* ontology (https://w3id.org/seas/), (Lefrançois, 2017), designed as a set of simple core ontology patterns modeling multiple engineering-related concepts and properties of the energy ecosystem (concepts in violet); *procedure execution ontology* (PEP) (https://w3id.org/pep/) including properties used to describe procedures, outputs, and results related to metering activities; (Lehmann et al., 2015) *DBpedia* resources (DBR) and *GeoNames (GN)* ontology (https://www.geonames.org/ontology/) have been exploited to model elements of the IT network and relationships about asset location (concepts in yellow). In Figs. 20.2 and 20.3, entities and relationships belonging to each ontology are denoted by their respective RDF namespace prefix, and they are shown in a different color. In addition, a custom *Energy Digital Twin (EDT)* vocabulary has been defined to model domain entities (reported in gray) and properties not included within the aforementioned ontologies.

20.3.2 Data flow and real-time monitoring

Data collected by sensing devices is stored through the information layer and exploited by data analysis tools to provide descriptive, predictive, and prescriptive analytics capabilities (Roy et al., 2022) in order to understand the operational behaviors of energy systems (Ahmad et al., 2022). The functions layer can also initiate data crawling processes, provided by the information layer, to extend the knowledge base with additional information obtained from cloud-based services and external resources. By means of data mining and ML algorithms, the system aims to identify actionable insights and extract useful knowledge from the vast volumes of data generated by the connected infrastructures. As highlighted in Alimi et al. (2020), potential applications include:

- *Fault detection and diagnosis*: identify possible malfunctions of the energy grid and monitor the health and performance of grid assets such as transformers, generators, or substations;
- *Preventive maintenance*: based on grid equipment structural and behavioral models, analyze status data series and predict when maintenance should be performed to prevent failures and minimize downtime;
- *Load forecasting and balancing*: predict future electricity demand based on historical data and balance the load across different parts of the grid to ensure stable operation and prevent overloading or congestion;
- *Demand response optimization*: optimize demand response programs to manage electricity consumption during peak periods or in response to pricing signals.

Depending on the timing of the response and the amount of data to be processed, the framework enables two main types of processing tasks: *batch processing* for large volumes of data and groups of transactions. Usually, data is collected, inserted, and processed to produce results automatically. Tasks are scheduled over even medium to long periods of time; *real-time processing* for tasks requiring a rapid response and performed by simple systems managing rapid transactions. Mainly used in environments where many events occur in a short time. Functions layer also enables the application of further techniques to simulate real-world operational scenarios according to the following analytical approaches: *what-if analysis* (Kegel et al., 2017): one or more parameters of the reference scenario are expected to be changed and, after the execution of multiple simulation sessions, results are stored to allow comparisons and sensitivity analysis; *goal-seeking analysis* (Nguyen & Gonzalez, 2020): A simulation objective is defined in terms of observable variables or key performance indicators (KPIs) and the system will provide support in defining scenarios aimed to identify and evaluate suitable ways to obtain the specific result. To translate these architectural concepts into a viable production environment, specific quantitative system budgets must be established. These budgets define the operational constraints for telemetry ingestion, processing latency, and model throughput, ensuring that the DT can maintain synchronization with the physical grid while supporting the computational overhead of Generative AI. Table 20.4 outlines the reference baseline targets for a standard EDT implementation.

20.3.3 Interfacing large language models with Digital Twin systems

The business layer plays a central role in coordinating the data-driven workflow. Positioned between user-facing services and backend cognitive components, it manages input interpretation, semantic retrieval, and task orchestration. By integrating DTs, this layer can simulate and optimize grid operations in real-time, enabling smarter decision-making and proactive system management. Within the reference architecture, LLMs can be interfaced with the DT infrastructure to enable advanced functionalities such as natural language interaction, semantic enrichment, and decision support. A key requirement in this context is the preservation of data privacy and operational autonomy. To meet this need, the architecture is configured to perform all processing locally, including analytics, ML, and data visualization, without transferring sensitive information to remote services (Gautam et al., 2025).

TABLE 20.4 Quantitative system budgets for LLM-enhanced Digital Twins.

System component	Metric	Example target budget/constraint
Telemetry ingestion	Ingestion rate	Handle 50 – 60 **Hz** sampling from Phasor Measurement Units (PMUs) and 1 – 5 **Hz** from SCADA; aggregate throughput target: >50k events/sec per cluster node.
Inference latency	Critical anomaly detection	< 100 ms end-to-end (hard real-time). Must bypass LLM reasoning paths to trigger immediate alerts.
	Interactive LLM response	< 200 ms time to first token (TTFT); <5 s total completion time for complex RAG queries (soft real-time).
Data throughput	RAG vector search	Retrieval of top- $k(k = 5)$ chunks from **1M** + vectors in < 50 **ms** (P95 latency) to avoid stalling the generation pipeline.
	Context window load	Support dynamic context loading of 4k–32k tokens per query without degrading prompt processing speed below 50 tokens /sec.

LLM, *large language models.*

TABLE 20.5 Operational architecture decision matrix: On-premises vs. cloud-based LLMs.

Criterion	On-premises (Edge/Local)	Cloud-based (Software as a Service/API)
Latency and SLA	**Deterministic/Low:** Essential for real-time control loops and safety-critical alerts.	**Variable:** Dependent on internet connectivity and provider load.
Data residency	**Strict control:** Data never leaves the physical asset perimeter. Complies with strict critical infrastructure regulations.	**Shared responsibility:** Data traverses' public networks; requires strict vendor agreements and potentially anonymization proxies.
Cost structure	**CapEx dominant:** High upfront cost for hardware and maintenance. Lower marginal cost per query.	**OpEx dominant:** Pay-per-token model. Low entry cost but scales poorly with high-frequency automated agent loops.
Model temporal validity	**Static:** Updates require manual intervention and downtime. Knowledge cutoff is fixed unless RAG is used.	**Continuous:** Provider handles updates. Access to the latest reasoning capabilities without infrastructure changes.
Hardware/Graphics Processing Unit (**GPU**)	**Constrained:** Limited by local power, cooling, and rack space. Often requires quantization to fit memory.	**Scalable:** Elastic access to high-end clusters for complex reasoning tasks on demand.

API, *application programming interface;* LLMs, *large language models;* RAG, *retrieval-augmented generation;* SLA, *service level agreements.*

Basically, both proprietary and open-source LLM backends can be exploited in smart grid applications. Open-source models, recommended for on-premises deployment, offer full control over data and system behavior. This is particularly important in scenarios requiring local processing or strict privacy controls. On the other hand, proprietary solutions, typically accessed via cloud-based APIs, may be considered when advanced model capabilities or greater scalability are required. Depending on the specific use case, organizations may adopt one approach or combine both to balance performance, cost, and control within the smart grid ecosystem. To address these deployment considerations systematically, two reference architectural patterns suitable for smart grid environments are proposed. Pattern A (hybrid approach) integrates a local model with an on-premises vector store for processing sensitive operational technology (OT) data, while offloading non-sensitive tasks to external, high-capacity cloud-based models. Pattern B implements a localized inference gateway entirely within the critical infrastructure perimeter, thereby ensuring data locality. The specific trade-offs governing the selection between these architectures, ranging from latency constraints to data residency compliance, are outlined in Table 20.5 (Chen et al., 2025).

However, general-purpose LLMs rely on a fixed knowledge base derived from their training data, which means they cannot access new or evolving information after their release. Moreover, these models often struggle with specific or highly technical domains, such as smart grid, where specialized knowledge and up-to-date context are essential for accurate support. To enhance the relevance and factual grounding of generated outputs, the proposed architecture integrates an RAG mechanism enabling the system to handle specialized questions without retraining the entire model. As illustrated in Fig. 20.4, user queries submitted via DT platforms or virtual assistants are matched

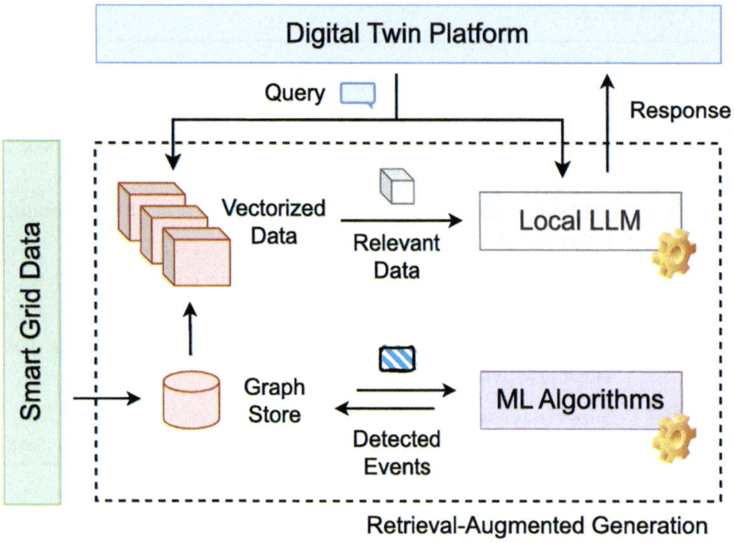

FIGURE 20.4 Retrieval-augmented generation framework for contextual LLM integration. *LLM*, Large language models.

against the system's KG. The graph store includes contextual information, such as infrastructure descriptors, detected events, and sensor measurements, converted into vectorized representations able to capture semantic meaning and relationships between terms and sentences. For example, in a smart grid context, the term *energy consumption* might be close to *power usage*. In a similar way, a document about *solar inverters* will be closer to one about *renewable generation* than to one about *billing policies*. The system then compares vectors using similarity measures to identify the most relevant pieces of information. Retrieved knowledge fragments are dynamically injected into the input prompt before being forwarded to the language model for response generation. This design improves not only the factual accuracy of outputs but also their contextual alignment and traceability (Ieva et al., 2024). In this way, RAG combines the reasoning power of LLMs with the precision and adaptability of information retrieval, making it a powerful solution for ensuring responses reflect both contextual data and up-to-date information.

20.3.4 AI-based predictive and prescriptive analytics

Beyond reactive and descriptive capabilities, the integration of LLMs within DT systems enables advanced forms of analytics, including predictive and prescriptive functionalities. These techniques exploit historical data, real-time sensor streams, semantic annotations, and contextual knowledge stored in the KG, transforming DT into intelligent systems capable of anticipating events and recommending optimal interventions (Motlagh et al., 2024; van Dinter et al., 2022). For *predictive analytics*, ML models are trained over operational data collected from distributed assets to forecast conditions such as equipment failure, energy demand, or anomalous behaviors. DTs provide a structured and continuous view of the monitored infrastructure, allowing these models to incorporate temporal, spatial, and relational dependencies that are often critical for accurate forecasting. Semantic enrichment further improves model interpretability by aligning raw data with domain-specific concepts and entities. *Prescriptive analytics* extends this paradigm by not only predicting future outcomes but also suggesting optimal decisions or corrective actions. This is achieved by combining predictive models with optimization algorithms, domain constraints, and feedback mechanisms that evaluate the effectiveness of executed actions. The resulting recommendations can support a wide range of operational goals, including performance optimization, fault mitigation, and resource allocation.

Within the proposed reference architecture, predictive and prescriptive capabilities are supported by a modular AI functionality interfaced with both the data management and user interaction components. Data originating from physical assets is continuously analyzed to detect early signs of degradation or risk. When potential issues are identified, inference pipelines activate reasoning mechanisms to identify possible causes and recommend suitable interventions. These may include scheduling of maintenance, adjustment of control parameters, or reassignment of operational priorities. LLMs contribute to this process by facilitating access to analytical results through natural language queries, summarizing findings, and translating insights into actionable knowledge. When combined with retrieval-based approaches, they can generate context-aware explanations grounded in the system's current state and historical patterns. This improves human interpretability and supports timely decision-making. As shown in Fig. 20.4, these insights are added to the KG to model the evolving representation of the smart grid. Completed interventions and operator feedback are also captured and semantically annotated, enabling continuous improvement of the predictive models and supporting knowledge-based reasoning. Over time, this loop contributes to a shift from data-centric monitoring to strategy-oriented control, where DT not only represents but also enhances the intelligence of the underlying systems.

20.4 Real-world applications for energy infrastructures

Modern electricity grid infrastructures, characterized by complex human-machine interactions and extensive data flows, serve as prime candidates for solutions powered by AI (Bayindir et al., 2016; Massaoudi et al., 2021). Specifically, LLMs are capable of handling unstructured data, reasoning through technical challenges, and communicating their findings in natural language. For example, a multi-step reasoning task for a CoT-based LLM can simulate the impact of taking a specific substation offline for maintenance. In this case, the model first analyzes which feeders and downtown loads are affected, then estimates how the load would be redistributed to neighboring substations, evaluating potential voltage drops or overloads. Finally, the associated risks are summarized, including possible service interruptions and mitigation strategies. This section explores several real-world applications of LLMs in smart grids, including predictive maintenance and fault detection, natural language interfaces for grid operators, and automated reporting with anomaly detection.

20.4.1 Predictive maintenance and fault detection

In a smart grid, predictive maintenance systems continuously monitor the condition of assets such as transformers, switchgear, and power lines. By analyzing sensor readings alongside historical failure data, AI models can detect early warning signs of malfunction and recommend maintenance before a fault occurs (Mahadik et al., 2025), which can significantly reduce operational expenses, improve service reliability, and enhance safety. Incorporating Internet of Things (IoT) devices represents a key element in this transition, establishing a widespread network of sensors that generate live data on the equipment status across the grid (Motlagh et al., 2020). These data streams offer significant input for ML models that are responsible for detecting degradation trends and predicting malfunctions. In addition to this, the 5G communication infrastructure ensures the low-latency and high-bandwidth transmission of sensor data necessary for time-critical decision-making (Ahmadzadeh et al., 2021). To further enhance responsiveness and scalability, edge computing is increasingly employed to process data close to the source. This reduces latency and minimizes dependency on centralized infrastructure, thereby enabling faster local decision-making and greater resilience in fault detection applications (Arcas et al., 2024; Cao et al., 2020; Minh et al., 2022). Smart nodes with edge AI functionality are capable of running lightweight ML models or federated learning (FL) algorithms (Loconte et al., 2024; Ruta et al., 2019), enabling predictive diagnostics in environments with constrained resources.

DT technologies offer an additional layer of insight by maintaining real-time, virtual representations of physical grid components (Singh et al., 2021). Specifically, a DT system is capable of combining sensor information, past records, and simulation outcomes to offer a detailed operational perspective of assets. This facilitates not only real-time anomaly detection but also long-term trend analysis and lifecycle management (Dietz & Pernul, 2020). Through enhanced situational awareness and collaborative interfaces (Liu et al., 2023), DTs empower operators and maintenance teams to proactively identify risks and coordinate interventions more effectively. LLMs do not directly model the statistical behavior of individual component failures. Instead, they enhance the capabilities of expert systems by integrating heterogeneous data sources, including numerical sensor data, maintenance logs, technical manuals, and real-time descriptions from field engineers. This integration allows the system to act as an informed assistant for fault diagnosis. Alongside more conventional ML models, LLMs can detect unusual sensor patterns and provide clear explanations of which component is likely malfunctioning and why, referencing relevant equipment documentation. Fine-tuned language models can even generate step-by-step maintenance strategies tailored to specific faults (Ieva et al., 2024). The main applications of this technology in smart grid operations are discussed in the following two sections.

20.4.2 Natural language user interfaces for grid operators

Traditionally, grid operators interact with energy management systems (EMS) through complex UIs. Conversational AI can be employed as an NLUI, enabling operators to query and control the grid using conversational inputs.

Rather than navigating the UI or recalling task-specific commands, an operator could simply pose questions or issue directives to an agent, like "Which station is currently overloaded?" or "Adjust the network to prioritize power for the hospital district." The LLM processes this high-level request and translates it into machine-understandable commands (Ieva et al., 2024). This interaction reduces the need for extensive operator training and reduces the chances of human mistakes. The LLM can also explain the decisions made, thereby improving user trust in the system. An LLM can also augment traditional expert systems by providing contextualized, narrative insights that combine real-time data with historical and technical knowledge. For instance, if a transformer sensor detects a voltage anomaly, the LLM can first identify the relevant sensor ID and query the associated KG. It retrieves the transformer's maintenance history, technical specifications from equipment manuals, and records of similar past incidents. The model then synthesizes this information into a concise, actionable summary for the engineer: *Transformer T-451 is experiencing a voltage spike similar to the event recorded on 12/05/2023, which was resolved by replacing the bushings. Current operating temperature is within specification. Recommend prioritizing the inspection of bushings to prevent recurrence.* This approach allows the engineer to quickly understand the context and recommended actions without manually cross-referencing multiple data sources.

Early implementations of such NLUIs are beginning to appear in grid control centers. In the United States, the National Renewable Energy Lab has prototyped an "eGridGPT" assistant that uses generative transformers to aid dispatchers in decision-making (Choi et al., 2024). China introduced an LLM-based assistant called GAIA to support power dispatch operations. GAIA was fine-tuned on power system data and can handle tasks like adjusting generator outputs, monitoring grid status, and even guiding operators through black-start procedures. GAIA significantly outperformed a base model (LLaMA-2) and enhanced human-machine interactions in dispatch control—operators using GAIA made decisions faster and with more confidence (Cheng et al., 2025). These results validate that LLMs can act as helpful assistants for grid operators. European pilot projects are likewise exploring conversational AI for control rooms. For example, researchers have proposed agent-based AI that would let operators interact with the grid, asking about anomalies or receiving real-time recommendations for network reconfiguration (European Commission, 2024). A common challenge across these implementations lies in ensuring that the LLM correctly interprets and applies highly domain-specific technical jargon, acronyms, and grid terminology. Without careful fine-tuning or deliberate prompt design, the model may misinterpret critical terms, leading to ambiguous or misleading outputs. For example, in the energy domain, DER should be understood as *distributed energy resources*, but a general-purpose LLM might instead interpret it as a *duplicate email remover,* resulting in irrelevant or confusing responses. Addressing this issue typically requires the integration of domain ontologies, supervised training data, or prompt engineering strategies tailored to the operational context of the smart grid.

20.4.3 Automated reporting and anomaly detection

Conventional systems frequently underperform in providing accurate results and user-friendly interpretability features that clarify prediction outcomes. On the other hand, LLMs can be combined with existing analytics tools to provide conversational agents that explain results in natural language. These models are ideal for smart grid anomaly detection, as they integrate data from multiple sources and can adapt to dynamic contexts. For instance, a conversational agent can determine whether a voltage spike is fault-related by examining sensor readings and log data. A particular case showed ChatGPT-4's substation monitoring system outperforming traditional methods by effectively analyzing complex communication patterns, achieving nearly 98% accuracy in detecting real anomalies (Ibrahim & Kashef, 2025).

LLMs are also increasingly used in DT-based smart grids as autonomous engines for anomaly detection and reporting. For instance, Zaboli et al. (2025) presented CyberGridToD, a system that employs an LLM to identify and react to anomalies in grid communications. The AI agent engages in dialogue, mimicking the thought process of an operator. When the DT identifies an irregularity, the LLM analyzes the situation and suggests a corrective measure. These systems have proven to be highly reliable in anomaly handling and provide enhanced scalability compared to traditional human-supervised methods (Ibrahim & Kashef, 2025).

LLMs are also used to automate reporting in energy infrastructures for compliance and auditing, as well as real-time alert summaries for operators. A fine-tuned LLM can synthesize outputs from a DT that includes sensor events, simulations, and control actions, into structured reports that clearly explain the system state and detected anomalies. For example, following a grid disturbance, an LLM can automatically generate an incident summary outlining what occurred, identifying likely causes, and suggesting remediation steps. Similar applications have been shown in other critical infrastructure domains. BERT-like models have been implemented to extract critical events from operator notes and incident logs, thereby enhancing the efficiency of post-event analysis (Yigit et al., 2025). During blackouts,

TABLE 20.6 Risk-control matrix.

Risk	Vector	Controls	Reference layer
Prompt injection and input manipulation	Manipulated LLM queries, malicious sensor data streams	Input anomaly detection, strict prompt templates, and sanitization policies	**Communication layer** (API Gateways) and input validator module
Cloud exposure and data leakage	Multi-territorial transfers, exposure of consumer patterns	On-premise inference, federated learning (local parameters), encryption	**Information layer** (local vector store) & edge nodes
Model reliability and hallucination	False assertions, reverse engineering training data	Confidence scoring, citation enforcement, and differential privacy	**Functions layer** (RAG pipeline anchored to KG)
Compromised infrastructure	Unauthorized actuator access, legacy protocol exploits	Zero-trust verification, network segmentation, and role-based access control (RBAC)	**Device layer** (hardware security modules)
Automation bias and over trust	Authoritative tone of text, "Black-box" opacity	eXplainable artificial intelligence (XAI) visuals, raw telemetry linking, and human-in-the-loop checks	**Business layer** (natural language UI & dashboards)
Data quality and drift	Sensor noise, heterogeneous formats, evolving loads	Automated extract, transform, and load (ETL) cleaning, drift detectors, and standardized ontologies	**Information layer** (data lake and preprocessing)

API, *application programming interface;* KG, *knowledge graph;* LLM, *large language models;* RAG, *retrieval-augmented generation;* UI, *user interface.*

an LLM can also look at streams of sensor and control data to make short reports (Huang et al., 2024). Furthermore, LLM-based early warning systems have improved prediction accuracy by revealing subtle consumption patterns preceding safety incidents (Wang et al., 2024).

20.5 Challenges and considerations in artificial intelligence integration

Incorporating ICT technologies and DTs into energy management presents a varicty of challenges and prospects that span both technical and human factors (Adel, 2023). Based on the insights gathered from the previous discussion, the subsequent sections will explore key challenges in DT adoption within energy systems: ensuring model interpretability and trustworthiness; addressing data quality, privacy, and security issues; and integrating AI with existing grid systems. Table 20.6 defines the security framework by systematically mapping each *risk* to its specific attack *vector* (the pathway of exploitation) and the corresponding *control* required for mitigation. Moreover, it identifies the specific *architectural layer* within the SGAM framework (see Section 20.3 and Fig. 20.1) responsible for enforcing these safeguards.

20.5.1 Data quality, privacy, and security concerns

DT-based systems rely on a continuous data stream between the physical asset and its virtual replica. The quality of this data is critical, as any flaws can significantly impact the accuracy of the models. Ensuring high data quality is particularly challenging due to the heterogeneity of data sources in the smart grid domain. Data may be generated in various formats and schemas (e.g., different IoT devices using custom JavaScript Object Notation (JSON) or extensible markup language (XML) structures), requiring extensive transformation and preprocessing before integration. Managing these data streams is non-trivial, as they often contain duplicate records, missing values, time misalignments, and other inconsistencies (Aghazadeh Ardebili et al., 2024).

Data governance is a critical aspect of a DT-based system. Smart grid systems process sensitive data, including consumer behavior, distributed energy resources, and grid operations, and the integration of LLMs introduces additional privacy concerns. Often, AI models are executed on centralized cloud-based infrastructure, meaning that information needs to be uploaded to external services, which may result in multi-territorial data transfers. Adherence to data protection laws becomes critical. The real-world application must therefore protect customer data using strict access controls, anonymization, and aggregation. Some solutions currently in research use data minimization, encryption, and FL, where data stays local, and only model parameters are shared (Mchirgui et al., 2024). On the data quality side, AI ethics research has noted that LLMs can generate "hallucinations," or plausible but false outputs, by combining information learned during training with inaccurate assertions (Mittelstadt et al., 2023). Because of this possibility, AI outputs need extensive human supervision before they are used for critical decision-making purposes.

LLM models remain exposed to potential vulnerabilities. Malicious sensor data or manipulated LLM queries could manipulate AI systems. Recent studies verify the danger posed by these attacks and underscore the importance of defensive strategies, such as anomaly detection. These findings highlight the importance of meticulous security measures and care in handling these systems to ensure the confidentiality, integrity, and availability of system data, as well as the outputs of LLM (Alsaigh et al., 2023).

20.5.2 Model interpretability and trustworthiness

Human factors, or the cognitive, social, and organizational components that influence how people interact with systems and make decisions, must be considered for the effective deployment of DT technology in energy management (Neumann et al., 2021). To optimize the potential of DTs, it is mandatory to analyze how these systems can be incorporated into human workflows. As these systems become more pervasive, non-expert end users are expected to engage collaboratively with DT applications, supervising their performance and outputs in specific tasks. Power grids are safety-critical systems, so any AI-generated outcome must be understandable to human operators and engineers. Models like deep neural networks and LLMs are often considered as "black boxes," since the method used to produce their results is not transparent to users. In contrast, traditional grid control algorithms are explicitly programmed and thus inherently interpretable. Grid operators are unlikely to act on an AI-generated recommendation without a clear rationale, given the liability and safety implications (Alsaigh et al., 2023).

Integrating LLMs into smart grids DT requires reliable XAI methods to ensure transparency and trust in decision-making (Adadi & Berrada, 2018). Techniques such as natural language justifications and highlighting can assist in elucidating which inputs affected the output of a model (Ieva et al., 2024). However, the reliability of LLMs is compromised by their tendency to hallucinate. Such errors could have severe operational consequences. Furthermore, operators may overtrust LLM-generated responses because of their tone, which is akin to that of an informed data source, even in cases where they are inaccurate. A further approach involves generating confidence scores for responses, allowing operators to gauge the reliability of recommendations before acting on them. Another strategy is to expose the retrieved context snippets used in a RAG pipeline, enabling engineers to verify which documents or data sources informed the model's reasoning. Additionally, for safety-critical sub-tasks, smaller and more interpretable models, such as rule-based classifiers or decision trees, can be deployed alongside the LLM to provide traceable, human-understandable justifications. Another aspect is the UI design, which plays a vital role in facilitating interaction with the platform itself. Interfaces must be intuitive and present relevant information and predictions in an accessible way, that is customized according to the user's needs and skill level (Nardo et al., 2020). This requires developing transparent validation mechanisms, clearly communicating uncertainties, and incorporating user feedback to improve model comprehensibility, particularly for non-expert users (Bonney et al., 2023). DT systems usually require extensive training and educational programs. Understanding the system's strengths and weaknesses is important for users who want to make good use of its features, solve problems, and make informed choices (Molino et al., 2020).

20.5.3 Legacy infrastructures

In the context of legacy grid infrastructure, there are obstacles to data interoperability, system architecture, security, and operational continuity. Legacy systems, including supervisory control and data acquisition (SCADA) systems and substation equipment, often rely on proprietary protocols, limited instrumentation, and fixed-function logic, producing often low-quality and heterogeneous data that is not suitable for ML analytics without significant data-preprocessing (Mchirgui et al., 2024). A "wrap-and-extend" strategy offers a sensible compromise, enabling the extension of the grid infrastructure without replacing existing hardware. The process involves the gradual integration with the current architecture by deploying supplementary sensors and actuators and developing layers of middleware and protocol translation (Veichtlbauer et al., 2017). This has the drawback of adding more complexity and requiring a significant amount of engineering work to safeguard against inaccurate data interpretation. As discussed

TABLE 20.7 Deployment readiness matrix: From implementation to research.

Readiness	Technology	Implementation focus
Ready (*deploy now*)	RAG with knowledge graphs	Integration of commercial LLMs with graph databases to query maintenance manuals and static asset data. Focus on reducing hallucinations.
Emerging (*pilot phase*)	Federated Digital Twins	Deploying local anomaly detection models on edge gateways. Aggregating weights, not data, to handle privacy across operator boundaries. Requires robust aggregation logic.
Emerging (*pilot phase*)	Multimodal LLMs	Extending diagnostics by ingesting heterogeneous data sources alongside standard telemetry and data streams.
Research (*3–5 years*)	Generative world models	End-to-end differentiable simulators that learn grid physics from raw observation. Used for autonomous "self-driving" grid control agents and complex counterfactual reasoning.

LLMs, *large language models;* RAG, *retrieval-augmented generation*

earlier in this section, this point is critical since processing inaccurate data can significantly impact the accuracy and effectiveness of predictive models and analytics.

Moreover, many legacy systems are constructed using outdated hardware and software, raising many security concerns (Amanlou et al., 2025). In the presence of discontinued or unmaintained software, systems are no longer protected against emerging threats (e.g., hardware trojan attacks (Massaro et al., 2025)) and zero-day vulnerabilities. Furthermore, outdated devices often implement obsolete authentication protocols, if any, exposing them to malicious actors. These are the ideal targets for malicious users, who are aware that any vulnerabilities that are discovered will likely remain unpatched. This presents the opportunity to gain unauthorized access to sensor data and, worse, actuators that can disrupt critical system functionality. System security is a critical concern, especially when dealing with grid infrastructure. The literature presents some strategies when dealing with this kind of challenge. The defense-in-depth approach uses many layers of security, like firewalls, intrusion detection systems, and fine-grained network segmentation, to ensure that even if one layer is compromised, the others can still maintain the whole system's safety (Mosteiro-Sanchez et al., 2020). On the other hand, zero-trust architectures operate under the principle that no user or device should be trusted by default, requiring continuous identity verification and access control for every interaction, regardless of network location (Teerakanok et al., 2021).

20.6 Future trends and research directions

The convergence of DT platforms with next-generation AI methods is expected to redefine the ecosystem of smart grid management. Table 20.7 categorizes the main technologies discussed in this chapter by operational maturity, outlining a roadmap from immediate implementation to long-term research directions. Emerging technologies are not only expanding the functional scope of DTs but also transforming their role from passive systems to proactive, context-aware reasoning systems. In this evolving scenario, research is progressively shifting from monolithic architectures toward more modular, adaptive, and learning-enabled infrastructures. Future developments will increasingly explore the integration of learning-based simulation models and decentralized cognitive agents to strengthen the decision-making capabilities of DTs in energy systems. On one side, extending LLMs to *large multimodal models* (Huang et al., 2024) and *world models* (Matsuo et al., 2022) can allow DTs to reason over heterogeneous data sources and simulate system evolution over time. Basically, a DT operates as a domain-specific world model, providing a structured digital representation of physical assets, processes, and their interactions. Traditionally, such models have been largely pre-programmed, designed to simulate expected system behavior under predefined scenarios. The current evolution, however, is to enhance these DTs with AI capabilities that make the model learnable and predictive. This shift allows the DT to move beyond static simulation toward adaptive reasoning and forecasting. On the other hand, collaborative AI approaches enable distributed agents to cooperate, adapt, and coordinate actions across complex energy infrastructures. The following sections examine these directions in depth, outlining their potential impact on autonomy, scalability, and responsiveness in next-generation smart grids.

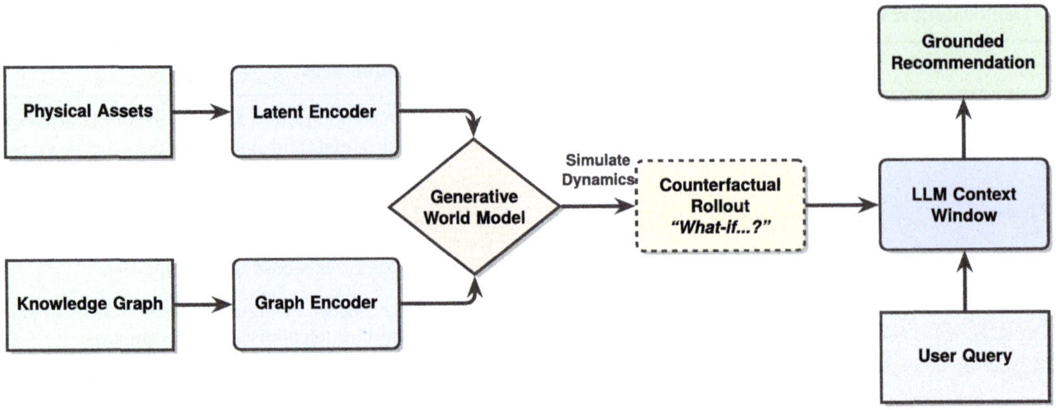

FIGURE 20.5 Conceptual architecture of a world model-enhanced Digital Twin. (A) Multimodal encoders process real-time sensor data to learn latent system dynamics. (B) These are aligned with the semantic knowledge graph. (C) The world model generates counterfactual rollouts that are injected as context into the LLM's pipeline. *LLM*, Large language models.

20.6.1 Multimodal models and world models for smart grids

In smart grid environments, the ability to predict, simulate, and understand system behavior is increasingly dependent on models that go beyond raw data capture. *World models* (Matsuo et al., 2022), originally introduced for reinforcement learning and autonomous control, represent an emerging direction for enabling predictive simulation and planning in complex, data-rich environments. These models learn latent representations of a system's dynamics and underlying physical laws from observed interactions, enabling agents to simulate possible futures in a coherent way without directly querying the environment. This principle, applied in agent-based planning frameworks such as *MuZero* (Schrittwieser et al., 2020), has strong potential for integration within DT systems.

Recent approaches extend the concept of world models to support multimodal input, allowing the integration of diverse data sources such as time-series measurements, geospatial information, maintenance logs, and visual feeds (Lin et al., 2024). These models offer a unified representation that enhances the interpretability of grid dynamics and supports holistic reasoning across different operational domains. By leveraging latent imagination and sequence modeling techniques (Hafner et al., 2020), world models enable DTs to anticipate future states, evaluate the outcomes of hypothetical actions, and support simulation-based analysis in scenarios with limited observability or high operational risk. When used in conjunction with semantic representations, world models can augment retrieval workflows by generating predictive or counterfactual knowledge that enriches user interaction. This information can be passed as contextual input to LLMs through RAG pipelines. This combination generates *predictive RAG architectures grounding their responses not only in static knowledge but also on dynamic data-driven inferences*, enabling the generation of explanatory and action-oriented responses aiming to anticipate system behavior. The integration of multimodal world models into DT architectures thus represents a promising avenue for proactive energy infrastructure management. It supports the development of predictive decision support tools, allows scenario-based training, and paves the way for cognitive control agents capable of operating within adaptive and uncertain environments. Fig. 20.5 illustrates a proposed pilot implementation framework for a world model-enhanced DT. This pipeline begins by learning latent system dynamics from high-frequency time-series data through the employment of multimodal encoders. These numerical representations are subsequently aligned with the grid KG to ensure interpretability. Once trained, the model functions as a simulation engine capable of generating counterfactual scenarios (e.g., "*What-if feeder x trips during peak load?*"). These outcomes are injected into the LLM's context window as retrieved artifacts. This integration enables the generative agent to reason about hypothetical future states, grounding its recommendations in physics-compliant simulations rather than relying solely on statistical language patterns.

20.6.2 Collaborative artificial intelligence systems for distributed energy networks

As energy infrastructures evolve toward decentralized cyber-physical architectures, there is growing interest in collaborative AI approaches capable of operating across distributed components of the grid. Unlike traditional centralized intelligence pipelines, collaborative systems enable local agents – embedded at the edge of the network – to perform learning, inference, and coordination tasks based on shared objectives and partial observability. This distributed paradigm aligns well with the needs of smart grids, where edge devices, substations, and microgrids must often make autonomous decisions while remaining aligned with overall energy goals (Munir et al., 2021). A key enabler of this shift is *FL*, which allows distributed nodes to collaboratively train shared models without exchanging raw data (Sun et al., 2020). This paradigm is particularly suited to DTs that integrate multiple stakeholders and heterogeneous devices while preserving privacy and reducing communication overhead (Mughal et al., 2024). These systems improve model personalization, scalability, and resilience to *data drift* and *concept drift* (Wahab, 2022). In particular, data drift refers to the change in the statistical properties of input data over time, while concept drift refers to changes in the underlying input-output relationships as systems evolve; both can lead to a degradation in the performance of ML models. Since most models are trained under the assumption that training and inference data follow the same distribution, any shift caused by evolving user behavior, external conditions, or changes in data collection processes may result in inaccurate predictions. To implement this collaborative approach in the real-world, federated DTs adopt hierarchical aggregation topologies where edge nodes (substations or microgrids) train local anomaly detection models and transmit only compressed weight updates to a central coordinator (the primary substation). This architecture explicitly addresses scenarios where data is non-independent and identically distributed – across varying locations and/or temporally due to drifts – by employing client re-weighting strategies and periodic teacher-guided distillation on proxy datasets (Moyeen et al., 2025). Furthermore, to account for intermittent connectivity and "straggler" devices, asynchronous aggregation protocols are employed to ensure global model convergence without stalling the network (Yang et al., 2023).

Beyond offline training, recent work proposes service-oriented architectures and dynamic edge orchestration mechanisms (Loseto et al., 2022) that distribute intelligence across nodes, enabling runtime task reallocation and adaptive model deployment (Loconte et al., 2024). Frameworks such as edge-as-a-service (EaaS) introduce modular coordination among agents, allowing for real-time adjustments to network load, topology, or operational constraints (Zhang et al., 2022). Multi-agent deep reinforcement learning also offers a promising strategy for decentralized energy management. In this context, agents collaboratively learn energy scheduling and demand response policies using shared reward functions that reflect grid-wide objectives (Munir et al., 2021). This approach supports long-term optimization without requiring centralized supervision. To realize these collaborative AI strategies, semantic and operational interoperability are crucial. Ontology-based task modeling and lightweight edge reasoning engines (Ruta et al., 2022) enable agents to dynamically discover, interpret, and execute actions based on current grid conditions. Recent frameworks stress the importance of scalable, secure coordination mechanisms to preserve user trust and support regulatory compliance (Abdelmoniem et al., 2024). Taken together, these trends signal a shift toward hybrid systems where localized autonomy is balanced by collective learning and decision-making, making distributed energy systems more adaptive, robust, and responsive to real-world variability.

20.7 Conclusion

This chapter has illustrated novel data-driven and knowledge-based energy DT trends. The discussion has been framed by means of a reference architecture model, integrating: (1) data stream gathering from sensing devices; (2) data stream mining and ML for predictive maintenance and analytics; (3) an ontology-based KG combining annotated data, background domain knowledge and ML outcomes; (4) RAG exploiting the KG to enrich a foundation LLM, leveraged to provide a conversational virtual agent for assisting users in making decisions on energy infrastructure performance and maintenance management; (5) a natural user interface combining conversational interaction with geospatial information presentation based on 3D models of infrastructure assets in a geographic information system. All these functions are supported by a cohesive, scalable cloud-edge microservice architecture, characterized by functional tiers and by cross-layers dedicated to device support, data integration, and cybersecurity. A prototype of the reference framework has been developed, leveraging commercial off-the-shelf components for a case study regarding the management of a section of the energy distribution infrastructure in central Italy. Currently, the platform only supports Italian as the reference language. The system is mainly conceived for Italian companies in the energy sector, and several documents and regulations referenced in the platform are only available in Italian. Using the local language facilitated easier text interpretation, also enhancing clarity and understanding of the platform functionalities. However, as part of future work, multi-language support is being considered and will be developed as an ongoing activity. This will enable the platform to accommodate users from diverse linguistic backgrounds, further expanding its accessibility and usability.

Future work also regards expanding the prototype into a complete implementation and using it for pilot trials in production environments. This will allow gathering feedback for the assessment of critical aspects of DT implementation, including: (1) performance and scalability; (2) ease of integration with a broad range of legacy systems; (3) user interaction effectiveness; (4) acceptability and trust of the system from operators and managers. Further investigation will concern improving and expanding ML methods and applications, with a focus on XAI approaches to further increase system trust, semantic representation refinement for improving RAG effectiveness for the conversational virtual assistant, and enhancing asset control and scenario simulation use cases within the framework. Another important direction for future work concerns the evaluation of the economic impact and return on investment (ROI) of implementing the proposed AI-enhanced DT systems. While the technical feasibility and effectiveness of the approach are crucial, industry adoption largely depends on the ability to demonstrate tangible economic benefits, such as cost reduction, performance optimization, or efficiency gains. Developing methodologies to systematically assess ROI through quantitative metrics, comparative case studies, or cost-benefit analyses would provide valuable insights for decision-makers and strengthen the case for large-scale deployment.

References

Abdelmoniem, A. M., Abdelmoniem, A. M., Jaber, M., Anwar, A., Zhang, Y., & Gao, M. (2024). Towards a decentralized collaborative framework for scalable edge AI. *Future Internet, 16*(11), 421.

Adadi, A., & Berrada, M. (2018). Peeking inside the black-box: A survey on explainable artificial intelligence (XAI). *IEEE Access, 6*, 52138–52160.

Adel, A. (2023). Unlocking the future: Fostering human-machine collaboration and driving intelligent automation through industry 5.0 in smart cities. *Smart Cities, 6*(5), 2742–2782.

Aghazadeh Ardebili, A., Zappatore, M., Ramadan, A. I. H. A., Longo, A., & Ficarella, A. (2024). Digital Twins of smart energy systems: A systematic literature review on enablers, design, management and computational challenges. *Energy Informatics, 7*(1), 94.

Ahmad, T., Madonski, R., Zhang, D., Huang, C., & Mujeeb, A. (2022). Data-driven probabilistic machine learning in sustainable smart energy/smart energy systems: Key developments, challenges, and future research opportunities in the context of smart grid paradigm. *Renewable and Sustainable Energy Reviews, 160*, 112128.

Ahmadzadeh, S., Parr, G., & Zhao, W. (2021). A review on communication aspects of demand response management for future 5G IoT-based smart grids. *IEEE Access, 9*, 77555–77571.

Alasali, F., Itradat, A., Abu Ghalyon, S., Abudayyeh, M., El-Naily, N., Hayajneh, A. M., & AlMajali, A. (2023). Smart grid resilience for grid-connected PV and protection systems under cyber threats. *Smart Cities, 7*(1), 51–77.

Alimi, O. A., Ouahada, K., & Abu-Mahfouz, A. M. (2020). A review of machine learning approaches to power system security and stability. *IEEE Access, 8*, 113512–113531.

Alotaibi, I., Abido, M. A., Khalid, M., & Savkin, A. V. (2020). A comprehensive review of recent advances in smart grids: A sustainable future with renewable energy resources. *Energies, 13*(23), 6269.

Alsaigh, R., Mehmood, R., & Katib, I. (2023). AI explainability and governance in smart energy systems: a review. *Frontiers in Energy Research, 11*, 1071291.

Amanlou, S., Hasan, M. K., Asma'Mokhtar, U., Malik, K. M., Islam, S., Khan, S., … Khan, M. A. (2025). Cybersecurity challenges in smart grid systems: Current and emerging attacks, opportunities, and recommendations. *IEEE Open Journal of the Communications Society, 6*, 1965–1997.

Arcas, G. I., Cioara, T., Anghel, I., Lazea, D., & Hangan, A. (2024). Edge offloading in smart grid. *Smart Cities, 7*(1), 680–711.

Bayindir, R., Colak, I., Fulli, G., & Demirtas, K. (2016). Smart grid technologies and applications. *Renewable and Sustainable Energy Reviews, 66*, 499–516.

Bizer, C., Heath, T., & Berners-Lee, T. (2023). Linked data-the story so far. *Linking the world's information: Essays on Tim Berners-Lee's invention of the world wide web* (pp. 115–143).

Bonney, M. S., de Angelis, M., Dal Borgo, M., & Wagg, D. J. (2023). Contextualisation of information in digital twin processes. *Mechanical Systems and Signal Processing, 184*, 109657.

Cao, K., Liu, Y., Meng, G., & Sun, Q. (2020). An overview on edge computing research. *IEEE Access, 8*, 85714–85728.

Chen, C., Zhao, K., Leng, J., Liu, C., Fan, J., & Zheng, P. (2025). Integrating large language model and digital twins in the context of industry 5.0: Framework, challenges and opportunities. *Robotics and Computer-Integrated Manufacturing, 94*, 102982.

Cheng, Y., Zhao, H., Zhou, X., Zhao, J., Cao, Y., Yang, C., & Cai, X. (2025). A large language model for advanced power dispatch. *Scientific Reports, 15*(1), 8925.

Choi, S.L., Jain, R., Feng, C., Emami, P., Zhang, H., Hong, J., Kim, T., Park, S., Ding, F., Baggu, M., & Kroposki, B. (2024). Generative AI for power grid operations. National Renewable Energy Laboratory (NREL). https://research-hub.nlr.gov/en/publications/generative-ai-for-power-grid-operations/.

Cronin, J., Anandarajah, G., & Dessens, O. (2018). Climate change impacts on the energy system: a review of trends and gaps. *Climatic Change, 151*, 79–93.

Cyganiak, R., Wood, D., & Lanthaler, M. (2014). RDF 1.1 concepts and abstract syntax. *W3C Recommendation*. https://www.w3.org/TR/rdf11concepts/.

Dietz, M., & Pernul, G. (2020). Digital twin: Empowering enterprises towards a system-of-systems approach. *Business & Information Systems Engineering, 62*, 179–184.

Dileep, G. (2020). A survey on smart grid technologies and applications. *Renewable Energy, 146*, 2589–2625.

Do Amaral, J. V. S., dos Santos, C. H., Montevechi, J. A. B., & De Queiroz, A. R. (2023). Energy Digital Twin applications: A review. *Renewable and Sustainable Energy Reviews, 188*, 113891.

European Commission. AI and generative AI: Transforming Europe's electricity grid for a sustainable future. https://digital-strategy.ec.europa.eu/en/library/ai-and-generative-ai-transformingeuropes-electricity-grid-sustainable-future. Accessed: 2025-05-22. 2024.

Gautam, A., Aryal, M. R., Deshpande, S., Padalkar, S., Nikolaenko, M., Tang, M., & Anand, S. (2025). IIoT-enabled digital twin for legacy and smart factory machines with LLM integration. *Journal of Manufacturing Systems, 80*, 511–523.

Gramegna, F., Bilenchi, I., Loseto, G., Manco, F., Mastrototaro, G., Scioscia, F., & Ruta, M. (2025). Integrating large language models into DataDriven frameworks for smart meter analytics. IEEE international conference on systems, man, and cybernetics (SMC). Vienna, Austria, 2025, pp. 6093-6096.IEEE, NY, USA.

Hafner, D., Lillicrap, T., Ba, J., & Norouzi, M. (2020). *Dream to control: Learning behaviors by latent imagination*. NY, USA: Curran Associates.

He, P., Liu, X., Gao, J., & Chen, W. (2021). DeBERTa: Decoding-enhanced BERT with Disentangled Attention. 9th international conference on learning representations, ICLR 2021, virtual event, Austria, online.

Huang, C., Li, S., Liu, R., Wang, H., & Chen, Y. (2024). Large foundation models for power systems. *IEEE power & energy society general meeting* (pp. 1–5). NY, USA: IEEE.

Huang, D., Yan, C., Li, Q., & Peng, X. (2024). From large language models to large multimodal models: A literature review. *Applied Sciences, 14*(12), 5068.

Ibrahim, N., & Kashef, R. (2025). Exploring the emerging role of large language models in smart grid cybersecurity: A survey of attacks, detection mechanisms, and mitigation strategies. *Frontiers in Energy Research, 13*, 1531655.

Ieva, S., Loconte, D., Loseto, G., Ruta, M., Scioscia, F., Marche, D., & Notarnicola, M. (2024). A retrieval-augmented generation approach for data-driven energy infrastructure digital twins. *Smart Cities, 7*(6), 3095–3120.

Ji, Z., Lee, N., Frieske, R., Yu, T., Su, D., Xu, Y., Ishii, E., Bang, Y.J., Madotto, A. & Fung, P. (2023). Survey of hallucination in natural language generation. ACM Computing Surveys, 55(12). doi:10.1145/3571730. https://doi.org/10.1145/3571730.

Joseph, A., & Balachandra, P. (2020). Smart grid to energy internet: A systematic review of transitioning electricity systems. *IEEE Access, 8*, 215787–215805.

Ke, Y. H, Jin, L., Elangovan, K., Abdullah, H. R., Liu, N., Sia, A. T. H., ... Kuo, C. (2025). Retrieval augmented generation for 10 large language models and its generalizability in assessing medical fitness. *npj Digital Medicine, 8*(1), 187.

Kegel, L., Hahmann, M., & Lehner, W. (2017). *Generating what-if scenarios for time series data* (pp. 1–12). NY, US: Association for Computing Machinery.

Lehmann, J., Isele, R., Jakob, M., Jentzsch, A., Kontokostas, D., Mendes, P. N., ... Auer, S. (2015). Dbpedia-a large-scale, multilingual knowledge base extracted from wikipedia. *Semantic Web, 6*(2), 167–195.

Lefrançois, M. (2017). Planned ETSI SAREF extensions based on the W3C&OGC SOSA/SSN-compatible SEAS ontology patterns. Proceedings of workshop on semantic interoperability and standardization in the IoT. SIS-IoT. CEUR Workshop Proceedings.

Lehner, D., Zhang, J., Pfeiffer, J., Sint, S., Splettstößer, A. K., Wimmer, M., & Wortmann, A. (2025). Model-driven engineering for digital twins: A systematic mapping study. *Software and systems modeling, 24*, 1339–1377.

Lewis, P., Perez, E., Piktus, A., Petroni, F., Karpukhin, V., Goyal, N., ... Rocktäschel, T. (2020). Retrieval-augmented generation for knowledge-intensive NLP tasks. *Advances in Neural Information Processing Systems, 33*, 9459–9474.

Lin, J., Du, Y., Watkins, O., Hafner, D., Abbeel, P., Klein, D., & Dragan, A. (2024). Learning to model the world with language. Proceedings of the 41st international conference on machine learning, ICML'24. Vienna, Austria: JMLR.org.

Liu, S., Lu, Y., Shen, X., & Bao, J. (2023). A digital thread-driven distributed collaboration mechanism between digital twin manufacturing units. *Journal of Manufacturing Systems, 68*, 145–159.

Loconte, D., Ieva, S., Pinto, A., Loseto, G., Scioscia, F., & Ruta, M. (2024). Expanding the cloud-to-edge continuum to the IoT in serverless federated learning. *Future Generation Computer Systems, 155*, 0167–739X.

Loseto, G., Scioscia, F., Ruta, M., Gramegna, F., Ieva, S., Fasciano, C., ... Loconte, D. (2022). Osmotic cloud-edge intelligence for IoT-based cyber-physical systems. *Sensors, 22*(6), 2166.

Mahadik, S., Gedam, M., & Shah, D. (2025). Environment sustainability with smart grid sensor. *Frontiers in Artificial Intelligence, 7*, 1510410.

Manickam, R., Vollmar, J., & Prabhakar, G. M. (2023). User experience–digital twin maturity model (UX-DTMM). *International conference on research into design* (pp. 877–889). Springer, Singapore.

Massaoudi, M., Abu-Rub, H., Refaat, S. S., Chihi, I., & Oueslati, F. S. (2021). Deep learning in smart grid technology: A review of recent advancements and future prospects. *IEEE Access, 9*, 54558–54578.

Massaro, A., Epicoco, N., Loseto, G., & Ardito, C. A. (2025). AI-based detection and simulation of hardware trojan attacks in industrial applications. *IEEE Access, 13*, 50967–50979.

Matsuo, Y., LeCun, Y., Sahani, M., Precup, D., Silver, D., Sugiyama, M., Uchibe, E. & Morimoto, J. (2022). Deep learning, reinforcement learning, and world models. Neural Networks, 152, 267–275.

Mchirgui, N., Quadar, N., Kraiem, H., & Lakhssassi, A. (2024). The applications and challenges of digital twin technology in smart grids: A comprehensive review. *Applied Sciences, 14*(23), 10933.

Minh, Q. N., Nguyen, V. H., Quy, V. K., Ngoc, L. A., Chehri, A., & Jeon, G. (2022). Edge computing for IoT-enabled smart grid: The future of energy. *Energies, 15*(17), 6140.

Mittelstadt, B., Wachter, S., & Russell, C. (2023). To protect science, we must use LLMs as zero-shot translators. *Nature Human Behaviour, 7*(11), 1830–1832.

Molino, M., Cortese, C. G., & Ghislieri, C. (2020). The promotion of technology acceptance and work engagement in Industry 4.0: From personal resources to information and training. *International Journal of Environmental Research and Public Health, 17*(7), 2438.

Mosteiro-Sanchez, A., Barcelo, M., Astorga, J., & Urbieta, A. (2020). Securing IIoT using defence-in-depth: towards an end-to-end secure industry 4.0. *Journal of Manufacturing Systems, 57*, 367–378.

Motlagh, N. H., Mohammadrezaei, M., Hunt, J., & Zakeri, B. (2020). Internet of Things (IoT) and the energy sector. *Energies, 13*(2), 494.

Motlagh, N. H., Zaidan, M. A., Lovén, L., Fung, P. L., Hänninen, T., Morabito, R., ... Tarkoma, S. (2024). Digital Twins for smart spaces-beyond IoT analytics. *IEEE Internet of Things Journal, 11*(1), 573–583. https://doi.org/10.1109/JIOT.2023.3287032.

Moyeen, M. A., Kaur, K., Agarwal, A., Manzano, S. R., Zaman, M., & Goel, N. (2025). Fed challenger: A robust challenge-response and aggregation strategy to defend poisoning attacks in federated learning. *IEEE Access, 13*, 153866–153885.

Mughal, F. R., He, J., Das, B., Dharejo, F. A., Zhu, N., Khan, S. B., & Alzahrani, S. (2024). Adaptive federated learning for resource-constrained IoT devices through edge intelligence and multi-edge clustering. *Scientific Reports, 14*, 28746.

Munir, M. S., Abedin, S. F., Tran, N. H., Han, Z., Huh, E. N., & Hong, C. S. (2021). Risk-aware energy scheduling for edge computing with microgrid: A multi-agent deep reinforcement learning approach. *IEEE Transactions on Network and Service Management, 18*(3), 3476–3497.

Nardo, M., Forino, D., & Murino, T. (2020). The evolution of man-machine interaction: The role of human in Industry 4.0 paradigm. *Production & Manufacturing Research, 8*(1), 20–34.

Neumann, W.P., Winkelhaus, S., Grosse, E.H., & Glock, C.H. (2021). Industry 4.0 and the human factor-A systems framework and analysis methodology for successful development. *International Journal of Production Economics, 233*, 107992.

Nguyen, T. N., & Gonzalez, C. (2020). *Effects of decision complexity in goal-seeking gridworlds: A comparison of instance-based learning and reinforcement learning agents*. PA: University Park.

Panda, D. K., & Das, S. (2021). Smart grid architecture model for control, optimization and data analytics of future power networks with more renewable energy. *Journal of Cleaner Production, 301*, 126877.

Patil, R., & Gudivada, V. (2024). A review of current trends, techniques, and challenges in large language models (llms). *Applied Sciences, 14*(5), 2074.

Pourpanah, F., Abdar, M., Luo, Y., Zhou, X., Wang, R., Lim, C. P., … Wu, Q. M. J. (2022). A review of generalized zero-shot learning methods. *IEEE Transactions on Pattern Analysis and Machine Intelligence, 45*(4), 4051–4070.

Statista. Smart grid market value worldwide 2022-2028. https://www.statista.com/study/111848/smart-grids-worldwide/. 2023.

Raza, M., Jahangir, Z., Riaz, M. B., Saeed, M. J., & Sattar, M. A. (2025). Industrial applications of large language models. *Scientific Reports, 15*(1), 13755.

Roy, D., Srivastava, R., & Jat, M. (2022). A complete overview of analytics techniques: Descriptive, predictive, and prescriptive. In Decision Intelligence Analytics and the Implementation of Strategic Business Management (pp. 15–30). Springer. https://doi.org/10.1007/978-3-030-82763-2_2.

Ruta, M., Scioscia, F., Bilenchi, I., Gramegna, F., Loseto, G., Ieva, S., & Pinto, A. (2022). A multiplatform reasoning engine for the semantic web of everything. *Journal of Web Semantics, 73*, 100709.

Ruta, M., Scioscia, F., Loseto, G., Pinto, A., & Di Sciascio, E. (2019). Machine learning in the Internet of Things: A semantic-enhanced approach. *Semantic Web, 9*(6), 183–204. https://doi.org/10.3233/SW-180314.

Schleussner, C. F., Rogelj, J., Schaeffer, M., Lissner, T., Licker, R., Fischer, E. M., … Hare, W. (2016). Science and policy characteristics of the Paris Agreement temperature goal. *Nature Climate Change, 6*(9), 827–835.

Schrittwieser, J., Antonoglou, I., Hubert, T., Simonyan, K., Sifre, L., Schmitt, S., … Silver, D. (2020). Mastering Atari, Go, chess and shogi by planning with a learned model. *Nature, 588*, 604–609.

Singh, M., Fuenmayor, E., Hinchy, E. P., Qiao, Y., Murray, N., & Devine, D. (2021). Digital Twin: Origin to future. *Applied System Innovation, 4*(2), 36.

Sun, W., Lei, S., Wang, L., Liu, Z., & Zhang, Y. (2020). Adaptive federated learning and digital Twin for industrial Internet of Things. *IEEE Transactions on Industrial Informatics, 17*(8), 5605–5614.

Syed, Z.H., Trabelsi, A., Helbert, E., Bailleau, V., & Muths, C. (2021). Question answering chatbot for troubleshooting queries based on transfer learning. *Procedia Computer Science, 192*, 941–950.

Teerakanok, S., Uehara, T., & Inomata, A. (2021). Migrating to zero trust architecture: Reviews and challenges. *Security and Communication Networks, 2021*, 1–10.

van Dinter, R., Tekinerdogan, B., & Catal, C. (2022). Predictive maintenance using digital twins: A systematic literature review. *Information and Software Technology, 151*, 107008.

Vaswani, A., Shazeer, N., Parmar, N., Uszkoreit, J., Jones, L., Gomez, A.N., Kaiser, Ł., & Polosukhin, I. (2017). Attention is all you need. In Advances in Neural Information Processing Systems 30 (NeurIPS 2017).

Veichtlbauer, A., Ortmayer, M., & Heistracher, T. (2017). OPC UA integration for field devices. *IEEE 15th international conference on industrial informatics (INDIN)* (pp. 419–424). IEEE, NY, USA.

Vidivelli, S., Ramachandran, M., & Dharunbalaji, A. (2024). Efficiency-driven custom chatbot development: Unleashing LangChain, RAG, and Performance-Optimized LLM Fusion. *Computers, Materials & Continua, 80*(2), 2423–2442.

Wahab, O. A. (2022). Intrusion detection in the IoT under data and concept drifts: Online deep learning approach. *IEEE Internet of Things Journal, 9*(20), 19706–19716.

Wang, R. G., Chuang, M. L., Ke, C. Y., Chien, Y. F., Ho, W. J., Chiang, K. C., … Chou, C. C. (2024). Predicting imminent electrical safety incidents using smart meter big data with large language models. *IEEE Access, 12*, 184940–184952.

Wang, Y., Yao, Q., Kwok, J. T., & Ni, L. M. (2020). Generalizing from a few examples: A survey on fewshot learning. *ACM Computing Surveys (CSUR), 53*(3), 134.

Wei, J., Wang, X., Schuurmans, D., Bosma, M., Xia, F., Chi, E., … Zhou, D. (2022). Chain-of-thought prompting elicits reasoning in large language models. *Advances in Neural Information Processing Systems, 35*, 24824–24837.

Wolf, T., Debut, L., Sanh, V., Chaumond, J., Delangue, C., Moi, A., … Funtowicz, M. (2020). *Transformers: State-of-the-art natural language processing. Proceedings of the 2020 conference on empirical methods in natural language processing: system demonstrations* (pp. 38–45). Stroudsburg, PA, USA: Association for Computational Linguistics.

Yang, C.-W., Dubinin, V., & Vyatkin, V. (2020). Automatic generation of control flow from requirements for distributed smart grid automation control. *IEEE Transactions on Industrial Informatics, 16*(1), 403–413.

Yang, J., Zhou, Y., Wen, W., Zhou, J., & Zhang, Q. (2023). Asynchronous hierarchical federated learning based on bandwidth allocation and client scheduling. *Applied Sciences, 13*(20), 11134.

Yigit, Y., Ferrag, M. A., Ghanem, M. C., Sarker, I. H., Maglaras, L. A., Chrysoulas, C., … Janicke, H. (2025). Generative AI and LLMs for critical infrastructure protection: Evaluation benchmarks, agentic AI, challenges, and opportunities. *Sensors, 25*(6), 1666.

Zaboli, A., Kim, Y.-H., & Hong, J. (2025). An advanced generative AI-based anomaly detection in IEC61850-based communication messages in smart grids, *IEEE Access* (13, pp. 89997–90016).

Zhang, M., Cao, J., Sahni, Y., Chen, Q., Jiang, S., & Wu, T. (2022, August). Eaas: A service-oriented edge computing framework towards distributed intelligence. In 2022 IEEE International Conference on Service-Oriented System Engineering (SOSE) (pp. 165-175). IEEE Computer Society, CA, USA.

Zhang, T., Kishore, V., Wu, F., Weinberger, K. Q., & Artzi, Y. (2020). BERTScore: Evaluating text generation with BERT. 8th International Conference on Learning Representations (ICLR 2020). International conference on learning representations (ICLR), NY, USA. *Curran Associates.*

CHAPTER

21

Artificial intelligence, Digital Twin and blockchain technology for secure smart grids

Burgess Lauren[a], *Hengshuo Liang*[b], *Milon Biswas*[a], *Qianlong Wang*[c], *Chao Lu*[a], *Wei Yu*[a]

[a]Department of Computer and Information Sciences, Towson University, Towson, MD, United States [b]School of Engineering & Technology, University of Washington Tacoma, WA, United States [c]Department of Electrical and Computer Engineering, Old Dominion University, VA, United States

21.1 Introduction

Cyber-physical systems (CPSs) are a vertical architecture or framework that combines physical components with sensing, communication, control, and computational technologies. This integration allows real-time data monitoring, control, and analytics, promoting accurate decision-making based on the information gathered from sensors deployed across various locations associated with physical objects/things and responsive actions given to actuators to control the physical systems to desired behaviors. Leveraging these interconnected devices, CPS can optimize performance, improve efficiency, and enhance safety across different applications, from industrial automation to smart infrastructure (Humayed et al., 2017; Kayamboo et al., 2022; Lin et al., 2017; Rajkumar et al., 2010; Stankovic, 2014).

Meanwhile, the global electricity demand has surged dramatically in recent decades due to rapid industrialization, urbanization, and the proliferation of digital technologies. Between 1950 and 2008, energy consumption in the United States alone outpaced production by a factor of 1.5, illustrating a persistent imbalance between supply and demand (Craig & Feng, 2017). Key contributors to this escalating energy demand include the residential, industrial, commercial, and public sectors (Borenstein & Bushnell, 2015). In response to these pressures, the energy sector is undergoing a paradigm shift from conventional power grids to more intelligent, adaptive infrastructures—commonly referred to as smart grids, as an essential CPS in the energy domain (Liu et al., 2019a; Xu et al., 2017).

Generally speaking, the smart grid systems aim to modernize the electric power system by incorporating two-way communication, distributed energy resources (DERs), automation, and advanced analytics (Abdelwahab et al., 2019; Moulema et al., 2015; Venayagamoorthy et al., 2016). These systems offer improved operational efficiency, sustainability, and consumer engagement. However, the same features that make smart grids innovative—ubiquitous connectivity, real-time data exchange, and cyber-physical integration—also introduce significant vulnerabilities (Aloul et al., 2012; Liu et al., 2011; Lin et al., 2012; Yang et al., 2014, 2017). For instance, a distributed denial-of-service (DDoS) attack on a data aggregator node can disrupt grid communication and potentially trigger cascading failures in energy distribution (Huseinović et al., 2020). Such incidents underscore the urgency of developing resilient, secure smart grid architectures.

Modernizing electrical infrastructure through smart grid technologies has introduced transformative benefits, including real-time monitoring, decentralized energy integration, predictive maintenance, and enhanced energy efficiency. However, this transformation also significantly increases system complexity by interconnecting millions of heterogeneous components—smart meters, DERs, control systems, communication networks, and edge-computing nodes. This growing interdependence between cyber and physical elements has escalated the risk surface for both cyberattacks and system-level failures. Recent incidents, such as ransomware attacks on energy providers and

Digital Twin Technology and Smart Grid.
DOI: https://doi.org/10.1016/B978-0-443-36664-2.00013-7

coordinated data-spoofing campaigns, have demonstrated that legacy security protocols are inadequate for today's highly dynamic and distributed grid environments. Traditional perimeter-based defenses are insufficient to cope with threats like false data injection attacks (FDIAs), DDoS attacks, and insider manipulation, all of which can propagate through grid layers undetected. One notable example is the Ukraine Power Grid Cyberattack in December 2015 (Lee et al., 2016), which was part of a coordinated cyberattack on Ukraine's power grid that resulted in outages affecting over 230,000 people. Attackers breached internal systems via phishing, launched FDIA against *Supervisory control and data acquisition* (SCADA) controls, and employed DoS attacks to disrupt the service availability of call centers. As the multilayered attack went undetected, it highlights the failure of traditional perimeter defenses (such as firewalls and intrusion detection systems [IDS]) and the urgent need for the design and deployment of intelligent, layered security to protect critical infrastructure systems, such as the smart grid system. Thus ensuring security and resilience requires a shift toward intelligent, adaptive, and tamper-resistant architectures capable of proactively identifying, simulating, and mitigating multivector threats in real time.

This chapter aims to design and validate a unified smart grid architecture that synergistically integrates three cutting-edge technologies: artificial intelligence (AI), Digital Twin (DT), and blockchain (BC). Each addresses distinct yet complementary dimensions of grid security and intelligence in the following three aspects. First, AI contributes advanced analytics, pattern recognition, and predictive control through deep learning (DL) models, enabling early detection of anomalies and intelligent response strategies based on real-time data or historical data, depending on the application. Second, DT offers a virtual replica of physical grid components and systems in real-time, enabling high-fidelity simulations for fault diagnosis, scenario testing, and operational optimization without disrupting live systems. Third, BC ensures immutable and decentralized data logging, secure peer-to-peer energy transactions, and tamper-proof audit trails via cryptographic mechanisms and smart contracts.

To this end, this chapter proposes a secure, scalable, and fault-tolerant smart grid framework by embedding these technologies into a layer architecture from physical infrastructure to application-level intelligence. The objective extends beyond conceptual design: through a DL-based case study using phasor measurement unit (PMU) data. We demonstrate the feasibility and efficacy of our approach in a real-world scenario. Ultimately, this unified framework aims to lay the groundwork for self-adaptive, cyber-aware, and resilient energy systems. We have also reviewed several recent studies that address smart grid security through AI, DT, and BC to position our work within the existing research landscape. Our proposed architecture uniquely unifies all three components to deliver an intelligent, secure, and modular framework for smart grid operations.

To summarize, this chapter makes the following key contributions toward advancing secure and intelligent smart grid systems:

- **Foundational overview:** We provide a detailed review of AI, DT, and BC technologies and examine their individual and synergistic roles in securing smart grids based on existing literature.
- **Proposed architecture:** We design a four-layer framework building upon the original smart grid layer by integrating AI for intelligent decision-making, DT for real-time system emulation, and BC for secure data exchange and transparency.
- **Empirical evaluation:** To validate our architecture, we introduce a case study using DL models on PMU data for fault detection. The evaluation demonstrates the viability of popular AI models in real-time grid applications.
- **Research outlook:** We conclude by identifying open challenges and future directions, including studying security challenges in new layers and cross-layer interactions, developing adversarially robust AI systems, secure DT synchronization mechanisms, and privacy-preserving BC protocols.

The remainder of this chapter is structured as follows: Section 21.2 presents the technological foundations of smart grids, AI, DT, and BC, and some examples of related works in those areas. Section 21.3 describes the threat model, mitigation strategies, our proposed architecture, and new challenges in the model. Section 21.3.4 discusses security challenges and implications. Section 21.4 presents our case study and evaluation methodology. Section 21.5 outlines promising research avenues, and Section 21.6 concludes the chapter.

21.2 Background and related works

This section provides a foundational overview of the smart grid, AI, DT, and BC.

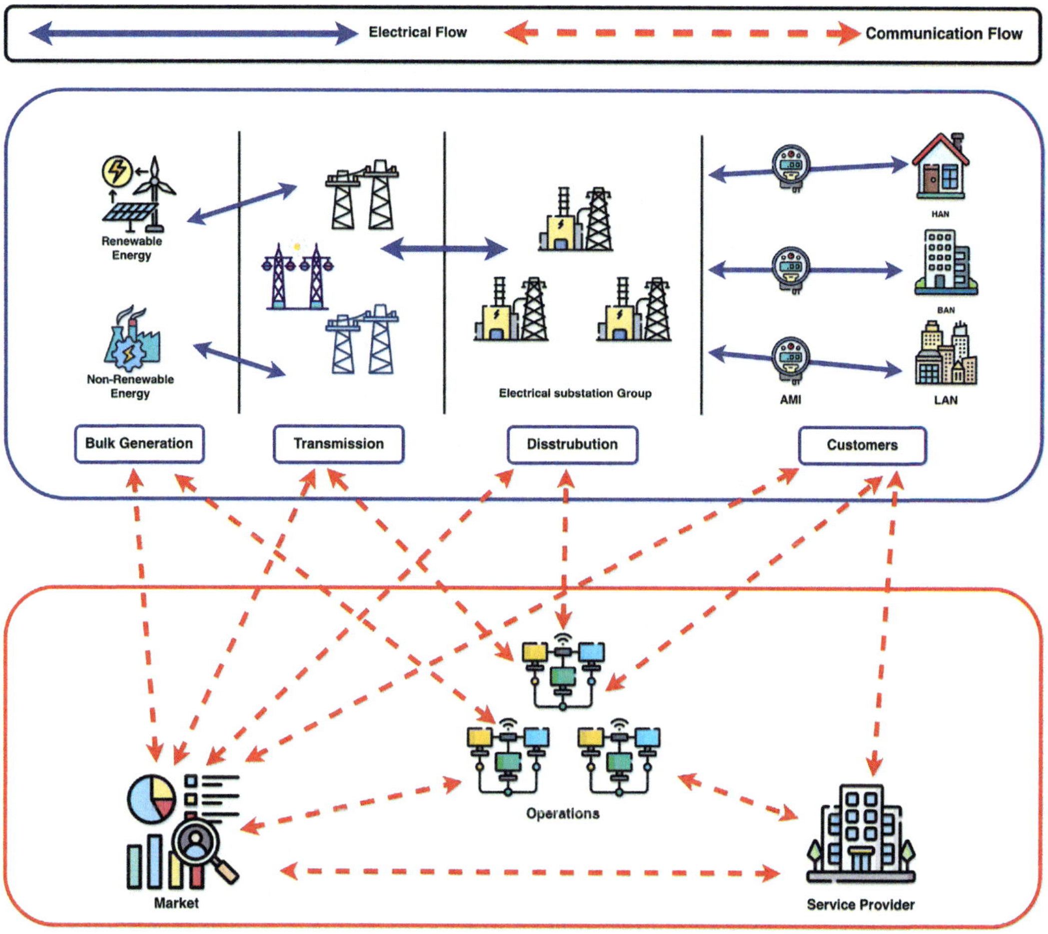

FIGURE 21.1 Conceptual architecture of a smart grid with electrical and communication flows.

21.2.1 Smart grid

CPS are physical devices integrating computing, networks, and physical environments with computing and communication algorithms. CPS provides the vertical structure of application, network communication, sensing, and physical systems. By leveraging information communication technology, the vision of CPS enables the efficient monitoring and control capability of various critical infrastructure systems across numerous application domains like energy, transportation, healthcare, and agriculture, among others (Kim et al., 2022; Levytskyi et al., 2024; Lin et al., 2015; Mulo et al., 2025; Rajkumar et al., 2010; Qian et al., 2024a). A smart grid is an energy CPS that leverages information and communications technology (sensing, control, communication, computing, and AI-driven data analytics) to enable the monitoring and control of physical grid infrastructure (Mahmood et al., 2015; Xu et al., 2016, 2017).

The smart grid's design and implementation models are still developing, with a key reference model from the US National Institute of Standards and Technology (NIST) (Greer et al., 2014) comprising seven domains as shown in Fig. 21.1, including electricity generation, transmission, distribution, customers, markets, operations, and service providers. Unlike traditional grid architecture, electrical power flows move bidirectionally in the first four domains, while the latter three manage these flows through communication channels. Customer categories include various networking systems covering homes, buildings, and industrial plants. Also, the advanced metering infrastructure (AMI) plays a crucial role in this system, facilitating the data collection of the physical grid and supporting monitoring and management capability, ensuring the system's efficiency and reliability. The bottom layer highlights smart grid operations, where communication flows connect market operators, grid control centers, and service providers. Red dashed arrows represent these bidirectional data pathways. This model illustrates energy flow from renewable and nonrenewable sources to residential, commercial, and industrial consumers, enabling effective data exchange and control.

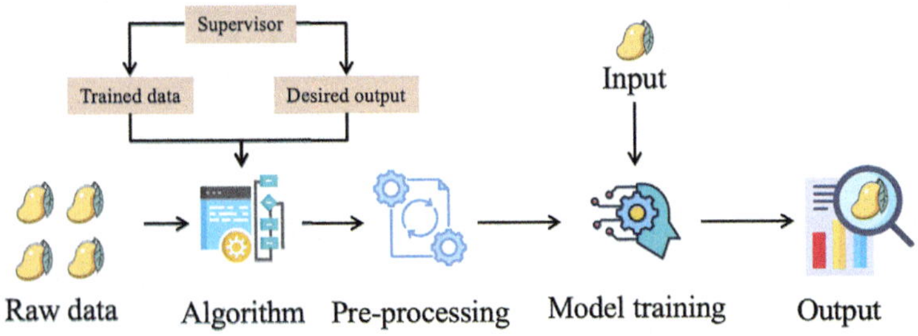

FIGURE 21.2 Supervised learning.

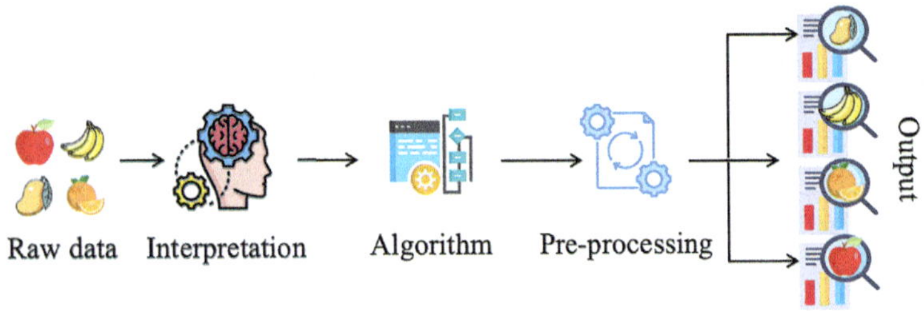

FIGURE 21.3 Unsupervised learning.

The key subsystem of a smart grid comprises: (1) *AMI:* As one key component of the smart grid, AMI consists of smart meters, communication infrastructure, and data management components, aiming to build the link of connection between consumers and system operators (Ghosal & Conti, 2019). (2) *SCADA:* It is responsible for conducting real-time monitoring and control of power transmission and distribution based on collected real-time and system-wide data (Raghunandan, 2022). With SCADA, the control centers can collect various measurements from the physical power grid. Such measurements can be further used to support multiple key functional components, including state estimation, bad/false data recognition, contingency analysis, and optimal power flow control. (3) *Wide area measurement system (WAMS):* It is deployed to enable the monitoring and control of power grid systems across wide geographical areas. As an example of WAMS, GPS satellite signals are used to synchronize the timing of PMUs deployed at critical locations in the power grid system. With PMU, real-time angle and magnitude status can be collected, transmitted to a control center, and further processed by the control center so that a greater monitoring and control capability based on the rich data received can be realized.

21.2.2 Artificial intelligence

AI, even though this is not a new idea and has been around for many years, has been reenergized by a few key factors, such as computing power, large data sets, and algorithmic advancements (Garg & Agarwal, 2019; Hatcher & Yu, 2018). Business and marketing strategies use AI to make forecasts and predictions. However, how these techniques can be used in CPS, like the smart grid, remains an open and challenging problem. We provide some fundamentals before we look into how AI can support smart grids.

Taking a closer look at the key factors, the increase in computing power through cloud services has made it feasible for individuals and organizations to access advanced machine learning (ML) systems and model training at affordable rates (Garg & Agarwal, 2019). The growth in large data sets has surpassed all previously recorded information, showing that these data sets frequently yield a higher value than sophisticated algorithms (Garg & Agarwal, 2019; Liang et al., 2019). From an algorithmic standpoint, the field has made significant advances by addressing historical challenges while strengthening the foundations and applications of ML (Garg & Agarwal, 2019; Hatcher & Yu, 2018).

AI, a term encompassing various approaches and algorithms, often brings to mind ML and DL. ML enables computers to learn from data, providing the ability to perform tasks by extracting insightful information from the data through a training process. This practicality leads to a broad range of applications across various CPS (Chen et al., 2021a, 2021b, 2024; Hatcher & Yu, 2018; Liang et al., 2019, 2020, 2022; Mohammadi et al., 2018; Musa et al., 2023a; Tian et al., 2022; Xu et al., 2020). DL focuses on neural networks that process information in layers, automatically learning a function and transforming inputs into the necessary outputs with minimal error (Esenogho et al., 2022; Hatcher & Yu, 2018).

We can categorize AI in several ways: supervised, unsupervised, and reinforcement learning (RL), as illustrated in Figs. 21.2–21.4. Supervised learning involves algorithms that can learn functions by mapping inputs to outputs from training data that is labeled (Sen et al., 2020). Taking a closer look at the various models of AI, artificial neural networks

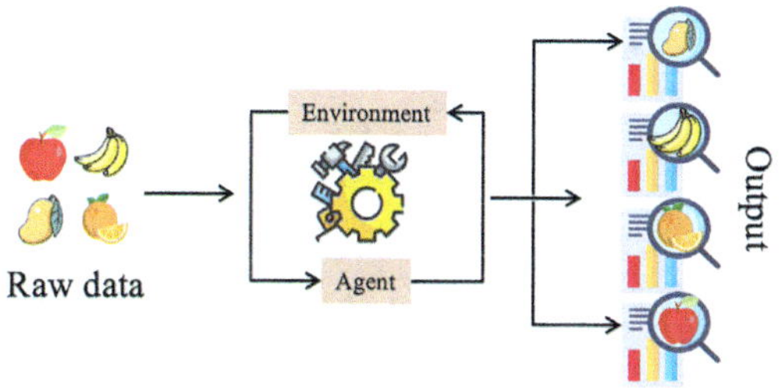

FIGURE 21.4 Reinforcement learning.

(ANNs) are often used across multiple research areas because they are based on algorithms or rules, reducing the need for substantial programming to make predictions from datasets (Bailly et al., 2022). ANNs work well for pattern recognition and image processing, enhancing security by analyzing data within the grid and triggering alerts when established patterns deviate (Shih, 2010). In addition to ANNs, some examples of DL models, *convolutional neural networks (CNNs)* is effective in dealing with spatial data such as image/video, *Recurrent neural networks (RNN)* that is efficient in processing temporal/sequential data like time-series data (long short-term memory [LSTM] is one type of RNN that is widely used in the field), *deep/stacked autoencoder* that works for visualization or mapping of input and conduct compression to maintain key features (Esenogho et al., 2022; Garg & Agarwal, 2019; Liang et al., 2019).

Unlike supervised learning, unsupervised learning focuses on the absence of labeled input samples, allowing for broader goals versus specific output classes to identify the similarities and differences in the data. Unsupervised learning techniques are well known for extracting and uncovering hidden patterns from the data (Johnston et al., 2019). Three key tasks within unsupervised learning include the reduction of input data complexity using techniques like autoencoders, clustering based on the data groups using statistical methods like fuzzy learning and image registration, and these statistical methods provide estimation of data distribution to include noise reduction in audio and analyzing traffic density with CNNs (Nubert et al., 2018).

RL (Liu et al., 2019, 2023) is different from supervised learning as there are no labels on the inputs, can often be thought of as a perception-action learning loop, and utilizes reward values to enhance decision-making even with limited knowledge or unforeseen circumstances (Arulkumaran et al., 2017). There are two examples of methods in RL: one is *Policy Search*—finding the optimal policy using gradient-based or gradient-free techniques, and the other is *Value Function Approximation*—estimating the expected rewards from action to optimize the learning process (Liu et al., 2019; Xu et al., 2014). There are numerous applications for AI within a smart grid, including the prediction of demand response (DR), anomaly detection, and grid stability (Wang et al., 2024a). For instance, forecasting is one of the key applications of AI methods, particularly in predicting power prices for various loads in DR environments. Accurate forecasts can significantly influence short-term power scheduling, enhancing services provided by aggregators and enabling customers to respond more effectively to DR signals. Furthermore, precise long-term estimates allow service providers and operators to make informed decisions regarding flexibility offerings, customer prioritization for DR, and appropriate compensation or charges (Gill et al., 2023; Liu et al., 2023).

21.2.3 Digital Twin

The DT (Singh et al., 2021) is referred to as a digital or virtual replical one based on the physical entity that can map the complex physical system via a network connected to the real-life system. This kind of network can provide a way to control, monitor, and optimize the physical entity through data dissemination. DT technology originated in aerospace, notably within NASA's Apollo program, where engineers developed physical twin systems to mirror the behavior of actual mission modules. One twin would remain on Earth as a high-fidelity simulation environment while its counterpart embarked on space missions (Barbie & Hasselbring, 2024; Gao et al., 2022; Mihai et al., 2022; Sun et al., 2023). This setup enabled engineers to diagnose, replicate, and mitigate system anomalies in real-time. However, this physical replication model was costly and logistically impractical for broader industrial deployment (Abideen et al., 2021). Despite early computational limitations, this laid the groundwork for the current generation of DTs, which rely on advances in data analytics, cloud infrastructure, IoT sensors, AI/ML algorithms, and the application in CPS. Today, DTs are regarded not only as simulation tools but as intelligent, real-time operational mirrors of physical assets, capable of interacting with human operators and autonomous systems, closely related to CPS (Musa et al., 2023b; Qian et al., 2022, 2024).

Three fundamental components (Barricelli et al., 2019) describe the DT technology in Fig. 21.5 as follows: (1) *DT Prototype:* It is used for carrying out the model illustration of the physical twin and can be considered as the design

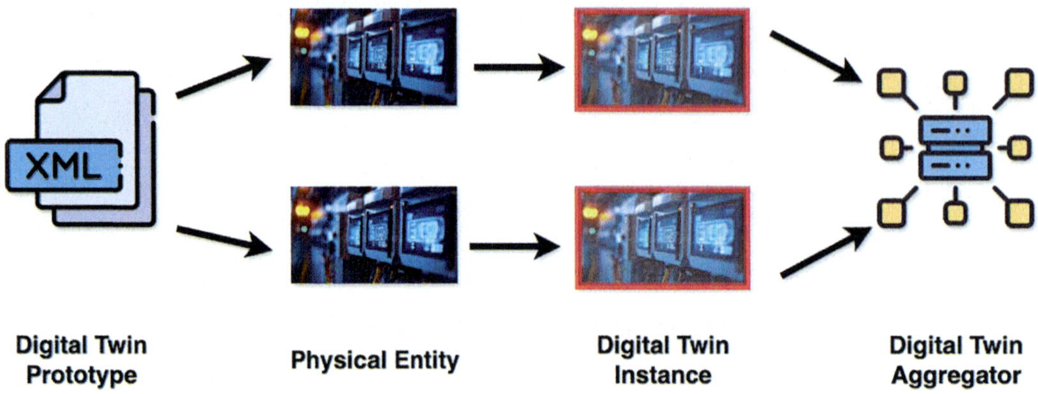

FIGURE 21.5 Correlation between the DT concepts.

interpretation of a specific entity model. For example, we can consider using the DT prototype of a smart meter as a visualized representation model for one specific smart meter design and model. (2) *DT instance:* It is considered an instance of a physical entity based on the DT prototype. The standard smart meter prototype of the smart meter model will represent the DT instance of the smart meter. (3) *DT aggregator:* It is a collection of DT instances that aggregate a collection of DTs. By definition, it represents a larger, overarching DT that integrates information from various components, systems, or processes within a defined scope. It is important to note that DT prototypes and instances can be represented as files in the form of Extensible Markup Language and JavaScript Object Notation (JSON) for enhanced data. This offers a convenient way to build and manage the DT in CPS.

Within the research side of DT, there are three models based on its integration with the physical entity at various levels; these are known as: (1) *Digital Model*—changes in either the physical or digital model do not impact the other without human intervention, as there is a manual data exchange between the two; this setup is primarily utilized for development purposes. (2) *Digital Shadow*—a unidirectional relationship between physical and digital models, where alterations in the physical model impact the digital counterpart. Nonetheless, any modification in the digital model will not impact the physical model, and (3) *Digital Twin*—enables a two-way data flow between the physical and digital realms, ensuring that any alterations made to one side are automatically reflected on the other (Barbie & Hasselbring, 2024; Fuller et al., 2020; Somma et al., 2023).

Part of the DT architecture consists of three primary pillars. The first pillar is the *Actual*, which refers to the physical smart devices currently operational and in use. The second pillar is the *Virtual*, which mirrors the configurations and settings of the Actual but exists virtually; this layer facilitates data analysis and testing and incorporates state-of-the-art technologies, including AI methodologies such as ML and DL. The third pillar, termed *Relationship or Link*, represents the connections between networks, which can be manual or automated. This component is essential for information sharing, configuration updates, and the provision of operational data for the virtualized network, with the connection type depending on the chosen model (Barbie & Hasselbring, 2024; Fuller et al., 2020; Jafari et al., 2023).

There are some existing research efforts on the research side of DT applications in smart grids. For example, Lei et al. (2023) worked to develop a DT-structured framework comprising the physical DC-DC converters and their DT replica. Their research focused on a hybrid twin modeling approach and considered cases like incorporating reference tuning, system model variation, and controller failure. One related use case is that the DT updates the model and parameters received in real time using live data to track and monitor the physical converters, ensuring that if there is a controller fault in the physical, the DT system can replace the controller to maintain normal operations instantly. Concerning fault detection, Jain et al. (2020) designed a model-based DT estimator prototype for real-time estimation of photovoltaic (PV) energy conversion unit output by comparing the outputs with the actual measurements to create error residual vectors to enable fault diagnosis. Their research focused on leveraging distributed power electronics to enable real-time fault detection and diagnosis. Using a prototype power converter to validate outdoor conditions demonstrated successful real-time analysis of various fault types, showing that their approach offers quick fault detection, is independent of installation type, adds minimal complexity, and can be applied to other energy battery-like systems.

Also, Saad et al. (2020) focused on creating an IoT-enabled DT model for voltage regulation in CPS, interfacing with control systems to enhance functionality. Their research featured cybersecurity frameworks, where the model demonstrated the capability to address isolated and synchronized cyber threats, and the NETEM tool for network

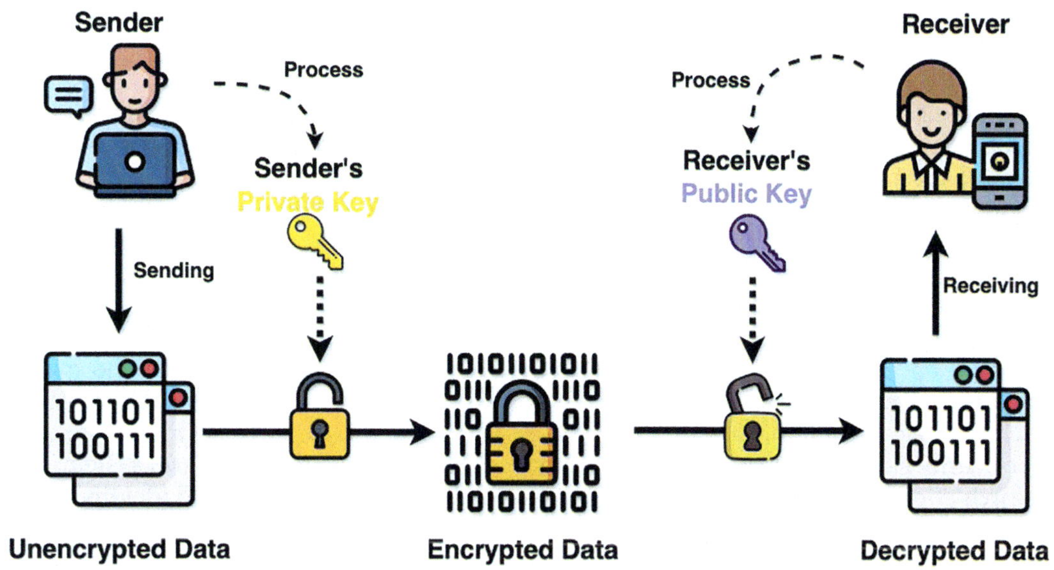

FIGURE 21.6 Asymmetric key encryption and decryption.

emulation to simulate DoS, network delays, and packet loss. Testing showed that the DT effectively mitigated the coordinated FDIs and DoS cyberattacks. Likewise, Gui et al. (2023) proposed a hierarchical control strategy for PV inverters to stabilize voltage in low-voltage electrical distribution networks featuring three tiers, aiming to enhance and optimize reactive power management from PV inverters using DT engagement with control systems. Their case studies showed improved voltage quality and increased capacity for hosting from coordinated inverter contributions within a Danish grid. Validation of their strategy was completed using a DT of the system.

21.2.4 Blockchains

BCs are a distributed infrastructure in which resources and data are distributed across multiple nodes. This allows for the integration of distributed data storage that contains transactions, known as ledgers (Varlamis et al., 2025), that are organized into blocks (Gao et al., 2018; Wang et al., 2024). These blocks then become arranged into logical chains, known as BCs, which continue to grow and are indisputable. In the decentralized world of BCs, all nodes are equal, and data is maintained collectively across the system, ensuring everyone is part of the process (Abaas et al., 2021; Chen et al., 2021c). When a node in the system becomes unavailable, the overall functionality remains intact (Mollah et al., 2020). The system uses cryptographic verification to establish trust, allowing transactions without third-party involvement. BCs are tamper-proof; they cannot be altered once data is recorded. Each new block is shared with all nodes to maintain data synchronization, enabling transactions to be tracked. This process enhances security, ensuring all nodes have the latest data, making attacks from malicious nodes difficult.

BC ensures integrity and credibility through cryptography and various consensus mechanisms. One key structure is the Merkle Tree (Mohan & Gladston, 2020), which verifies data integrity and summarizes large datasets. Consensus algorithms like Proof of Work (PoW) (Gervais et al., 2016) require significant computing power and are common in cryptocurrency, while Proof of Stake (PoS) (Saad & Radzi, 2020) reduces the time to reach consensus. Delegated Proof of Stake (DPoS) (Fan & Chai, 2018) improves efficiency by allowing selected delegates for transaction validation and new block creation. Additionally, Practical Byzantine Fault Tolerance (PBFT) (Onireti et al., 2019) is a fault-tolerant algorithm for permissioned BCs that involves known nodes and includes three phases: preprepared, prepared, and commit, to achieve consensus.

BC utilizes smart contracts, which are specified agreements defined digitally automatically, without the need for a third party to execute the terms of the contract, such as setting privacy protection to ensure user privacy, security, and trustworthiness, while being in a decentralized network (Cong & He, 2019). Smart contracts are encrypted by asymmetric encryption (Zhai et al., 2019) during data dissemination, also known as public-key cryptography, shown in Fig. 21.6. These security methods are key-based and have two paired keys: a public key and a private key, where the data encrypted by a public key can only be decrypted by its corresponding private key. For BCs, the transaction signatures are uniquely tied to transaction content, ensuring that even the same private key yields different signatures for different contents, providing an advantage over manual signing methods.

There are some existing research efforts on applying BC to the smart grid, including Abaas et al. (2021) and Kumar and Chittora (2024). For example, on the data management side, data is secured and tracked through the consensus models (Abaas et al., 2021). Fan and Zhang (2019) investigated a consortium BC-based data aggregation and regulation mechanism for the smart grid system for the preservation of user information to allow for monitoring, DR, management of data congestion, and size (Fan & Zhang, 2019). Their study showed that, compared to other schemes, their design can enable multidimensional data aggregation, facilitate multiparty reception, and demonstrate that it is well-suited for smart grid applications. Gao et al. (2022) showed in their research that integrating BC technology into the smart grid data traceability system has enhanced the data traceability capabilities of the smart grid dynamic data traceability model on the internal platform. In contrast, energy transactions are streamlined and strengthen grid operations with BC, providing peer-to-peer trading for consumers who own solar panels, empowering consumers, enabling energy balancing, and promoting renewable energy adoption (Abaas et al., 2021; Chen et al., 2021c; Kumar & Chittora, 2024).

On the cybersecurity side, BC is being explored to enhance the security of the smart grid system. Some examples of existing research efforts in this direction include integrating BC with smart meters to identify attack points and monitor cyber-physical assets without the need for third parties. For example, Liang et al. (2019) designed a self-defense framework for power systems using BC's consensus algorithms to protect data flows in the central power grid's communication and control layers, allowing smart meters to act as autonomous nodes that can reject malicious data entries. Their research showed that the distributed BC-based framework enhanced the data security in the power systems against cyberattacks due to BCs' self-defense features, higher transaction speeds, and lower investment risks.

Table 21.1 provides a comparative summary of these efforts, highlighting the technologies used, the datasets or case studies evaluated, and the key limitations of each approach. While most works focus on one or two technologies, few comprehensively integrate AI, DT, and BC.

21.3 Application-driven DT and blockchain-based smart grid

In this section, we first build a threat model to analyze potential threats in the smart grid, introduce our proposed countermeasure, and discuss potential threats to our proposed structure.

21.2.1 Threat model for smart grid

As an energy CPS, the smart grid introduces new attack surfaces because it integrates digital communication networks with physical electrical infrastructure. This interconnectivity enhances efficiency and real-time responsiveness but also exposes the grid to various vulnerabilities from both cyber and physical domains. As shown in Fig. 21.7, a comprehensive threat model is essential for understanding, assessing, and defending against potential security breaches that can disrupt operations, compromise sensitive data, or even cause widespread outages.

1. *Cybersecurity threats:* Cyber attacks target the digital components of the smart grid, such as communication links, control centers, and metering systems, by exploiting vulnerabilities in software, protocols, or network access. Common cyberattack types include:
   a. *DoS:* Attackers flood communication or control nodes with illegitimate traffic, rendering them unavailable. For example, a DDoS attack on a smart meter aggregator can prevent critical data from being transmitted, delaying response times and disrupting demand forecasting.
   b. *FDI:* Malicious actors inject deceptive sensor or metering data to mislead grid operators or AI decision engines. This may cause the grid to respond to non-existent faults, misallocate energy resources, or manipulate market pricing.
   c. *Man-in-the-middle (MITM):* Intercepting communication between components, adversaries alter or block legitimate messages. For example, an MITM attack between a substation and control center may manipulate setpoints or override safety thresholds.
   d. *Replay attacks:* Attackers capture legitimate commands or data and replay them later to disrupt operations. For instance, replaying an old load-shedding command could trigger unnecessary blackouts.

TABLE 21.1 Comparison of existing works with our proposed approach.

References	Focus area	AI used	DT used	Blockchain used	Dataset/case study	Key limitation
He et al. (2017)	Anomaly detection	CDBN	No	No	Synthetic grid data	Focused only on cyber intrusion detection
Jain et al. (2020)	PV dault detection (DT)	No	Yes	No	Real-time PV data	Lacks integration with AI or BC
Liang et al. (2019)	BC-based grid protection	No	No	Yes	Simulated drid	Limited to static threat scenarios
An et al. (2019)	DRL-based grid Protection	DRL	No	No	IEEE bus systems	Time overhead in large node systems
Sharma and Tiwari (2024)	Anomaly detection	Optimized XGBoost	No	No	Real-world smart grid data	Lacks integration with DT and BC for comprehensive security
Yılmaz and Martinez-Morales (2023)	PV fault detection	No	Yes	No	Real-time PV panel data	Does not incorporate AI for predictive analytics or BC for data integrity
Liu and Ajay (2024)	Secure communication in smart grids	No	No	Yes	Simulated smart grid environment	Focused solely on key exchange; lacks AI and DT integration
Wei et al. (2023)	Load-shedding strategy	DRL	No	No	IEEE 14-bus system	Does not address data security or real-time system modeling
Ness (2025)	PMU fault detection	VGG16, CNN	No	No	PMU image data	Limited to image-based analysis; lacks integration with DT and BC

(continued on next page)

TABLE 21.1 Comparison of existing works with our proposed approach—cont'd

References	Focus area	AI used	DT used	Blockchain used	Dataset/case study	Key limitation
Din and Su (2024)	Energy billing and trading	No	No	Yes	Simulated smart grid transactions	Does not incorporate AI for demand forecasting or DT for system modeling
Salazar-Pena et al. (2024)	PV system fault detection	ANN	No	No	Synthetic PV system data	Lacks real-time system modeling and secure data handling
Hajian and Chang (2022)	Cybersecurity in smart grids	No	No	Yes	Conceptual framework	Does not integrate AI for threat detection or DT for system simulation
Bailey et al. (2021)	Cybersecurity in distribution systems	DRL	No	No	IEEE 13-bus system	Focused on cybersecurity; lacks integration with DT and BC
Kolahi et al. (2024)	PV system inspection	No	Yes	No	Real-time PV system data	Does not incorporate AI for fault prediction or BC for data security
Ours	Unified AI-DT-BC architecture	CNN, ViT, etc.	Yes	Yes	PMU fault detection	Integrated, real-time, and modular design

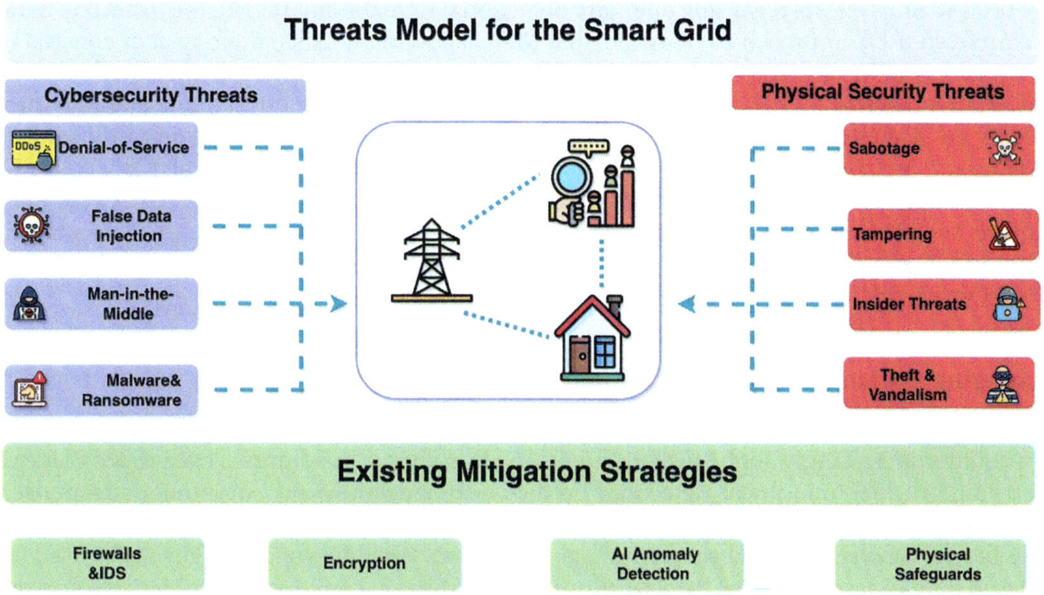

FIGURE 21.7 Threat model for smart grid.

   e. *Malware & ransomware:* Malicious code introduced via compromised firmware or third-party software can gain persistent control over grid devices. Stuxnet is a historical example of targeted malware affecting industrial control systems.
2. *Physical security threats:* In addition to cyber threats (Mahboubi et al., 2024), physical threats involve unauthorized access or sabotage of critical infrastructure, including transformers, substations, and distribution centers. These attacks may include:
   a. *Theft or vandalism of remote substations or smart meters:* Smart meters and remote substations are often installed in unguarded or poorly monitored environments, making them easy targets for physical attacks. Steal hardware (e.g., copper wiring, transformers, or meter units) for resale, disrupting local grid operations—bypass metering systems to illegally access electricity without billing, contributing to revenue loss. Destroy equipment deliberately (vandalism), leading to localized power outages or requiring costly replacements and downtime for restoration. These attacks affect operational continuity and pose significant safety hazards (e.g., electrical fires or electrocution).
   b. *Tampering with power lines or communication cables:* Attackers may physically access and compromise overhead or underground power lines or communication lines, especially in rural or less-monitored urban areas. Signal jamming or line-cutting disrupts real-time telemetry and control signal transmission. Interception of fiber-optic or copper lines to eavesdrop on grid control commands or customer usage patterns. Such actions undermine the security objectives (availability, integrity, etc.) of the system and can cascade into broader grid instability if not detected promptly.
   c. *Insider threats:* Inside threats can come from Authorized Personnel Employees or contractors with legitimate access to sensitive infrastructure, systems, or credentials may: (i) Leak confidential system configurations or passwords to external actors (e.g., cybercriminals or foreign intelligence), (ii) Override safety protocols or inject malicious scripts directly into grid controllers or SCADA systems, (iii) Sabotage software or firmware updates, introducing backdoors or zero-day vulnerabilities for later exploitation.
3. *Existing mitigation strategies:* Several mitigation techniques are currently employed to counter these threats:
   a. Firewalls and IDS protect the system against unauthorized access and monitor suspicious traffic patterns.
   b. Public key infrastructure and asymmetric encryption are used to secure communication, especially in smart meter and substation networks.
   c. Redundancy and failover mechanisms help maintain operations during component failures or attacks.
   d. AI-driven anomaly detection systems, such as DL models, are increasingly used to identify out-of-pattern behavior indicative of FDI or DoS attacks.
   e. BC is emerging as a tool for secure, tamper-proof logging of transactions and communications in the grid.

There have been various efforts regarding anomaly detection within the smart grid. For example, Jiang, Wu, Wang, & Xin, 2018 employed a DL approach to detect virtual MAC spoofing attacks; their system consists of three main components: an access point hosted on a workstation and two laptops—one as a benign node and the other as a malicious one. Bontemps et al. (2017) worked to develop an LSTMmodel for continuous anomaly monitoring and detection in cyber environments, where the mechanism recalled recent detection results and utilized time-series data and a circular array to recognize collective anomalies based on error thresholds. He et al. (2017) used DL to provide early security warnings for the smart grid system, allowing the prevention of such occurrences and minimizing potential damage and downtime when the system is under attack, like FDIAs. Also, for data integrity detection, An et al. (2019) designed a deep RL-based solution to deal with data integrity attacks within the AC power grid by utilizing a deep-Q-network as a detection engine.

21.3.2 Leveraging artificial intelligence for defense

As smart grids grow in complexity and scale, conventional rule-based or reactive defense mechanisms become insufficient to counter increasingly sophisticated threats. AI introduces an adaptive, data-driven layer of protection capable of carrying out real-time threat detection, predictive diagnostics, and autonomous response. By learning from vast operational data streams, AI can proactively identify abnormal behaviors like: (1) FDIA: Via spotting inconsistencies between expected and observed voltage or frequency patterns (e.g., the key states used for monitoring and control of grid operations). (2) DDoS: By observing sudden spikes in communication traffic from external sources. (3) Insider threats: Such threats may be detected via checking unusual access patterns or unauthorized configuration changes in control systems. AI can also flag malware activity by tracking abnormal process executions or command sequences within grid controllers. Moreover, predictive models can identify early signs of equipment failure by learning normal operational baselines and detecting deviations in sensor readings, such as overheating transformers or irregular load fluctuations, thereby enabling timely maintenance before outages occur. It can also forecast potential failures, making it a crucial component for enhancing the cyber-physical resilience of next-generation energy systems.

- ***AI-IDS:*** AI-powered IDSs utilize advanced ML or DL techniques to monitor and analyze network traffic across the smart grid continuously. These systems can detect signs of malicious activity that traditional firewalls or static thresholds may overlook. For instance, a CNN or LSTM model can be trained on grid telemetry data to identify anomalies such as latency spikes, abnormal packet loss, or irregular communication patterns, common indicators of DoS or MITM attacks. These AI-IDS frameworks provide real-time alerts, enabling rapid mitigation before the threat impacts grid operations.
- ***FDI attack detection:*** FDIA are among the most dangerous in smart grids, as they can mislead control centers by subtly corrupting measurement data. AI models like autoencoders or deep belief networks are particularly effective in this domain. They can learn the normal behavioral patterns of power grid metrics such as voltage, frequency, or load profiles and detect deviations that suggest the presence of falsified data. By flagging inconsistencies in real time, these models help grid operators maintain situational awareness and prevent malicious data from influencing operational decisions or triggering cascading failures.
- ***Preventive maintenance:*** AI techniques are also highly valuable in identifying physical infrastructure vulnerabilities before they lead to critical failures. Predictive maintenance systems leverage historical and real-time data, such as temperature, vibration, and electrical loads, to forecast the likelihood of component degradation or malfunction. ML models can detect early warning signs of transformer fatigue, cable overheating, or sensor miscalibration. This allows utilities to prioritize maintenance schedules, reduce downtime, and prevent physical sabotage or wear-related malfunctions that could otherwise compromise grid reliability.
- ***AI in load forecasting and resilience planning:*** Accurate load forecasting is essential to ensuring grid stability, especially during emergencies or cyberattacks. AI models such as RNNs or hybrid ensemble methods can predict short and long-term electricity demand patterns by analyzing historical usage, weather conditions, and real-time consumption behaviors. When a potential overload, outage, or attack is anticipated, these models can help balance supply and demand by triggering DR mechanisms, rerouting power, or advising preventive shutdowns. In doing so, AI enhances the grid's ability to recover from disruptions and maintain uninterrupted service to critical sectors.

21.3.3 Our proposed architecture

Based on what we summarize in Sections 21.2 and 21.3.1, we propose a four-layer architecture to secure the smart grid via AI, BC, and DT, as shown in Fig. 21.8. Its four layers are described as follows:

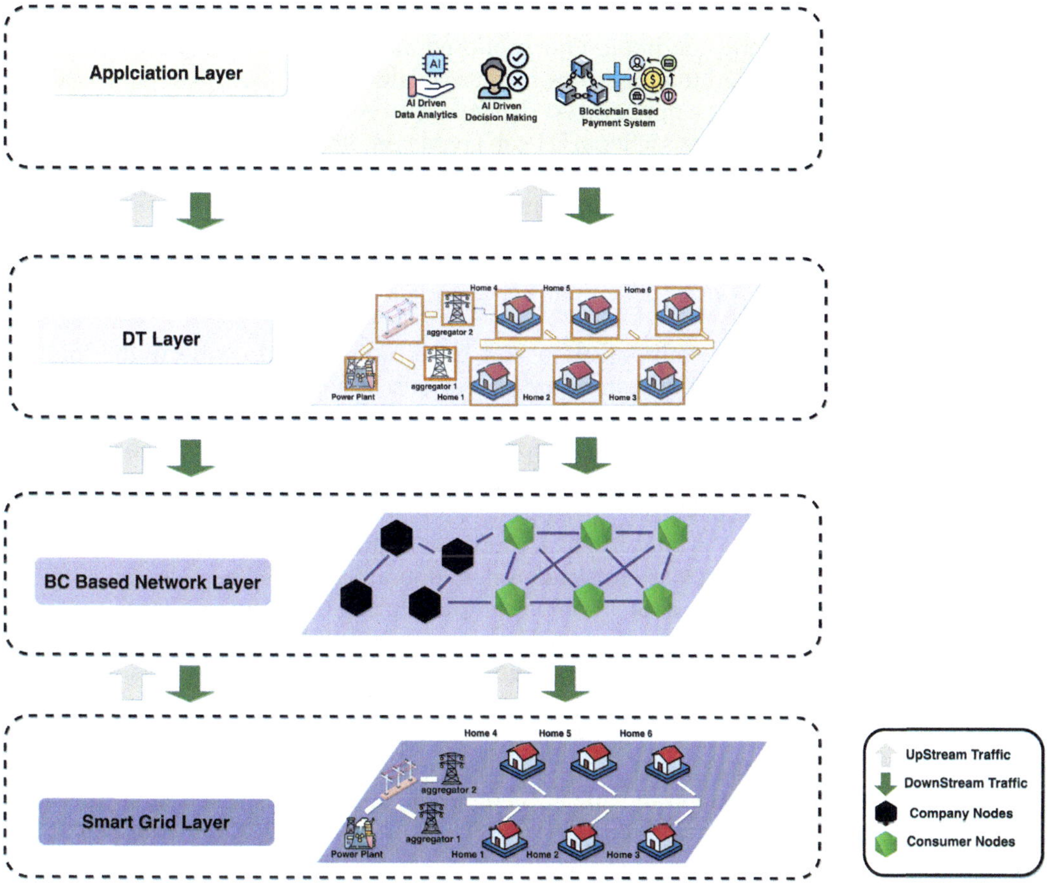

FIGURE 21.8 AI application-driven DT and blockchain-based smart grid architecture.

1. ***Smart grid layer:*** The smart grid layer consists of real-world infrastructure for delivering power services, including smart meters, substations, transmission lines, and consumer-end devices. Unlike conventional systems that collect data in isolation, this layer is tightly coupled with upper layers to enable real-time data collection and edge-assisted decision support. This integration ensures that anomalies in power flow or equipment status are detected at the source and forwarded upward for analysis and further action.
2. ***BC-based network layer:*** The BC-based network layer aims to secure all transmitted data by validating and recording it using tamper-resistant BC ledgers. While numerous existing solutions focus on BC for smart grid applications, such as energy trading, our design utilizes BC to secure smart grid communications. It prevents data tampering, ensures auditability, and builds trust among stakeholders through consensus mechanisms such as PoS or PBFT. This drives the grid's communication layer to be resilient to cyber threats and insider manipulation.
3. ***DT layer:*** The DT layer creates a real-time virtual replica of the physical grid, continuously synchronized with BC-validated data. This allows the system to run simulations, predict faults, and test and validate new operational strategies in a safe virtual environment. Unlike traditional DTs that are used solely for monitoring, our implementation supports interactive and predictive modeling guided by AI, making it an intelligent middle layer between physical infrastructure and decision-making.
4. ***Application layer:*** Finally, the application layer hosts AI-driven and BC-enabled services. AI modules perform predictive analytics, detect threats such as FDI or DDoS attacks, and optimize grid performance using insights from the data. BC smart contracts also manage tasks like secure billing, access control, and automated energy transactions. What makes this layer unique is its feedback mechanism: decisions made by AI are first validated in the DT environment before being deployed to the physical grid. This way can ensure safer, more reliable automation for grid operations.

In this proposed architecture, we use BC as the pipeline to secure the network security and transmit data integrity, DT as middleware to provide close-range data collection for serving higher layers, and to serve as a simulation for further test decision performance, and AI as the core to create different kinds of applications to better manage and support the smart grid in good working condition. Our proposed architecture stands out by unifying AI, DT, and BC in a tightly integrated, layered design that precedes further isolated applications. Unlike traditional systems, it offers real-time fault simulation, secure data flow, and AI-driven decision-making via coordinated feedback across all layers. BC ensures data integrity beyond transactions; DT can enable predictive simulation; and AI serves as the core engine for threat monitoring, detection, control, and response. This synergy creates a modular, intelligent ecosystem that not only secures but also enhances smart grid resilience.

21.3.4 New security challenges

Since smart grids rely on electronic communication and have bidirectional connectivity, they are inherently vulnerable and can expose the electricity infrastructure to various security risks (Khurana et al., 2010). These internet-based attacks on these systems can lead to serious repercussions, including data theft, cascading failures, complete blackouts, loss of synchronization, power supply disruptions, and substantial commercial and cultural losses (Gill et al., 2023). Although the technologies described earlier can improve the security of smart grid systems, they also introduce some challenges to consider.

With the introduction of AI, the smart grid needs to account for frequent attack techniques, including DDoS and FDIA, which aim to manipulate original data to mislead system operators as well as poisoning attacks that seek to contaminate the training data by injecting carefully crafted samples during the training phase; this tactic deteriorates the model's performance (Liang et al., 2019). Another attack to keep in mind with the introduction of AI is the evasion attack; adversaries manipulate test samples to evade detection during the system test process, compromising the ML model's accuracy without changing the training data (Sanjab & Saad, 2016). Timely and precise identification of these security vulnerabilities or attacks is crucial for maintaining the stable operation of smart grids (Gill et al., 2023).

Integrating DT technology into smart grids improves monitoring and operational efficiency, but also raises cybersecurity risks. DoS attacks (Hussaini et al., 2022; Zheng et al., 2022) can disrupt communication, leading to potential physical damage and outages. Data injection attacks (Hussaini et al., 2022; Jafari et al., 2023) can manipulate metering data, resulting in overcharging or resource misallocation. MITM attacks (Ali et al., 2023; Hussaini et al., 2022) can alter communications between DTs and system components, causing discrepancies in system status. Additionally, the supporting infrastructure, including IoT devices and cloud services (Saad et al., 2020), may have vulnerabilities that attackers can exploit.

Integrating BC technology into smart grids offers enhanced security and transparency. However, it also introduces specific cybersecurity challenges that shall be addressed to make sure the resilience and reliability of the power infrastructure. BC's transparency (Erturk et al., 2019) may expose sensitive user data, raising privacy concerns. If not properly coded and audited, smart contracts can be exploited (Muzumdar et al., 2021), leading to unauthorized access to grid operations.

21.4 Case study: AI-driven smart grid fault detection from phasor measurement unit

This section uses a special scenario to show how AI can help maintain and secure the smart grid's operational status by detecting and avoiding error signals. As part of our proposed architecture in Section III-C, this shows how AI can secure the smart grid.

21.4.1 Overview

PMUs (Ghosh et al., 2013) play an important role in the smart grid's wide area monitoring, protection, and control systems (Terzija, 2007), enabling real-time observation and response to power system disturbances. However, these systems are subject to cyber threats, particularly PMU data manipulation attacks (Wang et al., 2019). Such attacks, often a form of FDIAs (Zhang et al., 2020), can alter PMU readings stealthily, leading to incorrect decisions or failure to act on actual faults, potentially causing cascading failures. Despite data integrity checks (Li and Luo, 2012), well-crafted manipulations can bypass them, misleading control centers and undermining grid resilience. AI, particularly DL, has shown promise to enhance grid security. In this case study, we aim to deploy several DL models to demonstrate how well they detect those errors or attacks.

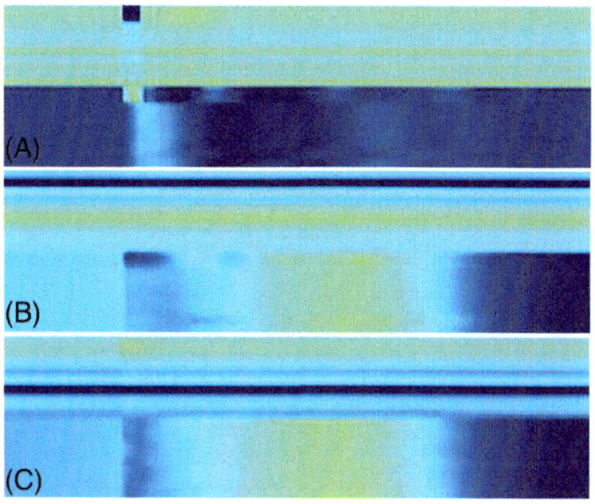

FIGURE 21.9 Example dataset. (A) DB FLT (B) DB GNL (C) DB SMS.

In this case study, we consider the task of supervised classification of short PMU segments into discrete disturbance classes. Each instance consists of a five-second window of synchronized voltage and frequency measurements, which we encode as a pseudo-color image $x \in \mathbb{R}^{300\times20\times3}$. The goal is to learn a mapping f_θ: $\mathbb{R}^{300\times20\times3} \rightarrow Y$, where Y = {DB FLT, DB GNL, DB SMS} denotes the set of event labels corresponding to faults, generation losses, and synchronous motor switching events. Training is performed on a finite collection of labeled pairs $(x_i, y_i)^{N_i}$ and the learned classifier's performance is evaluated on an independent test set to gauge its ability to generalize to unseen data.

21.4.2 Data collection and scenarios

The dataset (Dataset Smart Grid Phasor Measurement Unit Fault Images by PythonAfroz on Kaggle, available at https://www.kaggle.com/datasets/pythonafroz/smart-grid-phasor-measurement-unit-faulty-data.) utilized in the case study (Biswal et al., 2020) includes pseudo-color images derived from PMU time-series data. Each image measures 300 × 20 × 3 pixels, where 300 denotes the number of time points (i.e., 5 seconds of data at 60 samples per second), 20 represents the combined voltage and frequency measurements from 10 PMUs (10 voltage + 10 frequency channels), and 3 corresponds to the RGB color channels used to encode the data as a pseudo-color image.

To prepare the data, raw voltage and frequency signals were first synchronized and segmented into 5-second windows, including 0.5 seconds before the disturbance trigger and 4.5 seconds after. Normalization was applied to scale all values into a uniform range, ensuring consistency across measurements and improving model convergence. Missing values, if any, were handled using interpolation or masking to maintain temporal continuity. RGB encoding was chosen to convert the time-series data into a visual format that preserves both temporal and multichannel structure. By mapping signal variations to color intensities in the RGB space, pseudo-color images enable CNNs to learn spatial and temporal features for fault classification effectively. This approach leverages pretrained image-based models for tasks such as anomaly detection and disturbance classification in the smart grid. The dataset comprises three types of events: 344 instances of faults labeled as "DB FLT," 140 instances of loss of generation labeled as "DB GNL," and 21 instances of synchronous motor switching events labeled as "DB SMS." Fig. 21.9 provides examples of these three types.

In this study, we utilize several pretrained models of varying sizes to train and detect this dataset offline, assessing the DL model's successful detection rate.

21.4.3 Performance evaluation

We evaluate the effectiveness of DL models in classifying smart grid disturbances using a representative PMU-based dataset. For our evaluation environment, all experiments were conducted using PyTorch (Imambi et al., 2021) on a server equipped with two high-end NVIDIA RTX 4090 GPUs. To leverage the advantages of transfer learning, we selected four widely used pretrained models with varying architectural complexity: VGG16 (Theckedath & Sedamkar, 2020), ResNet50 (Mukti & Biswas, 2019), EfficientNet-B3 (Kumar et al., 2024), and Vision Transformer (ViT) (Yuan et al., 2021). These models were chosen for their proven success in image-based classification tasks and their ability to generalize well on new datasets. Only the final classification layer was fine-tuned using the PMU disturbance data, while the rest of the model weights remained fixed. Each model is fine-tuned with the Adam optimizer at a fixed learning rate of 0.001 for 50 epochs. The dataset was split 70% for training and 30% for testing in

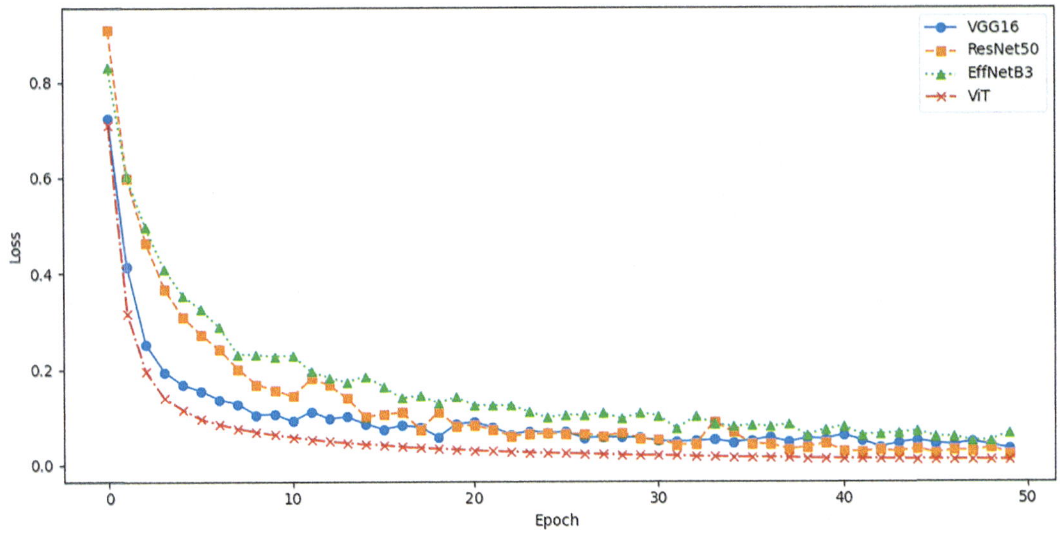

FIGURE 21.10 Training loss over epochs.

batches of 32 with shuffling on the training set. Each model was trained for 50 epochs. These experiments concentrate on selecting models under ideal server conditions; however, deploying at the edge would require lightweight networks such as MobileNet or SqueezeNet, combined with quantization and pruning, to fit within the resource limits of a PMU gateway.

Accuracy and cross-entropy loss were used as primary performance metrics, with accuracy reflecting the proportion of correctly classified events and cross-entropy loss indicating prediction confidence and convergence behavior. Although VGG16 and ResNet50 are not lightweight architectures, they were prioritized due to their stability, ease of interpretation, and established baselines in the DL community. We acknowledge that lighter models, such as MobileNet or SqueezeNet, would be better suited for edge deployment. However, the goal of this case study is to evaluate classification potential under ideal (nonconstrained) conditions. Future work can further explore these edge-optimized models by incorporating quantization and pruning to reduce memory and computational overhead.

This trade-off between classification performance and deployability is important for real-time grid systems deployed on constrained IoT devices.

Performance results show consistent trends across all models. As seen in Fig. 21.10, the loss value decreases steadily with epochs, and ViT exhibits the fastest convergence and lowest training loss. Despite its architectural complexity and fast convergence, the ViT model produces only marginal improvements in accuracy. We found that the training and test accuracy across all models exceeds 95%, as seen in Figs. 21.11 and 21.12. Notably, VGG16, as one of the earliest and representative CNN architectures, achieves comparable results, highlighting that even simpler models can be effective in smart grid applications, particularly where inference speed and resource constraints are a concern, as seen in Fig. 21.13. While accuracy remains the central metric in this case study, we recognize that security-critical applications in smart grids require defensive robustness against adversarial threats. FDIAs, for example, can be specifically crafted to manipulate PMU signals and bypass detection. To improve the validity of our approach, future work shall incorporate adversarial robustness evaluations using techniques such as the projected gradient descent (PGD) or the fast gradient sign method (FGSM). Quantifying how each model withstands adversarial perturbations would provide a more holistic assessment of its practical utility in a threat-prone smart grid environment.

21.4.4 Case study summary

We believe that future work involving adversarial robustness is crucial for ensuring the reliability of AI models used in safety-critical smart grid systems, particularly when working with PMU data streams. Since the data support essential grid controls, it is vital to evaluate model performance under bounded perturbations, as mentioned earlier, such as FGSM and PGD, as well as simpler, label-preserving corruptions, such as PMU channel dropout and clock drift. These stress tests simulate potential real-world disruptions to assess whether a model's reported accuracy of over 95% holds up under challenging, security-relevant conditions. We believe that by testing these models in

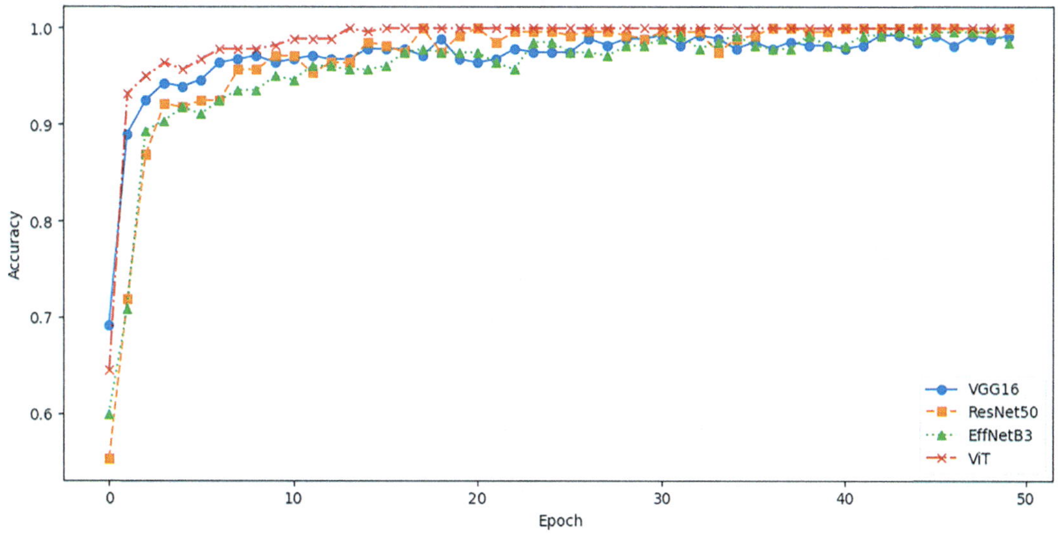

FIGURE 21.11 Training accuracy over epochs.

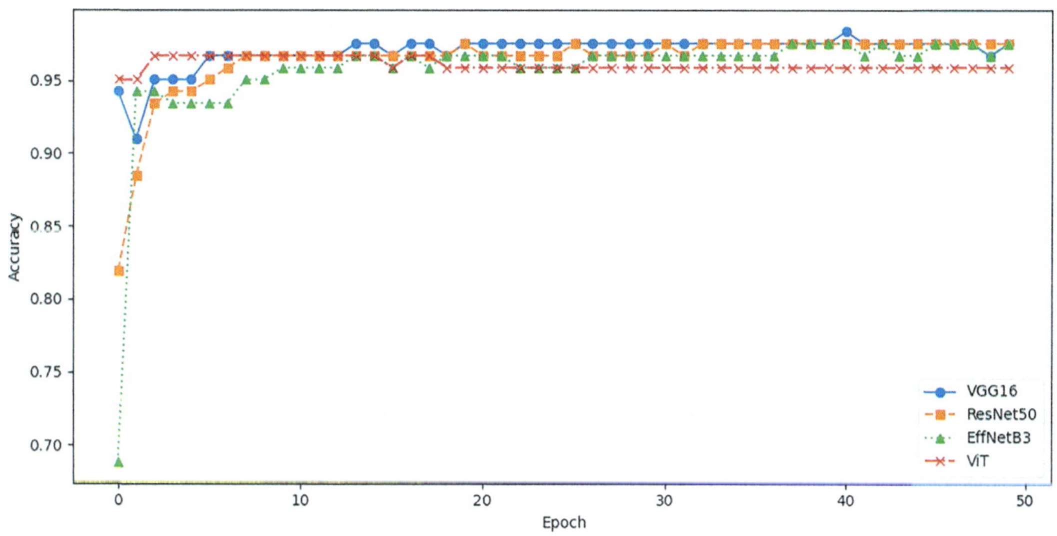

FIGURE 21.12 Testing loss over epochs.

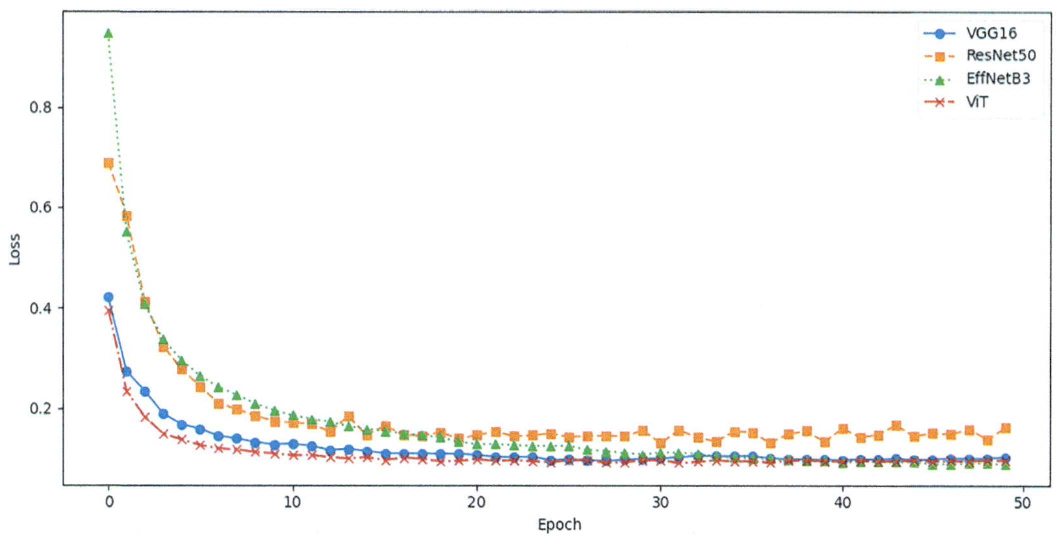

FIGURE 21.13 Testing accuracy over epochs.

such scenarios, future work can ensure they remain robust and effective, even in the presence of adversarial or nonmalicious data corruptions.

This case study demonstrates that by integrating BC, DT, and AI, the security and operational efficiency of a smart grid can be significantly enhanced. Utilizing BC ensures data integrity and traceability, while DT enables real-time monitoring and simulation of grid conditions. AI, through DL techniques, can detect fault signals from PMUs, enabling proactive fault detection and decision-making. Employing a cross-layer security framework across devices, middleware, and application layers addresses critical security concerns such as data spoofing, unauthorized access, and model reliability, ultimately supporting a more resilient and intelligent smart grid system.

21.5 Future research directions

Our current study emphasizes accuracy under edge-computing constraints, and we acknowledge the need for robustness analysis. In future work, we plan to incorporate adversarial testing techniques, specifically PGD, to assess model vulnerability and resilience systematically. This addition will enhance the practical reliability of our approach in adversarial settings. Here are some potential future research directions based on our proposed structure.

Security challenges in new layers—A research imperative: As smart grids become more layered and interconnected, future work must address the complex security dynamics introduced by individual new layers, as we introduce in Fig. 21.8. The interfaces between BC logging, DT synchronization, and AI-driven control represent new vectors for cyber-physical attacks. Research is needed to define formal threat models and design modeling, simulation, and testbeds to investigate these interfaces, identify new security risks, and develop lightweight, explainable intrusion detection and prevention systems that operate across multiple layers. Furthermore, new trust frameworks, which can be employed by using distributed learning or BC-consensus mechanisms, could ensure that data transitions between layers remain tamper-proof and verifiable. Addressing these emerging vulnerabilities will be critical for building resilient grid architectures that can withstand cyberattacks and cascading system failures.

Security challenges arising from layered interactions: The layered structure introduces critical interlayer communication points that also serve as new attack surfaces. For instance, the interface between the Smart Grid and BC Layer, as shown in Fig. 21.8, where raw sensor data is transmitted, may be vulnerable to integrity attacks if the sensors are compromised or to spoofing or replay attacks if communication is not authenticated. Similarly, the DT layer, which relies on data integrity from the BC layer, can propagate faulty simulations if the input data is compromised. Cross-layer dependencies, such as AI decision-making based on DT predictions, amplify risks, as a successful attack on lower layers can cascade into false optimizations or incorrect control signals at the application level. Moreover, synchronization delays or mismatches between layers can lead to real-time misalignment, which affects DR accuracy and resilience. Thus the architecture shall include multilayered security mechanisms and protocols (e.g., mutual authentication, data provenance mechanisms, and interlayer anomaly detection filters, among others).

Leveraging AI for resilient and efficient smart grids: AI will continue to be a cornerstone of smart grid automation, but future research must shift toward more adaptive, privacy-aware, and resource-efficient AI solutions. Lightweight DL models optimized for edge deployment (e.g., quantized CNNs or TinyML frameworks) can ensure real-time performance under constrained compute environments. Moreover, researchers should explore hybrid AI systems that combine symbolic reasoning with neural networks to improve explainability in critical grid decisions. Future directions also include the use of generative models and RL agents for predictive simulation within DTs, enabling proactive fault avoidance and autonomous grid recovery. Ultimately, integrating AI with secure, BC-based smart contracts offers new ways for decentralized energy trading and DR strategies.

BC overheads in real-time smart grid logging: While BC ensures data integrity and tamper-resistance in smart grid systems, it introduces significant computational and energy overhead, especially for high-frequency data sources, such as PMUs. Traditional consensus mechanisms, such as PoW, are computationally intensive and unsuitable for time-critical grid operations due to latency and energy consumption. Even PoS, though more efficient, may struggle to scale with the volume and velocity of PMU data streams. To address this, lightweight consensus protocols (e.g., PBFT or DPoS) or hybrid on-chain/off-chain architectures should be utilized for balancing security, scalability, and latency in real-time smart grid applications.

Secure and scalable synchronization of DT: The implementation of DT enables functionalities (e.g., real-time monitoring, simulation, emulation, and decision-making) for smart grids. The synchronization between physical grid components and their digital replicas faces difficulties when communication latency is high or during cyber incidents. Future research shall design secure synchronization protocols that employ predictive modeling and redundant data

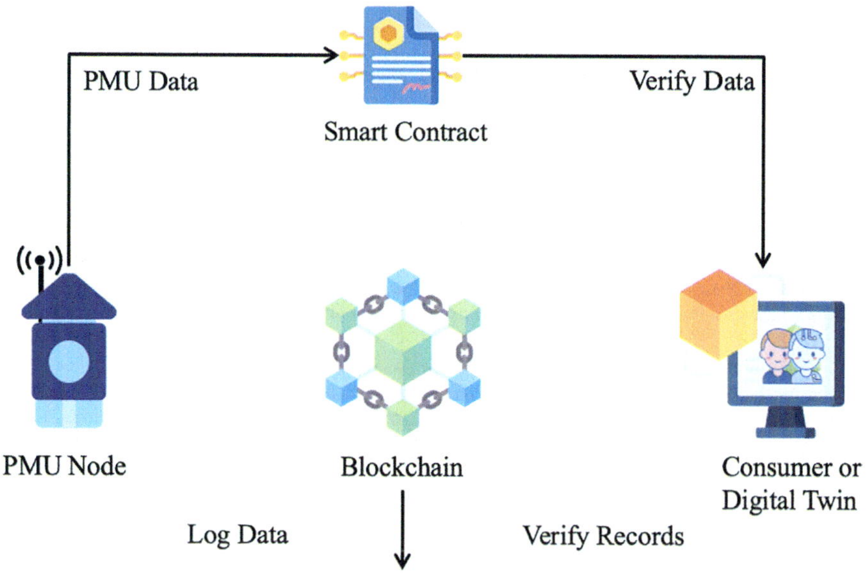

FIGURE 21.14 Smart contract: Grid data integrity.

channels to maintain operational coherence. The study should explore lightweight cryptographic techniques and federated update mechanisms to enable the scalable deployment of DTs across DERs and edge environments. A smart contract for grid data integrity ensures that each time-stamped data packet from grid devices such as smart meters or PMUs is authenticated, verified, and immutably logged. As illustrated in Fig. 21.14, the process begins when a device signs its data (e.g., voltage, frequency, timestamp) using a private key. The smart contract receives the signed data, verifies the source via its public key, checks the timestamp validity, and ensures the data hash matches the attached digital signature. Once all checks pass, the hash, not the raw data, is stored on-chain. This enables tamper-proof and efficient logging. This mechanism enables lightweight future auditing, in which off-chain records can be verified against the BC hash, leading to a balance between security and scalability.

Privacy-preserving and efficient BC protocols for energy transactions: The current BC implementations provide transparency and tamper resistance for smart grid data and transactions, but create privacy issues, scalability problems, and energy consumption challenges. Future research should investigate new consensus algorithms, such as Proof-of-Authority and zero-knowledge proofs, to develop BC systems that protect privacy while maintaining low latency and energy efficiency. Integrating off-chain computation with smart contract auditing tools helps reduce attack exposure and supports regulatory compliance, thereby building trust in BC-based smart energy systems.

21.5 Final remarks

This chapter proposes an AI application-driven architecture leveraging DT and BC technologies to enhance the security and efficiency of smart grids. In our proposed framework, BC ensures the integrity and traceability of transmitted data, DT facilitates real-time data collection and system simulation/emulation, and AI provides intelligent analytics and decision-making capabilities to support smart grid operations. We also analyzed the potential challenges of each technology, particularly concerning new security vulnerabilities and system complexity introduced by their integration. To validate the efficacy of our approach, we conducted a case study using DL techniques to detect fault signals in smart grid data collected from PMUs. This case study shows how AI assists the smart grid in detecting operational fault signals and demonstrates AI's potential to secure the smart grid. Lastly, we outlined some future research directions based on our proposed architecture.

References

Abaas, M., Singh, P., & Lee, R. (2021). Blockchain applications in smart grid systems. In *2020 52nd North American Power Symposium (NAPS)* (pp. 1–6). USA: Tempe AZ.

Abdelwahab, R. H., El-Habrouk, M., & Abdelhamid, T. H. (2019). Survey of communication techniques in smart grids. In *2019 21st International Middle East Power Systems Conference (MEPCON)* (pp. 550–555). Cairo, Egypt: IEEE.

Abideen, A. Z., Sundram, V. P. K., Pyeman, J., Othman, A. K., & Sorooshian, S. (2021). Digital twin integrated reinforced learning in supply chain and logistics. *Logistics, 5*(4), 84.

Ali, M., Kaddoum, G., Li, W.-T., Yuen, C., Tariq, M., & Poor, H. V. (2023). A smart digital twin enabled security framework for vehicle-to-grid cyber-physical systems. *IEEE Transactions on Information Forensics and Security, 18*, 5258–5271.

Aloul, F., Al-Ali, A., Al-Dalky, R., Al-Mardini, M., & El-Hajj, W. (2012). Smart grid security: Threats, vulnerabilities and solutions. *International Journal of Smart Grid and Clean Energy, 1*(1), 1–6.

An, D., Yang, Q., Liu, W., & Zhang, Y. (2019). Defending against data integrity attacks in smart grid: A deep reinforcement learning-based approach. *IEEE Access, 7*, 110835–110845.

Arulkumaran, K., Deisenroth, M. P., Brundage, M., & Bharath, A. A. (2017). Deep reinforcement learning: A brief survey. *IEEE Signal Processing Magazine, 34*(6), 26–38.

Bailey, T., Johnson, J., & Levin, D. (2021). Deep reinforcement learning for online distribution power system cybersecurity protection. In *2021 IEEE International Conference on Communications, Control, and Computing Technologies for Smart Grids (SmartGridComm)* (pp. 227–232). Aachen, Germany: IEEE.

Bailly, A., Blanc, C., Francis, E´., Guillotin, T., Jamal, F., Wakim, B., & Roy, P. (2022). Effects of dataset size and interactions on the prediction performance of logistic regression and deep learning models. *Computer Methods and Programs in Biomedicine, 213*, 106504.

Barbie, A., & Hasselbring, W. (2024). From digital twins to digital twin prototypes: Concepts, formalization, and applications. *IEEE Access, 12*, 75 337–75 365.

Barricelli, B. R., Casiraghi, E., & Fogli, D. (2019). A survey on digital twin: Definitions, characteristics, applications, and design implications. *IEEE Access, 7*, 167653–167671.

Biswal, M., Misra, S., & Tayeen, A. S. (2020). Black box attack on machine learning assisted wide area monitoring and protection systems. In *2020 IEEE Power & Energy Society Innovative Smart Grid Technologies Conference (ISGT)* (pp. 1–5). Washington, DC, USA: IEEE.

Bontemps L., Cao V. L., McDermott J., and Le-Khac N.-A., "Collective anomaly detection based on long short term memory recurrent neural network," 2017. https://arxiv.org/abs/1703.09752

Borenstein, S., & Bushnell, J. (2015). The us electricity industry after 20 years of restructuring. *Annual Review of Economics, 7*(1), 437–463.

Chen, Z., Liao, W., Hua, K., Lu, C., & Yu, W. (2021a). Towards asynchronous federated learning for heterogeneous edge-powered internet of things. *Digital Communications and Networks, 7*(3), 317–326. https://www.sciencedirect.com/science/article/pii/S2352864821000195.

Chen, Z., Tian, P., Liao, W., & Yu, W. (2021b). Zero knowledge clustering based adversarial mitigation in heterogeneous federated learning. *IEEE Transactions on Network Science and Engineering, 8*(2), 1070–1083.

Chen, G., He, M., Gao, J., Liu, C., Yin, Y., & Li, Q. (2021c). Blockchain-based cyber security and advanced distribution in smart grid. In 2021 IEEE 4th International Conference on Electronics Technology (ICET), Chengdu, China (pp. 1077-1080). IEEE.

Chen, Z., Tian, P., Qian, C., Liao, W., Hussaini, A., & Yu, W. (2024). Chapter 9- Machine learning frameworks in iot systems: A survey, case study, and future research directions. In Z. Lyu (Ed.), Smart spaces, *ser. intelligent data-centric systems* (pp. 197–218). Academic Press. https://www.sciencedirect.com/science/article/pii/B9780443134623000157.

Cong, L. W., & He, Z. (2019). Blockchain disruption and smart contracts. *The Review of Financial Studies, 32*(5), 1754–1797.

Craig, C. A., & Feng, S. (2017). Exploring utility organization electricity generation, residential electricity consumption, and energy efficiency: A climatic approach. *Applied Energy, 185*, 779–790.

Din, J., & Su, H. (2024). Blockchain-enabled smart grids for optimized electrical billing and peer-to-peer energy trading. *Energies, 17*(22), 5744.

Erturk, E., Lopez, D., & Yu, W. Y. (2019). Benefits and risks of using blockchain in smart energy: A literature review. *Contemporary Management Research, 15*(3), 205–225.

Esenogho, E., Djouani, K., & Kurien, A. M. (2022). Integrating artificial intelligence internet of things and 5g for next-generation smartgrid: A survey of trends challenges and prospect. *IEEE Access, 10*, 4794–4831.

Fan, M., & Zhang, X. (2019). Consortium blockchain based data aggregation and regulation mechanism for smart grid. *IEEE Access, 7*, 35929–35940.

Fan, X., & Chai, Q. (2018). Roll-DPoS: a randomized delegated proof of stake scheme for scalable blockchain-based internet of things systems. In *Proceedings of the 15th EAI International Conference on Mobile and Ubiquitous Systems: Computing, Networking and Services* (pp. 482–484). New York, NY, USA: Association for Computing Machinery (ACM).

Fuller, A., Fan, Z., Day, C., & Barlow, C. (2020). Digital twin: Enabling technologies, challenges and open research. *IEEE Access, 8*, 108952–108971.

Gao, B., Liu, H., Su, H., Hong, Z., Ji, C., & Zhao, M. (2022). Design and typical application of digital twin architecture in smart grid. In *2022 7th Asia Conference on Power and Electrical Engineering (ACPEE)* (pp. 2140–2144). Hangzhou, China: IEEE.

Gao, L., Pi, Z., Wang, J., & Sun, J. (2022). Smart grid data traceability system based on blockchain technologies. In *2022 8th Annual International Conference on Network and Information Systems for Computers (ICNISC)* (pp. 721–726). Guilin, China: IEEE.

Gao, W., Hatcher, W. G., & Yu, W. (2018). A survey of blockchain: Techniques, applications, and challenges. In *2018 27th International Conference on Computer Communication and Networks (ICCCN)* (pp. 1–11). Hangzhou, China: IEEE.

Garg, V., & Agarwal, B. (2019). *Survey of literature on machine intelligence and deep learning for smart grid applications* (pp. 1–5). Jaipur, India: IEEE(pp.

Gervais, A., Karame, G. O., Wu¨st, K., Glykantzis, V., Ritzdorf, H., & Capkun, S. (2016). On the security and performance of proof of work blockchains. In *Proceedings of the 2016 ACM SIGSAC Conference on Computer and Communications Security (CCS '16)* (pp. 3–16). New York, NY, USA: Association for Computing Machinery (ACM).

Ghosal, A., & Conti, M. (2019). Key management systems for smart grid advanced metering infrastructure: A survey. *IEEE Communications Surveys & Tutorials, 21*(3), 2831–2848.

Ghosh, D., Ghose, T., & Mohanta, D. K. (2013). Communication feasibility analysis for smart grid with phasor measurement units. *IEEE Transactions on Industrial Informatics, 9*(3), 1486–1496.

Gill, A., Khalid, A., & M. , R (2023). A crucial review on AI use in hybrid grids: An ultra-smart way towards microgrid. In *2023 International Conference on Power Energy, Environment Intelligent Control (PEEIC)* (pp. 1364–1368). Greater Noida, India: IEEE.

Greer, C., Wollman, D. A., Prochaska, D., Boynton, P. A., Mazer, J. A., Nguyen, C., FitzPatrick, G., Nelson, T. L., Koepke, G. H., Hefner, A. R., Jr., Pilliteri, V., Brewer, T., Golmie, N., Su, D., Eustis, A., Holmberg, D., & Bushby, S. T. (2014). *NIST framework and roadmap for smart grid interoperability standards, release 3.0.*

Gui, Y., Nainar, K., Ciontea, C.-I., Bendtsen, J. D., Iov, F., Shahid, K., & Stoustrup, J. (2023). Automatic voltage regulation application for pv inverters in low-voltage distribution grids – A digital twin approach. *International Journal of Electrical Power Energy Systems, 149*, 109022. https://www.sciencedirect.com/science/article/pii/S0142061523000790.

Hajian, A., & Chang, H.-C. (2022). A blockchain-based smart grid to build resilience through zero-trust cybersecurity. In *Handbook of smart energy systems* (pp. 1–19). Springer.

Hatcher, W. G., & Yu, W. (2018). A survey of deep learning: Platforms, applications and emerging research trends. *IEEE Access, 6*, 24411–24432.

He, Y., Mendis, G. J., & Wei, J. (2017). Real-time detection of false data injection attacks in smart grid: A deep learning-based intelligent mechanism. *IEEE Transactions on Smart Grid, 8*(5), 2505–2516.

Humayed, A., Lin, J., Li, F., & Luo, B. (2017). Cyber-physical systems security—a survey. *IEEE Internet of Things Journal, 4*(6).

Huseinovic´, A., Mrdovic´, S., Bicakci, K., & Uludag, S. (2020). A survey of denial-of-service attacks and solutions in the smart grid. *IEEE Access, 8*, 177447–177470.

Hussaini, A., Qian, C., Liao, W., & Yu, W. (2022). A taxonomy of security and defense mechanisms in digital twins-based cyber-physical systems. In *2022 IEEE International Conferences on Internet of Things (iThings), Green Computing and Communications (GreenCom), Cyber, Physical Social Computing (CPSCom), Smart Data (SmartData), and Cybermatics*, (pp. 597–604). Espoo, Finland: IEEE.

Imambi, S., Prakash, K. B., & Kanagachidambaresan, G. (2021). Pytorch. In *Programming with TensorFlow: Solution for edge computing applications* (pp. 87–104). Springer.

Jafari, M., Kavousi-Fard, A., Chen, T., & Karimi, M. (2023). A review on digital twin technology in smart grid, transportation system and smart city: Challenges and future. *IEEE Access, 11*, 17471–17484.

Jain, P., Poon, J., Singh, J. P., Spanos, C., Sanders, S. R., & Panda, S. K. (2020). A digital twin approach for fault diagnosis in distributed photovoltaic systems. *IEEE Transactions on Power Electronics, 35*(1), 940–956.

Johnston, B., Jones, A., & Kruger, C. (2019). *Applied Unsupervised Learning with Python: Discover hidden patterns and relationships in unstructured data with Python*. Packt Publishing Ltd.

Kayamboo, S., Ray, B., Das, N., & Tom, M. (2022). Iot-based cyber-physical distribution system planning. In *2022 IEEE International IOT, Electronics and Mechatronics Conference (IEMTRONICS)* (pp. 1–6). Toronto, Ontario, Canada: IEEE.

Khurana, H., Hadley, M., Lu, N., & Frincke, D. A. (2010). Smart-grid security issues. *IEEE Security & Privacy, 8*(1), 81–85.

Kim, S., Park, K.-J., & Lu, C. (2022). A survey on network security for cyber–physical systems: From threats to resilient design. *IEEE Communications Surveys Tutorials, 24*(3), 1534–1573.

Kolahi, M., Esmailifar, S., Sizkouhi, A. M., & Aghaei, M. (2024). Digital-PV: A digital twin-based platform for autonomous aerial monitoring of large-scale photovoltaic power plants. *Energy Conversion and Management, 321*, 118963.

Kumar, C., & Chittora, P. (2024). Integration of blockchain technology in smart grids for enhancing security, efficiency, and resilience: A comprehensive survey. In *2024 IEEE Third International Conference on Power Electronics, Intelligent Control and Energy Systems (ICPEICES)* (pp. 577–582). Delhi, India: IEEE.

Kumar, V., Prabha, C., Sharma, P., Mittal, N., Askar, S. S., & Abouhawwash, M. (2024). Unified deep learning models for enhanced lung cancer prediction with resnet-50–101 and efficientnet-b3 using dicom images. *BMC Medical Imaging, 24*(1), 63.

Jiang, P., Wu, H., Wang, C., & Xin, C. (2018). Virtual mac spoofing detection through deep learning. 2018 IEEE International Conference on Communications (ICC) Kansas City, MO, USA IEEE(pp. 1–6).

Lee R. M., Assante M. J., and Conway T., "Analysis of the cyber attack on the ukrainian power grid," SANS Industrial Control Systems, Tech. Rep., 2016. https://www.sans.org/white-papers/36297

Lei, Z., Zhou, H., Dai, X., Hu, W., & Liu, G. (2023). Digital twin based monitoring and control for DC-DC converters. *Nature Communications, 14*(09).

Levytskyi, V., Kruk, P., Lopuha, O., Sereda, D., Sapaiev, V., & Matsiievskyi, O. (2024). Use of deep learning methodologies in combination with reinforcement techniques within autonomous mobile cyber-physical systems. In *2024 IEEE 4th International Conference on Smart Information Systems and Technologies (SIST)* (pp. 455–460). Astana, Kazakhstan: IEEE.

Li, F., & Luo, B. (2012). Preserving data integrity for smart grid data aggregation. In *2012 IEEE Third International Conference on Smart Grid Communications (SmartGridComm)* (pp. 366–371). Tainan City, Taiwan: IEEE.

Liang, F., Hatcher, W. G., Liao, W., Gao, W., & Yu, W. (2019). Machine learning for security and the internet of things: The good, the bad, and the ugly. *IEEE Access, 7*, 158126–158147.

Liang, F., Qian, C., Yu, W., Griffith, D., & Golmie, N. (2022). Survey of graph neural networks and applications. *Wireless Communications and Mobile Computing, 2022*(1), 9261537.

Liang, F., Yu, W., Liu, X., Griffith, D., & Golmie, N. (2020). Toward edge-based deep learning in industrial internet of things. *IEEE Internet of Things Journal, 7*(5), 4329–4341.

Liang, G., Weller, S. R., Luo, F., Zhao, J., & Dong, Z. Y. (2019). Distributed blockchain-based data protection framework for modern power systems against cyber attacks. *IEEE Transactions on Smart Grid, 10*(3), 3162–3173.

Lin, J., Yu, W., Yang, X., Xu, G., & Zhao, W. (2012). On false data injection attacks against distributed energy routing in smart grid. In *in 2012 IEEE/ACM Third International Conference on Cyber-Physical Systems (ICCPS)* (pp. 183–192). Beijing, China: IEEE Computer Society.

Lin, J., Yu, W., Yang, X., Yang, Q., Fu, X., & Zhao, W. (2015). *A novel dynamic en-route decision real-time route guidance scheme in intelligent transportation systems* (pp. 61–72). Columbus, OH, USA: IEEE Computer Society(pp.

Lin, J., Yu, W., Zhang, N., Yang, X., Zhang, H., & Zhao, W. (2017). A survey on internet of things: Architecture, enabling technologies, security and privacy, and applications. *IEEE Internet of Things Journal, 4*(5), 1125–1142.

Liu, W., & Ajay, P. (2024). Smart grid environment using block chain-based key agreements. *Measurement: Sensors, 31*, 100992.

Liu, X., Qian, C., Hatcher, W. G., Xu, H., Liao, W., & Yu, W. (2019a). Secure internet of things (IoT)-based smart-world critical infrastructures: Survey, case study and research opportunities. *IEEE Access, 7*, 79523–79544.

Liu, X., Qian, C., Yu, W., Griffith, D., Gopstein, A., & Golmie, N. (2023). Using deep reinforcement learning to automate network configurations for internet of vehicles. *IEEE Transactions on Intelligent Transportation Systems, 24*(12), 15948–15958.

Liu, X., Xu, H., Liao, W., & Yu, W. (2019b). Reinforcement learning for cyber-physical systems. In *2019 IEEE International Conference on Industrial Internet (ICII)* (pp. 318–327).

Liu, X., Yue, G., & Shan, B. (2023). Research on sustainable management of grid companies based on artificial intelligence. In *2023 International Conference on Distributed Computing and Electrical Circuits and Electronics (ICDCECE)* (pp. 1–5). Ballari, India: IEEE.

Liu, Y., Ning, P., & Reiter, M. K. (2011). False data injection attacks against state estimation in electric power grids. *ACM Transactions in Information Systems Security, 14*(1). https://doi.org/10.1145/1952982.1952995.

Mahboubi, A., Luong, K., Aboutorab, H., Bui, H. T., Jarrad, G., Bahutair, M., … Gately, H. (2024). Evolving techniques in cyber threat hunting: A systematic review. *Journal of Network and Computer Applications, 234*, 104004.

Mahmood, A., Javaid, N., & Razzaq, S. (2015). A review of wireless communications for smart grid. *Renewable and Sustainable Energy Reviews, 41*, 248–260.

Mihai, S., Yaqoob, M., Hung, D. V., Davis, W., Towakel, P., Raza, M., Karamanoglu, M., Barn, B., Shetve, D., Prasad, R. V., Venkataraman, H., Trestian, R., & Nguyen, H. X. (2022). Digital twins: A survey on enabling technologies, challenges, trends and future prospects. *IEEE Communications Surveys Tutorials, 24*(4), 2255–2291.

Mohammadi, M., Al-Fuqaha, A., Sorour, S., & Guizani, M. (2018). Deep learning for iot big data and streaming analytics: A survey. *IEEE Communications Surveys Tutorials, 20*(4), 2923–2960.

Mohan, A. P., Gladston, A., et al. (2020). Merkle tree and blockchain-based cloud data auditing. *International Journal of Cloud Applications and Computing (IJCAC), 10*(3), 54–66.

Mollah, M. B., Zhao, J., Niyato, D., Lam, K.-Y., Zhang, X., Ghias, A. M., Koh, L. H., & Yang, L. (2020). *Blockchain for future smart grid: A comprehensive survey: 8* (pp. 18–43).

Moulema, P., Yu, W., Griffith, D., & Golmie, N. (2015). On effectiveness of smart grid applications using co-simulation. In *2015 24th International Conference on Computer Communication and Networks (ICCCN)* (pp. 1–8). Las Vegas, NV, USA: IEEE.

Mukti, I. Z., & Biswas, D. (2019). Transfer learning based plant diseases detection using ResNet50. In *2019 4th International conference on electrical information and communication technology (EICT)* (pp. 1–6). IEEE.

Mulo, J., Liang, H., Qian, M., Biswas, M., Rawal, B., Guo, Y., & Yu, W. (2025). Navigating challenges and harnessing opportunities: Deep learning applications in internet of medical things. *Future Internet, 17*(3). https://www.mdpi.com/1999-5903/17/3/107.

Musa, A. A., Hussaini, A., Liao, W., Liang, F., & Yu, W. (2023a). Deep neural networks for spatial-temporal cyber-physical systems: A survey. *Future Internet, 15*(6). https://www.mdpi.com/1999-5903/15/6/199.

Musa, U.S., Guo, Y., Qian, C., & Yu, W. (2023b). Optimal sampling for moving object trajectory tracking in smart transportation systems: A transformer-based approach. In 2023 IEEE International Conference on Big Data (BigData), Sorrento, Italy (pp. 1230-1235). IEEE.

Muzumdar, A., Modi, C., Vyjayanthi, C., et al. (2021). A trustworthy and incentivized smart grid energy trading framework using distributed ledger and smart contracts. *Journal of Network and Computer Applications, 183*, 103074.

Ness, S. (2025). Enhancing smart grid reliability: Fault detection in phasor measurement unit images with deep learning. *IEEE Access, 13*, 69257–69268.

Nubert, J., Truong, N. G., Lim, A., Tanujaya, H. I., Lim, L., & Vu, M. A. (2018). Traffic density estimation using a convolutional neural network. *arXiv preprint arXiv*, 1809.01564.

Onireti, O., Zhang, L., & Imran, M. A. (2019). On the viable area of wireless practical byzantine fault tolerance (pbft) blockchain networks. In *2019 IEEE global communications conference (GLOBECOM)* (pp. 1–6). Waikoloa, HI, USA: IEEE.

Qian, C., Guo, Y., Hussaini, A., Musa, A., Sai, A., & Yu, W. (2024a). A new layer structure of cyber-physical systems under the era of digital twin. *ACM Trans.actions onInternet Technology*. https://doi.org/10.1145/3674974. Accepted.

Qian, C., Liu, X., Ripley, C., Qian, M., Liang, F., & Yu, W. (2022). Digital twin—cyber replica of physical things: Architecture, applications and future research directions. *Future Internet, 14*(2). https://www.mdpi.com/1999-5903/14/2/64.

Qian, M., Qian, C., Xu, G., Tian, P., & Yu, W. (2024a). Smart irrigation systems from cyber–physical perspective: State of art and future directions. *Future Internet, 16*(7). https://www.mdpi.com/1999-5903/16/7/234.

Raghunandan, K. (2022). Supervisory control and data acquisition (SCADA). In *Introduction to wireless communications and networks: A practical perspective* (pp. 321–337). Springer.

Rajkumar, R. R., Lee, I., Sha, L., & Stankovic, J. (2010). Cyber-physical systems: The next computing revolution. In *Proceedings of the 47th Design Automation Conference, ser. DAC '10* (pp. 731–736). Association for Computing Machinery. https://doi.org/10.1145/1837274.1837461.

Saad, A., Faddel, S., Youssef, T., & Mohammed, O. A. (2020). On the implementation of IoT-based digital twin for networked microgrids resiliency against cyber attacks. *IEEE Transactions on Smart Grid, 11*(6), 5138–5150.

Saad, S. M. S., & Radzi, R. Z. R. M. (2020). Comparative review of the blockchain consensus algorithm between proof of stake (PoS) and delegated proof of stake (dpos). *International Journal of Innovative Computing, 10*(2).

Salazar-Pena, N., Tabares, A., & Gonzalez-Mancera, A. (2024). AI-powered dynamic fault detection and performance assessment in photovoltaic systems. *arXiv preprint arXiv*, 2409.00052.

Sanjab, A., & Saad, W. (2016). Data injection attacks on smart grids with multiple adversaries: A game-theoretic perspective. *IEEE Transactions on Smart Grid, 7*(4), 2038–2049.

Sen, P. C., Hajra, M., & Ghosh, M. (2020). Supervised classification algorithms in machine learning: A survey and review. In *Emerging technology in modelling and graphics: Proceedings of IEM graph 2018* (pp. 99–111). Springer.

Sharma, A., & Tiwari, R. (2024). Anomaly detection in smart grid using optimized extreme gradient boosting with SCADA system. *Electric Power Systems Research, 235*, 110876.

Shih, F. Y. (2010). *Image processing and pattern recognition: Fundamentals and techniques*. John Wiley & Sons.

Singh, M., Fuenmayor, E., Hinchy, E. P., Qiao, Y., Murray, N., & Devine, D. (2021). Digital twin: Origin to future. *Applied System Innovation, 4*(2), 36.

Somma, A., Benedictis, A. D., Zappatore, M., Martella, C., Martella, A., & Longo, A. (2023). Digital twin space: The integration of digital twins and data spaces. In *2023 IEEE International Conference on Big Data (BigData)* (pp. 4017–4025).

Stankovic, J. A. (2014). Research directions for the internet of things. *IEEE Internet of Things Journal, 1*(1), 3–9.

Sun, L., Shang, Y., Ma, Z., Zhou, H., & Cui, H. (2023). Power distribution safety assessment based on the digital twin. In *2023 4th International Conference on Smart Grid and Energy Engineering (SGEE)* (pp. 71–74). New York, NY, USA: Association for Computing Machinery (ACM).

Terzija, V. (2007). Wide area monitoring protection and control (WAMPAC). In *IET-UK International Conference on Information and Communication Technology in Electrical Sciences (ICTES 2007)* (pp. 1119–1126). Chennai, India: Institution of Engineering and Technology (IET).

Theckedath, D., & Sedamkar, R. (2020). Detecting affect states using VGG16, ResNet50 and SE-ResNet50 networks. *SN Computer Science, 1*(2), 79.

Tian P., Liao W., Yu W., and Blasch E., "WSCC: A weight-similarity-based client clustering approach for non-iid federated learning," IEEE Internet of Things Journal, vol. 9, no. 20, pp. 20243–20256, 2022.

Varlamis, I., Himeur, Y., Chronis, C., & Sardianos, C. (2025). Blockchain technology for secure digital twin data management. In *Blockchain and digital twin for smart healthcare* (pp. 439–452). Elsevier.

Venayagamoorthy, G. K., Sharma, R. K., Gautam, P. K., & Ahmadi, A. (2016). Dynamic energy management system for a smart microgrid. *IEEE Transactions on Neural Networks and Learning Systems, 27*(8), 1643–1656.

Wang, Q., Liao, W., Guo, Y., McGuire, M., & Yu, W. (2024a). Blockchain-empowered federated learning through model and feature calibration. *IEEE Internet of Things Journal, 11*(4), 5770–5780.

Wang, X., Shi, D., Wang, J., Yu, Z., & Wang, Z. (2019). Online identification and data recovery for PMU data manipulation attack. *IEEE Transactions on Smart Grid, 10*(6), 5889–5898.

Wang, X., Wang, H., Bhandari, B., & Cheng, L. (2024a). AI-empowered methods for smart energy consumption: A review of load forecasting, anomaly detection and demand response. *International Journal of Precision Engineering and Manufacturing-Green Technology, 11*(3), 963–993.

Wei, Y., Tian, A., Jiang, Y., Zhang, W., Ma, L., Ma, L., Sun, C., & Sun, J. (2023). A DRL-based load shedding strategy considering communication delay for mitigating power grid cascading failure. *Electronics, 12*(14), 3024.

Xu, G., Moulema, P., Ge, L., Song, H., & Yu, W. (2016). A unified framework for secured energy resource management in smart grid. In *Smart grid: Networking, data management, and business models: 2016* (pp. 73–96). CRC Press.

Xu, G., Yu, W., Griffith, D., Golmie, N., & Moulema, P. (2017). Toward integrating distributed energy resources and storage devices in smart grid. *IEEE Internet of Things Journal, 4*(1), 192–201.

Xu, H., Liu, X., Yu, W., Griffith, D., & Golmie, N. (2020). Reinforcement learning-based control and networking co-design for industrial internet of things. *IEEE Journal on Selected Areas in Communications, 38*(5), 885–898.

Xu, X., Zuo, L., & Huang, Z. (2014). Reinforcement learning algorithms with function approximation: Recent advances and applications. *Information Sciences, 261*, 1–31.

Yang, Q., An, D., Min, R., Yu, W., Yang, X., & Zhao, W. (2017). On optimal PMU placement-based defense against data integrity attacks in smart grid. *IEEE Transactions on Information Forensics and Security, 12*(7), 1735–1750.

Yang, Q., Yang, J., Yu, W., An, D., Zhang, N., & Zhao, W. (2014). On false data-injection attacks against power system state estimation: Modeling and countermeasures. *IEEE Transactions on Parallel and Distributed Systems, 25*(3), 717–729.

Yılmaz, M., & Martinez-Morales, A. A. (2023). Fault detection in photovoltaic panels using digital twin technology: A comprehensive study. *The Journal of Cognitive Systems, 8*(2), 28–32.

Yuan, L., Chen, Y., Wang, T., Yu, W., Shi, Y., Jiang, Z.-H., … Yan, S. (2021). Tokens-to-token vit: Training vision transformers from scratch on imagenet. In *Proceedings of the IEEE/CVF international conference on computer vision (ICCV)* (pp. 558–567). Montreal, QC, Canada: IEEE Computer Society.

Zhai, S., Yang, Y., Li, J., Qiu, C., & Zhao, J. (2019). Research on the application of cryptography on the blockchain. *Journal of Physics: Conference Series, 1168*, 032077.

Zhang, Y., Wang, J., & Chen, B. (2020). Detecting false data injection attacks in smart grids: A semi-supervised deep learning approach. *IEEE Transactions on Smart Grid, 12*(1), 623–634.

Zheng T., Liu M., Puthal D., Yi P., Wu Y., and He X., "Smart grid: Cyber attacks, critical defense approaches, and digital twin," arXiv preprint arXiv:2205.11783, 2022.

CHAPTER

22

Future perspectives for Digital Twin–based power grids

Pooja Kapoor[a], *Chitranjanjit Kaur*[b], *Ajeet Kumar Sharma*[c], *Avinash Kumar Sharma*[d]

[a]Noida Institute of Engineering & Technology, Greater Noida, Uttar Pradesh, India [b]GNA University, Punjab, India [c]School of Computer Science & Engineering, IILM University, Greater Noida, Uttar Pradesh, India [d]Department of CSE, Sharda School of Computing Science & Engineering (SSCSE), Sharda University, Greater Noida, Uttar Pradesh, India

22.1 Introduction

The global energy sector is undergoing a significant transformation driven by demands for reliability, resilience, and digitalization. The old power grid, traditionally designed to serve in centralized energy generation and one-way direction flow of energy, has become insufficient to answer the increasing requirements of the contemporary electricity networks. The emergence of renewable energy sources, electric vehicles (EVs), energy storage systems, and distributed energy resources (DERs) has brought modeling complexity in grid management, operation, and planning. A modular "Digital Twin System (DTS)," is proposed with the concept of intrinsic and external services. The architecture, the extension, and the generic data structure are applicable to universal use cases. The framework allows for generic implementation for future power systems (Song et al., 2023).

To solve these drawbacks, the idea of the Digital Twin (DT) has been brought out as yet another revolutionary technology in the power systems sector. A DT is a virtual tool of real assets, systems, or processes that are continually updated with real-time data. proposed a framework to implement a DT in electrical grid management. The concept of automation in electrical energy has transformed into a smart grid to utilize for the transmission and distribution of electrical power with dynamic scheduling. Further, this approach is used for all key components in the smart grid to reduce downtime and enhance the reliability of the entire system (Khan et al., 2023).

Srivastava et al. identified and examined the future advantages of DTs to enable cybersecurity and a resilient power grid to explore various use cases, while another study was performed to develop a digital twin power model from various dimensions like business, data, shape, time, and flow of energy (Srivastava et al., 2024; Zhang et al., 2022). DTs can assist in visualizing real-time grid as well as perform predictive maintenance, identify faults, optimize asset performance as well as risks DTs can change power grids to intelligent, autonomously learning systems and enable them to respond to needs and emergencies by connecting with other technologies, including the Internet of Things (IoT), artificial intelligence (AI), cloud computing, and big data analytics.

A collection of emerging technologies is highly contributing to the efficiency of DT applications in modern power systems by providing the opportunity to exchange data in real time, ensure safe transactions, analyze large quantities of data, and decentralize intelligence. These enabling technologies build the data foundation of a digitally enabled, scalable, intelligent, and resilient energy system:

- **IoT and edge computing**: The IoT and edge computing provide the foundation of both data acquisition and local intelligence in the power systems. IoT-based smart meters, intelligent sensors, and connected actuators are distributed in parallel throughout the grid to monitor physical parameters such as voltage, current, frequency, temperature, and asset testing at all times. Edge computing is used to solve the problems of latency and bandwidth

Digital Twin Technology and Smart Grid.
DOI: https://doi.org/10.1016/B978-0-443-36664-2.00014-9

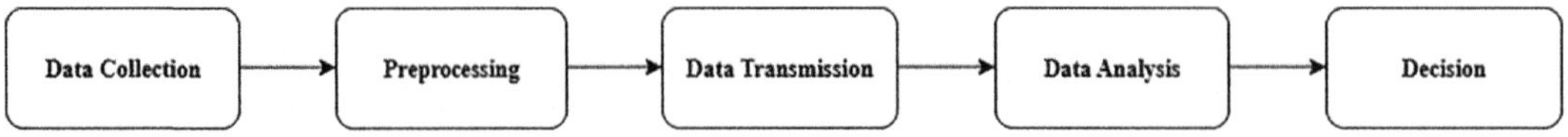

FIGURE 22.1 Illustrates the structured process for decision-making. Architecture of data acquisition.

by performing local data processing beside the source of information, complementing IoT. Rather than send all raw information to some central server providing it, the edge could do preanalysis, filtering, and decision-making locally. This will enable quicker responses, minimize communication overhead, and provide higher reliability in the application, such as fault detection, load balancing, and microgrid control. Combined with IoT, edge computing will facilitate contextually aware, autonomous operations on distributed energy systems in real time.

- **Big data analytics and cloud computing**: The scale of computation and storage space required to handle the sheer amounts of data generated by DTs is a level that can be met by cloud computing. Services such as Infrastructure as a Service (IaaS) and Platform as a Service (PaaS) could allow utilities to implement massive simulation, apply AI/machine learning (ML) models, and get access to data analytics features without the need to uphold physical infrastructure. Simultaneously, the structured and unstructured data are turned into usable insights using frameworks of big data analytics (e.g., Apache Hadoop and Spark). These are trend analysis, predictive maintenance, consumption forecasting, and asset performance benchmarking. The predicted result will be to enable decision-making that is evidence-based and smarter, and result in better system efficiency, reliability, and customer service because of the possibility to analyze historical and real-time data streams. Cloud and big data technologies are also compatible with multiuser collaborations, data visualization dashboards, and interfacing with external systems through application programming interfaces (APIs), and are rapidly becoming a fundamental component in a modern grid intelligence platform.
- **Blockchain:** Blockchain technology offers an effective solution for enabling secure, transparent, and tamper-resistant transactions in decentralized power systems. It allows grid participants to trade and share energy through a trusted, distributed ledger while supporting the execution of smart contracts. These capabilities ensure permanent and immutable records of transactions, fostering trust and accountability among stakeholders. In addition, smart contracts automate key processes such as energy billing, renewable energy credit trading, and demand-response payouts. By relying on decentralized authentication mechanisms, blockchain also reduces the risk of cyberattacks and data manipulation, thereby enhancing the overall security and reliability of decentralized energy markets. In a local energy market, as an example, the prosumers fitted with solar panels may safely resell the surplus to their neighbors, all the operations being confirmed and recorded in a blockchain ledger in real time. Blockchain has the potential to support the verification of the authenticity of data published on digital assets, provide grid components with digital identity, and exchange updates of models securely throughout the grid when connected to DTs (Sharma & Kumar, 2024).

The chapter presents the background knowledge of DT power grids as well as their increased role in current energy sectors. It also points to the possibilities of DTs in the future to solve such imminent issues as renewable integration, system resilience, and decentralized energy management. This chapter gives a broad view of the role of DTs in creating the next generation of smart and sustainable power grids through exposition of important technologies, applications, and expected developments, applying a panoramic approach to the topic.

22.2 Digital Twin-based power grid architecture

22.2.1 Layer of data acquisition

Data acquisition layer acts as a bridge between the actual facility of the power system and the virtual picture of the same. It is a layer with a broad interface of smart sensors, smart meters, and IoT-edge devices installed throughout the grid. These constituents facilitate real-time gathering of data on essential resources, including the substations, transformers, power lines, and DERs. Fig. 22.1 illustrates the process to transform the raw data into valuable insights to reach a decision.

Parameters that are usually obtained in the data gathered for the experiment:

- Electrical instruments (volts, amps, hertz)
- Conditions of the environment (environmental temperature, humidity, speed of wind)

TABLE 22.1 Integration of various technologies and their applications.

Aspect	Applications
Integration of AI & ML	Integration of both technologies support to predict anomaly and grid behavior in real time.
Edge Computing	It reduces latency and support for real time decision-making.
5G/6G Networks	These technologies are required for fast and synchronized data transfer between grid assets and control centers.
Blockchain	It supports for secure and transparent transaction records for decentralized energy trading.
IoT Expansion	It helps to monitor and control through smart sensors and devices across grid infrastructure.

- Asset performance indicators (equipment health and status, signs of wear)
- External factors (weather conditions)

This layer allows constant tracking and ensures the unprocessed information required by advanced analytics, forecasting, and decision-making of DTs. The data from this layer must be accurate, high-resolution data, to constitute the digital thread that connects physical components to their virtual twins to guarantee synchronization, situational awareness, and responsive control.

22.2.2 Infrastructure communications

The communication infrastructure is a fundamental part of the DT ecosystem, as it acts as a guarantor to the faultless and safe flow of information between field devices and centralized or distributed computing platforms. After sensors, smart meters, and IoT interventions receive the data, it is propagated along a mix of high-speed, low-latency communications networks.

Typical technologies applied in this level will be:

- 5G networks, which deliver low-latency capabilities and high bandwidth to time-sensitive applications.
- Low Power Wide Area Networks: An application that works well over long distances with power-constrained endpoints.
- Fiber optic networks have high-speed and highly reliable data transmission features.

Consistent communication is needed to ensure that the physical grid and its simulated version remain in real-time synchronization. Delays or any data losses might affect the applications of real-time monitoring, predictive analytics, and automated controls. Thus, the communication level should be redundant, scalable, and cybersecure, and it should support such industrial and utility-focused protocols as MQTT, IEC 61850, and DNP3 (Idrisov et al., 2023; Zomerdijk et al., 2024). Table 26.1 shows the integration of various technologies and their applications.

22.2.3 Other components

Modeling and simulation engine: This is the core of the DT; it is an engine that simulates the performance of the grid in terms of AI models, ML algorithms, and physics-based simulations. It assists in predicting the load demand, modeling faults, and optimizing the energy flows.

Visualization and control layer: This comprises dashboards and computer interfaces that display data in easily comprehensible forms. Operators will be able to visualize grid conditions and performance, and be able to intervene where necessary.

Decision-making systems: These systems can give insights and recommendations based on the real-time inputs and past trends, as well as automate (part of) the control functions to improve operational efficiency.

22.3 Power systems applications

22.3.1 Digital Twins

DTs are a major factor in transforming the work of the grid and maintaining strategies. DTs allow transitioning to condition-based monitoring and predictive maintenance, which reduces the reliance on scheduled/time-based maintenance (Idrisov et al., 2023; Kamyabi et al., 2022). Here, maintenance scheduling, extension of asset life, unexpected equipment failures, and unscheduled outages will be reduced. Operators are able to find problems before they blow up, which would improve grid reliability, safety, and efficiency by integrating data in real time and simulating it.

22.3.2 Integration of renewable energy

The growing integration of intermittent renewable energy resources like solar and wind poses significant problems for grid stability and reliability. DTs address these challenges by developing simulations of the renewable sources and environment. All these models use real-time weather cases, satellite-engineered data, and pre-existing results to imitate and forecast the energy production patterns of owners with unprecedented accuracy.

This way, grid operators will be able to foresee the changes in power delivery and start making preparations about switching to backup power generation, activating energy storage, or adjusting the demand side. An example is that a DT of a solar farm can predict the level of irradiance as well as panel performance and enable automated curtailment, energy trading decisions, or best times to charge a battery. Additionally, DTs assist in determining the effects of renewable penetration on grid inertia and frequency stability, thus they can anticipate actions to ensure the quality of the power is maintained.

DTS are also used to enable peer-to-peer (P2P) energy trading in microgrids and distributed energy systems and integration of EVs as flexible load/generation units, opening the path to a more decentralized and intelligent energy ecosystem.

22.3.3 Forecasting load in real time

Effective operation of a power system requires accurate prediction of the demand for electricity. DTs with AI and ML algorithms are able to consume big volumes of real-time information (such as temperature, occupancy patterns, industrial loads, and even the social aspect, such as holidays or large-scale activities) and provide extremely accurate load forecasts both short into the future and far into the future.

In contrast to classical forecasting models, which highly rely on historical averages, AI-driven DTs can learn to build prediction models by learning anomalies and updating them each time new data is acquired. This dynamic forecasting assists the grid operators in prioritizing the unit commitment, spinning reserves, reduction of fuel cost, and the efficiency of energy.

To give an example, during operation of an urban energy system, DT systems can predict residential peak loads on the basis of weather forecasts and customer behavior, and consequently, demand-response mechanisms, including smart thermostat control or time-of-use charges, can be used to balance supply and demand proactively.

22.3.4 Fault repair and self-healing

The second most important use of DTs in the power system is reinforcement of system resilience via early fault detection, predictive maintenance, and self-healing networks. It is expected that DTs would monitor the state of grid equipment, e.g., transformers, lines, switchgear, and relays, with sensor readings, historical trends, and anomaly detectors. As a fault happens, the DT can model various fault scenarios to identify the fault location and advise isolation measures. This highly mitigates the mean time to repair and promotes reliability KPI such as System Average Interruption Duration Index and System Average Interruption Frequency Index. DTs cooperate with automated reclosers, smart switches, and intelligent electronic devices to support self-healing capabilities with smart grids. In a distribution network, when a fault happens on a feeder, the digital twin may be used to determine the most advantageous reconfiguration to restore power to unaffected areas and isolate the fault; many such restorations can be done in a few seconds, and no human interaction is necessary (Gao et al., 2024; Hoffman, 2024).

22.3.5 Risk management and Cybersecurity

With the power grid growing more and more digital and interconnected, the attack surface is expanding as well. DTs offer a software platform to emulate a range of cyber-physical attack scenarios so that actual infrastructure remains untouched (Fotopoulou et al., 2023). The existence of DTs enables the cybersecurity departments to test the resilience of any SCADA, smart meters, and edge devices built on IoT against various types of intrusion, including denial-of-service attacks, malware injections, or data spoofing, without risk for the actual grid. Moreover, the use of DTs assists utility firms in undertaking risk analysis through a simulation of cascading failures, dependency, and system vulnerabilities. As an example, the ripple effects of a successful ransomware attack on a control center can be captured in a DT and can be used by the decision-makers to define business continuity plans, backup restoration, and multilayered security protocols. Additionally, combining DTs with the blockchain is also a new direction in ensuring the security of energy transactions and the integrity of the data throughout the decentralized energy networks.

A practical illustration of this capability can be seen in the following DT–based cybersecurity assessment workflow:

- **Fault injection and attack simulation:** The DT is used to emulate a ransomware attack by injecting faults such as the encryption or disabling of critical SCADA files within a virtual control center, allowing threat scenarios to be tested without endangering real assets.
- **Observation of cascading effects:** Building on this simulation, the DT reveals how the attack propagates across the cyber-physical system, disrupting control signals between SCADA systems, smart meters, and downstream components, and exposing interdependencies and failure points.
- **Risk and impact analysis:** Finally, the outcomes of the simulated attack are analyzed to quantify economic losses, operational downtime, and customer impact, supporting the development of business continuity plans, backup strategies, and multilayered cybersecurity defenses.

22.4 Future perspectives

22.4.1 AI-driven autonomous grids

The adoption of AI in power systems is generating the transformation of self-regulating, self-optimizing, and self-healing smart grids. They use real-time data of DTs and have super-powerful AI-enabled algorithms like ML, reinforcement learning, and neural networks that make efficient operational decisions with little or no human involvement.

Some of the most significant abilities of AI-powered autonomous grids comprise the following:

- Self-regulation: Automatic protection and real-time control of the grid stability through manipulation of voltage, frequency, and load flow.
- Self-optimization: The dynamic control of energy dispatch, storage, and demand-side resources to achieve the maximum efficiency with minimum expenditures.
- Self-healing: Automatic detection of faults, isolation of the involved portions, and redirecting power automatically restores service without the occurrence of human intervention.

Resilience, reliability, and sustainability can also be optimized because these AI-enabled grids could foresee equipment failures, identify cyber threats, and adapt to the abrupt variations in supply or demand. With increasingly complex and decentralized power systems, as renewables, electric vehicle networks, and prosumers increasingly enter the power grid, AI will become the enabler that constructs the orchestrated simulator and the intelligent grid.

22.4.2 Energy distributed markets

The move to decentralized energy is reshaping the established power markets, and the main contributor to it is the increased number of prosumers, or the combination of consumers and producers of electricity. DTs also operate in this changing environment, where they ensure P2P energy trading and real-time market functioning.

DTs can also help energy providers create precise models of energy generation, usage, and storage with the physical simulation of DERs with rooftop solar panels, home batteries, and EVs. In these models, it is possible for:

- Modeling of the energy transfer on the local grid
- Checking of transactions between peers
- Micro-market supply and demand forecasting

And automated settlement in a smart contract.

Added to blockchain technology or any other distributed ledger, DTs provide transparency, security, and trust to decentralized transactions. To provide another example, the energy output of a prosumer selling extra solar energy to a fellow could be modeled, checked, and sold in the DTS in real-time.

Moreover, DTs also contribute to a better market design because they can be used to analyze the consequences of diverse pricing methodologies, tariffs, and decision-making policies without risk in the real world. This aids utilities and regulators in making the informed decisions and makes consumers active in energy markets.

DTs will be used as the intelligent layer of secure, efficient, and scalable decentralized energy trading platforms as energy systems become more localized and democratized.

22.4.3 Robust grid infrastructure

It is important to construct a robust grid infrastructure to contend with rising climate variability, cyber threats and aging electricity infrastructure. DTs can significantly increase the resilience of a grid through complex simulations, real-time modeling and predictive analytics.

DTs are able to simulate those extreme events, which include:

- Extreme weather (e.g. thunderstorms, heat, floods),
- Defects of physical assets (product explosion, faults in the line),
- Man-made interferences (e.g., cyberattacks, sabotage).

In these simulations, utility operators and planners are able to assess vulnerability scenarios, define important weak points, and assess the efficiency of mitigation plans without risking real infrastructure. Minimize cascading failure The DT will be capable of predicting cascading failure during a virtual storm event, recommending the most effective feeder reconfiguration and deployment crew.

In an imaginary situation of cybersecurity, the twin will be used to simulate malware deployment in control systems and sample response procedures prior to an attack.

As power systems become more demanding when it comes to the flexibility and stability factors, DTs become one of the most valuable tools that can be used when designing, testing, and keeping the up-to-date grids resilient.

22.4.4 Novelties in regulations and policies

Current trends and the extensive deployment of DT technologies in the energy industry require parallel advancements in regulatory practices and policy formulation. As utility companies and grid operators adopt DTs for predictive analysis, automated operations, and real-time scenario simulation, governments and regulators must develop new standards. These are essential to ensure the technology is used responsibly, securely, and equitably. The following sections outline key areas where policy innovation is required.

22.4.4.1 *Privacy and ownership of data*

DTs rely heavily on the real-time transmission of data from smart meters, sensors, and user devices. Therefore, clear policies must be established to address data governance. Regulations need to explicitly define data ownership, clarifying whether rights belong to the consumer or the utility. Furthermore, guidelines must be established regarding the usage and sharing of this data, alongside strict requirements for anonymization and obtaining user consent to protect privacy.

22.4.4.2 *Standardization interoperability*

To ensure that DTs can interact seamlessly with other platforms and devices across multiple application domains, it is necessary to mandate standardized data formats, APIs, and communication protocols. Regulatory bodies should actively promote open architecture-based frameworks and support the development of generic representations for assets and the grid. These measures are vital for ensuring cross-vendor compatibility and seamless integration across various grid layers.

22.4.4.3 *Cybersecurity norms*

Because DTs replicate vital infrastructure and operate in real-time, they introduce new cybersecurity vulnerabilities. To mitigate these risks, regulatory bodies must mandate strict requirements, including regular cyber-resilience audits and the adoption of robust security standards. Policies should also enforce rigorous access controls and continuous monitoring of the DT environment. Ultimately, cybersecurity frameworks must treat DTs with the same gravity as national critical infrastructure.

22.4.4.4 *Validations and compliance*

For DTs to play a reliable role in guiding grid activity and market decisions, regulators will likely require independent model validation and certification of simulation accuracy. It is essential to establish clear, systematized processes for verifying real-time decision support tools. These measures will ensure that decisions relying on DTs remain transparent, verifiable, and fully aligned with regulatory objectives.

22.5 Limitations of Digital Twins

While DT technology offers transformative potential for the energy sector, its practical implementation is accompanied by significant technical, financial, and operational hurdles. Moving from theoretical frameworks to fully functional, real-time grid replicas requires overcoming complex challenges related to data integrity, computational demands, model accuracy, and resource allocation.

The efficacy of a DT is fundamentally limited by the quality of the data it receives, following the principle of "garbage in, garbage out." Power grids are often characterized by heterogeneous environments where modern smart sensors coexist with legacy analog infrastructure. Integrating these diverse data sources into a unified model is a complex engineering challenge. Information often resides in isolated silos—such as separate databases for SCADA, GIS, and metering—making it difficult to create a holistic view. Furthermore, issues regarding data granularity and synchronization are prevalent; if sensors do not capture high-frequency transients or if there is a lag in data transmission, the DT cannot accurately predict rapid stability issues or maintain real-time synchronization with the physical assets.

Creating a high-fidelity replica of a large-scale power grid involves solving complex differential-algebraic equations and processing massive streams of IoT data simultaneously. To be effective for critical applications like fault detection or frequency regulation, the DT must process data and run simulations within milliseconds. Current cloud-based architectures often suffer from latency that renders them too slow for these critical real-time control loops. Additionally, scalability becomes a major concern; as the number of DERs and EVs increases, the computational burden grows exponentially. Maintaining a twin that encompasses millions of active endpoints requires immense processing power, often necessitating expensive high-performance computing or advanced edge-computing solutions.

A DT is, by definition, an approximation of reality, creating an inevitable "reality gap" between the physical asset and its mathematical representation. Physical parameters drift over time due to environmental factors and equipment aging—for instance, the changing resistance of a cable or the degradation of a battery. If the DT does not autonomously update its internal parameters to match these physical changes, the simulation results will diverge from reality. Moreover, the stochastic nature of renewable energy generation and consumer behavior adds a layer of uncertainty. While AI helps predict these patterns, unexpected "Black Swan" events can lead to model failure, where the DT recommends actions based on incorrect assumptions or training data that does not reflect the current anomaly.

The deployment of a full-scale DTS represents a substantial capital expenditure (CAPEX) and operating expenditure (OPEX). Upgrading a grid with the necessary density of sensors, high-speed communication networks (such as 5G or fiber), and server infrastructure is financially prohibitive for many utilities, particularly in developing regions. Beyond the initial hardware, the DT itself requires ongoing maintenance; software updates, model recalibration, and cybersecurity patching create a continuous operational burden. Finally, there is a critical workforce skills gap. The successful operation of these systems requires professionals with interdisciplinary expertise in both electrical power systems and advanced data science, and training or hiring such personnel adds significantly to the long-term costs.

22.6 Challenges and research gaps

Despite the transformative potential of DTs in power systems, several challenges and unresolved issues remain. These technical, operational, and institutional limitations must be addressed through targeted research and policy innovations to ensure successful and scalable deployment. The key challenges and research gaps are outlined in the following section.

22.6.1 Data security and privacy

The implementation of DTSs relies heavily on a continuous stream of real-time data transmission between different segments of the power grid. This connectivity predisposes the grid to cyberattacks, data breaches, and unauthorized entry. The proliferation of open communication protocols, cloud platforms, and IoT devices expands the attack surface, creating multiple entry points for malicious actors to access sensitive infrastructure data. Potential risks include the illegal manipulation of grid resources and the compromise of consumer privacy, such as usage patterns and location data. Furthermore, existing access control and monitoring systems are often inadequate for these highly distributed environments.

Future research directions in this domain include:

- Advanced encryption & zero-trust architectures: Research is needed into end-to-end encryption technologies and zero-trust authentication mechanisms specifically adapted for low-latency, real-time distributed environments.
- Privacy-preserving analytics: Development of federated learning techniques that allow DTs to learn from local data without transmitting sensitive raw data to a central server.
- Cyber-physical resilience: Investigation into AI-driven intrusion detection systems that can distinguish between physical grid faults and cyberattacks in real-time.

22.6.2 Model accuracy and fidelity

A DT is only as effective as its ability to accurately mirror the physical system. Poor or outdated models can lead to ineffective decision-making, particularly during critical real-time operations. Ensuring high fidelity requires addressing issues related to consistent data synchronization, recalibration, and the modeling of nonlinear stochastic behaviors. Challenges include simulating the complex interactions between physical and cyber layers, ensuring the quality of data from sensors and field devices, and accounting for the degradation, aging, and unexpected behavior of assets over time.

Future research may focus on:

- Self-adaptive DTs: Research is needed into self-learning systems that can automatically recalibrate their parameters on the fly using real-time data streams and reinforcement learning, minimizing the "reality gap."
- Uncertainty quantification: Development of robust statistical methods to quantify model uncertainty, providing operators with a confidence interval for every prediction made by the DT.
- Hybrid modeling: Exploration of hybrid models that combine physics-based equations with data-driven ML to improve accuracy in scenarios where physical laws are complex or partially unknown.

22.6.3 Computational and infrastructure challenges

DTs require immense computational power to process massive datasets and run complex simulations in real-time. Current infrastructure often struggles with the latency and bandwidth requirements needed for immediate grid control. The sheer volume of data generated by millions of smart meters and sensors can overwhelm centralized cloud servers, leading to bottlenecks that hinder timely decision-making.

To address these computational bottlenecks, research must prioritize the development of distributed computing architectures, particularly edge-cloud orchestration. By investigating methods to intelligently share data processing tasks between local edge devices and centralized cloud servers, the industry can significantly reduce latency and bandwidth usage. Concurrently, significant effort is needed in the field of Model Order Reduction. This involves developing advanced mathematical techniques capable of simplifying complex, high-fidelity models into lighter versions that execute in real-time without sacrificing critical accuracy. Finally, the evolution of communication networks remains a priority; exploring the integration of 6G technologies will be essential to support the ultra-reliable low-latency communication required for the next generation of highly responsive DTs.

22.6.4 Regulatory and standardization hurdles

The rapid evolution of DT technology has outpaced the development of regulatory frameworks and industry standards. There is currently a lack of universal standards for data interoperability, meaning DTs from different vendors often cannot communicate effectively. Additionally, regulatory policies regarding data ownership, liability in case of AI-driven errors, and cross-border data sharing remain undefined or fragmented.

Overcoming these institutional barriers requires a dedicated focus on establishing universal interoperability standards. Researchers and industry bodies must collaborate to develop comprehensive data ontologies and communication protocols—expanding upon existing standards like IEC 61850—to ensure seamless integration across diverse platforms. In parallel, the establishment of regulatory sandboxes is crucial. These controlled testing environments would allow policymakers and researchers to experiment with and refine new regulations for AI and DTs without jeopardizing grid stability. Furthermore, as automation increases, there is a pressing need to formulate ethical AI frameworks that clearly define liability and governance structures for automated decisions made by DTS.

22.7 Conclusions

DT technology stands at the forefront of the energy sector's digital transformation, offering a paradigm shift in how power grids are designed, monitored, and operated. As explored in this chapter, the ability to create dynamic, virtual representations of physical assets enables utilities to move beyond traditional, reactive management toward a proactive, data-driven approach. By integrating the IoT, AI, and big data analytics, DTs provide the granular visibility needed to manage the increasing complexity of modern power systems.

The applications discussed—ranging from predictive maintenance and self-healing networks to the seamless integration of renewable energy and EVs—demonstrate the versatility of this technology. DTs are not merely simulation tools; they are active enablers of grid resilience, efficiency, and sustainability. They allow operators to test scenarios safely, optimize asset performance in real time, and facilitate the decentralized energy markets of the future.

However, realizing the full potential of DTs requires navigating significant hurdles. As highlighted in the previous sections, challenges regarding data quality, computational latency, cybersecurity vulnerabilities, and regulatory interoperability must be addressed. The transition from pilot projects to wide-scale deployment will demand a concerted effort from researchers, policymakers, and industry leaders to build robust, secure, and standardized frameworks.

Looking ahead, the role of the DT' will only grow in importance. As the global demand for clean, reliable energy accelerates, the power grid must become smarter and more autonomous. DTs will serve as the "digital brain" of this future infrastructure, bridging the physical and virtual worlds to create an energy system that is not only efficient and resilient but also adaptable to the needs of future generations.

References

Fotopoulou, M., Skopetou, N., Rakopoulos, D., & Iliadis, P. (2023). Digital Twin architecture for smart grids: The case of Ikaria Island in Greece. *International Conference on Electrical, Computer and Energy Technologies, ICECET 2023*. Greece: Institute of Electrical and Electronics Engineers Inc http://ieeexplore.ieee.org/xpl/mostRecentIssue.jsp?punumber=10389177. doi:10.1109/ICECET58911.2023.10389442.

Gao, S., Li, Z., Wang, Z., Song, C., Peng, L., & Ou, Z. (2024). Construction method and application of digital twin for power grid equipment based on electrical topology. In *Proceedings - 2024 International Conference on Artificial Intelligence and Power Systems, AIPS 2024* 1 2024/01. China: Institute of Electrical and Electronics Engineers Inc http://ieeexplore.ieee.org/xpl/mostRecentIssue.jsp?punumber=10710003. doi:10.1109/AIPS64124.2024.00112.

Hoffman, P. (2024). Realizing the digitalization and Digital Twin vision [Technology Leaders]. *IEEE Electrification Magazine, 12*(3), 5–8. http://ieeexplore.ieee.org/servlet/opac?punumber=6508996. doi:10.1109/MELE.2024.3423069.

Idrisov, I., Veretennikov, I., Vasilev, S., Gutierrez, S., & Ibanez, F. (2023). Russian Federation microgrid Digital Twin application for future virtual power plants. In *IECON Proceedings (Industrial Electronics Conference)*. IEEE Computer Society. doi:10.1109/IECON51785.2023.10311709.

Kamyabi, L., Lie, T. T, & Madanian, S. (2022). Applications of Digital Twins in ;power systems: A perspective. *Transactions on Energy Systems and Engineering Applications, 3*(2), 1–9. doi:10.32397/tesea.vol3.n2.484.

Khan, S. A., Rehman, H. Z. U., Waqar, A., Khan, Z. H., Hussain, M., & Masud, U. (2023). Digital Twin for advanced automation of future smart grid. In *1st International Conference in Advanced Innovation on Smart City, ICAISC 2023 - Proceedings*. Pakistan: Institute of Electrical and Electronics Engineers Inc http://ieeexplore.ieee.org/xpl/mostRecentIssue.jsp?punumber=10084917. doi:10.1109/ICAISC56366.2023.10085428.

Sharma, A. K., & Kumar, R. (2024). Introduction to smart cities: An overview of blockchain, AI, and advanced computing. In *Smart cities: Blockchain, AI, and advanced computing* (pp. 1–16). CRC Press. https://www.routledge.com/Smart-Cities-Blockchain-AI-and-Advanced-Computing/Sharma-Gupta-Jeon/p/book/9781032579276. doi:10.1201/9781003442660-1.

Song, Z., Hackl, C. M., Anand, A., Thommessen, A., Petzschmann, J., Kamel, O., Braunbehrens, R., Kaifel, A., Roos, C., & Hauptmann, S. (2023). Digital Twins for the future power system: An overview and a future perspective. *Sustainability, 15*(6), 5259. doi:10.3390/su15065259.

Srivastava, A., Liu, C. C., Stefanov, A., Basumallik, S., Hussain, M. M., Somda, B., & Rajkumar, V. S. (2024). Digital Twins serving cybersecurity: More than a model: Cybersecurity as a future benefit of Digital Twins 2. *IEEE Power and Energy Magazine, 22*(1), 61–71. http://www.ieee.org. doi:10.1109/MPE.2023.3325196.

Zhang, Y., Ma, C., Luo, J., Zhu, W., Wu, Y., & Zhang, X. (2022). Application of Digital Twins in smart grids. *IEEE 4th International Conference on Power, Intelligent Computing and Systems, ICPICS 2022*. China: Institute of Electrical and Electronics Engineers Inc http://ieeexplore.ieee.org/xpl/mostRecentIssue.jsp?punumber=9872525. doi:10.1109/ICPICS55264.2022.9873758.

Zomerdijk, W., Palensky, P., AlSkaif, T., & Vergara, P. P. (2024). On future power system digital twins: A vision towards a standard architecture. *arXiv*. https://arxiv.org. doi:10.48550/arXiv.2404.02568.

Index

Page numbers followed by "*f*" and "*t*" indicate, figures and tables respectively.

T

U

V

W

X

Printed and bound by CPI Group (UK) Ltd, Croydon, CR0 4YY
11/09/2026
02215335-0005